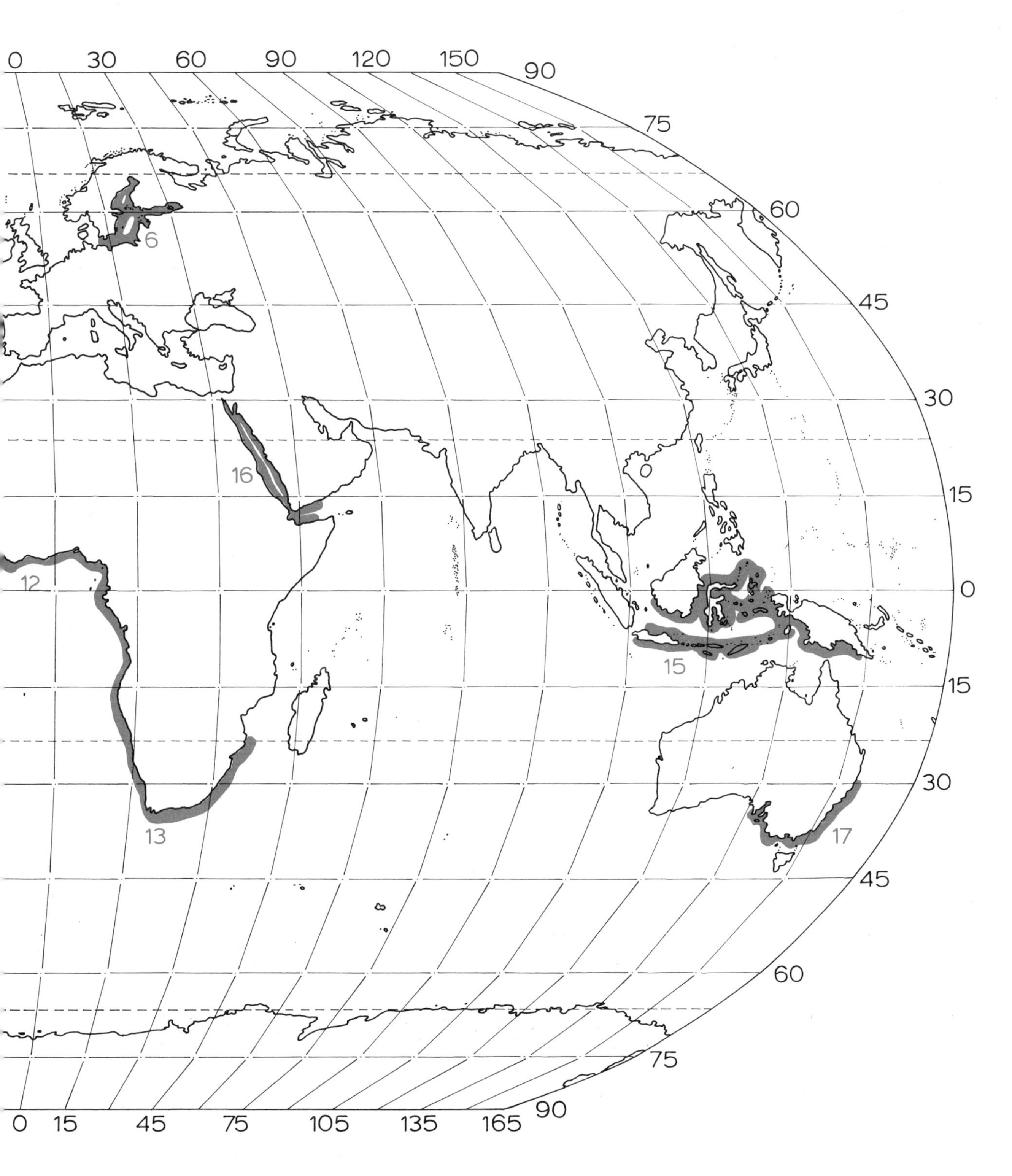

al ecosystems described in this volume.
e chapters in this book.

ECOSYSTEMS OF THE WORLD 24

INTERTIDAL AND LITTORAL ECOSYSTEMS

ECOSYSTEMS OF THE WORLD

Editor in Chief:

David W. Goodall

CSIRO Division of Wildlife and Ecology, Private Bag 4, Midland, W.A. 6056 (Australia)

I. TERRESTRIAL ECOSYSTEMS

A. Natural Terrestrial Ecosystems

1. Wet Coastal Ecosystems
2. Dry Coastal Ecosystems
3. Polar and Alpine Tundra
4. Mires: Swamp, Bog, Fen and Moor
5. Temperate Deserts and Semi-Deserts
6. Coniferous Forest
7. Temperate Deciduous Forest
8. Natural Grassland
9. Heathlands and Related Shrublands
10. Temperate Broad-Leaved Evergreen Forests
11. Mediterranean-Type Shrublands
12. Hot Deserts and Arid Shrublands
13. Tropical Savannas
14. Tropical Rain Forest Ecosystems
15. Forested Wetlands
16. Ecosystems of Disturbed Ground

B. Managed Terrestrial Ecosystems

17. Managed Grassland
18. Field-Crop Ecosystems
19. Tree-Crop Ecosystems
20. Greenhouse Ecosystems
21. Bioindustrial Ecosystems

II. AQUATIC ECOSYSTEMS

A. Inland Aquatic Ecosystems

22. River and Stream Ecosystems
23. Lakes and Reservoirs

B. Marine Ecosystems

24. Intertidal and Littoral Ecosystems
25. Coral Reefs
26. Estuaries and Enclosed Seas
27. Continental Shelves
28. Ecosystems of the Deep Ocean

C. Managed Aquatic Ecosystems

29. Managed Aquatic Ecosystems

ECOSYSTEMS OF THE WORLD 24

INTERTIDAL AND LITTORAL ECOSYSTEMS

Edited by

A.C. Mathieson

Department of Plant Biology and
Jackson Estuarine Laboratory
University of New Hampshire
Durham, NH 03824 (U.S.A.)

and

P.H. Nienhuis

Delta Institute for Hydrobiological Research
Royal Netherlands Academy of Arts and Sciences
4401 EA Yerseke (The Netherlands)

ELSEVIER

Amsterdam—London—New York—Tokyo 1991

ELSEVIER SCIENCE PUBLISHERS B.V.
Sara Burgerhartstraat 25
P.O. Box 211, 1000 AE Amsterdam, The Netherlands

Distributors for the United States and Canada:

ELSEVIER SCIENCE PUBLISHING COMPANY INC.
655 Avenue of the Americas
New York, N.Y. 10010, U.S.A.

Library of Congress Cataloging-in-Publication Data

Intertidal and littoral ecosystems / edited by A.C. Mathieson and P.H. Nienhuis.
p. cm. -- (Ecosystems of the world ; 24)
Includes bibliographical references and index.
ISBN 0-444-87409-7
1. Seashore ecology. 2. Intertidal ecology. I. Mathieson, Arthur C. II. Nienhuis, P. H. III. Series.
QH541.5.S35I58 1990
574.5'2638--dc20 90-42853
CIP

ISBN 0-444-87409-7 (Vol. 24)

Printed in The Netherlands

PREFACE

Intertidal and littoral environments are complex and productive ecosystems, which fascinate many of us who love the seashore and living things. The importance and beauty of such ecosystems require our best efforts to understand, conserve and manage them, lest the quality of our lives be diminished. Hopefully the present volume will help guide future studies and be of interest to senior students, teachers, researchers, as well as those involved with implementing coastal zone management/policy.

Many colleagues have contributed to the completion of this volume. Foremost we thank each of the authors and co-authors for their fine contributions, as well as their ability to synthesize diverse information. Their responsiveness to our editorial suggestions is appreciated. Ultimately each of the authors has been the primary judge of appropriate coverage of materials and illustrations; many of the latter are originals, while some are redrawn or reprinted with permission from various publishers. We are particularly grateful to those authors who completed and submitted their manuscripts on time. As with may cooperative efforts, we have encountered unavoidable delays, some of which have postponed the volume's scheduled publication. In addition, a few chapters originally envisioned did not materialize, resulting in a somewhat different volume.

David Goodall, Editor in Chief of the series, is acknowledged for his constructive review and editorial comments of each chapter. His insights and experience are impressive and very much appreciated.

Preparation of the indices and the Systematic List of Genera has presented a substantial challenge to one of us (ACM), particularly the animal portions. Several associates have helped with these synopses, providing proper names and authorities as well as details of classification for the diverse organisms within this volume (i.e., the division Bacteria to phylum Chordata). Garrett Crow, Department of Plant Biology, University of New Hampshire, is thanked for providing valuable insights regarding several taxonomic and nomenclatural problems. We gratefully acknowledge the extensive efforts of Duane Hope, Department of Invertebrate Zoology at the Smithsonian's Natural History Museum, who 'orchestrated' the circulation of an original draft of generic names and their classification (animals) through the Departments of Invertebrate Zoology, Vertebrate Zoology, Entomology and Paleobiology. He also solicited assistance from other colleagues when specific expertise regarding some animal groups was unavailable at the Smithsonian. The following individuals provided a variety of factual details for these synopses: Kay Stroemer Baker, Jay Grimes, Steven Jones, Richard Langan, Clayton Penniman, and Fred Short, Jackson Estuarine Laboratory, University of New Hampshire (UNH); Lawrence Abele, Department of Biological Sciences, Florida State University; Hans Bertsch, Center for Marine Ecosystem Studies, St. John, U.S. Virgin Islands; Donald Chandler and Marcel Reeves, Department of Entomology, UNH; Bruce C. Coull, Belle W. Baruch Institute for Marine Biology and Coastal Research, Unversity of South Carolina; Ray Grizzle, Department of Biology, Livingston University; Thomas Harrington, Department of Plant Biology, UNH; Arthur Borror, Larry Harris, Hunt Howell and Peter Sale, Department of Zoology, UNH; Max Hommersand, Department of Biology, University of North Carolina, Chapel Hill; Gerd

Hartmann, Zoologisches Institut and Zoologische Museum, University of Hamburg, David John, Department of Botany, British Museum of Natural History; Galen Jones, Department of Microbiology, UNH; Kenneth J. Boss and Herbert W. Levi, Museum of Comparative Zoology, Harvard University; Paul Silva, Herbarium, University of California, Berkeley; Les Kaufman, New England Aquarium; Mason Hale and James Norris, Department of Botany, Natural History Museum (NHM), Smithsonian Institution; Terry Erwin, Oliver S. Flint, Richard C. Froeschner and Ronald W. McGinely, Department of Entomology, NHM; J. Laurens Barnard, Fredrick M. Bayer, Thomas E. Bowman, Steven D. Cairns, Fenner A. Chace, Robert F. Cressey, Kristian Fauchald, Ray Germon, Richard Grant, M.G. Harasewych, Robert Hershler, Richard S. Houbrick, Brian Kensley, Raymond B. Manning, David L. Pawson, Marian H. Pettibone, J.W. Reid, Klaus Ruetzler, Victor Springeer, and W.J. Sweeney, Department of Invertebrate Zoology, NHM; Alan Cheetham, Department of Paleobiology, NHM; James Mead, Victor Springer, Richard W. Thorington, and George Zug, Department of Vertebrate Zoology, NHM; Jon Norenburg, Smithsonian Oceanographic Sorting Center; Mary E. Rice, Smithsonian Marine Station, Link Port.

Two staff members at the Jackson Estuarine Laboratory (Irene Hutchinson and Heather Talbot) deserve special thanks for their help with typing correspondences, selected manuscripts, and indices. The library staff at the University of New Hampshire (UNH), particularly Karen Fagerberg, David Lane, Francis Hallahan, Connie Stone, and Debbie Watson, is acknowledged for facilitating several searches and interlibrary loans, many of which were invaluable. Joseph Danahy and Terri Winters of University Computing (UNH) are also thanked for their invaluable help with the preparation of several indices for this volume.

Both editors appreciate the support and encouragement of their parent Institutions — i.e. the University of New Hampshire, including the Jackson Estuarine Laboratory and the Department of Plant Biology (ACM), and the Delta Institute for Hydrobiological Research (PHN). One of us (ACM) acknowledges a Faculty Development Grant that allowed a semester's release time from teaching during 1986 and a sabbatical leave (1987) at the Port Erin Marine Laboratory of the University of Liverpool (Isle of Man) where some of the indices were prepared. The help of various faculty, staff, and students at Port Erin is gratefully acknowledged, including Trevor A. Norton, Joana (Kain) Jones and Sigmar A. Steingrimmson. The latter individual helped with the translation and editing of several Icelandic terms and names.

Two wives, Myla J. Mathieson and Arine Nienhuis, deserve special thanks for their help extending from critical reading of manuscripts and proofs to the preparation of indices. Without their love, support and encouragement the volume would never have been finished.

A.C. MATHIESON
Durham, N.H.

P.H. NIENHUIS
Yerseke

LIST OF CONTRIBUTORS

R. ASMUS
Biologische Anstalt Helgoland
Litoralstation
D-2282 List/Sylt (F.R.G.)

J.J.W.M. BROUNS
Research Institute for Nature Management
P.O. Box 59
1790 AB Den Burg (Texel) (The Netherlands)

C.J. DAWES
Department of Biology
University of South Florida
Tampa, FL 33620 (U.S.A.)

A.P. DE VOGELAERE
Institute of Marine Sciences
University of California
Santa Cruz, CA 95064 (U.S.A.)

R.J. DIAZ
Virginia Institute of Marine
Science
School of Marine Sciences
College of William and Mary
Gloucester Point, VA 23062
(U.S.A.)

J.G. FIELD
Department of Zoology
University of Cape Town
Rondebosch 7700
Cape Town (South Africa)

M.S. FOSTER
Moss Landing Marine Laboratories
P.O. Box 450
Moss Landing, CA 95039 (U.S.A.)

C.L. GRIFFITHS
Department of Zoology
University of Cape Town
Rondebosch 7700
Cape Town (South Africa)

L.G. HARRIS
Department of Zoology
University of New Hampshire
Durham, NH 03824 (U.S.A.)

C. HARROLD
Monterey Bay Aquarium
886 Cannery Row
Montery, CA 93940 (U.S.A.)

K.L. HECK JR.
Marine Environmental Sciences
Consortium
Dauphin Island Sea Lab
Dauphin Island, AL 36528 (U.S.A.)

F.M.L. HEIJS
Science Department
Ministery of Science and Education
P.O. Box 25000
2700 LZ Zoetermeer (The Netherlands)

P.A. HUTCHINGS
Division of Invertebrate Zoology
Australian Museum
6-8 College Street
P.O. Box A285
Sydney South, N.S.W. 2000 (Australia)

D.M. JOHN
Department of Botany
British Museum (Natural History)
Cromwell Road
London SW7 5BD (U.K.)

M.G. KELLY
Department of Environmental Sciences
University of Virginia
Charlottesville, VA 22903 (U.S.A.)

R.J. KING
School of Botany
The University of New South Wales
P.O. Box 1
Kensington, N.S.W. 2033 (Australia)

A.W.D. LARKUM
School of Biological Sciences
University of Sydney
Macleay Building A12
Sydney, N.S.W. 2006 (Australia)

G.W. LAWSON
Department of Biological Sciences
Bayero University
PMB 3011
Kano (Nigeria)

Y. LIPKIN
Department of Botany
George S. Wise Faculty of Science
Tel Aviv University
Ramat Aviv 69 978
Tel Aviv (Israel)

D.S. LITTLER
Department of Botany
National Museum of Natural History
Smithsonian Institution
Washington, DC 20560 (U.S.A.)

M.M. LITTLER
Department of Botany
National Museum of Natural History
Smithsonian Institution
Washington, DC 20560 (U.S.A.)

K. LÜNING
Biologische Anstalt Helgoland, Zentrale
Notkestrasse 31
D-2000 Hamburg 52 (F.R.G.)

A.C. MATHIESON
Department of Plant Biology and Jackson Estuarine Laboratory
University of New Hampshire
Durham, NH 03824 (U.S.A.)

E.D. McCOY
Department of Biology
University of South Florida
Tampa, FL 33620 (U.S.A.)

I.M. MUNDA
Centre of Scientific Research of the Slovene Academy of Science and Arts
Novi trg 5
61000 Ljubljana (Yugoslavia)

S.N. MURRAY
Department of Biological Science
California State University
Fullerton, CA 92634 (U.S.A.)

P.H. NIENHUIS
Delta Institute for Hydrobiological Research
Royal Netherlands Academy of Arts and Sciences
Vierstraat 28
4401 EA Yerseke (The Netherlands)

J.S. OLIVER
Moss Landing Marine Laboratories
P.O. Box 450
Moss Landing, CA 95039 (U.S.A.)

R.J. ORTH
School of Marine Science
Virginia Institute of Marine Science
College of William and Mary
Gloucester Point, VA 23062 (U.S.A.)

J.S. PEARSE
Institute of Marine Sciences
University of California
Santa Cruz, CA 95064 (U.S.A.)

C.A. PENNIMAN
Narragansett Bay Project
291 Promenade Street
Providence, RI 02940 (U.S.A.)

G. RUSSELL
Department of Environmental and Evolutionary Biology
University of Liverpool
P.O. Box 147
Liverpool L69 3BX (U.K.)

B. SANTELICES
Facultad de Ciencias Biologicas
Departemento de Ecologia
Universidad Catolica de Chile
Grupo de Ecologia Marina
Casilla 114-D
Santiago (Chile)

W. SCHRAMM
Institut für Meereskunde
Abt. Meeresbotanie
Universität Kiel
Düsternbrooker Weg 20
D-2300 Kiel 1 (F.R.G.)

R.R. SEAPY
Department of Biological Science
California State University
Fullerton, CA 92634 (U.S.A.)

I. WALLENTINUS
Department of Marine Botany
Gothenburg University
Carl Skottsbergs Gata 22
S-41319 Gothenburg (Sweden)

R.J. WEST
The University of New South Wales
P.O. Box 1
Kensington, N.S.W. 2033 (Australia)

CONTENTS

Chapter 1

INTRODUCTION

P.H. NIENHUIS and A.C. MATHIESON

This volume fills the niche between wet coastal ecosystems dominated by terrestrial angiosperms (Volume 1: Chapman, 1977) and dry coastal ecosystems (above high-water level, mainly sand dunes but also cliffs; Volume 2: Van der Maarel, in prep.). The book covers the intertidal belt (or littoral zone of non-tidal coasts) down to the limits of macrophyte growth or the approximate 1% light level. The ecosystems comprise oceanic and coastal rocky and sandy shores, estuarine hard substrates and sand and mud flats, including seagrass communities. Mangroves and coral reefs are subjects treated respectively in Volume 1 (Chapman, 1977) and Volume 25 (Dubinsky, 1990). This volume provides a comprehensive description of a large part of the present day knowledge of the world's intertidal and littoral ecosystems, although some vital parts of the global survey are lacking (see figure on inside cover).

Intertidal and littoral ecosystems, as defined in this book, comprise a large variation in their physical and chemical properties over a world-wide geographical and climatic range. Water temperature is the major physical factor governing the geographical distribution of marine organisms. The latitudinal thermal gradient ranges from a surface maximum of 26 to 30°C in the tropics (35°C in shallow, enclosed areas like the Persian Gulf) to a minimum at lower latitudes, reaching the freezing point of seawater at −1.9°C. On a geological scale water temperature is also a highly significant factor. The evolution of cold-adapted species and new genera occurred mainly in response to temperature drops in the Miocene and Pliocene. The greatest diversity of cold-water biota in the Northern Hemisphere is found in the North Pacific coastal habitats. Biota in tropical areas were never extinguished by dramatic drops in water temperature. Consequently, the upper temperature limit of marine tropical plants and animals is approximately 35°C.

The availability of light under water governs the depth distribution of marine plants. A depth of 200 m in extremely clear water has commonly been accepted as the lower limit of the deepest representatives of multicellular marine algae (crustose calcified coralline red algae); 0.01% of the surface light intensity reaches that level. Along European coasts the deepest growing kelp individuals occur at a water depth of 100 m in the Mediterranean Sea or at 25 m on the turbid coast of Brittany.

Water temperature and light availability fix the geographical distribution of marine biota on a world-wide scale. Tidal amplitude, water movements, exposure to wave action and desiccation, and sediment stability structure, the benthic littoral communities on a regional or local scale. The last mentioned physical factors are of paramount importance for local zonation and distribution patterns of shore organisms (Lüning and Asmus, Ch. 2 this Volume).

The impact of chemical factors on the coastal zone is far more difficult to generalize than the influence of physical factors. Pronounced temporal variations, either long- or short-term, periodical or non-periodical, are typical for shallow-water systems. Generally, littoral communities do not suffer from nutrient depletion; open oceans contain more "deserts" (regions of low nutrient availability) than are found in coastal habitats. Even non-polluted, littoral, hard-substrate communities in the open ocean have higher nutrient concentrations than the surrounding sea, owing to runoff of freshwater and seepage. Coastal intertidal ecosystems are the main recipients of the major portion of riverine and land runoff, often loaded with high concentrations of nutrients. More than any other marine system, intertidal and littoral ecosystems are exposed to human impact, such as pollution and eutrophication (Schramm, Ch. 3 this Volume).

Along soft-substrate coasts there is often an intensive chemical and physical interaction between the water column and the sediment. Moreover, the regeneration of nutrients is enhanced by bioturbation or by the filtering capacity of bivalve molluscs (Schramm, Ch. 3; Wallentinus, Ch. 6).

Among all intertidal and littoral habitats the rocky shores are the best studied. Some older, but outstanding textbooks (Lewis, 1964; Stephenson and Stephenson, 1972) exist on the world-wide structural aspects of hard-substrate coastal ecosystems and the universal features of zonation patterns. Sediment coasts are ecologically far less known. Integrated studies on (exposed) sandy beaches started only recently (Field and Griffiths, Ch. 13; King et al., Ch. 17). An "Ecology of Sediment Shores", a world-wide, integrated survey of the soft-substrate shores, including estuarine, sheltered animal communities and macrophyte-dominated communities, does not exist, although the reviews by Wilkinson (1980) and Wolff (1983) cover a considerable part of the estuarine habitat.

The comprehensive book of T.A. and Anne Stephenson (1972) on life between tidemarks on rocky shores is, besides a compilation of existing literature, mainly based on their original papers written between 1930 and 1960. It contains a meticulous naturalist's description of the geographical distribution of shore biota under physical constraints. Moreover, it focuses on the world-wide features of vertical zonation in terms of distribution of organisms, related to tidal levels.

Lewis (1964) described mainly rocky shores in Britain, but his book has a far wider impact. Twenty years after Lewis' (1964) book on the distribution and zonation of shore organisms, and the factors influencing these phenomena, Moore and Seed (1985) dedicated a series of essays to J.R. Lewis on the same subject. Rocky shore ecology became more quantitative and experimental. Studies on autecology (life tables) and population dynamics, work on constancy and persistence of rocky shore communities, and experimental investigations concerning processes that govern distribution and abundance of littoral species, have deepened our views on rocky shore ecosystems. Lewis (1964) stressed the main control of shore phenomena by physical factors. Recent trends focus on biologically mediated processes (competition, predation) as the cause of distribution and zonation patterns. The truth of course lies in the blend of ideas (see also King et al., Ch. 17).

As several chapters in this book bear witness (Munda, Ch. 5; John and Lawson, Ch. 12; Santelices, Ch. 14; Lipkin, Ch. 16), in extensive regions of the earth the descriptive approach of the Stephensons is still in full use, simply because taxonomical and ecological explorations have just started in those areas. Owing to the tremendous heterogeneity and complexity of littoral and intertidal habitats and communities, both in time and in space, we are far from "generalizations", a "synthesis" or a mass balance approach, such as has been attempted for coral reefs by Smith (1988).

Several more or less "closed" intertidal or littoral ecosystems seem rather consistent in the use of their energy and nutrients. An example is the food chain dominated by grazing molluscs and benthic algae in the high intertidal zone of rocky shores. The herbivores can control the algal biomass in these habitats (Dawes et al., Ch. 9). Another example might be the self-sustaining nature of constant tropical seagrass beds, with a very low export of organic material and an assumed rapid turnover of nutrients (Nienhuis et al., 1989). In many cases, however, the intertidal and littoral ecosystems are "open" systems under physical constraints and subsidized by energy from the adjacent pelagic system (the case of the benthic filter feeders; Wallentinus, Ch. 6; Field and Griffiths, Ch. 13) or exporting energy (the case of the temperate communities dominated by macrophytes either kelps or seagrasses; Foster et al., Ch. 10; Field and Griffiths, Ch. 13).

A most remarkable feature in intertidal and littoral areas is the vertical zonation pattern of the biota. To avoid confusing terminology, the zonation in this book comprises three primary zones: (1) supralittoral zone — is nearly permanently emerged; (2) eulittoral zone — is subject to periodic immersion by tidal or other hydrological events; (3) sublittoral zone — is nearly permanently submerged. As to the causes of zonation, the upper limits of organisms are determined mainly by tolerances to environmental extremes (for example, desiccation) and the lower limits to the effects of competition (Russell, Ch. 4).

The geographical distribution of dominant intertidal organisms shows some significant features. In Arctic and Antarctic regions, where there is ice action, usually a definite belt around sea level is scraped bare of all organisms, except those living in crevices. Rocks that have been denuded by ice may be repopulated in summer with annual algae; fucoids, barnacles and mussels cannot reach their full development in one summer. Below the ice sheet a permanent growth of algae and associated animals is allowed.

On cold-temperate shores, at low water of spring-tides, generally, a heavy forest-like growth of large brown algae occurs, commonly laminarians, many of them of very large size. Warm-temperate coasts lack the giant kelp growth but fair-sized brown algae may be quite common. Moreover, there may be a short turf of algae among which red algae are prominent.

Tropical shores differ markedly in two respects from cold and temperate shores: (a) mangroves fringe many tropical coasts, occupying both the open sheltered beach and estuaries, inhabiting the upper intertidal region and supporting a characteristic fauna; (b) coral reefs form under stable conditions on the lower part of the shore, in areas approximately bordered by the 20°C winter isotherm of seawater. Both ecosystems are dealt with in separate volumes (respectively, Vol. 1: Chapman, 1977; Vol. 25: Dubinsky, 1990). In many places, however, mangroves and coral reefs proper are interspersed with rocky outcrops, harbouring intertidal hard-substrate communities. In other localities there is a wide, flat area in the lee of the coral reef, dominated by seagrass beds, beyond which there is a sandy beach.

The Baltic Sea is one of the world's larger brackish water areas, with stable, low salinities and without tides. It originated between 10 000 and 3000 years ago. The ecology of the Baltic littoral ecosystems is very well studied. In contrast to earlier ideas, the Baltic Sea harbours a low degree of endemism. Most littoral species that were previously considered endemic or specific for brackish areas are on a subspecific level either ecotypes or morphological forms. The development of ecotypes is ongoing in the Baltic, which makes this area interesting for studies in evolutionary biology. Another feature, studied in full detail in the Baltic Sea is the fact that the littoral ecosystem forms the margin of a much larger system. In terms of energy turnover the littoral zone does not function as an independent unit. There is an important coupling between the pelagic system and the littoral zone. Owing to the filter-feeding activities of bivalves (mussels), which constitute up to 90% of the animal biomass on hard substrates, the larger part of the annual pelagic phytoplankton production is assimilated by these invertebrates. The littoral mussel population functions as an important nutrient regenerator for the pelagic system (Wallentinus, Ch. 6).

Chapters 7 through 11 cover almost the entire coastline of North and Central America. Mathieson et al. (Ch. 7 this Volume) describe the ecology of northwest Atlantic rocky shores mainly in terms of patterns of community distribution and zonation. The overall species richness of this area is low compared with the northeast Atlantic coastline. Cold-temperate western and eastern North Atlantic algal floras have a high degree of similarity (50%) and an approximate equality of species numbers. In contrast, the warm-temperate flora of Northwest Europe has a much greater species richness than that of Northeast America, due to the dominant influence of the Gulf Stream. In general terms the North Atlantic rocky shore flora and fauna is poor compared to the North Pacific, due to repeated and drastic cooling during Pleistocene glaciations; the Bering land bridge acted in the past as a barrier for cold water flowing in the direction of the Pacific.

The mid-Atlantic region of the U.S.A. is characterized by a notable lack of natural hard substrates. Soft substrate habitats and seagrass beds are the main intertidal features in this area. Data on primary and secondary productivity are available from selected areas. The secondary production of bare mud and sand flats is much less than that recorded from vegetated flats (Orth et al., Ch. 8).

The tropical west Atlantic, including the Caribbean Sea, have the richest algal flora of the entire Atlantic Ocean, with a high degree (55%) of endemism. Although less well known, the fauna is also remarkably rich. Community structure in the littoral zone is well defined but it has to be kept in mind that the high grazing pressure by sea urchins and herbivorous fish is largely responsible for the distribution pattern of benthic algae. Only scanty data on primary and secondary productivity are available (Dawes et al., Ch. 9).

The west coast of North America has been described in Chapters 10 and 11. Although their surveys are mainly based on community structure, Foster et al. (Ch. 10) and Littler et al. (Ch. 11) offer a substantial amount of quantitative data on the functioning of rocky coasts. The intertidal zone may represent the only habitat where there is a consistent, energetically based, grazer–algae interaction, controlling the structure of the algal community. Herbivorous limpets and littorinids can control algal biomass in this habitat. The interplay between grazing intensity and seasonal, abiotically induced changes in primary production, accounts for much of the observed community structure (Foster et al., Ch. 9).

Littler et al. (Ch. 11) stress the effects of human impact on the functioning of intertidal ecosystems along southern California. They distinguish, besides a macrophyte based grazing food web, also a particulate matter based food web near sewage outfalls.

The tropical west coast of Africa has been extensively dealt with by John and Lawson (Ch. 12). The coastline is dominated by sandy beaches, but rocky outcrops, river deltas, mangroves and lagoons occur also. The western coast of Africa shares with the western shores of other continents, such as the Americas, the absence of coral reefs and consequently the rich and varied life associated with these biogenic structures. Similarly lacking are the extensive meadows of seagrasses, which often occur in the lee of fringing reefs. The marine flora and fauna on Africa's tropical west coast is relatively poor, owing to ecological (cold upwelling, lack of firm substrates) and historical (elimination of tropical marine biota during Pleistocene) reasons. On the basis of sound taxonomic and descriptive ecological work, more functional and experimental investigations of intertidal ecosystems may start in the years to come (John and Lawson, Ch. 12). The study of coastal ecology has already reached a mature level in southern Africa where the cold-temperate ecosystems on the west coast and the warm-temperate ones on the east coast offer steep biological gradients. Biogeographic (historical) factors determine the species composition of the rocky shore communities, but the degree of exposure to wave action determines the trophic structure. Just as has been investigated for the brackish Baltic Sea (Wallentinus, Ch. 6), a tight relation exists between the pelagic and the benthic tidal system. The littoral kelp beds together with the communities of filter-feeders on the west coast form an open system governed by upwelling and downwelling, depending on weather conditions. The sandy beaches receive an enormous amount of kelp detritus resulting in high macrofaunal biomasses. The subtidal filter-feeders suffer from a shortage of food under conditions of continuous upwelling when detritus is exported from the kelp beds. Under downwelling conditions, the large reservoir of phytoplankton from offshore provides a rich source of organic matter for nearshore benthic filter-feeders (Field and Griffiths, Ch. 13).

Knowledge of intertidal and littoral ecosystems along the temperate coasts of South America is incomplete, dealing mainly with distribution patterns of biota and community structure. The marine flora and fauna along the coast of Chile shows a high degree of endemism, but species richness both for algae and invertebrates is relatively low compared to eastern Pacific temperate communities in the Northern Hemisphere (California) (Santelices, Ch. 14).

Soft-substrate benthic communities occurring in the intertidal and upper subtidal area of the tropical West Pacific (Indonesia), have been described by Brouns and Heijs (Ch. 15). Monospecific and mixed seagrass beds contain only 12 flowering plants and many more marine algae. Much work remains to be done on trophic relations and production and decomposition characteristics of these communities.

Lipkin (Ch. 16) offers a colourful description of biota, community structures and zonation patterns in the Red Sea, lined with coral reefs and interspersed with rocky and sandy coasts. The Red Sea gradient shows a decreasing species richness from south to north. Many tropical genera and species living in the southern part do not occur farther north. Quantitative and functional data are scarce.

The marine biota on the temperate coast of southeastern Australia is characterized by high species richness and a high degree of endemism. Early accounts of shoreline ecology were largely descriptive and restricted to rocky shores, but studies are now quantitative and experimental and often concerned with sheltered shores and estuaries (King et al., Ch. 17).

The final chapter in this volume deals with remote sensing (Kelly, Ch. 18), started in the early sixties with aerial photography, but nowadays using satellite and multispectral scanner imageries. The use of aerial photography and satellite imagery for the analysis of coastal community structures and biomass assessment appears to be a rapidly developing and useful research application (Klemas et al., 1987; Meulstee et al., 1988).

Appendix

In this volume the following abbreviations for tidal levels have been used:

HHWS	Highest high water spring tide
HWN	High water neap tides
HWS	High water spring tides
LLWS	Lowest low water spring tides
LWM	Low water mean
LWN	Low water neap tides
LWOST, LWS	Low water of spring tides
MHHW	Mean higher high water
MHW	Mean high water
MHWS	Mean high water of spring tides
MLLW	Mean lowest low water
MLW, MLWL	Mean low water level
MLWS	Mean low water of spring tides
MSL	Mean sea level

References

Chapman, V.J. (Editor), 1977. *Wet Coastal Ecosystems, Ecosystems of the World, Vol. 1.* Elsevier, Amsterdam, 428 pp.

Dubinsky, Z. (Editor), 1990. *Coral Reefs, Ecosystems of the World, Vol. 25.* Elsevier, Amsterdam, 550 pp.

Klemas, V., Thomas, J.P. and Zaitzeff, J.B. (Editors), 1987. *Proc. Workshop on Remote Sensing of Estuaries.* U.S. Department of Commerce, National Oceanic and Atmospheric Administration, Washington DC, 245 pp.

Lewis, J.R., 1964. *The Ecology of Rocky Shores.* The English University Press, London, 323 pp.

Meulstee, C., Nienhuis, P.H. and Van Stokkom, H.T.C., 1988. Aerial photography for biomass assessment in the intertidal zone. *Int. J. Remote Sensing*, 9(10/11): 1859–1867.

Moore, P.G. and Seed, R. (Editors), 1985. *The Ecology of Rocky Coasts.* Hodder and Stoughton, London, 467 pp.

Nienhuis, P.H., Coosen, J. and Kiswara, W., 1989. Community structure and biomass distribution of seagrasses and macrofauna in the Flores Sea, Indonesia. *Neth. J. Sea Res.*, 23(2): 197–214.

Smith, S.V., 1988. Mass balance in coral reef-dominated areas. In: B.-O. Jansson (Editor), *Coastal-Offshore Ecosystem Interactions. Lecture Notes on Coastal and Estuarine Studies. Vol. 22.* Springer, Berlin, pp. 209–226.

Stephenson, T.A. and Stephenson, A., 1972. *Life Between Tidemarks on Rocky Shores.* Freeman, San Francisco, 245 pp.

Van der Maarel, E., in prep. *Dry Coastal Ecosystems, Ecosystems of the World, Vol. 2.* Elsevier, Amsterdam.

Wilkinson, M., 1980. Estuarine benthic algae and their environment: a review. In: J.H. Price, D.E.G. Irvine and W.F. Farnham (Editors), *The Shore Environment, Vol. 2. Ecosystems.* Academic Press, London, pp. 425–486.

Wolff, W.J., 1983. Estuarine benthos. In: B.H. Ketchum (Editor), *Estuaries and Enclosed Seas, Ecosystems of the World, Vol. 26.* Elsevier, Amsterdam, pp. 151–182.

Chapter 2

PHYSICAL CHARACTERISTICS OF LITTORAL ECOSYSTEMS, WITH SPECIAL REFERENCE TO MARINE PLANTS

K. LÜNING and R. ASMUS

TEMPERATURE

Water temperature is the major factor governing the geographical distribution of marine organisms. The latitudinal thermal gradient ranges from a surface maximum of 26 to 30°C in the tropics (35°C in shallow, enclosed areas like the Persian Gulf) to a minimum at lower latitudes, reaching the freezing point of seawater at −1.9°C.

Paleo-temperatures

The steep temperature gradient during the Recent period is atypical for most of the earth's history. Since the Late Palaeozoic, only two relatively short-lived major glaciations with ice formation near the poles occurred. The first started about 300 million years ago, near the Carboniferous/Permian boundary; and the second 2 million years ago, at the beginning of the Pleistocene. During non-glaciated eras, surface temperatures of 10°C may have been characteristic in polar waters, persisting into the early Tertiary (Savin, 1977; Frakes, 1979; Shackleton and Boersma, 1981).

The picture obtained from oxygen isotopic data from planktonic Foraminifera indicates that the pole-to-equator temperature gradient was only around 5°C in the Early Tertiary (Frakes, 1979). Through the course of the earth's history, biota at lower latitudes were thus exposed to contrasting temperature conditions or even driven towards the equator by permanent ice, while the tropical regions were narrowed during glaciations. However, during non-glaciated eras since the Late Precambrian, the tropical surface seawater temperatures near the equator probably never surpassed a temperature of 33°C (Schopf, 1980), and tropical biota were never extinguished within the tropical regions.

Consequently, the upper temperature limit of marine-tropical animals and plants is around 35°C (see below), near to the long-term seawater maximum. Moreover, one may assume that the present-day cold-water biota are geologically "young" and have developed only within the Tertiary. The greatest diversity of cold-water biota of the Northern Hemisphere is found in the North Pacific, as well as the majority of the Northern Hemisphere's cold-water genera, some of which also occur in the cold North Atlantic and its "appendage", the Arctic Ocean (Briggs, 1974; Van den Hoek, 1984).

In the Eocene, 50 million years ago, surface-water temperature was probably still as high as 18°C in the North Pacific (Savin, 1977; Frakes, 1979). World-wide, two major temperature drops occurred, the first near the Eocene/Oligocene boundary, 38 million years ago (Antarctica ice-free), and the second in the early middle Miocene, 10 million years ago. It is conceivable that the evolution of cold-adapted species and of new cold-water genera took place mainly in response to these temperature drops (Van den Hoek, 1984).

The last major temperature decline in the late Tertiary incorporated the glaciation of Antarctica, with the first glaciers at sea-level appearing 14 million years ago (Grant-Mackie, 1979; Hempel, 1985). In contrast, the permanent ice cover of the Arctic Ocean is thought to have existed for only the last 700 000 years (Herman and Hopkins, 1980).

Isotherm patterns and biogeographical consequences

The major ocean currents (Fig. 2.1) move clockwise in circular gyres in the Northern Hemisphere (or counter-clockwise in the Southern Hemisphere). This is caused by the zonal winds, which initiate the major ocean currents, which are deflected to the right (or left in the Southern Hemisphere) due to the Coriolis effect (see, for instance, Brown and Gibson, 1983).

The major gyres, as well as local currents, markedly affect the latitudinal gradient of seawater temperatures, as shown by the world-wide patterns of seawater isotherms in Fig. 2.2. Since, in the Northern Hemisphere, along the western shores of the Atlantic and Pacific Oceans cold water flows from the north and warm water from the south, the isotherms are compressed along the western shores. Put another way, the annual temperature range at a given location is much wider on the American than on the European side of the North Atlantic; for instance, 0 to 20°C at Cape Cod, but only 3 to 10°C in northern Norway, 9 to 17°C in Brittany, or 14 to 19°C in mid-Portugal.

Consequently, eurythermal species (with a wide range of temperature tolerance) prevail along the western shores of the North Atlantic and North Pacific Oceans. These are complemented by many stenothermal species (with a narrow range of temperature tolerance) along the eastern shores (Joosten and Van den Hoek, 1986; Yarish et al., 1986.)

Northern and southern biogeographical boundaries

The relationships between surface seawater isotherms (February or August isotherms as annual minima and maxima, respectively, for the Northern Hemisphere) and the geographical distribution of individual species have been analyzed, for instance by Hutchins (1947) for marine animals

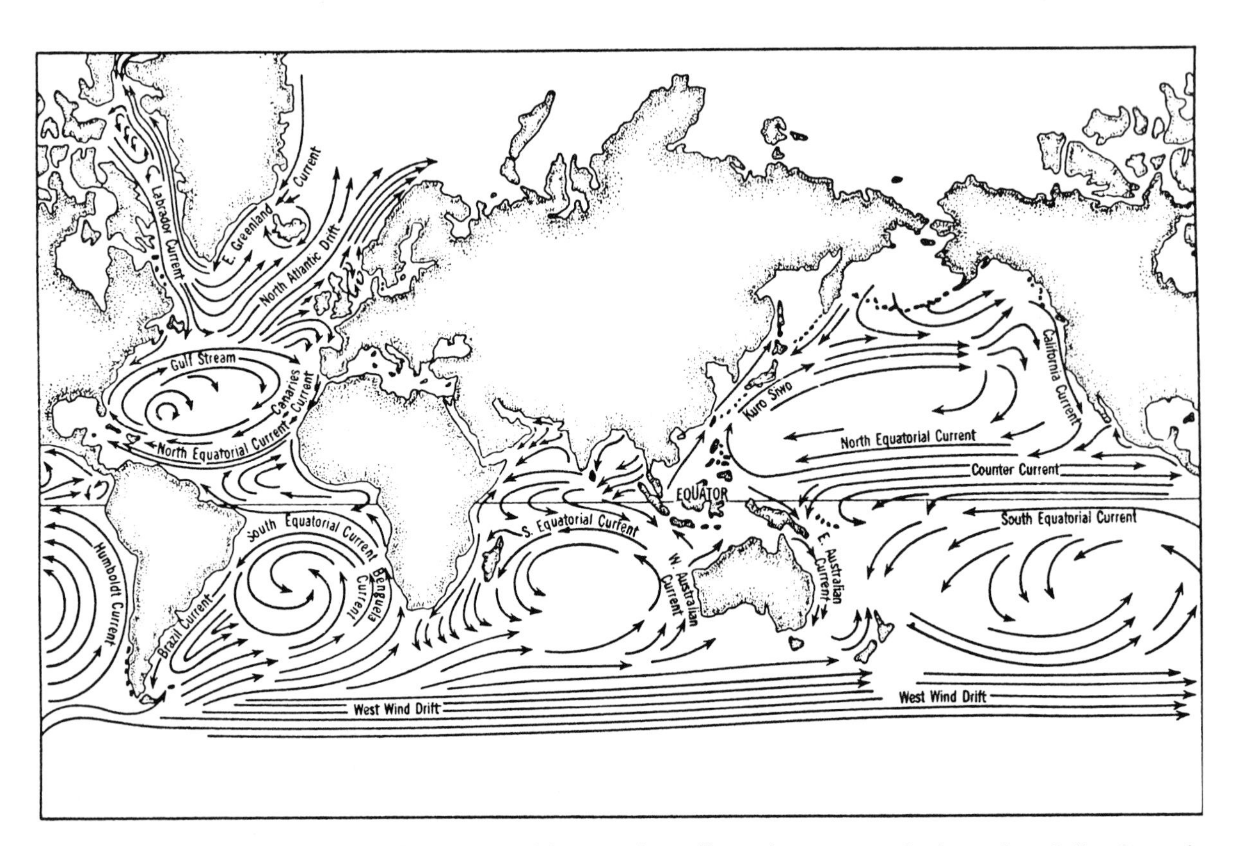

Fig. 2.1. Prevailing winds and major surface currents, with zones of upwelling and convergence. In the northern Indian Ocean the summer circulation is shown. (From MacArthur and Connell, 1966.)

and by Van den Hoek (1975, 1982a, b, 1984) for marine macroalgae. In principle, these authors set up the following scheme:

A species cannot surpass a certain isotherm, because at higher or lower temperatures

(a) lethal damage occurs, implying

(a1) a southern lethal limit for cold-adapted species, corresponding to the August isotherm,

(a2) a northern lethal limit for warm-adapted species, corresponding to the February isotherm; or

(b) growth and/or reproduction are impeded, implying

(b1) a southern growth and/or reproduction boundary, according to the February isotherm, and

(b2) a northern growth and/or reproduction boundary, according to the August isotherm.

The isotherms are based on averages of many years, and an exceptionally warm summer or cold winter, with temperatures deviating by a few degrees from the average isotherm value, may eradicate the representatives of a species at a particular location. To cope with such situations, Van den Hoek (1982a, b) proposed to use a safety margin of 1°C on isotherms for temperatures below 10°C, of 2°C for temperatures above 10°C, and of 3°C in case of the southern lethal limit.

The distribution of the warm-temperate kelp *Saccorhiza polyschides* is used here as an example for demonstrating the correlation between seawater isotherms and an alga's temperature limits (Fig. 2.3). The sporophytes of this alga die at 3°C and at 25°C, and its gametophytes become fertile to produce a new sporophyte generation at a maximum temperature of 17°C. Hence, the coastline inhabited by this alga is bordered by the 4°C February isotherm (lethal boundary, with a safety margin of 1°C) in the north, and the 22°C August isotherm (lethal boundary, with a safety margin of 3°C) in the south. These conditions do not exist in any locality along the North American coast, as the two isotherms cross when passing the North Atlantic (Fig. 2.3).

In this way a species with a relatively narrow temperature survival range is excluded from the western side of the Atlantic. In contrast, on the eastern side of the North Atlantic the two isotherms are separated by a long stretch of coastline. Here *Saccorhiza polyschides* can grow from mid-Norway (a yearly range from 5 to 12°C) to Morocco (a yearly range from 15 to 22°C), although only on open coasts. On the coasts of the North Sea temperatures are too low in winter and too high in summer, so that the species is absent. In the Mediterranean the species occurs only locally, where summer temperatures do not surpass 22°C.

Biogeographical regions

These are inhabited by groups of organisms with similar temperature requirements. More or less conspicuous changes in floral and faunal composition should occur at the boundary between two biogeographical regions.

A rather uniform system of marine biogeographical regions, as represented below, evolved in marine zoogeography (Ekman, 1953; Hedgpeth, 1957; Briggs, 1974; Vermeij, 1978), and marine phycogeography (Van den Hoek, 1975, 1984; Michanek, 1979, 1983). Basically, one may distinguish seven latitudinal groups (Fig. 2.2), each containing one or more biogeographical regions (numbers of regions according to Briggs, 1974):

(1) Arctic group (1 region).

(2) Cold-temperate group, Northern Hemisphere (3 regions).

(3) Warm-temperate group, Northern Hemisphere (4 regions).

(4) Tropical group (4 regions).

(5) Warm-temperate group, Southern Hemisphere (5 regions).

(6) Cold-temperate group, Southern Hemisphere (5 regions).

(7) Antarctic group (1 region).

The geographical boundaries separating these seven groups have been extensively discussed by Briggs (1974), and also follow more or less closely the scheme of critical temperatures proposed by Stephenson (1948) and Van den Hoek (1975). This scheme (Fig. 2.4) may look arbitrary, as it is based on 5°C intervals, and might provoke endless disputes — for instance, whether "warm" temperate conditions should begin above 15°C in summer. On the other hand, it is well-established that coral reefs, the main indicators of tropical regions, require temperatures higher than 20°C, so that tropical regions should be bordered by the 20°C winter isotherm (Fig. 2.4).

Another example, which may help one to accept

"rounded-off numbers", is furnished by the distribution of kelps (order Laminariales). They live in cold- and warm-temperate regions, but do not enter tropical regions. Experimentally determined temperature requirements for gametophytes of warm-temperate kelp species, which thrive in temperatures up to 25°C and reproduce sexually at somewhat lower temperatures, are in accordance with the summer and winter isotherms separating warm-temperate from tropical regions, as indicated in Fig. 2.4. Examples of warm-temperate kelp species with the given characteristics are *Saccorhiza polyschides* from European coasts (Norton, 1977), *Ecklonia maxima* and *Laminaria pallida* from South Africa (Branch, 1974), *Ecklonia radiata* from New Zealand (Novaczek, 1984), and *Undaria pinnatifida* from Japan (Akiyama, 1965).

Seven genera of seagrasses are characteristic for tropical seas: *Cymodocea*, *Enhalus*, *Halodule*, *Halophila*, *Syringodium*, *Thalassia* and *Thalassodendron*, according to Den Hartog (1970). This author also mentions that some species of these genera have extended their distribution into warm-temperate areas (e.g. *Halophila ovalis*, *Syringodium isoetifolium*).

Five genera are typical for temperate and cold areas: *Amphibolis*, *Heterozostera*, *Phyllospadix*, *Posidonia* and *Zostera*. Some species of *Zostera* subgen. *Zosterella* are still living in the tropics. *Posidonia* and *Zostera* have a bipolar distribution, whereas *Phyllospadix* is restricted to the northern Pacific, and *Heterozostera* and *Amphibolis* are living in the temperate zone of the Southern Hemisphere (Den Hartog, 1970). The fundamental work of Den Hartog has lately been broadened concerning geographical distribution in relation to temperature (De Oliveira et al., 1983; Phillips et al., 1983; McMillan, 1984). McMillan (1984) found that tropical seagrass species differ in their tolerance of high temperatures. Seagrasses growing in shallow sites tolerate higher temperatures than those living constantly submerged in the tropics.

In contrast to macroalgae and seagrasses, the geographical distribution of diatom species does not show a clear relationship with temperature. The great majority of benthic marine diatoms of

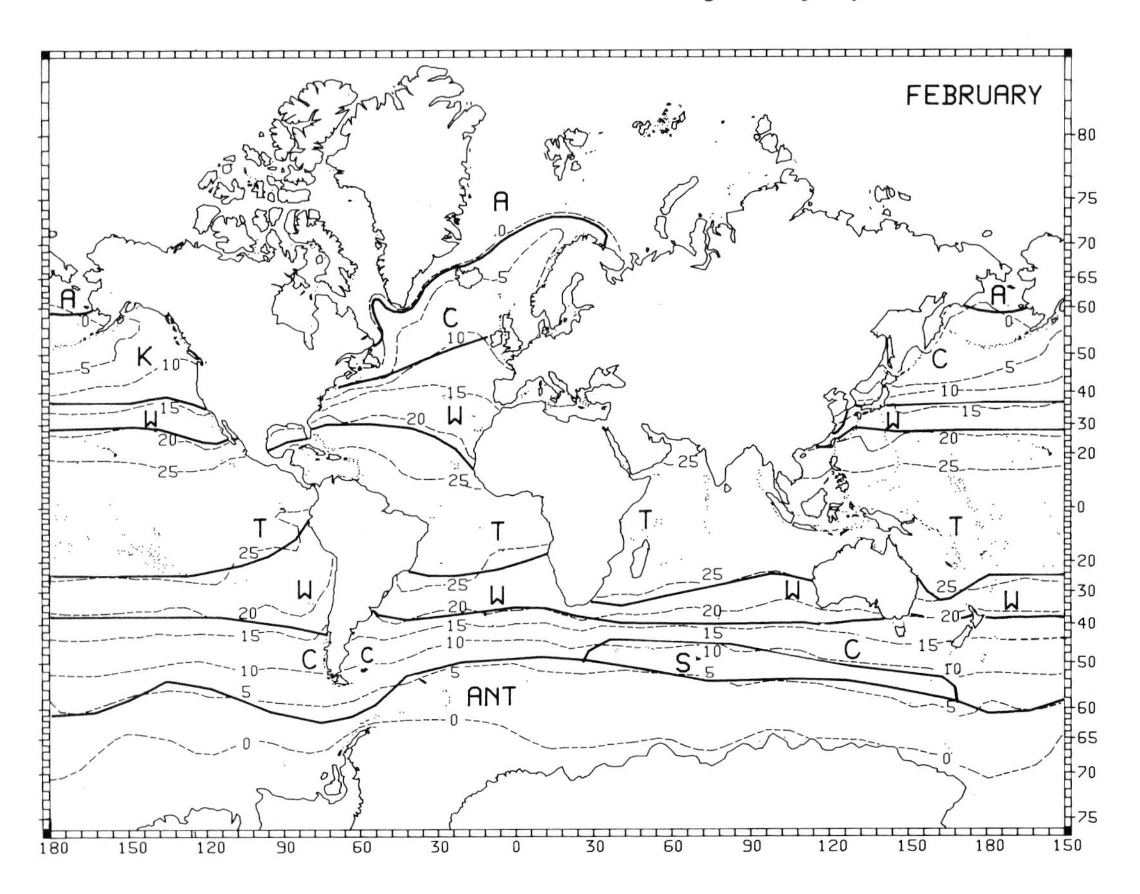

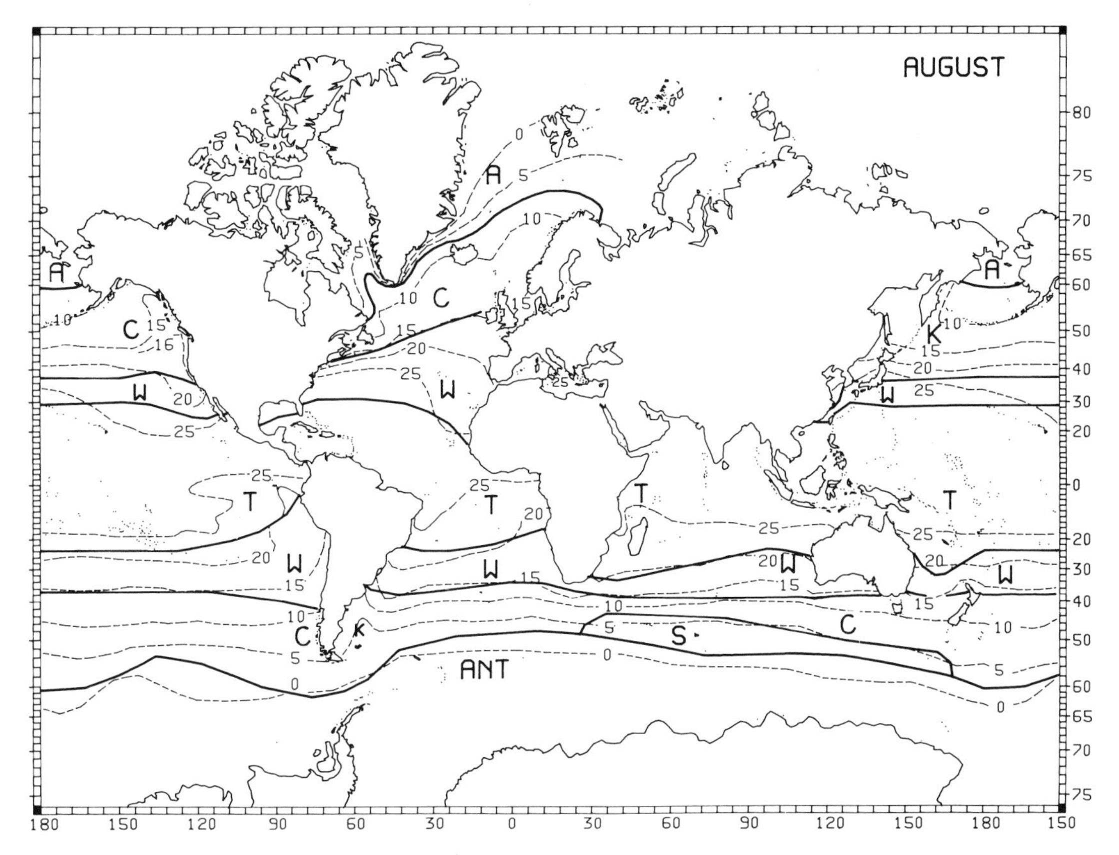

Fig. 2.2. Seawater surface isotherms (dashed lines) and boundaries between groups of biogeographical regions (solid lines) given for February and August. (A) Arctic region; (C) cold-temperate and, (W) warm-temperate regions of the Northern and Southern Hemisphere; (T) tropical regions; (ANT) Antarctic region; (S) subantarctic (cold-temperate) islands region. (After Briggs, 1974; from Lüning, 1985.)

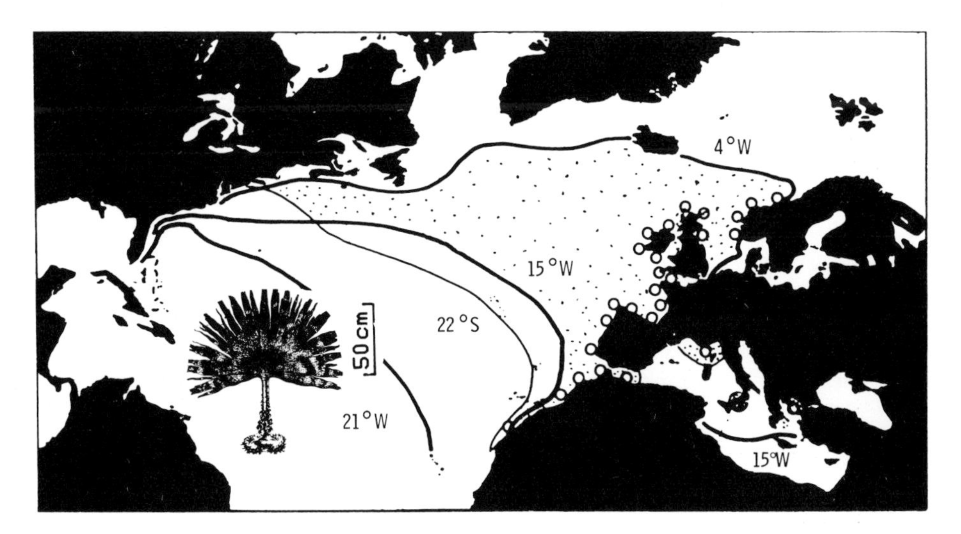

Fig. 2.3. Distribution of the kelp *Saccorhiza polyschides*. Circles represent occurrence (species is absent on North American coasts). 4°C winter isotherm = northern lethal boundary; 15°C winter isotherm = southern reproduction boundary; 22°C summer isotherm = southern lethal boundary; 21°C winter isotherm = southern growth boundary (not reached). (From Van den Hoek, 1982b; habit drawing from Sauvageau, 1918.)

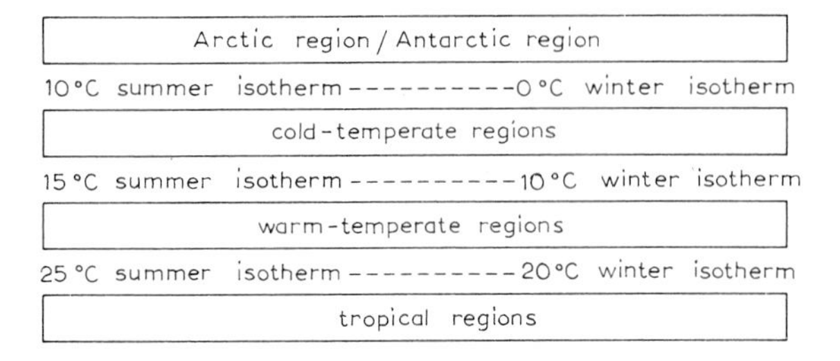

Fig. 2.4. Scheme of isotherms separating groups of marine biogeographical regions. (After Stephenson, 1948; from Lüning, 1985.)

cold and temperate waters are found over a broad geographical range (Castenholz, 1967; Kennett and Hargraves, 1984). These findings from the west and east coasts of North America, Europe and South Africa are based on valve morphology. Thus it is not known if physiological differences exist. Regrettably there is not enough information from warm-temperate and tropical regions in order to make a comparison of the species composition of benthic diatoms. The other members of the microphytobenthos, such as blue-green algae, dinoflagellates and other phototrophic flagellates and bacteria, are even less investigated world-wide.

Temperature tolerance in different biogeographical groups

Based on the results obtained for various seaweed species investigated by Biebl (1958, 1962, 1968), which are shown in Table 2.1, the temperature tolerance of different biogeographical groups may be summarized:

(1) Eulittoral species. These exhibit a wider tolerance range than sublittoral species. Arctic and cold-temperate species of the eulittoral zone survive temperatures below the freezing point of seawater.

(2) Tropical sublittoral species. These exhibit only limited tolerance of temperatures below 10°C, as they probably are lacking the capacity to form cold-adapted membranes and enzymes. Upper lethal limits of 33 to 35°C are quite common in tropical sublittoral species (Table 2.1). For example, tropical species of the red-algal genus *Gracilaria* survive 34°C, but not 36°C (McLachlan and Bird, 1984). In this respect it is interesting to remember the hypothesis of Schopf (1980) regarding the stable maximum of 33°C in tropical regions since the Precambrian, as mentioned above.

(3) Cold-temperate and polar sublittoral species. These do not survive freezing in seawater (Table 2.1), which is conceivable, as sublittoral species are *never* exposed to freezing conditions in polar regions, since in winter they are living below the ice cover, and in summer in an ice-free habitat.

Species of cold-temperate regions and of the Arctic may suffer lethal damage at temperatures around 20°C, for instance, North Atlantic species of the brown algae *Chorda*, *Desmarestia* and *Laminaria* (Lüning, 1984). The findings that the cold-temperate brown alga *Durvillaea antarctica* from the Southern Hemisphere dies at 14°C (Delépine and Asensi, 1976), and the gametophytes of the Arctic-endemic kelp *Laminaria solidungula* at temperatures higher than 18°C (Bolton and Lüning, 1982), point to the considerably older history of cold-water biota in the Southern, compared to the Northern Hemisphere. Many benthic invertebrates on Antarctic coasts are extremely stenothermal (Hempel, 1985; Picken, 1985).

TABLE 2.1

Temperature survival ranges of eulittoral and sublittoral species from different biogeographical regions after a 12 h exposure in seawater. (Compiled after Biebl, 1958, 1962, 1968; from Lüning, 1985.)

Type of region	Coast	Annual temperature span (°C)	Temperature survival range	
			eulittoral species (°C)	sublittoral species (°C)
Arctic	W. Greenland	0 to 6	−10 to 28	−1 to 22 (24)
Warm-temperate	Brittany	10 to 16	−8 to 30 (35)	−1 (0) to 25 (30)
Warm-temperate	Naples	14 to 24	−7 to 35	1 (2) to 27 (30)
Tropical	Puerto Rico	26 to 28	−2 to 35 (40)	+14 (5) to 35 (32)

Temperature requirements for growth and metabolism

Growth optima are around 10°C for polar seaweed species, 10 to 15°C for various cold-temperate species, 10 to 20°C for warm-temperate species, and 15 to 30°C in tropical species (see, for instance, the compilation of data by Lüning, 1985). Geographical isolates of the brown alga *Ectocarpus siliculosus* exhibit differences of up to 10°C with regard to optimal growth temperature, and upper survival limits; for instance, isolates from Texas died at 35°C, and from the Canadian Arctic at 25°C (Bolton, 1983). Similarly, the growth optima of the tropical to temperate red algal genus *Gracilaria* range from 20 to 30°C (McLachlan and Bird, 1984).

The relationship between photosynthesis and temperature is represented by an optimum curve. Due to the fact that the enzymatic reactions limiting photosynthesis at saturating irradiances, or being involved in respiration, are temperature-dependent, the rates of photosynthesis or respiration, within limits, roughly double if temperature increases by 10°C ($Q_{10}=2$).

In most poikilothermal animals, Q_{10} ranges from 2 to 3 for oxygen consumption, and in many species acclimatization to seasonal temperature change adjusts them to the new condition (reviews: Bayne, 1985; Newell, 1976; Levinton, 1982). Little is known in this respect for seaweeds.

The temperature optimum for photosynthesis is 20°C for Arctic *Fucus distichus* (Healey, 1972), and 15°C for the Antarctic red alga *Leptosomia simplex*, or Antarctic brown alga *Himanthothallus grandifolius* (Drew, 1977, 1983). Also terrestrial Arctic plants exhibit photosynthetic optima near 20°C, hence well above the environmental temperatures (Berry and Raison, 1981). The reason for this may be that shifting of the photosynthetic temperature optimum to the cold side requires raising of the enzyme content in order to balance the adverse effect of low temperatures on chemical reactions (Berry and Björkman, 1980). A low temperature optimum of photosynthesis, for instance at 5°C, might not be reached without high energy cost for the plant in terms of enzyme synthesis, and hence may lie outside of the plant's adaptive capacity.

Light-saturated net photosynthesis of seagrasses increases with increasing temperature up to an optimum temperature (Marsh et al., 1986; Bulthuis, 1987). Above the optimum temperature the photosynthetic rate decreases due to a disruption of the functional integrity of chloroplasts (Berry and Björkman, 1980). Various seagrasses have an optimum temperature of 30 to 35°C (Bulthuis, 1987). However, the respiration of seagrasses increases too with rising temperature (Marsh et al., 1986). At higher temperatures a higher light intensity is necessary to maintain a positive carbon balance than at lower temperatures. This may lead to unfavourable low or negative carbon balances of seagrasses living in deeper water at low light intensities and higher temperatures, as in tropical waters or in summer in temperate regions.

In winter, seagrasses can have a potentially higher photosynthetic rate than in summer because the optimum temperature is lower in seagrasses living in dim light (Bulthuis, 1987). However, the actual growth rate is not only determined by the photosynthetic rate. Other physiological processes and environmental conditions are important in addition. Some measurements of the growth rates of seagrasses showed that temperature is of minor importance (Sand-Jensen and Borum, 1983; Wium-Andersen and Borum, 1984), whereas light was the most important factor regulating the growth of seagrass. Despite the proved effect of temperature on the photosynthesis of seagrasses, it is still under discussion how important temperature is for growth in different natural environments.

Benthic microalgae are capable of growth in all temperature regions of the oceans from polar to tropical regions provided other biological, chemical and physical factors are sufficient for photosynthesis. Despite very low temperatures, benthic microalgae can build up a large standing stock. This occurs in the Arctic, where benthic microalgae bloom after the breakup of the shorefast ice. The benthic microalgae became a more important source of primary productivity than phytoplankton or algae living on ice in shallow areas (Matheke and Horner, 1974). In the Antarctic too, benthic microalgae can develop a considerable biomass.

The benthic microalgae seem to be an important food source for benthic invertebrates (Dayton et al., 1986), but they cannot grow till the shading ice cover breaks up or thaws. Ice algae can bloom before benthic microalgae as soon as light is sufficient. These ice algae can grow at very low temperatures [around −1.8°C, and even lower temperatures in hypersaline brine pockets (Palmisano et al., 1987)]. They are psychrophilic, with low temperature optima for light-saturated photosynthesis, although temperature optima are about 8°C higher than the ambient temperature (Palmisano et al., 1987). It is in accordance with earlier investigations that in general the optimal temperature is somewhat higher than the temperature in nature (Smayda, 1969; Gessner, 1970; Durbin, 1974; Admiraal, 1977; Rasmussen et al., 1983). Growth rates in terms of cell divisions in relation to increasing temperature have been investigated only for a few species of benthic diatoms (*Amphiprora* cf. *paludosa*, *Gyrosigma spenceri*, *Navicula arenaria*, *Nitzschia dissipata*, *N. sigma*) (Admiraal, 1977; Admiraal and Peletier, 1980). Species-specific temperature optima could be determined but the temperature range of growth is broad. Thus the tested species are eurythermal (Admiraal and Peletier, 1980). The relationship between growth and temperature is almost linear up to the optimum temperature (Admiraal, 1977). The relation between cell division and temperature is so close that, under light-saturated conditions as on high-level tidal flats, the cell-division rates are primarily regulated by the temperature throughout the year. High temperatures above 30°C have an inhibitory and detrimental effect only in continuous long-term incubations. A short-term rise of temperature above 30°C does not inhibit cell division (Admiraal, 1980). These investigations of Admiraal were conducted in northern Europe. Studies from other areas are lacking. Especially the investigation of benthic microalgae in tropical areas is still in its initial phase. Thus all questions concerning benthic microalgae living constantly under high temperature and high light intensities are unresolved.

The relation between growth of benthic microalgae and temperature could be confirmed not only in laboratory experiments but also in field measurements. The primary productivity of benthic microalgae increases with increasing temperature *in situ* (Pomeroy, 1959; Pamatmat, 1968; Cadée and Hegeman, 1974; Rasmussen et al., 1983; Asmus and Asmus, 1985).

Because in the field many factors act at the same time (in addition to temperature, they may include light, sediment stability, salinity fluctuations, grazing pressure, and chemical factors such as concentrations of carbon dioxide, sulphide or nutrients), the relation between primary productivity and temperature is sometimes concealed by relations with factors of more importance in the specific situation (Davis and McIntire, 1983; Colijn and De Jonge, 1984; Varela and Penas, 1985).

Temperature as an indicator of season

Reproduction is limited to a narrow temperature interval in various species; in these cases, the seasonal course of water temperature may be taken as an indicator of season. Gametophytes of the brown alga *Chorda tomentosa* arise in late spring from the zoospores of the macroscopic sporophytes. Temperatures above 8°C (summer) prevent the day-neutral gametophytes from becoming fertile (Lüning, 1980). In this way it is secured that only one sporophyte generation per year is produced, and a similar temperature-controlled response has been ascertained for *Desmarestia viridis* (Lüning, 1980).

Müller (1962) demonstrated a temperature-regulated switching response in the Mediterranean *Ectocarpus siliculosus*, with the formation of mainly unilocular sporangia at temperatures around 10°C, therewith initiating sexual reproduction, and plurilocular sporangia close to 19°C.

The green alga *Urospora wormskioldii* is an interesting example for the regulation of its morphogenesis and chemical cell-wall composition by temperature (Bachmann et al., 1976). Three morphologically different stages may be formed, all connected by zoospores (Fig. 2.5), namely a filamentous stage (at temperatures below 5°C), a dwarf stage (around 10°C), and a saccate *Codiolum*-stage (above 14°C). The temperature which acts on the zoospores decides which morphological pathway will be entered, and which pattern of chemical cell-wall composition will be realized.

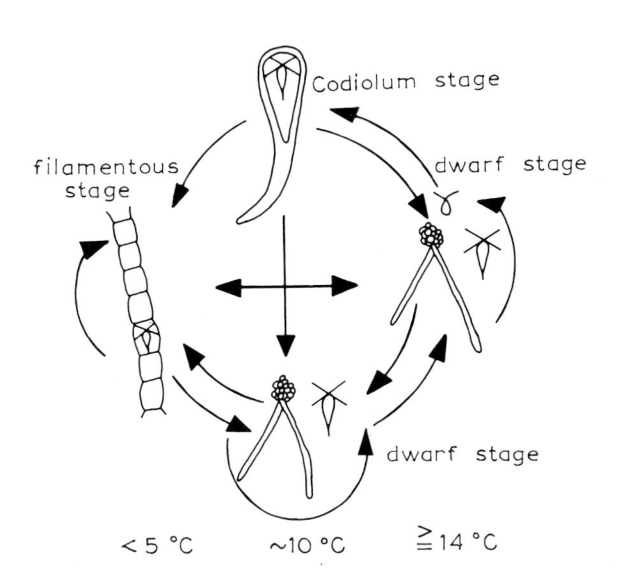

Fig. 2.5. Temperature regulation of the life history of the eulittoral green alga *Urospora wormskioldii*. All stages connected by zoospores (no sexuality involved). (From Bachmann et al., 1976.)

LIGHT

Spectral distribution in the sea and vertical distribution of seaweeds

Light is attenuated in the water due to absorption and scattering. In the clearest water ["oceanic water type I", Fig. 2.6; see Jerlov (1976, 1978)] maximum transmittance occurs in the blue range (400–500 nm), and only this spectral range remains at a water depth of 110 m (Fig. 2.7).

Turning to somewhat less clear water, such as "oceanic water type III", the majority of quanta at greater depths occur in the blue-green range (450–550 nm; Fig. 2.7). In green and brown algae, blue light is well absorbed by chlorophyll *a* (serving photosystem I), and by accessory pigments (serving photosystem II) – chlorophyll *b* and siphonaxanthin in green, and fucoxanthin in brown algae. In effect, both photosystems "run at full speed", which is not the case in red algae in short-wave blue light, because their short-wave accessory pigments, the phycoerythrins, absorb only slightly from about 450 nm and attain their maximum absorption in the green range, roughly at 500 to 600 nm.

Nevertheless, red algae are abundant also in tropical, clear and "blue" waters at depth, together with brown and green algae. Even here red algae, as crustose calcified Corallinales, represent the deepest-occurring multicellular algae at the lower algal limit. Their photosynthetic "handicap" in short-wave blue light may be balanced by the fact that the calcified crustose life form, which one does

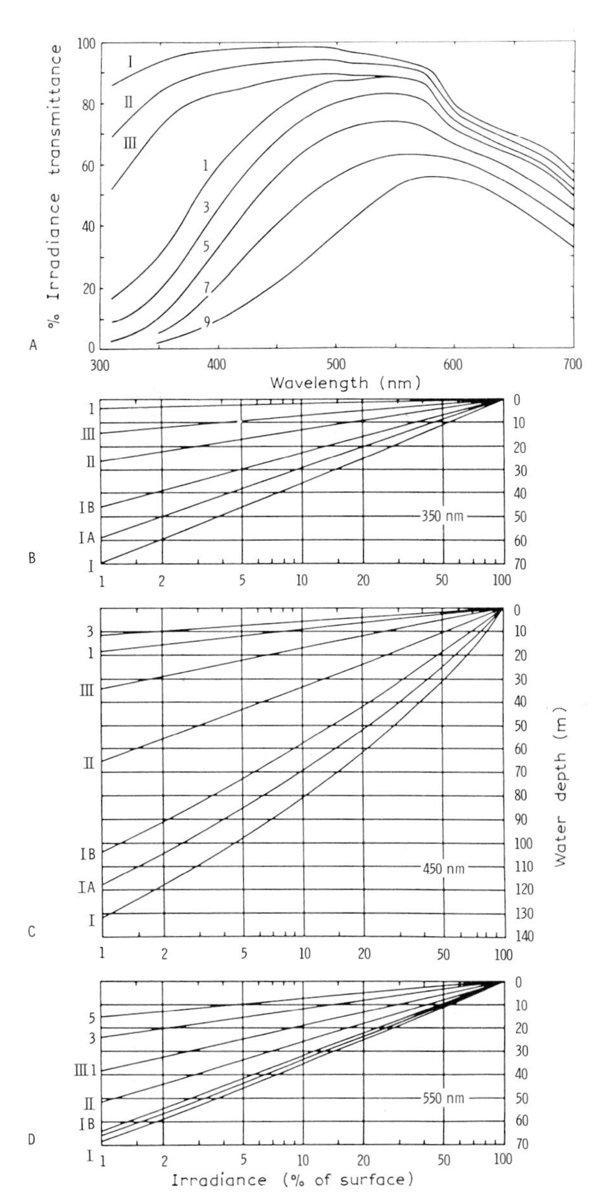

Fig. 2.6. (A) Transmittance per metre of downward irradiance in oceanic (I–III) and coastal (1–9) Jerlov water types. (B–D) Percentage of irradiance in different water depths for three selected wavelengths, in relation to Jerlov water types. (A from Jerlov, 1976; B–D from Jerlov, 1978.)

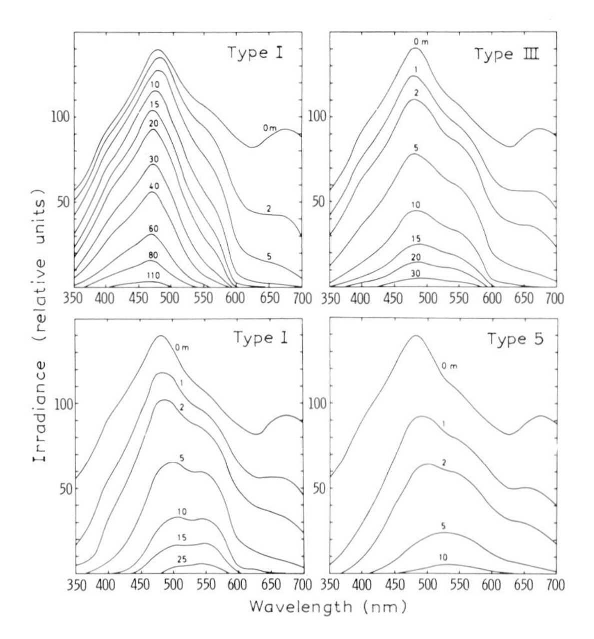

Fig. 2.7. Spectral irradiance of daylight in the sea in oceanic water types I and III, and coastal water types 1 and 5. (From Jerlov, 1978.)

not find in green or brown algae, effectively protects the algae from grazers. Longevity may be an important pre-condition for survival near the lower algal limit, where photosynthesis may attain a positive balance only during a few months or weeks of the summer.

In more turbid waters (Jerlov's "coastal water types 1–9" in Fig. 2.6), green light prevails at depth. Only the wavelength range 480 to 560 nm remains below a water depth of 25 m, if "coastal water type 1" prevails, and 460 to 620 nm at 10 m depth in "coastal water type 5" (Fig. 2.7). Red algae possess the ideal pigment for absorption of green light. The well-known fact that close to the lower algal limit in "green waters" all conspicuous seaweeds belong to the red algae, fits Engelmann's hypothesis (1883) that phycobilins evolved for optimal light absorption in predominantly green underwater light.

It is true that thick green and brown algae also fill the "absorption window" of chlorophyll *a* in the green part of the spectrum by capturing green light in multiple chlorophyll layers (Ramus, 1981). However, Kirk (1983) pointed out that this uneconomic way of multiplying a basically unsuitable pigment for absorption of green light is hardly likely to work at depths, where light quantity becomes the minimum factor, and energetic cost for forming an upright thallus must be kept at a minimum.

Today, in view of the symbiont theory on the origin of chloroplasts, one is inclined to shift the emphasis of Engelmann's phylogenetic hypothesis from the red to the blue-green algae. The scenario would have to be set up at some point in the Precambrian, when phycobilins evolved in prokaryotes which were forced to live at greater water depths because of high intensity of ultraviolet radiation in the upper water layers. As may be seen from Fig. 2.6, UV-A (330–400 nm) is relatively well transmitted in clear water. This is valid to a somewhat lesser extent also for the more dangerous UV-B (300–330 nm), which is absorbed by proteins and DNA, thus causing lethal damage. It is only the ozone and oxygen content of the atmosphere (not existing during the greater part of the Precambrian) which today keeps away dangerously high levels of ultraviolet radiation from the surface of the earth.

In an experiment, benthic microalgae developed the lowest chlorophyll *a* content and cell numbers under blue light, and the highest under red light (Antoine and Benson-Evans, 1983). Diatoms grew better under blue light than green algae or blue-green algae. Antoine and Benson-Evans suggested that accessory phytopigments (carotenoids) act in diatoms as protective substances against harmful effects of blue light.

Quantitative light levels in the sea

Adaptations of deep-water algae to low light are similar to those in higher, terrestrial shade plants, for example, low metabolic activity and reduced respiration rates, a low content of photosynthetic enzymes, since photosynthesis is light-restricted, and a high content of photosynthetic pigments in order to increase the chances for photons to be absorbed (see Ramus, 1981; Dring, 1982; Kirk, 1983; Lobban et al., 1985).

Maximum sunlight levels (400–700 nm) above water in warm-temperate and tropical regions are of the order of 600 W m^{-2} for irradiance or 3000 µmol m^{-2} s^{-1} (= 3000 µE m^{-2} s^{-1}) for photon fluence rate. In clear oceanic water (type I) a reduction of surface irradiance by a factor of ten occurs roughly after every 50 m. This means that levels of 10, 1, 0.1, and 0.01% of surface light are reached at water depths of roughly 50, 100, 150, and 200 m respectively (see Fig. 2.8, neglecting the fact that this figure is based on the waveband 350–400 nm, instead of 400–700 nm).

The depth of 200 m in very clear water has commonly been accepted as the lower limit in "blue" as well as in "green" waters, of the deepest representatives of multicellular marine algae, both represented by crustose calcified corallines. In absolute terms, these deepest macroalgae obtain irradiance levels in the order of 0.3 μmol m^{-2} s^{-1} at 200 m water depth. This level may be in accordance with experiences from cultivation work in the laboratory, where one obtains growth in deep-water red algae at least at 1 μmol m^{-2} s^{-1}. The light minimum at which terrestrial unicellular

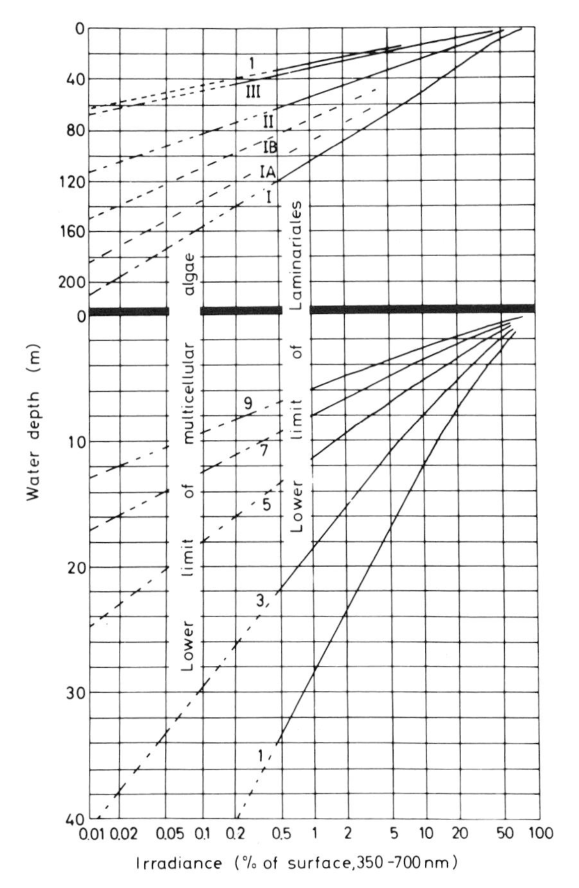

Fig. 2.8. Depth profiles of total irradiance (350–700 nm), and critical lower depth limits of marine algae in temperate regions. Top: oceanic water types I to III, and coastal water type 1. Bottom: coastal water types 1 to 9. (After Jerlov, 1976; from Lüning and Dring, 1979.)

algae are found growing within caves is of the order of 0.1 μmol m^{-2} s^{-1} (Leclerc et al., 1983).

At 250 m water depth a light level of 0.001% of the surface light or a photon fluence rate of 0.03 μmol m^{-2} s^{-1} is reached in clearest water. As Bünning (1973) states, in moonlight one measures a value of 0.02 μmol m^{-2} s^{-1} in the tropics, or half this value in temperate regions (equivalent to an illuminance of 0.5 lux in the latter case). Exactly to this "moonlight level" has thinking about irradiance at the lower algal limit been pushed by Littler et al. (1985, 1986). These authors discovered, from a submersible, crustose-calcified corallines at 268 m in clear water near the Bahamas. Higher up, the depth records fitted phycologists' preconceptions better. The lower depth limit for the boring green alga *Ostreobium* was at 210 m (roughly 0.01% of surface light), for the uncalcified red alga *Peyssonnelia* at 189 m, and for the deepest upright alga, namely the green alga *Johnson-sea-linkia profunda*, at 157 m (0.1% surface light).

This example shows that increase in morphological diversity, that is, the change from a crustose to a small upright thallus, requires an increase of light input up to one hundred times (from 0.001 to 0.1% surface light). The kelp growth form requires increase by a further factor of ten (1% surface light). This can be inferred from Fig. 2.8, which summarizes data for temperate regions, where the deepest crustose algae are found at 0.05 to 0.1% of surface light. The deepest kelp individuals live at a depth of 100 m in the Mediterranean (*Laminaria rodriguezii*), or at 25 m on the coast of Brittany (*Laminaria hyperborea*). In both cases surface light is reduced to about 1% at the depths indicated, the difference in absolute depth being due to the clearer water in the Mediterranean (water type IA) compared to the coast of Brittany [water type III; see Lüning and Dring (1979) for further details].

Some seagrass species are well suited to live in dim light (*Posidonia oceanica, Zostera marina*) (Libes, 1986; Mazzella and Alberte, 1986). Undersaturation with light is reported from deeper waters, especially in winter and bad weather (Libes, 1986). By transplanting eelgrass from lower to deeper stations and vice versa, Dennison and Alberte (1986) proved that the lower limit of eelgrass distribution is determined by light. Not only irradiance is reduced in deeper water but also the daily period of compensating and saturating light intensity decreases. The survival of the plants is secured as long as the relation between the reduced photosynthetic capacity at greater depths and the plant respiration allows a positive carbon balance. In shallow waters growth of seagrasses can become light-limited through self-shading (Short, 1975, 1980; Dennison, 1979; Sand-Jensen and Borum, 1983). Photosynthesis increases with rising light intensity up to saturation, as in other plants.

A seasonal change in the photosynthesis–light relationship was observed in *Zostera marina*, so that the low light intensities in winter could be utilized effectively. Photoinhibition was measured in *Posidonia oceanica* in late spring and summer (Libes, 1986). Primary productivity was highest in the morning, and the photoinhibiting effect, beginning at mid-day, lasted till the end of daylight. It was not neutralized in the afternoon. The reactions of other seagrass species may be similar, or quite different as shown for *Zostera marina*. This species developed no photoinhibition under natural conditions (Mazzella and Alberte, 1986).

Benthic microalgae experience a very steep gradient in light intensity within the upper few millimetres of the sediment. Fine-grained sediments attenuate light more than coarse sand. One millimetre of moist mud reduces the surface light intensity to 2% (Perkins, 1963). While rinsed quartz sand is more transparent, Fenchel and Straarup (1971) showed that 1% of red and infrared light penetrates down to a depth of 4.8 mm. These authors also found that light penetrated undisturbed estuarine sediment down to 3 mm.

Turbid waters above the sediment may reduce the light available for photosynthesis considerably. In very turbid estuaries this may have the consequence that most of the photosynthetic activity takes place during periods of exposure (Colijn, 1982). In sublittoral areas of the temperate zone, light-saturated photosynthesis of benthic microalgae occurs down to a depth of approximately 5 m from April to September. Below 5 m, light is a limiting factor all the year round in a nontidal area in the Kattegat (Sundbäck and Jönsson, 1988). In tropical areas, a considerable primary productivity (60 g C m^{-2} yr^{-1}) is measured in water depths of 5 m down to 60 m (Plante-Cuny, 1984).

A quantification of the light level to which benthic microalgae can grow is given by an investigation of the ice algae in the Antarctic. Algal populations increase *in situ* at light levels lower than 0.4% of surface intensities (Garrison et al., 1986). These algae are shade-adapted. Shade adaptation could be confirmed for sublittoral benthic microalgae in the Baltic (Gargas, 1971), and is supposed to be a common phenomenon in benthic microalgae existing in dim light.

On tidal flats in the European Wadden Sea, benthic diatoms may be light-limited in winter when short periods of daylight, cloudy weather and turbid water conditions coincide (Admiraal, 1980). With seasonally increasing light intensity the benthic microalgae change their phase of adaptation towards higher light intensity (Gargas, 1971). Many *in situ* measurements of the seasonal trend of primary productivity of microphytobenthos have shown a statistical correlation between light intensity and increasing primary productivity (Pomeroy, 1959; Pamatmat, 1968; Gargas, 1970; Leach, 1970; Cadée and Hegeman, 1974; Asmus and Asmus, 1985). Although light is a very important environmental factor in primary productivity some authors did not find this relationship with *in situ* measurements (Colijn and De Jonge, 1984; Varela and Penas, 1985). In this case other factors, as for instance very high disturbance of the sediment (Varela and Penas, 1985), could mask this basic relationship.

Benthic microalgae can generally get on well with rapidly increasing light intensity as it usually occurs in tidal flats from high tide to low tide or in laboratory experiments. Benthic intertidal diatoms are not very sensitive to high light intensities and do not tend to show strong photoinhibition (Taylor, 1964; Gargas, 1971; Cadée and Hegeman, 1974; Colijn and Van Buurt, 1975; Admiraal, 1980; Rasmussen et al., 1983). However, there are some indications of photoinhibition if algal cultures (of *Amphiprora alata*) are not adapted to high light intensities (Colijn and Van Buurt, 1975). Until now, differences between species in their reactions to changes in the light climate and to high light intensities have seldom been investigated. Most investigations are concerned with mixed populations with *in situ* experiments. In one investigation, strong photoinhibition was found in exposed populations of benthic microalgae under full sunlight in a salt-marsh in Georgia, U.S.A.

(Whitney and Darley, 1983). In this study, the authors succeeded in measuring primary productivity of benthic microalgae in air, whereas all other authors used smaller or larger water-filled incubation chambers.

Most studies were conducted in cold- and warm-temperate regions; there are no data available concerning photoinhibition of benthic microalgae in tropical areas. Primary productivity of benthic microalgae is high in tropical areas. Plante-Cuny (1984) showed that primary productivity values decrease with increasing latitude or water depth.

Light as an indicator of season

A considerable number of seaweed species with seasonally occurring generations or phases, or distinct seasonal growth cycles, use daylength as an indicator of season [reviews on photoperiodism in seaweeds: Dring (1984) and Dring and Lüning (1983)]. As for higher plants, daylength is more reliable in this respect than the seasonal course of temperature, which is nevertheless used by other seaweed species for the same purpose (see above).

The kelp *Laminaria hyperborea*, which dominates the mid-sublittoral zone from northern Norway to mid-Portugal, offers an example. A new frond is formed in the field from December to June, but in the laboratory only in short days, up to daylengths of 12 h light per day, or in continuous darkness. Long days prevent induction and growth of the new frond, and this result is also obtained if one cultivates experimental individuals at 8 h light per day, but with 1 h of light in the middle of the long night ("night-break regime"). This result indicates the presence of a genuine photoperiodic response, and shows that long nights are required for induction of the new frond. The environmental signal "short day = long night" is obtained by *L. hyperborea* in late autumn, so that the new frond can be formed during winter well ahead of the time, when it is finally required to serve as the "light antenna of the new year" in summer, when the bulk of the yearly light supply arrives in deeper water.

A total of 37 photoperiodic responses in seaweeds have been listed by Dring (1984), all discovered since 1965. Figure 2.9 shows four examples of short-day reactions in some detail. Basically, cryptic, oversummering stages or generations receive the autumnal short-day signal, so that the macroscopic stage or generation is initiated to be formed in winter and early spring.

WATER MOTION AND OTHER PHYSICAL FACTORS

Water motion and its vertical aspect

Wave action or water motion is a complex factor, and difficult to measure, which also applies to its biological effects (see reviews by: Riedl, 1971; Dring, 1982; Wheeler and Neushul, 1981; Norton et al., 1982; Hiscock, 1983; Lobban et al., 1985). To cite some extreme values from Riedl (1971), maximum current speeds in shallow or narrow coastal areas may amount to 5 m s^{-1}, particle speed in rocky surf areas to 15 m s^{-1}, depth ranges of waves to 500 m, surf pressures to 100 t m^{-2}, wave heights to 45 m, and spray heights to 50 m.

In the vertical dimension on the shore, one may distinguish three bathymetric zones in regard to qualitative and quantitative changes in water movement (Riedl, 1964, 1971): (1) a surf zone (down to 2 or 4 m water depth) with omnidirectional particle movement and high wave pressures acting from all directions on the benthic organism; (2) an oscillating zone (down to 10 or 20 m depth) with a reduced, mainly two-directional mechanical stress; and (3) deeper down, a flowing zone with the main flow moving parallel to the coast and with slight mechanical stress on the organisms. Barnacles and calcified corals show the characteristic adaptations to zone (1), the arrangement of planar passive filter-feeders parallel to the coast is typical for the oscillating zone (2), and the arrangement of such filter-feeders at right angles to the coast for the flowing zone (3) in unidimensionally flowing water, or the occurrence of radial growth forms in two-dimensionally flowing water, again in the flowing zone (3) (Riedl, 1971). Such relationships for seaweeds have still to be worked out by phycologists.

Tides

Most shore areas of the world, except for large inner areas of the Baltic, Mediterranean, Black and Caspian Seas, experience tides. Semidiurnal tides that have two high and two low waters of almost

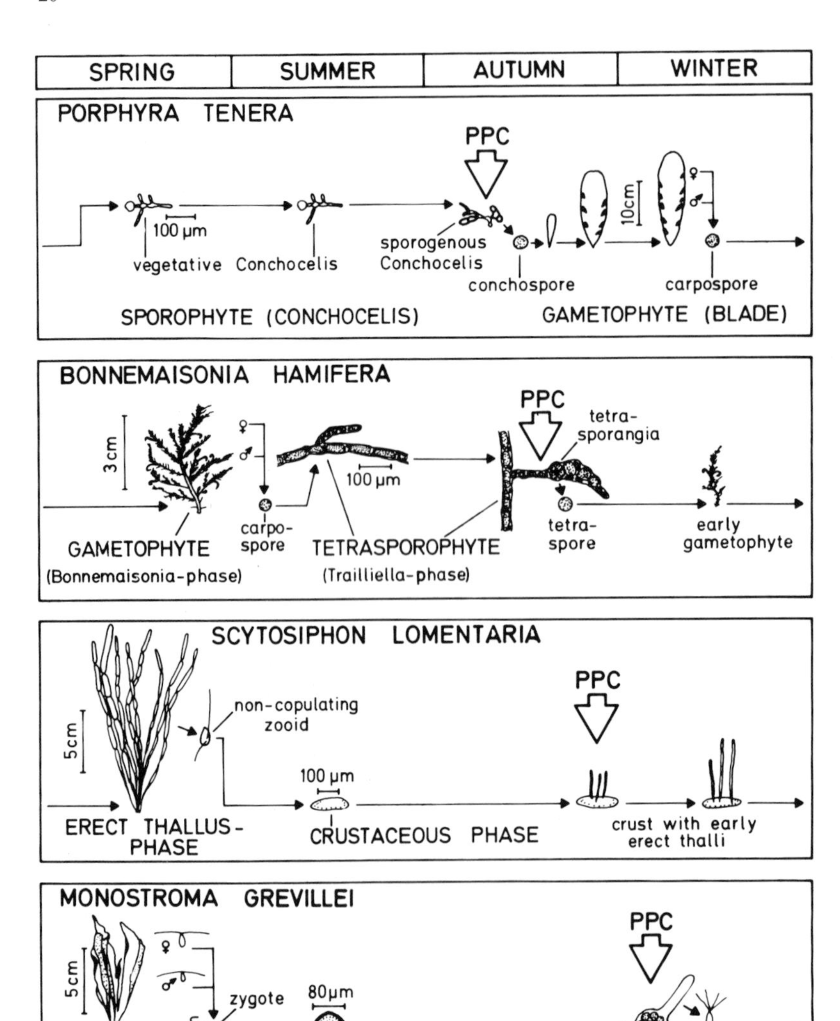

Fig. 2.9. Seasonal development in four short-day algae. PPC : photoperiodic control signal. Photoperiodic reactions: Red alga, *Porphyra tenera*: formation of conchosporangia by the microscopic Conchocelis-phase, leading to the macroscopic *Porphyra*-phase. Red alga, *Bonnemaisonia hamifera*: formation of tetrasporangia by the filamentous, prostrate Trailliella-phase, leading to the macroscopic, upright *Bonnemaisonia*-phase. Brown alga, *Scytosiphon lomentaria*: vegetative formation of upright thalli from the prostrate crust. Green alga, *Monostroma grevillei*: formation of zoospores in the microscopic Codiolum-phase, leading to the macroscopic, upright *Monostroma*-phase. (From Dring and Lüning, 1983.)

equal height per lunar day are well-known from the Atlantic. The periodic exposure to air has led to a vertical zonation in which the upper laminarians may be regarded as biologically characterizing the upper limit of the sublittoral zone in the North Atlantic (Lewis, 1964). Biologically important "critical tide levels", which occur as crests and troughs in daily or long-term tidal cycles, indicate predictable levels at which the duration of exposure or submergence markedly increases (Doty, 1946; Swinbanks, 1982).

On other shores, and well-known for marine biologists from the North American coast of the Pacific, mixed semidiurnal tides occur, with the two daily low waters being considerably unequal in height (and correspondingly the two high waters).

In consequence, the highest-growing representatives of the Laminariales are exposed to the air only once per day. Hence, there are species of the Laminariales growing "in the lower intertidal" along the North American coast of the Pacific (see Abbott and Hollenberg, 1976), although the order is strictly sublittoral in the North Atlantic.

The periodic exposure of tidal flats inhabited by benthic microalgae has the positive effect of a better light supply at low tide. At the same time several factors inhibiting the primary productivity of benthic microalgae are connected with periods of exposure. Salinity may vary strongly, the pH increases with increasing primary productivity leading to carbon dioxide limitation, and desiccation is an additional stress factor (Rasmussen et al., 1983). However, many species of benthic microalgae have evolved strategies to resist these harmful effects, and grow intensively on tidal flats.

During submersion benthic microalgae benefit by transport processes from tidal water movement. Substances necessary for primary productivity such as nutrients and gaseous carbon dioxide can easily reach benthic microalgae, while other substances which may be detrimental to the algae like sulfide and other metabolic products from bacteria, algae and animals are diluted and removed from the system.

Diatom cells living in mucus films are not as easily resuspended as those without microbial binding (Grant et al., 1986). Vertical migration of several species is another factor reducing the loss of cells due to resuspension. At times, resuspension is a very important ecological process (Grøntved, 1962; Varela and Penas, 1985; De Jonge, 1985; Lukatelich and McComb, 1986). Losses of cells due to resuspension are greatest for unattached or loosely attached diatoms, and they are greater for diatoms attached to mud particles than for those attached to sand grains (De Jonge, 1985). However, resuspension may also have a positive effect on the community of benthic microalgae. The competition for space will be reduced when some microalgae are harvested from time to time.

Finally, it must be added that due to characteristics of the various ocean basins diurnal tides are also possible, with a single low and high water per day, for instance in the Gulf of Mexico. Other aspects of tides and tidal currents have been treated in more detail in this series by Ketchum (1983).

Horizontal gradients of wave action

The drastic changes of algal vegetation and faunal composition along a horizontal gradient have long been recognized. Exposed and sheltered shores are inhabited by completely different biota, as illustrated by numerous examples in the present Volume.

One of the first authors who thoroughly described the impact of wave action on the composition of the seaweed vegetation was Börgesen (1903–1908), in his famous work on the algal vegetation of the Faeröes. On exposed, rocky coasts, or at the entrance of fjords, laminarian species with wave-resistant and wave-dependent thalli prevail, such as *Alaria esculenta* and *Laminaria digitata*. The total frond area may be ten times as great as the ground area ("thallus area index" = 10), and only the continuous movement of the thalli ensures that each thallus may momentarily capture light and readily exchange gases and ions with the water.

Survival at locations with strong wave action is accomplished by different strategies. *Durvillaea antarctica*, a belt-forming massive brown alga in central and southern Chile, is pulled and deformed by the moving water, since it has an elastic stipe and extensible blade (Koehl, 1982). In contrast, *Lessonia nigrescens*, a companion seaweed, has a stiff stipe and bends with the flow. For firm fixation of seaweeds on the substrate, the texture of the substrate is important, since smooth substrata cannot be colonized by larger seaweeds (see review by Hartnoll, 1983).

The beneficial effects of wave action are absent in the still water of interior parts of a fjord. Here, species dominate which by morphological adaptations make use of minimal turbulence in the water in order to lift the thallus from the bottom. An example is provided by *Laminaria saccharina* with its curled thallus margins, each curl working like a ship's propeller to move the thallus and to protect the vegetation from death caused by settling of the thalli too thickly on the bottom.

Sediment stability

On exposed beaches wave action may lead to mixing, redistribution and sorting of particle size of the sand, and may thus inhibit the formation of surface mats of algae (Steele et al., 1970). Primary productivity was extremely low in this locality ($5 \, g \, C \, m^{-2} \, yr^{-1}$: Steele and Baird, 1968). The vertical distribution of algae in the sediment is strongly determined by water movement, not only by the light gradient. In calm weather a high chlorophyll concentration in the surface layer, a steep vertical gradient and an accumulation of degradation products at slightly greater depth were found (Plante et al., 1986). Rough weather led to mixing of the sediment, and a slow decline in chlorophyll *a* concentration with depth (Plante et al., 1986).

The associations of benthic diatoms living in moderate exposed localities differ from those of sheltered areas. Exposure to water movement restricts the amount of protruding life-forms (Sundbäck, 1984). Epipsammic (attached to sediment) species are more dominant in exposed localities, whereas epipelic (not attached to sediment) species need a certain degree of shelter in order to build up a large standing stock (Taasen and Høisaeter, 1981). Wave-induced sediment instability can lead to a homogeneous epipsammic community of benthic diatoms (Amspoker, 1977).

In calm weather benthic microalgae may accumulate in the troughs or on the slopes of ripple marks. Ripple marks, however, often are a very unstable environment. Ripple marks with a wave length of 3 to 10 cm were found only to last some hours in one example from the Mediterranean (Plante et al., 1986). Thus a small-scale patchiness was present, but it was not correlated with the actual ripple marks (Plante et al., 1986). A patchy accumulation of plant material may persist even when the ripple-mark profiles are erased (Hogue and Miller, 1981). A film of benthic diatoms can be affected by a migrating ripple mark in different ways; it may be eroded, buried or exposed (Grant et al., 1986). However, benthic diatoms are not only passively affected by sediment instability, they can stabilize the sand by producing mucus films.

Boundary layer and turbulence

The exchange of ions and gas molecules is impeded by the boundary layer, that is, the layer of slowly moving fluid surrounding the alga surfaces (Koehl, 1982). Where there is little water motion, the boundary layer has a thickness of a few millimetres. Crustaceous algae and algal sporelings live within the boundary layer. Water turbulence decreases the thickness of the boundary layer and speeds up the rate of ion and gas exchange. Increasing the current velocity from 0 to $4 \, cm \, s^{-1}$ results in an increase of photosynthesis by a factor of 3, and that of nitrate uptake by 5 (Gerard, 1982; Wheeler, 1982). Much of the morphological diversity in seaweeds serves the purpose of creating turbulence and enhancing gas and ion exchange in slowly flowing water; for instance, the formation of spines, hairs and protuberances (see Norton et al., 1982).

In seagrass meadows, current flow may enhance nutrient uptake at the leaf surface too by reducing the diffusion boundary layer and presenting more nutrients to the leaf (Fonseca and Kenworthy, 1987). The current speed of waters flowing over a seagrass meadow will be reduced by the seagrass leaves, and thus sedimentation of organic and inorganic particles is increased in seagrass meadows in comparison with unvegetated sediments (Short and Short, 1984). Upper limits of current speed beyond which the meadows of different seagrass species decrease are 120 to 150 $cm \, s^{-1}$ (Scoffin, 1970; Fonseca et al., 1983).

Desiccation tolerance

The highest fucalean species growing on European shores, i.e. *Pelvetia canaliculata*, may survive an emersion period of 4 to 6 days, the deeper-growing *Fucus spiralis* 1 to 2 days, and the bathymetrically following species, *Ascophyllum nodosum, F. vesiculosus, and F. serratus*, only an emersion period in the range of hours. It is the critical water content, to which fucalean species may be dried and retain their full photosynthetic rate after re-immersion, which decides, as a physical factor, to which upper level a seaweed species may grow on the shore (Dring, 1982). Thus, the water content may be reduced to 4% in *Pelvetia canaliculata*, to 10% in *Fucus spiralis*, to 30% in *F. vesiculosus*, and still the photosynthetic rate 2 h after return in water recovers to 100% of its normal value in the submerged state. In the uppermost representative of the Laminariales in the upper sublittoral zone on European shores, that is, *Laminaria digitata*, the value for this critical water content is as high as 45%. In other words, if particular individuals lose too much water, their photosynthesis will never recover to normal values, and these individuals will be irreversibly damaged to death, as demonstrated by Schramm (1968) for Baltic *F. vesiculosus*.

Microbenthic algae too are endangered by desiccation (Wulff and McIntire, 1972). Differences between species of benthic diatoms in the resistance against desiccation could be revealed (Moore and McIntire, 1977).

The differential tolerances of intertidal algal species in regard to recovery of their metabolism after desiccation illustrate the commonly accepted rule that in the eulittoral zone physical factors often set the upper limits of occurrence for a particular species, whereas the lower limits are determined by biological competition (for examples see Chs. 5 and 6).

REFERENCES

Abbott, I.A. and Hollenberg, G.J., 1976. *Marine Algae of California*. Stanford University Press, Stanford, Calif., 827 pp.

Admiraal, W., 1977. Influence of light and temperature on the growth rate of estuarine benthic diatoms in culture. *Mar. Biol.*, 39: 1–9.

Admiraal, W., 1980. Experiments on the ecology of benthic diatoms in the Eems–Dollard estuary. *BOEDE, Publicaties en Verslagen*, 3: 1–125.

Admiraal, W. and Peletier, H., 1980. Influence of seasonal variations of temperature and light on the growth rate of cultures and natural populations of intertidal diatoms. *Mar. Ecol. Prog. Ser.*, 2: 35–43.

Akiyama, K., 1965. Studies of ecology and culture of *Undaria pinnatifida* (Harv.) Sur. II. Environmental factors affecting the growth and maturation of gametophyte. *Bull. Tohoku Reg. Fish. Res. Lab.*, 25: 143–170.

Amspoker, M.C., 1977. The distribution of intertidal epipsammic diatoms on Scripps Beach, La Jolla, Calif., USA. *Bot. Mar.*, 20: 227–232.

Antoine, S.E. and Benson-Evans, K., 1983. The effect of light intensity and quality on the growth of benthic algae. *Arch. Hydrobiol.*, 99: 118–128.

Asmus, H. and Asmus, R., 1985. The importance of grazing food chain for energy flow and production in three intertidal sand bottom communities of the northern Wadden Sea. *Helgol Wiss. Meeresunters.*, 39: 273–301.

Bachmann, P., Kornmann, P. and Zetsche, K., 1976. Regulation der Entwicklung und des Stoffwechsels der Grünalge *Urospora* durch die Temperatur. *Planta*, 128: 241–245.

Bayne, B.L., 1985. Responses to environmental stress: tolerance, resistance and adaptation. In: J.S. Gray and M.E. Christiansen (Editors), *Marine Biology of Polar Regions and Effects of Stress on Marine Organisms*. Wiley, Chichester, pp. 331–349.

Berry, J. and Björkman, O., 1980. Photosynthetic response and adaptation to temperature in higher plants. *Ann. Rev. Plant Physiol.*, 31: 491–543.

Berry, J.A. and Raison, J.K., 1981. Responses of macrophytes to temperature. In: O.L. Lange, P.S. Nobel, C.B. Osmond and H. Ziegler (Editors), *Encyclopedia of Plant Physiology, New Series, Vol. 12 A*. Springer, Berlin, pp. 277–338.

Biebl, R., 1958. Temperatur- und osmotische Resistenz von Meeresalgen der bretonischen Küste. *Protoplasma*, 50: 217–242.

Biebl, R., 1962. Temperaturresistenz tropischer Meeresalgen. (Verglichen mit jener von Algen in temperierten Meeresgebieten). *Bot. Mar.*, 4: 241–254.

Biebl, R., 1968. Über Wärmehaushalt und Temperaturresistenz arktischer Pflanzen in Westgrönland. *Flora, Abt. B*, 157: 327–354.

Bolton, J.J., 1983. Ecoclinal variation in *Ectocarpus siliculosus* (Phaeophyceae) with respect to temperature growth optima and survival limits. *Mar. Biol.*, 73: 131–138.

Bolton, J.J. and Lüning, K., 1982. Optimal growth and maximal survival temperatures of Atlantic *Laminaria* species (Phaeophyta) in culture. *Mar. Biol.*, 66: 89–94.

Börgesen, F., 1903–1908. The algae-vegetation of the Faeröese coasts with remarks on the phyto-geography. In: *Botany of the Faeröes based upon Danish investigations*. Det Nordiske Forlag, Kopenhagen, pp. 683–834.

Branch, M.L., 1974. Limiting factors for the gametophytes of three South African Laminariales. *Investigational Report Sea Fisheries Branch South Africa*, 104: 1–38.

Briggs, J.C., 1974. *Marine Zoogeography*. McGraw-Hill, New York, 475 pp.

Brown, J.H. and Gibson, A.C., 1983. *Biogeography*. Mosby Company, St. Louis, 643 pp.

Bulthuis, D.A., 1987. Effects of temperature on photosynthesis and growth of seagrasses. *Aquat. Bot.*, 27: 27–40.

Bünning, E., 1973. *The Physiological Clock*. Springer, Berlin.

Cadée, G.C. and Hegeman, J., 1974. Primary production of the benthic microflora living on tidal flats in the Dutch Wadden Sea. *Neth. J. Sea Res.*, 8: 260–291.

Castenholz, R.W., 1967. Seasonal ecology of non-planktonic marine diatoms on the western coast of Norway. *Sarsia*, 29: 237–256.

Colijn, F., 1982. Light absorption in the waters of the Ems–Dollard estuary and its consequences for the growth of phytoplankton and microphytobenthos. *Neth. J. Sea Res.*, 15: 196–216.

Colijn, F. and De Jonge, V.N., 1984. Primary production of microphytobenthos in the Eems–Dollard estuary. *Mar. Ecol. Prog. Ser.*, 14: 185–196.

Colijn, F. and Van Buurt, G., 1975. Influence of light and temperature on the photosynthetic rate of marine benthic diatoms. *Mar. Biol.*, 31: 209–214.

Davis, M.W. and McIntire, C.D., 1983. Effects of physical gradients on the production dynamics of sediment-associated algae. *Mar. Ecol. Prog. Ser.*, 13: 103–114.

Dayton, P.K., Watson, D., Palmisano, A., Barry, J.P., Oliver, J.S. and Rivera, D., 1986. Distribution patterns of benthic microalgal standing stock at McMurdo Sound, Antarctica. *Polar Biol.*, 6: 207–213.

De Jonge, V.N., 1985. The occurrence of "epipsammic" diatom

populations: a result of interaction between physical sorting of sediment and certain properties of diatoms species. *Estuar. Coast. Shelf Sci.*, 21: 607–622.

Delépine, R. and Asensi, A., 1976. Quelques données expérimentales sur l'écophysiologie de *Durvillaea antarctica* (Cham.) Hariot (Phéophycées). *Bull. Soc. Phycol. France*, 21: 65–80.

Den Hartog, C., 1970. *The Seagrasses of the World.* Verh. K. Ned. Ak. Wet. Afd. Natuurkd., 59(1): 1–275.

Dennison, W.C., 1979. *Light adaptions of plants: A model based on the seagrass Zostera marina* L. M.S. thesis, University of Alaska, Fairbanks, 70 pp.

Dennison, W.C. and Alberte, R.S., 1986. Photoadaptation and growth of *Zostera marina* L. (eelgrass) transplants along a depth gradient. *J. Exp. Mar. Biol. Ecol.*, 98: 265–282.

De Oliveira, E.F., Pirani, J.R. and Giulietti, A.M., 1983. The Brazilian seagrasses. *Aquat. Bot.*, 16: 251–267.

Doty, M.S., 1946. Critical tide factors that are correlated with the vertical distribution of marine algae and other organisms along the Pacific coast. *Ecology*, 27: 315–328.

Drew, E.A., 1977. The physiology of photosynthesis and respiration in some Antarctic marine algae. *Br. Antarct. Surv. Bull.*, 46: 59–76.

Drew, E.A., 1983. Light. In: R. Earll, and D.G. Erwin (Editors), *Sublittoral Ecology. The Ecology of the Shallow Sublittoral Benthos.* Clarendon Press, Oxford, pp. 10–57.

Dring, M.J., 1982. *The Biology of Marine Plants.* Edward Arnold, London, 199 pp.

Dring, M.J., 1984. Photoperiodism and phycology. *Progr. Phycol. Res.*, 3: 159–162.

Dring, M.J. and Lüning, K., 1983. Photomorphogenesis of marine macroalgae. In: W. Shropshire and H. Mohr (Editors), *Encyclopedia of Plant Physiology, New Series, Vol. 16B.* Springer, Berlin, pp. 545–568.

Durbin, E.G., 1974. Studies on the autecology of the marine diatom *Thalassiosira nordenskiöldii* Cleve. I. The influence of daylength, light intensity, and temperature on growth. *J. Phycol.*, 10: 220–225.

Ekman, S., 1953. *Zoogeography of the Sea.* Sidgwick and Jackson, London, 417 pp.

Engelmann, T.W., 1883. Farbe und Assimilation. *Bot. Ztg.*, 41: 1–13, 17–29.

Fenchel, T. and Straarup, B.J., 1971. Vertical distribution of photosynthetic pigments and the penetration of light in marine sediments. *Oikos*, 22: 172–268.

Fonseca, M. and Kenworthy, W.J., 1987. Effects of current on photosynthesis and distribution of seagrasses. *Aquat. Bot.*, 27: 59–78.

Fonseca, M.S., Zieman, J.C., Thayer, G.W. and Fisher, J.S., 1983. The role of current velocity in structuring seagrass meadows. *Estuar. Coast. Shelf Sci.*, 17: 367–380.

Frakes, L.A., 1979. *Climates Throughout Geologic Time.* Elsevier, Amsterdam, 310 pp.

Gargas, E., 1970. Measurements of primary production, dark fixation and vertical distribution of the microbenthic algae in the Øresund. *Ophelia*, 8: 231–253.

Gargas, E., 1971. "Sun–shade" adaptation in microbenthic algae from the Øresund. *Ophelia*, 9: 107–112.

Garrison, D.L., Sullivan, C.W. and Ackley, S.F., 1986. Sea ice microbial communities in Antarctica. *Biol. Sci.*, 36: 243–250.

Gerard, V.A., 1982. In situ water motion and nutrient uptake by the giant kelp *Macrocystis pyrifera. Mar. Biol.* 69: 51–54.

Gessner, F., 1970. Temperature: plants. In: O. Kinne (Editor), *Marine Ecology, Vol. I, Part 1. Environmental Factors.* Wiley, Chichester, pp. 363–406.

Grant, J., Bathmann, U.V. and Mills, E.L., 1986. The interaction between benthic diatom films and sediment transport. *Estuar. Coast. Shelf Sci.*, 23: 225–238.

Grant-Mackie, J.A., 1979. Cretaceous–Recent plate tectonic history and paleoceanographic development of the Southern Hemisphere. In: *Proc. Int. Symp. on Marine Biogeography and Evolution in the Southern Hemisphere, Auckland, New Zealand, July 1978. N.Z. DSIR Information Series 137, Vol. 1,* pp. 27–42.

Grøntved, J., 1962. Preliminary report on the productivity of microbenthos and phytoplankton in the Danish Wadden Sea. *Medd. Dan. Fisk. Havunders.*, 3: 347–378.

Hartnoll, R.G., 1983. Substratum. In: R. Earll and D.G. Erwin (Editors), *Sublittoral Ecology. The Ecology of the Shallow Sublittoral Benthos.* Clarendon Press, Oxford, pp. 97–124.

Healey, F.P., 1972. Photosynthesis and respiration of some Arctic seaweeds. *Phycologia*, 11: 267–271.

Hedgpeth, J.W., 1957. Marine biogeography. *Geol. Soc. Amer. Mem.*, 67: 359–382.

Hempel, G., 1985. On the biology of polar seas, particularly the Southern Ocean. In: J.S. Gray and M.E. Christiansen (Editors), *Marine Biology of Polar Regions and Effects of Stress on Marine Organisms.* Wiley, Chichester, pp. 3–33.

Herman, Y. and Hopkins, D.M., 1980. Arctic Oceanic climate in Late Cenozoic time. *Science*, 209: 557–562.

Hiscock, K., 1983. Water movement. In: R. Earll and D.G. Erwin (Editors), *Sublittoral Ecology. The Ecology of the Shallow Sublittoral Benthos.* Clarendon Press, Oxford, pp. 58–95.

Hogue, E.W. and Miller, C.B., 1981. Effects of sediment microtopography on small-scale distributions of meiobenthic nematodes. *J. Exp. Mar. Biol. Ecol.*, 53: 181–191.

Hutchins, L.W., 1947. The bases for temperature zonation in geographical distribution. *Ecol. Monogr.*, 17: 325–335.

Jerlov, N.G., 1976. *Marine Optics.* Elsevier, Amsterdam, 231 pp.

Jerlov, N.G., 1978. The optical classification of sea water in the euphotic zone. *Rep. Kjob. Univ. Inst. Fys. Oceanogr.*, 36: 1–46.

Joosten, A.M.T. and Van den Hoek, C., 1986. World-wide relationships between red seaweed floras. *Bot. Mar.*, 29: 195–214.

Kennett, D.M. and Hargraves, P.E., 1984. Subtidal benthic diatoms from a stratified estuarine basin. *Bot. Mar.*, 27: 169–183.

Ketchum, B.H. (Editor), 1983. *Estuaries and Enclosed Seas.* Elsevier, Amsterdam, 500 pp.

Kirk, J.T.O., 1983. *Light and Photosynthesis in Aquatic Ecosystems.* University Press, Cambridge, 401 pp.

Koehl, M.A.R., 1982. The interaction of moving water and sessile organisms. *Sci. Am.*, 247: 124–132.

Leach, J.H., 1970. Epibenthic algal production in an intertidal mudflat. *Limnol. Oceanogr.,*, 15: 514–521.

Leclerc, J.C., Couté, A. and Dupuy, P., 1983. Le climat annuel de deux grottes et d'une église du Poitou, ou vivent des colonies pures d'algues sciaphiles. *Cryptogamie, Algologie*, 4: 1–19.

Levinton, J.S., 1982. *Marine Ecology*. Prentice-Hall, Englewood Cliffs, N.J., 526 pp.

Lewis, J.R., 1964. *The Ecology of Rocky Shores*. English Universities Press, London, 323 pp.

Libes, M., 1986. Productivity–irradiance relationship of *Posidonia oceanica* and its epiphytes. *Aquat. Bot.*, 26. 285–306.

Littler, M.M., Littler, D.S., Blair, S.M. and Norris, J.N., 1985. Deepest known plant life discovered on an uncharted seamount. *Science*, 227: 57–69.

Littler, M.M., Littler, D.S., Blair, S.M. and Norris, J.N., 1986. Deep-water plant communities from an uncharted seamount off San Salvador Island, Bahamas: distribution, abundance, and primary productivity. *Deep-Sea Res.*, 33: 881–892.

Lobban, C.S., Harrison, P.J. and Duncan, M.J., 1985. *The Physiological Ecology of Seaweeds*. University Press, Cambridge, 242 pp.

Lukatelich, R.J. and McComb, A.J., 1986. Distribution and abundance of benthic microalgae in a shallow southwestern Australian estuarine system. *Mar. Ecol. Prog. Ser.*, 27: 287–297.

Lüning, K., 1980. Control of algal life-history by day length and temperature. In: J.H. Price, D.E.G. Irvine and W.F. Farnham (Editors), *The Shore Environment, Vol. 2. Ecosystems*. Academic Press, London, pp. 915–945.

Lüning, K., 1984. Temperature resistance and biogeography of seaweeds: the marine algal flora of Helgoland, North Sea, as an example. *Helgol. Wiss. Meeresunters.*, 38: 305–317.

Lüning, K., 1985. *Meeresbotanik. Verbreitung, Ökophysiologie und Nutzung der marinen Makroalgen*. Georg Thieme Verlag, Stuttgart, New York, 375 pp.

Lüning, K. and Dring, M.J., 1979. Continuous underwater light measurement near Helgoland (North Sea) and its significance for characteristic light limits in the sublittoral region. *Helgol. Wiss. Meeresunters.*, 32: 403–424.

MacArthur, R.H. and Connell, J.H., 1966. *The Biology of Populations*. Wiley, New York, 200 pp.

Marsh, J.A., Dennison, W.C. and Alberte, R.S., 1986. Effects of temperature on photosynthesis and respiration in eelgrass (*Zostera marina* L.). *J. Exp. Mar. Biol. Ecol.*, 101: 257–267.

Matheke, G.E.M. and Horner, R., 1974. Primary productivity of the benthic microalgae in the Chukchi Sea near Barrow, Alaska. *J. Fish. Res. Board. Can.*, 31: 1779–1786.

Mazzella, L. and Alberte, R.S., 1986. Light adaptation and the role of autotrophic epiphytes in primary production of the temperate seagrass, *Zostera marina* L. *J. Exp. Mar. Biol. Ecol.*, 100: 165–180.

McLachlan, J. and Bird, C.J., 1984. Geographical and experimental assessment of the distribution of species of *Gracilaria* in relation to temperature. *Helgol. Wiss. Meeresunters.*, 38: 319–334.

McMillan, C., 1984. The distribution of tropical seagrasses with relation to their tolerance of high temperatures. *Aquat. Bot.*, 19: 369–379.

Michanek, G., 1979. Phytogeographic provinces and seaweed distribution. *Bot. Mar.*, 22: 375–391.

Michanek, G., 1983. World resources of marine plants. In: O. Kinne (Editor), *Marine Ecology, Vol. V, Part 2*. Wiley, New York, pp. 795–837.

Moore, W.W. and McIntire, C.D., 1977. Spatial and seasonal distribution of littoral diatoms in Yaquina estuary, Oregon (U.S.A.). *Bot. Mar.*, 20: 99–109.

Müller, D.G., 1962. Über jahres- und lunarperiodische Erscheinungen bei einigen Braunalgen. *Bot. Mar.*, 4: 140–155.

Newell, R.C. (Editor), 1976. *Adaptation to the Environment: Essays on the Physiology of Marine Animals*. Butterworths, London, 539 pp.

Norton, T.A., 1977. Experiments on the factors influencing the geographical distributions of *Saccorhiza polyschides* and *Saccorhiza dermatodea*. *New Phytol.*, 78: 625–635.

Norton, T.A., Mathieson, A.C. and Neushul, M., 1982. A review of some aspects of form and function in seaweeds. *Bot. Mar.*, 25: 501–510.

Novaczek, I., 1984. Response of gametophytes of *Ecklonia radiata* (Laminariales) to temperature in saturating light. *Mar. Biol.*, 82: 241–246.

Odum, E.P., 1980. *Grundlagen der Ökologie*. Thieme, Stuttgart, 836 pp.

Palmisano, A.C., Beeler Soo Hoo, J. and Sullivan, C.W., 1987. Effects of four environmental variables on photosynthesis–irradiance relationships in Antarctic sea-ice microalgae. *Mar. Biol.*, 94: 299–306.

Pamatmat, M.M., 1968. Ecology and metabolism of a benthic community on an intertidal sand flat. *Int. Rev. Gesamten Hydrobiol.*, 53: 211–298.

Perkins, E.J., 1963. Penetration of light into littoral soils. *J. Ecol.*, 51: 687–692.

Phillips, R.C., McMillan, C. and Bridges, K.W., 1983. Phenology of eelgrass, *Zostera marina* L., along latitudinal gradients on North America. *Aquat. Bot.*, 15: 145–146.

Picken, G.B., 1985. Benthic research in Antarctica: past, present and future. In: J.S. Gray and M.E. Christiansen (Editors), *Marine Biology of Polar Regions and Effects of Stress on Marine Organisms*. Wiley, Chichester, pp. 167–184.

Plante, R., Plante-Cuny, M.-R. and Reys, J.-P., 1986. Photosynthetic pigments of sandy sediments on the north Mediterranean coast: their spatial distribution and its effect on sampling strategies. *Mar. Ecol. Prog. Ser.*, 34: 133–141.

Plante-Cuny, M.-R., 1984. Le microphytobenthos et son rôle a l'échelon primaire dans le milieu marin. *Oceanis*, 10: 417–427.

Pomeroy, L.R., 1959. Algal productivity in salt marshes of Georgia. *Limnol. Oceanogr.*, 4: 386–397.

Ramus, J., 1981. The capture and transduction of light energy. In: C.S. Lobban and M.J. Wynne (Editors), *The Biology of Seaweeds*. Blackwell, Oxford, pp. 458–492.

Rasmussen, M.B., Henriksen, K. and Jensen, A., 1983. Possible causes of temporal fluctuations in primary production of the microphytobenthos in the Danish Wadden Sea. *Mar. Biol.*, 73: 109–114.

Riedl, R. 1964. Die Erscheinungen der Wasserbewegung und ihre Wirkung auf Sedentarier im mediterranen Felslitoral. *Helgol. Wiss. Meeresunters.*, 10: 155–186.

Riedl, R., 1971. Water movement. Animals. In: O. Kinne (Editor), *Marine Ecology, Vol. 1, Part 2*. Wiley–Interscience, London, pp. 1123–1156.

Sand-Jensen, K. and Borum, J., 1983. Regulation of growth of eelgrass (*Zostera marina* L.) in Danish coastal waters. *Mar. Technol. Soc. J.*, 17: 15–21.

Sauvageau, C., 1918. Recherches sur les laminaires des côtes de France. *Mém. Acad. Sci. Inst. Fr.*, 56: 1–240.

Savin, S.M., 1977. The history of the earth's surface temperature during the past 100 million years. *Ann. Rev. Earth Planet. Sci.*, 319–355.

Schopf, T.J.M., 1980. *Paleoceanography*. Harvard University Press, Cambridge, Mass., 341 pp.

Schramm, W., 1968. Ökologisch–physiologische Untersuchungen zur Austrocknungs- und Temperaturresistenz an *Fucus vesiculosus* L. der westlichen Ostsee. *Int. Rev. Gesamten Hydrobiol.*, 53: 469–510.

Scoffin, T.P., 1970. The trapping and binding of subtidal carbonate sediments by marine vegetation in Bimini Lagoon, Bahamas. *J. Sediment. Petrol.*, 40: 249–273.

Shackleton, N. and Boersma, A., 1981. The climate of the Eocene ocean. *J. Geol. Soc. London*, 138: 153–157.

Short, F.T., 1975. *Eelgrass production in Charlestown Pond: An ecological analysis and simulation model*. M.S. Thesis, University of Rhode Island, Kingston, 180 pp.

Short, F.T., 1980. A simulation model of the seagrass production system. In: R.C. Phillips and C.P. McRoy (Editors), *Handbook of Seagrass Biology: An Ecosystem Perspective*. Garland STPM Press, New York, pp. 275–295.

Short, F.T. and Short, C.A., 1984. The seagrass filter: Purification of estuarine and coastal waters. In: V.S. Kennedy (Editor), *The Estuary as a Filter*. Academic Press, New York, 395–413.

Smayda, T.J., 1969. Experimental observations on the influence of temperature, light, and salinity on cell division of the marine diatom *Detonula confervacea* (Cleve) Gran. *J. Phycol.*, 5: 150–157.

Steele, J.H. and Baird, T.E., 1968. Production ecology of a sandy beach. *Limnol. Oceanogr.*, 13: 14–25.

Steele, J.H., Munro, A.L.S. and Giese, G.S., 1970. Environmental factors controlling the epipsammic flora on beach and sublittoral sands. *J. Mar. Biol. Ass. U.K.*, 50: 907–918.

Stephenson, T.A., 1948. The constitution of the intertidal fauna and flora of South Africa. Part III. *Ann. Natal Mus.*, 11: 207–324.

Sundbäck, K., 1984. Distribution of microbenthic chlorophyll *a* and diatom species related to sediment characteristics. *Ophelia*, Suppl., 3: 229–246.

Sundbäck, K., 1986. What are the benthic microalgae doing on the bottom of Laholm Bay? *Ophelia*, Suppl., 4: 273–286.

Sundbäck, K. and Jönsson, B., 1988. Microphytobentic productivity and biomass in sublittoral sediments of a stratified bay, southeastern Kattegat. *J. Exp. Mar. Biol. Ecol.*, 122: 63–81.

Swinbanks, D.D., 1982. Intertidal exposure zones: a way to subdivide the shore. *J. Exp. Mar. Biol. Ecol.*, 62: 69–86.

Taasen, J.P. and Høisaeter, T., 1981. The shallow-water soft-bottom benthos in Lindaspollene, western Norway. 4. Benthic marine diatoms, seasonal density fluctuations. *Sarsia*, 66: 293–316.

Taylor, W. Roland, 1964. Light and photosynthesis in intertidal benthic diatoms. *Helgol. Wiss. Meeresunters.*, 10: 29–37.

Van den Hoek, C., 1975. Phytogeographic provinces along the coasts of the northern Atlantic Ocean. *Phycologia*, 14: 317–330.

Van den Hoek, C., 1982a. Phytogeographic distribution groups of benthic marine algae in the North Atlantic Ocean. A review of experimental evidence from life history studies. *Helgol. Wiss. Meeresunters.*, 35: 153–214.

Van den Hoek, C., 1982b. The distribution of benthic marine algae in relation to the temperature regulation of their life histories. *Biol. J. Linn. Soc.*, 18: 81–144.

Van den Hoek, C., 1984. World-wide longitudinal seaweed distribution patterns and their possible causes, as illustrated by the distribution of rhodophytan genera. *Helgol. Wiss. Meeresunters.*, 38: 227–257.

Varela, M. and Penas, E., 1985. Primary production of benthic microalgae in an intertidal sand flat of the Ria de Arosa, NW Spain. *Mar. Ecol. Prog. Ser.*, 25: 111–119.

Vermeij, G.J., 1978. *Biogeography and Adaptation. Patterns of Marine Life*. Harvard University Press, Cambridge, Mass., 332 pp.

Wheeler, W.N., 1982. Nitrogen nutrition of *Macrocystis*. In: L.M. Srivastava (Editor), *Synthetic and Degradative Processes in Marine Macrophytes*, Walter de Gruyter, Berlin, pp. 121–135.

Wheeler, W.N. and Neushul, M., 1981. The aquatic environment. In: O.L. Lange, P.S. Nobel, C.B. Osmond and H. Ziegler (Editors), *Encyclopedia of Plant Physiology, New Series, Vol. 12A*. Springer, Berlin, pp. 229–247.

Whitney, D.E. and Darley, W.M., 1983. Effect of light intensity upon salt marsh benthic microalgal photosynthesis. *Mar. Biol.*, 75: 249–252.

Wium-Andersen, S. and Borum, J., 1984. Biomass variation and autotrophic production of an epiphyte–macrophyte community in a coastal Danish area: I. Eelgrass (*Zostera marina* L.) biomass and net production. *Ophelia*, 23: 33–46.

Wulff, B.L. and McIntire, C.D., 1972. Laboratory studies of assemblages of attached estuarine diatoms. *Limnol. Oceanogr.*, 17: 200–214.

Yarish, C., Breeman, A.M. and Van den Hoek, C., 1986. Survival strategies and temperature responses belonging to different biogeographic distribution groups. *Bot. Mar.*, 24: 215–230.

Chapter 3

CHEMICAL CHARACTERISTICS OF MARINE LITTORAL ECOSYSTEMS

W. SCHRAMM

INTRODUCTION

Intertidal and littoral ecosystems as defined in this book comprise, over a world-wide geographical and climatic range, vast variation in their physical, chemical and biological properties (see, for instance, Odum et al., 1974; Mann, 1982). Because of this diversity of coastal ecosystems, the attempt to generalize their chemical characteristics and to classify them according to these characteristics, necessarily leads to the question of which factors or processes common to all types of systems primarily control chemical parameters.

Intertidal and littoral ecosystems are areas of transition between the terrestrial and oceanic environment and thus to a lesser or greater extent the recipients of terrigenous materials. Input from these sources as well as exchange with the bordering off-shore or oceanic systems are of great importance. Usually nearshore or coastal waters, in particular tidal areas, are sites of high energy inputs in the form of tidal or wind forced water motion, promoting lateral and vertical mixing, increased rates of particle flux and intensive resuspension of superficial sediments.

Intensive mixing will often increase nutrient levels and enhance primary production, favouring other biological activities that in turn modify and control to a great extent the chemical parameters of the water column as well as the bottom systems.

The relatively shallow water depths, together with usually intensive vertical mixing, allow a strong influence of benthic processes, both biological and non-biological. Benthic primary production is often dominating over other biological activities, as for example in seagrass meadows, kelp beds and algal mats, in salt marshes or mangroves. In contrast to deep oceanic systems with their high physical as well as chemical buffer capacities, pronounced temporal variations, either long- or short-term, periodical or non-periodical, are typical for most of the comparatively shallow coastal systems under consideration.

On the other hand, the above mentioned direct and intensive interactions between the water column and the sediment may become an effective buffer mechanism; for instance, through release or uptake of nutrients, deposition of pollutants etc. Finally, more than any other marine system, intertidal and littoral ecosystems are often exposed to human impact. Pollution, or eutrophication through discharge of industrial, agricultural or domestic wastes into rivers or directly into the sea, may drastically influence the chemical characteristics of coastal waters.

Within the limited frame of this general chapter, out of the large number of chemical components only the biologically most relevant key parameters will be discussed, such as salinity, the inorganic plant nutrients, dissolved and particulate organic matter of natural origin and the dissolved metabolic gases oxygen, carbon dioxide or hydrogen sulphide. Trace elements such as heavy metals will only be touched upon, for example in connection with chemical speciation or estuarine mixing processes.

It is important to realize that, because of the multiplicity of types of intertidal and littoral ecosystems, it is not possible to set up a rigid classification based on the chemical characteristics. An attempt will only be made to illustrate and to exemplify the influence of the above mentioned controlling factors and processes on selected chemical parameters in some contrasting types of intertidal and littoral systems. Using this approach, these systems may be grouped according to the following classificational concepts:

(1) Systems with pronounced temporal (diurnal, seasonal) variation in contrast to those with little temporal variation.
(2) High-energy systems with intensive mixing (tidal, wind-forced, thermal) in contrast to low-energy (stratified) systems.
(3) Open oceanic or estuarine (brackish), rocky or sandy shores in contrast with semi-enclosed estuaries and mud flats or compared to closed-off systems such as lagoons, fjords and land-locked coastal salt lakes of maritime origin (thalassohaline waters).
(4) Sandy or muddy soft-bottom systems without macrophytes in contrast with macrophyte-based systems such as seagrass meadows, kelp beds, algal turfs, salt marshes or mangroves.

Coral reefs and mangroves are the subjects of separate volumes within this series and thus will be largely excluded, except for cases where they interact with other adjacent intertidal or littoral systems as, for example, mangroves or coral reef flats may control the chemistry of lagoons (Mee, 1978). Estuaries and enclosed seas are also treated in a separate volume of the series; however, they will not be completely excluded as far as general processes of land run-off or estuarine mixing are concerned.

IMPORT AND EXPORT PROCESSES AND CHEMICAL CHARACTERISTICS

Besides processes, either biological or non-biological, within intertidal and littoral ecosystems, external sources or sinks may considerably influence the chemical characteristics. Riverine land run-off flowing worldwide to the oceans is estimated to carry approximately 9×10^9 t of suspended particulate matter and 4×10^9 t of dissolved materials annually (Wedepohl, 1984). Other fluxes are estimated to include 2×10^9 t yr^{-1} materials from polar ice and 6×10^8 t yr^{-1} atmospheric input into the ocean (Garrels and Mackenzie, 1971). Intertidal and littoral ecosystems occupying the transition zone between land and the oceans are naturally the main recipients of the major portion of these inputs.

In river water the major elements of seawater are usually present at much lower concentrations, while the biologically important nutrient elements nitrogen, phosphorus and silicon, certain trace elements and sometimes organic materials may be considerably more concentrated, very often as a result of human activities. Hence, especially in estuaries the mixing of fresh water and seawater produces a continuous spectrum of chemical and physio-chemical gradients, showing a complexity far greater than that of the contributing water bodies. For example, in 1980 shallow coastal waters of the North Sea received from land run-off total inputs of 127×10^3 t yr^{-1} phosphorus and 1100×10^3 t yr^{-1} nitrogen, of which only 18×10^3 t phosphorus and 273×10^3 t nitrogen were estimated to be natural loads, that is, originating from non-anthropogenic sources. Based on existing input figures, phosphorus and nitrogen concentrations have been computed to be 2 to 5 times higher in the nearshore areas of the North Sea compared to the offshore central parts (Anonymous, 1985). In estuaries the nutrient gradients may be even greater and pronounced spatial variations, sometimes over some orders of magnitude, are typical [see, for instance data given by Biggs and Cronin (1981) reproduced in Table 3.1].

Whereas import of dissolved organic matter through land run-off is probably small in proportion to the pool of dissolved organic carbon (DOC) on an oceanic scale, the contribution through great rivers, which are estimated to carry an average of 5 mg C l^{-1} to estuaries, can be significant (Williams, 1975). In samples from the Wadden Sea (The Netherlands), for example, the DOC content was inversely correlated with salinity, suggesting that the organic carbon may originate from riverine input (Duursma, 1961).

In some cases, the input of organic matter by rivers may even exceed total primary production of a system as, for example, has been shown for a tidal mud-flat estuary from Canada where annually almost 200 g DOC and 56 g particulate organic carbon (POC) were supplied per square metre of mud flat (Naiman and Sibert, 1979). Although the major portion of terrigenous inputs into coastal waters is carried by rivers, "diffuse" land run-off can be significant, particularly in lagoons, sheltered bays or fjords. Mee (1978), for example, observed in a lagoon (Mexico) that

TABLE 3.1

Nutrient gradients in selected estuaries. (From Biggs and Cronin, 1981.)

Estuary	S (‰)	NO_3^- (µmol l^{-1})	NH_4^+ (µmol l^{-1})	$Si(OH)_4$ (µmol l^{-1})	PO_4^{3-} (µmol l^{-1})
Delaware (USA)	0–30	150– 1	50 – 1	110– 7	2 –0.3
Zaire (Africa)	0–35	8– 0	0.5– 0	160– 0	1.2–0
Magdalena (S. America)	0–35	17– 0	–	225– 0	3 –0.2
Scheldt (Europe)	0–30	0–30	600 –40	230–10	40 –2
Potomac (USA)	0–10	110– 1	200 – 1	–	32 –0.2

the first rains of the rainy season caused flash-flooding and run-off from the bordering land providing an important source of inorganic nutrients. In salt marshes it could be shown that seepage of groundwater may carry substantial amounts of inorganic nutrients that are more than enough to sustain plant production in the marsh (Valiela et al., 1978; Teal et al., 1979).

Significant increase in inorganic nutrients has also been observed where nitrogen cycle waste products from bird or seal colonies are washed into the sea (Myrcha et al., 1985; Dawson et al., 1985). Dawson and co-workers considered also the possible effects of freshwater contribution from ice-melts, often carrying high loads of nutrients and particulate matter which may be important in the coastal systems of high latitudes, although little information is available until now. Coastal land erosion is another source of particulates and mineral elements, which can significantly alter the chemical characteristics of near-shore waters. For the shallow Kiel Bight (western Baltic), Gerlach (1986) estimated annual phosphorus inputs, as a result of cliff erosion, in the range of 250 t, which is about 15 to 20% of the annual riverine inputs or of the total amount of phosphorus dissolved in the Kiel Bight during winter.

Just like import from land, the input from offshore or deep water into intertidal and littoral systems affects the chemical characteristics. Local wind-induced upwelling is probably a regular feature in locations adjacent to deep water and where land winds prevail. In the fjords of Kiel Bight (Baltic), for example, driven by strong winds, colder and nutrient rich water, occasionally oxygen depleted and enriched in hydrogen sulfide, reaches the surface, indicating upwelling processes (Ehrhardt and Wenck, 1984). The effects of upwelling water under the lee of Marion Island (Antarctica) on the coastal hydrogeographical and chemical parameters as well as on plankton composition have recently been reported by Grindley and David (1985).

THE INFLUENCE OF NON-BIOLOGICAL PROCESSES ON CHEMICAL CHARACTERISTICS

A general characterization of the chemical parameters of marine intertidal and littoral systems is hardly possible without consideration of physical, physico-chemical or chemical factors and processes involved. Particularly, water motion determines to a great extent the hydrographical or chemical situation in coastal waters. In principle, two different types of water movement may be distinguished, either turbulent mixing or mass transport generated by tidal or wind forces and by density gradients. In intertidal and exposed littoral systems the tidal rise and fall, wave action or wind-forced currents usually produce intensive mixing of the entire water body, preventing pronounced physical or chemical gradients. On the other hand, low energy systems, such as sheltered bays, fjords or lagoons, are characterized by such gradients. Especially in higher latitudes, seasonal fluctuations of the physical parameters light, temperature, wind, precipitation or land run-off, and temporal as well as spatial variations of chemical parameters are typical.

Density stratification, as a result of vertical salinity or temperature gradients, is a common phenomenon in nearshore low-energy systems and may be of particular influence because vertical transport is reduced or sometimes entirely interrupted.

Haline stratification is known for all climatic zones where fresh water from rivers, glaciers or rain spreads over saltier water, or where saltier water intrudes into the bottom layers, while thermal stratification is typical for tropical coastal waters and for many polar or temperate nearshore waters during the summer months. Where evaporation is extreme, as in some tropical lagoons, even reverse temperature gradients may occur.

As thermal stabilization of the water masses is common to coastal low-energy systems, convective, thermal or wind-forced mixing, increasing with latitude and in the direction of shallow areas, characterizes the winter situation.

Other processes that considerably influence the chemical characteristics of intertidal areas, and closed-off or semi-closed-off waters in particular, are concentration changes due to dilution with fresh water or evaporation. In tropical lagoons salinity may vary from virtually zero to five times the normal seawater concentration between the rainy and the dry season when evaporation dominates (Mee, 1978; Copeland and Nixon, 1974). Extreme values up to 300‰ salinity, for example, have been reported for a Mexican lagoon (Copeland, 1967). But also in temperate zones significant increases in salinity due to evaporation can be observed, for instance, in tidal or rock pools during summer. Extreme evaporation and hence concentration of dissolved materials may result in precipitation of inorganic as well as organic substances in isolated shallow water bodies, such as lagoons or tide pools (Mee, 1978).

In mats of blue-green algae, where evaporation may concentrate dissolved organic material up to 100 mg l^{-1} (Wilson, 1963), co-precipitation with inorganic salts (for example, microcrystalline aragonite) may form a distinctive grain type termed micrite, leading to considerable deposits of organic materials (Birke, 1974).

No information was found as to what extent freezing of seawater would affect the chemistry of shallow waters, for example, by concentrating dissolved substances.

As mentioned before, strong water movement causes and supports the suspension and resuspension of particulate materials, which can greatly influence chemical properties in the water as a result of physico-chemical adsorption and desorption processes on the particle surface.

The role of particle adsorption has been extensively studied in connection with estuarine mixing processes which have been repeatedly described (Aston, 1978; Duinker, 1980; Neilson and Cronin, 1981; Kennedy, 1984).

Scavenging of dissolved or colloidal substances through adsorption is most effective where concentrations of suspended particulate matter are high as, for example, in the turbulent waters of intertidals or littorals. Particularly estuaries may be rich in suspended particles, either through river inputs or as a result of specific estuarine circulation patterns with zones of increased concentrations of suspended solids. In addition, in estuaries the mixing of fresh water and seawater causes changes in ionic strength or electric charges, which strongly influence the interaction between solid and solute phases.

It is unclear how important is the transformation of dissolved organic material into particulate forms, which, according to Wangersky (1978), is one of the major mechanisms of forming organic particles in the sea. Since surfaces of inorganic particles or air bubbles are considered as the aggregation sites of surface active molecules, this mechanism may be particularly effective in the particle loaded surf zone in the intertidal or littoral where air bubbles are continuously generated by the crashing waves.

Due to these processes, the influx of riverine elements to the sea can be drastically modified. It has been estimated that more than 90% of the suspended inorganic or organic materials are deposited in most estuaries when river water mixes with seawater (Martin and Whittfield, 1983; Biggs and Howell, 1984).

Likewise, dissolved materials carried into the estuary may be removed from the dissolved phase and deposited, depending on their reactivity. Those substances which are only diluted while mixing with the seawater are described as having conservative properties. Others show non-conservative behaviour, which may be either positive when they are removed from the water through, for example, flocculation or adsorption and deposition, or negative when, for example, substances adsorbed to suspended particles are mobilized under estuarine conditions. Silicon, for instance, which in river water is on average six times more concentrated than in seawater, is probably non-biologically removed for 10 to 20% through combined effects of seawater electrolytes and particle adsorption (Aston, 1978). Similar physico- or geochemical "filter effects" in estuaries have been suggested for humic acids, iron, phosphate or aluminium, among other substances (Morris et al., 1981; Sharp et al., 1984).

The role of physico-chemical particle interactions and the distribution and cycling of trace metals in nearshore environments have been reviewed by Duinker (1980) and recently discussed by Kremling (1987). Trace metals in shallow coastal waters may undergo varying degrees of internal cycling before they are ultimately trapped in the deeper layers of the sediment or exported to off-shore regions. The recycling can involve non-biological as well as biological processes, such as flocculation, adsorption and precipitation, biological incorporation and accumulation and biological or chemical remobilization, either in the water column or within the superficial sea-bottom layers. Until recently little has been known of the extent to which biological activities control observed temporal and spatial variations of trace metals in coastal waters (Danielson and Westerlund, 1984; Osterroth et al., 1985), although metal uptake rates and contents of planktonic as well as benthic organisms suggest the quantitative significance of such processes (e.g. Bryan, 1984).

One of the more important non-biological mechanisms in controlling trace metals in coastal environments seems to be the partitioning between particles and seawater. It appears that the affinity for the solid phase (expressed as the partition coefficient K_d) decreases with increasing electronegativity, ranging from the lanthanide elements to the halide ions (Martin and Whittfield, 1983). The gradual increase of seasonal differences in trace metal concentration from deeper to shallower coastal waters, where vertical mixing becomes more effective (see Kremling, 1987), points to the influence of benthic processes, in particular resuspension of superficial sediments and/or diagenetic remobilization with subsequent transport into the water column.

During the last years, investigations on benthic chemical fluxes and on pore water chemistry in sediments of various coastal areas revealed the importance of biologically mediated redox reactions and of the oxidation kinetics of the elements involved (see Balzer et al., 1983; Brumsack and Gieskes, 1983; and the review by Wong et al., 1983). In intertidal and littoral waters, where the depth of the redox layer is usually small and where steep concentration gradients between the pore water system and the overlying bottom water may occur, trapping or remobilization at the sediment–seawater interface certainly play a major role in trace metal cycling.

Chemical speciation is another non-biological mechanism, although to a larger extent controlled by biological processes, which is essential for the understanding of the chemical characteristics of trace metals in marine systems. Trace metals, particularly heavy metals, are often bound in chemical complexes by various naturally occurring organic or inorganic ligands in seawater. The biological availability of the metals and their bioreactivity, for example toxicity, depends widely on the nature and concentration of complexing compounds (Stumm and Brauner, 1975; Leppard, 1983; Kramer and Duinker, 1984).

There is evidence that, particularly in nearshore waters where concentrations of trace metals as well as of dissolved organic compounds are usually higher than in the open ocean, these mechanisms may become effective. Organic copper concentration in coastal waters of the Baltic Sea were correlated with patterns of seasonal variations in dissolved organic carbon, but not with phytoplankton activity (Osterroth et al., 1985). Besides a variable fraction of organic copper, presumably formed of primary algal exudates and readily degradable by bacteria, a more stable background fraction was found, probably consisting of material of the humic or fulvic acid type that is known for its great chelating capacity (Mantoura et al., 1978).

Gillespie and Vaccaro (1978), who employed a bacterial bioassay technique to determine the complexation capacity of various types of natural seawater, found the copper binding capacity of coastal and salt-marsh waters to be 2 to 11 times higher than that of seawater from the Sargasso Sea. However, the complexation values when expressed per unit DOC, as measured in various near-shore waters ("DOC-normalized"), were only half of the values obtained for open ocean water. This difference may mean that the highly refractory DOC fraction prevailing in offshore waters, although present in relatively low concentration, has more effective chelation properties than the biologically more unstable, dissolved organic materials produced in nearshore waters.

BIOLOGICAL ACTIVITY AND CHEMICAL PARAMETERS

Variations of chemical parameters in the marine environment are primarily driven by physical processes. Within these constraints, however, biological activity can shape the chemical characteristics to various extents. In the relatively small water bodies of shallow intertidal and littoral systems, biological processes and activities, such as uptake, incorporation or accumulation, transformation, excretion and exudation, remobilization through degradation, remineralization or bioturbation, are of more importance than in deep offshore or oceanic systems. In fact, biological activity may even surpass the effects of physical, physico-chemical or chemical processes. Because of low depths and usually intensive mixing, the influence of benthic activity may dominate that of pelagic activity. In addition, primary production, often favoured by increased nutrient levels and radiant energy inputs unlimited by depth, may control biologically relevant chemical parameters to a great extent. This applies especially to the plant nutrient elements phosphorus, nitrogen and silica as shown in numerous investigations on various types of coastal ecosystems (see, for example, McRoy et al., 1972; Mann, 1979; Von Bodungen, 1985) and to a certain extent also to trace metals as recently discussed by Kremling (1987).

Plant nutrients are used by primary producers to build up organic material. On average, 180 moles of hydrogen, 106 moles of carbon, 46 moles of oxygen, 16 moles of nitrogen and 1 mole of phosphorus (plus other mineral elements) are needed to produce 3258 g dry weight of plant material, taking into account that the overall photosynthetic product is not simply sugar but a mixture of various organic compounds (Odum, 1971).

Since intertidal and littoral systems are known to range among the most productive ecosystems (see Lieth and Whittaker, 1975; Turner, 1976; McRoy and McMillan, 1977; Mann, 1982), considerable amounts of nutrient elements can be bound in the form of plant biomass. Primary production of salt marshes, kelp beds or seagrass meadows is second to that of tropical rain forests, when agricultural systems are discounted. Surveys of phytoplankton productivity in the sea have shown that the greater part of the ocean fixes less than 50 g C m^{-2} yr^{-1}, whereas coastal waters have productivities up to five times higher than this level and may be even higher in coastal lagoons or upwelling areas (Koblentz-Mishke et al., 1970).

In addition, benthic microphytes significantly contribute to primary production in most of the intertidal or littoral ecosystems (Leach, 1970; Cadée and Hegeman, 1977; Sundbäck, 1983). The actual contents or uptake rates of nutrients may vary considerably depending on the external supply of nutrients as well as on the physiological state of the plants (Wallentinus, 1985; Schramm et al., 1988).

Marine plants are often capable of so-called surplus uptake of plant nutrient elements. Phosphorus, in particular, may be taken up in excess far beyond the levels needed to satisfy maximum production (Kuhl, 1974; Schramm and Booth, 1981). In this context, the influence of bioaccumulation on the chemical characteristics of nearshore waters is of interest. It is a well-known fact that many marine organisms are able to take up and accumulate trace elements, for instance heavy metals, up to concentrations several orders of magnitude beyond that of the surrounding medium (Bryan, 1984). Hence, on one side, certain elements may be scavenged and removed from seawater and bound by the organisms. Where these organic materials are deposited and accumulated, as for example decomposing plankton on the sea bottom or macrophytic materials in depressions, release of accumulated elements in the course of decomposition may lead to significant increase in the concentration of these elements.

Another process related to primary production that may significantly alter the nitrogen budget of shallow coastal systems is nitrogen fixation, mainly by blue-green algae or Cyanobacteria. In shallow brackish areas of the Baltic Sea (Hiddensee, western Baltic), for example, nitrogen fixation by the benthic blue-green alga *Calothrix scopulorum* ranged between 2 and 87 mg N m^{-2} d^{-1}. Maximum uptake rates of 9.3 mg N m^{-2} h^{-1} were determined in July, and the annual nitrogen fixation was estimated to be 7.6 g N m^{-2} (Savela, 1983).

In *Spartina* salt marshes of Nova Scotia, nitrogen fixation was 2.2 g N m^{-2} yr^{-1} on the mud surface and 9.3 g N m^{-2} yr^{-1} in the rhizosphere of *Spartina* (Patriquin and McLung, 1978). This was estimated to be almost sufficient to provide for the nitrogen requirements of the marsh (Patriquin, 1978). Even higher fixation rates (20 g N m^{-2} yr^{-1} and twice that level in marsh pools) were determined in *Puccinella–Festuca* marshes of northwestern England (Jones, 1974).

Removal of nutrients and other elements by primary producers are counteracted by numerous biological processes of regeneration or transformation in both pelagic and benthic environments. Whereas in deep water systems, primary production is confined to the euphotic layer and regeneration processes on the sea-floor are spatially separated from primary production, in the shallow intertidal and littoral the sea-floor is included within the euphotic zone, or sufficiently close to it so that these processes are very often closely linked with each other. In this context, the contribution of benthic organisms, in particular that of microbes, is of great importance (see Johannes, 1968; Nixon et al., 1976; Fenchel and Blackburn, 1979; Zeitzschel, 1980; Rumohr et al., 1987. The principal pathways of benthic regeneration processes have been described in detail by Fenchel (1972) and are summarized below.

Except for beaches exposed to strong wave action or sediments with much bioturbation, marine sediments are anaerobic beneath a thin oxidized layer. Hence, the initial steps of breakdown of organic compounds are anaerobic fermentative processes, giving rise to low molecular organic compounds. If these reach the oxidized surface layers, they will be aerobically oxidized by chemoheterotrophic or photoautotrophic bacteria. Alternatively, further decomposition may be anaerobic due to micro-organisms that use inorganic hydrogen acceptors (SO_4^{2-}, CO_2, NO_3^-). The end-products of these anaerobic processes are either H_2S, CH_4 and NH_3 or N_2 (denitrification). Whereas the reduced inorganic compounds in the presence of oxygen may be further used by photosynthetic sulfur bacteria or by chemoautotrophic organisms to form elemental sulphur or sulphate, carbon dioxide and nitrate respectively, denitrification to N_2 could lead to a net loss of nitrogen from the system.

Earlier investigations (e.g. Kuenzler, 1961; Johannes, 1965) suggested that, besides micro-organisms, benthic meio- and macrofauna may notably contribute to nutrient regeneration (Rowe et al., 1975; Kautsky and Wallentinus, 1980; Raine and Patching, 1980; Jordan and Valiela, 1982). According to Kautsky and Wallentinus (1980), release of total inorganic nitrogen and phosphorus from *Mytilus edulis* populations in the northern Baltic proper ranged from 1.4 to 75.5 µg N and 0.4 to 24.1 µg P h^{-1} g^{-1} shell-free dry weight of mussels, respectively, the release rates being highest in summer. Budget estimates for the investigated area near Askö (160 km^2) showed that the released nutrients would by far exceed the nutrient requirements of the associated macroalgal communities. The remaining nutrients were estimated to be sufficient to supply nearly 6% of the nitrogen and 17% of the phosphorus demands of the pelagic system. Estimates for the annual regeneration by mussels in the entire northern Baltic proper amounted to 250×10^3 t inorganic nitrogen, 97×10^3 t amino-nitrogen and 77×10^3 t inorganic phosphorus, which would be of the same order of magnitude as natural and man-made terrigenous inputs (Larsson et al., 1985).

Fluxes of nutrients and other dissolved materials across the sediment–water interface are controlled by various physical and geochemical processes, such as diffusion along concentration gradients, resuspension, density displacement, or mobilization of pore water as a result of wave action or current or gaseous percolation (e.g. methane bubbles). In addition to these, biological activity in the form of bioturbation is an essential factor and may influence the chemical characteristics both of the sediment and the overlying water column to a great extent (Hylleberg and Henriksen, 1980; Dicke, 1986). Besides direct effects of pumping or movements of animals on the flow of water in the pores and across the sediment–water interface (Luedtke and Bender, 1979), bioturbation also includes effects of the organisms' activity (for example, effects of sorting and burrowing) on the physical structure of the sediments or on the chemical characteristics of the interstitial water through selective uptake and transport of substances and metabolic activities such as excretion, respiration, production of faecal pellets etc. (Rhoads, 1973, 1974; Aller, 1982, 1984).

Whereas the flux of inorganic nitrogen compounds or silica from the sediments into the water column seems to be closely related to diffusive or bioturbative transport (Dicke, 1986), the recycling of phosphorus depends to a great extent on redox conditions in the sediment (Balzer et al., 1983). Phosphate becomes adsorbed to ferric compounds of the sediment under oxic conditions and is released when the environment becomes anoxic. Hence, bioturbative transport of oxygen-rich seawater into the sediment will rather decrease than increase phosphorus recycling (Dicke, 1986). A special mechanism, which may also be considered as bioturbation, is the active transport of substances in rooting marine vascular plants (McRoy and McMillan, 1977). *Zostera*, for example, can "pump" phosphate and ammonia from the sediment and release it to the ambient water, as shown for phosphate in a lagoonal system in Alaska (McRoy et al., 1972; McRoy and Goering, 1974).

Total nutrient fluxes, as determined for various intertidal and littoral systems, may vary considerably depending on climatic zone, sediment type, depth and other factors. Often these variations reflect the seasonality of organic inputs from the pelagic (plankton blooms) or from other sources (for example, lateral inputs of macrophytic materials), as shown in a comprehensive review by Nixon (1981) and in later investigations (Pomroy et al., 1983; Balzer, 1984; Balzer et al., 1986; Dicke, 1986).

Generally, it can be concluded that, in intertidal and littoral systems, regeneration processes in the sediments are the major source of nutrient replenishment of the water column. In stratified systems, however, where vertical exchange is interrupted, or in areas with sediments poor in organic carbon content and thus little benthic activity, regeneration of nutrients in the pelagic may be important and may even exceed benthic recycling (Nixon, 1981). In coastal waters of southern California, for example, 92% of the nitrogen required for annual primary production was recycled in the water column (Harrison, 1978). Hartwig (1976) estimated that nutrient release from a siliceous sediment with low organic content contributed only 5% of the requirement for nitrogen and 10% for phosphorus.

Generally speaking there are two sources of regenerated nutrients in the water column: those regenerated by micro-organisms and those by zooplankton. The relative importance of each may vary considerably, as emphasized in earlier reviews (Corner and Davies, 1971; Mann, 1982). In waters rich in organic carbon sources, bacteria may even compete with phytoplankton for nutrient uptake. In this case, however, those organisms which feed on bacteria (for example Ciliata) probably contribute significantly (Johannes, 1968). More important than regeneration by bacteria is probably zooplankton excretion, which is in the average range of 1 to 10 μmol N mg^{-1} dry weight per day and usually one order of magnitude lower for phosphorus (Corner and Davies, 1971).

An interesting aspect of biological nutrient cycling in nearshore waters has been discussed by Nixon (1981). Whereas the ratio of nitrogen to phosphorus fixed in phytoplankton is on average 16:1 (by atoms) over much of the ocean and also of most coastal waters (Redfield, 1934; Goldmann et al., 1978), the N/P ratio in coastal water as well as of the net sediment–water nutrient flux is often below 10. Nixon (1981) excluded the possibility of additional phosphorus inputs, export of nitrogen (for example in the form of nitrogen-rich animal tissue), deposition or burial of nitrogen compounds in the sediment, or release of bigger amounts of organic nitrogen from the sediments. His final conclusion was that gaseous nitrogen might be lost from the sediments through denitrification, which is favoured in coastal waters where a greater portion of the organic production is remineralized in the sediment under anaerobic conditions supporting denitrification processes. The role of denitrification in marine nitrogen cycles has been discussed by various authors in recent years (see, for example, Carpenter and Capone, 1983). Denitrification rates as determined for various types of coastal sediments are in the range from 0 to 3 mmol m^{-2} day^{-1} (equivalent to 14 g N m^{-2} yr^{-1} at maximum), and show pronounced seasonal variations (see, for instance, Knowles,

1982; Seitzinger et al., 1980, 1984; Jenkins and Kemp, 1984).

Photosynthesis and respiration are the two biological processes that predominantly control the dissolved gases oxygen, carbon dioxide, and indirectly also hydrogen sulfide, in aquatic environments. Although low depths and usually thorough mixing of the intertidal and littoral waters generally allow rapid exchange and equilibrium between the seawater and the atmosphere, pronounced variations from saturation are typical for shallow coastal waters.

Particularly in macrophyte-dominated systems, oxygen concentration may vary from far below saturation, sometimes down to nearly zero due to respiratory activity during the night-time, up to several hundred percent oversaturation due to photosynthesis during daytime. In extreme cases, where water exchange is obstructed — for example by stable thermal stratification or in dense algal mats — oxygen may be entirely depleted even in very shallow waters. Under these conditions, anaerobic production of free hydrogen sulfide may occur, as for example was observed in thick, loose-lying *Cladophora* mats in the Bermuda inshore waters (Schramm and Booth, 1981), in blue-green algal mats (Dalrymple, 1965) or in bottom waters of lagoons (Carrada and Rigillo-Troncone, 1975).

Oxygen which enters the sediment from the overlying water, supported by bioturbative activity, is consumed by biological oxidation of organic materials. Even in rather coarse sediments, free oxygen is restricted to the upper surface layer in the range of a few millimetres, except for canals and cavities produced by burrowing activity of benthic animals (Revsbech et al., 1980). Below this oxygenated layer the sediment becomes anaerobic. Sometimes, for example in intertidal mud flats, different zones can be distinguished which are characterized by their typical discoloration (Reise, 1985). In an upper oxygenated brownish layer stained by ferric hydroxides, oxic decomposition of organic matter prevails. Below this horizon the sediment is entirely anaerobic. A black layer stained by ferrous sulfide (FeS) and comparatively rich in organic material is followed by a deeper greyish sediment containing pyrite (FeS_2) and being poor in organic substrates.

Variations in carbon dioxide, as a result of metabolic activity, are significant in nearshore waters, although less pronounced compared to oxygen because of the great buffer capacity of the seawater carbonate system. The amount of CO_2 needed to saturate seawater is 100 to 200 times that of the physically dissolved carbon dioxide (Dietrich et al., 1980). One of the major effects of variations of the carbon dioxide–carbonate system lies in the related changes of pH, which may influence dissolution and precipitation processes particularly in the bottom layer or in macrophyte stands where the pH can considerably diverge from the normal seawater pH of around 8.3 (Skirrow, 1975). The interaction between biological carbonate deposition and the equilibrium of the carbonate system, which is important in tropical intertidal or littoral coral reefs, will be treated in a special volume of this series.

Intertidal and littoral systems, as pointed out earlier, are usually areas of considerably greater productivity than the open oceans. Hence, the content of organic materials, either particulate (POM) or dissolved (DOM), in the seawater is comparatively high. The usually quoted average range of dissolved organic material in the euphotic layers of the oceans is 0.5 to 2 mg C l^{-1}, equivalent to approximately 1 to 4 mg DOM l^{-1} (Williams, 1975; Wangersky, 1978; Duursma and Dawson, 1981). In coastal waters, DOM concentrations are normally two to five times higher. Extreme values up to 100 mg DOM l^{-1} have been reported for littoral depressions (Erokhin, 1972) or over blue-green algal mats (Wilson, 1963).

Similarly, the figures for suspended organic matter range from 10 to 300 μg POC l^{-1} in the surface layers of the oceans, while they reach the milligram range in coastal waters (Parsons, 1975). Whereas land run-off of organic material seems to be less important, except for certain river estuaries, the major portion of dissolved as well as particulate organic matter is produced in the coastal zone itself. The great bulk originates from attached macrophytes which are often the major producers of plant biomass (see Table 3.2). Marine algae as well as vascular plants release varying amounts of organic substances into the surrounding water, either by exudation or by leaching, depending on their physiological state and on the environmental conditions. For seaweeds, the reported release

TABLE 3.2

Contribution of phytoplankton, attached microalgae and macrophytes to total primary production in some selected coastal ecosystems. (Values expressed in g C m^{-2} yr^{-1}) (Field, 1983)

System	Phytoplankton	Attached microalgae	Macrophytes	Total
Narragansett Bay	98 (100%)	–	–	98
Lynher mud flat (Cornwall)	82 (36%)	143 (64%)	–	225
Askö (Baltic Sea)	191 (72%)	?	75 (28%)	266
Mangrove forest (Florida)	–	–	890 (100%)	890
Marsh/estuary (Sapelo Island)	79 (6%)	150 (10%)	1216 (84%)	1445
Kelp bed (Nova Scotia)	226 (11%)	–	1750 (89%)	1976
Kelp bed (S. Africa)	502 (40%)	–	767 (60%)	1269

ranges from almost zero up to 40% of the net fixation of carbon (Sieburth, 1969; Moebus et al., 1974; Fankboner and de Burgh, 1977; Schramm et al., 1984).

Existing data suggest that, in the absence of stress, exudation from seaweeds will not exceed 2 to 5% of net primary production. It must be considered, however, that even in the normal situation plants of the intertidal or shallow waters are subjected to extreme temperature and osmotic conditions. Stress release, as observed for intertidal plants, may significantly increase DOC content in the seawater (Sieburth, 1969; Schramm et al., 1984; Moebus and Johnson, 1974).

Less information is available about the production of DOM by salt-marsh plants or by seagrasses, although there is evidence that these contribute significantly to the pool of dissolved organics in coastal waters (e.g. Gallagher et al., 1976; Kirkman and Reid, 1979).

The role of DOM production by phytoplankton, which is on average 3 to 4 times more productive in near shore waters compared to the major part of the ocean, has been reviewed by Fogg (1983). He concluded that, when various sources of error in determination of release rates and activity of heterotrophs are taken into account, approximately 5% of total carbon fixed in eutrophic waters, rising to 40% in oligotrophic waters, is released from physiologically intact phytoplankton. These rates will probably increase during the end phase of a bloom when the physiological state of the organisms weakens.

Relatively little is also known about the contribution of sources other than primary producers to the DOM pool of intertidal and littoral ecosystems. Excretion of organic nitrogen, mostly in the form of amino acids, together with ammonia from animals, such as zooplankton, can be significant (Webb and Johannes, 1967; Jawed, 1969).

Just as for DOM production, macrophytes are similarly important for the production of nonliving particulate organic matter (POM) in intertidal and littoral ecosystems. In many cases, as much as nine tenths of the benthic primary production enters the food web as organic detritus (Melchiorri-Santolini and Hopton, 1972; Pomeroy, 1980). Particularly the larger brown seaweeds or kelps, such as *Laminaria*, *Ecklonia*, *Himantothallus* or *Fucus*, contribute the greater portion of their annual production to the detritus pool through thallus erosion, fragmentation, "leaf fall" or through the activity of shredders (Mann, 1972; Field et al., 1977; Dieckmann et al., 1985). Johnston et al. (1977), for example, estimated that 64% of the annual net production of a *Laminaria* population from a Scottish sea-loch was passed to the heterotrophs as particulate material. In *Fucus vesiculosus* communities of the Baltic (Kiel Bight), the total production of POM in the form of fragmented material or receptacles cast off in late summer accounts for 90% of the annual net production (Grützmacher, 1983).

The role of detritus production by salt-marsh plants or seagrasses has been repeatedly studied. Early investigations had already shown that only a small fraction of the salt-marsh plants were grazed while alive (Teal, 1962). Around 90% of salt-marsh primary production, which is estimated to be in the average range of 200 to 400 g C m^{-2} yr^{-1}, is utilized as dead particulate material (Odum and De la Cruz, 1967; Hatcher and Mann, 1975; Ribelin and Collier, 1979).

Another important source of particulate organic material is the seagrasses, the average net production of which ranges between 120 and 320 g C m^{-2} yr^{-1} (McRoy and Helfferich, 1977), although maximum production around 1000 g C m^{-2} yr^{-1} has been reported for tropical *Thalassia* or *Cymodocea* meadows (Jones, 1968; Quasim and Bhattathiri, 1971). The conspicuous amounts of drifting dead seagrass leaves that can be observed in most coastal areas of the world after storms or where seasonal "leaf fall" occurs (for example *Zostera*) indicate the importance of this detrital source. Kirkman and Reid (1979). For example, estimated that approximately half of the annual production of a *Posidonia* bed in South Australia was utilized as dead particulate matter, while the remaining portion was either grazed alive by herbivores (3%) or released as dissolved organic material.

Production of dissolved or particulate organic materials and external inputs are counteracted by various non-biological and particularly by biological transformation and removal processes. Organic materials originating from natural sources comprise a vast variety of chemical compounds. The major fraction of the identified dissolved organics consist of primary photosynthetic products such as low-molecular carbohydrates or free sugars, and free or combined amino acids (Williams, 1975; Wangersky, 1978; Duursma and Dawson, 1981). Most of these compounds are readily available for heterotrophic utilization and thus undergo a rapid turnover (Andrews and Williams, 1971).

The importance of microbial heterotrophy in coastal systems has been often emphasized (Pomeroy, 1974; Morita, 1977). In nearshore waters the numbers of free bacteria are normally one to two orders of magnitude higher (10^5 to 10^7 cells ml^{-1}) than in offshore waters and may be considerably greater in nearshore sediments (Palumbo and Ferguson, 1978; Hoppe, 1978; Meyer-Reil, 1978).

It appears that, particularly in intertidal and littoral systems where suspended solids and detrital material are abundant, bacteria attached to particles account for a considerable portion of heterotrophic activity (Hanson and Wiebe, 1977; Wangersky, 1978; Hoppe, 1984). The quantitative role of bacteria in organic-matter transformation and degradation is not yet fully understood. Estimates ranging between 1 and 50% of phytoplankton net production (Andrews and Williams, 1971; Derenbach and Williams, 1974) are usually based on uptake rates of labelled low-molecular organics such as monosaccharides or amino acids, and do not consider the higher molecular fractions of the naturally occurring dissolved materials. A greater still uncharacterized portion of the total DOC pool, often referred to as marine humus, seems to be highly refractory to biological transformation or degradation. To this fraction also belong substances termed "Gelbstoff", the concentration of which is generally much higher in nearshore waters than in offshore waters.

Early investigations by Craigie and McLachlan (1964), Sieburth and Jensen (1968, 1969) or Khailov and Burlakova (1969) already suggested that most of these materials originate from marine plants, particularly from seaweeds, rather than from terrestrial sources as thought earlier (see also Moebus et al., 1974; Fankboner and de Burgh, 1977; Schramm et al., 1984). Besides bacteria, other groups of organisms, including phytoplankton, zooplankton or benthic animals, are capable of heterotrophic uptake of dissolved organic materials, sometimes in considerable amounts (see review by Wangersky, 1978). The quantitative significance of such activities for intertidal and littoral ecosystems is not yet clear, although the existing data suggest that bottom living animals, particularly molluscs, may be important.

It appears that in intertidal and littoral systems biologically mediated transformation and removal of particulate organic material is at least as important as physical and physico-chemical processes. A major role is played by microbes, for which organic particles are an excellent substrate.

Through their activity dead plant or animal matter is broken down into simpler dissolved forms of organic material, and eventually decomposed to its mineral constituents. An important function is also performed by those animals that may be functionally grouped as shredders; they reduce the particle size, thus increasing the surface/volume ratio. As particle size decreases, both the number of micro-organisms per unit weight in particulate organic matter and the microbial activity increase considerably (Fenchel, 1970). As a result of increasing bacterial numbers the quality of detritus changes markedly. Whereas C/N ratios of macrophytic detrital material range between 8 and 40, that of attached bacteria is approximately 6, thus increasing the food value for filter or deposit feeders (Harrison and Mann, 1975).

Various investigations have shown that planktonic and benthic detritivores are often incapable of digesting the structural material of macrophytes and that they utilize, in the first place, the microbes attached to the plant particles rather than the plant material itself (Kristensen, 1972; Fenchel, 1977; Heinle et al., 1977; Moriarty, 1977).

In nearshore waters there is a large number of planktonic and benthic groups of organisms, including sponges, coelenterates, polychaetes, crustaceans, molluscs, ascidians and holothurians that very efficiently remove suspended particles through filter feeding or browsing. For example, in Delaware Bay (U.S.A.), zooplankton alone has been estimated to be capable of filtering the entire water volume of the bay 3 to 5 times a year, and would be able to deposit 200 times the annual fluvial input of suspended material to the bay (Biggs and Howell, 1984). Similarly high were the estimates for filtration rates of the benthic filter feeders such as mussels, oysters or barnacles in that area. These estimates are impressive, although the authors point out that this example represents only the lower end of the scale of biodeposition when comparing the population numbers of filter feeders in other coastal systems.

A major effect of marine animals on the chemical properties of particulate organic matter is the packaging and pelletization of organic and inorganic materials into agglomerated forms, which exhibit substantially different physical and chemical properties compared to their composite materials (Biggs and Howell, 1984; Reise, 1985).

REFERENCES

Aller, R.C., 1982. The effects of macrobenthos on chemical properties of marine sediments and over-lying water. In: P.L. McCall and M.J.S. Tevesz (Editors), *Animal-Sediment Relations*. Plenum Press, New York, pp. 53–102.

Aller, R.C., 1984. The importance of relict burrow structure and burrow irrigation in controlling sedimentary solute distributions. *Geochim. Cosmochim. Acta*, 28: 1929–1934.

Andrews, P. and Williams le B., P.J., 1971. Heterotrophic utilization of dissolved organic compounds in the sea. III. Measurement of oxidation rates and concentration of glucose and amino acids in seawater. *J. Mar. Biol. Ass. U.K.*, 51: 111–125.

Anonymous, 1985. *Harmonisatie Noordzeebeleid, Water-kwaliteitsplan Noordzee. Achtergronddocument 2. De ecologie van de Noordzee.* Rijkswaterstaat and Waterloopkundig Laboratorium, Den Haag.

Aston S.R., 1978. Estuarine chemistry. In: J.P. Riley and R. Chester (Editors), *Chemical Oceanography*. Vol. 7 (2nd ed.). Academic Press, London, pp. 362–440.

Balzer, W., 1984. Organic matter degradation and biogenic element cycling in nearshore sediments (Kiel Bight). *Limnol. Oceanogr.*, 29: 1231–1246.

Balzer, W., Grasshoff, K., Dieckmann, P., Haardt, H. and Petersohn, U., 1983. Redox turnover at the sediment–water interface studied in a large bell jar system. *Oceanol. Acta*, 6: 337–344.

Balzer, W., Erlenkeuser, H., Hartmann, M., Müller, P.J. and Pollehne, F., 1986. Diagenesis and exchange processes at the benthic boundary. In: E. Walger, B. Zeitzschel and J. Rumohr (Editors), *Seawater Sediment Interactions in Coastal Waters: An Interdisciplinary Approach*. Springer, Berlin.

Biggs, R.B. and Cronin, L.E., 1981. Special characteristics of estuaries. In: B.J. Neilson and L.E. Cronin (Editors), *Estuaries and Nutrients*. Humana Press, Clifton, N.J., pp. 3–24.

Biggs, R.B. and Howell, B.A., 1984. The estuary as a sediment trap: Alternate approaches to estimating its filtering efficiency. In: V.S. Kennedy (Editor), *The Estuary as a Filter*. Academic Press, Orlando, pp. 107–130.

Birke, L., 1974. Marine blue-green algal mats. In: H.T. Odum, B.J. Copeland and E.A. McMahan (Editors), *Coastal Ecological Systems of the United States, Vol. 1.* The Conservation Foundation, Washington, D.C., pp. 331–345.

Brumsack, H.J. and Gieskes, J.M., 1983. Interstitial water trace-metal chemistry of laminated sediments from the Gulf of California, *Mex. Mar. Chem.*, 14: 89–106.

Bryan, G.W., 1984. Pollution due to heavy metals and their compounds. In: O.Kinne (Editor), *Mar. Ecol.*, Vol. 5, Part 3. Wiley, Chichester, pp. 1289–1432.

Cadée, G.C. and Hegeman, J., 1977. Distribution of primary production of the benthic microflora and accumulation of organic matter on tidal flat area, Balgzand, Dutch Wadden Sea. *Neth. J. Sea, Res.*, 14: 305–322.

Carpenter, J. and Capone, D.G., 1983. *Nitrogen in the Marine Environment*. Academic Press, New York, 900 pp.

Carrada, G.C. and Rigillo-Troncone, M., 1971. Presence of "red water" and environmental condition in some meromictic brackish-water lagoons of the Pontine region. *Rapp. P.-V. Réun. Comm. Int. Explor. Rapp. Sci. Mer Mediterr., Monaco*, 22: 33–35.
Copeland, B.J., 1967. Environmental characteristics of hypersaline lagoons. *Univ. Tex. Contr. Mar. Sci.*, 12: 207–218.
Copeland, B.J. and Nixon, S.W., 1974. Hypersaline lagoons. In: H.T. Odum, B.J. Copeland and E.A. Mc Mahan, (Editors), *Coastal Ecological Systems of the United States, Vol. 1*. The Conservation Foundation, Washington, D.C., pp. 312–345.
Corner, E.D.S. and Davies, A.G., 1971. Plankton as a factor in the nitrogen and phosphorus cycles in the sea. *Adv. Mar. Biol.*, 9: 101–204.
Craigie, J.S. and McLachlan, J., 1964. Excretion of colored ultraviolet-absorbing substances by marine algae. *Can. J. Bot.*, 42: 23–33.
Dalrymple, D.W., 1965. Calcium carbonate deposition associated with blue-green algal mats, Baffin Bay, Texas. *Univ. Texas Inst. Mar. Sci. Publ.* 10: 187–200.
Danielson, L.G. and Westerlund, S., 1984. Short term variations in trace metal concentrations in the Baltic. *Mar. Chem.*, 15: 273–277.
Dawson, R., Schramm, W. and Bölter, M., 1985. Factors influencing the production, decomposition and distribution of organic and inorganic matter in Admiralty Bay, King George Island. In: W.R. Siegfried, P.R. Condy and R.M. Laws (Editors), *Antarctic Nutrient Cycles and Food Webs, Proc. IVth SCAR Symp. Antarct. Biol. Wilderness, S.A., 1983*. Springer, Berlin, pp. 109–114.
Derenbach, J.B. and Williams le B., P.J., 1974. Autotrophic and bacterial production: fractionation of samples from the English Channel. *Mar. Biol.*, 25: 263–269.
Dicke, M., 1986. *Vertikale Austauschkoeffizienten und Porenwasserfluss an der Sediment–Wasser Grenzfläche*. Diss. Univ. Kiel, 224 pp.
Dieckmann, G.S., Reichardt, W. and Zielinski, K., 1985. Growth and production of the seaweed, *Himantothallus grandifolius* at King George Island. In: W.R. Siegfried, P.R. Condy and R.M. Laws (Editors), *Antarctic Nutrient Cycles and Food Webs*, Springer, Berlin, pp. 104–108.
Dietrich, G., Kalle, K., Krauss, W. and Siedler, G., 1980. *General Oceanography*. Wiley, New York, 626 pp.
Duinker, J.C., 1980. Suspended matter in estuaries: adsorption and desorption processes. In: E. Olausson and I. Cato (Editors), *Chemistry and Biogeochemistry of Estuaries*. Wiley, New York, pp. 71–117.
Duursma, E.K., 1961. Dissolved organic carbon and phosphorus in the sea. *Neth. J. Sea Res.*, 1: 1–148.
Duursma, E.K. and Dawson, R., 1981. *Marine Organic Chemistry*. Elsevier, Amsterdam, 521 pp.
Ehrhardt, M. and Wenck, A., 1984. Wind pattern and hydrogen sulfide in shallow waters of the Western Baltic Sea, a cause and effect relationship? *Ber. Dtsch. Wiss. Komm. Meeresforsch.*, 30: 101–110.
Erokhin, V.B., 1972. Dissolved carbohydrates of some biotopes in the littoral zone of the sea *Okeanol.*, 12: 291–298 [in Russian, English Abstract].
Fankboner, P.V. and de Burgh, M.E., 1977. Diurnal exudation of ^{14}C- labelled compounds by the large kelp *Macrocystis integrifolis Bory*. *J. Exp. Mar. Biol. Ecol.*, 28: 151–162.
Fenchel, T., 1970. Studies on decomposition of organic detritus derived from turtle grass *Thalassia testudinum*. *Limnol. Oceanogr.*, 15: 14–20.
Fenchel, T., 1972. Aspects of decomposer food chains in marine benthos. *Verhandl. Deutsch. Zool. Ges.*, 14: 14–22.
Fenchel, T., 1977. Aspects of decomposition of seagrasses. In: C.P. McRoy and C. Helfferich (Editors), *Seagrass Systems: A Scientific Perspective*. Marcel Dekker, New York, pp. 123–145.
Fenchel, T. and Blackburn, T.H., 1979. *Bacteria and Mineral Cycling*. Academic Press, London.
Field, J.G., 1983. Flow patterns of energy and matter. In: O. Kinne (Editor), *Marine Ecology, Vol. V, Part 2*. John Wiley, Chichester, pp. 758–794.
Field, J.G., Jarman, N.G., Dieckmann, G.S., Griffiths, C.L., Velimirov, B. and Zoutendyk, P., 1977. Sun, waves, seaweeds and lobsters: the dynamics of a west coast kelp bed. *S. Afr. Sci.*, 73: 7–10.
Fogg, G.E., 1983. The ecological significance of extra-cellular products of phytoplankton photosynthesis. *Bot. Mar.*, 26: 3–14.
Gallagher, J.L., Pfeiffer, W.J. and Pomeroy, L.R., 1976. Leaching and microbial utilization of dissolved organic carbon from leaves of *Spartina alterniflora*. *Estuarine Coast. Mar. Sci.*, 4: 467–471.
Garrels, R.M. and Mackenzie, F.T., 1971. *Evolution of the Sedimentary Rocks*. Norton, New York, 317 pp.
Gerlach, S., 1986. Langfristige Trends bei den Nährstoffkonzentrationen im Winterwasser und Daten für eine Bilanzierung der Nährstoffe in der Kieler Bucht. *Meeresforsch.*, 31: 153–174.
Gillespie, P.A. and Vaccaro, R.F., 1978. A bacterial bioassay for measuring the copper-chelation capacity of seawater. *Limnol. Oceanogr.*, 23: 543–548.
Goldmann, J.C., McCarthy, J.J. and Peavey, D.G., 1978. Growth rate influence on the chemical composition of phytoplankton in oceanic waters. *Nature*, 279: 210–214.
Grindley, J.R. and David, P., 1985. Nutrient upwelling and its effects in the lee of Marion Island. In: W.R. Siegfried, P.R. Condy and R.M. Laws (Editors), *Antarctic Nutrient Cycles and Food Webs*. Springer, Berlin, pp. 46–51.
Grützmacher, M., 1983. *Produktionsökologische Untersuchungen an Fucus-Beständen der Kieler Bucht (Westliche Ostsee)*. Diss. Univ. Kiel, 108 pp.
Hanson, R.B. and Wiebe, W.J., 1977. Heterotrophic activity associated with particulate size fractions in Spartina saltmarsh estuary, Sapelo Island, Georgia, USA, and the continental shelf waters. *Mar. Biol.*, 42: 321–330.
Harrison, W.G., 1978. Experimental measurements of nitrogen remineralization in coastal waters. *Limnol. Oceanogr.*, 23: 649–694.
Harrison, W.G. and Mann, K.H., 1975. Chemical changes during the seasonal cycle of growth and decay of eelgrass (*Zostera marina*) on the Atlantic coast of Canada. *J. Fish. Res. Bd. Can.*, 32: 615–621.
Hartwig, E.O., 1976. The impact of nitrogen and phosphorus release from a siliceous sediment on the overlying water. In: M. Wiley (Editor), *Estuarine Processes*. Academic Press, New York, pp. 103–117.

Hatcher, B.G. and Mann. K.H., 1975. Above ground production of marsh cord grass (*Spartina alterniflora*) near the northern end of its range. *J. Fish. Res. Bd. Can.*, 32: 83–87.

Heinle, D.R., Harris, R.P., Ustach, J.F. and Flemer, D.A., 1977. Detritus as food for estuarine copepods. *Mar. Biol.*, 40: 341–353.

Hoppe, H.G., 1978. Relations between active bacteria and heterotrophic potential in the sea. *Neth. J. Sea Res.*, 12: 78–98.

Hoppe, H.G., 1984. Attachment of bacteria: advantage or disadvantage for survival in the aquatic environment. In: K.C. Marshall (Editor), *Microbial Adhesion and Aggregation. Dahlem Konf. 1984.* Springer, Berlin, pp. 283–301.

Hylleberg, J. and Henriksen, K., 1980. The central role of bioturbation in sediment remineralization element recycling. *Ophelia*, Suppl. 1: 1–16.

Jawed, M., 1969. Body nitrogen and nitrogenous excretion in *Neomysis rayii Murdoch* and *Euiphausia pacifica Hansen. Limnol. Oceanogr.*, 14: 748–754.

Jenkins, M.S. and Kemp, W.M., 1984. The coupling of nitrification and denitrification. *Adv. Microbiol. Ecol.*, 1: 135–214.

Johannes, R.E., 1965. Influence of marine protozoa on nutrient regeneration. *Limnol. Oceanogr.*, 10: 434–442.

Johannes, R.E., 1968. Nutrient regeneration in lakes and oceans. *Adv. Microbiol. Sea*, 1: 203–213.

Johnston, C.S., Jones, R.G. and Bunt, R.D., 1977. A seasonal carbon budget for a *Laminaria* population in a Scottish sea loch. *Helgoländer Wiss. Meeresunters.*, 30: 527–545.

Jones, J.A., 1968. *Primary productivity of the tropical marine turtle grass Thalassia testudinum König and its epiphytes.* Thesis, Univ. Miami, Fla.

Jones, K., 1974. Nitrogen fixation in a salt marsh *J. Ecol.*, 62: 553–563.

Jordan, T.E. and Valiela, I., 1982. A nitrogen budget of the ribbed mussel. *Geukensia demissus*, and its significance in nitrogen flow in a New England salt marsh. *Limnol. Oceanogr.*, 27: 75–90.

Kautsky, N. and Wallentinus, I., 1980. Nutrient release from a Baltic *Mytilus* red algal community and its role in benthic and pelagic productivity. *Ophelia*, Suppl., 1: 17–30.

Kennedy, V.S. (Editor), 1984. *The Estuary as a Filter*. Academic Press, London, 511 pp.

Khailov, K.M. and Burlakova, Z.P., 1969. Release of dissolved organic matter by marine seaweeds and distribution of their total organic production to inshore communities. *Limnol. Oceanogr.*, 14: 521–527.

Kirkman, H. and Reid, D.D., 1979. A study of the role of seagrass *Posidonia australis* in the carbon budget of an estuary. *Aquat. Bot.*, 7: 173–183.

Knowles, R.C., 1982. Denitrification. *Microbiol. Rev.*, 46: 43–70.

Koblentz-Mishke, O.J., Volkovinsky, V.V. and Kabanova, J.G., 1970. Plankton primary production in the world ocean. In: W.S. Wooster (Editor), *Scientific Exploration of the South Pacific.* National Academy of Sciences, Washington, D.C., pp. 183–193.

Kramer, C.J.M. and Duinker, J.C. (Editors), 1984. *Complexation of Trace Metals in Natural Waters.* Martinus Nijhoff/ Dr. W. Junk Publishers, Dordrecht, 448 pp.

Kremling, K., 1987. Metal cycles in coastal environments. In: L.D. Lacerda and U. Seeliger (Editors), *Metals in Coastal Environments of Latin America.* Springer, Berlin.

Kristensen, J.H.M., 1972. Carbohydrates of some invertebrates with notes on their food and on the natural occurrence of the carbohydrates studied. *Mar. Biol.*, 14: 130–142.

Kuenzler, E.J., 1961. Phosphorus budget of a mussel population. *Limnol. Oceanogr.*, 6: 400–415.

Kuhl, A., 1974. Phosphorus. In: W.D.P. Stewart (Editor), *Algal Physiology and Biochemistry.* Blackwell, Oxford, pp. 636–654.

Larsson, U., Elmgren, R. and Wulff, F., 1985. Eutrophication and the Baltic Sea: causes and consequences. *Ambio*, 14: 9–14.

Leach, J.H., 1970. Epibenthic algal production in intertidal mudflat. *Limnol. Oceanogr.*, 15: 514–521.

Leppard, G.G. (Editor), 1983. *Trace Element Speciation in Surface Waters and its Ecological Implications. NATO Conf. Ser. I. Ecology, Vol. 6.* Plenum Press, New York, 320 pp.

Lieth, H. and Whittaker, R.H. (Editors), 1975. *Primary Productivity of the Biosphere.* Springer, Berlin, 339 pp.

Luedtke, N.A. and Bender, M.L., 1979. Tracer study of sediment–water interactions in estuaries. *Estuar. Coast. Mar. Sci.*, 9: 643–651.

Mann, K.H., 1972. Macrophyte production and detritus food chains in coastal waters. *Mem. Ist. Ital. Idrobiol.*, 29: 353–384.

Mann, K.H., 1979. Nitrogen limitations on the productivity of *Spartina* marshes, *Laminaria* kelp beds and higher trophic levels. In: R.L. Jefferies and A.J. Davy (Editors), *Ecological Processes in Coastal Environments.* Blackwell, Oxford, pp. 363–370.

Mann, K.H., 1982. *Ecology of Coastal Waters.* Blackwell, Oxford, 321 pp.

Mantoura, R.F.C., Dickson, A. and Riley, J.P., 1978. The complexation of metals with humic materials in natural waters. *Estuar. Coast. Mar. Sci.*, 6: 387–408.

Martin, J.M. and Whittfield, M., 1983. The significance of river input of chemical elements to the ocean. In: C.S. Wong, E. Boyle, K.W. Bruland, J.D. Burton and E.D. Goldberg (Editors), *Trace Metals in Seawater, NATO Conf. Ser. IV. Mar. Sci.* Plenum Press, New York, pp. 265–296.

McRoy, C.P. and Goering, J.J., 1974. Nutrient transfer between the seagrass *Zostera marina* and its epiphytes. *Nature*, 248: 173–174.

McRoy, C.P. and Helfferich, C. (Editors), 1977. *Seagrass Ecosystems: A Scientific Perspective.* Marcel Dekker, New York.

McRoy, C.P. and McMillan, C., 1977. Production ecology and physiology of seagrasses. In: C.P. McRoy and C. Helfferich (Editors), *Seagrass Ecosystems: A Scientific Perspective.* Marcel Dekker, New York, pp. 53–88.

McRoy, C.P., Barsdate, R.J. and Nebert, M., 1972. Phosphorus cycling in eelgrass (*Zostera marina L.*) ecosystem. *Limnol. Oceanogr.*, 17: 58–67.

Mee, L.D., 1978. Coastal lagoons. In: J.P. Riley and R. Chester (Editors), *Chemical Oceanography, Vol. 7* (2nd ed.). Academic Press, London, pp. 441–490.

Melchiorri-Santolini, U. and Hopton, J.W., 1972. Detritus and

its role in aquatic ecosystems. *Mem. Istr. Ital. Idrobiol.*, 29, Suppl.: 1–280.
Meyer-Reil, L.A., 1978. Uptake of glucose by bacteria in the sediment. *Mar. Biol.*, 44: 293–298.
Moebus, K. and Johnson, K.M., 1974. Exudation of dissolved organic carbon by brown algae. *Mar. Biol.*, 26: 117–125.
Moebus, K., Johnson, K.M. and Sieburth, J. McN., 1974. Rehydration of desiccated intertidal brown algae: release of dissolved organic carbon and water uptake. *Mar. Biol.*, 26: 127–134.
Moriarty, D.J.W., 1977. Quantification of carbon, nitrogen, and bacterial biomass in the food of some penaeid prawns. *Austr. J. Mar. Freshwat. Res.*, 28: 113–118.
Morita, R.Y., 1977. The role of microorganisms in the environment. In: N.R. Andersen and B.J. Zahuranec (Editors), *Oceanic Sound Scattering Prediction*. Plenum Press, New York, pp. 445–455.
Morris, A.W., Bale, A.J. and Howland, R.J.M., 1981. Nutrient distribution in an estuary: evidence of chemical precipitation of dissolved silicate and phosphate. *Estuar. Coast. Shelf Sci.*, 12: 205–216.
Myrcha, A., Pietr, S.J. and Tatur, A., 1985. The role of pyoscelid penguin rookeries in nutrient cycles of Admiralty Bay, King George Island. In: W.R. Siegfried, P.R. Condy and R.M. Laws (Editors), *Antarctic Nutrient Cycles and Food Webs*. Springer, Berlin, pp. 156–162.
Naiman, R.J. and Sibert, J.R., 1979. Detritus and juvenile salmon production in the Nanaimo Estuary. III. Importance of detrital carbon to the estuarine ecosystem. *J. Fish. Res. Board Can.*, 36: 504–520.
Neilson, B.J. and Cronin, L.E. (Editors), 1981. *Estuaries and Nutrients*. Humana Press, Clifton, N.J., 643 pp.
Nixon, S.W., 1981. Remineralization and nutrient cycling in coastal marine ecosystems. In: B.J. Neilson and L.E. Cronin (Editors), *Estuaries and Nutrients*. Humana Press, Clifton, N.J., pp. 111–138.
Nixon, S.W., Oviatt, C.A. and Hale, S.S., 1976. Nitrogen regeneration and the metabolism of coastal marine bottom communities. In: J.M. Anderson and A. Macfaden (Editors), *The Role of Terrestrial and Aquatic Organisms in Decomposition Processes*. Blackwell, Oxford, pp. 269–283.
Odum, E.P., 1971. *Principles of Ecology*. Saunders, Philadelphia, Phil., 574 pp.
Odum, E.P. and De la Cruz, A.A., 1967. Particulate organic detritus in a Georgia salt-marsh estuarine ecosystem. In: G.H. Lauff (Editor), *Estuaries*. Publ. 83, Am. Ass. Advance Mar. Sci., Washington, DC., pp. 383–388.
Odum, H.T., Copeland, B.J. and McMahan, E.A. (Editors), 1974. *Coastal Ecological Systems of the United States. Vols. 1–4*. The Conservation Foundation, Washington, D.C.
Osterroht, C., Wenck, A., Kremling, K. and Gocke, K., 1985. Concentration of dissolved organic copper in relation to other chemical and biological parameters in coastal Baltic waters. *Mar. Ecol. Prog. Ser.*, 22: 273–279.
Palumbo, A.V. and Ferguson, R.L., 1978. Distribution of suspended bacteria in the Newport river estuary. *Estuar. Coast. Mar. Sci.*, 7: 521–529.
Parsons, T.R., 1975. Particulate organic carbon in the sea. In: J.P. Riley and G. Skirrow (Editors), *Chemical Oceanography, Vol. 2* (2nd ed.). Academic Press, London, pp. 365–383.
Patriquin, D.G., 1978. Nitrogen fixation (acetylene reduction) associated with cord grass *Spartina alterniflora Loisel. Ecol. Bull.* (Stockholm), 26: 20–27.
Patriquin, D.G. and McLung, C.R., 1978. Nitrogen accretion, and the nature and possible significance of N_2-fixation (acetylene reduction) in a Nova Scotian *Spartina alterniflora* stand. *Mar. Biol.*, 47: 227–242.
Pomeroy, L.R., 1974. The oceans food web, a changing paradigm. *BioScience*, 24: 499–504.
Pomeroy, L.R., 1980. Detritus and its role as a food source. In: R.K. Barnes and K.H. Mann (Editors), *Fundamentals of Aquatic Ecosystems*. Blackwell, Oxford, pp. 84–102.
Pomroy, A.J., Joint, I.R. and Clarke, K.R., 1983. Benthic nutrient flux in a shallow coastal environment. *Oecologia*, 60: 306–312.
Quasim, S.Z. and Bhattathiri, P.M.A., 1971. Primary production of a seagrass bed on Kavaratti Atoll (Lacadives). *Hydrobiologia*, 38: 29–38.
Raine, R.C.T. and Patching, J.W., 1980. Aspects of carbon and nitrogen cycling in a shallow marine environment. *J. Exp. Mar. Biol. Ecol.*, 47: 127–139.
Redfield, A.C., 1934. On the proportions of organic derivatives in sea water — their relation to the composition of the plankton. In: *James Johnstone Memorial Volume*. Liverpool Univ. Press, Liverpool, pp. 176–192.
Reise, K., 1985. *Tidal Flat Ecology. Ecological Studies, Vol. 54*, Springer, Berlin, 191 pp.
Revsbech, N.P., Sørensen, J. and Blackburn, T.H., 1980. Distribution of oxygen in marine sediments measured with microelectrodes. *Limnol. Oceanogr.*, 25: 403–411.
Rhoads, D.C., 1973. The influence of deposit-feeding benthos on water turbidity and nutrient cycling. *Am. J. Sci.*, 273: 1–22.
Rhoads, D.C., 1974. Organism–sediment relations on the muddy sea floor. *Oceanogr. Mar. Biol. Ann. Rev.*, 12: 263–300.
Ribelin, B.W. and Collier, A.W., 1979. Ecological considerations of detrital aggregates in the salt-marsh. In: R.J. Livingston (Editor), *Ecological Processes in Coastal and Marine Systems*. Plenum Press, New York, pp. 47–68.
Rowe, G.T., Clifford, C.H., Smith, K.L. and Hamilton, P.L., 1975.Benthic nutrient regeneration and its coupling to primary productivity in coastal waters. *Nature*, 225: 215–217.
Rumohr, J., Walger, E. and Zeitzschel, B. (Editors), 1987. *Seawater–Sediment Interactions in Coastal Waters. Lecture Notes on Coastal and Estuarine Studies, Vol. 13*. Springer, Berlin, 338 pp.
Savela, K., 1983. Stickstoff-Fixierung der Blaualge *Calothrix scopulorum* in Küstengewässern der Ostsee. *Ann. Bot. Fennici*, 20: 399–405.
Schramm, W. and Booth, W., 1981. Mass bloom of the alga *Cladophora prolifera* in Bermuda: Productivity and phosphorus accumulation. *Bot. Mar.*, 24: 419–426.
Schramm, W., Gualberto, E. and Orosco, C., 1984. Release of dissolved organic carbon from marine tropical reef plants: temperature and desiccation effects. *Bot. Mar.*, 27: 71–77.
Schramm, W., Abele, D. and Breuer, G., 1988. Nitrogen and

phosphorus nutrition and productivity of two community forming seaweeds (*Fucus vesiculorus, Phycodrys rubens*) from the western Baltic (Kiel Bight) in the light of eutrophication processes. *Kieler Meeresforsch. Sonderh.*, 2: 221–240.

Seitzinger, S., Nixon, S.W., Pilson, M. and Burke, S., 1980. Denitrification and N_2O production in nearshore marine sediments. *Geochim. Cosmochim. Acta*, 44: 1853–1860.

Seitzinger, S., Nixon, S.W. and Pilson, M.E.Q., 1984. Denitrification and nitrous oxide production in a coastal marine ecosystem. *Limnol. Oceanogr.*, 29: 73–83.

Sharp, J.H., Pennock, J.R., Church, T.M., Tramentano, J.M. and Cifuentes, L.A., 1984. The estuarine interaction of nutrients, organics and metals: A case study in the Delaware estuary. In: V.S. Kennedy (Editor), *The Estuary as a Filter*, Academic Press, London, pp. 241–260.

Sieburth, J.M., 1969. Studies on algal substances in the sea. III. The production of extracellular organic matter by littoral marine algae. *J. Exp. Mar. Biol. Ecol.*, 3: 290–309.

Sieburth, J.M. and Jensen, A., 1968. Studies on algal substances in the sea. I. Gelbstoff (humic material) in terrestrial and marine waters. *J. Exp. Mar. Biol. Ecol.*, 2: 174–189.

Sieburth, J.M. and Jensen, A., 1969. Studies on algal substances in the sea. II. The formation of gelbstoff (humic material) by exudates of Phaeophyta. *J. Exp. Mar. Biol. Ecol.*, 3: 275–289.

Skirrow, G., 1975. The dissolved gases — carbon dioxide. In: J.P. Riley and G. Skirrow (Editors), *Chemical Oceanography. Vol. 2* (2nd ed.). Academic Press, London, pp. 1–192.

Stumm, W. and Brauner, P.A., 1975. Chemical speciation. In: J.P. Riley and G. Skirrow (Editors), *Chemical Oceanography. Vol. 1* (2nd ed.). Academic Press, London, pp. 173–240.

Sundbäck, K., 1983. *Microphytobenthos on sand in shallow brackish water. Öresund, Sweden.* Diss. Univ. Lund, Sweden, 209 pp.

Teal, J.M., 1962. Energy flow in the salt marsh ecosystem of Georgia. *Ecology*, 43: 614–624.

Teal, J.M., Valiela, I. and Berlo, D., 1979. Nitrogen fixation by rhizosphere and free-living bacteria in salt-marsh sediments. *Limnol. Oceanogr.*, 24: 126–132.

Turner, R.E., 1976. Geographic variations in salt marsh macrophyte production: a review. *Contr. Mar. Sci. Univ. Texas*, 20: 47–68.

Valiela, I., Teal, J.M., Volkman, S., Shafer, D. and Carpenter, E.J., 1978. Nutrient and particulate fluxes in salt-marsh ecosystems: tidal exchange and inputs by precipitation and ground water. *Limnol. Oceanogr.*, 23: 798–812.

Von Bodungen, B. 1985. Annual cycles of nutrients in a shallow inshore area, Kiel Bight — variability and trends. *Sonderforschungsbereich 95, Univ. Kiel Contr. No. 453.*

Wallentinus, I., 1985. Partitioning of nutrient uptake between annual and perennial seaweeds in a Baltic archipelago area. In: C.J. Bird and M.A. Ragan (Editors), *Proc. XI. Int. Seaweed Symp. China, 1983.* Junk, Dordrecht, pp. 363–370.

Wangersky, P.J., 1978. Production of dissolved organic matter. In: O.Kinne (Editor), *Marine Ecology. Vol. 4.* Wiley, London, pp. 115–200.

Webb, K.L. and Johannes, R.E., 1967. Studies of the release of dissolved free amino acids by marine zooplankton. *Limnol. Oceanogr.*, 12: 376–382.

Wedepohl, K.H., 1984. Die Zusammensetzung der oberen Erdkruste und der natürliche Kreisläufe ausgewählter Metalle. In: E. Merian (Editor), *Metalle in der Umwelt*, Verlag Chemie, Weinheim, pp.1–10.

Williams, le B., P.J., 1975. Biological and chemical aspects of dissolved organic material in seawater. In: J.P. Riley and G. Skirrow (Editors), *Chemical Oceanography, Vol. 2* (2nd ed.). Academic Press, London, pp. 301–364.

Wilson, R.F., 1963. Organic carbon levels in some aquatic ecosystems. *Univ. Tex. Inst. Mar. Sci. Publ.*, 9: 64–76.

Wong, C.S., Boyle, E., Burton, J.D. and Goldberg, E.D. (Editors), 1983. *Trace Metals in Seawater. NATO Conf. Ser. IV. Mar. Sci.* Plenum Press, New York, 920 pp.

Zeitzschel, B., 1980. Sediment–water interactions in nutrient dynamics. In: K.R. Tenore and B.C. Coull (Editors), *Marine Benthic Dynamics.* Univ. South Carolina Press, pp. 195–218.

Chapter 4

VERTICAL DISTRIBUTION

G. RUSSELL

ZONATION

The vertical distribution of benthic marine plants and animals is rarely, if ever, random. Non-randomness may be manifest in a variety of ways, but its most frequent expression is zonation; an arrangement of organisms in series of horizontal bands. Zonation is not an exclusive property of marine ecosystems: it is common and widespread in terrestrial habitats where there is a transition from one kind of environment to another, as at the margins of freshwater lakes or on mountain slopes. It is on marine rocky shores, however, that zonation is particularly sharply delineated and spatially condensed — as Fig. 4.1, drawn by Lewis (1964) of a rocky shore in the British Isles, very clearly illustrates. Marine zonation is also a world-wide phenomenon; few, if any, rocky shores have been found to lack some kind of banding pattern in the disposition of species.

The volume of literature on marine zonation is vast, but an excellent introduction to the descriptive aspects of shore ecology may be obtained from the works by Stephenson and Stephenson (1949, 1972), Lewis (1961, 1964), Chapman (1974) and Pérès (1982 a, b). Marine ecologists have recognized the existence of zonation since the beginning of the 19th century, and the history of these early investigations has been ably recounted by Doty (1957), Lewis (1964) and Ricketts and Calvin

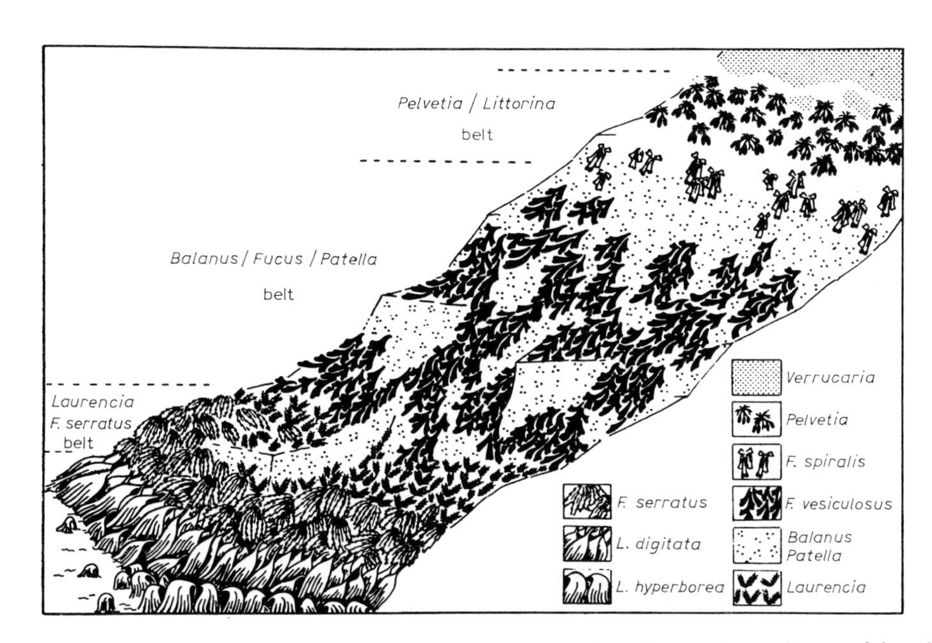

Fig. 4.1. Profile of a rocky shore in the British Isles. The three belts illustrated are subzones of the eulittoral zone. Also shown are the lower supralittoral (*Verrucaria*) and upper sublittoral (*Laminaria*) belts. Note patchiness of *Fucus vesiculosus*. (From Lewis, 1964.)

(1968). No further comment on the historical aspects of this subject is therefore required, and none will be given here. Nevertheless it is appropriate to consider briefly how some ecologists have perceived and analyzed benthic communities, because their approaches and methods have in turn influenced ideas on the ways in which rocky-shore communities are organized.

Zonal systems

Zonation, as Lewis (1961) has pointed out, is primarily a biological phenomenon, whatever importance the physical conditions of the environment may have had in bringing it about. Through time, ecologists have come to recognize that locally-observed patterns tend to recur in other, often very distant and different, habitats. From this observation the idea was born that there are certain universal, or at least general, features of rocky-shore zonation that provide a framework for the comparative description of shores throughout the world.

The great exponents of this approach are Stephenson and Stephenson (1949, 1972) and Lewis (1961, 1964). These workers proposed that there are three primary zones on marine rocky shores each characterized by particular kinds of organisms. When these zones have been recognized by an ecologist in the field, it is a simple matter for him then to record their species composition and the possible presence of any secondary distribution patterns.

It is evident from Fig. 4.1 that identification of the primary zones by inspection of a shore is a process which is necessarily influenced by the species composition of the topmost layer of the community. This point is illustrated in Fig. 4.2, which shows the stratification of the algal vegetation of the eulittoral zone on a Netherlands dyke as described by Den Hartog (1959). It may be seen that at the rock surface the entire extent of the zone is vegetated by the crustose red alga *Hildenbrandia rubra*. The middle stratum consists also of red algae but with an up-slope belt of *Catenella caespitosa* and a down-slope belt of *Mastocarpus stellatus*. Finally, the canopy layer consists of large brown (fucoid) algae in four conspicuous belts with *Pelvetia canaliculata* up-slope, followed successively by *Fucus spiralis, Ascophyllum nodosum* and *Fucus serratus*. A three-layered community of this kind is structurally quite simple, others having been described with four or more component strata. Nevertheless this diagram demonstrates that zonation is a three-dimensional construction and hence that zones defined by the species of the topmost stratum conceal a number of other patterns.

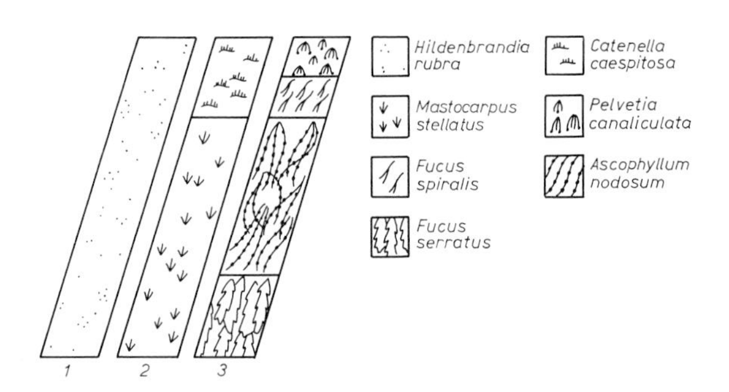

Fig. 4.2. Stratification of vegetation in the eulittoral zone of a dyke in the Netherlands. The rock surface (1) bears the encrusting red alga *Hildenbrandia rubra*; the second stratum (2) consists of *Catenella caespitosa* and *Mastocarpus stellatus* and the canopy (3) comprises, in descending order, *Pelvetia canaliculata, Fucus spiralis, Ascophyllum nodosum* and *Fucus serratus*. Original drawing based on a diagram in Den Hartog (1959).

The Stephenson–Lewis approach to shore ecology has provided us with a practical and simple method, and one which has been used with success by ecologists throughout the world. Reference should be made, however, to the Scandinavian approach. This school recognizes only one primary zonal boundary, the "litus line" which lies at the lower limit of the supralittoral zone. This system has the advantage that it makes only one prior assumption about patterns of vertical distribution, and at the same time it provides a bench mark against which the belt heights of the various shore organisms can be recorded (Fig. 4.3). The Scandinavian tradition owes much to the work by Sjöstedt (1928), but more recent examples of its applicability are given by Jorde and Klavestad (1963) and by Jorde (1966).

Numerical analysis

The increased availability of computers during the last three decades has revolutionized descriptive ecology of rocky shores by making it possible to analyze rapidly very large amounts of data.

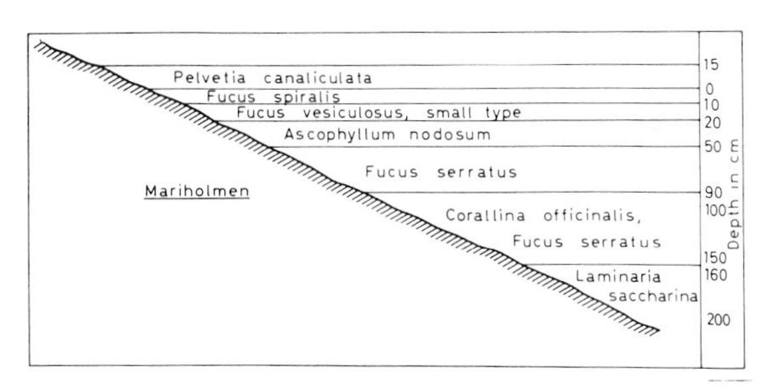

Fig. 4.3. Zonal analysis of the algal vegetation of a Norwegian rocky shore by Jorde (1966). The zero point on the vertical scale denotes the position of the "litus line" (i.e. the boundary between eulittoral and supralittoral zones). All species distributions are measured in relation to this (biological) level.

Ecologists have thus been able to give more attention to quantifying the various elements of community structure, resulting in more refined analyses than was possible hitherto.

As with zonal systems, numerical analysis has developed its own set of ground rules. These have been discussed by Russell and Fielding (1981) and by Lobban et al. (1985) and will not be considered in detail here. Numerical analysis depends upon a sampling programme which must be unbiased. Samples, whether area (quadrat) or plotless (point), may be located randomly or in some predetermined array, but their siting ought not to be tainted with any suspicion of subjectivity. The organisms sampled may then be quantified, either by counting numbers of individuals, or by means of some measurable attribute such as cover or biomass, or by some expression of incidence such as frequency. The measurement of species abundance is usually effected without regard to stratification. Thus, if the shore depicted in Fig. 4.2 were to be sampled, and the cover values of every species summed, then the total cover in each quadrat would probably exceed 100%. It would also follow that, in numerical terms, the encrusting alga *Hildenbrandia* would be as important as the canopy species. Such a conclusion might not reflect the respective biological roles of these organisms.

The techniques available for the analysis of the assembled data are numerous and various but they usually involve calculation of some measure (coefficient) of similarity. Samples may then be classified, that is, grouped together on the basis of similarity in species composition and abundance. Alternatively, they may be ordinated, ordination being the location of samples as mathematical points in one-, two- or multi-dimensional space, interpoint distance expressing the similarities of the samples involved.

Comparative evaluation of numerical techniques by Boudouresque (1971a) and by Lindstrom and Foreman (1978) do not clearly indicate a "best method". Indeed there is evidence of rather contradictory findings; an ordination method found unsatisfactory by Lindstrom and Foreman gave acceptable results when used by Russell (1980). Practitioners of numerical analysis are now required to apply several contrasting techniques to their data. Thus, Field and McFarlane (1968) who analyzed the zoobenthic community structure at False Bay, South Africa, noted that groupings of species were fairly constant whatever clustering technique was adopted. Consistency of results inspires more confidence that the patterns detected reflect the realities of species distributions in the field.

Phytosociology

Phytosociology has evolved as a group of somewhat related classificatory techniques for the analysis of terrestrial vegetation. Marine phytosociologists have usually adopted the method of the Braun–Blanquet (Zürich–Montpellier) school, or some minor variant of it. The Z–M approach has been summarized by Westhoff and Van der Maarel (1978) and needs no additional exposition. Some comments on its applicability to marine communities may be pertinent, however. First, Z–M practitioners are required to sample from homogeneous stands of vegetation, examples of which may not prove easy to find. Spatial heterogeneity in marine communities is too widespread and too common to be regarded simply as a structural aberration, and therefore to be ignored. Second, the sampling procedure in Z–M phytosociology is openly subjective, the particular locations for quadrats being selected with considerable care. This invalidates any subsequent statistical treatment of data; and even the application of non-statistical analytical techniques to such data might be construed as questionable.

The primary unit of Z–M phytosociology is the association, a rather abstract concept based upon certain diagnostic (characteristic) species. The characteristic species of one association properly ought not to occur in another, but in practice such rigorous segregation of associations may prove impracticable. It has been argued by Feldmann (1951) that marine algal associations may not have quite the same ecological status as their terrestrial counterparts, and other workers have preferred to use the term in much looser senses (Jorde and Klavestad, 1963; Jorde, 1966; Tittley et al., 1976).

Zürich–Montpellier phytosociology nevertheless has enthusiastic followers. These are strongly concentrated along the shores of the Mediterranean Sea, and a notable exponent from this region is Boudouresque (1969,1971b). However, the marine vegetation of Japanese coastal waters has also been investigated using phytosociological techniques by Taniguti (1962).

Costs and benefits

Stephenson and Stephenson (1972) have extolled the advantages of their approach to community description, emphasizing the ability of the trained human eye to integrate the complexities of natural communities rapidly and accurately. Also, the fact that so much of the work discussed by them was carried out using the same eyes gives their observations an enhanced comparative value. The Stephensons did not disparage more labour-intensive analytical techniques, but suggested that the virtues of these lie chiefly in the study of local communities. This comment is true insofar as comparative numerical studies of a variety of shores, on the geographical scale achieved by the Stephensons, simply do not exist. Nevertheless, the application of the precepts of numerical analysis can give greater insight into the details of community structure than is possible by the Stephenson approach. It may also reveal more clearly ways in which communities may change through time or in response to changing environmental conditions. Zürich–Montpellier phytosociology is not really comparable with either of these approaches. Its sampling procedure is unusual and its hierarchy of vegetational units, and the criteria for their determination, are unique. It may be argued, as Stephenson and Stephenson (1972) have contended, that to consider community structure strictly in terms of plants and to ignore the animals (or vice versa) is bound to lead to an ecologically unsatisfactory analysis.

No absolute verdict on these several approaches to descriptive ecology is possible, however. Each can be judged only according to the objectives of the investigator.

The problems of scale

Quadrat users among shore ecologists have often attached considerable importance to the question of sample size; and, in particular, to establishing the minimal area of substrate that needs be sampled to be representative of the community as a whole. The simplest procedure for obtaining minimal size is to record species increment with increase in sample area, minimal area being reached when no additional species are recorded. This and similar methods have been adopted by Russell (1972), Boudouresque (1974) and Niell (1974). However in many cases it may prove impossible to establish a minimal area, either because rearrangement of the sequence of samples gives rise to a curve of different form, or because sampling continues to add new species. In such cases it is better to pool all sample data and then to calculate the predicted rate of increase of species number with area, as demonstrated by Hawkins and Hartnoll (1980).

The question of scale in marine ecological studies, recently discussed by Dayton et al. (1984), applies to the problem of delimiting zones. By emphasizing that the upper boundary of the eulittoral zone on certain British rocky shores is determined by the upper limit of barnacles "in quantity", Stephenson and Stephenson (1949) simply make the point that zonal frontiers are sometimes fuzzy. In that event, sample size and sampling interval may have a considerable influence on the pattern detected.

Sampling along the line of a zone is less likely to reveal disjunctions in community structures, and a pattern of continuous change (continuum) is frequently obtained. A continuum of this sort is usually associated with differences in wave action (see Fig. 4.6) but may also be expressed on a geographical scale. Figure 4.4 illustrates differences in distribution of barnacles, *Patella* species and fucoid algae at selected sites in Western Europe, and shows how geographical differences interact with wave action (Ballantine, 1961). A more detailed appraisal of continuum structure on a

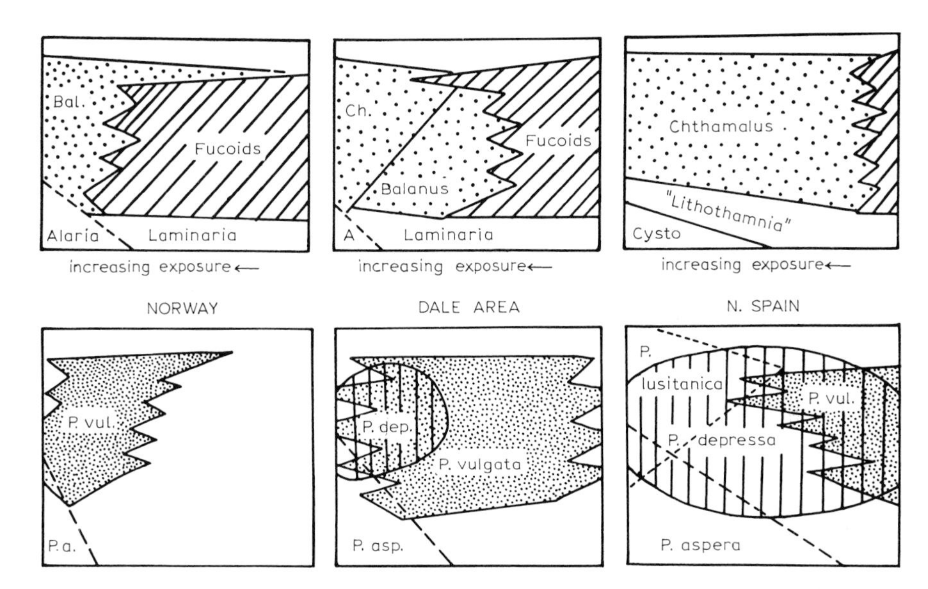

Fig. 4.4. Diagram showing distribution of eulittoral-zone organisms at three geographically distinct localities (Norway, England, Spain; Ballantine, 1961). Note geographical and wave-action components of distributions of barnacles (*Chthamalus*, *Balanus* (= *Semibalanus*)), limpets (*Patella vulgata*, *P. aspera*, *P. depressa*, *P. lusitanica*) and fucoids (Cysto = *Cystoseira*).

rocky shore has been carried out by Boudouresque (1970).

Finally, the problem of scale also applies to the organisms themselves. The existence of zonation on a microscale has been demonstrated, for example, among epiphyte communities of large perennial algae (Tokida, 1960; Markham, 1969; Russell, 1983; Novak, 1984).

Zonation and dominance

Zones defined by the presence of certain important species form the basis of the Stephenson–Lewis approach to shore ecology. It is also very much in the tradition of some terrestrial plant ecologists such as Whittaker (1978), who has characterized communities by the most important species forming the uppermost stratum of the vegetation. The communities defined in this way he refers to as "dominance-types". Dominance is an equally common term in the literature on shore ecology, although it is used in two rather different senses. For some authors it evidently indicates simply a great abundance of certain species. Others use it to describe particular species which have a profound influence on the structure of the community irrespective of their abundance. Thus a herbivore may not account for much of the total community biomass, but may have a very great impact upon its structure. This functional interpretation of dominance has been favoured by Dayton (1972, 1975a, b), who has coined the term "foundation species" for these influential organisms. This emphasis on dynamic aspects of shore ecology is also evident in an important paper by Connell (1972), which dealt principally with zoobenthos, and in an excellent review by Chapman (1986), which covers recent research on algae.

The principal zones

The zonation system adopted in this chapter is depicted in Fig. 4.5, which shows three primary zones. The uppermost band is the supralittoral zone; its upper boundary is determined by the upper limit of sea spray and its lower boundary (the litus line) marks the upper limit of submergence either by waves or tides or both. The middle eulittoral zone is subject to periodic immersion by tidal or other hydrological events, while the sublittoral zone is permanently or nearly permanently submerged. Each of the three zones is populated by a particular assemblage of organisms, and it is these which give the shore its characteristic structure.

Hatched lines in Fig. 4.5 indicate the possible

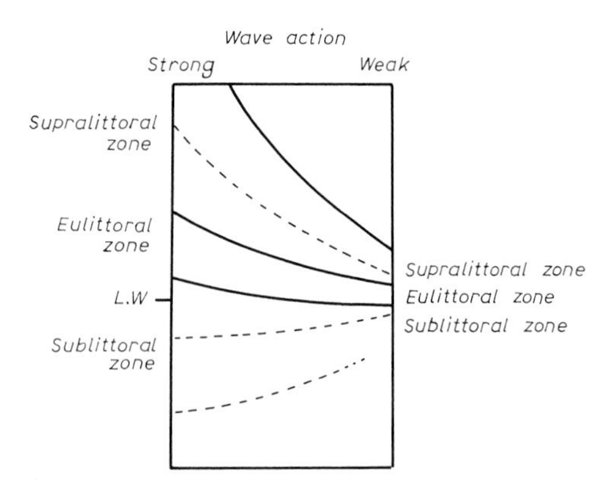

Fig. 4.5. Explanatory diagram of zonal terminology adopted in text. Three primary zones are recognized and their boundaries indicated by continuous lines. Dashed lines denote possible subzonal boundaries in supralittoral and sublittoral zones. Absence of dashed lines from eulittoral zones should not be construed as evidence of absence of subzones. Note zonal enlargement and displacement due to effects of wave action. L.W. indicates approximate position of low water on both tidal and non-tidal shores. Original drawing based upon Lewis (1961), Shepherd and Womersley (1970) and others.

location of secondary zones. The lower belt of the supralittoral zone is characterized by the presence of blackening organisms, usually lichens or Cyanobacteria. Division of the sublittoral zone into three subsidiary belts is in accordance with the observations of Neushul (1967), Shepherd and Womersley (1970) and Edgar (1984), but it is conceded that other patterns are obtainable. The absence of secondary zones from the eulittoral zone is not intended to indicate structural uniformity. Rather, the complexity of eulittoral zone organization is too variable to be resolved even tentatively into a consistent pattern. The zonal scheme in Fig. 4.5 follows that of Nienhuis (1975), and is close in concept to those of Stephenson and Stephenson (1949) and of Lewis (1961), although differing in terminology.

Figure 4.5 also incorporates the expansion of zones, which takes place as wave action increases, as shown by Burrows et al. (1954), Jones and Demetropoulos (1968), Shepherd and Womersley (1970) and Edgar (1984).

Zone convergence and divergence

Marine ecological literature contains numerous references to convergence in community structure. Here again, the term is used in somewhat different senses. According to some, this phenomenon is evident at localities where, for example, great regional biotas converge. This may be seen at the southern tip of South Africa which has been the subject of detailed study by Stephenson and Stephenson (1972). Divergent changes may also be seen in species composition along continental coast-lines; such taxonomic disjunctions are evident in the floristic works of Druehl (1970), Van den Hoek (1975, 1982) and Santelices (1980).

Other authors have used the term convergence in a way which suggests analogy with evolutionary change. Thus Thorson (1957) has drawn attention to the convergent features of the bivalve mollusc faunas of certain types of sediments throughout the world. Thorson's ideas have been taken up by Kussakin (1977), who has applied them to rocky shores. Kussakin distinguished four community types: (1) uniform communities (of the same species, e.g. *Fucus evanescens*); (2) parallel communities (of closely related species, for example, *F. distichus* and *F. evanescens*; (3) isomorphic or convergent communities (of distantly or unrelated species, but of similar life-form, for example, *F. serratus* and *Hedophyllum sessile*); and (4) anisomorphic communities (of taxa which share only functional properties, as perhaps in the case of sublittoral kelp forests and reef-forming corals). Parallelism and convergence are to be found in each of the primary zones of marine benthos, as the following examples will indicate.

In the following pages, attention is given primarily to the vertical distribution of plants and primary consumers. The importance of predators in determining community structure is great, but could not be dealt with effectively in a short account. The author also regrets any nomenclatural errors relating to unfamiliar taxa.

THE SUPRALITTORAL ZONE

Principal features

The upper boundary of the supralittoral zone is the position beyond which sea spray ceases to be an effective force in determining community structure. On coasts subject to strong wave action, its influence may be detected several kilometres inland (Stephenson and Stephenson, 1972); precise measurements of the relationship between spray droplet size and mobility have been given by Boyce (1954).

The lichen vegetation of the upper supralittoral (Fig. 4.5) has been divided by Fletcher (1980) into four bands. These are (1) the mesic supralittoral, (2) the sub-mesic supralittoral, (3) the xeric supralittoral, and (uppermost) (4) the terrestrial-halophilic. These bands are perhaps not always sharply delimited, especially when the lichens are obscured by halophilic flowering plants or mosses. The upper supralittoral zone is unquestionably maritime rather than marine in character. Its vegetation consists almost entirely of salt-tolerant species with marked terrestrial affinities; its ecology is therefore a little outside the scope of this chapter (see Van der Maarel, in prep.).

The lower supralittoral zone (= littoral fringe, *sensu* Lewis, 1961) is characteristically black in appearance, although colour variants are known. The blackening is due chiefly to the presence of lichens, usually species of *Verrucaria*, and/or of Cyanobacteria. Fletcher (1980) has commented on the problems of drawing ecological conclusions from information published about this zone because misidentification of lichens is common; the same taxonomic difficulties, it might be added, surround the Cyanobacteria. The lower supralittoral is more distinctly marine in character than the upper supralittoral, and several examples of parallelism are to be found among its biota.

Lichens

The characteristic lichen of the lower supralittoral on North Atlantic coasts is *Verrucaria maura*, although other species may be common, such as *V. amphibia* and *V. ditmarsica* (Fletcher, 1980) and *V. microspora* (Lewis, 1964). *Verrucaria maura* has been reported from Spitsbergen (Svalbard) (Svendsen, 1959), from the Baltic (Hällfors et al., 1981), from the North American Arctic and Subarctic (Ellis and Wilce, 1961) and from Newfoundland (Bolton, 1981). On Mediterranean shores *V. maura* may be replaced by *V. antricola* or *V. symbalana* (Ballesteros, 1984a,b) and on South Australian shores the genus is represented by *V. microsporoides* (Womersley and Edmonds, 1958). *Verrucaria* is also present in the lower supralittoral of Subantarctic and Antarctic shores (Heywood and Whitaker, 1984) and on those of Pacific North America (Carefoot, 1977). It is frequently accompanied by other lichens, notably of the genus *Lichina, L. confinis* having been reported from the coasts both of Britain (Fletcher, 1980) and South Australia (King et al., 1971; King, 1973).

Cyanobacteria

Cyanobacteria are common associates of black lichens on temperate coasts; indeed they may replace lichens altogether, as on certain British shores (Lewis, 1964). Their importance seems to be somewhat diminished in subpolar regions, however, and they are evidently absent altogether from the Antarctic supralittoral (Heywood and Whitaker, 1984). Nevertheless, they may be quite abundant on coasts of Labrador and northwestern Newfoundland (Wilce, 1959). Places where the lower supralittoral is vegetated exclusively by Cyanobacteria are to be found mainly in warm temperate, subtropical and tropical regions. Examples are to be found in New Zealand (Chapman, 1950; Dellow, 1950), in South Australia (May et al., 1970), in Queensland, Australia (Cribb, 1965a), on Indo-Pacific atolls (Taylor, 1971; Russell, 1981), on Caribbean shores (Brattström, 1980), on West and East African coasts (Lawson, 1966, 1969), on those of the Canary Islands (Lawson and Norton, 1971), in southwestern France (Renoux-Meunier, 1965) and on the Mediterranean coast of Israel (Lipkin and Safriel, 1971). The species recorded belong to numerous genera including *Calothrix, Entophysalis, Hydrocoleus, Hyella, Lyngbya, Oscillatoria, Plectonema, Rivularia, Schizothrix* and *Scytonema*. In view of the considerable taxonomic uncertainties in Cyanobacteria, it would be a little rash to discuss their vertical distributions in terms of parallelism or convergence.

Cyanobacterial taxa may occupy different bands within the lower supralittoral. Examples of this kind of patterning are to be found in the Mediterranean (Ballesteros, 1984a) and at Aldabra in the Indian Ocean (Potts and Whitton, 1980).

Other algae

On chalk cliffs of the English Channel, the lower supralittoral may be vegetated by microscopic green algae (*Endoderma* spp.) and by benthic Chrysophyceae (Anand, 1937a,b; Tittley and Shaw, 1980). These algae give this subzone a green and/or brown colour, replacing the more familiar blackening organisms. The latter may also be overgrown, on many shores, by populations of larger and more conspicuous algae. These are usually annuals, and their development is often confined to certain favourable seasons. They include members of the following rather cosmopolitan genera: *Bangia, Porphyra* (red algae), *Ectocarpus* (brown algae), *Prasiola, Blidingia, Ulothrix, Urospora* (green algae). The shaded and humid conditions of caves promote the development of another type of supralittoral flora which includes species of *Audouinella* (= *Rhodochorton*) and the curious ectocarpoid alga *Waerniella*, which has been found in caves both of northern Europe (Waern, 1952) and of Queensland, Australia (Cribb, 1965b). Under dense shade from overhanging trees, tropical shores may be vegetated by a short turf of *Polysiphonia howei* (Russell, 1981).

Truly perennial seaweeds are relatively rare in the supralittoral zone, and their presence seems to be confined to shores where wave action is strong, and the atmosphere humid. For example, in the cool, damp climate of northern Scotland, the exposed-shore form of *Fucus distichus* is present (Lewis, 1964). *Pelvetia canaliculata* (North Atlantic) and *P. wrightii* (U.S.S.R., Sea of Japan) may also be common in this zone; they also provide an example of ecological parallelism in fucoids (Kussakin, 1977). Figure 4.6 illustrates lower-supralittoral algae at a site in northern Scotland, and the

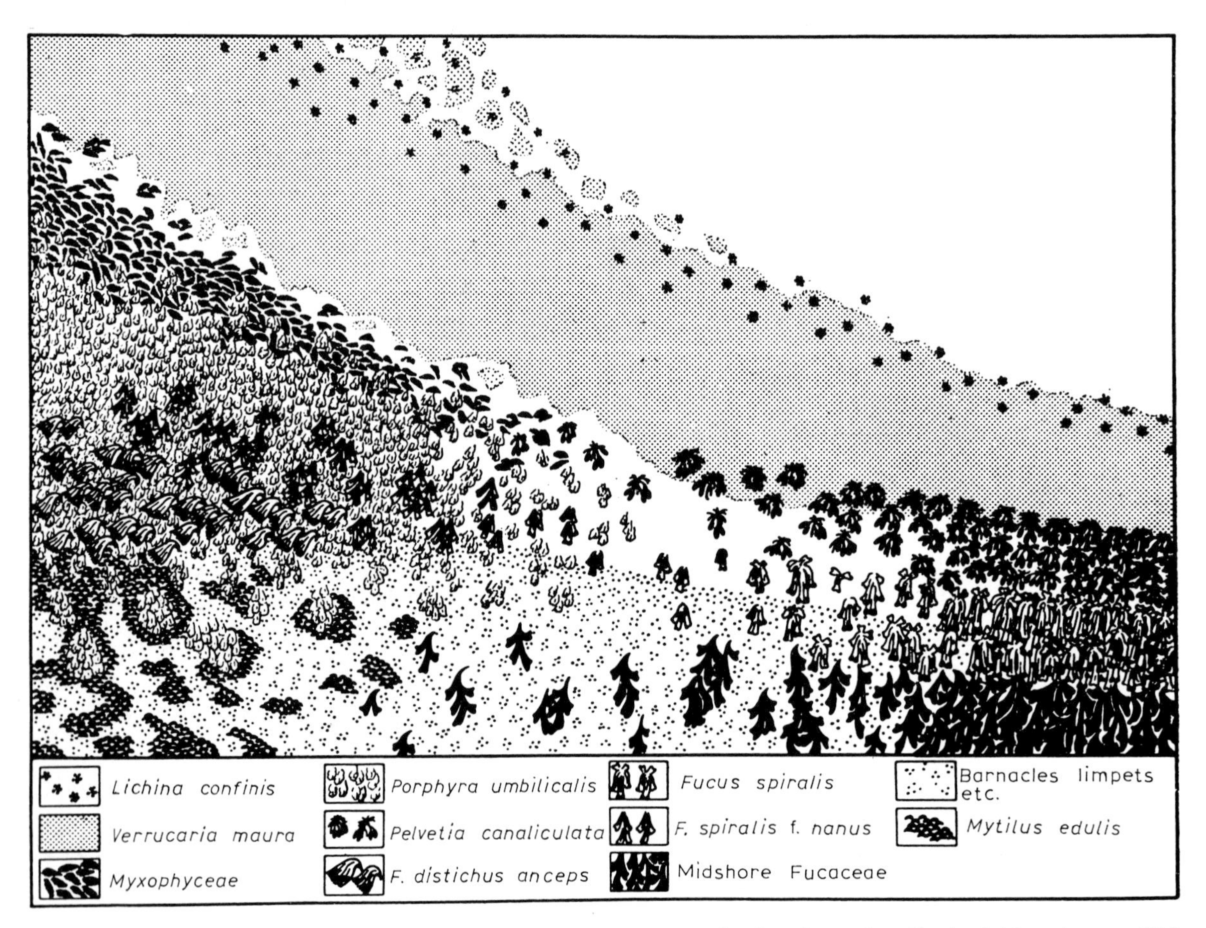

Fig. 4.6. Lower supralittoral zone (and upper part of eulittoral zone) on a rocky shore in northern Scotland. Note absence of lichen *Verrucaria maura* from the lowest levels of the subzone and its replacement by various annual and perennial algae. Note also widening and elevating effects of wave action (horizontal axis) on zones as well as vegetation continuum in response to this factor. (From Lewis, 1964.)

kind of vegetation continuum associated with wave action.

Littorinid snails

Littorinid gastropods are probably the most widespread and characteristic animals of the lower supralittoral zone, the most common being species of *Littorina* (or *Melarhaphe*). *Littorina saxatilis* and, at higher levels, *L. neritoides* are frequent on North Atlantic shores (Lewis, 1964). The latter is also found at the Canary Islands (Lawson and Norton, 1971) and in the Mediterranean (Lipkin and Safriel, 1971; Ballesteros, 1984b). In New Zealand and South Australia, the common supralittoral snails include *Melanerita atramento*, *Melarhaphe oliveri*, *M. paludinella*, *M. praetermissa*, *M. unifasciata* and *Nodilittorina pyramidalis* (Dellow, 1950; Womersley and Edmonds, 1958; King et al., 1971; King, 1973; May et al., 1970). *Littorina punctata* is characteristic of West African shores (Lawson, 1966) and, at Aldabra in the Indian Ocean, *L. glabrata* and *L. scabra* are common, together with *Nerita albicilla*, *N. plicata*, *N. textilis* and *N. undata* (Taylor, 1971). *Littorina angustior*, *L. ziczac* and *Nodilittorina tuberculata* occur in the lower supralittoral of the Caribbean coast of Colombia (Brattström, 1980). Other supralittoral snails include species of *Peasiella*, *Risella* and *Tectarius* (Pérès, 1982b).

These animals are herbivores and their grazing may have a marked effect upon the vegetation structure of the community. The supralittoral zone of the Vishákhapatnam coast (India) has been reported devoid of algae, yet with a *Littorina* population present (Umamaheswararao and Sreeramulu, 1964). It is possible that the absence of algae was due to climatic severity but it is also possible that grazing was responsible. Figure 4.7 demonstrates the effect of grazing by *L. saxatilis* on the vegetation of the lower supralittoral zone at the Isle of Man, British Isles. Black spots represent small cavities occupied by snails, while clear areas have been grazed free of algae; the presence of Cyanobacteria and the green alga *Blidingia minima* is indicated by dotted and squared areas, respectively. After making this observation, Hawkins and Hartnoll (1983) removed the snails and filled in the holes, an operation which resulted in continuous algal cover in a few months.

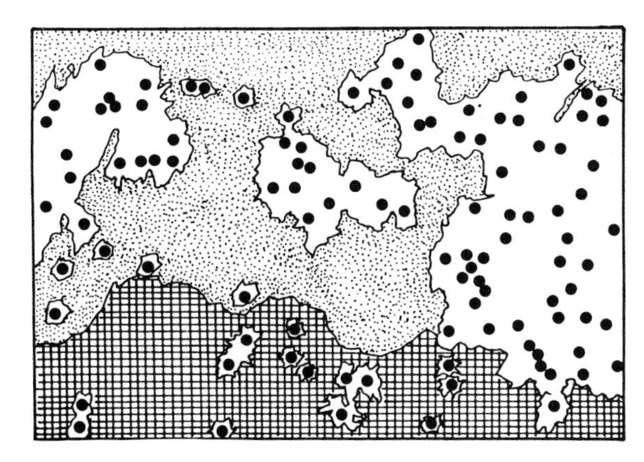

Fig. 4.7. Spatial heterogeneity (patchiness) of algal vegetation in lower supralittoral zone on a shore of the Isle of Man. Dotted areas indicate cover by Cyanobacteria, squared areas cover by *Blidingia minima*, and white areas bare rock. The black spots mark positions of pits inhabited by the snail *Littorina saxatilis*. Subsequent removal of animals and sealing up of pits led to complete vegetation cover in several months. (From Hawkins and Hartnoll, 1983.)

Other fauna

Amphipod crustaceans belonging to the genera *Ligia* and *Megaligia* are common residents of the lower supralittoral and, as with the littorinids, some parallelism is evident. The common sea slater of Pacific American coasts is *L. pallasii* (Carefoot, 1977); *L. oceanica* is characteristic of the Atlantic coasts of Europe (Lewis, 1964), *L. italica* of the Mediterranean Sea (Lipkin and Safriel, 1971; Ballesteros, 1984b) and *L. australiensis* of South Australia (Womersley and Edmonds, 1958). These are detritus eaters (Pérès, 1982b), although *L. pallasii* has a particular preference for drift algae (Carefoot, 1977).

Supralittoral crevices are sometimes populated by barnacles, animals which are more characteristic of the eulittoral zone, such as *Chthamalus depressus* and *C. stellatus* (Lewis, 1964; Ballesteros, 1984b). The lower supralittoral may also be visited by animals that do not normally reside there. On tropical shores, grapsid and hermit crabs (*Grapsus* and *Calcinus* spp.) may occur in very large numbers. Other opportunistic visitors include insects (Formicidae), arachnids, birds, and small rodents.

THE EULITTORAL ZONE

Principal features

Plants and animals of the eulittoral zone are subject to periodic total immersion. Few are well adapted to very prolonged emergence, however, and some are ill-adapted to permanent submergence. On tidal shores, the lower boundary of this zone lies at, or a little above, mean low water of spring tides, but the limits of the eulittoral, like those of the supralittoral, are raised and enlarged with increasing wave action (Fig. 4.5). Subzones are variable in number. On certain tropical shores, the zone is rather uniform in structure (Umamaheswararao and Sreeramulu, 1964) but other examples are readily divisible into 2, 3, or more subzones (Figs. 4.1 and 4.2). According to Pérès (1982b), two subzones are usually distinguishable, while the present account suggests that three may be recognized. In practice, neither of these somewhat arbitrary alternatives can adequately convey the immense structural and environmental complexity of this zone.

Algae

On Arctic and Antarctic shores, the upper part of the eulittoral zone is barren for most of the year, the lack of life being due to the grinding action of ice (Ellis and Wilce, 1961; Heywood and Whitaker, 1984). During summer ice-melt periods, however, this subzone may become vegetated by benthic diatoms and ephemeral algae such as *Ulothrix* (*U. pseudoflacca* and *U. australis* in Arctic and Antarctic, respectively). Rock crevices in the barren area may escape ice abrasion and harbour a few of the tougher perennial algae such as the crustose *Hildenbrandia*.

In lower latitudes, where ice action is less severe or absent, the upper eulittoral may contain upright perennial forms, but this part of the zone is never rich in species. On northern and Pacific coasts of the U.S.S.R., species of *Pelvetia* (*P. babingtoni*, *P. wrightii*: Gurjanova, 1968; Yamada, 1980) are present, accompanied by *Fucus distichus* and/or *F. evanescens*. *Fucus distichus* also occurs in the upper eulittoral at Spitsbergen (Svalbard) (Svendsen, 1959), but in Iceland *F. spiralis* and *Pelvetia canaliculata* are present (Munda, 1972). These last two species are common in the upper eulittoral zone of many North Atlantic shores. In Newfoundland, *F. spiralis* is accompanied by *F. distichus* (Bolton, 1981), as may be the case on certain very exposed shores in northern Scotland (Lewis, 1964). A degree of parallelism is therefore evident among the upper eulittoral fucoids of the Northern Hemisphere.

These fucoids often have an undergrowth of small red algae such as *Catenella* (Fig. 4.2) or *Gelidium* spp. (Renoux-Meunier, 1965). On the Pacific coast of North America, this kind of association is represented by *Pelvetiopsis limitata* (Carefoot, 1977) or *Pelvetia fastigiata* (Gunnill, 1982) with the red alga *Endocladia muricata*. A ground layer of crustose algae such as *Hildenbrandia* (Fig. 4.2) or, on Atlantic coasts, the amphibious lichen *Verrucaria mucosa* is usually present (Munda, 1972). The fruticose lichen *Lichina pygmaea* may also be found at this level (Lewis, 1964). These perennial plants are frequently associated with annual ephemeral forms, some of which (*Bangia*, *Ulothrix*, *Porphyra*, *Prasiola*) are found also in the lower supralittoral zone.

Fucus spiralis has been reported from the Canary Islands (Lawson and Norton, 1971) and from the Atlantic coast of Morocco (Gayral, 1958), but these localities lie close to its southern limit. Further south, the upper eulittoral assumes a different character with a greater proportion of annual species. On the West African coast, these include species of *Bachelotia*, *Bangia*, *Enteromorpha*, *Porphyra* and *Ulva* (Lawson, 1966), while on the East African coast *Bostrychia* spp. are common (Lawson, 1969). The importance of Cyanobacteria also increases when the tropics are reached (Van den Hoek et al., 1978; Potts and Whitton, 1980; Littler and Littler, 1984), although these are usually accompanied by eukaryotic algae. The species diversity of the eulittoral zone in the tropics is greatly increased by wave action. Exposed shores of an Indian Ocean atoll and of a section of the Great Barrier Reef have been reported to have species of lithothamnia, and also *Boodlea*, *Centroceras*, *Cladophora*, *Ectocarpus*, *Enteromorpha*, *Gelidiella*, *Lophosiphonia*, *Sphacelaria* and *Ulva* (Russell, 1981; Cribb, 1965a). Severe climatic conditions may, however, result in the formation of a barren upper eulittoral, analogous to that of polar regions (Brattström, 1980).

On coasts of South Australia and New Zealand, the upper eulittoral zone contains lithothamnia, *Bangia, Cladophora, Ectocarpus, Gelidium, Scytosiphon* and *Ulva* (Dellow, 1950; May et al., 1970; King, 1973); the common *Hildenbrandia* of North Atlantic coasts (*H. rubra*) has its counterpart here in *H. occidentalis* (May et al., 1970).

The middle and lower levels of the eulittoral zone throughout the world support richer floras than the upper subzone. In temperate conditions the vegetation may have a dense canopy of fucoid algae including species of *Ascophyllum*, *Fucus* and *Himanthalia* (Northern Hemisphere) or *Hormosira* (Southern Hemisphere) (Womersley and Edmonds, 1958; Lewis, 1964; Carefoot, 1977). Middle and low eulittoral subzones in Pacific North America may also have kelp canopies formed from *Hedophyllum sessile* (Carefoot, 1977) or *Postelsia palmaeformis* (Stephenson and Stephenson, 1972); the latter occurs in very characteristic clumps, as is depicted in Fig. 4.8.

The middle and low subzones on Arctic and Subarctic coasts contain *Fucus* spp. (*F. distichus* and *F. evanescens*), but with numerous smaller algae including species of *Blidingia*, *Chordaria*, *Pilayella*, *Prasiola*, *Ptilota*, *Rhodomela*, *Sphacelaria*, *Stictyosiphon* and *Ulothrix* (Ellis and Wilce, 1961). The same levels on Antarctic and Subantarctic coasts share a few taxa such as *Blidingia*, but have a high proportion of endemic species. The vegetation contains *Adenocystis*, *Chaetomorpha*, *Curdiea*, *Gigartina*, *Iridaea*, *Leptosomia*, *Rhodoglossum* and *Urospora* spp. (Price and Redfearn, 1968; Heywood and Whitaker, 1984).

Vegetational differences between middle and low subzones are not always distinct, but the latter frequently has a short and floristically complex turf. Though containing species from all major algal groups, the turf is composed mainly of red algae and the term "red algal turf" is common in the literature. On temperate North Atlantic shores, this turf includes species of *Corallina*, *Gigartina*, *Laurencia*, *Lomentaria* and *Palmaria*, all of which have fairly robust thalli; but mixed with these are numerous more delicate forms of *Membranoptera*, *Plumaria*, *Polysiphonia* and *Ptilota* (Lewis, 1964).

Fig. 4.8. Drawing of the eulittoral and upper sublittoral zones at Mission Point, Pacific Grove, California by T.A. Stephenson. Note distribution of barnacles in the upper, and mussels in the lower, parts of the eulittoral zone. Note also the patchiness of the lower eulittoral introduced by clumps of the kelp *Postelsia palmaeformis*. (From Stephenson and Stephenson, 1972.)

Corallina officinalis, a common turf alga of the North Atlantic, is also present in South Australia (King et al., 1971), but some parallelism also occurs insofar as the species of *Laurencia* differ between the two localities. Other *Laurencia* species characterize tropical examples of the algal turf (Cribb, 1965a; Van den Hoek et al., 1978), which contain, in addition, species of *Acanthophora*, *Amphiroa*, *Bryopsis*, *Centroceras*, *Codium*, *Corallina*, *Dictyota*, *Ectocarpus*, *Enteromorpha*, *Gelidiella*, *Giffordia*, *Gracilaria*, *Hypnea*, *Jania*, *Sphacelaria* and *Ulva* (Lawson, 1966, 1969; Van den Hoek et al., 1978; Brattström, 1980; Russell, 1981).

Benthic plankton feeders

The eulittoral zone contains a large number of benthic animals which obtain their food by removing suspended particles and planktonic organisms from seawater. These include cirriped crustaceans (barnacles) and bivalve molluscs (mussels and oysters). As with the algae, some parallelism is evident among these.

Barnacles tend to be most successful in the upper half of the eulittoral zone (Fig. 4.8). The very widespread genus *Chthamalus* is represented in South Australian waters by *C. antennatus* (Womersley and Edmonds, 1958; King et al., 1971; King, 1973), on Pacific North American coasts by *C. dalli* (Carefoot, 1977), on northern Pacific coasts of the U.S.S.R. by *C. challengeri* (Gurjanova, 1968), on Caribbean shores by *C. angustitergum* (Brattström, 1980) and, on temperate and warm Atlantic shores by *C. stellatus* (Lewis, 1964; Lawson, 1966; Lawson and Norton, 1971).

In the North Atlantic, *Semibalanus (Balanus) balanoides* is an extremely abundant component of the eulittoral zone (Lewis, 1964; Bolton, 1981) but in Autralasian waters it is replaced by *Chamaesipho columna, Elminius modestus* and *E. plicatus* (Chapman, 1950; Dellow, 1950; Womersley and Edmonds, 1958; May et al., 1970; King et al., 1971). The genus *Tetraclita* is widespread on tropical shores with, for instance, *T. squamosa* in Brazil (Pérès, 1982b), and *T. stalactifera* in Colombia (Brattström, 1980).

Mussels, when present in abundance, tend to occupy lower levels of the eulittoral zone than barnacles (Fig. 4.8). They likewise provide examples of eulittoral-zone parallelism. The common eulittoral mussel of North Atlantic coasts is *Mytilus edulis* (Lewis, 1964; Bolton, 1981), but its counterpart on American North Pacific coasts is *M. perna* (Lawson, 1966) and, in the Mediterranean, *M. galloprovincialis* (Ballesteros, 1984b).

In warm seas, the common bivalves of the low eulittoral zone may be oysters such as *Crassostrea cucullata*, found in the Indian Ocean (Taylor, 1971), and *C. glomerata* in New Zealand (Dellow, 1950).

Herbivores

Littorinid and other snails are prominent among eulittoral herbivores. *Littorina littorea* is abundant on American and European Altantic coasts (Lewis, 1964; Lubchenco, 1978), but *L. obtusata* (*L. littoralis*) is also common (Lewis, 1964). Pacific North American species include *L. scutulata* and *L. sitkana* (Carefoot, 1977), while *L. glabrata* is frequent in the Indian Ocean eulittoral (Taylor, 1971). Other genera which appear to be widespread in the Southern Hemisphere include *Bembicium, Melanerita* and *Nerita*.

Perhaps more characteristic of the eulittoral zone than these, however, are the limpets or limpet-like pulmonate molluscs such as *Fissurella* and *Siphonaria* (Pérès, 1982b). On shores of northwestern Europe the commonest limpets are species of *Patella (P. aspera, P. vulgata*: Lewis, 1964); while on those of Pacific North America several species of *Acmaea* are the principal limpet herbivores (Carefoot, 1977). *Patelloida* and *Siphonaria* are frequent in Australian waters (Womersley and Edmonds, 1958; May et al., 1970; King, 1973), as they are in other warm southern seas (Pérès, 1982b).

Species interactions

Species interactions on rocky shores are now known to be numerous and diverse; to review the very considerable body of literature on the subject is outside the scope of this chapter. Nevertheless, it is pertinent to refer briefly to several specific interactions which are important determinants of pattern in certain eulittoral communities.

The low-eulittoral kelp *Hedophyllum sessile* of the Pacific coasts of North America (see subsection on Algae above) has been shown by Dayton

(1975a, b) to be the competitive dominant ("foundation species") alga of its community. Experimental removal of *Hedophyllum* led to a die-back of its understorey algae and a temporary increase in ephemeral "fugitive species", before *Hedophyllum* recovery took place.

The mussel *Mytilus californianus* is the dominant organism on certain exposed Pacific American shores, but its cover may be interrupted by clumps of the annual kelp *Postelsia palmaeformis* (Fig. 4.8). It has been shown by Dayton (1973) and by Paine (1979) that the existence of *Postelsia* groves is maintained because diaspores settle and grow on encroaching mussels, the epizooic plants making them liable to removal by wave action. The kelp therefore requires recurrent wave disturbance for its persistence. A comparable example of interaction between benthic algae (fucoids) and animals (chiefly *Semibalanus balanoides*) is illustrated in Fig. 4.9. In this case, the species interaction revealed by Hartnoll and Hawkins (1985) is once again subject to the modifying effect of wave action. The *Mytilus–Postelsia* subzone depicted in Fig. 4.8 shows the kind of spatial heterogeneity that may occur in the eulittoral zone. A similar example of patchiness, among North Atlantic fucoids, has been found by Hawkins and Hartnoll (1983) to be related to the browsing activities of *Patella* spp.

The chief herbivore of the low eulittoral on New England shores is the snail *Littorina littorea*. Selective grazing by this animal determines the species composition of rock-pool vegetation (Lubchenco, 1978). The dominant low-eulittoral alga of open rock in this locality is *Chondrus crispus*, but experimental removal of this plant enables *Fucus* spp. to colonize the vacant space. However, *Fucus* establishment is most successful when *Littorina* is also excluded, and the latter has therefore an important secondary role in determining subzone structure (Lubchenco, 1980).

Finally, space competition may occur also between benthic animal species, as for example between the barnacles *Chthamalus stellatus* and *Semibalanus (Balanus) balanoides* (see Carefoot, 1977).

Non-tidal analogues

According to Pérès (1982b), the eulittoral zone on shores which lack, or have only very small, tides

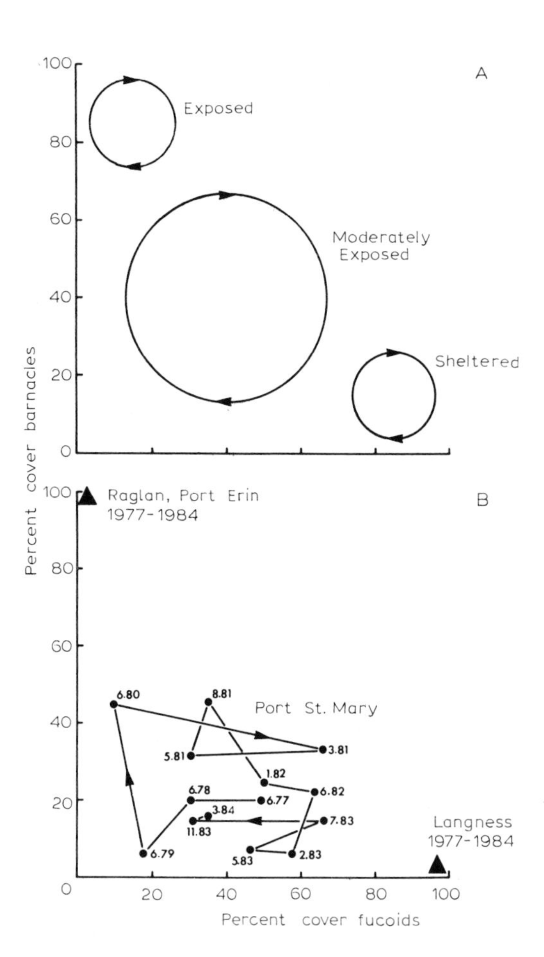

Fig. 4.9. (A) Hypothetical relationship between cover of fucoid algae and of barnacles in relation to wave action. (B) Actual cover-abundance data for fucoids and barnacles of three sites on the Isle of Man. The exposed (Raglan) and sheltered (Langness) shores lacked fucoids and barnacles respectively. The connected points indicate temporal changes in abundances of these organisms at Port St. Mary between June 1977 and March 1984. (From Hartnoll and Hawkins, 1985.)

is divisible into two distinct subzones. This analysis certainly holds for certain Mediterranean shores (Fig. 4.10). In this example, there is an upper subzone comprising the encrusting alga *Ralfsia verrucosa* and the barnacle *Chthamalus stellatus*, while the lower subzone consists of a platform of the encrusting coralline alga *Lithophyllum tortuosum* together with the mussel *Mytilus galloprovincialis* and the herbivores *Acanthochiton* spp. and *Patella* spp. (Ballesteros, 1984b).

Changes in seawater level in the Mediterranean are controlled largely by changes in barometric pressure and wind force (Colman and Stephenson, 1966). This is also true of the similarly "tideless"

ZABALA 82

Baltic Sea, but its eulittoral zone cannot readily be resolved into two subzones. In early spring, the eulittoral of the Baltic proper is often rather bare as a consequence of winter ice and subsequent exposure to air. As the water level rises, the eulittoral becomes vegetated by ephemeral algae such as *Dictyosiphon*, *Enteromorpha* and *Pilayella* spp. but including also the freshwater *Cladophora glomerata*. Marine crustaceans such as *Gammarus* species and *Idotea* graze these algae, but the herbivorous gastropods present include salt-adapted freshwater taxa such as *Lymnaea* and *Theodoxus* (Hällfors et al., 1981).

THE SUBLITTORAL ZONE

Principal features

The sublittoral zone extends downwards from its boundary with the eulittoral to the point at which growth of macroalgae ceases. In terms of depth, this position is very variable. A sublittoral zone may be absent altogether in turbid coastal waters (Russell, 1972), while the record depth for macroalgae stands presently at 268 m. The plants in the latter case were crustose corallines, collected from a submarine peak off the Bahamas by Littler et al. (1985).

Figure 4.5 indicates that the sublittoral zone is divisible into three subzones; however, as with the eulittoral zone, this number is liable to vary. The uppermost subzone is recognized by some authors as distinctive in composition (Stephenson and Stephenson, 1972) and referred to as the sublittoral fringe, while the lower sublittoral has been treated by others as a separate circalittoral zone (Pérès, 1982b). The widening and deepening effect of wave action upon subzones shown in Fig. 4.5 follows the observations of Shepherd and Womersley (1970) and Edgar (1984), but there is rather less evidence on the influence of wave action on sublittoral zonation than there is on the intertidal zones.

Structurally, the sublittoral communities can be single-layered or multilayered and, in the latter case, are sometimes of very considerable complexity. For the plants in this zone, light is probably the most important structuring agency.

Algae

Single-layered plant communities of the sublittoral zone are likely to consist of coralline encrusting red algae. They occur widely in the low sublittoral (see above), but are also to be found in shallower waters, especially if grazing is intense. Examples include *Porolithon* communities of coral-reef ridges (Littler and Littler, 1984), and *Lithophyllum* vegetation of Mediterranean shores (Pérès, 1982b).

Encrusting corallines may be accompanied by turf-forming (usually red) algae, giving rise to a two-layered community. Some authors distinguish shade- and light-loving turf species (Pérès, 1982b) but, in practice, the light optima of species are unlikely to be rigidly classifiable in this way. An example of a sciaphilic algal turf from the Mediterranean is given in Fig. 4.11. This diagram, by Boudouresque and Lück (1972) also illustrates very clearly the patchiness in species distribution which may be present in the sublittoral zone.

Examples of turf vegetation may occur locally in cool-water habitats; Lewis (1964) describes such a community in a sea loch of Western Scotland. Turves like this usually contain some species from the low-eulittoral zone, together with other taxa of a more truly sublittoral distribution (*Delesseria, Odonthalia, Phycodrys, Phyllophora, Plocamium* etc.). However, it is in the warm waters of tropical shores, and especially in the shallows formed to landward of coral-reef ridges (reef flats), that algal turves are routinely found. These are extremely complex floristically; examples described by Cribb

Fig. 4.10. Drawing of zonation on a Mediterranean shore (Illes Medes, Spain). Supralittoral zone (upper left) contains the lichen *Verrucaria symbalana* with barnacles (*Chthamalus depressus*) in crevices. The upper eulittoral zone (middle left) contains the encrusting brown alga *Ralfsia verrucosa* with barnacles (*Chthamalus stellatus*) and limpets (*Patella rustica*). The lower eulittoral zone (from lower left to top right) consists of a shelf of the encrusting coralline algae *Lithophyllum tortuosum* with the mussel *Mytilus galloprovincialis*. Herbivores present include *Patella caerulea, Acanthochiton fasicularis* and *Paracentrotus lividus*. The upper sublittoral zone (bottom) contains numerous algae, barnacles (*Balanus perforatus*) and sponges. (This drawing by M. Zabala is reproduced with the kind permission of the editors of *Els Sistemes Naturals de les Illes Medes*, Institut d'Estudis Catalans, Barcelona.)

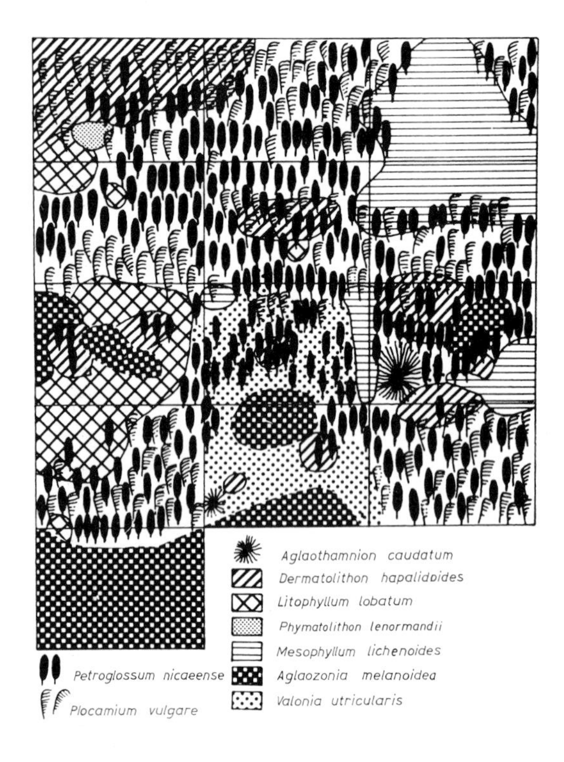

Fig. 4.11. Diagram of sciaphilic algal turves from the Mediterranean sublittoral zone, after Boudouresque and Lück (1972): note patchiness.

(1965a) and Russell (1981) contained members of the genera *Acanthophora*, *Boodlea*, *Caulerpa*, *Centroceras*, *Ceramium*, *Champia*, *Cladophoropsis*, *Dictyosphaeria*, *Dictyota*, *Ectocarpus*, *Gelidiella*, *Herposiphonia*, *Hypnea*, *Jania*, *Laurencia* and *Sphacelaria* (see Dubinsky, 1990).

Where the sublittoral vegetation comprises three or more layers, the higher strata usually consist of large brown algae. Three principal types of canopy may be obtained, dominated respectively by members of the orders Laminariales, Fucales and Desmarestiales. These types are not mutually exclusive, however, and mixed stands are commonplace.

Kelp forests are found in most of the cool- and cold-water regions of the world. The most complex structurally are probably located on the Pacific coast of North America, whereas floristically complex examples are to be found in Japan. Dayton et al. (1984) have described a five-layered Californian kelp community with *Macrocystis pyrifera* forming a floating canopy, *Eisenia* and *Pterygophora* a stipitate canopy and *Laminaria farlowii* a prostrate canopy. Beneath these lay an algal turf and an encrusting layer of calcified red algae. As an example of floristic complexity in the kelp community, the sublittoral vegetation of Hokkaido, Japan contains *Agarum cribrosum, Alaria crassifolia, A. praelongata, Laminaria angustata, L. diabolica, L. japonica, L. religiosa* and species of *Costaria, Ecklonia* and *Undaria* (Yamada, 1980).

North and mid-Atlantic kelp forests are simpler both structurally and floristically. On these coasts, the mid-sublittoral (Fig. 4.5) is usually dominated by a *Laminaria hyperborea* forest. The forest thins out to "parkland" in the lower sublittoral before ceasing altogether. Above *L. hyperborea*, the upper sublittoral usually contains different kelp species, *L. digitata* being particularly common (Lüning, 1970; Castric-Fey et al., 1973).

Fucoid-dominated sublittoral vegetation is often well-developed in the warm-water parts of the world, although a cold-water exception is found in the inner Baltic Sea where *Fucus vesiculosus* forms a canopy in the upper sublittoral (Hällfors et al., 1981). The most frequently encountered sublittoral fucoids of warm seas are *Cystoseira* and *Sargassum*, species of which are to be found on mid-Atlantic shores (Lawson, 1966; Donze, 1968; Lawson and Norton, 1971) and in the Indo-Pacific oceans (Umamaheswararao and Sreeramulu, 1964). Tropical and subtropical coasts may also carry stands of the fucoid *Turbinaria*, possibly in association with *Sargassum* (Taylor, 1965; Taylor, 1971).

South Australian coasts provide a number of good examples of kelps and fucoids occupying different subzones of the same sublittoral, although this kind of pattern is by no means exclusive to the Southern Hemisphere. Figure 4.12 shows a vegetation profile at Pearson Island, South Australia, with an upper belt of the fucoid *Cystophora* plus a red-algal turf, and a mid-sublittoral forest of the kelp *Ecklonia radiata* which thins into parkland in the lower sublittoral (Shepherd and Womersley, 1970).

Antarctica is unique among cold-water floras in that it contains no kelps (Moe and Silva, 1977). Among its large brown algae are certain kelp-like members of the Desmarestiales, notably *Himantothallus, Phaeoglossum* and *Phyllogigas* (Price and

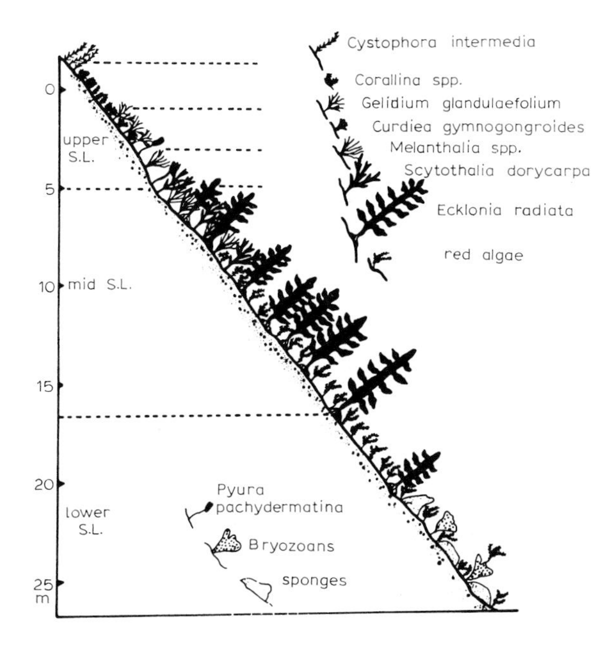

Fig. 4.12. Profile of sublittoral zone at West Island, South Australia, after Shepherd and Womersley (1970). The upper sublittoral is vegetated by the fucoid *Cystophora intermedia* and, at greater depths, by a red agal turf. The mid-sublittoral is dominated by a forest of the kelp *Ecklonia radiata*. The lower sublittoral contains a red-algal turf, sponges and bryozoans.

Redfearn, 1968; Richardson, 1979; Heywood and Whitaker, 1984) although these taxa are now considered congeneric (Moe and Silva, 1977). A similarity in appearance to kelps is possessed also by the fucoid *Durvillaea* which occupies the upper sublittoral of certain exposed shores of southern oceans (Womersley and Edmonds, 1958). The sublittoral zone may therefore be seen to contain examples of algal convergence and parallelism. It may also illustrate zonal divergence in that *Laminaria*, normally characteristic of upper and middle subzones, may occur only in the lower subzone, as in the case of *L. rodriguezii* on the Italian Mediterranean coast (Giaccone, 1967, 1969).

Competition and herbivory

Experiments by Harkin (1981), involving removal of the blades from *Laminaria hyperborea*, resulted in a significant increase in biomass of small algae on the remaining stipes. These algae were members of species which normally occur as stipe epiphytes and which may therefore be interpreted as obligate understorey species (*sensu* Dayton, 1975a). The absence of fugitive species of the kind that develop in intertidal clearance experiments (*Enteromorpha* spp., *Ulva* spp.) suggests a different pattern of revegetation in the sublittoral. There may also be a problem in applying to sublittoral macrophytes the functional terminology devised by Dayton because certain canopy species can evidently function also as fugitive species, for instance, *Alaria fistulosa* (Dayton, 1975b) and *Saccorhiza polyschides* (Kain, 1975).

Sublittoral vegetation is grazed by animals belonging to a variety of taxonomic groups, and their influence on community structure may be great. This has been demonstrated by Wanders (1977), who introduced experimental plates into the upper sublittoral zone at Curaçao, Netherlands Antilles. Plates protected by cages against the attention of herbivorous fish rapidly developed a dense vegetation of ephemeral algae, which were later succeeded by larger fleshy algae. Grazed plates bore only calcified encrusting algae such as *Porolithon* at the end of the experiment.

Preferential grazing by sea urchins is known also to exist (Vadas, 1977) and may produce similar effects. Likewise, these animals serve to illustrate parallelism among sublittoral animals. The common urchin of northwestern European waters is *Echinus esculentus*, but in Newfoundland the chief grazer is *Strongylocentrotus droebachiensis* (South, 1983), a species found also in Pacific North America (Carefoot, 1977) with *S. purpuratus*, the important herbivore of Californian shores (Dayton, 1975a). In warmer European waters *Paracentrotus lividus* is often abundant (see Fig. 4.10) while in the tropics *Diadema* species may be important grazers (Van den Hoek et al., 1978).

The impact of urchin grazing upon kelp vegetation has been demonstrated by Jones and Kain (1967), who systematically removed *Echinus* from a heavily-grazed area of sublittoral in the Isle of Man, British Isles. Continual removal of urchins from the area, which had previously lacked kelp vegetation, led to a restoration of *Laminaria* forest.

It has become evident from many animal removal and exclusion experiments that herbivores often preferentially graze soft algae. Encrusting calcified algae may therefore have selective advantage in heavily grazed ecosystems. It is thought that some algae have evolved chemical defences which make them unpalatable to herbivores (see Hawkins and Hartnoll, 1983). It is also possible that water movement causes sublittoral kelps physically to sweep herbivores away (Velimirov and Griffiths, 1979). Thus, organismic interactions of the kind that influence the community structure of the eulittoral zone evidently operate also in the sublittoral.

CAUSES OF ZONATION

The publication of *Life Between Tidemarks on Rocky Shores* marked the completion of the careers of T.A. and A. Stephenson; and it remains the outstanding achievement of a remarkable scientific and artistic partnership. Yet, even at the time of publication, the kind of ecology it represented had fallen into serious decline. It was felt that the numerous descriptions and illustrations of shores applied only to the particular moments of observation (Ricketts and Calvin, 1968). Conversely, interest was growing in the variations that might occur through time in species distributions at particular places. There was also a suspicion that the zones, associations and other units of descriptive ecology had more to do with the self-fulfilling nature of the systems that begat them than to the realities of life on the shore (Ricketts and Calvin, 1968; Russell, 1972). Rocky-shore ecology had, in the words of Lewis (1980), moved on from its descriptive to its dynamic phase, and the dwindling interest in zonation was directed more towards understanding its causes than to detailing its biological composition.

The quest for causes of zonation has concentrated primarily upon the physical and chemical conditions of the shore environment, although some have argued that zonation is mainly a consequence of interspecific competition (Chapman, 1973). Supralittoral marine plants and animals are subject to greater dehydration and to greater extremes of salinity than those in lower zones, and have correspondingly greater tolerances to these factors (Dring and Brown, 1982; Russell, 1987). The importance of biological interactions has not been neglected, however. It has been contended that upper limits of organisms are determined mainly by tolerances to environmental extremes ("stress") and lower limits to the effects of competition (Connell, 1972; Schonbeck and Norton, 1978, 1980). This view, though popular, may not be invariably correct; Hawkins and Hartnoll (1985) have, by experimental removal of competitors, enabled certain species of algae to extend their ranges upward as well as downward. In practice, vertical distributions of species are related to their physico-chemical tolerances, but boundaries between species are likely also to involve a degree of biological interaction: the former determines the potential range of a species, the latter its actual distribution (Connell, 1975). The causative relations between these factors and zonation have been discussed by Hawkins and Hartnoll (1983), from whose work Fig. 4.13 was drawn.

Tides have proved an irresistible attraction for those who seek to establish physical and biological correlation on rocky shores, despite the fact that they do not actually cause zonation (Stephenson and Stephenson, 1972). The idea of linking zonal and tidal patterns is an old one, going back to Vaillant (1891) and earlier (see Ricketts and Calvin, 1968). The term "critical tide levels", (CTL) was introduced by Colman (1933), who noted certain irregularities in the curve produced when percentage time of exposure to air was plotted against tide height. These irregularities in the emersion curve corresponded with major zonal boundaries on the shore at Wembury, southern England. Colman calculated his emersion curves from published tidal predictions, as has Underwood (1978), who was unable to detect any irregularities. Underwood therefore concluded that CTL, in Colman's sense, did not exist, although some confirmation that they might do so has been claimed by Hartnoll and Hawkins (1982) on the basis of actual (as opposed to predicted) tidal data.

The arguments over CTL are further complicated by the presence in the literature of a different system based, not upon mean tide levels, but upon the maximum period of continuous emersion (or submergence) experienced. This concept was introduced by Doty (1946) and has since been developed by Swinbanks (1982). The latter has concluded that variations in daily, monthly and annual tidal cycles are sufficient to generate many CTL, no matter what kind of tidal regime operates. As a result of his researches, Swinbanks has

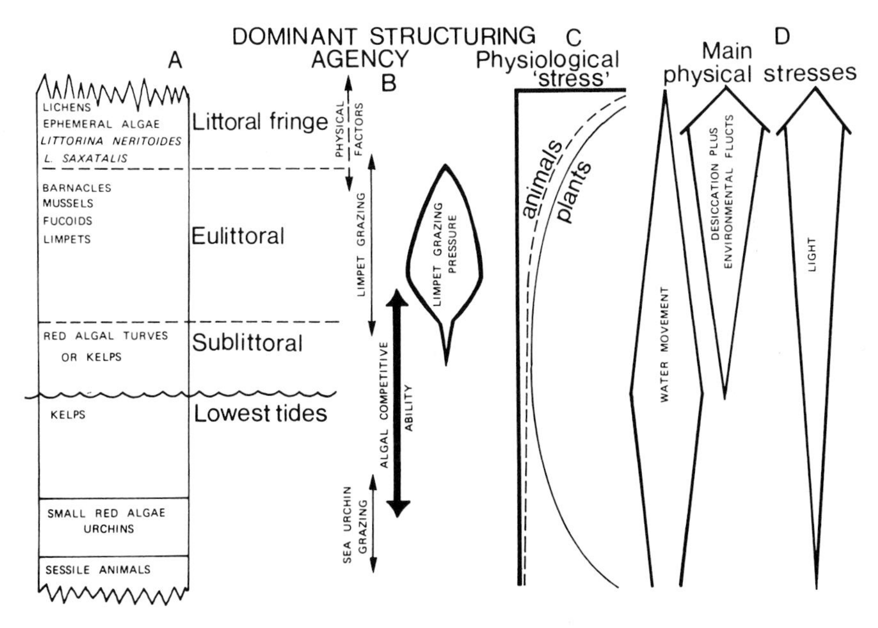

Fig. 4.13. Diagram showing how several important environmental forces may determine zonation on rocky shores, after Hawkins and Hartnoll (1983). (A) Zones with their principal animals and plants. (Note, littoral fringe = lower supralittoral zone in text.) (B) Respective importances of physical and biotic agencies operating at different shore heights. (C) Relative amounts of "stress" experienced by plants and animals at different shore heights. (Note, increased stress in sublittoral plants arises mainly from light deficiency.) (D) Principal physical stress involved in zonation.

proposed a new three-zone terminology for use on rocky shores (atmozone, amphizone, aquazone) but admits that causal relations with biological zonation remain to be established. If CTL do exist, then they may be expected to occur only on sheltered shores (Fig. 4.5); and even on these, the effects of biological interactions may nullify those of tides.

The dynamic phase of intertidal ecology has been especially marked by increasing application of the precepts of population ecology. From careful observation of small areas of substrate (permanent quadrats), details of species recruitment and survivorship have been obtained annually and seasonally. Analyses of this kind have been made both on benthic animals (see, for example, Kendall et al., 1985) and algae (see Chapman, 1986). These observations have demonstrated the presence of spatial and temporal heterogeneity among marine benthic organisms, which is also evident in Figs. 4.1, 4.7, 4.8 and 4.11. In one case at least, the dynamics of *Fucus*–barnacle–*Patella* patchiness have been modelled (Hartnoll and Hawkins, 1985). These findings evoke agreement with the comments of Den Boer (1968):

> "...After some time, I realized that heterogeneity and instability must not be considered as just a drawback of field data to be neglected (averaged away or seen through by intuition) or circumvented by retreating into the laboratory because they are mere deviations from the typical or representative case (or even noise). On the contrary, heterogeneity and/or instability must be recognised as fundamental features of a natural situation."

Most shore ecologists have accepted that zonation is a fact [but see Ricketts and Calvin (1968) for at least one heretical view]. However it may prove as fruitless to pursue a search for a fundamental cause as to generalize over the infinitely variable patterns of species. More facts on vertical distribution seem likely to be forthcoming by attempting to elucidate the forces that initiate, maintain and change patterns at particular sites. This information will be supplied by the field worker, the only one who can tell us what actually happens in nature (Den Boer, 1968). Those who wish to be introduced to these new and exciting developments are referred to the collection of essays dedicated to J.R. Lewis, edited by Moore and Seed (1985).

REFERENCES

Anand, P.L., 1937a. An ecological study of the algae of the British chalk cliffs, Part I. *J. Ecol.*, 25: 153–188.

Anand, P.L., 1937b. An ecological study of the algae of the British chalk cliffs. Part II. *J. Ecol.*, 25: 344–367.

Ballantine, W.J., 1961. A biologically-defined exposure scale for the comparative description of rocky shores. *Field Stud.*, 1: 1–19.

Ballesteros, E., 1984a. Contribucio al coneixement dels cianofits de les Illes Medes. In: J. Ros., I. Olivella, and J.M. Gili (Editors), *Els Sistemes Naturels de les Illes Medes. Arxius de la Seccio de Ciències*, 73: 321–332.

Ballesteros, E., 1984b. Els estatges supralitoral i mediolitoral de les Illes Medes. In: J. Ros, I. Olivella and J.M. Gili (Editors), *Els Sistemes Naturels de les Illes Medes. Arxius de la Seccio de Ciències*, 73: 647–657.

Bolton, J.J., 1981. Community analysis of vertical zonation patterns on a Newfoundland rocky shore. *Aquat. Bot.*, 10: 299–316.

Boudouresque, C.-F., 1969. Une nouvelle méthode d'analyse phytosociologique et son utilisation pour l'étude des phytocoenoses benthiques. *Tethys*, 1: 529–534.

Boudouresque, C.-F., 1970. Recherches sur les concepts de biocoenose et de continuum au niveau de peuplements benthiques sciaphiles. *Vie et Milieu, Ser.* B, 21: 103–136.

Boudouresque, C.-F., 1971a. Méthodes d'étude qualitative et quantitative du benthos (en particulier du phytobenthos). *Tethys*, 3: 79–104.

Boudouresque, C.-F., 1971b. Contribution à l'étude phytosociologique des peuplements algaux des côtes Varoises. *Vegetatio*, 22: 83–184.

Boudouresque, C.-F., 1974. Aire minima et peuplements algaux marins. *Bull. Soc. Phycol. France*, 19: 141–157.

Boudouresque, C.-F. and Lück, H.B., 1972. Recherches de bionomie structurale au niveau d'un peuplement benthique sciaphile. *J. Exp. Mar. Biol. Ecol.*, 8: 133–144.

Boyce, S.G., 1954. The salt spray community. *Ecol. Monog.*, 24: 29–67.

Brattström, H., 1980. Rocky-shore zonation in the Santa Marta area, Colombia. *Sarsia*, 65: 163–226.

Burrows, E.M., Conway, E., Lodge, S.M. and Powell, H.T., 1954. The raising of intertidal algal zones on Fair Isle. *J. Ecol.*, 42: 283–288.

Carefoot, T., 1977. *Pacific Seashores. A Guide to Intertidal Ecology*. J.J. Douglas, Vancouver, 208 pp.

Castric-Fey, A., Girard-Descatoire, A., Lafargue, F. and L'Hardy-Halos, M-T., 1973. Etagement des algues et des invertébrés sessiles dans l'Archipel de Glénans. *Helgol. Wissensch. Meeresunters.*, 24: 490–509.

Chapman, A.R.O., 1973. A critique of prevailing attitudes towards the control of seaweed zonation on the sea shore. *Bot. Mar.*, 16: 80–82.

Chapman, A.R.O., 1974. The ecology of macroscopic marine algae. *Ann. Rev. Ecol. Syst.*, 5: 65–80.

Chapman, A.R.O., 1986. Population and community ecology of seaweeds. *Adv. Mar. Biol.*, 23: 1–161.

Chapman, V.J., 1950. The marine algal communities of Stanmore Bay, New Zealand (Studies in Intertidal Zonation, 1). *Pac. Sci.*, 4: 63–68.

Colman, J., 1933. The nature of the intertidal zonation of plants and animals. *J. Mar. Biol. Ass. U.K.*, 18: 435–476.

Colman, J.S. and Stephenson, A., 1966. Aspects of the ecology of a "tideless" shore. In : H. Barnes (Editor), *Some Contemporary Studies in Marine Science*. Allen and Unwin, London, pp. 163–170.

Connell, J.H., 1972. Community interactions on marine rocky intertidal shores. *Ann. Rev. Ecol. and Syst.*, 3: 169–192.

Connell, J.H., 1975. Some mechanisms producing structure in natural communities: A model and evidence from field experiments. In: M.L. Cody and J.M. Diamond (Editors), *Ecology and Evolution of Communities*. Harvard University Press, Cambridge, Mass., pp. 460–490.

Cribb, A.B., 1965a. The marine and terrestrial vegetation of Wilson Island, Great Barrier Reef. *Proc. R. Soc. Queensl.*, 77: 53–62.

Cribb, A.B., 1965b. An ecological taxonomic account of the algae of a semi-marine cavern, Paradise Cave, Queensland. *Univ. Queensl. Pap. Dep. Bot.*, 4: 259–282.

Dayton, P.K., 1972. Towards an understanding of community resilience and the potential effects of enrichments to the benthos at McMurdo Sound, Antarctica. In: B.C. Parker (Editor), *Colloquium on Conservation Problems in Antarctica*. Allen Press, London, pp. 81–96.

Dayton, P.K., 1973. Dispersion, dispersal and persistence of the annual intertidal alga *Postelsia palmaeformis* Ruprecht. *Ecology*, 54: 433–438.

Dayton, P.K., 1975a. Experimental evaluation of ecological dominance in a rocky intertidal algal community. *Ecol. Monog.*, 45: 137–159.

Dayton, P.K., 1975b. Experimental studies of algal canopy interactions in a sea otter dominated kelp community at Amchitka Island, Alaska. *Fish. Bull.*, 73: 230–237.

Dayton, P.K. and Tegner, M., 1984. The importance of scale in community ecology: A kelp forest example with terrestrial analogs. In: P.W. Price, C.N. Slobodchikoff and W.S. Gand (Editors), *A New Ecology: Novel Approaches to Interactive Systems*. Wiley, Chichester, pp. 457–481.

Dayton, P.K., Currie, V., Gerrodette, T., Keller, B.D., Rosenthal, R. and VenTresca, D., 1984. Patch dynamics and stability of some Californian kelp communities. *Ecol. Monogr.*, 54: 253–289.

Dellow, V., 1950. Inter-tidal ecology at Narrow Neck Reef, New Zealand (Studies in Intertidal Zonation, 3). *Pac. Sci.*, 4: 355–374.

Den Boer, P.J., 1968. Spreading of risk and stabilization of animal numbers. *Acta Biotheor.*, 18: 165–194.

Den Hartog, C., 1959. The epilithic algal communities occurring along the coast of the Netherlands. *Wentia*, 1: 1–241.

Donze, M., 1968. The algal vegetation of the Ria de Arosa (N.W. Spain). *Blumea*, 16: 159–183.

Doty, M.S., 1946. Critical tide factors that are correlated with the vertical distribution of marine algae and other organisms along the Pacific coast. *Ecology*, 27: 315–328.

Doty, M.S., 1957. Rocky intertidal surfaces. *Geol. Soc. Am. Mem.*, 67: 535–585.

Dring, M.J. and Brown, F.A., 1982. Photosynthesis of intertidal brown algae during and after periods of emersion: A renewed search for physiological causes of zonation. *Mar. Ecol. Prog. Ser.*, 8: 301–308.

Druehl, L.D., 1970. The pattern of Laminariales distribution in the northeast Pacific. *Phycologia*, 9: 237–247.

Dubinsky, Z. (Editor), 1990. *Coral Reefs. Ecosystems of the World, Vol. 25.* Elsevier, Amsterdam, 550 pp.

Edgar, G.J., 1984. General features of the ecology and biogeography of Tasmanian subtidal rocky shores. *Pap. Proc. R. Soc. Tasmania*, 118: 173–186.

Ellis, D.V. and Wilce, R.T., 1961. Arctic and subarctic examples of intertidal zonation. *Arctic*, 14: 224–235.

Feldmann, J., 1951. Ecology of marine algae. In: G.M. Smith (Editor), *Manual of Phycology–An Introduction to the Algae and their Biology*. Chronica Botanica, Waltham, Mass, pp. 313–334.

Field, J.G. and McFarlane, G., 1968. Numerical methods in marine ecology. I. A quantitative "similarity" analysis of rocky shore samples in False Bay, South Africa. *Zool. Afr.*, 3: 119–137.

Fletcher, A., 1980. Marine and maritime lichens of rocky shores: Their ecology, physiology and biological interactions. In: J.H. Price, D.E.G. Irvine and W.F. Farnham (Editors), *The Shore Environment, Vol. 2. Ecosystems.* Academic Press, London, pp. 789–842.

Gayral, P., 1958. *Algues de la Côte Atlantique Marocaine.* Rabat, Morocco, 523 pp.

Giaccone, G., 1967. Populamenti a *Laminaria rodriguezii* Bornet sul Banco Apollo dell' Isola di Ustica (Mar Tirreno). *Nova Thalassia*, 3: 1–9.

Giaccone, G., 1969. Note sistematiche ed osservazioni fitosociologiche sulle Laminariales del Mediterraneo occidentale. *G. Bot. Ital.*, 103: 457.

Gunnill, F.C., 1982. Macroalgae as habitat patch islands for *Scutellidium lamellipes* (Copepoda: Harpacticoida) and *Ampithoe tea* (Amphipoda: Gammaridae). *Mar. Biol.*, 69: 103–116.

Gurjanova, E.F., 1968. The influence of water movements upon the species composition and distribution of the marine fauna and flora throughout the Arctic and north Pacific intertidal zones. *Sarsia*, 34: 83–94.

Hällfors, G., Niemi, Å., Ackefors, H., Lassig, J. and Leppäkoski, E., 1981. Biological oceanography. In: A. Voipio (Editor), *The Baltic Sea.* Elsevier, Amsterdam, pp. 219–274.

Harkin, E., 1981. Fluctuations in epiphyte biomass following *Laminaria hyperborea* canopy removal. *Proc. Int. Seaweed Symp.*, 10: 303–308.

Hartnoll, R.G. and Hawkins, S.J., 1982. The emersion curve in semidiurnal tidal regimes. *Estuarine Coastal Shelf Sci.*, 15: 365–371.

Hartnoll, R.G. and Hawkins, S.J., 1985. Patchiness and fluctuations on moderately exposed rocky shores. *Ophelia*, 24: 53–63.

Hawkins, S.J. and Hartnoll, R.G., 1980. A study of the small-scale relationship between species number and area on a rocky shore. *Estuarine Coastal Mar. Sci.*, 10: 201–214.

Hawkins, S.J. and Hartnoll, R.G., 1983. Grazing of intertidal algae by marine invertebrates. *Oceanogr. Mar. Biol.: Ann. Rev.*, 21: 195–282.

Hawkins, S.J. and Hartnoll, R.G., 1985. Factors determining the upper limits of intertidal canopy-forming algae. *Mar. Ecol. Prog. Ser.*, 20: 265–271.

Heywood, R.B. and Whitaker, T.M., 1984. The Antarctic marine flora. In: R.M. Laws (Editor), *Antarctic Ecology, Vol. 2.* Academic Press, London, pp. 373–419.

Jones, N.S. and Kain, J.M., 1967. Subtidal algal colonization following the removal of *Echinus*. *Helgol. Wissensch. Meeresunters.*, 15: 460–466.

Jones, W.E. and Demetropoulos, A., 1968. Exposure to wave action: Measurements of an important ecological parameter on rocky shores on Anglesey. *J. Exp. Mar. Biol. Ecol.*, 2: 46–63.

Jorde, I., 1966. Algal associations of a coastal area south of Bergen, Norway. *Sarsia*, 23: 1–52.

Jorde, I. and Klavestad, N., 1963. The natural history of the Hardangerfjord. 4. The benthonic algal vegetation. *Sarsia*, 9: 1–99.

Kain, J.M., 1975. Algal recolonization of some cleared subtidal areas. *J. Ecol.*, 63: 739–765.

Kendall, M.A., Bowman, R.S., Williamson, P. and Lewis, J.R., 1985. Annual variation in the recruitment of *Semibalanus balanoides* on the north Yorkshire coast 1969–1981. *J. Mar. Biol. Ass. U.K.*, 65: 1009–1030.

King. R.J., 1973. The distribution and zonation of intertidal organisms in Bass Strait. *Proc. R. Soc. Victoria.*, 85: 145–162.

King, R.J., Hope Black, J. and Ducker, S., 1971. Intertidal ecology of Port Phillip Bay with systematic lists of plants and animals. *Mem. Nat. Mus. Victoria*, 32: 93–128.

Kussakin, O.G., 1977. Intertidal ecosystems of the seas of the U.S.S.R. *Helgol. Wissensch. Meeresunters.*, 30: 243–262.

Lawson, G.W., 1966. The littoral ecology of West Africa. *Ocean. Mar. Biol.: Ann. Rev.*, 4: 405–448.

Lawson, G.W., 1969. Some observations on the littoral ecology of rocky shores in east Africa (Kenya and Tanzania). *Trans. R. Soc. S. Afr.*, 38: 329–339.

Lawson, G.W. and Norton, T.A., 1971. Some observations on littoral and sublittoral zonation at Teneriffe (Canary Isles). *Bot. Mar.*, 14: 116–120.

Lewis, J.R., 1961. The littoral zone on rocky shores — a biological or physical entity? *Oikos*, 12: 280–301.

Lewis, J.R., 1964. *The Ecology of Rocky Shores.* English Universities Press, London, 323 pp.

Lewis, J.R., 1980. Objectives in littoral ecology — a personal viewpoint. In: J.H. Price, D.E.G. Irvine and W.F. Farnham (Editors), *The Shore Environment, Vol. 1. Methods.* Academic Press, London, pp. 1–18.

Lindstrom, S.C. and Foreman, R.E., 1978. Seaweed associations of the Flat Top Islands, British Columbia: A comparison of community methods. *Syesis*, 11: 171–185.

Lipkin, Y. and Safriel, U., 1971. Intertidal zonation on rocky shores at Mikmoret (Mediterranean, Israel). *J. Ecol.*, 59: 1–30.

Littler, M.M. and Littler, D.S., 1984. Models of tropical reef biogenesis. The contribution of algae. *Prog. Phycol. Res.*, 3: 323–364.

Littler, M.M., Littler, D.S., Blair, S.M. and Norris, J.N., 1985. Deepest known plant life discovered on a unchanted seamount. *Science*, 227: 57–59.

Lobban, C.S., Harrison, P.J. and Duncan, M.J., 1985. *The*

Physiological Ecology of Seaweeds. Cambridge University Press, Cambridge, 242 pp.
Lubchenco, J., 1978. Plant species diversity in a marine intertidal community: Importance of herbivore food preference and algal competitive abilities. *Am. Nat.*, 112: 23–39.
Lubchenco, J., 1980. Algal zonation in the New England rocky intertidal community: An experimental analysis. *Ecology*, 61: 333–344.
Lüning, K., 1970. Tauchuntersuchungen zur Vertikalverteilung der sublitoralen Helgolander Algenvegetation. *Helgol. Wissensch. Meeresunters.*, 21: 271–291.
Markham, J.W., 1969. Vertical distribution of epiphytes on the stipe of *Nereocystis luetkeana* (Mertens) Postels and Ruprecht. *Syesis*, 2: 227–240.
May, V., Bennett, I. and Thompson, T.E., 1970. Herbivore–algal relationships on a coastal rock platform (Cape Banks, N.S.W.). *Oecologia (Berlin)*, 6: 1–14.
Moe, R.L. and Silva, P.C., 1977. Antarctic marine flora: Uniquely devoid of kelps. *Science*, 196: 1206–1208.
Moore, P.G. and Seed, R. (Editors), 1985. *The Ecology of Rocky Coasts*. Hodder and Stoughton, London, 467 pp.
Munda, I., 1972. General features of the benthic algal zonation around the Icelandic coast. *Acta Nat. Isl.*, 21: 1–34.
Neushul, M., 1967. Studies of subtidal marine vegetation in western Washington. *Ecology*, 48: 83–94.
Niell, X., 1974. Les applications de l'indice de Shannon à l'étude de la végétation intertidale. *Bull. Soc. Phycol. France*, 19: 238–254.
Nienhuis, P.H., 1975. *Biosystematics and ecology of Rhizoclonium riparium (Roth) Harv. (Chlorophyceae: Cladophorales) in the estuarine area of the rivers Rhine, Meuse and Scheldt*. Ph.D. Thesis, University of Groningen, Netherlands, 240 pp.
Novak, R., 1984. A study in ultra-ecology: Microorganisms on the seagrass *Posidonia oceanica* (L.) Delile. *Mar. Ecol.*, 5: 143–190.
Paine, R.T., 1979. Disaster, catastrophe and local persistence of the sea palm *Postelsia palmaeformis*. *Science*, 205: 685–687.
Pérès, J.M., 1982a. Zonations. In: O. Kinne (Editor), *Marine Ecology, Vol. 5, Part 1*. Wiley, Chichester, pp. 9–45.
Pérès, J.M., 1982b. Major benthic assemblages. In: O. Kinne (Editor), *Marine Ecology, Vol. 5, Part 1*. Wiley, Chichester, pp. 378–522.
Potts, M. and Whitton, B.A., 1980. Vegetation of the intertidal zone of the lagoon of Aldabra, with particular reference to the photosynthetic prokaryote communities. *Proc. R. Soc. London*, B208: 13–55.
Price, J.H. and Redfearn, P., 1968. The marine ecology of Signy Island, South Orkney Islands. In: *Symp. on Antarctic Oceanography* Santiago, Chile, 13–16 September 1966. Scott Polar Research Institute, Cambridge, pp. 163–164.
Renoux-Meunier, A., 1965. Etude de la végétation algale du cap Saint-Martin (Biarritz). *Bull. Cent. Etudes Rech. Sci. Biarritz*, 5: 378–564.
Richardson, M.G., 1979. The distribution of Antarctic marine macro-algae related to depth and substrate. *Brit. Antarct. Surv. Bull.*, 49: 1–13.
Ricketts, E.F. and Calvin, J., 1968. *Between Pacific Tides*, 4th ed. revised by J.W. Hedgpeth. Standford University Press, Stanford, Cal., 614 pp.
Russell, G., 1972. Phytosociological studies on a two-zone shore. I. Basic pattern. *J. Ecol.*, 60: 539–545.
Russell, G., 1980. Applications of simple numerical methods to the analysis of intertidal vegetation. In: J.H. Price, D.E.G. Irvine and W.F. Farnham (Editors), *The Shore Environment, Vol. 1. Methods*. Academic Press, London, pp. 171–192.
Russell, G., 1981. Report on the marine vegetation of Egmont Is., Chagos Bank (Indian Ocean). *Proc. Int. Seaweed Symp.*, 8: 464–468.
Russell, G., 1983. Formation of an ectocarpoid epiflora on blades of *Laminaria digitata*. *Mar. Ecol. Prog. Ser.*, 11: 181–187.
Russell, G., 1987. Salinity and seaweed vegetation. In: R.M. Crawford (Editor), *Physiological Ecology of Amphibious and Intertidal Plants*. Blackwell, Oxford, pp. 35–52.
Russell, G. and Fielding, A.H., 1981. Individuals, populations and communities. In: C.S. Lobban and M.J. Wynne (Editors), *The Biology of Seaweeds*. Blackwell, Oxford, pp. 393–420.
Santelices, B., 1980. Phytogeographic characterization of the temperate coast of Pacific South America. *Phycologia*, 19: 1–12.
Schonbeck, M.W. and Norton, T.A., 1978. Factors controlling the upper limits of fucoid algae on the shore. *J. Exp. Mar. Biol. Ecol.*, 31: 303–313.
Schonbeck, M.W. and Norton, T.A., 1980. Factors controlling the lower limits of fucoid algae on the shore. *J. Exp. Mar. Biol. Ecol.*, 43: 131–150.
Shepherd, S.A. and Womersley, H.B.S., 1970. The sublittoral ecology of West Island, South Australia: 1. Environmental features and algal ecology. *Trans. R. Soc. Austr.*, 94: 105–137.
Sjöstedt, L.J., 1928. Littoral and supralittoral studies on the Scanian shores. *Lunds Univ. Arsskr.*, 24: 1–36.
South, G.R., 1983. Benthic marine algae. In: G.R. South (Editor), *Biogeography and Ecology of the Island of Newfoundland*. W. Junk, The Hague, pp. 385–420.
Stephenson, T.A. and Stephenson, A., 1949. The universal features of zonation between tide-marks on rocky coasts. *J. Ecol.*, 37: 289–305.
Stephenson, T.A. and Stephenson, A., 1972. *Life Between Tidemarks on Rocky Shores*. Freeman, San Francisco, 425 pp.
Svendsen, P., 1959. The algal vegetation of Spitsbergen. *Norsk Polarinst. Skr.*, 116: 1–47.
Swinbanks, D.D., 1982. Intertidal exposure zones: A way to subdivide the shore. *J. Exp. Mar. Biol. Ecol.*, 62: 69–86.
Taniguti, M., 1962. *Phytosociological Study of Marine Algae in Japan*. Inoue, Tokyo, 129 pp.
Taylor, J.D., 1971. Intertidal zonation at Aldabra Atoll. *Philos. Trans. R. Soc. London*, B260: 173–213.
Taylor, W.R., 1965. The genus *Turbinaria* in eastern seas. *J. Linn. Soc. London Bot.*, 58: 475–487.
Thorson, G., 1957. Bottom communities (sublittoral or shallow shelf). *Geol. Soc. Am. Mem.*, 67: 461–534.
Tittley, I. and Shaw, K.M., 1980. Numerical and field methods

in the study of the marine flora of chalk cliffs. In: J.H. Price, D.E.G. Irvine, and W.F. Farnham (Editors), *The Shore Environment, Vol. 1. Methods*. Academic Press, London, pp. 213–240.

Tittley, I., Irvine, D.E.G. and Jephson, N.A., 1976. The infralittoral marine algae of Sullom Voe, Shetland. *Trans. Bot. Soc. Edinburgh*, 42: 397–419.

Tokida, J., 1960. Marine algae epiphytes on Laminariales plants. *Bull. Fac. Fish. Hokkaido Univ.*, 11: 72–105.

Umamaheswararao, M. and Sreeramulu, T., 1964. An ecological study of some intertidal algae on the Visakhapatnam coast. *J. Ecol.*, 52: 595–616.

Underwood, A.J., 1978. A refutation of critical tidal levels as determinants of the structure of intertidal communities on British shores. *J. Exp. Mar. Biol. Ecol.*, 33: 261–276.

Vadas, R.L., 1977. Preferential feeding: An optimization strategy in sea urchins. *Ecol. Monogr.*, 47: 337–371.

Vaillant, L., 1891. Zones littorales. *Ann. Sci. Nat. Zool.*, 7(12): 39–50.

Van den Hoek, C., 1975. Phytogeographic provinces along the coasts of the northern Atlantic Ocean. *Phycologia*, 14: 317–330.

Van den Hoek, C., 1982. The distribution of benthic marine algae in relation to the temperature regulation of their life histories. *Biol. J. Linn. Soc.*, 18: 81–144.

Van den Hoek, C., Breeman, A.M., Bak, R.P.M. and Van Buurt, G., 1978. The distribution of algae, corals and gorgonians in relation to depth, light attenuation, water movement and grazing pressure in the fringing coral reef of Curaçao, Netherlands Antilles. *Aquat. Bot.*, 5: 1–46.

Van der Maarel, E. (Editor), in prep. *Dry Coastal Ecosystems. Ecosystems of the World, Vol. 2*. Elsevier, Amsterdam.

Velimirov, B. and Griffiths, C.L., 1979. Wave-induced kelp movement and its importance for community structure. *Bot. Mar.*, 22: 169–172.

Waern, M., 1952. Rocky-shore algae in the Oregrund archipelago. *Acta Phytogeogr. Suec.*, 30: 1–298.

Wanders, J.B.W., 1977. The role of benthic algae in the shallow reef of Curaçao (Netherlands Antilles). III: The significance of grazing. *Aquat. Bot.*, 3: 357–390.

Westhoff, V. and Van der Maarel, E., 1978. The Braun–Blanquet approach. In: R.H. Whittaker (Editor), *Classification of Plant Communities*. W. Junk, The Hague, pp. 287–399.

Whittaker, R.H., 1978. Dominance types. In: R.H. Whittaker (Editor), *Classification of Plant Communities*. W. Junk, The Hague, pp. 65–79.

Wilce, R.T., 1959. The marine algae of the Labrador Peninsula and north-west Newfoundland (ecology and distribution). *Bull. Nat. Mus. Can.*, 19: 1–103.

Womersley, H.B.S. and Edmonds, S.J., 1958. A general account of the intertidal ecology of South Australian coasts. *Austr. J. Mar. Fresh. Res.*, 9: 217–260.

Yamada, I., 1980. Benthic marine algal vegetation along the coasts of Hokkaido, with special reference to the vertical distribution. *J. Fac. Sci. Hokkaido Univ.* Ser. 5 (*Botany*), 12: 11–98.

Chapter 5

SHORELINE ECOLOGY IN ICELAND, WITH SPECIAL EMPHASIS ON THE BENTHIC ALGAL VEGETATION

IVKA M. MUNDA

INTRODUCTION

Iceland, which is centrally positioned in the North Atlantic between the European and American continents, is the largest emerged part of the Mid-Atlantic Ridge. It was formed during the Miocene by volcanic activity (Figs. 5.1 and 5.2). A variety of vegetation patterns and biotopes occurs within relatively short distances in Iceland, because of extreme hydrographic variations caused by different water masses (conveyed by ocean currents), as well as climatic and geological changes during historical times.

GEOLOGICAL HISTORY

Stream conditions in the North Atlantic are dependent upon its complicated bottom topography. The Mid-Atlantic Ridge divides the North Atlantic into an eastern and a western basin. The western and eastern parts of the North Atlantic were connected by uninterrupted coastlines and archipelagos until the Eocene. The Scotland–Greenland Ridge (Wyville–Thompson Ridge) separates the North Atlantic from the Arctic Ocean and runs from northern Ireland to western Scotland over the Færøerne (Faeröes), Iceland, and Greenland to Disko Island in Davis Strait. Tertiary volcanic rocks occur along this chain, including the extensive Rockall Plateau, which is a possible remnant of a land area in the North Atlantic (Einarsson, 1961). During the Eocene the climate throughout the North Atlantic was tropical to warm-temperate (Frakes, 1979). Subsequent periods of cooling occurred during the Oligocene and early Miocene, when the Scotland–Greenland Ridge subsided (Thiede and Eldholm, 1983), allowing a cold-water intrusion from the Arctic Ocean (McKenna, 1983).

An exchange of species and genera was possible along the land connection between the eastern and western Atlantic coasts during the Eocene (Frakes, 1979). This floral exchange proceeded during periods of cooling in the Miocene and Oligocene. The steep temperature drops may have induced the evolution of cold-temperate from warm-temperate floras on both sides of the North Atlantic Ocean.

ABIOTIC ENVIRONMENTAL FACTORS

Iceland is surrounded by water masses of widely different origins and characters. Warm and saline Atlantic water surrounds southern Iceland (Krauss, 1958; Malmberg, 1962). The northwest-flowing Irminger Current is gradually cooled during its passage over the Icelandic shelf. It divides off Látrabjarg into two branches, one flowing into the Irminger Sea (Fig. 5.3) and the other farther northeastward, following the slopes of the North Icelandic shelf. The North Icelandic Irminger Current is formed mainly by water originating from the Irminger Sea off Breiðafjörður (Dietrich, 1957; Stefánsson, 1962). There are, however, wide seasonal and annual variations in the volume influx of Atlantic water flowing into the north Icelandic coastal area (Einarsson, 1949; Stefánsson, 1949, 1962). A conspicuous hydrographic limit is found in northwest Iceland, due to the diminished influx of Atlantic water and the admixing of Polar water from the East Greenland Current. The northeastward-flowing Irminger Current is gradually cooled and diluted during its

Fig. 5.1. Topographical (a) and orographical (b) map of Iceland. (After Stefánsson, 1961.)

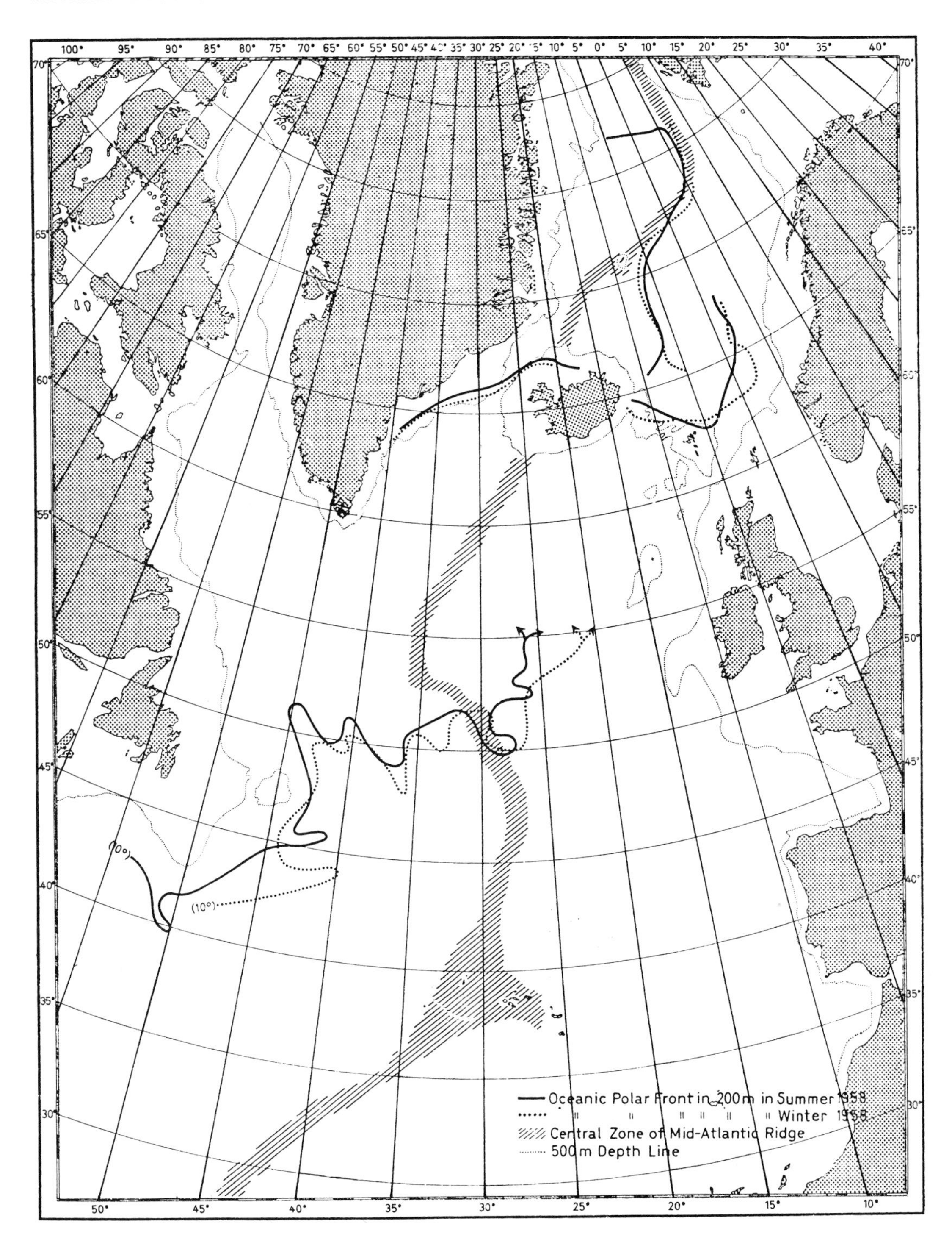

Fig. 5.2. Position of Oceanic Polar Front and the central zone of the Mid-Atlantic Ridge. (After Dietrich, 1957.)

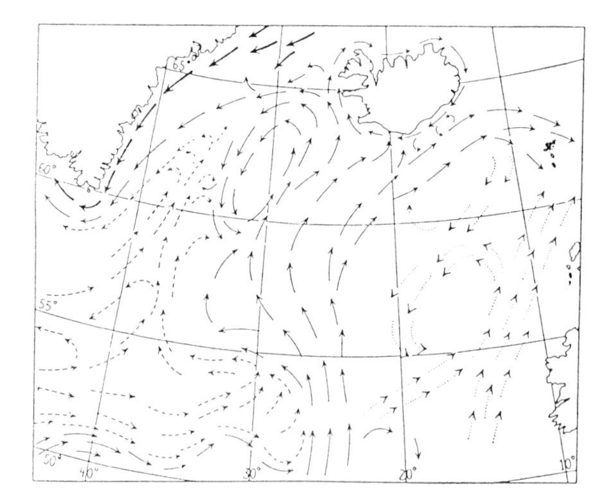

Fig. 5.3. Currents in the Irminger Sea and south of Iceland. (After Hermann and Thompsen, 1946.)

passage over the shelf. It is mixed with other primary and secondary water masses, such as Polar water, North Icelandic Winter water, Arctic water, Arctic Intermediate water, Arctic Bottom water and Coastal water (Stefánsson, 1962). The cold East Icelandic Current enters the Icelandic coastal area north of Melrakkaslétta. Its origin is south of the Scoresby Sound in Greenland. Between Iceland and Jan Mayen an anticlockwise circulation is formed (from waters of the East Greenland Current and the North Icelandic Irminger Current) which feeds the cold East Icelandic Current (Fig. 5.4). It follows the slopes of the east Icelandic coast and meets warm and saline Atlantic water in

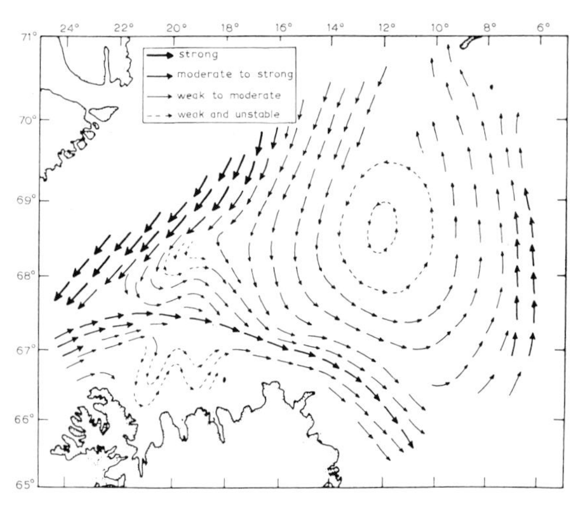

Fig. 5.4. Surface currents of the Iceland Sea. (After Stefánsson, 1962.)

the southeast. The frontal zone is subjected to seasonal and annual translocations. Notable changes in water masses, ice conditions and the location of the frontal zone have occurred during recent decades (Malmberg, 1984; Malmberg and Stefánsson, 1972; Stefánsson, 1972).

Stream conditions around Iceland are reflected in regional variations of temperature and salinity (Fig. 5.5). Seasonal, annual and long-term fluctuations in ice conditions (Fig. 5.6) and other hydrographic parameters occur in Iceland and the North Atlantic as a whole. Hence, cold-temperate and arctic conditions grade into each other, and a sharp delimitation on the basis of winter and summer isotherms is impossible. A striking example is the recent changes in ice conditions around Iceland, and of hydrographic conditions in the East Icelandic Current (Einarsson, 1969; Malmberg, 1984). This current changed from an ice-free Arctic Current into a Polar Current, transporting drift ice. Conditions in the East Icelandic Current are, however, one of the most important aspects in climatic changes on the Northern Hemisphere in recent and historical times (Lamb, 1979).

A further distinguishing character between coastal districts in Iceland is evident with respect to the tidal range. This varies from 3.82 m in southern Iceland (Reykjavík) to 1.23 m in eastern Iceland for the spring tides, and from 1.57 to 0.50 m for the neap tides (Stefánsson, 1961).

VEGETATION PATTERNS AND SPECIES DISTRIBUTION

The recent distribution of seaweeds in the North Atlantic and within Iceland is primarily a function of hydrographic conditions (that is, temperature, salinity, ice conditions and prevailing ocean currents). Although several floral districts have been delineated (for example, Börgesen and Jónsson, 1905; Van den Hoek, 1975, 1982, 1984; Van den Hoek and Donze, 1967; Michanek, 1979), they do not coincide with geographical latitudes. For example, in Iceland, which is centrally positioned in the North Atlantic (Fig 5.2), several vegetational types occur in a rather small area, due to varying hydrographic conditions. Most authors have neglected these small-scale hydrographic variations, which profoundly influence the benthic algal vegetation and create different vegetation types (cf. Strömfelt, 1886; Jónsson, 1910, 1912; Munda, 1975, 1976a, b, 1977a, 1978, 1983, 1984).

A general feature of cold-temperate areas like Iceland is the dominance of fucoid algae, which form distinct belts (Figs. 5.7–5.9). A complete fucoid zonation, ranging from *Pelvetia canaliculata* over *Fucus spiralis*, *F. vesiculosus*, *Ascophyllum nodosum* to *F. serratus* is characteristic of the relatively warm southern coast of Iceland (Munda, 1972b). *Pelvetia* spreads farthest to the southwest, while *Fucus serratus* is limited to the south. The number of fucoid belts is reduced in northwestern and northeastern Iceland. In these colder regions *Fucus distichus* is the main fucoid species, being represented by various subspecies and forms. The north and east coasts are only sporadically covered by ice, and attachment of fucoids and other algal belts is possible within the eulittoral zone. Ice conditions have been rather changeable during the last few years, depending upon ice coverage along east Greenland and prevailing wind directions. Masses of drift ice may be transported southwards to the Iceland–Greenland Ridge during the spring and early summer (Fig. 5.6).

Southern Iceland

The benthic algal vegetation of southern Iceland is of a warm-boreal type (Munda, 1976a). Here, lower eulittoral belts of *Mastocarpus stellatus* (Fig. 5.10) and *Corallina officinalis* (Fig. 5.11) are a characteristic feature, shared with the southwest and northwest coasts of Iceland (Munda, 1977b, c). In exposed sites, *Callithamnion sepositum* grows attached to small *Mytilus edulis*. A mixture of *Dilsea carnosa* and *Mastocarpus* is also conspicuous.

The Reykjanes Peninsula (Fig. 5.1) proceeds into an offshore submarine ridge, which restricts the extension of warm water masses from southern Iceland. The Peninsula represents a conspicuous floristic limit for several warm-boreal species, including *Asperococcus fistulosus*, *Desmarestia ligulata*, *Fucus serratus*, *Lomentaria clavellosa*, *L. orcadensis* and *Plocamium cartilagineum* (Munda, 1975).

Several tide-pool associations are conspicuous in southern Iceland, including *Ahnfeltia plicata*, *Asperococcus fistulosus*, *Ceramium* spp., *Chondrus crispus*, *Corallina officinalis*, *Cystoclonium purpureum* and *Dumontia contorta*. Another characteristic feature is a narrow and inconspicuous belt composed of *Porphyra umbilicalis* f. *umbilicalis*, *Urospora* spp. and *Ulothrix* spp.

Wide fields of fucoid algae dominate the area between the Ölfusá and Thjörsá Rivers in southern Iceland (Munda, 1964, 1976a). *Ascophyllum nodosum* forms extensive fields 1 to 2 km wide, with an average biomass of 3 to 18 kg m^{-2} (mean 6 kg m^{-2}). The chemical composition of *Ascophyllum* (i.e. contents of protein, ash, mannitol and alginic acid) from different habitats has been studied (Munda, 1964, 1972a). Within the upper sublittoral zone, two major associations dominate: *Laminaria digitata* f. *stenophylla* and *L. saccharina*, in exposed and protected sites respectively. *Laminaria hyperborea* also forms prolific forests within the lower eulittoral. Hence, southern Iceland is characterized by a high floristic diversity, a prolific vegetation within the eulittoral and upper sublittoral zones, and high biomass of most algal populations.

Southwestern Iceland

The benthic algal vegetation in southwestern Iceland is floristically impoverished, but still has several warm-boreal features. Low-level and tide-pool associations are similar to those in southern Iceland, except where soft substrata (Munda, 1980a) and freshwater dominate (Munda, 1980b). Fucoid stands are less prolific than in southern Iceland. The Snaefellsnes Peninsula represents a major floristic boundary (Munda, 1984, 1987). Northwards, *Alaria esculenta* dominates the upper sublittoral on exposed rocky sites. The Snaefellsnes Peninsula also represents the northern distributional limit of *Chondrus crispus*, *Dilsea carnosa*, *Gloiosiphonia capillaris* and *Pelvetia canaliculata* in Iceland. Around this peninsula, *Mastocarpus stellatus* forms an extensive zone, with its biomass greatly exceeding that found in southern Iceland (Munda, 1977c).

Northwestern Iceland

The benthic algal vegetation of southern and southwestern Iceland shows considerable similarities to that of the northwestern peninsula of

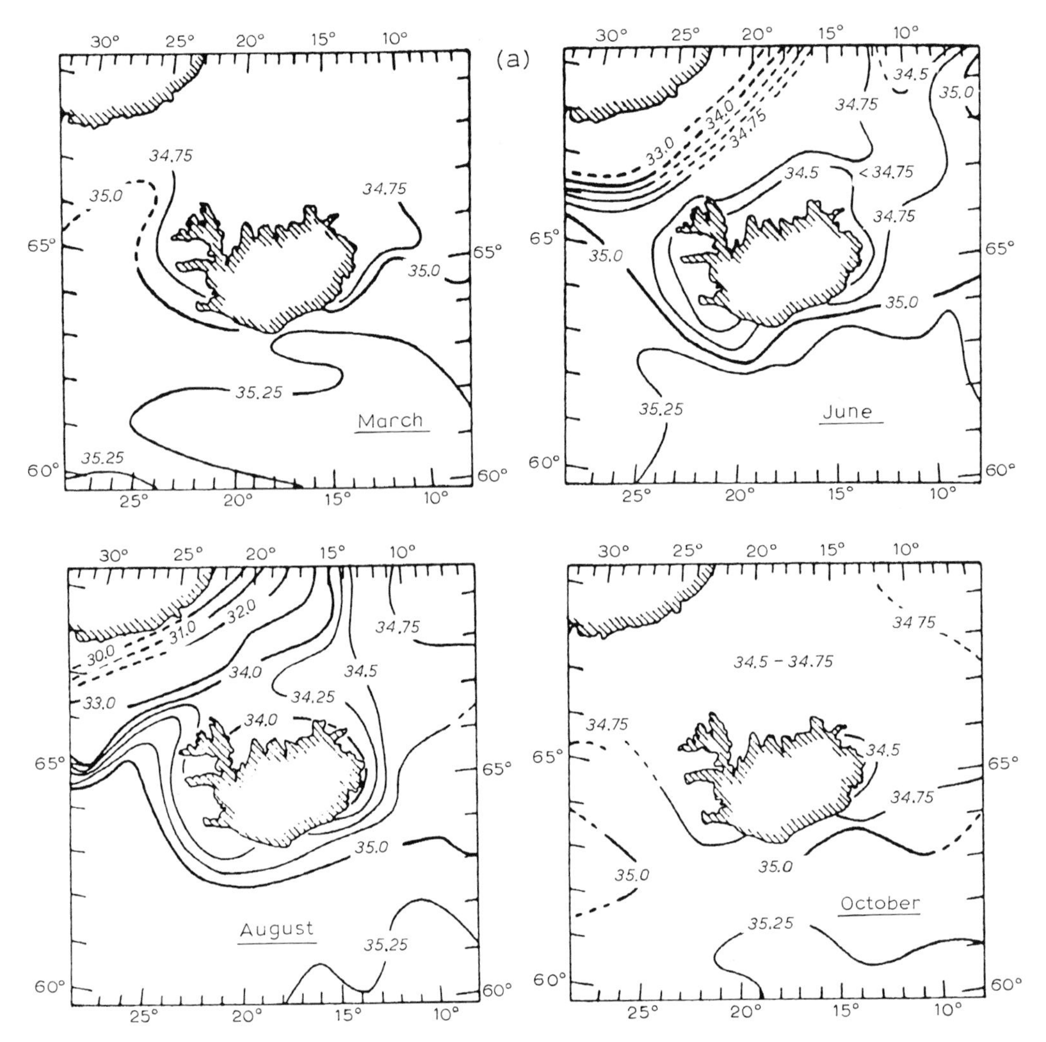

Fig. 5.5(a). For legend see p. 73.

Iceland, up to Hornbjarg (Fig. 5.1). Along this peninsula the flora is impoverished compared with the southwest and south coasts. The northern limit of distribution of *Ahnfeltia plicata, Phyllophora pseudoceranoides* and *Polysiphonia nigrescens* occurs in Breidafjördur, on the Bardaströnd Coast (Munda, 1985). Here the dominant low-eulittoral algal belts, which are characteristic of Atlantic waters in Iceland, are absent (for example, *Callithamnion sepositum, Corallina officinalis* and *Mastocarpus stellatus*). This impoverishment is primarily due to the occurrence of extensive sandy surfaces as well as changed hydrographic conditions. By contrast, in deep fjords of the northwestern peninsula these low-level associations are again prolific and dominant (e.g. Munda, 1978). In tide pools, warm- and cold-boreal components are intermixed. Atlantic floristic elements, such as *Ceramium* spp., *Corallina officinalis,* and *Cystoclonium purpureum* occur side by side with tide-pool associations of *Acrosiphonia* spp. and *Devaleraea ramentacea.* Filamentous brown algae, such as *Chordaria flagelliformis, Dictyosiphon foeniculaceus, Ectocarpus siliculosus, Eudesme virescens, Petalonia* spp. and *Stictyosiphon tortilis* are more frequent in these pools than farther south. Furthermore, the width of the *Ulothrix* spp.–*Urospora* spp. belts increases in a northwesterly direction. Cold-water species join the sublittoral vegetation here — for example, *Polysiphonia arctica, Ralfsia fungiformis* and *Turnerella pennyi.* However, the vegetation of the northwestern peninsula of Ice-

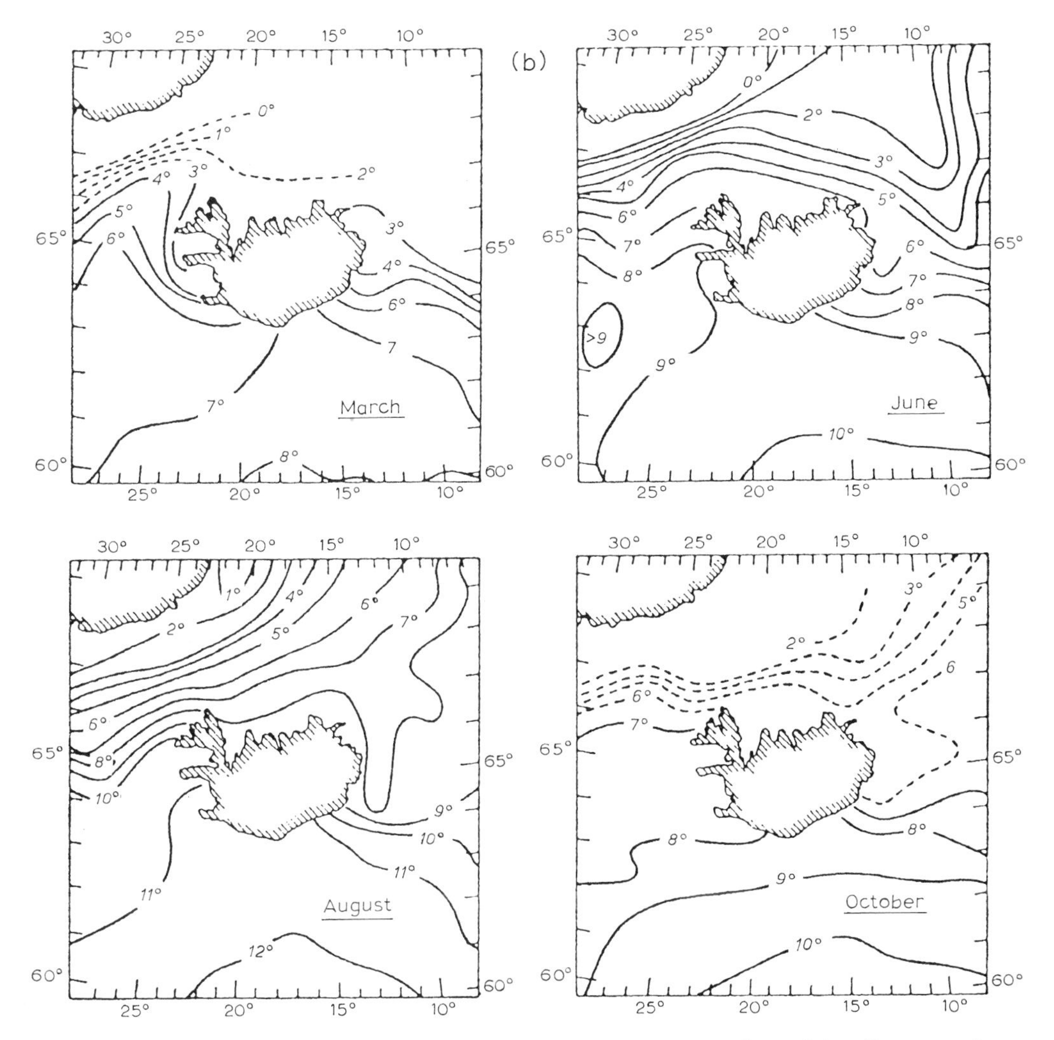

Fig. 5.5. Hydrographic conditions in Iceland (after Krauss, 1958). (a) Mean surface salinity; (b) mean surface temperature.

land is of a cold-boreal type. It exhibits a floristic impoverishment and admixture of "cold-water" associations.

Northern Iceland

Because of the conspicuous hydrographic changes in the northwestern part of Iceland, a sharp floristic discontinuity occurs around Hornbjarg in the extreme northwest (Munda, 1975, 1976b). Vegetational changes occur in the lower eulittoral, where belts of *Corallina officinalis* and *Mastocarpus stellatus* are replaced by *Acrosiphonia* species and *Devaleraea ramentacea*. Vegetational changes are less pronounced in tide pools, with *Ceramium* spp., *Corallina officinalis* and *Cystoclonium purpureum* occurring side by side with *Acrosiphonia* spp., *Devaleraea ramentacea* and diverse filamentous brown algae. Near Hornbjarg, *Callithamnion sepositum, Ceramium shuttleworthianum, Membranoptera alata* and *Plumaria elegans* reach their northern distributional limit, while *Mastocarpus stellatus* is uncommon.

In northern Iceland a gradient of floristic composition and vegetational features occurs from the northwest to the northeast, which is related to a gradual cooling of water masses along this coastline. The vegetation of northern Iceland is intermediate between the typical Atlantic vegetation of the south, southwest and northwest of Iceland and the subarctic eastern vegetation. It shows most similarity with the latter. The major algal associations

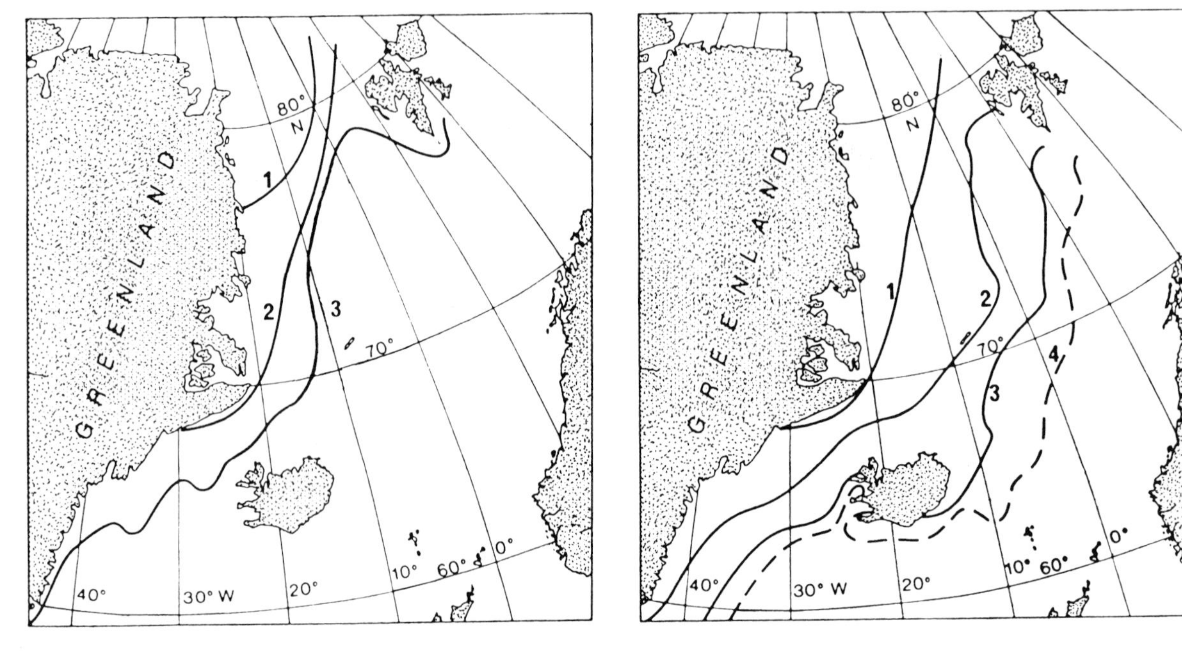

Fig. 5.6. Ice conditions between Iceland and Greenland (after Eythórsson and Sigtryggsson, 1971). Right: Ice limits during March through May : (1) recent minimum; (2) recent normal; (3) recent maximum; (4) estimated maximum in historical times. Left: Ice limits in early October : (1) recent minimum; (2) recent normal; (3) recent maximum.

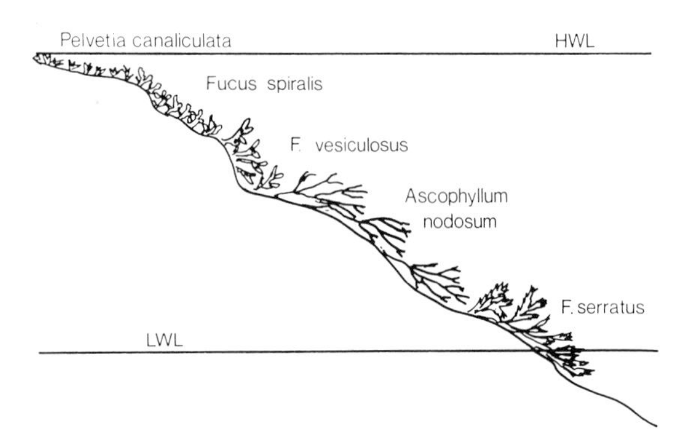

Fig. 5.7. Pattern of fucoid zonation in southern Iceland.

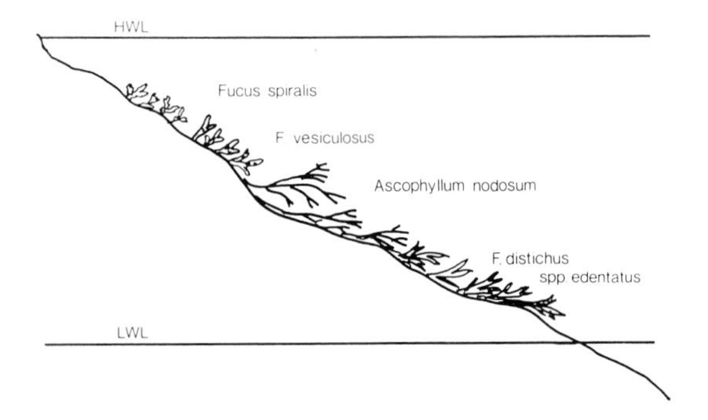

Fig. 5.8. Pattern of fucoid zonation in northwestern Iceland.

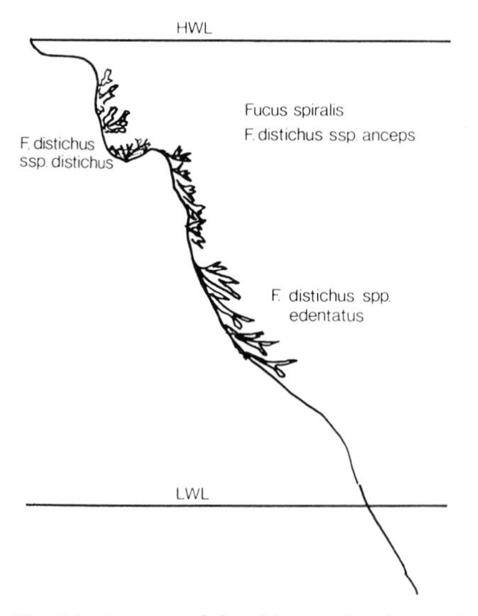

Fig. 5.9. Pattern of fucoid zonation in northern and eastern Iceland.

within the lower eulittoral zone and in tide pools consist of *Acrosiphonia* spp., *Chordaria flagelliformis, Devaleraea ramentacea, Petalonia fascia* and *P. zosterifolia*. Besides, tide-pool associations characteristic of Atlantic regions of Iceland also occur (e.g. *Ceramium* spp., *Corallina officinalis* and

Fig. 5.10. *Mastocarpus stellatus* belt in southwestern Iceland (Snaefellsnes).

Fig. 5.11. *Corallina officinalis* belt in southern Iceland (Reykjanes).

Cystoclonium purpureum). Locally, *Palmaria palmata* is prolific and belt-forming. Differences from the typical Atlantic vegetational pattern occur, as well as the depletion of undergrowth species beneath fucoid algae — e.g. *Cladophora rupestris, Membranoptera alata, Plumaria elegans*. Cold-water floristic elements such as *Clathromorphum circumscriptum* and *Ralfsia fungiformis* join the undergrowth. This "north Icelandic" vegetation gradually changes in the hydrographically mixed area between Melrakkaslétta and Vopnafjörður, which is probably due to the gradual cooling of this area by the cold East Icelandic Current. *Ceramium* spp., *Cladophora rupestris, Corallina officinalis, Cystoclonium purpureum* and *Mastocarpus stellatus* gradually disappear along this transitional area.

Eastern Iceland

The east Icelandic coast is exposed to cold water masses of Arctic origin, and is occasionally influenced by drift or compact ice. Ice conditions have become more severe during recent decades. The vegetation found in eastern Iceland (Munda, 1983) is "subarctic" in nature, since ice cover is infrequent as compared to the Arctic proper, allowing the attachment of a rather prolific eulittoral vegetation. The algal vegetation in eastern Iceland has several characteristic associations and zonation patterns. In the upper eulittoral, a vertical extension of the *Ulothrix* spp. –*Urospora* spp. belts occurs. On highly exposed rocky coasts or ice-scoured surfaces, this belt can extend over the entire eulittoral zone, that is, down to the *Alaria esculenta* belt. In the mid-eulittoral, different types of fucoid zonation occur, with *Fucus distichus* as the dominant species. Under extreme exposure *F. distichus* spp. *anceps* is the only fucoid present. Frequently both *F. distichus* spp. *anceps* and *F. distichus* spp. *edentatus* follow one another along the same transect. *Ascophyllum nodosum* and *Fucus vesiculosus* occur sporadically and in rather narrow belts, depending on the configuration of the coastal slope and exposure conditions. Within tide pools in eastern Iceland, *Fucus distichus* ssp. *distichus* dominates. A morphocline is observed from the upper to lowermost tide pools (cf. Sideman and Mathieson, 1983a, b). In splash pools, *F. distichus* spp. *anceps* and ssp. *distichus* seem to grade into each other. Other tide pools in the east are occupied by *Acrosiphonia* spp., *Devaleraea ramentacea* and diverse filamentous brown algae (*Chordaria flagelliformis, Dictyosiphon* spp., *Petalonia* spp., *Scytosiphon lomentaria, Stictyosiphon tortilis*.) A typical eastern association consists of *Coilodesme bulligera* and *Ralfsia fungiformis*. All typical Atlantic floristic elements are absent from the vegetation, which is joined by

Enteromorpha groenlandica, Monostroma arcticum, Porphyra thulaea and *Saccorhiza dermatodea* (Fig. 5.12). In the lower eulittoral *Devaleraea ramentacea* (Fig. 5.13) covers wide areas of moderately sloping rocky surfaces, mingled with extensive mats of *Acrosiphonia arcta, A. centralis, A. grandis* and *A. sonderi. Acrosiphonia* mats are also found in the upper eulittoral and within shallow lagoons. On steep slopes, *Chordaria flagelliformis, Palmaria palmata* and *Porphyra thulaea* occur. The number of eulittoral belts becomes reduced with increasing exposure. The upper sublittoral is occupied solely by *Alaria esculenta* (narrow growth form), while *Laminaria* species are restricted to fjords and inlets, for example, *Laminaria digitata, L. saccharina* and the rather rare *L. nigripes*. The lower eulittoral is occupied by prolific *Laminaria hyperborea* forests with a rich epiphytic and epilithic rhodophycean flora, including *Callophyllis cristata, Delesseria sanguinea, Fimbrifolium dichotomum, Odonthalia dentata, Phycodrys rubens, Polysiphonia urceolata, P. arctica, Porphyra miniata* f. *amplissima, Ptilota serrata, P. plumosa, Rhodomela confervoides* and *Turnerella pennyi*. This subarctic vegetation shows an abrupt termination within southeastern Iceland, coinciding with the maximum southwards extension of the frontal zone. Westwards, the south Icelandic coastal area is a sandy desert, bare of algal growth. Around the island of Hrollaugseyjar, off Hornafjördur, a prolific warm-boreal vegetation is again found, similar to that around the Reykjanes Peninsula.

Regional differences

Hydrographically conditioned differences between the main vegetation types around Iceland are primarily due to the effects of temperature, that is, its maximum and minimum values. Additional modifying factors determine the local distributional patterns of algal associations. Regional differences are most pronounced in the lower eulittoral zone and within tide pools (cf. Figs. 5.14

Fig. 5.12. *Porphyra thulaea* belt in eastern Iceland (Reydarfjördur).

Fig. 5.13. *Devaleraea ramentacea* population in northern Iceland (Húnaflóa).

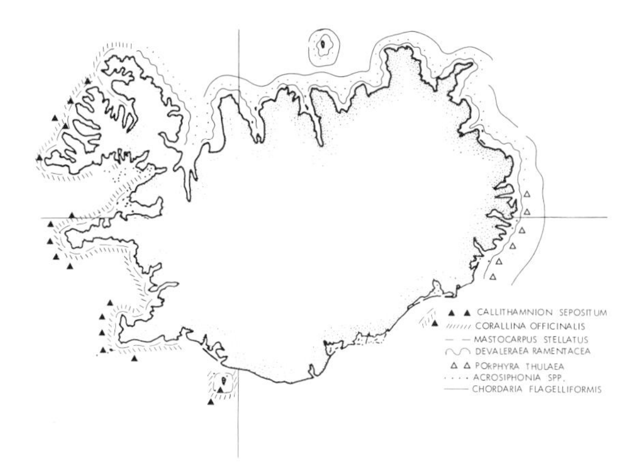

Fig. 5.14. Distribution of low-eulittoral associations in Iceland.

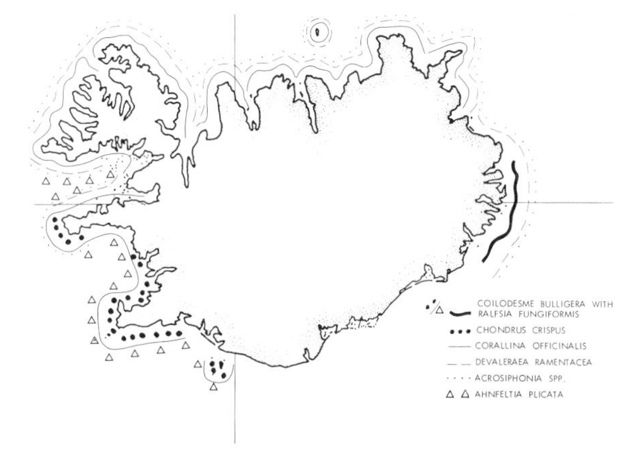

Fig. 5.15. Distribution of tide-pool associations in Iceland.

and 5.15). The *Mastocarpus stellatus* association (Munda, 1977b) is characteristic of warm- and cold-boreal vegetational types in Iceland. The same is true for the low-level *Corallina officinalis* association. Both associations are replaced by *Acrosiphonia* spp. and *Devaleraea ramentacea* in the north and east of Iceland. Tide-pool associations of *Corallina officinalis* and *Devaleraea ramentacea* have a wider distributional range around Iceland than their low-level associations (Munda, 1976b, 1977b). In eastern Iceland, coastal lagoons are dominated by *Saccorhiza dermatodea*, whereas in other regions of Iceland they are mainly overgrown by *Chorda filum*. Tide pools, although widely distributed around Iceland, are most abundant in the north and east, where they are dominated by *Acrosiphonia* spp., *Chordaria flagelliformis* and *Devaleraea ramentacea*.

The sequence and vertical extension of fucoid belts are determined by the inclination of coastal rocks and exposure conditions. Regional differences between fucacean associations are also obvious in their undergrowth and epiphytic covering. A striking example is the hemiparasite *Polysiphonia lanosa* on *Ascophyllum nodosum*, which is absent along the east Icelandic coast. Regional differences in the floristic composition of fucacean associations in Iceland have been described (Munda, 1976a, 1977a, 1978, 1983, 1987); they centre on the presence or absence of typical Atlantic-water floristic elements as undergrowth and subdominant species (e.g. *Ceramium* spp., *Cladophora rupestris, Corallina officinalis, Membranoptera alata, Phymatolithon lenormandii, P. polymorphum, Plumaria elegans*).

In the upper eulittoral zone a gradual extension of the *Ulothrix* spp.–*Urospora* spp. belts occurs in the northwesterly direction and further northeastward. The latter belts reach their maximum vertical extension in eastern Iceland. For the Icelandic subarctic vegetation type, a high-level belt of small filamentous brown algae is also characteristic (e.g. *Isthmoplea sphaerophora, Petalonia filiformis, Pilayella littoralis* f. *rupincola, Scytosiphon pygmaeus*). In southern Iceland well-developed *Fucus spiralis* and *Pelvetia canaliculata* belts prevent the development of other high-level associations. A *Porphyra umbilicalis* association is most prolific in northern Iceland. Minor regional differences are obvious within the sublittoral zone. The most outstanding feature is a change in the southeast, where *Alaria esculenta* replaces *Laminaria digitata* f. *stenophylla* in the upper sublittoral zone of exposed rocky sites.

On protected shores throughout Iceland *Blidingia* species and *Enteromorpha intestinalis* form conspicuous belts above the fucoids. Low-level associations dominated by *Acrosiphonia* spp. and *Pilayella littoralis* are also common on semi-exposed rocky shores.

Semibalanus balanoides forms a conspicuous belt in the upper eulittoral zone throughout Iceland. *Mytilus edulis* occurs as an undergrowth below the fucoids all around Iceland. Within the lower eulittoral, belts of small mussel shells are most frequent in the south and southwest.

COMPARISONS WITH OTHER AREAS IN THE NORTH ATLANTIC

The benthic algal vegetation of southern and southwestern Iceland shows considerable similarities to the south and east coasts of Norway, up to the middle of the western Norwegian coast (Printz, 1926; Rueness, 1977; Sundene, 1953), and with the Færøerne (Börgesen, 1905; Irvine, 1982; Price and Farnham, 1982). A similar vegetation occurs along the northwestern peninsula of Iceland, up to Hornbjarg. The vegetation of northern Iceland is different and more closely related to eastern Iceland than to the typical Atlantic vegetation of the south, southwest and northwest. Although floristically impoverished, the vegetation in northern Iceland resembles that in northern Norway (Jaasund, 1965). The major algal associations within the lower eulittoral zone and in tide pools consist of *Acrosiphonia* spp., *Chordaria flagelliformis, Devaleraea ramentacea, Petalonia fascia* and *P. zosterifolia*. In northern Norway, *Mastocarpus stellatus* is still found in the lower eulittoral zone, while *Chondrus crispus* occurs in tide pools.

As noted previously, the south, southwest and northwest coasts of Iceland show strong affinities with the Færøerne and southern Norway. By contrast, northern Iceland exhibits strong affinities with northern Norway and Newfoundland (e.g. Mathieson et al., 1969; South, 1970, 1971; South and Hooper, 1980).

The vegetation of eastern Iceland is subarctic in nature, since ice cover only occurs occasionally, unlike in the Arctic proper. The east Icelandic vegetation has no parallel in the northeast Atlantic, since the entire Norwegian coast is influenced by warm Atlantic water masses. It resembles Labrador's vegetation (Wilce, 1959; Hooper and Whittick, 1984) as far as the eulittoral is concerned.

Differences in vegetation are obvious in the sublittoral. In Newfoundland and Labrador the arctic *Laminaria solidungula* appears below the thermocline. *Agarum cribrosum* and *Laminaria saccharina* ssp. *longicruris* are present in the sublittoral vegetation (Mathieson et al., 1969; South and Hooper, 1980), while *Laminaria hyperborea* is absent. The algal vegetation of southernmost Greenland (Pedersen, 1976) also shows some affinities with the east Icelandic vegetation regarding vegetation patterns and floristic composition. However, fucoids are limited to protected shores and rocky fissures in southern Greenland. In sites exposed to ice scouring the eulittoral is occupied by diverse brown and green filamentous algae. *Devaleraea ramentacea* only occupies tide pools and is displaced to the sublittoral in Greenland (Lund, 1959; Pedersen, 1976).

Callithamnion sepositum is a characteristic lower-shoreline association of exposed rocky sites in western Norway, the Færøerne and the Atlantic regions of Iceland up to Hornbjarg. Apparently it does not spread into the northwest Atlantic. This dominant species is usually attached to small *Mytilus edulis*. *Mastocarpus stellatus* forms a characteristic association of warm and cold boreal vegetational types in Iceland. This low-level association reaches the middle of the Norwegian coast (Jaasund, 1965); it is prolific in the Færøerne (Irvine, 1982), in Shetland (Irvine et al., 1973) and reappears in Newfoundland (South, 1983). *Corallina officinalis* is association-forming in Atlantic water regions around Iceland, both in tide pools and on eulittoral slopes. It forms extensive mats lower down than *Mastocarpus stellatus* and protrudes locally into the sublittoral (Munda, 1977b). This species is common and widespread all throughout the North Atlantic. It is absent in colder regions, such as eastern Iceland, Greenland and Labrador.

Chordaria flagelliformis forms a conspicuous low-level association throughout the entire northern North Atlantic. It grows prolifically in northern and eastern Iceland, frequently being accompanied by *Chorda tomentosa, Petalonia* spp. and *Scytosiphon lomentaria*. The association is also common in northern Norway, the Færøerne, Newfoundland and Labrador, as well as in Greenland, where it is displaced into the sublittoral zone (Lund, 1959). It was observed in the eulittoral zone near Angmagsalik (Rosenvinge, 1898), as well as farther south (Pedersen, 1976). In eastern Iceland and southern Greenland, belts of *Porphyra thulaea* occur (Munda and Pedersen, 1978). Wide belts of *Acrosiphonia* spp. are a further characteristic feature of the lower eulittoral zone of colder regions in the North Atlantic. In northern and eastern Iceland they replace *Corallina* mats, which dominate in the south, southwest and northwest. A reliable measure of hydrographically-conditioned vegetational types is found also in tide pools. *Chondrus crispus* has a warm-boreal character and occupies mid-eulittoral tide pools in southern and southwestern Iceland (Munda, 1977c). It extends along the entire Norwegian coast, to the British Isles and Spain. Its transoceanic distribution reaches Newfoundland, but not Labrador. Other typical Atlantic tide-pool associations are represented by *Ahnfeltia plicata, Ceramium* spp., *Cladophora rupestris, Corallina officinalis, Cystoclonium purpureum, Dumontia contorta* and *Phyllophora pseudoceranoides*. In eastern Iceland a subarctic tide-pool association, that is, *Coilodesme bulligera* with *Ralfsia fungiformis*, is conspicuous. It also occurs in Labrador and southern Greenland.

The distribution of sublittoral crustose corallines around Iceland is determined by temperature minima and maxima in deep water layers (Adey, 1968). Hence, it reflects hydroclimatic conditions during many years, contrary to eulittoral algal belts.

Marine lichens (e.g. *Verrucaria maura, V. mucosa* — Fletcher, 1980), Cyanobacteria (e.g. *Calothrix crustacea*), together with *Prasiola* spp., *Rosenvingiella polyrhiza* and diverse *Ulothrix* and *Urospora* species, are common features of the littoral fringe zone in the northeast and northwest Atlantic. The latter species form conspicuous belts in northern and eastern Iceland, and may colonize the eulittoral zone. *Porphyra umbilicalis* f. *umbilicalis* forms a conspicuous belt above the fucoids on both sides of the North Atlantic. *Bangia atropurpurea* is a summer species in northern and eastern Iceland, but a winter species farther south.

Iceland represents the transoceanic distributional limit for several typical northeast-Atlantic species, such as *Delesseria sanguinea, Fucus ceranoides, Laminaria hyperborea* and *Pelvetia canaliculata*. *Fucus serratus*, which is limited to the warm boreal vegetation of southern Iceland, reappears again in the Canadian Maritime Provinces.

Several species that spread from the northern European mainland to the Færøerne (Börgesen, 1905; Irvine, 1982) do not reach Iceland, for instance, *Callithamnion corymbosum, Callophyllis laciniata, Furcellaria lumbricalis, Griffithsia corallinoides, G. flosculosa, Halidrys siliquosa, Himanthalia elongata, Laurencia pinnatifida, Lomentaria articulata, Polyides rotundus, Polysiphonia brodiaei, P. elongata, P. violaecea, Pterosiphonia parasitica.* Hence, vegetational differences centre on the absence of *Himanthalia* belts in Iceland as well as the physiognomy of the *Corallina* association, which mingles with *Lomentaria articulata* on the Faerøerne coasts. However, *Mastocarpus* belts are a common feature of both areas.

A floristic impoverishment is obvious from the northeast to the northwest Atlantic, that is from the northern European mainland, over the Shetlands (Irvine et al., 1975), Færøerne (Irvine, 1982) and Iceland (Munda, 1976a, 1977a, 1978, 1983) to Greenland (Lund, 1959; Wilce, 1964) and to Labrador (Wilce, 1959).

The similarity of the Færøerne, Shetlands, Iceland, Newfoundland and also Labrador in their vegetation patterns along with the transatlantic NE–NW distribution of most eulittoral species, is probably due to their common origin in the late Tertiary. The transoceanic ridge between Scotland and Greenland allowed the dispersal of species to both sides of the Atlantic. The transoceanic impoverishment in species diversity may be also explained by conditions in the late Tertiary, followed by a further loss of species and genera during Pleistocene glaciations.

REFERENCES

Adey, W.H., 1968. The distribution of crustose corallines on the Icelandic coast. *Soc. Sci. Islandica*, 1: 1–31.

Börgesen, F., 1905. The algae vegetation of the Faeröese coasts with remarks on the phytogeography. In: E. Warming (Editor), *Botany of the Faeröes, Vol. 3.* Det Nordiske Forlag, Copenhagen, pp. 683–834.

Börgesen, F. and Jónsson, H., 1905. The distribution of the marine algae in the Arctic Sea and in the northernmost part of the Atlantic. In: E. Warming (Editor), *Botany of the Faeröes, Vol. 3.* Det Nordiske Forlag, Copenhagen, Appendix, pp. 1–28.

Dietrich, G., 1957. Schichtung and Zirkulation der Irminger See im Juni 1955. *Ber. Deutsch Wiss. Komm. Meeresforsch.*, 4(4).

Einarsson, H., 1949. Influx of Atlantic water north of Iceland 1947. *Ann. Biol.*, 4: 10–11.

Einarsson, T., 1961. Pollenanalytische Untersuchungen zur spät- und postglazialen Klimageschichte Islands. *Sonderveröff. Geol. Inst. Univ. Köln*, 6: 1–52.

Einarsson, T., 1969. The ice in the North Polar Basin and the Greenland Sea and general causes of occasional approach of ice in the coast of Iceland. *Jökull*, 19: 2–6.

Eythórsson, J. and Sigtryggsson, H., 1971. The climate and weather of Iceland. *The Zoology of Iceland*, 1: 1–62.

Fletcher, A., 1980. Marine and maritime lichens of rocky shores: their ecology, physiology and biological interactions. In: J.H. Price, D.E.G. Irvine, and W.F. Farnham (Editors), *The Shore Environment, Vol. 2, Ecosystems.* Academic Press, London, pp. 789–842.

Frakes, L.A., 1979. *Climates Throughout Geologic Time.* Elsevier, Amsterdam, 310 pp.

Hermann, F. and Thompsen, H., 1946. Drift bottle experiments in the northern North Atlantic. *Medd. Komm. Havundersögelser. Hydrografi*, 3/4.

Hooper, R.G. and Whittick, A., 1984. The benthic marine algae of the Kaipokok bay, Makkovik bay and Big River Bay region of the central Labrador coast. *Nat. Can.*, 111: 131–138.

Irvine, D.E.G., 1982. Seaweeds of the Faeröes. I. The flora. *Bull. Br. Mus. Nat. Hist. Bot.*, 10: 109–131.

Irvine, D.E.G., Guiry, M.D. and Tittley, I., 1975. New and interesting marine algae from the Shetland Isles. *Br. Phycol. J.*, 10: 57–71.

Jaasund, E., 1965. Aspects of the marine algal vegetation of north Norway. *Bot. Gothob. Acta Univ. Gothob.*, 4: 1–174.

Jónsson, H., 1910. Om algevegetasjonen ved Islands Kyster. *Bot. Tidskr.*, 30: 223–328.

Jónsson, H., 1912. The marine algal vegetation. In: K.L. Rosenvinge and E. Warming (Editors), *The Botany of Iceland, Vol. 1*, pp. 1–196.

Krauss, W., 1958. Temperatur, Salzgehalt und Dichte an der Oberfläche des Atlantischen Ozeans. *Wiss. Ergeb. Deutsch. Atl. Exp. "Meteor"*, 5: 251–410.

Lamb, H.H., 1979. Climatic variations and changes in the wind and ocean circulation. The little ice age in the northeast Atlantic. *Quaternary Res.*, 11.

Lund, S., 1959. The marine algae of east Greenland. II. Geographic distribution. *Meddr. Grönland*, 156: 1–71.

Malmberg, S.A., 1962. Schichtung und Zirkulation in den südisländischen Gewässern. *Kieler Meeresforsch.*, 18: 3–28.

Malmberg, S.A., 1984. Hydrographic conditions in the East Icelandic Current and sea ice in north Icelandic waters, 1970–1980. *Rapp. Proc. Réun. Cons. Int. Explor. Mer.*, 185: 170–178.

Malmberg, S.A. and Stefánsson, U., 1972. Recent changes in the water masses of the East Icelandic Current. *Rapp. Proc. Verb. Con., Int. Explor. Mer.*, 162: 196–200.

Mathieson, A.C., Dawes, C.J. and Humm, H.J., 1969. Contributions to the marine algae of Newfoundland. *Rhodora*, 71: 110–159.

McKenna, M.C., 1983. Cenozoic paleogeography of North Atlantic land-bridges. In: M.H.P. Bott, S. Saxov, M. Talwani and J. Thiede (Editors), *Structure and Development of the Greenland Scotland Ridge, New Methods and Concepts.* Plenum Press, New York, pp. 351–399.

Michanek, G., 1979. Phytogeographic provinces and seaweed distribution. *Bot. Mar.*, 22: 375–391.

Munda, I.M., 1964. The quantity and chemical composition of *Ascophyllum nodosum* (L.) Le Jol. along the coast between the rivers Olfusa and Thjorsa (southern Iceland). *Bot. Mar.*, 7: 76–89.

Munda, I.M., 1972a. On the chemical composition, distribution and ecology of some common benthic marine algae from Iceland. *Bot. Mar.*, 15: 1–45.

Munda, I.M., 1972b. General features of benthic algal zonation around the Icelandic coast. *Acta Nat. Isl.*, 21: 1–34.

Munda, I.M., 1975. Hydrographically conditioned floristic and vegetation limits in Icelandic coastal waters. *Bot. Mar.*, 18: 223–235.

Munda, I.M., 1976a. Some aspects of the benthic algal vegetation of the south Icelandic coastal area. *Res. Inst. Nedri as Hveragerdi Icel. Rep.*, 25: 1–69.

Munda, I.M., 1976b. The distribution of the *Halosaccion ramentaceum* (L). J. Agardh associations in Icelandic waters and their hydrographically conditioned variations. *Bot. Mar.*, 19: 161–179.

Munda, I.M., 1977a. The benthic algal vegetation of the island of Grimsey (Eyjafjardarsysla, north Iceland). *Res. Inst. Nedri as Hveragerdi Icel. Rep.*, 28: 1–69.

Munda, I.M., 1977b. A comparison of the north and south European associations of *Corallina officinalis* L. *Hydrobiologia*, 52: 73–87.

Munda, I.M., 1977c. The structure and distribution of the *Gigartina stellata* (Stackh.) Batt. and *Chondrus crispus* Stackh. associations in Icelandic waters. *Bot. Mar.*, 20: 291–301.

Munda, I.M., 1978. Survey of the benthic algal vegetation of the Dryafjordur, Northwest Iceland. *Nova Hedwigia*, 29: 281–403.

Munda, I.M., 1980a. Contribution to the knowledge of the benthic algal vegetation of the Myrar area (Fayafloi, southwest Iceland). *Res. Inst. Nedri as Hveragerdi. Icel. Rep.*, 33: 1–49.

Munda, I.M., 1980b. Survey of the benthic algal vegetation of the Borgarfjördur, southwest Iceland. *Nova Hedwigia*, 31: 855–927.

Munda, I.M., 1983. Survey of the benthic algal vegetation of the Reydarfjördur as a typical example of the east Icelandic vegetation pattern. *Nova Hedwigia*, 37: 545–640.

Munda, I.M., 1984. The benthic algal vegetation of the Snaefellsnes Peninsula, southwest Iceland. *Hydrobiologia*, 116: 371–373.

Munda, I.M., 1985. General survey of the benthic algal vegetation along the Bardaströnd coast (Breidafjördur, West Iceland). *Res. Inst. Nedri as Hveragerdi Icel. Rep.*, 44: 1–62.

Munda, I.M., 1987. Characteristic features of the benthic algal vegetation along the Snaefellsnes peninsula (southwest Iceland). *Nova Hedwigia*, 44(3/4): 399–448.

Munda, I.M. and Pedersen, P.M., 1978. *Porphyra thulaea* sp. nov. (Rhodophyceae, Bangiales) from East Iceland and West Greenland. *Bot. Mar.*, 21: 283–288.

Pedersen, P.M., 1976. Marine benthic algae from southernmost Greenland. *Meddr. Grönland*, 199: 1–80.

Price, J.H. and Farnham, W.F., 1982. Seaweeds of the Faeröes. 3. Open shores. *Bull. Br. Mus. Nat. Hist. Bot.*, 10: 153–225.

Printz, H., 1926. Die Algenvegetation des Trondheimfjordes. *K. Norske Vidensk. Akad. Selsk. Skr.*, 5: 1–274.

Rosenvinge, K.L., 1898. Deuxième mémoire sur les algues marines du Groenlande. *Meddr. Grönland*, 20: 1–125.

Rueness, J., 1977. *Norsk Algeflora*. Universitetsforlaget, Oslo, 266 pp.

Sideman, E.J. and Mathieson, A.C., 1983a. Ecological and genecological distinctions of a high intertidal dwarf form of *Fucus distichus* (L.) Powell. *J. Exp. Mar. Biol. Ecol.*, 72: 171–188.

Sideman, E.J. and Mathieson, A.C., 1983b. The growth, reproductive phenology and longevity of non-tide pool *Fucus distichus* (L.) Powell. *J. Exp. Mar. Biol. Ecol.*, 68: 111–127.

South, G.R., 1970. *Checklist of marine algae from Newfoundland, Labrador and the French Islands of St. Pierre and Miquelon*. MSRL Techn. Rep. 2. Memorial Univ., Newfoundland, pp. 1–19.

South, G.R., 1971. Additions to the benthic marine algal flora of insular Newfoundland. *Nat. Can.*, 98: 1027–1031.

South, G.R., 1983. Benthic marine algae. In: G.R. South (Editor), *Biogeography and Ecology of the Island of Newfoundland*. Junk, Den Haag, pp. 385–420.

South, G.R. and Hooper, R.G., 1980. *A catalogue and atlas of the benthic marine algae of the island of Newfoundland*. Occasional Papers Biol. 3, 136 pp.

Stefánsson, U., 1949. Influx of Atlantic water north of Iceland 1948. *Ann. Biol.*, 5: 16–18.

Stefánsson, U., 1961. *Hafid*. Ed. Almena bokafélagid, Reykjavik, 299 pp.

Stefánsson, U., 1962. North Icelandic waters. *Rit Fiskideildar*, 3: 1–269.

Stefánsson, U., 1972. Near-shore fluctuations of the frontal zone southeast of Iceland. *Rapp. Proc. Réun. Cons. Int. Explor. Mer.*, 162: 201–205.

Strömfelt, H.P.G., 1886. Om algevegetationen vid Islands Kuster. *Akad. avh. Göteborg.*, 1–89.

Sundene, O., 1953. The algal vegetation of Oslofjord. *Kgl. Norske Visdensk. Akad. Skr., I. Mat.-Nat. Kl.*, 2: 1–224.

Thiede, J. and Eldholm, O., 1983. Speculations about the paleodepth of the Greenland–Scotland Ridge during late Mesozoic and Cenozoic. In: M.H.P. Bott, S. Saxov, M. Talwani and J. Thiede (Editors), *Structure and Development of the Greenland–Scotland Ridge*. Plenum Press, New York, pp. 351–399.

Van den Hoek, C., 1975. Phytogeographic provinces along the coasts of the northern Atlantic Ocean. *Phycologia*, 14: 317–330.

Van den Hoek, C., 1982. Phytogeographic distribution groups of benthic algae in the North Atlantic Ocean. *Helgol. Wiss. Meeresunters.*, 35: 153–214.

Van den Hoek, C., 1984. World-wide latitudinal and longitudinal seaweed distribution patterns and their possible causes, as illustrated by the distribution of Rhodophytan genera. *Helgol. Wiss. Meeresunters.*, 38: 227–257.

Van den Hoek, C. and Donze, M., 1967. Algal phytogeography of the European Atlantic coast. *Blumea*, 15: 63–89.

Wilce, R.T., 1959. The marine algae of the Labrador Peninsula and northwest Newfoundland (ecology and distribution). *Bull. Nat. Mus. Can.*, 158: 1–103.

Wilce, R.T., 1964. Studies on benthic marine algae in northwest Greenland. In: A. Davy de Virville and J. Feldmann (Editors), *Proc. 4th Int. Seaweed Symp., Biarritz, France*. Macmillan, New York, pp. 280–287.

Chapter 6

THE BALTIC SEA GRADIENT

I. WALLENTINUS

INTRODUCTION

The Baltic Sea (Fig. 6.1), one of the world's largest brackish water areas, has during recent decades attracted much attention from marine scientists. There are many recent reviews (for example, Elmgren, 1978, 1984; Jansson, 1978, 1980, 1981, 1984, 1989; Melvasalo et al., 1981; Voipio, 1981; Müller, 1982; Jansson et al., 1984; Wulff et al., 1986), including a full account in the volume in this series dealing with enclosed seas (Kullenberg, 1983). Hence, only a short summary of the hydrographical and geological development will be given here, and the biological communities will not be described in detail. The emphasis will be on the function and ecological performance of the biota in the Baltic Sea. However, since the geological development after the last glaciation in the Quaternary is not only of historical interest, but is highly relevant to the present situation, some brief comments will be included.

About 15 000 years B.P. the Baltic and the surrounding landmasses were still covered by ice. When, about 5000 years later, the ice had reached the coast of present Finland (Ignatius et al., 1981), it left, as a result, an ice lake in about the same position as the Baltic proper of today. During the following periods "the Baltic" underwent several stages, being both an isolated lake and connected to the ocean, inhabited by Arctic biota. Some of these organisms are still present in the Baltic today as glacial relicts, having invaded from marine or limnic environments. About 7000 to 8000 years ago (Ignatius et al., 1981) it reached its final stage, the Littorina Sea, with the sounds between the Danish islands as a transitional zone to the ocean. The higher temperature and salinity during the first part of this stage can be witnessed by records of both marine diatoms and invertebrates found further in the Baltic than today (Ignatius et al., 1981).

The Baltic Sea, as one knows it today, is only about 3000 years old (Ignatius et al., 1981) and now covers an area of *c.* 370 000 km^2. At that time the salinity had again decreased, since the connections to the ocean became shallower and narrower through sedimentation and land uplift. This uplift is still going on in most areas around the Baltic, reaching about 10 mm yr^{-1} in the northernmost parts, which continuously influences the succession of the terrestrial vegetation on the shores. This geologically and evolutionarily very short existence has had a considerable impact on the development of the Baltic biota. The connections to the oceans brought in organisms of marine origin, and these still dominate in all but the northern and innermost areas of the Baltic.

The Baltic Sea can be regarded as a gigantic threshold fjord (see also Waern, 1965), separated by sills from the ocean. Sills also divide the Baltic into several basins (Fig. 6.1) which, besides those of the Baltic proper, also include the Bothnian Sea and Bothnian Bay in the north, together named the Gulf of Bothnia, and the Gulfs of Finland and of Riga in the east. From a geographical, hydrographical and biological point of view, the natural outer border of the Baltic is inside the Danish sounds. However, when dealt with in political contexts, for instance by the Helsinki Commission for the Protection of the Baltic Marine Environment, the adjacent areas of the Belt Sea and the Kattegat are also included, being hydrographically much influenced by the Baltic.

The surrounding land-masses and the many

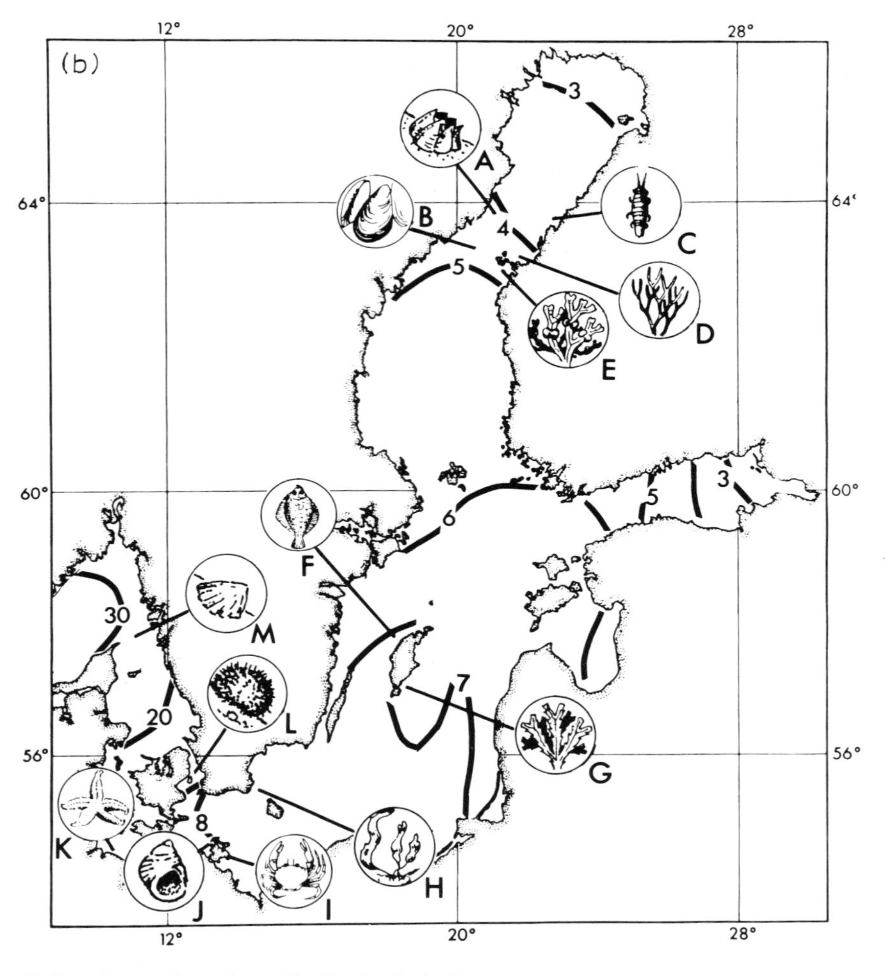

Fig. 6.1. The Baltic Sea and the Skagerrak with (a) subareas; (b) surface isohalines (‰S) and approximate inner distribution limits for some common marine species of importance for the littoral system. (A) *Balanus improvisus*, (B) *Mytilus edulis*, (C) *Idothea balthica*, (D) *Furcellaria lumbricalis*, (E) *Fucus vesiculosus*, (F) *Pleuronectes platessa*, (G) *Fucus serratus*, (H) *Laminaria saccharina*, (I) *Carcinus maenas*, (J) *Littorina littorea*, (K) *Asterias rubens*, (L) *Strongylocentrotus droebachiensis*, (M) *Patella vulgata*. Compiled from various sources.

wide archipelagos reduce the exposure to the oceanic swell considerably, and the mean depth is as shallow as *c*. 60 m. The proximity to land and a drainage area more than four times its surface also make the Baltic highly influenced by nutrient enrichment from land (Larsson et al., 1985) and by other anthropogenic discharges (Melvasalo et al., 1981; Voipio, 1981) as well as by a high turbidity. A long water residence time in the Baltic proper of tens of years and the weak average current velocities, contribute to the trapping of hazardous compounds as well as nutrients. The topography and salinity regimes further strengthen the development of a permanent primary halocline, which in the Baltic proper is located around 50 to 70 m depth, while in the Kattegat it is fluctuating around 15 m, being as shallow as 10 to 15 m in the Sound. In the northern Gulf of Bothnia a halocline is almost absent. Frequent upwelling in many areas still permits nutrient-rich bottom water to enter the autotrophic zone and thus contributes to the eutrophication. A seasonal thermocline develops in spring, but breaks up in autumn, which enhances the mixing in the upper water column.

The differences between the Baltic and true estuaries are as important as the similarities. Conspicuous differences include the almost complete absence of tides and the occurrence of stable, low salinities inside the transitional area. The largest parts of the Baltic can be classified as β-mesohaline (cf. Fig. 6.1b), with an annual salinity range offshore of only 0.5 to 1‰. However, the inner parts of the archipelagos, influenced by high river run-off and meltwater during spring, often experience salinity declines of 3 to 5‰. The effect of this can be witnessed by a strong reduction in the number of marine species in those areas.

Although geographically the Baltic is located in the high northern latitudes, equivalent to that of the Hudson Bay (also in size), the influence of the Gulf Stream makes the environment more benign for the flora and fauna. However, the northern latitudes do imply large seasonal fluctuations of both light regime and temperature.

The range in temperatures extends from below zero during winter in the north and in the Baltic proper (i.e. regular ice cover) to greater than 20°C in surface coastal waters during summer. The length of the period favourable for growth is about 9 to 10 months in the south, but only 4 to 5 months in the northernmost part (Hällfors et al., 1981).

Discharges from the many rivers contribute, to a large extent, to a high light attenuation in most of the area due to both eroded sediment particles and humic substances. This is especially noticeable in spring, when meltwater is discharged. Phytoplankton blooms further reduce the light intensities reaching the benthic primary producers. In the outer coastal areas of the Baltic proper macroalgal vegetation can be found down to a depth of *c*. 20 to 25 m. This compares well with the annual quantum flux densities received at maximum depth of light-limited growth in other areas (Table 6.1). Further into the archipelagos and in the Bothnian Bay macroalgal vegetation is restricted to even shallower depths — in areas close to rivers and urban centres even as shallow as less than 5 m (Pekkari, 1973; Wallentinus, 1976; Kautsky et al., 1981). Light limitation also sets the lower distribution limits for *Fucus* species, which in many areas have moved closer towards the surface during recent decades (Kautsky et al., 1986; Breuer and Schramm, 1988).

The nutrient concentrations are seasonally pulsed too, governed mainly by the activities in the pelagic system. Thus, the spring phytoplankton bloom almost totally strips the water of available nutrients, especially of the inorganic nitrogen compounds. The main parts of the Baltic have inorganic N:P ratios below the Redfield ratios of 16:1 (Fonselius, 1978), caused by upwelling of phosphorus-rich bottom water. The Bothnian Sea, and especially the Bothnian Bay, have lower nutrient concentrations and higher N:P ratios, which in the Bothnian Bay can even exceed 50:1 (by atoms: Fonselius, 1978). In the Baltic proper and the outer areas of the Sound and the Kattegat the production of both the phytoplankton (Granéli, 1981; Larsson, 1984; Granéli and Sundbäck, 1985; Granéli et al., 1986) and macroalgae (Feldner, 1976; Wallentinus, 1981a, 1983; Schramm et al., 1988) is mainly limited by nitrogen, except in inner coastal areas influenced by land run-off and rivers.

Although most of today's flora and fauna in the Baltic have existed there during the last 3000 years, new species are still invading, sometimes introduced by man (Elmgren, 1984; Leppäkoski, 1984). The degree of endemism in the Baltic is very low, in

TABLE 6.1

Annual quantum irradiance received at different depths in the outer archipelago at Askö, northern Baltic proper. Calculated from average global insolation (A), assuming PAR is 45%, and an average annual light extinction coefficient for the area (B) of 0.317 m^{-1} (5 years observations). Surface reflection is assumed to be 15%. Lower limit of attached macroalgae observed by diving: *c.* 20 to 25 metres. I = *Fucus vesiculosus* belt; II = Red algal belt; III = Single algal specimens

	Depth (m)	$E\ m^{-1}\ yr^{-1}$ Askö	Helgoland[1]	
I	1	4523		
	2	3294	1037	L. digitata
	4	1748	388	
	6	927		
II	8	492	71	L. hyperborea
	10	261	33	
	15	53	6	Lower limit of macroalgae
III	20	11		
	22	6		Lower limit of macroalgae
	25	2		

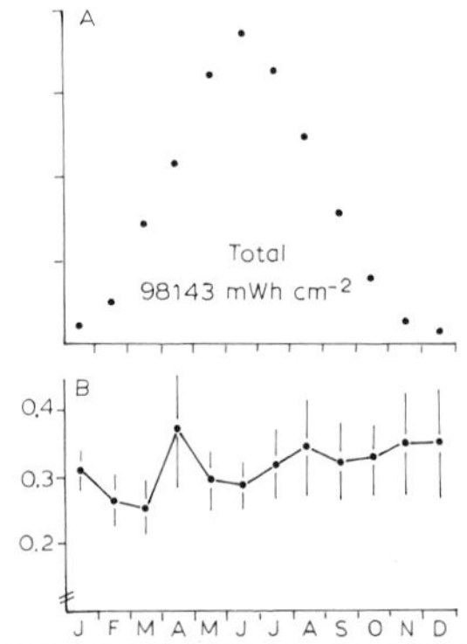

[1]From Lüning and Dring (1979).

contrast to the ideas earlier this century. Most species previously considered endemic or specific for brackish areas are either not true species, or have been found in culture experiments to be able to hybridize with marine relatives. In many cases, however, they have specific eco-physiological characters, which justifies their classification, on a subspecific level, as ecotypes, ecoclines or physiological clines (Rueness, 1978; Russell, 1985a; Tedengren and Kautsky, 1986), or as subspecies (Russell, 1988). Some algal species also show a trend towards more anatomical and morphological specialization (Russell, 1985b). The ongoing development of Baltic ecotypes and ecoclines emphasizes the theory that the Baltic Sea is a highly interesting area for studies in evolutionary biology, but hitherto there have been surprisingly few studies of that kind.

The historical development of the Baltic flora and fauna, with their high dominance of marine species and very few brackish water species, led Remane more than forty years ago to propose the theory of minimum number of species in brackish water (Remane, 1940), which has since been confirmed in many Baltic studies. For macroalgae this pattern is also revealed in Fig. 6.2, where the gradually decreasing number of species in common with the true marine environment of the North Sea is very obvious. Very few algal species which do not occur in the North Sea are found in the Baltic. In contrast to similar analyses for the floral provinces along the Atlantic European coasts (Van den Hoek, 1975), the gradient from the North Sea to the innermost parts of the Baltic does not involve any pattern of floral discontinuity between areas of different homogeneous floras, but merely a more or less stepwise decline. An additional example can also be found in the reduction of the number of physiognomically important large brown seaweeds (Fig. 6.3), which occupy the most important niches in many marine littoral systems.

It is also a well-known fact that the taxonomic groups of macroalgae react differently, the decline being more pronounced in the red than in the green algae (Fig. 6.4). The percentage of the latter even increases inwards, being due both to more euryhaline species and to the incoming freshwater species, thus highly affecting the ratios between the taxonomic groups (see, for instance, Wallentinus, 1979).

Most of the Baltic organisms are living close to their limit of tolerance to low salinity. Generally they are found at lower salinities in the Baltic than in estuaries, thanks to the low fluctuation in

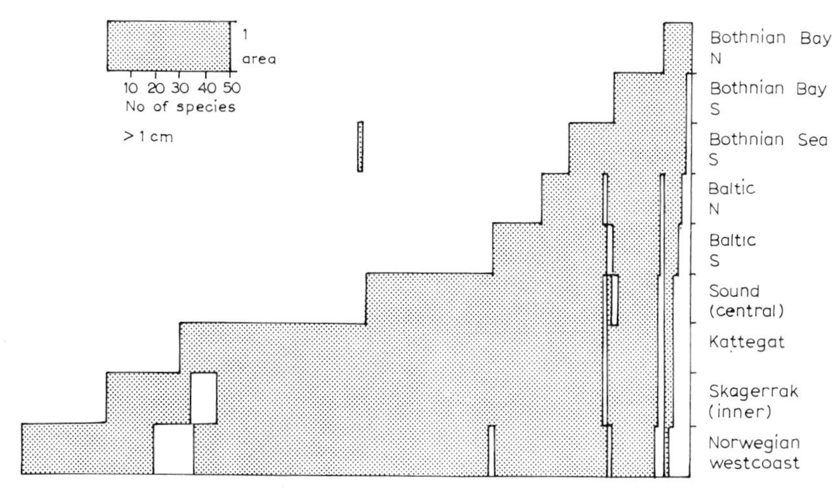

Fig. 6.2. Total number of macroalgal species (> 1 cm) in the different regions in the gradient from the North Sea to the innermost area of the Baltic. A continuous shading shows that the species occur in all areas, and a break indicates that they are absent in some areas.

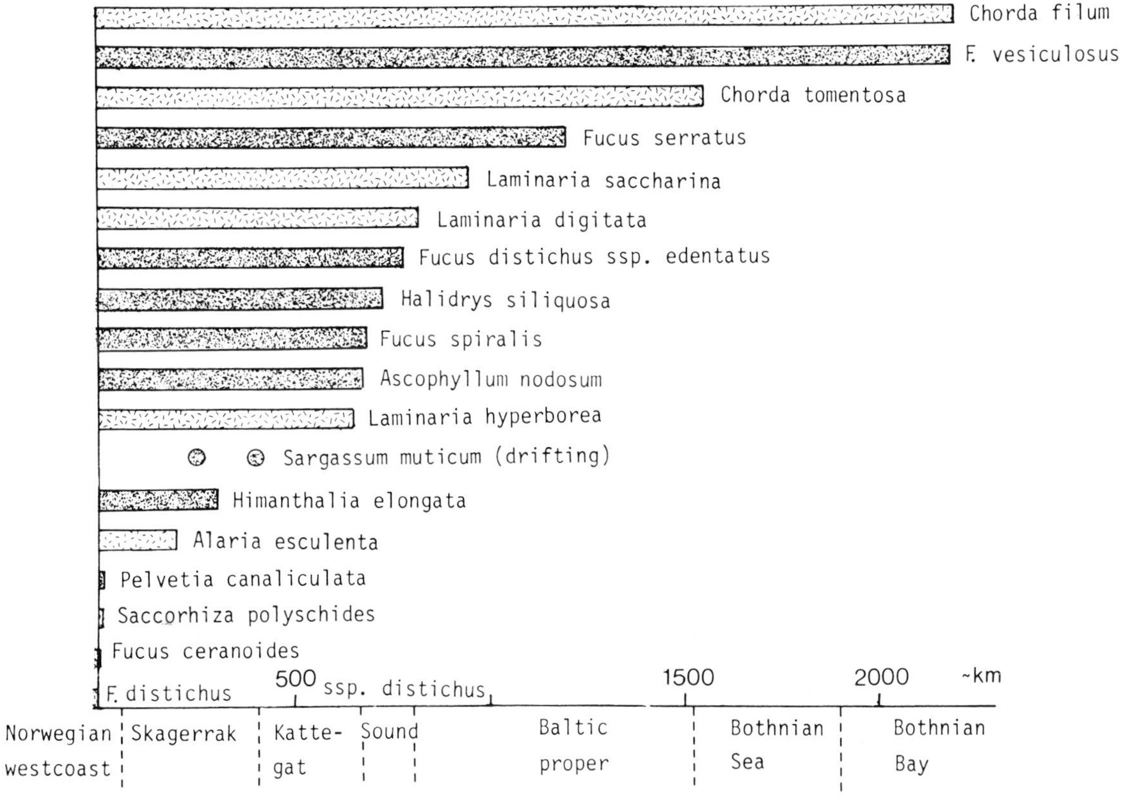

Fig. 6.3. Inner distribution limits of the *Fucales* and *Laminariales* along the gradient from the North Sea. *Sargassum muticum* has since 1987 been found attached on the Skagerrak coast.

salinity. The rather few Baltic representatives of many functional groups imply, however, that further stress on the ecosystem is crucial, since there are only few alternative species present to replace them.

A BRIEF QUALITATIVE DESCRIPTION OF THE LITTORAL ECOSYSTEM

Generally, the Baltic ecosystem is characterized by a low diversity of all macrocomponents, while

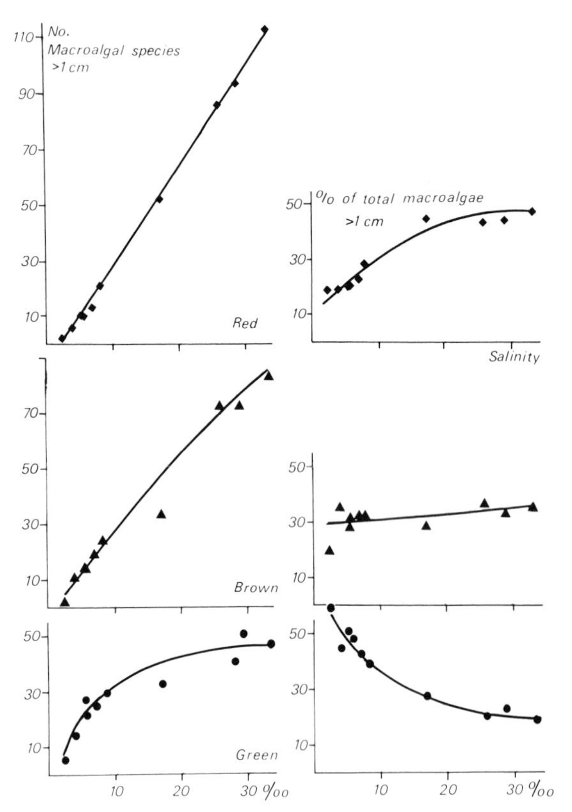

Fig. 6.4. Numbers (left) and percentage (right) of red, brown and green macroalgae (> 1 cm) versus average salinities in the Baltic.

the diversity increases in the adjacent outer areas. To understand the structural and functional aspects of the Baltic littoral ecosystem, it must be remembered, however, that the changes in species composition of both the flora and the fauna even within the Baltic subareas are almost as large as between the North Sea and the southern parts of the Baltic. To these are added the local differences, as experienced along all marine coasts, which depend mainly on the morphology of the coastline and the degree of wave exposure. Many littoral organisms, both animals and plants, also occupy broader niches in the Baltic than in the true marine environment, where biological competition is stronger.

Figure 6.5 summarizes the structurally most important communities of the littoral ecosystem in the Baltic proper, together with their major forcing functions. Naturally, these subsystems are not self-supporting, but have a continuous exchange (import and export) of organic material and nutrients, as well as of many organisms inhabiting these communities. This applies both between the various benthic subsystems and between the pelagic and benthic systems.

COMMUNITIES ON ROCKY SUBSTRATES

Eulittoral

Although the tidal range in most of the Baltic is negligible, meteorologically induced fluctuations in mean water level (which is normally low during several weeks in spring and high in autumn), together with ice-scouring in winter, form a community which in many respects compares with the eulittoral zone of tidal areas.

Above this "eulittoral", the supralittoral zone is, as on most marine rocky coasts, characterized by lichens, though *Lichina* species are only rarely met with. Apart from blue-greens, algae are sparse, but the green algae *Prasiola* spp. are often conspicuous among the yellowish lichens on the many skerries and rocks occupied by gulls and other birds. Also, the fauna is strikingly impoverished, and the periwinkles, *Littorina* spp., only reach the southern part of the Baltic proper.

The upper part of the "eulittoral", in Baltic literature often called the geolittoral zone, also maintains very few algae apart from the blue-green, nitrogen-fixing *Calothrix* belt on the southerly inclined rocks, while lichens dominate the northerly inclined ones. In spring and autumn, however, belts of the green algae *Urospora penicilliformis* and *Ulothrix* spp. extend from the zone below, in some areas also accompanied by the brownish red filaments of *Bangia atropurpurea* during the cold season.

The lower part of the "eulittoral" (the hydrolittoral) is, in contrast to the Atlantic coasts, dominated by opportunistic, mostly filamentous, annual macroalgae with their epiphytic microflora and -fauna. Because of the irregularities in water level and the mechanical disturbances by ice, this community is almost devoid of true perennials, including the barnacles, except for those confined to crevices in the rocks (for example, crusts of brown and red algae), which are regularly splashed by waves and protected from the drifting ice. Thus, the litus line is rather inconspicuous. The algal

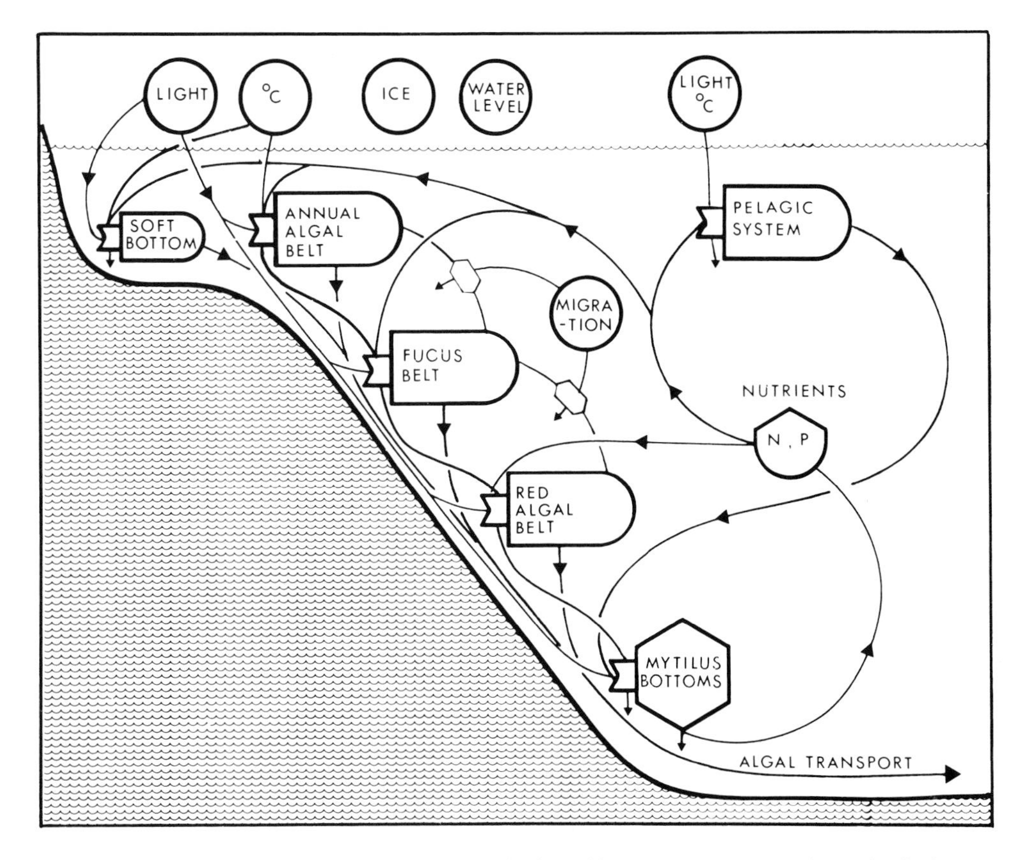

Fig. 6.5. Structurally important littoral communities in the Baltic proper versus approximate depths (0–25 m) and their main forcing functions. Symbols used are according to the energy circuit language by Odum (1972).

belts in this part of the "eulittoral" shift with season [for references to algae in this zone and in the sublittoral see, for instance, Levring (1940), Waern (1952, 1965), Ravanko (1968), Wallentinus (1979), Hällfors et al. (1981)]. In the Baltic proper and adjacent inner areas the brown alga *Pilayella littoralis* dominates the lower part of the belt in spring, together with tube-forming diatoms (Kautsky et al., 1984), the green algae *Monostroma grevillei* and *Acrosiphonia centralis*. In late spring the light green fringe of the freshwater species *Cladophora glomerata* begins to appear with concomitant brown (*Eudesme virescens, Scytosiphon lomentaria, Dictyosiphon* spp.) and green algae (*Enteromorpha* spp.). Towards the end of summer (the time depending on the persistence of the water level when the last generation of *Cladophora* spores settles) the zone is gradually taken over by the red alga *Ceramium tenuicorne* (for taxonomical affinity see also Rueness, 1978). Both in summer and autumn the blue-green alga *Rivularia*, with nitrogen-fixation capacity, is frequent on the basal parts of the algae and on the rocks.

The fauna of the "eulittoral" (Jansson, 1974; Hällfors et al., 1975, 1981) is especially rich in summer and autumn, when notably numerous juvenile crustaceans (*Jaera* spp., *Idothea* spp., *Gammarus* spp.), molluscs (*Cardium* spp., *Hydrobia* spp., *Mytilus edulis*, *Theodoxus fluviatilis*), insect larvae (for example, chironomids) and turbellarians utilize the rich food supply of epiphytic microflora and -fauna, and of organic matter trapped among the algal filaments. During winter most of these species migrate down into the sublittoral.

Further into the Baltic, several of these marine species disappear. The more severe ice-scouring and prolonged winter period, with an ice-cover lasting even up to 5 to 6 months in the very north, mainly restrict the "eulittoral" vegetation to summer and autumn with *Ceramium tenuicorne*,

Cladophora spp., *Rivularia atra* and *Ulothrix* spp. as the main components (Hällfors, 1976; Kangas, 1976; Hällfors et al., 1981; Kautsky et al., 1981; H. Kautsky, 1988).

Outwards, already in the Belt Sea and the Sound, the zonation becomes more similar to that of the Atlantic coasts. Below the litus line of *Semibalanus balanoides* a true eulittoral belt of *Fucus vesiculosus* occurs. Also, more annuals appear, such as species of *Dumontia, Petalonia* and *Porphyra* and the *Littorina* species occur regularly (Schwenke, 1974; Von Wachenfeldt, 1975). In the Kattegat the differences from the Atlantic coasts are mainly to be found among the Fucales (Fig. 6.3).

Sublittoral

In the upper sublittoral in the Baltic, including the Bothnian Sea and the Gulf of Finland, one finds the only architecturally important and true canopy-forming seaweed *Fucus vesiculosus*. During periods of extremely low water levels, the upper plants in this belt are also affected by desiccation and ice-scouring. Due to the many rocky archipelagos, and the extension of the belt downwards to about 5 to 8 m, *Fucus vesiculosus* also comprises the dominant biomass in most of these areas. In the southern parts (Fig. 6.3) *Fucus serratus* is a codominant species (Levring, 1940; Waern, 1965; Jansson et al., 1982; H. Kautsky, 1988). Their epiphytes are mainly marine, including also the host specific brown alga *Elachista fucicola*. As on most marine coasts, however, both species avoid the most exposed localities where they are replaced by filamentous "eulittoral" species and by perennials from the deeper part of the sublittoral.

The cause of the sublittoral occurrence of bladder-wrack in the Baltic has been the subject of many theories. However, as pointed out in several papers (Wallentinus, 1979; Hällfors et al., 1981; Rönnberg, 1981), the physical disturbance in the "eulittoral", and the absence of strong perennial competitors in the upper sublittoral, are the major responsible factors. A similar sublittoral *Fucus* belt is found in the Gulf of the St. Lawrence, Canada (Bird et al., 1983), where the ice-scouring is extensive. Recently, there have been many reports on a decline of the *Fucus* community in the Baltic. This has happened in many areas (see, for instance, references in Wallentinus, 1981b; Lindvall, 1984; Plínski and Florczyk, 1984; Wennberg, 1987) as a consequence of pollution, but also in nonpolluted areas, especially along the southern coast of Finland. There the decline is not due to direct influence of pollution, but rather to large scale fluctuations of nutrients in the Baltic and indirectly to increased grazing (Kangas et al., 1982; Haahtela, 1984; Hällfors et al., 1984; Mäkinen et al., 1984; Rönnberg, 1984; Salemaa, 1987). In some areas the communities have started to recover (Kangas and Niemi, 1985; Rönnberg et al., 1985).

The *Fucus* plants, together with their epiphytes and understorey species, being in both cases a seasonally varying complex of mostly filamentous algae, constitute the basis of the most diverse subsystem in the Baltic proper, with a variety of mobile and sessile macro- and meiofauna (Haage, 1975, 1976; Skult, 1977; Kangas, 1978; Jansson et al., 1982; H. Kautsky, 1988). About 55 taxa of macrofauna are included in the *Fucus* fauna in the Baltic proper, the most richly represented being crustaceans (for example, *Balanus improvisus, Gammarus* spp., *Idothea* spp., Mysidae, *Palaemon* spp.), molluscs (such as *Cardium* spp., *Hydrobia* spp., *Lymnaea* spp., *Mytilus edulis, Theodoxus fluviatilis*), insect larvae (such as Chironomidae, Trichoptera), the bryozoan *Electra crustulenta* and small demersal fishes (such as sticklebacks, gobiids). This rich fauna is in turn preyed upon by larger fishes such as perch (*Perca fluviatilis*) and pike (*Esox lucius*). Herring roe is frequently deposited in the *Fucus* community, which as a whole is a mixture of limnic and marine species. Still more species are represented in the transitional areas (Hagerman, 1966; Anders and Möller, 1983), and in the sublittoral *Fucus serratus* belt in the Kattegat.

In the northern and inner archipelago areas, where *Fucus* does not occur, it is replaced by perennials such as the freshwater species *Cladophora aegagropila* and the water moss *Fontinalis* spp. down to a maximum of about 10 m depth. This community has a much impoverished fauna of mainly limnic invertebrates (Kangas, 1976; Hällfors et al., 1981; Kautsky et al., 1981; H. Kautsky, 1988). However, there are still many demersal fish species, including also whitefish (*Coregonus lavaretus*) and vendace (*C. albula*).

Below the *Fucus* belt (i.e. down to about 25 m

depth depending both on the light attenuation and the availability of rocky substrate), a community composed mainly of red algae remains. The algal biomass is dominated by late successional perennials such as *Furcellaria lumbricalis* and *Phyllophora* spp., the latter mostly with highly reduced morphology. Filamentous perennial algae are also common (for example, *Ceramium* spp., *Polysiphonia* spp., *Rhodomela confervoides*, and the glacial relict *Sphacelaria arctica*). To a great extent all these algae are not permanently attached but occur as loose specimens entangled among the byssus threads of the blue mussel, *Mytilus edulis*. As pointed out already in the early studies on the Baltic, many of these red algae have lost their fertility and, like many other macroalgal species, they often have a reduced size (Fig. 6.6). It should also be emphasized that the calcareous red algae are found only in the southern parts of the Baltic and outside, and thus sublittoral crusts are represented only by a few brown and red algae in most of the Baltic proper and in its gulfs.

The fauna of the deeper sublittoral is totally dominated by mats of *Mytilus edulis*, which below 10 to 15 m more and more dominate the community, consisting of small-sized individuals up to *c.* 4 cm. This high abundance is possible because of the lack of most of its normal invertebrate predators, such as the starfish or sea star *Asterias rubens* (and in fact all echinoderms are absent), predatory snails, and the crab *Carcinus maenas* (and all the large decapods). There is also very little interspecific competition for food, since the other main marine filter-feeders, the ascidians, are also excluded due to the low salinities. Thus, the mussel populations usually reach carrying capacity, with natural death as the main mortality factor (Kautsky, 1981a). The absence of the blue mussel in the Bothnian Bay and inner Gulf of Finland highly reduces the importance of the deeper sublittorals for the energy flow in those areas (Jansson et al., 1980; Elmgren, 1984; H. Kautsky, 1988). Together with small crustaceans (*Gammarus* spp., *Mesidothea entomon*) and other molluscs (e.g. *Hydrobia* spp., *Macoma balthica*), several demersal fish also frequently occupy the deeper sublittoral in the Baltic, such as flounder, turbot, cod, lump-sucker, eelpout, gobiids and many others (Jansson et al., 1985).

Already in the Belt Sea the fauna has a more marine appearance (Lüthje, 1978). In the Kattegat, where salinities in the deeper sublittoral zone approach those of the North Sea, the algal communities on rocky substrates mainly have the same species composition as on the Norwegian Atlantic coast (Lüning, 1985; see also Fig. 6.2). That includes both the three *Laminaria* species below the sublittoral *Fucus serratus* belt and a variety of red algae, where many species — in contrast to those in the Baltic proper — have a leaf-like morphology. However, even in this transitional area many algal species are still reduced in

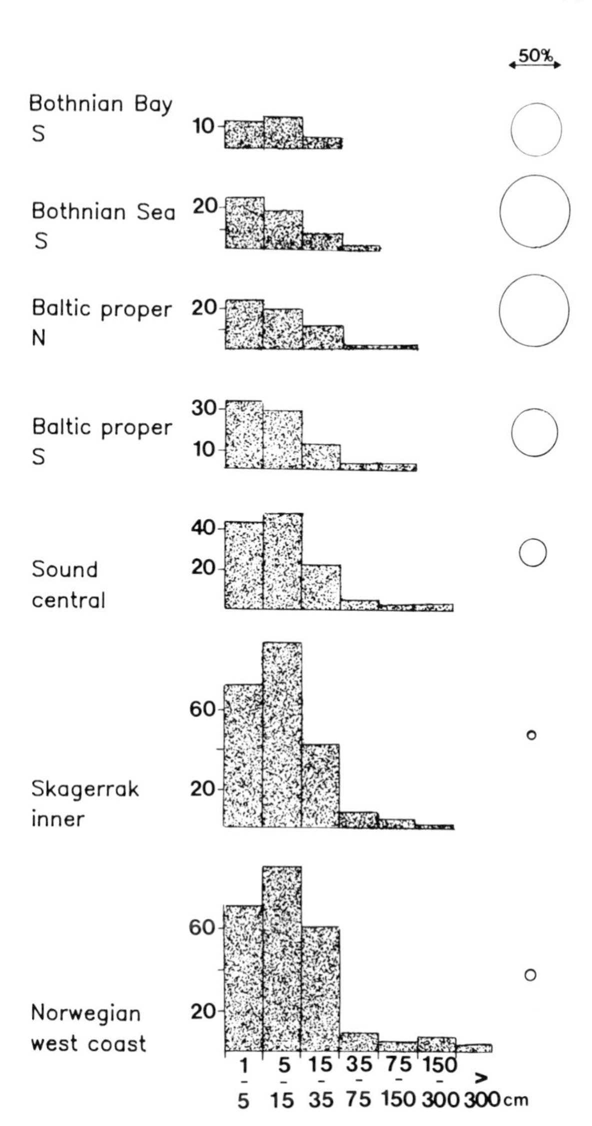

Fig. 6.6. Number of macroalgal species in different size-classes and the percentage of red algae with disturbed reproduction along the North Sea–Baltic gradient.

TABLE 6.2

Population and community attributes in environmentally constant and environmentally disturbed habitats according to Littler and Littler (1981). The disturbed habitats correspond to the "eulittoral" zone and the constant to the sublittoral zone in the Baltic proper. Reprinted with permission of *Marine Ecology Progress Series*

Environmentally constant habitat	Environmentally disturbed habitat
Populational attributes	
(1) Thalli larger and morphologically complex; low surface/volume ratios; high structural/photosynthetic tissue ratios	(1) Small and simple thalli; high surface/volume ratios; low structural/photosynthetic tissue ratios
(2) Long-lived perennials	(2) Opportunistic ephemerals
(3) Low reproductive output	(3) High reproductive output
(4) Low productivity	(4) High productivity
(5) Stenotopic forms[1]	(5) Eurytopic forms
(6) Mostly late successional	(6) Mostly early successional
Community attributes	
(1) Structurally heterogeneous; three-dimensional	(1) Structurally simple; two-dimensional
(2) Mosaic pattern of successional seres	(2) Homogeneous patterns
(3) Diversity, evenness and richness high	(3) Diversity, evenness and richness low
(4) More biotic interactions (e.g. crowding, layering of canopies)	(4) Fewer biotic interactions
(5) Biologically accommodated	(5) Physically accommodated

[1]This does not apply to Baltic macroalgae which are all eurytopic.

size (cf. Fig. 6.6). *Furcellaria lumbricalis* is still one of the biomass dominants (Kornfeldt, 1979), and red calcareous crusts are common on rocks and boulders. The fauna also has a much more marine appearance: ascidians, echinoderms, large decapods, a great variety of molluscs and bryozoans, the soft coral *Alcyonium digitatum*, the sponge *Halichondria panicea* and marine fish species are very common.

A description of population and community attributes for environmentally fluctuating versus constant systems has been presented by Littler and Littler (1981) (Table 6.2). Originally these attributes were based on intertidal communities in the Gulf of California and the Pacific. However, they fit surprisingly well on the "eulittoral" and sublittoral subsystems of the Baltic proper for environmentally fluctuating and stable conditions, respectively. This further strengthens the concept described above of a physical control of the "eulittoral" zone in the Baltic, in contrast to the often biologically controlled eulittoral zones in many true marine habitats (cf. Ch. 4).

If the functional-form model approach (Littler and Littler, 1980) is applied instead of taxonomical affinities to the macroalgal composition in the Baltic, there is a surprisingly high similarity in the percentage of species in the same form groups all along the gradient from the North Sea (Fig. 6.7). The filamentous group has by far the most representatives, with the jointed calcareous algae as the least common group, being almost totally absent from the Baltic proper.

LITTORAL COMMUNITIES ON SANDY AND MUDDY SUBSTRATES

These communities are usually dominated by freshwater phanerogams and charophytes. However, the marine sea-grass *Zostera marina* also occurs as far into the Baltic as the southernmost part of the Bothnian Sea and the Helsinki archipelago in the Gulf of Finland. This Baltic ecocline of *Zostera*, with narrower leaves than on the Atlantic coasts, did not show a decline during the 1930s because of its low reproductive potential and ability to spread mostly by vegetative diaspores.

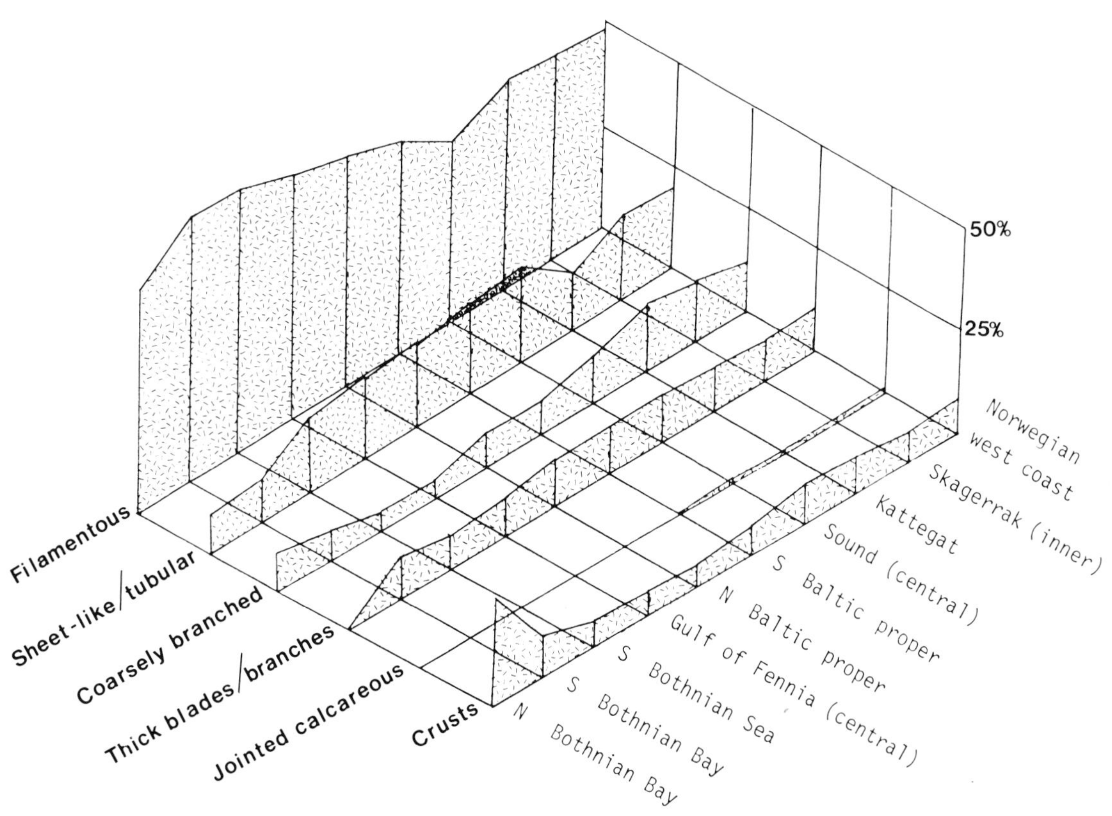

Fig. 6.7. Percentage of number of macroalgal species (> 1 cm) belonging to different functional-form groups along the North Sea–Baltic gradient.

Throughout the area, sheltered bays are covered by extensive reed beds, composed of the emergent species *Phragmites australis* and *Scirpus* spp. Epiphytes of macro- and microalgae and sessile animals are common on the stems. These reed beds, occupying depths down to *c.* 2 m, are also frequently found along moderately exposed shores in less dense stands. They are comparable to the *Spartina* salt marshes in the tidal areas of the Atlantic coasts. Since the reed beds are not flushed by tidal currents, their ability to trap sediment among the rhizomes considerably contributes to decreased water movements, especially in sheltered sounds and bays. Thus, anaerobic conditions in the sediments are also widespread in such areas. Finally, the reduced mixing can affect even the rocky shores through increased sedimentation, thus converting communities previously occupied by macroalgae into soft bottoms. The reed beds benefit from the ongoing nutrient enrichment of the Baltic and from the decreased grazing by cattle in recent decades.

The fauna of the reed beds is a mixture of marine and freshwater species. The latter can be exemplified by the many species of insect larvae of Odonata, Trichoptera and Chironomidae, of gastropods such as species of *Bithynia*, *Lymnaea* and *Theodoxus*, the small isopod *Asellus aquaticus*, and many fishes [such as bream (*Abramis brama*), rudd (*Scardinius erythrophthalmus*), roach (*Rutilus rutilus*), bleak (*Alburnus alburnus*), perch and pike, see for instance Blomqvist (1984)]. Bays with reed beds consititute important spawning areas for many of these fishes. The emergent plants have a significant structural function, too, in providing shelter for various nesting birds in early summer, and also in providing a rich variety of insects and bottom fauna as food. Birds and fish are preyed upon in turn by top carnivores [such as osprey (*Pandion haliaetus*) and white-tailed eagle (*Haliaetus albicilla*)].

The muddy substrates outside the reed beds mostly have a patchy distribution of submerged phanerogams (for example, *Myriophyllum* spp.,

Potamogeton spp., *Ranunculus* spp., *Ruppia* spp., *Zannichellia* spp., and in areas with low salinities also the quillwort, *Isoëtes* spp.; Luther, 1951; Wallentinus, 1979; H. Kautsky, 1988; L. Kautsky, 1988). Many of these species are also physiognomically important, offering shelter for a fauna similar to both that of the reed beds and the *Fucus vesiculosus* belt. In the sediment between the plants, molluscs (such as *Cardium* spp., Hydrobiidae, *Macoma balthica, Mya arenaria*), crustaceans (such as *Corophium volutator*) and polychaetes (such as *Nereis diversicolor*) are common. Charophytes, mainly *Chara* spp., can form extensive stands in very sheltered bays. During the last decades many of these *Chara* communities have disappeared, due to increased boat movement and dredging for marinas (cf. also Trei, 1985).

Sandy substrates and pebbles and shells on mud are frequently colonized by the marine brown alga *Chorda filum*. Although reduced in size in comparison to the marine specimens, these often metre-long annuals develop into dense meadows. Together with the sea-grass *Zostera*, they play a large role along the sandy coasts in the southern and southeastern parts of the Baltic, in the Baltic proper accompanied by the charophyte *Tolypella nidifica*, and along the sandy coasts of the Kattegat. In the many rocky archipelagos along the Swedish and Finnish coasts the role of the *Zostera* community is diminished. This is particularly obvious when compared to the rich fauna of commercially important species in the marine *Zostera* meadows. However, they still fulfill the function of feeding and nursery grounds for fish such as flounder, gobiids, sand-eel and turbot, as well as spawning areas for herring. The fauna is mainly composed of the same species as among the other submerged phanerogams (Lappalainen et al., 1977; Hällfors et al., 1981), but also includes the crustacean *Crangon crangon*, although not commonly enough to be harvested. Outwards the number of marine species increases (Gründel, 1976; Asmus et al., 1980).

The phanerogams are normally restricted to the upper six to eight metres of the seabed. Deeper along the sandy bottoms, communities of loose-lying red algae usually replace them, the species being the same as on deeper sublittoral rocky substrates. On mud the bottoms are normally devoid of macrovegetation from about five metres downwards.

Benthic microalgae, mainly diatoms, are efficient primary producers on the sediments (Sundbäck, 1983). In spring tube-forming diatoms may build up a dense cover on hibernating phanerogams and on the bottom. During calm weather their high photosynthetic activity creates gas bubbles that lift them off the substrate, giving rise to floating mats that may cover several square metres before they disperse.

BIOMASS DISTRIBUTION

The variation in species composition along the salinity gradient from the Kattegat inwards, with the decreasing number of canopy species (Fig. 6.3), is also reflected in the biomass distribution (Figs. 6.8 and 6.9). Within a geographical region the biomasses are also highly dependent on the type of bottom substrate and the communities they support. The high abundance of annual species among the macroalgae further implies quite substantial seasonal fluctuations (Fig. 6.10).

The number of investigations on biomass distribution is rather limited, and strict comparisons are further hampered by the fact that the calculations are based on different units of area (averages over total sea area or per area of harvested bottom substrate).

Plants

In the outer areas of the Kattegat and the Sound (Fig. 6.8), the presence of many perennial species results in biomasses in the sublittoral that are only slightly lower than those in fully marine environments (Von Wachenfelt, 1975; Kornfeldt, 1979, 1984). The eulittoral zone was not included in these studies, but can be expected to have high values too, since perennial species occur frequently there. In the northern part of the Sound, where the algal communities investigated were more typical for the area than along the exposed coast of the Kattegat, *Fucus serratus* comprised 56%, *Laminaria* spp. 11%, *Furcellaria lumbricalis* 14%, *Phyllophora* spp. 5%, *Chondrus crispus* 3% and thin filamentous and sheet-like algae the remaining 11% of the total mean biomass over the seasons (400 g dry weight per square metre of harvested rock) between 2 and 10 m depth. The highest single

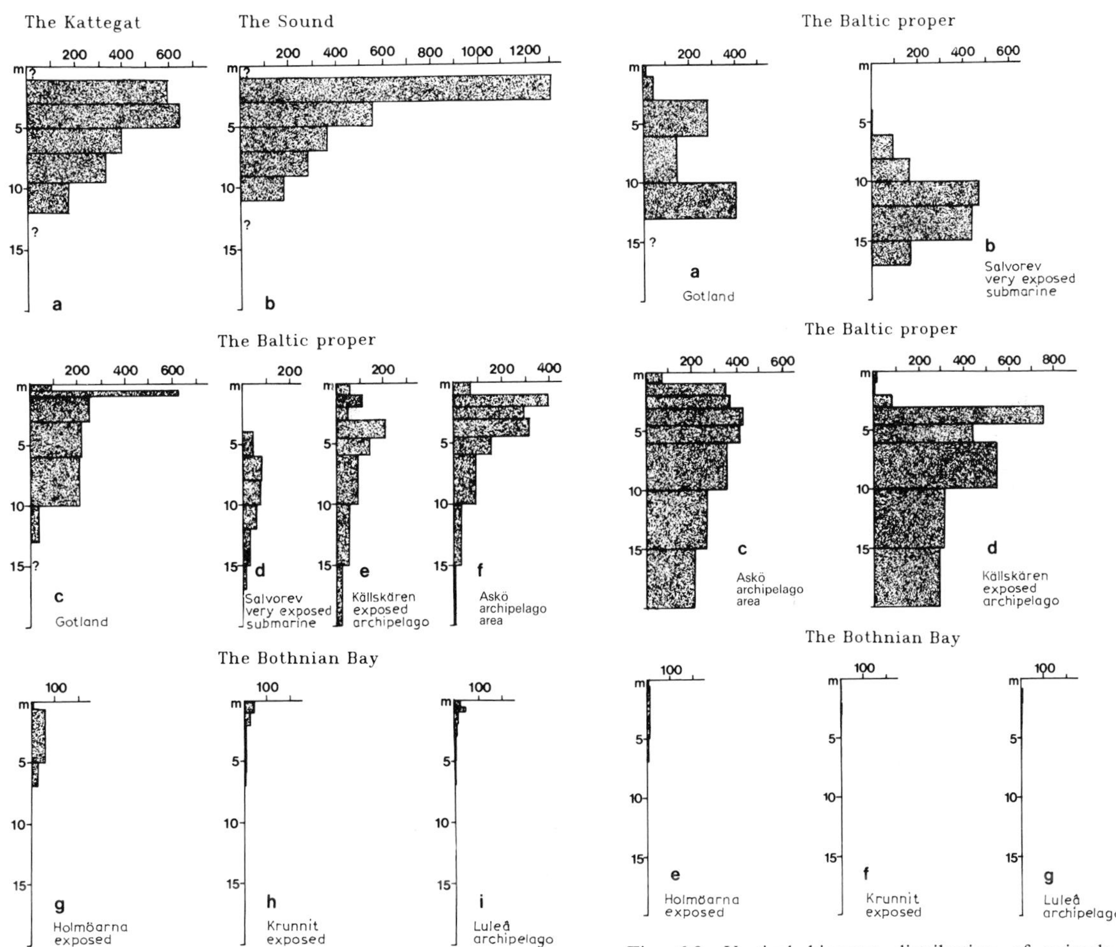

Fig. 6.8. Vertical plant biomass distribution (g dry wt m^{-2}) (biomass = horizontal axis; water-depth = vertical axis) in different regions of the Baltic. Values are from (a) Kornfeldt (1979); (b) Kornfeldt (1984); (c, d, e) H. Kautsky (1988); (f) Jansson and Kautsky (1977); (g) H. Kautsky (1988); (h) Hällfors in Kangas (1976); (i) Kautsky et al. (1981).

Fig. 6.9. Vertical biomass distribution of animals (g dry wt m^{-2}, incl. shells) in different regions of the Baltic. Values are from (a, b) H. Kautsky (1988); (c) Jansson and Kautsky (1977); (d, e) H. Kautsky (1988); (f) Kangas (1976); (g) Kautsky et al. (1981).

biomass values found there were for *F. serratus* 2.1 kg dry wt m^{-2} (2 m), for *F. lumbricalis* 0.5 (6 m) and for *L. digitata* 0.5 (8 m).

In the Baltic proper (Fig. 6.8), the existence of the sublittoral *Fucus vesiculosus* belt generates the highest biomasses in the interval between 1 and 5 m water depth (Jansson and Kautsky, 1977; H. Kautsky, 1988), while the values were lower both in the "eulittoral" zone and deeper down. In the first investigation *F. vesiculosus* made up 33% of the total average for the whole archipelago area (corresponding to 21.5 g dry wt m^{-2} of hard substrate available between 0 and 25 m), *F. lumbricalis* 15% (9.5 g), *Phyllophora* spp. 11% (6.9 g), *Pilayella littoralis* 11% (6.8 g) and *Ceramium tenuicorne* 10% (6.7 g), with a total average biomass in summer of 64.2 g dry wt m^{-2}. Individual maximum biomass values per m^2 area of the rocks were 2.3 kg dry wt for *F. vesiculosus*, 1.0 for *F. lumbricalis*, 0.4 for *Cladophora glomerata*, and 0.3 for *P. littoralis* (Jansson and Kautsky, unpubl.). In the outer part of the Gulf of Finland the algal biomass distribution is quite similar, having about the same proportions between the species (Hällfors et al., 1975; Lindgren, 1978). Offshore areas in the Baltic proper have a lower biomass of algae than the archipelagos (Jansson and Kautsky, 1977; H. Kautsky, 1988), since the *Fucus* belts are

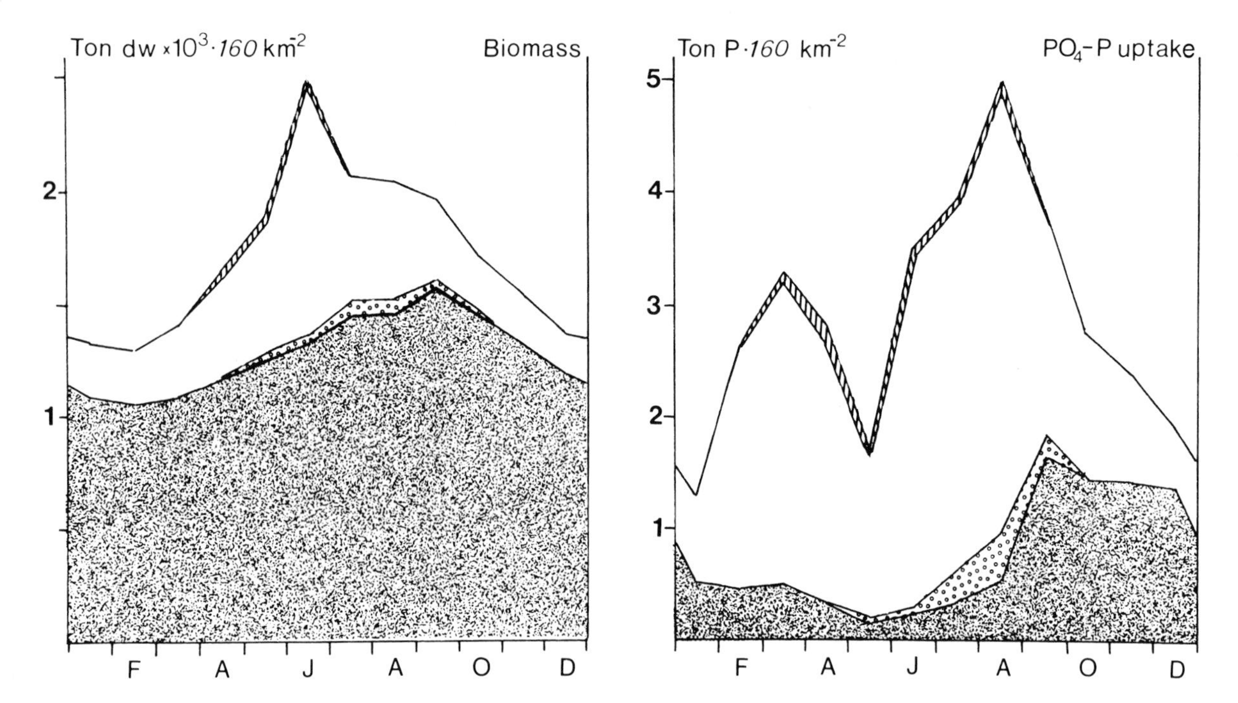

Fig. 6.10. Biomass (left) and nutrient uptake (right) of macroalgae belonging to different functional-form groups in a Baltic archipelago area. Sheet-like/thin tubular algae (stripes), thin filamentous algae (blank), coarsely branched algae (dots) and algae with thick leathery blades or branches (shaded). Based on values given by Wallentinus (1984b).

often missing or occur only sparsely. The proportions between the other dominants are about the same, except that the biomass of perennials is lower in the most exposed areas, due to the mobility of the substrate (boulders).

In the northern Bothnian Sea and Bothnian Bay, the biomasses are still lower (Fig. 6.8), and more than 3/4 of the plant biomass on hard substrates consists of green algae (Hällfors, 1976; Kautsky et al., 1981; H. Kautsky, 1988).

Calculations, based on these investigations, of algal biomass per metre of shorelength reveal large variations. These are primarily due to inclinations of the rocks in the study area rather than to regional differences. For instance, the large areas of rocky substrate available in the upper sublittoral zone around the island of Gotland (H. Kautsky, 1988) support an extensive growth of *Fucus vesiculosus*, with total biomass being on average *c.* 110 kg dry wt m^{-1}. The corresponding value from the archipelago in the northern Baltic proper was *c.* 8 kg, and 5 to 10 kg in the Finnish archipelagos and *c.* 1 kg in the Bothnian Bay (Wallentinus, 1983, fig. 10). The high value from Gotland was due to a mean transect length of *c.* 290 m, while those in the archipelagos were about 1/10 to 1/3 of that in length. However, the mean transect in the Bothnian Bay was almost as long, and thus reflected true biomass differences due to the low salinity.

On shallow soft substrates the extensive reed beds build up considerable above ground biomasses, usually around 1 to 2 kg dry wt m^{-2} bottom area (Wallentinus et al., 1973; Krisch et al., 1979; Granéli, 1980). Hardly any Baltic studies include the below-ground biomass, but in lakes the biomass of the widespread rhizomes are even higher than those above-ground (Schierup, 1978). During winter, nutrients are translocated into the rhizomes and stored, and then the next year used to support the growth of the emergent stems. This transport reduces the amount of nitrogen and phosphorus in the dead overwintering culms to about 1/4 and 1/10, respectively (Granéli, 1980).

The patchy distribution of most of the submerged phanerogams results in rather low overall biomass. An extensive review of these values was given by Luther (1983). Since most of these plants, except *Ruppia cirrhosa, Zannichellia major* and *Zostera marina*, are either annuals or hibernate without winter-green leaves, the biomasses fluctuate widely over the seasons. In the outer transitional areas of the Sound and the Belt Sea, *Zostera marina* meadows can reach a total maximum biomass of about 440 to 620 g dry wt m^{-2} bottom area (Sand-Jensen, 1975; Wium-Andersen and Borum, 1984), with the below-ground fraction being more stable over the seasons than that of the above-ground. The maximum biomasses in the northern Baltic proper and Gulf of Finland are normally less than 10% of these values (L. Kautsky, 1988).

Animals

The blue mussel *Mytilus edulis* constitutes more than 90% of the biomass on rocky substrates (Fig. 6.9) in most areas of the northern Baltic proper (Jansson and Kautsky, 1977; H. Kautsky, 1988). In addition to its total dominance in the deeper sublittoral zone (maximum biomass value including shells 1.6 kg dry wt m^{-2} rock), it is usually also the biomass dominant in the *Fucus* belt in these areas (Haage, 1976). Also, in terms of ash-free dry weight, *Mytilus* dominates the biomass. For the whole Baltic inside the Sound and the Belt Sea, Kautsky and Wallentinus (1980) estimated 7.6 x 10^6 tonnes of *Mytilus* dry weight above 25 m (including shells).

In the "eulittoral" zone the small crustaceans together form the group with the highest biomass (Jansson, 1974; Hällfors et al., 1975; Jansson and Kautsky, 1977; H. Kautsky, 1988), reaching average biomasses of about 1 to 10 g dry wt m^{-2}, but with large seasonal fluctuations because of migration downwards during winter. In the upper sublittoral *Fucus* belt these animals are also important quantitatively, reaching about the same biomass as in higher regions (Haage, 1976; Jansson et al., 1982). Their abundance, and hence their functional role, in Finnish coastal areas has increased in the 1980s (see p. 101).

Equally important in both communities mentioned are the gastropods, which in the *Fucus* belt even build up total dry weights about five to ten times as high as those of the small crustaceans, but have about the same ash-free dry weights as the crustaceans. In the Bothnian Bay they form the most significant group, with average biomasses of *c.* 1 to 5 g dry wt m^{-2} (Kangas, 1976; Kautsky et al., 1981).

The biomass of insect larvae in the "eulittoral" zone may in summer reach about 0.5 g dry wt m^{-2} (Jansson, 1974; Hällfors et al., 1975), while their share of the *Fucus* belt biomass is usually negligible.

In sediment communities in the Baltic, bivalves other than *Mytilus* normally dominate the biomass, although the latter is still significant on sandy substrates. *Macoma balthica* usually dominates on all kind of soft substrates with biomasses of about 2 to 15 g ash-free dry wt m^{-2} (Ankar and Elmgren, 1976; Lappalainen et al., 1977; Lappalainen, 1979, 1980a, b), with Hydrobiidae equally well spread, having biomasses up to 5 g. Also *Cardium* spp. and *Mya arenaria* are found in about the same amounts on more sandy substrates, whereas slightly lower values are found for *Nereis diversicolor*, mainly in sediments with a high organic content. A review of biomass and production of the macrofauna (mostly the mobile epibenthic species) and fish in several communities on shallow soft substrates along the Swedish coast is given in the publication edited by Rosenberg (1984).

FUNCTIONAL ASPECTS OF THE LITTORAL SYSTEM

Plants

Annual production values for plant communities are scarce. Guterstam (1979) measured the oxygen production of *Fucus vesiculosus* plants in the northern Baltic proper and calculated an annual production of 350 g C m^{-2} corresponding to an annual P/B ratio of 3.3 and an energy yield of 1.3% of that available *in situ*. For annual macroalgae only short-term diurnal measurements have been performed. They resulted in turnover times for *Cladophora* of 12 days and *Pilayella* 52 days (Wallentinus, 1978), which can be compared to 110 days for *Fucus* in Guterstam's study. For perennials such as *Furcellaria* and *Phyllophora* Wallentinus (1978) calculated 77 and 107 days, respectively, but considered that on an annual basis the turnover times must be longer, since the measurements were performed during early summer when light climate is much more favourable than during most of their growth period. The different macroalgae and submerged phanerogams in the Baltic were grouped according to their approximate production rates by Wallentinus (1979, based also on literature data). The opportunistic species (only algae) mainly had rates higher than 5 mg C (g dry wt)$^{-1}$ h^{-1}, with a large group, including some phanerogams, having rates between 1 and 5 mg C (g dry wt)$^{-1}$ h^{-1}, while the lowest rates, below 2, were found among the rest of the phanerogams and slow-growing perennials. Rates up to 5 to 7 mg C (g dry wt)$^{-1}$ h^{-1} were measured by L. Kautsky (1988) for phanerogams with ruderal and competitive strategies, while stunted and biomass-storing plants generally had rates below 2.

For submerged phanerogams very few annual production values have been published, that are not based on harvested biomasses alone. For *Zostera marina* from the Sound, Wium-Andersen and Borum (1984) used the leaf marking technique and calculated a total annual net production (including the below-ground fraction) of 815 g C m^{-2}, about five times the net biomass increase. This corresponds to an annual P/B ratio of 5.5 or a turnover time of 66 days. The epiphytes in their system contributed only *c*. 70 g C m^{-2} yr^{-1} (Borum et al., 1984). Using the same technique Sand-Jensen (1975) estimated a total annual net production for *Zostera* in the Belt Sea of 415 g C m^{-2}, or four times the biomass net increase, and with a measured life-time of leaves of 56 days.

The functional-form model approach to the ecological performance of the macroalgae (Littler and Littler, 1980) can considerably facilitate attempts to estimate their role for the energy flows and nutrient cycling of the Baltic ecosystem in the future. Productivity and nutrient uptake rates have been measured for the dominant Baltic macroalgae (King and Schramm, 1976a,b; Wallentinus, 1978, 1984a; Schramm et al., 1988) and the total annual nutrient uptake has been calculated for one Baltic archipelago area on the Swedish east coast (Wallentinus, 1984b).

As anticipated, both productivity and nutrient uptake rates were highest for algae with high ratios of surface to volume. As shown in similar studies in other areas, algae of the filamentous and sheet-like/tubular form groups have the highest rates, and also here the ranges within each group are quite wide (including intraspecific variations) and agree well with those from the same form groups in other areas. The rates of the coarsely branched and thick leathery algae were considerably lower. Although in an overall Baltic littoral system the filamentous algae do not compose the highest biomasses, their much higher metabolic rates made them the most important group for the total annual nutrient uptake by macroalgae in the study area (Fig. 6.10). The much lower uptake rates of the thick leathery forms, despite their dominating biomasses, reduce their influence on the total uptake, except during winter. Then the biomasses of the filamentous algae are low, and the leathery forms can utilize the high winter nutrient concentrations in the water.

Perennial seaweeds, found mainly among the leathery forms, thus highly contribute to the annual persistence and stability of the littoral ecosystem through their ability to bind large amounts of nitrogen and phosphorus in their biomass. This nutrient storage strategy also implies that they have a nutrient supply that can be used in the period after the phytoplankton spring bloom, when nutrient concentrations in the water mass are low. The same strategy is known from perennials in other areas (for instance, Chapman and Craigie, 1977; Gagné and Mann, 1981; Gerard, 1982). With this strategy the bladder-wrack can also endure the heavy epiphytic growth of opportunistic macroalgae in spring. When present, these epiphytes, with prior claim to the nutrients in the surrounding media, can efficiently strip the water of nutrients, both at low and higher concentrations (Fig. 6.11). Compared to the total annual nutrient uptake by phytoplankton in an archipelago area, however, the macroalgae can only secure about 15% (Wallentinus, 1984b).

This budget calculation also showed that, of the total nitrogen uptake by macroalgae over the year, two-thirds were in the form of nitrate and one-third ammonium. Analogous to the concept of "new" and "old" production by phytoplankton (Dugdale and Goering, 1967), on an annual basis most of the macroalgal production could be considered as new production based on nitrate. Still, during the period June to October, around 50% or more of the nitrogen utilization was in the form of ammonium (Wallentinus, 1984b).

There are hardly any published quantitative studies on nutrient uptake by submerged phanerogams in the Baltic. However, the high concentra-

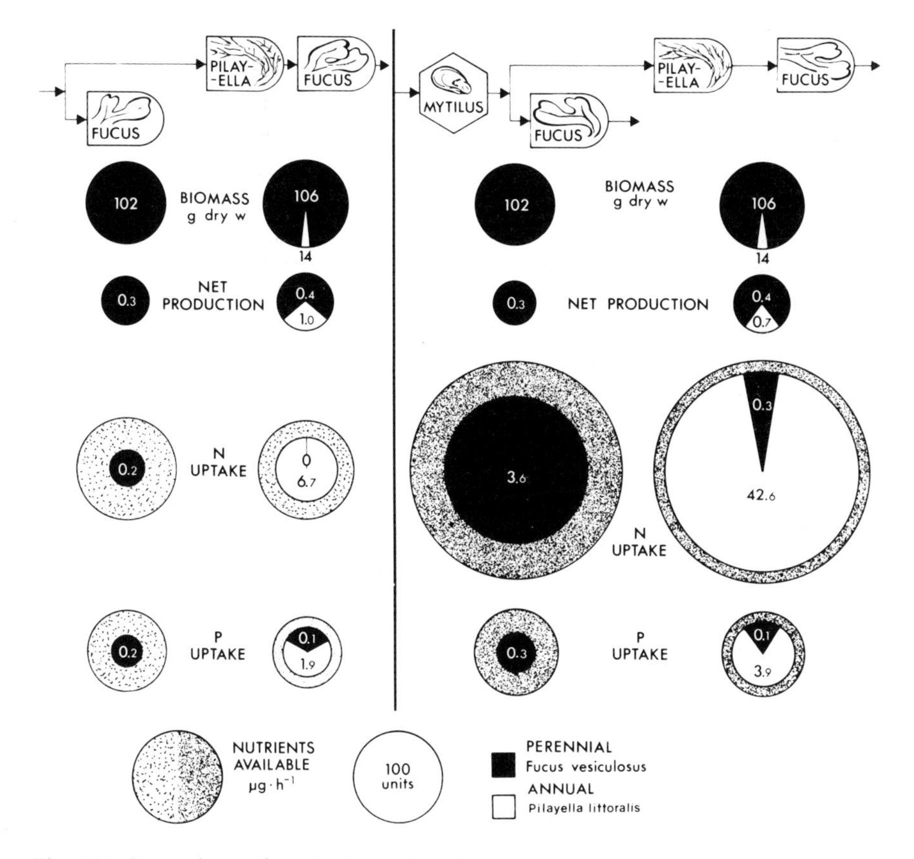

Fig. 6.11. Comparisons of production and nutrient uptake between the epiphytic, filamentous *Pilayella littoralis* and the perennial *Fucus vesiculosus* in a flow-through apparatus (Kautsky, 1984) with non-enriched seawater (right) and seawater enriched by nutrients excreted from the blue mussel, *Mytilus edulis*, (left). The organisms were kept in separate cylinders, coupled (arrows) in series to simulate *Fucus* covered by epiphytes, versus the parallel system of *Fucus* free from epiphytes. The areas of the circles indicate the rates per total biomass in the two parallel systems; the figures indicate the rates per gram dry weight of alga. Note the differences in nutrient uptake between *Fucus* with and without "epiphytes" and the increased efficiency in nutrient uptake (larger circles) of the combination *Pilayella* and *Fucus* versus *Fucus* alone. Measurements *in situ* at Askö, northern Baltic proper (Jansson, B-O., Kautsky, N. and Wallentinus, I., unpubl.)

tions of nutrients in the sediment pore-water (Engvall, 1978; Holm, 1978) make it plausible to hypothesize that their main uptake is through the roots, as described for limnic phanerogams (Carignan, 1982): over 80% of the total phosphorus uptake of the plants was carried out by the roots, if 20 times more phosphate was available in the sediments than in the water, and almost 100% if the sediment concentration was 200 times greater.

Little is known of the role of the benthic microalgae for the nutrient turnover in the Baltic proper. However, for the Sound and the Kattegat, measurements both in shallow water (Granéli and Sundbäck, 1985) and on sediments in the sublittoral (Sundbäck, 1986; Granéli and Sundbäck, 1986; Sundbäck and Granéli, 1988) have shown that through algal uptake, as long as sufficient light is available, this community greatly reduces the leakage of nutrients from the sediments. Their annual productivity in shallow sheltered areas was as high as *c.* 110 g C m^{-2} (Sundbäck, 1983), and has been estimated on the sublittoral bottoms in the Kattegat to be around 10 to 15 g C m^{-2} yr^{-1} (Sundbäck, 1986).

Nitrogen fixation has been quite well studied in the pelagic system of the Baltic, where blue-green algal blooms occur regularly in summer, and comprises about 11% of the total annual nitrogen load (Larsson et al., 1985). Blue-green algae in the littoral, particularly in the "eulittoral" zone, have been shown to have high nitrogen-fixation rates (Hübel and Hübel, 1974). Thus, they can probably also support some of the nitrogen demands of the annual algal belts through leakage. However, since they only cover a small area of the rocks and normally have low biomass, their total influence on the Baltic nitrogen turnover is certainly low. There are no figures available for the nitrogen fixation carried out by micro-organisms among the roots and rhizomes of the phanerogams. In other areas this has proved to be substantial (Capone, 1983).

Several attempts have been made to estimate the total Baltic primary production, but because of lack of good quantitative data, especially for macrovegetation, all these estimates have been rather uncertain. Elmgren (1984) calculated a total primary productivity of 52×10^6 tonnes C per year for the area inside the Belt Sea and the Sound, based on values for the different basins (weighted average 139 g C m^{-2} yr^{-1}). Of these, phytoplankton production made up 50×10^6 tonnes C (134 g C m^{-2}) and the benthic primary production (including benthic microalgae) was estimated at 2.0×10^6 tonnes C (5 g C m^{-2}). The accuracy in this estimate is probably highest for the pelagic system, where the values to a large extent are based on real measurements in several areas spread over the year and also corrected for production of exudates. Even though there is a certain patchiness in the pelagic system, it is much less pronounced than among the mosaic communities on the bottoms. For the benthic system there are very few production studies carried out over the years for any of the communities, and there are also very few accurate biomass data available to serve as a basis for transforming the rates per area. Thus, the figures given for the benthic annual primary production have to be considered as a rough approximation, which is probably underestimated, since the P/B values used (Jansson and Kautsky, 1977) are very conservative and do not include exudates or thallus parts lost through erosion or fragmentation. This figure also assumes a rather inconspicuous role for the benthic microalgae.

Even though benthic primary productivity is most probably somewhat higher than estimated by Elmgren (1984), it will certainly not be as high as that of the phytoplankton. The area above 20 m water depth, available for primary benthic productivity, is only about 25% (Stigebrandt and Wulff, 1987) of the total area of the Baltic (that is, the area supporting phytoplankton production) and the production rates per biomass unit are also higher for the phytoplankton than for the average benthic primary producers. Anyhow, the importance of the macrophytes for the total Baltic ecosystem cannot be measured by the amounts of carbon fixed alone. It must also include the structural importance of the plant's architecture for the macrofauna, the surface enlargement and many niches provided for meio- and micro-organisms, the increased areas for absorbing nutrients from the land run-off, and the plant's ability to maximize the ecological performance of the littoral system by combining opportunistic ruderals and biologically competitive dominants with their respective strategies of rapid growth/uptake, as well as storages of energy and nutrients over the year.

Animals

Several studies have been published on respiration rates of the Baltic macrofaunal species. For species occurring in the *Fucus* belt (including many from soft substrates) some of these studies are summarized in the paper by Jansson et al. (1982, table 21), ranging from 8 to 51 ml O_2 g^{-1} dry wt (excluding shells) day^{-1} and corrected for their activity. Large intraspecific variations could be found, depending both on size and sex. Respiration rates in the field are also much affected by the seasonal variation in water temperature (see, for instance, Kautsky and Wallentinus, 1980).

Estimates of average P/B ratios for infauna and epibenthic fauna along the Swedish coast are summarized in the publication edited by Rosenberg (1984; see also Möller et al., 1985). They showed slightly lower values for species in the Baltic proper (1–3) than in the Sound (2–3) with the highest values being found along the Swedish west coast (4–11). However, there are large intraspecific variations and higher values have been measured also for Baltic species (Ankar, 1980; Lappalainen, 1980b). The generally low values for the Baltic are not surprising since many species in that area have a reduced size and spend much energy compensating for low salinities. Such energy could otherwise be turned into growth (see also Tedengren and Kautsky, 1986). Also the dominant macrofaunal species, *Mytilus edulis*, has a low annual P/B ratio of 0.6 (Kautsky, 1981b).

The seasonal variation in growth is often considerable (Jansson, 1974; Ankar, 1980; Lappalainen, 1980a, b; Kautsky, 1982), and suspension-feeders such as *Mytilus* reach their maximum growth and spawning when the spring phytoplankton bloom sinks out of the pelagic system. The amount of primary production also determines their individual growth rates (Kautsky, 1982).

The overall productivity of the trophic levels among the fauna in the Baltic and its major subareas was estimated by Elmgren (1984). As for his estimates of the production of the macroflora, the background values of animal production in the littoral zone are uncertain. The absolute dominance of the blue mussel, however, makes the species the single chief contributor to both production and nutrient regeneration, and its biomass is relatively easy to predict, since on rocky substrates the mussels normally reach carrying capacity. Thus his estimate for the benthic fauna is better than that for the benthic flora.

For the entire Baltic inside the Belt Sea and the Sound, Elmgren (1984) estimated the net production and respiration (from P/B and R/B ratios) for bivalves above 25 m water depth to be 0.64 and 3.20×10^6 tonnes of carbon per year, respectively (corresponding to 1.7 and 8.6 g C m^{-2} yr^{-1}). Below 25 m the figures were about one-quarter of these. For the remaining macrofauna only values covering the entire depth range are given, amounting to 0.72 and 2.17×10^6 tonnes of carbon per year, respectively (1.9 and 5.8 g C m^{-2} yr^{-1}). Similar estimates for the meiofauna are 0.55 and 1.47×10^6 tonnes of carbon (1.5 and 3.9 g C m^{-2} yr^{-1}). For all demersal fish in the same area, a total production of 0.36×10^6 tonnes of carbon per year (1 g C m^{-2} yr^{-1}) was estimated, of which the registered catch each year was 20%, with about 1% consumed by seals.

Plant–animal interactions

Generally the grazing pressure on macroalgae is low in the Baltic, thanks to the absence of macrograzers such as sea urchins, marine gastropods and chitons. Herbivorous fish occur, but these mainly limnic species prefer phanerogams, which are also consumed by some birds; none of them have a great impact on the vegetation. On the other hand, many small invertebrates, such as crustaceans and freshwater gastropods, are quite large consumers of benthic microflora, as well as epilithic (Skoog, 1978), epiphytic (Jansson, 1967) and epipsammic components (Sundbäck and Persson, 1981). In the 1980s however, there have been several reports of enhanced grazing pressure by the isopod *Idothea balthica* on *Fucus vesiculosus* (for example, Salemaa, 1979, 1987; Kangas et al., 1982; Haahtela, 1984; Hällfors et al., 1984; Kangas and Niemi, 1985), which was due to increased abundance of *Idothea* on *Fucus*, especially during winter and where *Fucus* plants were few. Changes in the age structure of the *Fucus* populations can be an additional reason, since the older plants are preferred (Salemaa and Autio-Ikkala, unpubl. data). The content of phenolic compounds in other brown algae has been shown to affect grazing pressure (Steinberg, 1984). Accumulation of phenolic compounds in *F. vesiculosis* correlated inversely with nitrogen (Ilvessalo and Tuomi, 1989). Thus, changes of chemical constituents in *Fucus* may occur depending upon nutrient conditions. Also for many terrestrial plants there is a negative correlation between nutrients and phenols (see, for instance, Jonasson et al., 1986). In the Kattegat, where several of the macrograzers are present, the grazing pressure on macroalgae is probably even higher. There are still no reports of extensive grazing by sea urchins, as on the Norwegian Atlantic coast (Hagen, 1983).

There are many examples of feed-back mechanisms in the Baltic ecosystem. The benefit to the fauna of the structure of the plants has already been mentioned. Positive impacts on the plants from the fauna are mainly generated through the animals' excretion of nutrients, which are readily taken up by the plants (Fig. 6.11). The excretion of nutrients (mainly ammonium and phosphate) from *Mytilus* has been measured over the year in a Baltic archipelago area (Kautsky and Wallentinus, 1980). Calculations for the Baltic proper and adjacent inner areas indicate an annual regeneration by *Mytilus* of *c.* 250 000 tonnes of inorganic nitrogen and *c.* 77 000 tonnes of inorganic phosphorus. On an annual basis this implies that mussels in an archipelago area can supply all the macroalgae with nitrogen and phosphorus, and still support about 2 and 12% respectively of the phytoplankton nutrient demand (Wallentinus, 1984b). If biodeposited material (Kautsky and Evans, 1987) is included it will support 6 and 16%, respectively. But since such a large part of the annual macroalgal uptake is in the form of nitrate (see above), the phytoplankton probably utilizes much more of the nitrogen excretion by *Mytilus* than these figures indicate. In addition, nutrients are also regenerated by other organisms and by processes in the sediment (cf. Fig. 6.12).

During early summer, fish roe deposited in the littoral system can also provide nutrients to their "host" plants during a period when nutrients are scarce. Herring roe occurs in large quantities, and in a spawning area of 4 km^2 in a 160 km^2 Baltic archipelago area these eggs were calculated to contain about 25 tonnes of nitrogen (Aneer, 1983). During June, the death of many eggs (among filamentous algae a mortality as high as 75% has been measured; Aneer, 1985) means that these nitrogen compounds are broken down and thus become available to the algae. Nutrients are further excreted from the roe during its development before hatching (cf. Paulson, 1980).

One of the quantitatively most significant mechanisms of coupling between the pelagic system and the littoral zone is the filter-feeding activity of the large mussel beds, which can assimilate about 1/3 of the phytoplankton production per year in an archipelago (Kautsky, 1981a). Their dispersal of larvae into the pelagic system, estimated to equal about half the annual zooplankton production (Kautsky, 1981a), is another substantial link. Complementary to their great role as nutrient regenerators for the plants, the mussels also provide benefit to the benthic primary producers through filtering suspended particles in the water mass, and thereby improving the light climate.

Competition for space between plants and sessile animals may not be as severe a factor in the Baltic as in many true marine areas. But there are reports (for example, Waern, 1952; Mäkinen et al., 1988) that the abundance of some red algae is lower in archipelago areas with large mussel populations than further offshore. In the Kattegat, where more sessile marine animals occur, this competition for space is probably very high.

During the last two decades many attempts have been made to analyze the energy and/or nutrient flows in various Baltic littoral communities for

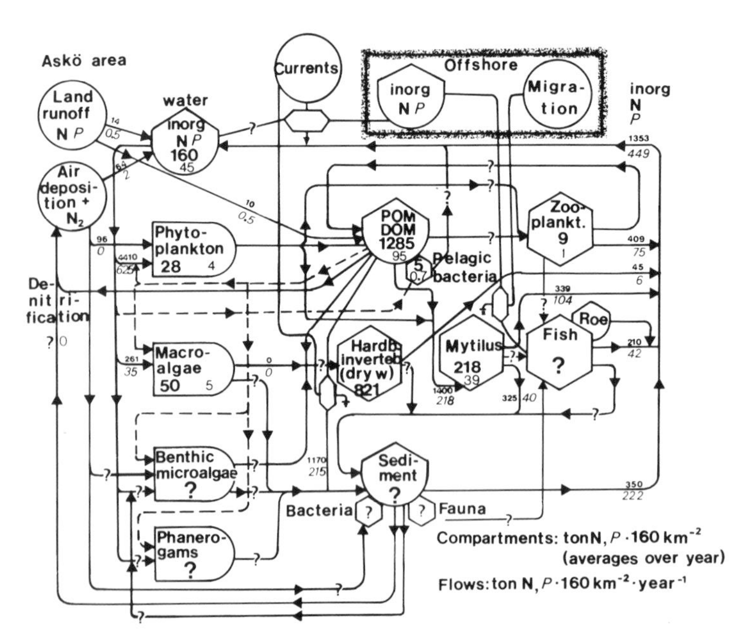

Fig. 6.12. An approximate budget model of annual fluxes and storages of nitrogen (bold figures) and phosphorus (italics) in an archipelago area of 160 km^2 at Askö, northern Baltic proper. Symbols used are according to the energy circuit language by Odum (1972). Values given are compiled from several published and unpublished studies in the area and are often approximate (see text), with storages as estimated averages over the year, and fluxes as annual rates. Question marks indicate major gaps of knowledge of rates or storages, and broken lines flows, the importance of which are not known. Major inputs from land, air and offshore areas are indicated, the uncertainty of the latter also emphasizing the non-steady state of the model (that is, the nutrient uptake by the primary producers exceeding the nutrient recirculation and known imports).

short periods or on an annual basis (for example, Jansson, 1972; Elmgren and Ganning, 1974; Jansson, 1974; Ankar and Elmgren, 1976; Gründel, 1976; Ankar, 1977; Jansson and Wulff, 1977; Arntz, 1978; Elmgren, 1978; Schramm, 1978; Guterstam, 1979; Asmus et al., 1980; Kautsky and Wallentinus, 1980; Kautsky, 1981a, b; Kautsky et al., 1981; Jansson et al., 1982; Arndt et al., 1984; Elmgren, 1984; Wallentinus 1984b, Zuchetto and Jansson, 1985; H. Kautsky, 1988; Wulff and Ulanowicz, 1989). It is not possible to summarize the results of these studies here; most of them, in any case, deal with only one community and are based on studies during one season only.

To try to summarize the community interactions, I will here instead present a very approximate annual budget of the most significant fluxes and storages of nitrogen and phosphorus (Fig. 6.12), as best we know them from an archipelago area of 160 km^2 in the northern Baltic proper. Many values are based only upon preliminary, unpublished results by several scientists at the Askö Laboratory, Sweden, or even on approximations from a few pilot studies. Thus they should be considered only as a rough estimate and not as a final quantitative analysis of the marine ecosystem in this area. The figures represent averages over the year in which large seasonal fluctuations are concealed.

Besides pointing out all still unknown data for flows and storages, the model emphasizes the importance of the offshore region for the coastal area. Thus the annual uptake of inorganic nitrogen and phosphorus is about 4 and 2 times greater respectively than that which reaches the nutrient pool through recirculation, from land-based inputs and air deposition direct to the area. The rest must be imported from the offshore areas. There are also indications (Shaffer and Rönner, 1984; Stigebrandt and Wulff, 1987) that denitrification is important in the Baltic system, which might contribute to low release rates from the sediment. The differences between the proportions of nitrogen and phosphorus recirculated, in comparison with that which is taken up, also show that nitrogen is more limiting for the primary producers than phosphorus. This has also been shown in nutrient enrichment experiments in the area (Larsson, 1984) and in comparisons of nitrogen and phosphorus surplus for macroalgae (Wallentinus, 1981a). The much higher turnover of nutrients by the planktonic primary producers than by the macroalgae is also evident, while more nutrients are bound by the macroalgae. The same relations can also be found for storages and nutrient excretions when comparing zooplankton and the macrofauna (including *Mytilus*) on rocky substrate.

The budget discussed above deals with one archipelago area in the Baltic. In his analysis of the trophic dynamics in the Baltic, Elmgren (1984) compared the overall pattern of energy flows for the different Baltic subareas with that of the marine ecosystem in the North Sea. His results showed that, despite the simplified food web in the Baltic ecosystem, the overall picture of its energy flow was very similar to that of the North Sea. This was true for most subareas, except for the northernmost Bothnian Bay, where the absence of bivalves (the main benthic suspension feeders) led to a qualitatively different benthic energy flow pattern, dominated instead by meiofauna. He also pointed out that the energy flow pattern in the Baltic is strongly influenced by man.

The effects on the Baltic biota of eutrophication and accumulation of hazardous substances are described in many of the references quoted in this chapter. In many coastal areas the structure and function of the Baltic subsystems have already changed considerably. There is also an obvious risk that several communities will irreversibly change into other subsystems more tolerant of the impact of human activities. Severe effects of harmful substances have also been recorded among several top carnivore species feeding on prey from the Baltic. If this human influence on the Baltic continues its adverse effect on the conditions, our unique chance in the future to follow the natural full-scale experiment of evolution, among species of both marine and limnic origin, might be lost. Still, however, the Baltic ecosystem is mainly intact, offering exclusive opportunities for studies on the ecology and taxonomy of both freshwater and marine species in a stable brackish water environment.

ACKNOWLEDGEMENTS

Much of the information presented I originally learned from my former colleagues at the Askö

Laboratory, University of Stockholm and from many other scientists all around the Baltic Sea. I am especially obliged to those who first taught me the Baltic seaweeds, Prof. M. Waern, Dr. S. Pekkari and the late Prof. H. Luther. I am also most grateful to Prof. B.-O. Jansson and Dr. F. Wulff, who many years ago at Askö introduced me into systems ecology. I also appreciate all the years I spent in the archipelago, benefitting from the many interdisciplinary studies in marine science. My special thanks go to all those who provided data from unpublished studies, to the previous phytobenthos group at Askö, and to Dr. R. Elmgren, who gave valuable comments to the manuscript.

Many of the results presented are from projects sponsored by the Swedish Natural Science Research Council and by the National Swedish Environmental Protection Board; the former unit also sponsored my participation in the 2nd International Phycological Congress in 1985 where some of these data were presented.

REFERENCES

Anders, K. and Möller, H., 1983. Seasonal fluctuations in macrobenthic fauna of the *Fucus* belt in Kiel Fjord (western Baltic Sea). *Helgol. Meeresunters.*, 36: 277–283.

Aneer, G., 1983. *Studie för klarläggande av sill/strömmingens val av lekplatser. Forskningsredogörelse för 1982. Bilaga till ansökan till Statens Naturvårdsverk.*, (Kontr. 5312011–9). Mimeo, Askö Laboratory, Univ. Stockholm, 9 pp [in Swedish].

Aneer, G., 1985. Some speculations about the Baltic herring (*Clupea harengus membras*) in connection with the eutrophication of the Baltic Sea. *Can. J. Fish. Aquat. Sci.*, 42 (Suppl. 1): 83–90.

Ankar, S., 1977. The soft bottom ecosystem of the northern Baltic proper with special reference to the macrofauna. *Contr. Askö Lab. Univ. Stockholm*, 19: 1–62.

Ankar, S., 1980. Growth and production of *Macoma balthica* (L.) in a northern Baltic soft bottom. *Ophelia* (Suppl.), 1: 31–48.

Ankar, S. and Elmgren, R., 1976. Benthic macro- and meiofauna of the Askö-Landsort area (northern Baltic proper) — A stratified random sampling survey. *Contr. Askö Lab. Univ. Stockholm*, 11: 1–115.

Arndt, H., Debus, L., Heerkloss, R. and Schneese, W., 1984. Diurnal changes in the matter flux of a shallow-water ecosystem in a Baltic inlet. *Ophelia* (Suppl.), 3: 1–9.

Arntz, W.E., 1978. The "upper part" of the benthic food web: the role of macrobenthos in the Western Baltic. *Rapp. P.-V. Réun. Cons. Int. Explor. Mer.*, 173: 152–169.

Asmus, H., Theede, H., Neuhoff, H.-G. and Schramm, W., 1980. The role of epibenthic macrofauna in the oxygen budget of *Zostera* communities from the Baltic Sea. *Ophelia* (Suppl.), 1: 99–111.

Bird, C.J., Greenwell, M. and McLachlan, J., 1983. Benthic marine algal flora of the north shore of Prince Edward Island (Gulf of St. Lawrence), Canada. *Aquat. Bot.*, 16: 315–335.

Blomqvist, E.M., 1984. Changes in fish community structure and migration activity in a brackish bay isolated by land upheaveal and reverted by dredging. *Ophelia*, (Suppl.), 3: 11–21.

Borum, J., Kaas, H. and Wium-Andersen, S., 1984. Biomass variation and autotrophic production of an epiphyte–macrophyte community in a coastal Danish area. II. Epiphyte species composition, biomass and production. *Ophelia*, 23: 165–179.

Breuer, G. and Schramm, W., 1988. Changes in macroalgal vegetation of Kiel Bight (Western Baltic Sea) during the past 20 years. *Kiel. Meeresforsch. Sonderh.*, 6: 241–255.

Capone, D.G., 1983. Benthic nitrogen fixation. In: E.J. Carpenter and D.G. Capone (Editors), *Nitrogen in the Marine Environment*. Academic Press, New York, pp. 105–137.

Carignan, R., 1982. An empirical model to estimate the relative importance of roots in phosphorus uptake by aquatic macrophytes. *Can. J. Fish. Aquat. Sci.*, 39: 243–247.

Chapman, A.R.O. and Craigie, J.S., 1977. Seasonal growth in *Laminaria longicruris*: Relations with dissolved inorganic nutrients and internal reserves of nitrogen. *Mar. Biol.*, 40: 197–205.

Dugdale, R.C. and Goering, J.J., 1967. Uptake of new and regenerated forms of nitrogen in primary productivity. *Limnol. Oceanogr.*, 12: 196–206.

Elmgren, R., 1978. Structure and dynamics of Baltic benthos communities with particular reference to the relationship between macro- and meiofauna. *Kiel. Meeresforsch. Sonderh.*, 4: 1–22.

Elmgren, R., 1984. Trophic dynamics in the enclosed, brackish Baltic Sea. *Rapp. P.-V. Réun. Cons. Int. Explor. Mer.*, 183: 152–169.

Elmgren, R. and Ganning, B., 1974. Ecological studies of two shallow, brackish water ecosystems. *Contr. Askö Lab. Univ. Stockholm*, 6: 1–56.

Engvall, A.-G., 1978. The fate of nitrogen in early diagenesis of Baltic sediments — A study of the sediment–water interface. *Contr. Microbial. Geochem. Dep. Geol. Univ. Stockholm*, 2: 1–103.

Feldner, R., 1976. *Untersuchungen über die eutrophierende Wirkung einiger Nährstoffkomponenten häuslicher Abwässer auf Benthosalgen der Kieler Bucht (westliche Ostsee)*. Ph.D. Thesis, Univ. Kiel, 134 pp [in German].

Fonselius, S., 1978. On the nutrients and their role as production limiting factors in the Baltic. *Acta Hydrochim. Hydrobiol.*, 6: 329–339.

Gagné, J.A. and Mann, K.H., 1981. Comparison of growth strategy in *Laminaria* populations living under differing seasonal patterns of nutrient availability. In: T. Levring (Editor), *Proc. Xth Int. Seaweed Symp., Göteborg, Sweden, August 11–15, 1980*. de Gruyter, Berlin, pp. 297–302.

Gerard, V.A., 1982. Growth and utilization of internal nitrogen reserves by the giant kelp *Macrocystis pyrifera* in low-nitrogen environment. *Mar. Biol.*, 66: 27–35.

Granéli, E., 1981. Biomass experiments in the Falsterbo Channel — nutrients added daily. *Kiel. Meeresforsch. Sonderh.*, 5: 82–90.

Granéli, E. and Sundbäck, K., 1985. The response of planktonic and microbenthic algal assemblages to nutrient enrichment in shallow coastal waters, southwest Sweden. *J. Exp. Mar. Biol. Ecol.*, 85: 253–268.

Granéli, E., Rydberg, L. and Granéli, W., 1986. Nutrient limitation at the ecosystem and the phytoplankton community level in the Laholm Bay, southeast Kattegat. *Ophelia*, 26: 181–194.

Granéli, W., 1980. *Energivass — Rapport Etapp II.* LUNBDS/(NBLI-3028)/1–37/(1980). Mimeo Dept. Limnol., Univ. Lund, 37 pp. [in Swedish].

Granéli, W. and Sundbäck, K., 1986. Can microphytobenthic photosynthesis influence below-halocline oxygen conditions in the Kattegat? *Ophelia*, 26: 195–206.

Gründel, E., 1976. Qualitative und quantitative Untersuchungen an einem Ökosystem "*Zostera* Wiese" vor Surendorf (Kieler Bucht, Westliche Ostsee). *Rep. Sonderforsch. 95, Univ. Kiel*, 18: 1–157 [in German].

Guterstam, B., 1979. *In situ-Untersuchungen über Sauerstoffumsatz und Energie-fluss in Fucus-Gemeinschaften der Ostsee.* Ph.D. Thesis, Univ. Kiel, 154 pp. [in German, with English summary].

Haage, P., 1975. Quantitative investigations of the Baltic *Fucus* belt macrofauna. 2. Quantitative seasonal fluctuations. *Contr. Askö Lab. Univ. Stockholm*, 9: 1–88.

Haage, P., 1976. Quantitative investigations of the Baltic *Fucus* belt macrofauna. 3. Seasonal variation in biomass, reproduction and population dynamics of the dominant taxa. *Contr. Askö Lab. Univ. Stockholm*, 10: 1–84.

Haahtela, I., 1984. A hypothesis of the decline of the bladder wrack (*Fucus vesiculosus* L.) in SW Finland, 1975–1981. *Limnologica* (Berlin), 15: 345–350.

Hagen, N.T., 1983. Destructive grazing of kelp beds by sea urchins in Vestfjorden, northern Norway. *Sarsia*, 68: 177–190.

Hagerman, L., 1966. The macro- and meiofauna associated with *Fucus serratus* L. with some ecological remarks. *Ophelia*, 3: 1–42.

Hällfors, G., 1976. The plant cover of some littoral biotopes at Krunnit (NE Bothnian Bay). *Acta Univ. Oul.*, A42: 87–95.

Hällfors, G., Kangas, P. and Lappalainen, A., 1975. Littoral benthos of the northern Baltic Sea. III. Macrobenthos of the hydrolittoral belt of filamentous algae on rocky shores in Tvärminne. *Int. Rev. Gesamten Hydrobiol.*, 60: 313–333.

Hällfors, G., Niemi, Å., Ackefors, H., Lassig, J. and Leppäkoski, E., 1981. Biological oceanography. In: A. Voipio (Editor), *The Baltic Sea.* Elsevier, Amsterdam, pp. 219–274.

Hällfors, G., Kangas, P. and Niemi, Å., 1984. Recent changes in the phytal at the south coast of Finland. *Ophelia* (Suppl.), 3: 51–59.

Holm, N., 1978. Phosphorus exchange through the sediment–water interface. Mechanism studies of dynamic processes in the Baltic Sea. *Contr. Microbial. Geochem. Dept. Geol. Univ. Stockholm*, 3: 1–149.

Hübel, H. and Hübel, M., 1974. *In situ*-Messungen der Stickstoff-Fixierung an Mikrobenthos der Ostseeküste. *Arch. Hydrobiol.* (Suppl.), 46: 39–54 [in German].

Ignatius, H., Axberg, S., Niemistö, L. and Winterhalter, B., 1981. Quaternary geology of the Baltic Sea. In: A. Voipio (Editor), *The Baltic Sea.* Elsevier, Amsterdam, pp. 54–104.

Ilvessalo, H. and Tuomi, J., 1989. Nutrient availability and accumulation of phenolic compounds in the brown alga *Fucus vesiculosus. Mar. Biol.*, 101: 115–120.

Jánsson, A.-M., 1967. The food web of the *Cladophora* belt fauna. *Helgol. Wiss. Meeresunters.*, 15: 574–588.

Jánsson, A.-M., 1974. Community structure, modelling and simulation of the *Cladophora* ecosystem in the Baltic Sea. *Contr. Askö Lab. Univ. Stockholm*, 5: 1–130.

Jánsson, A.-M. and Kautsky, N., 1977. Quantitative survey of hard bottom communities in a Baltic archipelago. In: B.F. Keegan, P.Ó. Céidigh and P.J.S. Boaden (Editors), *Biology of benthic organisms*, 11th Europ. Symp. Marine Biology, Galway, October, 1976. Pergamon Press, Oxford, pp. 359–366.

Jánsson, A.-M., Kautsky, N., Von Oertzen, J.-A., Schramm, W., Sjöstedt, B., Von Wachenfeldt, T. and Wallentinus, I., 1982. Structural and functional relationships in a southern Baltic *Fucus* ecosystem — A joint study by the BMB Phytobenthos Group. *Contr. Askö Lab. Univ. Stockholm*, 28: 1–95.

Jánsson, B.-O., 1972. Ecosystem approach to the Baltic problem. *Bull. Ecol. Res. Comm. Stockholm*, 16: 1–82.

Jánsson, B.-O., 1978. The Baltic — a system analysis of a semi-enclosed sea. In: H. Charnock and G. Deacon (Editors), *Advances in Oceanography.* Plenum, New York, pp. 131–183.

Jánsson, B.-O., 1980. Natural systems of the Baltic Sea. *Ambio*, 9: 128–136.

Jánsson, B.-O., 1981. Dynamics and energy flow in Baltic ecosystems. In: W.J. Mitsch, R.W. Bossermann and J.M. Klopatek (Editors), *Energy and Ecological Modelling.* Elsevier, Amsterdam, pp. 507–516.

Jánsson, B.-O., 1984. Baltic Sea ecosystem analysis: Critical areas for future research. *Limnologica* (Berlin), 15: 237–252.

Jánsson, B.-O., 1989. The Baltic Sea. In: L. Rey and V. Alexander (Editors), *Proc. 6th Conf. Comité Arctique Intern., 13–15 May 1985.* E.J. Brill, Leiden, pp. 283–326.

Jánsson, B.-O. and Wulff, F., 1977. Ecosystem analysis of a shallow sound in the northern Baltic proper — A joint study by the Askö group. *Contr. Askö Lab. Univ. Stockholm*, 18: 1–160.

Jánsson, B.-O., Wallentinus, I. and Wulff, F., 1980. Östersjön — en livsmiljö i förvandling. In: E. Söderberg (Editor), *Forska för livet. NFR Årsbok 1979–80.* Redaktionstjänsten NFR, Stockholm, pp. 181–218 [in Swedish].

Jánsson, B.-O., Wilmot, W. and Wulff, F., 1984. Coupling the sub-systems. — The Baltic Sea as a case study. In: M.J. Fasham (Editor), *Flows of Energy and Materials in Marine Ecosystems.* Plenum, New York, pp. 549–595.

Jánsson, B.-O., Aneer, G. and Nellbring, S., 1985. Spatial and temporal distribution of the demersal fish fauna in a Baltic archipelago as estimated by SCUBA census. *Mar. Ecol. Prog. Ser.*, 23: 31–43.

Jonasson, S., Bryant, J.P., Stuart Chapin III, F. and Andersson, M., 1986. Plant phenols and nutrients in relation to variations in climate and rodent grazing. *Am. Nat.*, 128: 394–408.

Kangas, P., 1976. Littoral stony-bottom invertebrates in the Krunnit area of the Bothnian Bay. *Acta Univ. Oul.*, A42: 97–106.

Kangas, P., 1978. On the quantity of meiofauna among the epiphytes of *Fucus vesiculosus* in the Askö area, northern Baltic Sea. *Contr. Askö Lab. Univ. Stockholm*, 24: 1–32.

Kangas, P. and Niemi, Å., 1985. Observations of recolonization by the bladder wrack *Fucus vesiculosus*, on the southern coast of Finland. *Aqua Fennica*, 15: 133–141.

Kangas, P., Autio, H., Hällfors, G., Luther, H., Niemi, Å. and Salemaa, H., 1982. A general model of the decline of *Fucus vesiculosus* at Tvärminne, south coast of Finland in 1977–81. *Acta Bot. Fennica*, 118: 1–27.

Kautsky, H., 1988. *Factors structuring phytobenthic communities in the Baltic Sea.* Ph.D. Thesis, Univ. Stockholm, Stockholm, 221 pp.

Kautsky, H., Widbom, B. and Wulff, F., 1981. Vegetation, macrofauna and benthic meiofauna in the phytal zone of the archipelago of Luleå, Bothnian Bay. *Ophelia*, 20: 53–77.

Kautsky, L., 1988. Life strategies of aquatic soft bottom macrophytes. *Oikos*, 53: 126–135.

Kautsky, N., 1981a. On the trophic role of the blue mussel (*Mytilus edulis* L.) in a Baltic coastal ecosystem and the fate of the organic matter produced by the mussels. *Kiel. Meeresforsch. Sonderh.*, 5: 454–461.

Kautsky, N., 1981b. *On the role of the blue mussel Mytilus edulis L. in the Baltic ecosystem.* Ph.D. Thesis, Univ. Stockholm, 22 pp.

Kautsky, N., 1982. Growth and size structure in Baltic *Mytilus edulis* populations. *Mar. Biol.*, 68: 117–133.

Kautsky, N., 1984. A battery-operated continuous-flow enclosure for metabolism studies in benthic communities. *Mar. Biol.*, 81: 47–52.

Kautsky, N. and Evans, S., 1987. Role of biodeposition by *Mytilus edulis* in the circulation of matter and nutrients in a Baltic coastal ecosystem. *Mar. Ecol. Prog. Ser.*, 38: 201–212.

Kautsky, N. and Wallentinus, I., 1980. Nutrient release from a Baltic *Mytilus*–red algal community and its role in benthic and pelagic productivity. *Ophelia* (Suppl.), 1: 17–30.

Kautsky, N., Kautsky, H., Kautsky, U. and Waern, M., 1986. Decreased depth penetration of *Fucus vesiculosus* (L.) since the 1940's indicates eutrophication of the Baltic Sea. *Mar. Ecol. Prog. Ser.*, 28: 1–8.

Kautsky, U., Wallentinus, I. and Kautsky, N., 1984. Spring bloom dynamics of an epilithic microphytobenthic community in the northern Baltic proper. *Ophelia* (Suppl.), 3: 89–99.

King, R.J. and Schramm, W., 1976a. Photosynthetic rates of benthic marine algae in relation to light intensity and seasonal variations. *Mar. Biol.*, 37: 215–222.

King, R.J. and Schramm, W., 1976b. Determination of photosynthetic rates for the marine algae *Fucus vesiculosus* and *Laminaria digitata*. *Mar. Biol.*, 37: 209–213.

Kornfeldt, R.-A., 1979. Makroalgers biomassa längs Hallandskusten. *Svensk Bot. Tidskr.*, 73: 131–138. [in Swedish, with English abstract].

Kornfeldt, R.-A., 1984. Variation in distribution and biomass of marine benthic algae off Kullen, S Sweden. *Nord. J. Bot.*, 4: 563–584.

Krisch, H., Krauss, N. and Kahl, M., 1979. Der Einfluss von Schnitt und Frost auf Entwicklung und Biomasseproduktion der *Phragmites*-Röhrichte am Greifswalder Bodden. *Folia Geobot. Phytotaxon.* (Praha), 14: 121–144 [in German].

Kullenberg, G., 1983. The Baltic Sea. In: B.H. Ketchum (Editor), *Ecosystems of the World 26. Estuaries and Enclosed Seas.* Elsevier, Amsterdam, pp. 309–335.

Lappalainen, A., 1979. Seasonal recruitment and population structure of coexisting mud snails (Hydrobiidae) in the Baltic Sea. In: E. Naylor and R.G. Hartnoll (Editors), *Cyclic Phenomena in Marine Plants and Animals.* Pergamon, Oxford, pp. 57–63.

Lappalainen, A., 1980a. *On the ecology of sandy bottoms in the Baltic Sea with special reference to mud snails (Hydrobiidae).* Ph.D. Thesis, Univ. Stockholm, 33 pp.

Lappalainen, A., 1980b. *Reproduction, growth and annual production of Hydrobia ulvae (Pennant) in the northern Baltic proper.* Mimeo, Askö Lab., Univ. Stockholm, 57 pp.

Lappalainen, A., Hällfors, G. and Kangas, P., 1977. Littoral benthos of the northern Baltic Sea. IV. Pattern and dynamics of macrobenthos in a sandy-bottom *Zostera marina* community in Tvärminne. *Int. Rev. Gesamten Hydrobiol.*, 62: 465–503.

Larsson U., 1984. Östersjön. In: R. Rosenberg (Editor), *Gödning av havsområden kring Sverige — En kunskapsöversikt.* SNV PM, 1808, pp. 17–73 [in Swedish].

Larsson, U., Elmgren, R. and Wulff, F., 1985. Eutrophication and the Baltic Sea. *Ambio*, 14: 9–14.

Levring, T., 1940. *Studien über die Algenvegetation von Blekinge, Südschweden.* Ph.D. Thesis, Univ. Lund, 178 pp.

Leppäkoski, E., 1984. Introduced species in the Baltic Sea and its coastal ecosystems. *Ophelia* (Suppl.), 3: 123–135.

Lindgren, L., 1978. *Algzonering på klippiga stränder i Porkala, Helsingfors och Sibbo som bas för fortsatt kontroll av föroreningsläget.* M. Sc. Thesis, Dept. Bot., Univ. Helsinki, 155 pp. [in Swedish, with English summary].

Lindvall, B., 1984. The condition of a *Fucus*-community in a polluted archipelago area on the east coast of Sweden. *Ophelia* (Suppl.), 3: 147–150.

Littler, M.M. and Littler, D.S., 1980. The evolution of thallus form and survival strategies in benthic marine macroalgae: Field and laboratory tests of a functional form model. *Am. Nat.*, 116: 25–44.

Littler, M.M. and Littler, D.S., 1981. Intertidal macrophyte communities from Pacific Baja California and the upper Gulf of California: Relatively constant vs environmentally fluctuating systems. *Mar. Ecol. Prog. Ser.*, 4: 145–158.

Lüning, K., 1985 *Meeresbotanik.* Thieme, Stuttgart, 375 pp.

Lüning, K. and Dring, M.J., 1979. Continuous underwater light measurement near Helgoland (North Sea) and its significance for characteristic light limits in the sublittoral region. *Helgol. Wiss. Meeresunters.*, 32: 403–424.

Luther, H., 1951. Verbreitung und Ökologie der höheren Wasserpflanzen im Brackwasser der Ekenäs-Gegend in Südfinnland. II. Specieller Teil. *Acta Bot. Fennica*, 50: 1–370.

Luther, H., 1983. On life forms, and above-ground and underground biomass of aquatic macrophytes. *Acta Bot. Fennica*, 123: 1–23.

Lüthje, H., 1978. The macrobenthos in the red algal zone of Kiel Bay (Western Baltic). *Kiel. Meeresforsch.* (Sonderh.), 4: 108–114.

Mäkinen, A., Haahtela, H., Ilvessalo, H. and Lehto, J., 1984. Changes in the littoral rocky shore vegetation in the Seili area, SW archipelago of Finland. *Ophelia* (Suppl.), 3: 157–166.

Mäkinen, A., Kääriä, J. and Rajasilta, M., 1988. Factors controlling the occurrence of *Furcellaria lumbricalis* (Huds.) Lamour. and *Phyllophora truncata* (Pallas) Zinova in the littoral of the archipelago of SW Finland. *Kiel. Meeresforsch. Sonderh.*, 6: 140–146.

Melvasalo, T., Pawlak, J., Grasshoff, K., Thorell, L. and Tsiban, A. (Editors), 1981. *Assessment of the Effects of Pollution on the Natural Resources of the Baltic Sea, 1980.* Baltic Sea Environment Proceedings, No. 5B, pp. 1–426.

Möller, P., Pihl, L. and Rosenberg, R., 1985. Benthic faunal energy flow and biological interaction in some shallow bottom habitats. *Mar. Ecol. Prog. Ser.*, 27: 109–121.

Müller, K. (Editor), 1982. *Coastal Research in the Gulf of Bothnia.* Junk, the Hague, 462 pp.

Odum, H.T., 1972. An energy circuit language for ecological and social systems: Its physical basis. In: B.C. Patten (Editor), *Systems Analysis and Simulation in Ecology, Vol. II.* Academic Press, New York, pp. 140–211.

Paulson, L.J., 1980. Models of ammonia excretion for brook trout (*Salvelinus fontinalis*) and rainbow trout (*Salmo gairdneri*). *Can. J. Fish. Aquat. Sci.*, 37: 1421–1425.

Pekkari, S., 1973. Effects of sewage water on benthic vegetation. Nutrients and their influence on the algae in the Stockholm Archipelago during 1970. No. 6. *Oikos Suppl.*, 15: 185–188.

Plínski, M. and Florczyk, I., 1984. Changes in the phytobenthos resulting from the eutrophication of the Puck Bay. *Limnologica* (Berlin), 15: 325–327.

Ravanko, O., 1968. Macroscopic green, brown, and red algae in the southwestern archipelago of Finland. *Act. Bot. Fennica*, 79: 1–50.

Remane, A., 1940. Einführung in die zoologische Ökologie der Nord- und Ostsee. In: G. Grimpe and E. Wagler (Editors), *Tierwelt der Nord- und Ostsee, I.a.* Akad. Verlagsges., Leipzig, 238 pp. [in German].

Rönnberg, O., 1981. Traffic effects on rocky-shore algae in the Archipelago Sea, SW Finland. *Acta Academ. Aboensis Ser. B.*, 41(3): 1–86.

Rönnberg, O., 1984. Recent changes in the distribution of *Fucus vesiculosus* L. around the Åland islands (N Baltic). *Ophelia* (Suppl.), 3: 189–193.

Rönnberg, O., Lehto, J. and Haahtela, I., 1985. Recent changes in the occurrence of *Fucus vesiculosus* in the Archipelago Sea, SW Finland. *Ann. Bot. Fennici*, 22: 231–244.

Rosenberg, R. (Editor), 1984. *Biologisk värdering av grunda svenska havsområden — Fisk och bottendjur.* SNV PM 1911: 1–384 [in Swedish, with English summary].

Rueness, J., 1978. Hybridization in red algae. In: D.E.G. Irvine and J.H. Price (Editors), *Modern Approaches to the Taxonomy of Red and Brown Algae.* Academic Press, London, pp. 247–262.

Russell, G., 1985a. Recent evolutionary changes in the algae of the Baltic Sea. *Br. Phycol. J.*, 20: 87–104.

Russell, G., 1985b. Some anatomical and physiological differences in *Chorda filum* from coastal waters of Finland and Great Britain. *J. Mar. Biol. Ass. U.K.*, 65: 343–349.

Russell, G., 1988. The seaweed flora of a young semi-enclosed sea: The Baltic. Salinity as a possible agent of flora divergence. *Helgol. Wiss. Meeresunters.*, 42: 243–250.

Salemaa, H., 1979. Ecology of *Idotea* spp. (Isopoda) in the northern Baltic. *Ophelia*, 18: 133–150.

Salemaa, H., 1987. Herbivory and micro-habitat preferences of *Idotea* spp. (Isopoda) in the northern Baltic Sea. *Ophelia*, 27: 1–15.

Sand-Jensen, K., 1975. Biomass, net production and growth dynamics in an eel-grass (*Zostera marina* L.) population in Vellerup Vig, Denmark. *Ophelia*, 14: 185–201.

Schierup, H.-H., 1978. Biomass and primary production in a *Phragmites communis* Trin. swamp in North Jutland, Denmark. *Verh. Int. Verein. Limnol.*, 20: 94–99.

Schramm, W., 1978. Die ökophysiologische Analyse bentischer mariner Ökosysteme: *in situ*-Methoden und interdisziplinäre Wege. *Verh. Ges. f. Ökol., Kiel 1977*, pp. 53–58.

Schramm, W., Abele, D. and Breuer, G., 1988. Nitrogen and phosphorus nutrition and productivity of two community forming seaweeds (*Fucus vesiculosus, Phycodrys rubens*) from the western Baltic (Kiel Bight) in the light of eutrophication processes. *Kiel. Meeresforsch. Sonderh.*, 6: 221–240.

Schwenke, H., 1974. Die Benthosvegetation. In: L. Magaard and G. Rheinheimer (Editors), *Meereskunde der Ostsee.* Springer, Berlin, pp. 131–146 [in German].

Shaffer, G. and Rönner, U., 1984. Denitrification of the Baltic proper deep water. *Deep-Sea Res.*, 31: 197–220.

Skoog, G., 1978. Influence of natural food items on growth and egg production in brackish water populations of *Lymnea peregra* and *Theodoxus fluviatilis* (Mollusca). *Oikos*, 31: 340–348.

Skult, P., 1977. Composition of phytal macrofauna communities on transects extending seawards from Helsinki. *Memoranda Soc. Fauna Flora Fenn.*, 53: 43–56.

Steinberg, P.D., 1984. Algal chemical defense against herbivores: Allocation of phenolic compounds in the kelp *Alaria marginata. Science*, 223: 405–407.

Stigebrandt, A. and Wulff, F., 1987. A model for the dynamics of nutrients and oxygen in the Baltic proper. *J. Mar. Res.*, 45: 729–759.

Sundbäck, K., 1983. *Microphytobenthos on sand in shallow brackish water, Öresund, Sweden.* Ph.D. Thesis, Univ. Lund, 209 pp.

Sundbäck, K., 1986. What are the benthic microalgae doing on the bottom of the Laholm Bay? *Ophelia* (Suppl.), 4: 273–286.

Sundbäck, K. and Granéli, W., 1988. The influence of microphytobenthos on the nutrient flux between sediment and water — a laboratory study. *Mar. Ecol. Prog. Ser.*, 43: 63–69.

Sundbäck, K. and Persson, L.-E., 1981. The effect of microbenthic grazing by an amphipod, *Bathyporeia pilosa*, Lindström. *Kiel. Meeresforsch.* (Sonderh.), 5: 573–575.

Tedengren, M. and Kautsky, N., 1986. Comparative study of the physiology and its probable effect on size in blue mussels (*Mytilus edulis* L.) from the North Sea and the northern Baltic proper. *Ophelia*, 25: 147–155.

Trei, T., 1985. Long-term changes in the bottom macroflora of the coastal waters of Estonia. In: T. Trei (Editor), *Problems Concerning Bioindication of the Ecological Condition of the Gulf of Finland.* Hydrobiological Research 15. Academy of Sciences of the Estonian SSR, Tallinn, pp. 117–122.

Van den Hoek, C., 1975. Phytogeographic provinces along the coasts of the northern Atlantic Ocean. *Phycologia*, 14: 317–330.

Voipio, A. (Editor), 1981. *The Baltic Sea.* Elsevier, Amsterdam, 418 pp.

Von Wachenfeldt, T., 1975. *Marine benthic algae and the environment in the Öresund. I–III.* Ph.D. Thesis, Univ. Lund, 328 pp.

Waern, M., 1952. Rocky-shore algae in the Öregrund archipelago. *Acta Phytogeogr. Suec.*, 30: 1–298.

Waern, M., 1965. A vista on the marine vegetation. *Acta Phytogeogr. Suec.*, 50: 13–27.

Wallentinus, H.-G., Gustafsson, K. and Söderström, B., 1973. Bladvassen, *Phragmites communis* Trin., i Brunnsviken, Stockholm 1971. *Svensk. Bot. Tidskr.*, 67: 81–96 [in Swedish, with English summary].

Wallentinus, I., 1976. Environmental influences on benthic macrovegetation in the Trosa–Askö area, northern Baltic proper. I. Hydrographical and chemical parameters, and the macrophytic communities. *Contr. Askö Lab. Univ. Stockholm*, 15: 1–138.

Wallentinus, I., 1978. Productivity studies on Baltic macroalgae. *Bot. Mar.*, 21: 365–380.

Wallentinus, I., 1979. Environmental influences on benthic macrovegetation in the Trosa–Askö area, northern Baltic proper. II. The ecology of macroalgae and submersed phanerogams. *Contr. Askö Lab. Univ. Stockholm*, 25: 1–210.

Wallentinus, I., 1981a. Chemical constituents of some Baltic macroalgae in relation to environmental conditions. In: T. Levring (Editor), *Proceedings Xth Int. Seaweed Symp., Göteborg, Sweden, August 11–15, 1980.* de Gruyter, Berlin, pp. 363–370.

Wallentinus, I., 1981b. Phytobenthos. In: T. Melvasalo, J. Pawlak, K. Grasshoff, L. Thorell and A. Tsiban (Editors), *Assessment on the Effects of Pollution on the Natural Resources of the Baltic Sea. Baltic Sea Environment Proc.*, 5B, pp. 322–342.

Wallentinus, I., 1983. Vegetationsklädda bottnar. In: *Eutrofiering i marin miljö — Östersjön, Delrapport V till Statens Naturvårdsverk.* Mimeo, Askö Lab. Univ. Stockholm, 57 pp.

Wallentinus, I., 1984a. Comparisons of nutrient uptake rates for Baltic macro-algae with different thallus morphologies. *Mar. Biol.*, 80: 215–225.

Wallentinus, I., 1984b. Partitioning of nutrient uptake between annual and perennial seaweeds in a Baltic archipelago area. *Hydrobiologia*, 116/117: 363–370.

Wennberg, T., 1987. *Långsiktiga förändringar av makroalgflorans sammansättning och utbredning i södra Laholmsbukten sedan 1950-talet.* Statens Naturvårdsverk Rapport 3290, 47 pp. [in Swedish, with English summary].

Wium-Andersen, S. and Borum, J., 1984. Biomass variation and autotrophic production of an epiphyte-macrophyte community in a coastal Danish area: I. Eel-grass (*Zostera marina* L.) biomass and net production. *Ophelia*, 23: 33–46.

Wulff, F. and Ulanowicz, R.E., 1989. A comparative anatomy of the Baltic Sea and Chesapeake Bay ecosystems. In: F. Wulff, J.G. Field and K.H. Mann (Editors), *Network Analysis in Marine Ecology. Methods and Applications. Coastal and Estuarine Studies, Vol. 32.* Springer, New York, pp. 232–256.

Wulff, F., Aertebjerg, G., Nicolaus, G., Niemi, Å., Ciszewski, P., Schultz, S. and Kaiser, W., 1986. The changing pelagic ecosystem of the Baltic Sea. *Ophelia* (Suppl.), 4: 299–319.

Zuchetto, J. and Jansson, A.-M., 1985. *Resources and Society — A Systems Ecology Study of the Island of Gotland, Sweden*, Ecological Studies 56. Springer, New York, 246 pp.

Chapter 7

NORTHWEST ATLANTIC ROCKY SHORE ECOLOGY

ARTHUR C. MATHIESON, CLAYTON A. PENNIMAN and LARRY G. HARRIS

INTRODUCTION

The northwest Atlantic area between Cape Cod/Nantucket Shoals and Newfoundland (Fig. 7.1) exhibits conspicuous regional differences in temperature, tidal fluctuation, ice scouring, wave exposure and nutrient enrichment. However, the dominant substratum is hard rock, ranging from granitic headlands to sandstone benches. Major fluctuations in sea levels and temperatures have occurred during recent geological times (Pielou, 1979). The wide range of present hydrographic parameters, combined with extremes of abiotic conditions imposed over time, have resulted in complex distributional patterns for many species. Overall, species diversity is low compared with the northeast Atlantic or the northeast Pacific. Patterns of community structure and zonation, and the abiotic and biotic mechanisms controlling them, are consistent with other north temperate regions.

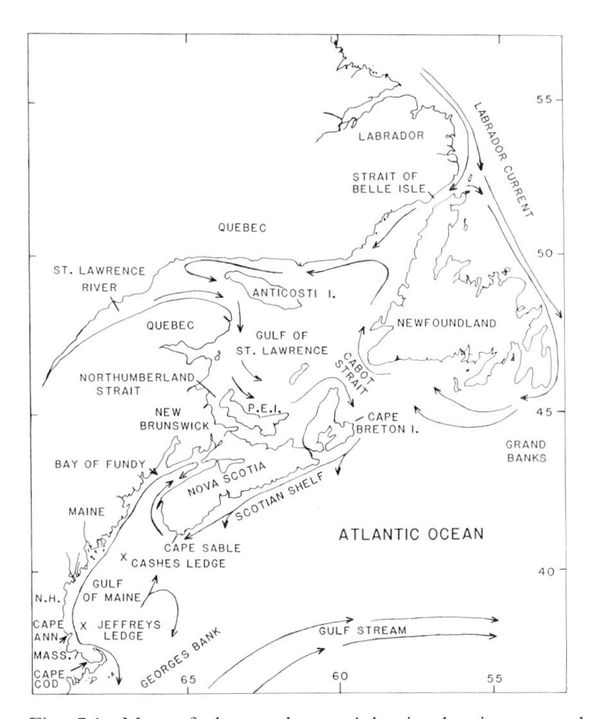

Fig. 7.1. Map of the northwest Atlantic showing general circulation patterns and five major hydrographic regions: (1) Gulf of Maine, (2) Bay of Fundy, (3) Atlantic Coast of Nova Scotia, (4) Gulf of St. Lawrence, and (5) Newfoundland.

The present chapter is limited to populations inhabiting open coastal rocky substrata, which extend from the littoral fringe to approximately −45 m where animals replace seaweeds on all surfaces. The primary focus is upon community organization and zonation, and structural mechanisms involved. Literature reviewed encompasses more than the species, populations and processes identified within this geography. Thus, we have illustrated the universality of patterns by comparing them with systems having higher species diversities and complexities.

ABIOTIC ENVIRONMENT

Geographical context

The northwest Atlantic area between Nantucket Shoals/Cape Cod and Newfoundland is an extensive and diverse coastline containing five major hydrographic regions (Appendices I–V, Fig. 7.1): (1) Gulf of Maine, (2) Bay of Fundy, (3) Atlantic Coast of Nova Scotia, (4) Gulf of St. Lawrence, and (5) Newfoundland. The southernmost region, the Gulf of Maine, is a partially enclosed sea that extends from Nantucket Shoals/Cape Cod to Cape Sable, Nova Scotia; approximately 30% of the Gulf is exposed to the open Atlantic. The Bay of Fundy is located in the northeastern corner of the Gulf of Maine, with Grand Manan, New Brunswick, being its southern boundary. The highly indented eastern Atlantic coastline of Nova Scotia generally faces southeast. The Gulf of St. Lawrence is an enclosed (inland) sea connected to the Atlantic Ocean by three narrow straits. Newfoundland, which is the world's sixteenth largest island, is comparable to Cuba and Iceland in size (South, 1983). Overall, the coastline within this northwest Atlantic area contains enormous irregularities. For example, the landward length of the Gulf of Maine is approximately 966 km, but if all bays and tidal rivers are traced, it is nearly 8050 km or 8.3 times greater (Apollonio, 1979). Specific details regarding the geography, as well as geology and hydrographic parameters within each of the five regions are summarized in Appendices I–V, along with many pertinent references.

Geology

Unlike the mid-Atlantic coast of the United States, which is primarily characterized by a lack of natural hard substrata (Orth et al., Ch. 8), regions north of Cape Cod, Massachusetts, have an abundance of rocky substrata, particularly in more northerly portions of the Gulf of Maine and within the Canadian Atlantic coastline (Appendices I–V). For example, steep granitic headlands dominate the coastline in the northern Gulf of Maine as well as on the Atlantic coast of Nova Scotia (Anonymous, 1974a–c). Barrier beaches and lower shore profiles are most conspicuous in the central and southern portions of the Gulf of Maine. Within the Bay of Fundy the coastline is steeply to moderately cliffed and primarily composed of massive granitic outcrops or friable shale-like materials (Stephenson and Stephenson, 1954a, b, 1972). These irregular rocky platforms and masses may also alternate with large and small boulders in the Bay of Fundy, as well as in scattered areas along the Atlantic coast of Nova Scotia and farther south in New England (Larsen and Doggett, 1981). The coastline in the Gulf of St. Lawrence is strikingly different from the other four regions, as its shorelines are composed of low profiles which are dominated by red sandstone (Bird et al., 1983; Pringle and Semple, 1983, 1984). Even so, a mixture of harder boulders and pebbles can occur, plus some stretches of sandy beaches alternating with rocky headlands. The coastline of Newfoundland is rocky and precipitous, with few extensively eroded platforms and sandy beaches (South, 1983; Steele, 1983). The island is highest in the west due to post-glacial rebound since the Wisconsin Period (approximately 10 000 years ago). The crustal rebound (that is, after the Pleistocene glaciation) occurred approximately 8500 years ago in the Gulf of Maine (Schnitker, 1974).

Circulation

As outlined by Apollonio (1979) and Sutcliffe et al. (1976), the Labrador Current and the Gulf Stream are the two major current systems within the northwest Atlantic (Fig. 7.1). The south-flowing Labrador Current moves along the coast of Labrador and the eastern shore of Newfoundland to the Grand Banks where it converges with the Gulf Stream flowing to the east-northeast. A small branch of the Labrador Current enters the Strait of Belle Isle north of Newfoundland and then moves along the Anticosti side of the Gaspé Passage into the Gulf of St. Lawrence. On the Scotian Shelf, a southwesterly flowing current occurs because of extensive fresh-water outflow from the Gulf of St. Lawrence. Neither the Labrador Current nor the Gulf Stream flows into the Gulf of Maine and only rarely does the Gulf Stream spill onto Georges Bank. The fresh-water drainage from Cape Elizabeth, Maine, to the Bay of Fundy, plus the strong tidal power of the latter region, provides much of the energy that generates a counter-clockwise circulation within the Gulf of Maine. Off the southwestern coastline of Nova Scotia a major dichotomy of this counterclockwise gyre occurs, with a northward flow into the Bay of Fundy and a southwestward flow towards coastal Maine. In the southern half of the Gulf of Maine an easterly flow skirts the northern edge of Georges Bank, while to the south of Cape Cod there is a sluggish southward movement of water. Of the five hydrographic regions in the northwest

Atlantic, the Bay of Fundy has the strongest tidal currents (mean maximum ebb = 3.9 knots or 201 cm s^{-1}). Tidal currents are also a characteristic feature of the Northumberland Strait near Prince Edward Island, but they are weak (2.3 knots or 118 cm s^{-1} maximum) in comparison to those in the Bay of Fundy.

Tides, waves and upwelling

Tidal amplitudes are extremely variable geographically, as well as being greater in open coastal than adjacent embayment–estuarine areas (Appendices I–V). The world's largest tidal amplitudes occur within the Minas Basin of the Bay of Fundy (Stephenson and Stephenson, 1954a, b, 1972; Anonymous,1981b). Within the Gulf of Maine, tides increase in a northerly and easterly direction, primarily because of resonance of the tidal wave in the constricted eastern Gulf and within the Bay of Fundy (Apollonio, 1979). Average tidal ranges within the Gulf of Maine vary from 2.5 to 5.6 m (mean spring tides = 2.9 to 6.4 m), while those within the other four geographic regions vary from 2.7 to 11.7 m (mean spring tides = 3.1 to 13.3 m) in the Bay of Fundy, 0.7 to 2.2 m (mean spring tides = 0.9 to 2.6 m) on the Atlantic coast of Nova Scotia, 0.4 to 2.0 m (mean spring tides = 0.6 to 2.6 m) in the Gulf of St. Lawrence and 0.6 to 1.5 m (mean spring tides = 0.8 to 1.9 m) in Newfoundland (Anonymous, 1984a). Typically, tides throughout this area are equal and semidiurnal, although the complicated topography of some areas (for example, the southern Gulf of St. Lawrence) can cause pronounced diurnal inequalities (Appendix IV).

A wide range of exposure conditions (that is, wave action) occurs (Appendices I–V), ranging from high-energy headlands to low-energy shores with low profiles. Similarly, strong gradients of wave action exist within short distances (for example, the highly indented shoreline of the Atlantic coast of Nova Scotia). Coastal upwelling is extensive within the northeastern sections of the Gulf of Maine (Anonymous, 1974a–c) and along the southwestern shore of Nova Scotia (Garrett and Loucks, 1976). Within central portions of the Gulf of Maine localized upwelling occurs either seasonally or more consistently within channel areas (Hulburt, 1968; Hulburt and Corwin, 1970). Wind-driven upwelling and pronounced turbidity occur in the Gulf of St. Lawrence due to its shallowness (Trites, 1979a). The eastern shoreline of Newfoundland (the Avalon Peninsula) is particularly exposed to the open Atlantic, and often receives upwelled cold intermediate waters (Steele, 1983; Appendix V).

Temperature, salinity and nutrients

Hydrographic patterns are extremely variable because of diverse topographic features and variability of horizontal and vertical mixing (Appendices I–V, Fig. 7.2). Surface water temperatures show major discontinuities during summer, but are generally uniform over the whole area during winter (Steele, 1975). Within the Gulf of Maine and the Bay of Fundy, the least stable areas occur off southwestern Nova Scotia and eastern Maine (Garrett and Loucks, 1976; Apollonio, 1979) where surface waters in summer usually are the coolest and saltiest, and vertical differences are minimal because of upwelling and tidal mixing. During summer, a maximum temperature difference of 10°C occurs within the Gulf of Maine, while larger differences are noted when comparing sites to the south of Cape Cod. Limited upwelling and warmer summer temperatures occur along the Atlantic coast of Nova Scotia, and the warmest summer temperatures north of Cape Cod are recorded from the Gulf of St. Lawrence (mean 17°C and maximum 23°C). On the other hand, winter temperatures within the Gulf of St. Lawrence are cooler than most coastal areas in the Gulf of Maine, the Bay of Fundy and the Atlantic coast of Nova Scotia, as it is ice-bound between December and mid-April (Stephenson and Stephenson, 1954a, b, 1972). In Newfoundland, the south-flowing Labrador Current reduces water temperatures and differentially affects patterns of vertical stratification (Appendix V). Exposed open coastal sites in Newfoundland like those on the Avalon Peninsula tend to have unstable thermoclines due to mixing, while deep embayments on the south and west coasts have stable summer–early fall thermoclines, with warmer waters overlying the essentially arctic water below (Hooper et al., 1980; South, 1983).

In a detailed synopsis of salinity patterns, Bugden et al. (1982) stated that 96% of the total

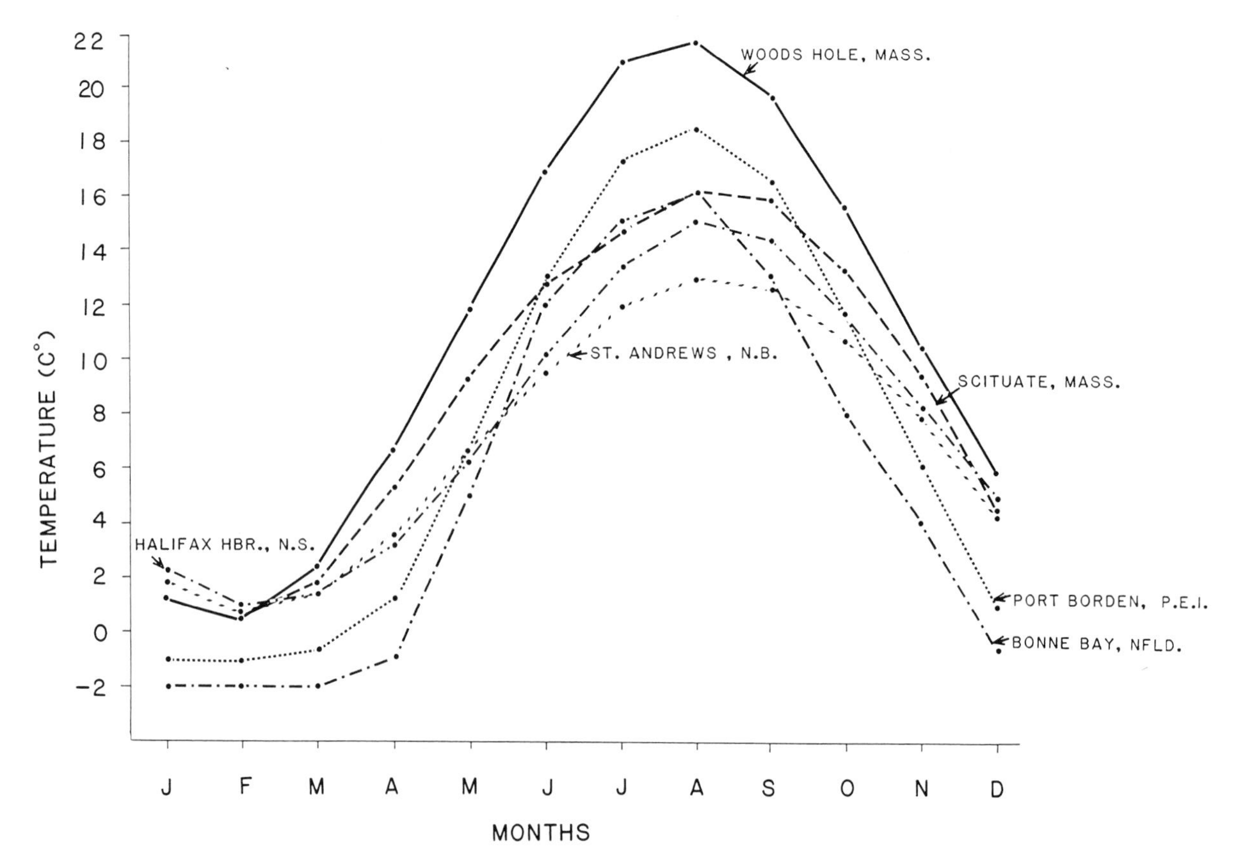

Fig. 7.2. Seasonal variations of surface-water temperatures at six northwest Atlantic sites ranging from Woods Hole, Massachusetts to Bonne Bay, Newfoundland.

global fresh-water input [3 to 5.1 x 10^3 km^3 based upon data of Wright and Worthington (1970)] is derived from northern latitudes, principally because of river runoff and excess precipitation in the north polar sea. This non-homogeneous distribution of fresh water is a major feature of the northwest Atlantic, and ultimately influences stratification, transport, and mixing processes (Appendices I–V). In general, salinity increases in an offshore direction to the Grand Banks (30.5 to 34.0‰), with an incremental increase near the edge of the continental shelf (that is, between coastal and slope water) and at the edge of the Gulf Stream. The lower salinity water (32 to 33‰) of the Labrador Current flows south around Newfoundland to the Grand Banks, the outer regions of the Scotian Shelf and Georges Bank. Water of even lower salinity (32‰), which is derived from runoff from the Gulf of St. Lawrence, covers inner portions of the Scotian Shelf. Sutcliffe et al. (1976) estimated that the annual discharge of fresh water from the St. Lawrence Estuary alone (423 km^3 yr^{-1}) exceeds that from the entire eastern United States. Considerable amounts of fresh water enter the Gulf of Maine in its northwestern sector, reducing salinities in the Gulf to 28–32‰ and implying that the highest salinities in spring, the time of maximum runoff, occur in the eastern sector (Appendix I). Salinities in estuarine habitats and open coastal areas near river mouths are often reduced and extremely variable throughout the northwest Atlantic (Appendices I–V).

As noted by Bugden et al. (1982) and others, fresh-water runoff and upwelling are primary factors influencing nutrient regimes. Typically, nutrient enrichment of coastal waters occurs where fresh and salt water interface at frontal zones, causing entrainment and mixing of nutrient-rich waters. Settling of biological detritus from surface waters introduces nutrients into deeper waters.

Subsequently, upwelling, entrainment or other modes of vertical mixing return nutrients to the surface where they again become involved in biological cycling. As outlined in Appendices I–V, pronounced differences of nutrient regimes occur within the northwest Atlantic. For example, there is a clear gradient of increasing nutrients (that is, nitrate) from the Atlantic coast of Nova Scotia into the Bay of Fundy due to varying patterns of upwelling (Gagné et al., 1982), which causes nutrient-rich waters year-round in some sites and varying periods of nutrient depletion in others. Hydrographically discrete regions like the Gulf of St. Lawrence (Trites, 1972), which trap nutrients by a combination of water circulation and regeneration processes, have very high levels (Coote and Yeats, 1979) that are approximately three times those on the Scotian Shelf.

Ice

The southernmost effects of the Arctic ice pack occur in the north coastal areas of Newfoundland (South, 1983). Fast ice also occurs in sheltered bays along the south, west, and especially the north coasts of Newfoundland (Steele, 1983). Icebergs drifting with the south-flowing waters reach Newfoundland (Appendix V). The Gulf of St. Lawrence is more or less covered with ice in winter (Fig. 7.3). Some ice is formed *in situ* by freezing of the surface water (fast ice), but a larger amount is brought in from the St. Lawrence River (40 to 50%) and through the Strait of Belle Isle by the Labrador Current (Stephenson and Stephenson, 1954a, b, 1972). Ice accumulated in the Gulf of St. Lawrence is discharged into the Atlantic during February and March, from which point it spreads 121 km or more from Cape Breton Island. East and south winds sometimes bring ice as far west as Halifax along the Atlantic coast of Nova Scotia. Ice forms *in situ* only within sheltered inlets and harbors on the Atlantic coast of Nova Scotia, the Bay of Fundy and the Gulf of Maine (Fig. 7.3). Depending upon the severity of winter conditions, these habitats may be covered with a continuous ice sheet of 0.9 to 2.7 m (Anonymous, 1974b; Fefer and Schettig, 1980a; Mathieson et al., 1982; Appendices I–V).

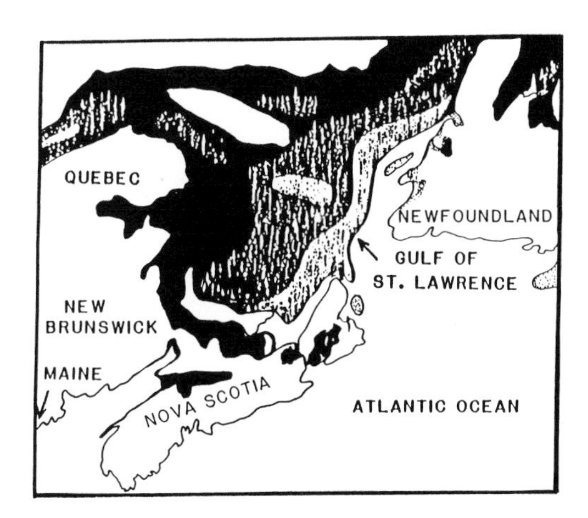

Fig. 7.3. Average sea ice conditions in the northwest Atlantic during January, 1901–1937. (Modified from O'Neill, 1979.)

BIOGEOGRAPHY

Because of the apparent correlation between water temperature and distribution of organisms (Setchell, 1915, 1917, 1920; Hutchins, 1947; Van den Hoek, 1975; Steele, 1975, 1983; Earll and Farnham, 1983), biogeographic regions (provinces) within the northwest Atlantic and other geographies are often delimited by temperature isotherms (Dana, 1852; Van Hofsten, 1915; Runnstrom, 1928; Ekman, 1953; Hedgpeth, 1957; Van den Hoek, 1975; Michanek, 1979). For example, in Briggs' (1974) comprehensive survey of zoogeography he divided the world's oceans latitudinally into various zones (polar, cold temperate and warm temperate of each hemisphere, plus tropical) with these being divided longitudinally into biogeographic regions (separated by geographical barriers) and then ultimately into provinces. Based upon extensive data on seaweed distribution Van den Hoek (1975) delineated five analogous geographic regions [arctic, cold temperate (eastern and western), warm temperate and tropical], with the northwest Atlantic area being designated as a cold-temperate (western) Atlantic region. It is bounded in the north (the Arctic) by the 10°C summer isotherm and to the south (the warm temperate) by either the 10°C winter isotherm or the 15°C summer isotherm, which represents the temperature tolerance of seaweeds in this region (Van den Hoek, 1982a, b). The boundary between cold temperate (or boreal) and arctic is marked by a 10°C summer isotherm. Temperate regions in the northeast Atlantic occupy a wider range of latitudes (particularly the warm temperate) than on the northwestern side, because the Gulf Stream carries relatively warm waters towards northwest Europe, while the cold Labrador Current dominates the American side (Appendices I–V).

Cape Cod, Massachusetts (Fig. 7.1), has long been considered a major biogeographic boundary within the northwest Atlantic (Dana, 1852, 1853; Harvey, 1852; Farlow, 1881), delimiting distinctive northern and southern biotas (Setchell, 1920; Hutchins, 1947; Abbott, 1968; Hecht, 1969; Humm, 1969; Van den Hoek, 1975). Among others, Dana (1852) recognized the uniqueness of Cape Cod and emphasized the relative uniformity of cold-water biota between Newfoundland and Cape Cod (that is, his Nova Scotian Province). Setchell (1922), in an early assessment of critical temperatures and phytogeography of seaweeds, based upon 5°C isotherms of maximum summer water temperatures, noted that Cape Cod was a dividing line between his 15 and 20°C isotherms (Appendix I), and that cold- and warm-water floras were to be distinguished north and south of the Cape. Although Cape Cod is a distinctive biogeographic boundary, several investigators (for example, Taylor, 1957; Briggs, 1974; Sears and Wilce, 1975; Steele, 1983) have recorded "cold-water" biotas south of this geographical feature. For example, Taylor (1957) stated that the flora as far south as New Jersey (where the sandy coastline begins) is boreal rather than tropical in its affinities. He recognized two assemblages with northern, subarctic centers of distribution; one extended as far south as Cape Cod, the other into Long Island Sound. Humm (1969) stated that the long coastline between Cape Cod, Massachusetts, and Cape Canaveral, Florida, is a vast transition zone occupied by two major geographic units. One originated and is centered in the tropics, with its major inshore northern boundary being Cape Canaveral. The other originated and is centered in the colder waters of the North Atlantic, with its major inshore southern boundary being Cape Cod. In discussing the boundary between cold- and warm-temperate regions in northeastern America, Searles (1984) suggested that, although many investigators have had difficulties in recognizing it (Stephenson and Stephenson, 1972; Van den Hoek, 1975; Kapraun, 1980), recent phycological and zoological evidence indicates its occurrence at Cape Hatteras, North Carolina.

Steele (1975, 1983) outlined several zoogeographic accounts of the northwest Atlantic. In an early synthesis, Ganong (1891) divided the waters of eastern North America into four Provinces: (1) Arctic; extending south to Labrador but not the Strait of Belle Isle; (2) Syrtensian; the northern Gulf of St. Lawrence, the St. Lawrence Estuary, all deeper portions of the Gulf of St. Lawrence, coastal waters of Newfoundland, except for localized areas on its south coast, and offshore banks south to Georges Bank; (3) Acadian; shallow southwestern portions of the Gulf of St. Lawrence, localized areas on the south coast of Newfoundland, coastal waters of Nova Scotia south to Massachusetts Bay and deeper parts of Long Island Sound; (4) Virginian; from Massachusetts Bay to Cape Hatteras, North Carolina, with outliers in the Gulf of St. Lawrence, at Sable Island, Nova Scotia, and in Casco Bay, Maine. The three northern provinces belong to Briggs's (1974) Arctic, sub-Arctic and boreal (cold temperate) climatic regions. While there is general agreement with these overall patterns (Leim, 1956; Hedgpeth, 1957; Coomans, 1962; Hall, 1964; Bousfield and Laubitz, 1972; Bousfield and Thomas, 1975; Watling, 1979; Lubinsky, 1980), there is no consensus regarding their boundaries because of the region's complex hydrography (Appendices I–V) and inadequate distributional data (i.e. horizontally and vertically).

Steele (1975, 1983) considered Newfoundland to have an impoverished boreal (cold-temperate) fauna. The sub-Arctic boundary of this fauna is at the Strait of Belle Isle where pronounced changes in fauna and flora occur (Hooper et al., 1980; Hooper, 1982; South, 1983). The north shore of the Gulf of St. Lawrence and the St. Lawrence Estuary are sub-Arctic areas, due to tidal upwelling of cold water. However, the southern Gulf of St. Lawrence (near Northumberland Strait and Prince Edward Island) is not a typical sub-Arctic area (Stephenson and Stephenson, 1972; Bird et al., 1983) as its summer temperatures are comparatively warm (Appendices IV and V). Thus, despite cold winters, a large number of "southerly" organisms survive among the "northerly" majority. The Atlantic coast of Nova Scotia, Bay of Fundy, and Gulf of Maine are boreal (cold temperate) in nature (Stephenson and Stephenson, 1954a, b, 1972; Anonymous, 1974a–c; Apollonio, 1979). However, some warm-temperate areas occur in the Gulf of Maine, particularly in warm estuarine embayments (Mathieson and Hehre, 1986).

The floristic affinity ratios (Cheney, 1977) for several North Atlantic and adjacent regions are presented in Table 7.1. These figures represent the total number of rhodophycean (red algal) plus chlorophycean (green algal) taxa divided by the number of phaeophycean (brown algal) taxa, this ratio being adapted from Feldmann's ratio (1937). The cold-water affinities of Newfoundland's flora are emphasized by the low index (1.8) compared with sites farther south and from the tropics, while comparable values are found in Iceland, Scandinavian countries, etc. (South, 1983). Cold-water affinities are also evident in the Gulf of St. Lawrence, Bay of Fundy and Gulf of Maine (Bird et al., 1983; Mathieson and Penniman, 1986a), whereas warm temperate habitats occur in estuar-

TABLE 7.1

Floristic affinities in the North Atlantic and adjacent regions[1]

Location	(R+C)/P[2]	References
Iceland	1.8	Caram and Jónsson (1972) Jónsson and Gunnarsson (1978)
Greenland	1.1	Lund (1959)
Norway	1.6	Rueness (1977)
Baltic Sea	2.0	Pankow (1971)
Helgoland, Germany	2.2	Kornmann and Sahling (1977)
Netherlands	2.2	Den Hartog (1959)
Great Britain	2.3	Parke and Dixon (1976)
Ireland	2.3	Guiry (1978)
Roscoff (NW France)	2.4	Feldmann (1954) Feldmann and Magne (1964)
Portugal	3.1	Ardré (1970)
Albères, France (Mediterranean)	3.8	Feldmann (1937)
Canary Islands	4.9	Börgesen, cited in Feldmann (1937)
Canadian Arctic	1.7	Lee (1980)
Labrador	1.4	Wilce (1959)
Newfoundland	1.8	South (1976a)
Eastern Canada	1.7	South (1976b)
Magdalen Islands, Quebec (Gulf of St. Lawrence)	1.7	De Seve et al. (1979)
Prince Edward Island, North Shore (Gulf of St. Lawrence)	1.5	Bird et al. (1983)
Bay of Fundy	1.8	Wilson et al. (1979)
Gulf of Maine and outer Bay of Fundy (open coastal)	2.3	Mathieson and Penniman (1986a)
Gulf of Maine (estuarine)	3.1–3.3	Mathieson and Penniman (1986b)
Rhode Island	2.1	Wood and Villalard-Bohnsack (1974)
Connecticut	2.8	Schneider et al. (1979)
Maryland and Virginia	3.4	Ott (1973)
North Carolina	4.1	Searles and Schneider (1978)
Bermuda	5.3	Taylor and Bernatowicz (1969)
Western Florida	5.2	Dawes (1974)

[1]Original data in part from Bird et al. (1983), Earll and Farnham (1983), and South (1983).
[2]Proportion of algal species from different classes (Cheney, 1977): (Rhodophyceae + Chlorophyceae)/Phaeophyceae.

ies in the Gulf of Maine (Mathieson and Hehre, 1986; Mathieson and Penniman, 1986b) and south of New England (floristic affinity ratios of 3.4–4.1, Maryland, Virginia and North Carolina). Similar patterns occur in the northeastern Atlantic, with Great Britain having a cold-temperate flora (2.3); by contrast the Mediterranean is warm temperate (i.e., 3.8, Albères, France) as are the Canary Islands (4.9).

Pielou (1979) described two patterns of disjunct distributions of invertebrate species in New England and the Maritime Provinces of Canada based upon studies by Bousfield and Thomas (1975). One set of "warm-water" animals (the three-lined snail *Triphora perversa* and an amphipod *Haustorius canadensis*) has its major center of distribution south of Cape Cod, but is also found in the southwestern Gulf of St. Lawrence. The other cold-water (boreal) set (a whelk, *Buccinum undatum*, and a fairy shrimp, *Mysis gaspensis*) is absent from the Gulf of St. Lawrence, but reappears farther south between Bar Harbor, Maine, and St. John, New Brunswick, Canada, near the mouth of the Bay of Fundy.

Pielou (1979) speculated that both sets of disjunct species resulted from changes in tidal amplitudes within the Bay of Fundy, which are now 17 m maximally (Appendix II). Approximately 6000 to 10 000 years ago these coastal waters were shallower (Kranck, 1972), tidal amplitudes were slight, and the world was warmer. During the warmest part of this "Hypsithermal Period" (7000 years ago), average water temperatures were approximately 2.5°C warmer than today, and cold-water species survived only below the summer thermocline. The coastline between Bar Harbor and St. John, where these species still survive, emulates this condition. As the Bay of Fundy became deeper and more turbulent, a warm, stratified surface layer was prevented, excluding the warm-water species. The latter survive in the southwestern Gulf of St. Lawrence because of its shallowness, small tides and pronounced summer stratification (Appendix IV). McAlice (1981) and Mathieson and Hehre (1986) have given a similar explanation for the post-glacial history of disjunct patterns for the copepod *Acartia tonsa* and several seaweeds in the Gulfs of Maine and St. Lawrence. For example, several seaweed taxa, which are most abundant south of Cape Cod, primarily occur in shallow embayments or protected habitats such as within the Great Bay Estuary System of New Hampshire and Maine, the Damariscotta River in central Maine, or the Northumberland Strait between Prince Edward Island and New Brunswick. The phenologies of these "southerly" seaweeds are conspicuously different in northern estuaries or embayments from those in southern, open coastal habitats (Hehre and Mathieson, 1970; Mathieson and Dawes, 1975; Bird et al., 1976; Mathieson and Hehre, 1986; Novaczek et al., 1987), with some only reproducing vegetatively in northern latitudes.

Most early studies dealing with temperature and biogeography were correlative assessments of species lists against latitude and/or temperature patterns (Stewart, 1984). Recently many investigations of thermal optima/tolerances for growth and reproduction of individual species have been conducted (Vernberg and Vernberg, 1970; Van den Hoek, 1982a, b; Cambridge et al., 1984; Rietema and Van den Hoek, 1984; Stewart 1984; Yarish et al., 1984, 1986) as well as detailed computer assessments (cluster, correspondence and factor analyses, reciprocal averaging techniques, etc.) of faunistic and floristic distributional records (Van den Hoek and Donze, 1967; Van den Hoek, 1975, 1984; Lawson, 1978; Joosten and Van den Hoek, 1986; Mathieson and Penniman, 1986a). Several experimental studies dealing with North Atlantic animals suggest the potential for more extensive distributions than exist in nature. Steele (1983) stated that such discrepancies may be explained by the fact that reproductive and young stages (eggs and/or larvae) often have more restrictive temperature tolerances than adult maintenance or vegetative activities (Orton, 1920; Runnstrom, 1928, 1930, 1936). Based upon an assessment of early biogeographic studies, Hutchins (1947) concluded that distributional limits of many animals of the Northern Hemisphere were determined either by adult survival or reproductive requirements in winter or summer. Hence, he delineated four distributional patterns, based on: (1) minimum temperature for survival; (2) minimum temperature for reproduction; (3) maximum temperature for survival; or (4) maximum temperature for reproduction.

In a series of detailed distributional and experimental studies, Van den Hoek and his colleagues

(Van den Hoek, 1979, 1982a, b, 1984; Cambridge et al., 1984; Rietema and Van den Hoek, 1984; Yarish et al., 1984, 1986; Joosten and Van den Hoek, 1986) extended Hutchins's (1947) hypotheses regarding temperature requirements for growth, reproduction and survival of North Atlantic seaweeds. For example, during a phytogeographic study of 42 species of the chlorophycean genus *Cladophora*, Van den Hoek (1982c) outlined ten phytogeographic distributional groups of wide applicability to other seaweeds. He subsequently described critical temperatures limiting growth and/or reproduction in diverse seaweeds (Van den Hoek, 1982a, b). Some distributional limits appear to be of a composite nature (Hutchins, 1947). For example, the southern limit of the brown alga *Laminaria digitata* follows the 10°C winter isotherm in Europe (i.e., it has a "southern reproductive boundary"), while in northeastern America it is a 19°C summer isotherm (a "southern lethal boundary"). Van den Hoek (1982a) also found that several species with restricted distributional limits have comparatively narrow temperature ranges for growth and/or survival (for example, the tropical western Atlantic, the warm-temperate Mediterranean–Atlantic and the Arctic groups), while other more widespread species have broader temperature spans (such as the amphiatlantic tropical-to-warm temperate groups with northeastern or northwestern extensions, the amphiatlantic tropical-to-temperate group, amphiatlantic temperate group and the northeast American tropical-to-temperate group). He speculated that many seaweeds found in the eastern North Atlantic are absent in the west because they cannot cope with annual fluctuations of more than 20°C (Appendices I–V). This and other reasons (such as recent glacial histories: see Pielou, 1979) may partially explain the low frequency of endemic northwest Atlantic seaweeds, as well as animals. Studies by Adey (1971), Guiry (1984), Lüning (1984) and McLachlan and Bird (1984), together with those of Van den Hoek and colleagues outlined above, have also correlated the growth and/or survival of North Atlantic seaweeds with their geographic distributions.

In discussing the North Atlantic, Van den Hoek (1975) noted that the cold temperate western and eastern floras have a high degree of similarity (>50%) and an approximate equality of species numbers (160 versus 140 taxa). In contrast, the warm-temperate flora of northwest Europe has a much greater diversity than that of northeastern America (7.4 times as great) due to the dominant influence of the Gulf Stream. Thus, temperate algal propagules can be transported from America to Europe, but are less likely to survive a westward flow through tropical or polar waters. Van den Hoek (1975, 1982a) suggested that the large hydrographic variability of temperature in the western North Atlantic (20°C, Appendices I–V) may restrict many stenothermal species, while the limited extent of rocky substrata in the warm temperate portion of this coastline is also important.

Comparing the biota of the North Atlantic and North Pacific, Pielou (1979) stated that the former is depauperate. The North Atlantic was repeatedly and drastically cooled during the Pleistocene glaciation, while the Pacific was exposed to less variability. The Bering land bridge was in existence during much of the Pleistocene, and acted as a barrier to the Pacific, whereas there was a direct flow of cold water from the Arctic into the Atlantic during this period. Thus, surface water temperatures within the North Atlantic fluctuated widely, forcing its biota to undergo vast latitudinal shifts, causing impoverishment (Van den Hoek, 1975; Joosten and Van den Hoek, 1986). Pielou (1977, 1978) noted that the geographical ranges ("spans") of individual littoral species are much greater on the Atlantic than on the Pacific coasts of the Americas. Historical reasons for this may have been the selective Pleistocene extinction described above, which eliminated species with narrow ecological tolerances (Van den Hoek, 1982a; Yarish et al., 1986), while the "modern" cause may be wider temperature variability in the Atlantic — approximately twice as great (Sverdrup et al., 1942).

COMMUNITY PATTERNS OF OPEN COASTAL ROCKY ECOSYSTEMS: LITTORAL AND SUBLITTORAL ZONES

General features of intertidal zonation on rocky shorelines within the northwest Atlantic (Table 7.2, Figs. 7.4 and 7.5) are relatively well-known and analogous to the "universal features" described by Lewis (1964) and by Stephenson and

TABLE 7.2

Patterns of littoral zonation[1]

Lewis (1964)	Zottoli (1978) General: northwest Atlantic	Bolton (1981) Bay Bulls, Newfoundland	Lobban and Hanic (1984) North Rustico, Prince Edward Island
Littoral fringe (extent of marine influence)	Lichen zone (*Verrucaria maura*) or Blue green algal zone *Bangia atropurpurea* *Blidingia minima* *Calothrix scopulorum* *Littorina saxatilis* *Porphyra linearis* *Ulothrix flacca*	Littoral fringe *Bangia atropurpurea* *Fucus spiralis* *Littorina saxatilis* *Urospora penicilliformis* *Verrucaria maura*	Intertidal Black zone (HHWS) (blue-green algae) *Calothrix scopulorum* *Rivularia* sp.
Eulittoral zone (upper limits of barnacles)	Barnacle zone *Fucus spiralis* *Mytilus edulis* *Nucella lapillus* *Semibalanus balanoides*	Eulittoral zone (+1.0 m) *Fucus distichus* ssp. *edentatus* *Mytilus edulis* *Pilayella littoralis* *Semibalanus balanoides*	Barnacle zone (MHHW) *Semibalanus balanoides*
Shores dominated by barnacles Shores dominated by mussels Shores dominated by fucoids	Brown algal zone *Ascophyllum nodosum* *Fucus distichus* ssp. *edentatus* *F. spiralis* *F. vesiculosus* *Hildenbrandia rubra* *Littorina littorea* *Littorina obtusata* *Monostroma* spp. *Mytilus edulis* *Nucella lapillus* *Polysiphonia lanosa* *Semibalanus balanoides* *Ulva lactuca* *Ulvaria obscura* Red algal zone *Chondrus crispus* *Mastocarpus stellatus*		*Fucus* zone (MSL) *Littorina littorea* *Littorina saxatilis* *Mytilus edulis*
Sublittoral zone (upper limits of kelps) Maximum exposure: *Alaria* shores Average exposure: *Laminaria* shores	Sublittoral zone *Alaria esculenta* *Homarus americanus* *Laminaria* spp. *Palmaria palmata* *Strongylocentrotus droebachiensis*	Sublittoral zone *Alaria esculenta* *Devaleraea ramentaceum*	*Enteromorpha/Ulva* zone[2] (sand-impacted rocks) *Chordaria zone*[3] (MLLW–LLWS) (sand-impacted rocks) *Ceramium* spp. *Dictyosiphon foeniculaceus* *Polysiphonia* spp. *Rhodomela confervoides* Subtidal zone (LLWS) *Chondrus crispus* *Corallina officinalis* *Fucus vesiculosus* *Laminaria saccharina* *Mytilus edulis*

[1] See p. 5 for abbreviations of tidal levels.
[2] Replaced by *F. vesiculosus* in non-sandy areas.
[3] Replaced by *Chondrus*/*Fucus serratus* in non-sandy areas.

Fefer and Schettig (1980b) Maine coast	Johnson and Skutch (1928 a, b, c) Mount Desert Island, Maine	Femino and Mathieson (1980) Bald Head Cliffs, Maine	Hardwick-Witman and Mathieson (1983) Mathieson et al. (1981) Jaffrey Point, New Hampshire
High intertidal *Anurida maritima* *Littorina saxatilis*	Supralittoral (+4.3 to +9.2 m) (*Plantago–Porphyridium* belt) *Aira flexuosa* *Blidingia minima* *Calothrix fasiculata* *Mougeotia* spp. *Oscillatoria* spp. *Plantago decipiens* *Porphyridium cruentum*	Littoral fringe (+3.4 to +4.6 m) *Bangia atropurpurea* *Littorina saxatilis* *Ulothrix flacca* *Urospora* spp.	Littoral fringe (+2.7 to +3.8 m) *Bangia atropurpurea* *Calothrix scopulorum* *Codiolum pusillum* *Prasiola stipitata* *Urospora speciosa* *Verrucaria maura*
Barnacle *Mytilus edulis* with the associated animals: *Hyale nilsonni* oligochaetes nematodes	Upper littoral (+2.1 to +4.3 m) (*Fucus–Ascophyllum–Calothrix–Verrucaria* belt) *Ascophyllum nodosum* *Bangia atropurpurea* *Blidingia minima* *Calothrix scopulorum* *Codiolum pusillum* *Fucus vesiculosus* *Semibalanus balanoides* *Ulothrix flacca* *Urospora penicilliformis* *Verrucaria maura*	Eulittoral zone (+0.1 to +3.4 m) Upper *Enteromorpha* spp. *Mytilus edulis* *Semibalanus balanoides* Middle *Ascophyllum nodosum* *Fucus vesiculosus* *Littorina littorea* *Monostroma grevillei* *M. pulchrum* *Mytilus edulis* *Rhizoclonium tortuosum* *Ulvaria obscura*	Eulittoral zone (+0.1 to +2.7 m) Upper *Fucus spiralis* *Verrucaria maura* Middle *Ascophyllum nodosum* *Fucus vesiculosus* *Hildenbrandia rubra* *Littorina littorea* *Littorina obtusata* *Mastocarpus stellatus* *Verrucaria maura*
Rockweed *Acmaea testudinalis* *Littorina littorea* *Littorina obtusata* *Modiolus modiolus* *Mytilus edulis* with nematodes oligochaetes *Nucella lapillus* *Semibalanus balanoides*	Lower littoral (+0.6 to 2.1 m) (*Porphyra–Fucus–Spongomorpha* belt) *Ascophyllum nodosum* *Fucus distichus* spp. *edentatus* *Palmaria palmata* *Porphyra umbilicalis* *Spongomorpha* spp.	Lower *Acmaea testudinalis* *Chondrus crispus* *Gammarus oceanicus* *Halichondria panicea* *Hildenbrandia rubra* *Littorina littorea* *Mastocarpus stellatus* *Mytilus edulis* *Nucella lapillus* *Spongomorpha* spp. *Ulva lactuca*	Lower *Acmaea testudinalis* *Chondrus crispus* *Littorina littorea* *Mastocarpus stellatus* *Mytilus edulis* *Nucella lapillus* *Petrocelis cruenta*
Irish moss *Acmaea testudinalis* *Asterias vulgaris* *Carcinus maenas* *Gammarellus angulosus* encrusting bryozoans *Halichondria panicea* *Hiatella artica* *Lacuna vincta* *Lepidonotus squamatus* *Littorina littorea* *Modiolus modiolus* *Mytilus edulis* nematodes oligochaetes *Semibalanus balanoides* *Tonicella ruber*	Sublittoral zone (to +0.6 m) (*Alaria–Devaleraea–Phymatolithon* belt) *Alaria esculenta* *Devaleraea ramentaceum* *Phymatolithon lenormandii* *Saccorhiza dermatodea* *Ulva lactuca*	Sublittoral zone (to +0.1 m) *Acmaea testudinalis* *Alaria esculenta* *Antithamnionella floccosa* *Chondrus crispus* *Devaleraea ramentaceum* *Gammarus oceanicus* *Halichondria panicea* *Hildenbrandia rubra* *Modiolus modiolus* *Mastocarpus stellatus* *Mytilus edulis* *Petrocelis cruenta* *Strongylocentrotus droebachiensis*	Sublittoral (to +0.1 m) see Table 7.3

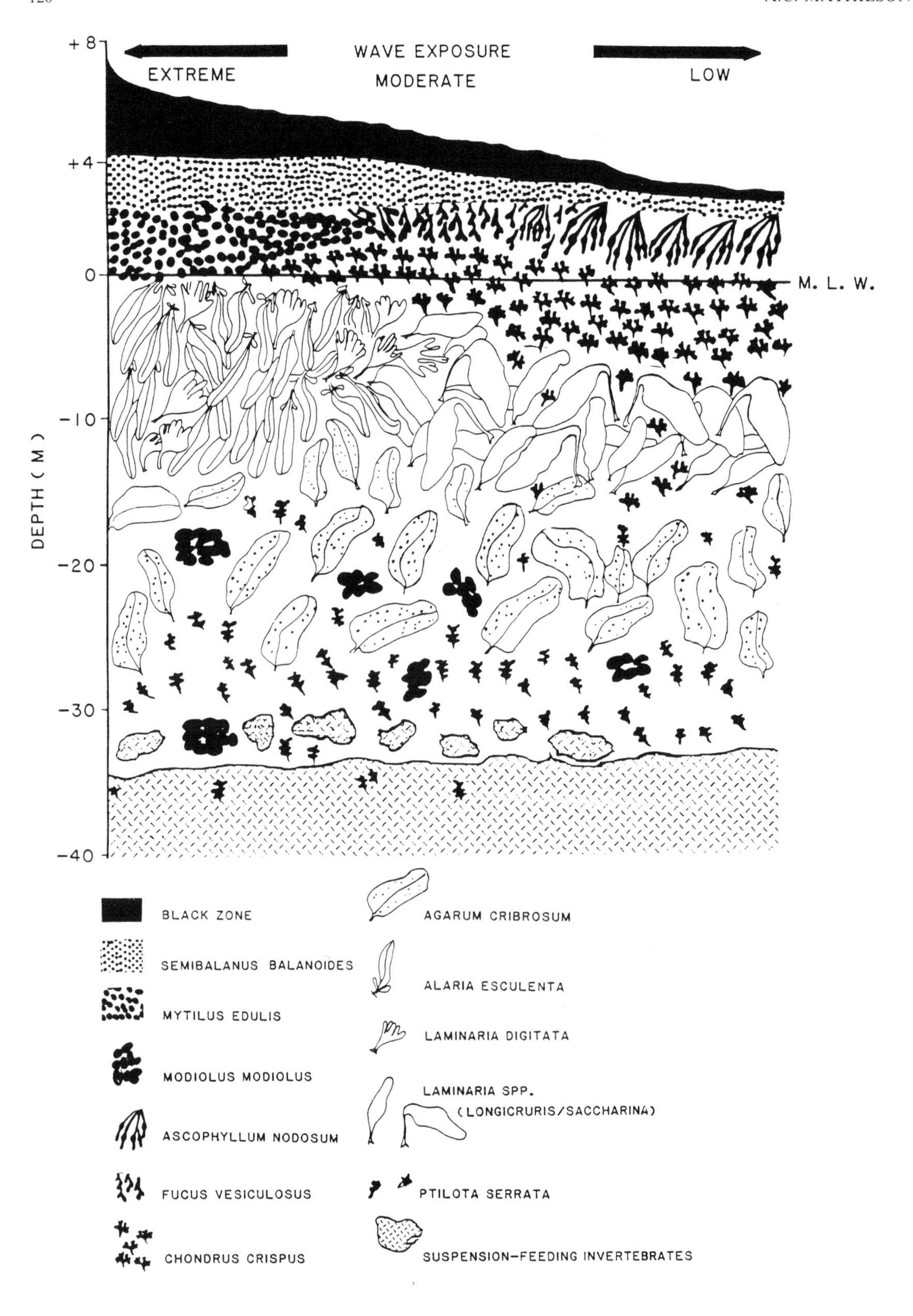

+8
+4
0
-10
-20
-30
-40
DEPTH (M)
WAVE EXPOSURE
EXTREME
MODERATE
LOW
M. L. W.
BLACK ZONE
SEMIBALANUS BALANOIDES
MYTILUS EDULIS
MODIOLUS MODIOLUS
ASCOPHYLLUM NODOSUM
FUCUS VESICULOSUS
CHONDRUS CRISPUS
AGARUM CRIBROSUM
ALARIA ESCULENTA
LAMINARIA DIGITATA
LAMINARIA SPP.
(LONGICRURIS/SACCHARINA)
PTILOTA SERRATA
SUSPENSION-FEEDING INVERTEBRATES

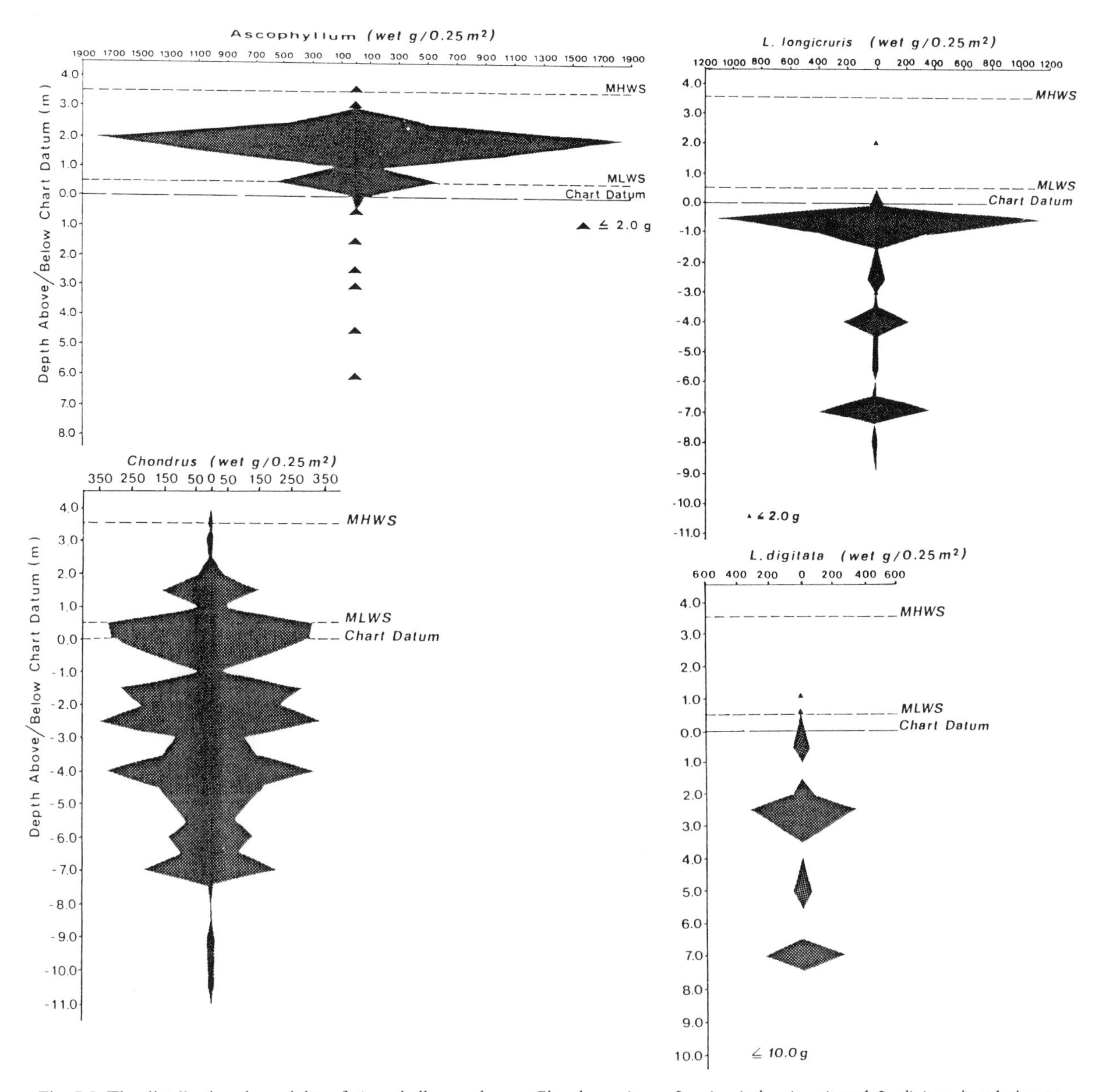

Fig. 7.5. The distribution, by weight, of *Ascophyllum nodosum*, *Chondrus crispus*, *Laminaria longicruris* and *L. digitata* in relation to chart datum along the West Pubnico Peninsula, Nova Scotia. (From Pringle and Semple, 1980.)

Stephenson (1972). Local habitat conditions (degree of wave exposure, disturbance due to ice and sand scouring, instability and discontinuity of rocky substrata, etc.) modify many of these general patterns (Sousa, 1984). While several contrasting schemes have been devised to describe intertidal zonation, Lewis' classification (Lewis, 1964), which emphasizes the functional differences between intertidal (i.e. between the tides) and littoral (i.e. coastal) as well as the modifying role of wave

Fig. 7.4. Schematic sketch showing the distribution of major benthic organisms in the northwest Atlantic in relation to exposure and depth.

action, is used in the following synopsis. The littoral zone (*sensu* Lewis) is divided into the littoral fringe, which extends from the upper limits of marine organisms down to the upper limits of barnacles, and the eulittoral zone that lies below the littoral fringe and includes the shore down to the upper limits of kelps. The sublittoral zone (*sensu* Lewis) extends from the eulittoral downward to the lower distributional limits of vegetation.

Littoral fringe

The littoral fringe, which is also referred to as a black or spray zone (Table 7.2, Fig. 7.4), comprises a distinct band throughout the northwest Atlantic, as well as the Arctic (Ellis and Wilce, 1961). The zone is characterized by blue-green algae (*Calothrix, Lyngbya, Rivularia*, etc.) and ephemeral macrophytes (such as *Bangia, Blidingia, Codiolum, Porphyra, Prasiola, Ulothrix*, and *Urospora*), lichens (such as *Verrucaria maura*), and a periwinkle (*Littorina saxatilis*). The zone (Fig. 7.4) is particularly sensitive to wave action (Johnson and Skutch, 1928a, b; Kingsbury, 1976) and can be dilated (i.e. expanded in vertical dimensions) and/or uplifted (i.e. vertically displaced) up to 20 m above high tide levels in extremely exposed situations (South, 1983). In sheltered open coastal sites an overlapping of freshwater algae, terrestrial algae (*Porphyridium cruentum*) and spermatophytes can occur within this complex transition zone (Johnson and Skutch, 1928c). Numerical clustering analysis of intertidal communities confirms a major discontinuity in species content at the junction of the littoral fringe and eulittoral zone (Bolton, 1981).

Eulittoral zone

On a "typical" semi-exposed rocky shoreline, three major zones occur within the eulittoral region (Kingsbury, 1976; Zottoli, 1978; Lubchenco, 1980; Pringle and Semple, 1980): (1) an upper barnacle zone with *Semibalanus balanoides* dominating; (2) a mid-shore brown algal zone with *Ascophyllum nodosum* or *Fucus* spp. and (3) a lower red algal zone with *Chondrus crispus* (Irish moss) and *Mastocarpus stellatus*[1] (Table 7.2, Figs. 7.4 and 7.6). Kingsbury (1976) and Lubchenco (1980) have described abundance patterns of major species in New England under various exposure regimes. The *S. balanoides* zone exhibits a conspicuous uplifting and dilation with increased wave action, while the brown and red algal zones are compressed and displaced downward. Thus, barnacles usually dominate upper shorelines in exposed and moderately exposed sites, while a greater number of species occur at these elevations in more sheltered habitats. Barnacles may also extend into the lower eulittoral zone (Hardwick-Witman and Mathieson, 1983), particularly in extremely exposed habitats. Depending upon wave action and other associated physical and biological factors, either *A. nodosum* or *Fucus* spp. will dominate the mid-shore in New England (Lubchenco, 1980). As in Europe, *A. nodosum* is most abundant in sheltered sites (Baardseth, 1970) and is replaced by *F. vesiculosus* and *F. distichus* ssp. *edentatus* with increasing exposure (Fig. 7.7). Under extreme wave action the latter fucoids are limited and *Mytilus edulis* becomes the major occupier of space within the mid-shore (Sideman and Mathieson, 1983a, b). In the lower eulittoral zone, *C. crispus* and/or *Mastocarpus stellatus* dominate at all but the most exposed sites, where mussels are most abundant. *Chondrus crispus* is mainly found on shelving and horizontal surfaces, whereas *Mastocarpus stellatus* dominates vertical ones which dry out more quickly and are exposed to stronger wave action (Pringle and Mathieson, 1987). Substrata with intermediate slope are populated by a mixture of both red algae (Burns and Mathieson, 1972b; Mathieson and Burns, 1975; Green and Cheney, 1981; Green, 1983).

Besides *Mytilus edulis* and *Semibalanus balanoides*, numerous other invertebrate species, both sessile and motile, are characteristic of the eulittoral zone (Table 7.2). Several herbivorous crustaceans and gastropods are common (Vadas, 1985), including amphipods (such as *Hyale nilssoni*), snails (*Littorina littorea, L. obtusata, L. saxatilis*, and *Lacuna vincta*) and limpets (*Acmaea testudinalis*). The chiton *Tonicella ruber* and the sea urchin *Strongylocentrotus droebachiensis* forage

---

[1]Guiry et al. (1984) reinstated the genus *Mastocarpus* Kützing for four widely distributed species of *Gigartina*, including *Mastocarpus stellatus* (= *Gigartina stellata*) (C. Agardh) Kützing, which alternates with a crustose sporophytic stage (*Petrocelis cruenta*).

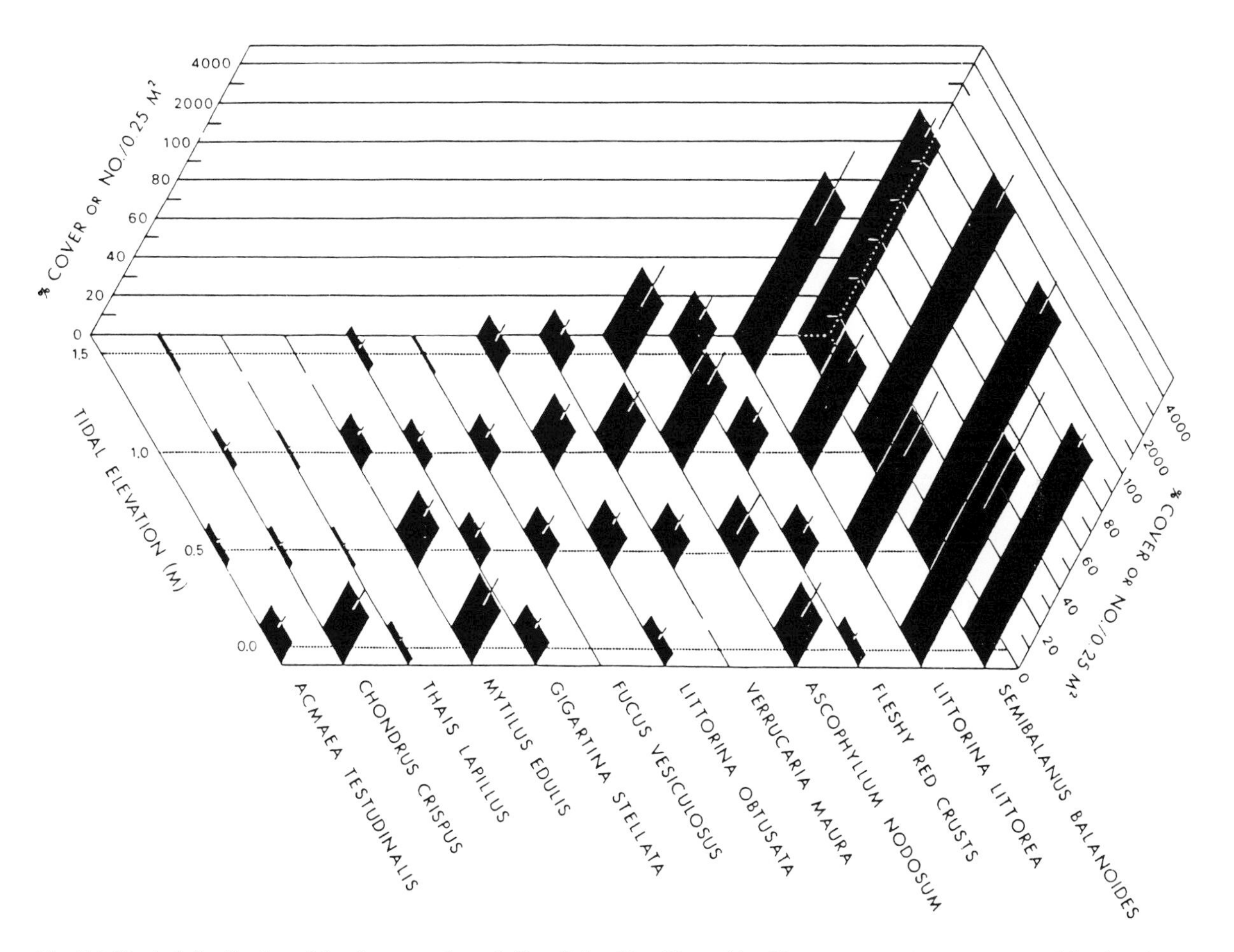

Fig. 7.6. Vertical distribution of dominant species at Jaffrey Point, New Hampshire. Data are mean percentage cover (algae) and mean number/0.25 m^2 (invertebrates) $\pm$95% confidence intervals. Fleshy red crusts are primarily *Hildenbrandia rubra* with some *Petrocelis cruenta* in the lower elevations. See previous comments regarding *Mastocarpus stellatus* (= *Gigartina stellata*), its crustose sporophyte (*Petrocelis cruenta*) and *Nucella* (= *Thais*) *lapillus*. (From Hardwick-Witman and Mathieson, 1983.)

within the lower eulittoral and sublittoral zones. The whelk *Nucella (Thais) lapillus*, two crab species, *Carcinus maenas* and *Cancer irroratus*, and a starfish, *Asterias vulgaris*, are important predators of rocky shorelines, particularly within the lower eulittoral and sublittoral zones. With increasing wave exposure the abundances of herbivores and predators decrease.

These patterns of littoral zonation are fairly typical throughout the northwest Atlantic (Table 7.2). However, modifications occur in more northerly areas, due primarily to ice scouring (p. 124,

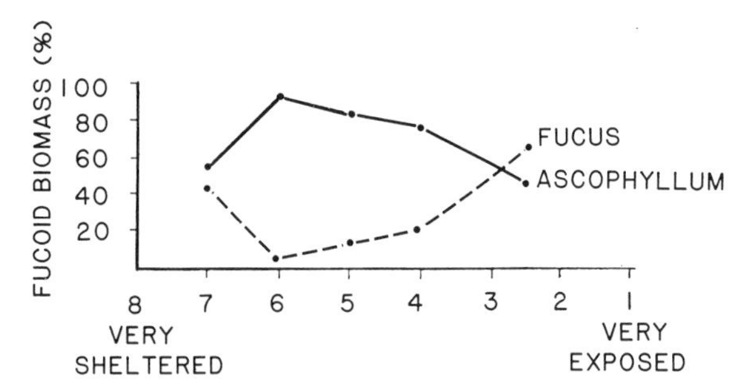

Fig. 7.7. Distribution of fucoid biomass with wave exposure in Maine. (Modified from Topinka et al., 1981.)

143, 145, 157). For example, *Fucus vesiculosus* may be restricted to the upper sublittoral zone in ice-impacted areas like the Gulf of St. Lawrence (Taylor, 1975; Bird et al., 1983; Lobban and Hanic, 1984; McLachlan et al., 1987). Ice limits the development of perennial zone-forming taxa, and only ephemeral organisms (both plants and animals) or individuals restricted to rock crevices are present (Wilce, 1959). Intertidal habitats not considered in detail in this chapter (such as gravel, sand and mud) generally have lower diversity than more stable rocky shores, and dissimilar zonation patterns (Dring, 1982).

Fucoid algae and Irish moss (*Chondrus crispus*) are important structural elements, providing important habitats for many plants and animals (Table 7.2). *Ascophyllum nodosum* is host for several epiphytic (such as *Elachista fucicola* and *Polysiphonia lanosa*) and epifaunal populations (for example, *Spirorbis* spp., hydrozoans, bryozoans), as well as amphipods (such as *Hyale nilssoni*) which grow amongst the epiphytes (Baardseth, 1970; McBane, 1981; McBane and Croker, 1983). Fucoids serve a major functional role by providing abundant canopy within the eulittoral zone (see pp. 122, 153, 157). Wave action, vertical position, and scouring by ice or sand affect the dominant organisms as well as influence the development of epiphytic and epifaunal communities (South and Hill, 1976; Pringle and Mathieson, 1987; Kingsbury, 1976). For example, the distribution of *P. lanosa* differs from that of its obligate host *A. nodosum*, as it is absent or sparse in extremely sheltered sites with regular ice scouring. Distributional patterns of littorinid snails (e.g. *Littorina obtusata*) and the major canopy-forming species *Ascophyllum nodosum* and *Fucus vesiculosus* are very similar (Van Dongen, 1955; Kingsbury, 1976; Hardwick-Witman and Mathieson, 1983): all three taxa diminish under extreme exposure.

Based upon a series of long-term transect studies on the Isles of Shoals (Gulf of Maine), Kingsbury (1976) noted a relatively consistent ratio of invertebrates/algal species (1.5 to 2.0) throughout the eulittoral, even though there are simultaneous patterns of maximum species richness in the lower eulittoral/upper sublittoral and at moderately exposed sites (Daly and Mathieson, 1977; Mathieson et al., 1981; Mathieson and Penniman, 1986a). A mean number of 17.4 macroscopic organisms per quadrat (20 cm^2) was recorded within the combined mid-eulittoral and upper sublittoral zones (Kingsbury, 1976).

Spatial variability of plant types (i.e. canopy, understorey, crusts and algal mats) occurs on rocky shorelines, depending upon wave exposure, substratum stability, etc. (Den Hartog, 1959; Hay, 1981; Hardwick-Witman and Mathieson, 1983; Taylor and Hay, 1984). For example, at a semi-exposed site like Jaffrey Point, New Hampshire, canopy, crustose and understorey species dominate, with the total coverage primarily reflecting that of the canopy species (Fig. 7.8). Nearby estuarine sites show reduced total and canopy coverage due to a dominance of mud, and an increased abundance of salt-marsh species and algal mats in the upper shoreline.

Sublittoral zone (nearshore)

The sublittoral zone within the northwest Atlantic (Table 7.3; Fig.7 4) has a relatively uniform cold-water biota (Taylor, 1957; Humm, 1969; Steele, 1975, 1983). A gradual change in species dominance occurs, especially in the northernmost sections (p. 114) and from the upper to lower sublittoral zones. In contrast, sublittoral biota south of Cape Cod have a greater number of warm temperate species and enhanced species richness in the deep sublittoral zone (Coleman and Mathieson, 1974, 1975; Sears and Wilce, 1975). Even so, some southern Cape Cod sites with strong tidal currents and lower summer temperatures have many cold temperate species (Farlow, 1881; Sears and Wilce, 1975).

Various schemes have been used to characterize

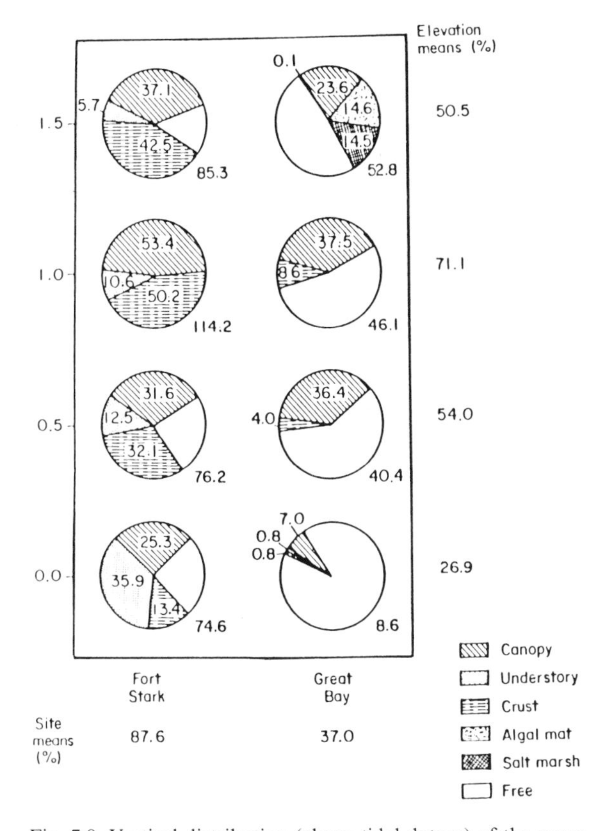

Fig. 7.8. Vertical distribution (above tidal datum) of the mean percentage cover of plant types at Fort Stark (Jaffrey Point), New Hampshire and within the adjacent Great Bay Estuary System, New Hampshire–Maine. Open areas in circles represent the percentage of free substratum. (From Hardwick-Witman and Mathieson, 1983.)

sublittoral biota (Table 7.3) depending upon the site and/or the investigators' interpretation (Golikov and Scarlato, 1973; Erwin, 1983; Hiscock, 1986). A fringe may be designated in the upper sublittoral zone if a distinct belt of species occurs which lacks a continuous extension downwards (Lewis, 1964). Numerical clustering analysis showed no dramatic difference in species composition between the lower eulittoral zone and the sublittoral fringe (Russell, 1973; Bolton, 1981); rather, many species are common on the lower shoreline but absent in the mid-shore. Similar evaluations within the sublittoral zone have demonstrated shallow and deep-water communities (Hooper et al., 1980). Variable numbers of associations have been described depending upon depth-related factors (that is, light, wave action, salinity, temperature and oxygen stratification), type and degree of substratum stability, and phenological patterns (Edelstein et al., 1969; Mann, 1972a; Sears and Wilce, 1975; Hooper et al., 1980; Erwin, 1983; Davis, 1984; Hiscock, 1986).

Zonation and biomass patterns of sublittoral seaweeds in St. Margaret's Bay, Nova Scotia, typify many of the conspicuous patterns on rocky shorelines in northern New England, Nova Scotia and Newfoundland (Mann, 1972a). Four major belts occur with increasing depth: (1) *Chondrus crispus*/*Chorda* spp.; (2) *Laminaria* spp.; (3) *Agarum cribrosum*; and (4) *Ptilota serrata*. The last zone reaches a lower limit around −20 to −30 m; *P. serrata* is often a conspicuous understorey species at shallower depths (Table 7.4, Fig. 7.4). Kelps (*L. digitata*, *L. longicruris* and *A. cribrosum*) account for approximately 83% of total biomass. A wide range of sublittoral habitats occurs within the Bay, ranging from very exposed sites dominated by *Alaria esculenta* to sheltered shorelines characterized by *Zostera marina*. With decreasing exposure *A. esculenta* is replaced by *L. digitata* within the shallow sublittoral zone, while strap-shaped laminarians (such as *L. longicruris*) ultimately replace the digitate ones in more sheltered sites (Lewis, 1964; Smith, 1985, 1986; Fig. 7.4). Various patterns of uplifting or vertical displacement occur. For example, where *L. digitata* replaces *Alaria esculenta* in shallow, moderately exposed sites, other non-digitate kelps move in at lower levels. The presence of an *Agarum cribrosum* zone below *Laminaria* spp. is a common feature of the northwest Atlantic (cf. Lamb and Zimmerman, 1964; Edelstein et al., 1969; Mathieson, 1979; Mathieson et al., 1981), while it is apparently absent on the northwest coastline of Europe (Mann, 1972a). Sea urchins (*Strongylocentrotus droebachiensis*) dominate many upper and mid-sublittoral areas and their grazing activities produce an alternate community state dominated by crustose coralline algae. Removal of urchins results in a return to the characteristic foliose algal community typical of that depth (p. 148).

The sublittoral zone along the Atlantic coast of Nova Scotia and within the Gulf of Maine is more diverse than in Newfoundland and the southern Gulf of St. Lawrence. Edelstein et al. (1969) described a rich and productive sublittoral flora near Halifax, with three major perennial macrophyte associations (Table 7.3). Analogous tripar-

TABLE 7.3

Patterns of sublittoral zonation (nearshore)

Bird et al. (1983); McLachlan et al. (1987) Prince Edward Island North Shore (Gulf of St. Lawrence)	Edelstein et al. (1969) Halifax County, Nova Scotia	Mathieson et al. (1981) Jaffrey Point, N.H.
Sublittoral fringe (0.0 m) *Fucus vesiculosus* belt (0.0 to −2.5 m) *Chordaria flagelliformis* *Dictyosiphon foeniculaceus* *Fucus serratus* *Fucus vesiculosus* *Scytosiphon lomentaria*	Sublittoral fringe (0.0 to −10(15) m) *Laminaria/Desmarestia* Association (> 50 species) *Ceramium rubrum* *Chondrus crispus* *Corallina officinalis* *Cystoclonium purpureum* *Desmarestia* spp. *Laminaria* spp. *Palmaria palmata* *Polysiphonia* spp. *Phyllophora* spp. *Saccorhiza dermatodea*	Upper (sublittoral fringe) (+0.1 to −2.0 m) *Chondrus crispus* *Chordaria flagelliformis* *Devaleraea ramentaceum* *Fucus distichus* ssp. *edentatus* *Fucus distichus* ssp. *evanescens* *Hildenbrandia rubra* *Laminaria* spp. *Leathesia difformis* *Petrocelis cruenta* *Spongomorpha arcta*
Chondrus crispus belt (−2.5 to −5.0 m) *Chondrus crispus* *Chordaria flagelliformis* *Dictyosiphon foeniculaceus* *Scytosiphon lomentaria*		
Chondrus/Furcellaria belt (−5.0 to −7.5 m) *Chondrus crispus* *Furcellaria lumbricalis* *Furcellaria* belt (−7.5 to −10.0 m) *Furcellaria lumbricalis* *Laminaria* spp.	*Agarum/Ptilota* Association (−10 to −30 m) *Agarum cribrosum* *Alaria esculenta* *Callophyllis cristata* *Fimbrifolium dichotomum* *Membranoptera alata* *Odonthalia dentata* *Peyssonnelia rosenvingii* *Turnerella pennyi*	Middle (−2.0 to −12.5 m) *Agarum cribrosum* *Ahnfeltia plicata* *Antithamnionella floccosa* *Ceramium rubrum* *Chondrus crispus* *Gymnogongrus crenulatus* *Laminaria* spp. *Membranoptera alata* *Phycodrys rubens* *Polyides rotundus* *Polysiphonia* spp.
Phyllophora belt (−10.0 to −15(20) m) *Laminaria* spp. *Phyllophora pseudoceranoides* *Phyllophora truncata* *Agarum/Polysiphonia/* crustose coralline belt (−20 to −25 m) *Agarum cribrosum* crustose coralline algae *Polysiphonia* spp.	*Phyllophora/Polysiphonia* Association (−30 to −40 m) *Agarum cribrosum* crustose coralline algae *Neodilsea integra* *Phycodrys rubens* *Phyllophora* spp. *Polysiphonia* spp. *Ptilota serrata*	Lower (−12.5 to −19.0 m) *Agarum cribrosum* *Lithothamnion glaciale* *Phyllophora* spp. *Phymatolithon rugulosum* *Rhodophysema elegans*

tite zonation patterns have been recorded at several sites within the Gulf of Maine (Lamb and Zimmerman, 1964; Harris and Mathieson, 1976; Hulbert et al., 1976, 1981; Mathieson, 1979; Hulbert, 1980; Mathieson et al., 1981). For example, at a semi-exposed site like Jaffrey Point, New Hampshire, three zones (upper, middle and lower) were differentiated based upon the occurrence of dominant species and biomass patterns (Mathieson et al., 1981). A sublittoral fringe was recognized because of the presence of a cluster of species (*Leathesia difformis*, *Spongomorpha* spp., etc.; Table 7.3) lacking a continuous extension downwards (Lewis, 1964), a recognizable stand of kelps (*Laminaria* spp., *Saccorhiza dermatodea*), a large number of red algae, few green algae and

Hulbert (1980); Hulbert et al. (1976)
Star Island, Isles of Shoals, N.H.

Zone I
(0 to −12 m)
Asterias forbesi
Asterias vulgaris
Buccinum undatum
Cancer borealis
Cancer irroratus
Carcinus maenas
Chondrus crispus
Henricia sanguinolenta
Homarus americanus
Laminaria digitata
Laminaria saccharina
many ephemeral seaweeds
Pseudopleuronectes americanus
Strongylocentrotus droebachiensis
Tautogolabrus adspersus

Zone II
(−13 to −30 m)
Acmaea testudinalis
Agarum cribrosum
Amphipholis squamata
Amphiporus angulatus
Amphitrite spp.
Asterias vulgaris
Boltenia echinata
Boltenia ovifera
Buccinum undatum
Cancer borealis
Cancer irroratus
crustose coralline algae
Cucumaria frondosa
Harmothoe imbricata
Henricia sanguinolenta
Hiatella arctica
Homarus americanus
Hyas araneus
Ischnochiton alba
Lepidonotus squamatus
Leptasterias spp.
Modiolus modiolus
Nereis pelagica
Onchidorus fusca
Ophiopholis aculeata
Pectinaria gouldii
Psolus fabricii
Ptilota serrata
Strongylocentrotus droebachiensis
Tautogolabrus adspersus

Zone III
(−31 to −45 m)
Asterias vulgaris
Boltenia ovifera
Buccinum undatum
Cancer borealis
Crossaster papposus
crustose coralline algae
Henricia sanguinolenta
Homarus americanus
Hyas araneus
Modiolus modiolus
Pseudopleuronectes americanus
Psolus fabricii
Ptilota serrata
Strongylocentrotus droebachiensis
Tautogolabrus adspersus

Lamb and Zimmerman (1964)
Cape Ann, Massachusetts

Sublittoral fringe
(0.0 to −2.0 m)
Alaria esculenta
Chondrus crispus
Chorda filum
Devaleraea ramentaceum
Fucus distichus ssp. *edentatus*
Mastocarpus stellatum
Saccorhiza dermatodea

Laminaria belt
(−2.0 to −12 m)
Ahnfeltia plicata
Callophyllis cristata
Ceramium rubrum
Chondrus crispus
Corallina officinalis
Laminaria spp.
Pantoneura baerii
Phyllophora spp.

Agarum belt
(−12 to −20 m)
Agarum cribrosum
Antithamnionella floccosa
Chondrus crispus
crustose coralline algae
Modiolus modiolus
Phyllophora spp.
Polyides rotundus
Ptilota serrata
Rhodomela confervoides
Scagelia pylaisaei

Lower limits of vegetation
(−20 m)

TABLE 7.4

Zonation and biomass (fresh weight) of seaweeds in St. Margaret's Bay, Nova Scotia, Canada[1]

Zone	Average width (m)	Biomass (kg m^{-2})	Biomass per metre of shoreline (kg)	Percent of total biomass
Fucus and *Ascophyllum*	15.5	10.7	124.9	8.7
Chorda and fine Phaeophyceae	87.9	1.1	95.3	6.5
Chondrus crispus	6.0	3.5	20.9	1.4
Zostera marina	4.9	1.0	5.0	0.3
Laminaria digitata and *L. longicruris*	22.7	16.0	363.5	25.0
L. longicruris	46.5	11.5	534.6	35.8
L. longicruris and *Agarum cribrosum*	36.7	4.9	179.2	11.6
A. cribrosum and *Ptilota serrata*	86.3	1.8	158.1	10.7

[1]Averaged from 24 transects (modified from Mann, 1972a).

many annuals. By contrast, the mid-sublittoral zone has a more abundant kelp forest (*Agarum cribrosum* and *Laminaria* spp.), fewer total species, a conspicuous understorey of foliose and turf-forming red algae (Gislèn, 1930), and many epiphytes. The lower sublittoral zone has a paucity of plant species (biomass and total) with locally abundant *Agarum cribrosum* and *Phyllophora* spp., plus crustose coralline algae.

Although not explicit in the above descriptions, it should be emphasized that crustose corallines are major occupiers of primary substrata within the algal-dominated sublittoral zones (Steneck, 1985, 1986). Other organisms may be more conspicuous and overgrow crustose coralline algae, but corallines survive and persist in a variety of habitats (Adey, 1964, 1965, 1966a–c; Hulbert et al., 1976, 1981; Garwood et al., 1985; Keats, 1985; see also p. 131, 144, 154). In particular, they thrive in, and commonly dominate, environments of intense herbivory (Adey, 1973; Steneck, 1977, 1978, 1982a, b, 1985, 1986).

Hulbert (1980) and Hulbert et al. (1976, 1981) have described three characteristic sublittoral communities within the Gulf of Maine (Table 7.3). The shallowest community (Zone I, 0 to −12 m), which represents an extension of the lower eulittoral, is dominated by *Chondrus crispus*, *Laminaria* spp. and several ephemeral algae. Seaweeds provide physical shelter and refuge for a large number of animals (Littler and Littler, 1986; Johns and Mann, 1987; Keats et al., 1987), as well as secondary substrata for various epifaunal species (*Electra pilosa*, *Spirorbis* spp., etc.). Crabs (*Cancer borealis*, *C. irroratus* and *Carcinus maenas*), lobsters (*Homarus americanus*), urchins (*Strongylocentrotus droebachiensis*), whelks (*Buccinum undatum*), sea stars (*Asterias forbesi*, *A. vulgaris* and *Henricia sanguinolenta*) and fish (*Pseudopleuronectes americanus* and *Tautogolabrus adspersus*) dominate this shallow zone. A sharp discontinuity occurs at approximately −12 m, which is the upper limit of urchin grazing on the lower limit of *Laminaria* spp. (cf. Prentice and Kain, 1976). Below this depth, a community dominated by crustose coralline algae occurs (Zone II, −13 to −30 m), covering approximately 85% of available rocky substrata. Scattered clumps of *Modiolus modiolus* occupy the remainder of primary substrata (Witman, 1984, 1985; p. 129); they trap sediments and provide refuge and secondary substrata for a wide range of infaunal and epifaunal species, including amphipods, asteroids, chitons, sea cucumbers, spider crabs, tunicates, polychaetes, urchins and many other invertebrates (Table 7.3, Fig. 7.4). Large urchins, snails and sea stars occur on the open substrata between mussel clumps. *Agarum cribrosum* and *Ptilota serrata* are the only conspicuous erect seaweeds at these depths. A community dominated by sessile suspension-feeding invertebrates occurs between −31 to −45 m (Zone III), with only a few residual algae (*P. serrata* and crustose corallines) occurring in the upper part. Many animals found here are restricted to depths below the summer thermocline (−10 to −25 m).

In contrast to the algal-dominated horizontal

surfaces (from the eulittoral zone to -30 m) in the Gulf of Maine, vertical and undercut rock faces have a complex of sessile suspension-feeding invertebrates, including sponges, tunicates, brachiopods, etc. (Hulbert et al., 1976; Hulbert, 1980; Witman, 1982), which are often termed fouling communities (Lundalv, 1971; Sutherland, 1974; Gulliksen, 1978; Harris and Irons, 1982). Sebens (1986a, b) described these communities in shallow habitats (p. 150); below -30 m they begin to dominate upper horizontal surfaces (Harris and Mathieson, 1976; Hulbert et al., 1976; Witman, 1984). Although some red algae occur (crustose corallines, *Ptilota serrata*, etc.), their densities decrease with depth, and they are replaced by a matrix of polychaete tubes and sessile invertebrates, including sponges, hydroids, anemones, soft corals, brachiopods, ectoprocts, and tunicates. This complex of attached forms traps sediments, which in turn provides a thin habitat for several motile invertebrates, such as ophiuroids, gastropods, polychaetes, and crustaceans. Unlike fouling communities on vertical faces, which are dominated by sheet-like growth forms (Woodin and Jackson, 1979; Jackson and Hughes, 1985), most sessile invertebrates on upper surfaces have upright and mounding shapes, presumably to avoid smothering by detrital accumulation.

Near vertical and undercut surfaces the most abundant predatory species include large winter flounder (*Pseudopleuronectes americanus*), haddock (*Melanogrammus aeglefinus*), wrasse (*Tautogolabrus adspersus*), eelpout (*Macrozoarces americanus*), and wolffish (*Anarhichas lúpus*) (Wheeler, 1980). A greater diversity of sea stars occurs here than at shallower depths (Hulbert, 1980). The horse mussel *Modiolus modiolus*, which forms conspicuous clumps and is an important refuge for many small invertebrates in shallower zones (see p. 140 and Table 7.3), decreases in abundance with depth. Below -40 m, *M. modiolus* is primarily limited to cracks and crevices. Dominance of upright algae on shallower (horizontal) surfaces suggests that competition limits fouling communities to vertical and undercut surfaces (that is, refuges) where light is inadequate for upright seaweeds (Sebens, 1986a, b).

Although upright algae dominate upper horizontal surfaces (to -30 m) in the Gulf of Maine, numerous sessile and motile invertebrates also occur. Foliose red algal canopies (to -30 m) serve as refuges for complexes of sponges, hydroids, ectoprocts, and tunicates, as well as mats of tubicolous amphipods (for example, *Jassa falcata* and *Corophium bonnellii*). These complexes are major occupiers of secondary space, growing upon a primary coverage of crustose coralline algae. Another suite of sessile invertebrates grows on foliose algae. Thus, kelps often have colonies of hydroids and ectoprocts (*Electra pilosa* and *Membranipora membranacea*), while red algae may be covered with ectoprocts, hydroids and tunicates, as well as spat of *Mytilus edulis* and *Modiolus modiolus*. Abundance of sessile invertebrates increases during fall and winter, and decreases in late winter and spring, in part because of predation by nudibranchs (Harris, 1973) and urchins (see p. 148). During spring and summer, red algae are often heavily covered with epiphytic algae; subsequently the coverage decreases because of heavy grazing by juvenile *Lacuna vincta* and herbivorous amphipods (Hulbert et al., 1976; Hulbert, 1980; Witman, 1984; Sebens, 1986b; Harris, unpubl. data). The red algal understorey also provides refuge and feeding ground for numerous motile invertebrates, particularly polychaetes and small crustaceans such as shrimps, amphipods, isopods, and young crabs.

Numerous herbivores occur in the algal-dominated sublittoral zone in the Gulf of Maine (Vadas, 1985). The most abundant species is the gastropod *Lacuna vincta*, which feeds on kelp, filamentous red algae and diatom films (Fralick et al., 1974). *Littorina littorea*, the dominant littoral herbivore within this area (Lubchenco, 1978; Lubchenco and Menge, 1978; Petraitis, 1979, 1987; see also p. 139, 141), extends to -10 m in scattered, protected embayments (for example, Casco Bay, Maine and Gosport Harbor, Isles of Shoals, Maine and New Hampshire). Chitons (*Tonicella marmorea* and *T. ruber*) and the limpet *Acmaea testudinalis* are common within urchin barrens and amongst *Ptilota serrata*. Few grazing molluscs occur amongst dense stands of *Laminaria* spp. or *Chondrus crispus* (Hulbert et al., 1976; Witman, 1984). In beds of kelp and *C. crispus*, *Strongylocentrotus droebachiensis* is uncommon and limited to cracks and crevices (Witman, 1984). Outside algal forests, urchins are the most important sublittoral herbivores. Their grazing can cause barrens to occur where abundance of foliose algae is severely reduced (p. 146).

Shallow sublittoral populations in Newfoundland and the Gulf of St. Lawrence often differ from the general northwest Atlantic pattern because of extreme ice abrasion, the occurrence of warm summer temperatures, and stable thermoclines (Appendices IV and V, Figs. 7.2 and 7.3). In Newfoundland extremes of ice disturbance occur on the southern and eastern coastlines, resulting in patterns of maximum species diversity and numbers of perennials in the former, and a diminutive and ephemeral biota in the latter (South, 1983). The open Atlantic coastline of Newfoundland, which is exposed to irregular ice-scouring, has a bimodal distribution of seaweeds. A shallow zone of high biomass occurs, followed by an intermediate zone with low algal productivity and numerous sea urchins (*Strongylocentrotus droebachiensis*), and a deeper zone of high algal biomass dominated by *Agarum cribrosum* and *Ptilota serrata* (Himmelman, 1980, 1985; Keats et al., 1982b, 1985; Steele, 1983; Keats, 1985). The width of the shallow zone varies, depending upon the periodic occurrence of ice-scouring and the effects of wave action, limiting urchin grazing. Hooper et al. (1980) distinguished two major sublittoral populations throughout Newfoundland: (1) a shallow community dominated by southern species (particularly on the west and south coasts); and (2) a deep-water community (below −25 m) dominated by northern species. Many northern species are at their southern limits in Newfoundland (for example, *Fimbrifolium dichotomum, Laminaria solidungula, Lithothamnion tophiforme, Neodilsea integra, Odonthalia dentata* and *Polysiphonia arctica*), while some southern ones are at their northern limits (for example, *Bryopsis plumosa, Chondrus crispus, Cystoclonium purpureum, Leathesia difformis* and *Phyllophora pseudoceranoides*). According to Bird et al. (1983), the sublittoral zone within the southern Gulf of St. Lawrence is unique and impoverished, as *Fucus serratus* has been recently introduced (Dale, 1979), many green algae and canopy-forming kelps (such as *Laminaria* spp.) are absent, and some typical zone-forming intertidal species (*Chondrus crispus* and *Fucus vesiculosus*) flourish in the upper sublittoral zone (Table 7.3). Below the *C. crispus* belt four deeper ones were delineated, based upon the dominance of perennial red and brown algae.

Several patterns are apparent when comparing these sublittoral studies, as well as those by Adey (1966a), Colinvaux (1966, 1970), Edelstein et al. (1970), Stone et al. (1970), Pringle and Semple (1980) and Norall et al. (1981). Foremost, a progressive surfacing of deep-water (i.e. "northern") forms occurs in cooler and less insolated (e.g. foggy) sites like the Bay of Fundy (Bousfield and Thomas, 1975), as compared to their occurrence in deeper waters in more southerly latitudes. For example, the sea stars *Crossaster papposus* and *Solaster endeca* are limited to −25 m and deeper at the Isles of Shoals (Gulf of Maine), while they, and many other deep-water forms in New England, are found at −1 m in Passamaquoddy Bay, Maine–New Brunswick (Meinkoth, pers. commun.). In addition, many plants and animals are considerably larger in the Bay of Fundy than at other locales (for example, New England and the Atlantic coast of Nova Scotia). Edelstein et al. (1970) emphasized that extremely low tides within the Bay of Fundy (Appendix II) occur during early morning and evening, particularly in winter. Light penetration is reduced by turbidity and the prevalence of fog in summer.

Patterns of seaweed species diversity are maximal within the upper sublittoral zone (Fig. 7.9), except in areas of extreme ice and/or wave action (Hooper et al., 1980; Bird et al., 1983; McLachlan et al., 1987). Usually the numbers of green algae decrease with depth more rapidly than those of brown and red algae. Red algae dominate the deepest sublittoral zone (Mathieson, 1979). Similar patterns of ratios of Chlorophyceae/Rhodophyceae and numbers of species decreasing with depth occur in northwestern Europe (Levring, 1960; Smith, 1967; Lüning, 1970; Norton and Milburn, 1972; Dring, 1982; Drew, 1983), with many authors emphasizing the significant role of light. This decrease in algal diversity with depth is inversely related to an increase in animal diversity (Hulbert et al., 1976); many invertebrates are limited to depths below the summer thermocline (Earll and Erwin, 1979; Earll and Farnham, 1983).

Another major pattern is a dominance of perennial taxa within the deep sublittoral, the ratios of perennials/annuals increasing with depth (Sundene, 1953; Ernst, 1968). Shallow-water habitats with extreme variability support a greater number of annuals, while perennials dominate

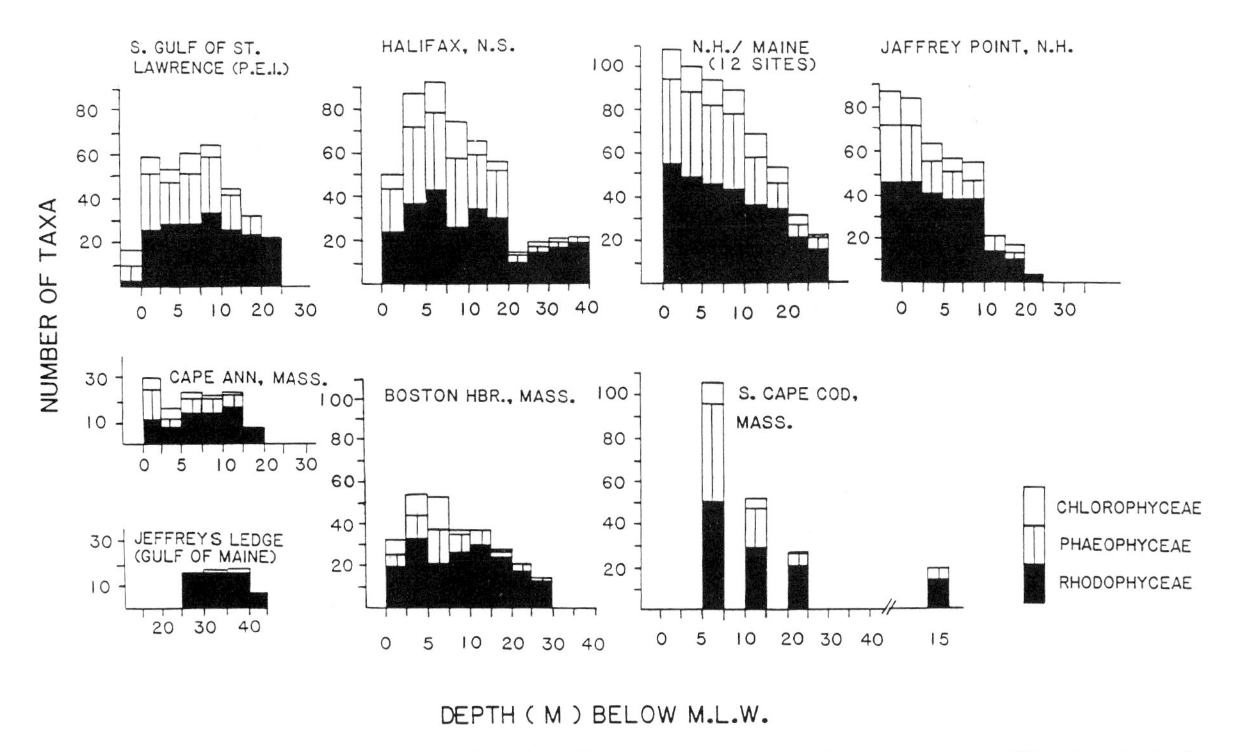

Fig. 7.9. Vertical distribution of seaweed taxa within the sublittoral zone at seven nearshore sites and one offshore location in the northwest Atlantic.

deeper and more stable habitats (Lamb and Zimmerman, 1964; Edelstein et al., 1969; Coleman and Mathieson, 1974, 1975; Hooper, 1975; Sears and Wilce, 1975; Mathieson, 1979; Hooper et al., 1980; Mathieson et al., 1981; Bird et al., 1983; South, 1983; McLachlan et al., 1987). The occurrence of southern annual and northern perennial taxa above and below the stable summer thermocline in Newfoundland supports this pattern (South, 1983).

A comparison of the distribution of some individual species shows several unique patterns. Few plants are restricted to the deepest sublittoral zone, and the majority penetrates to the lower limits from shallower depths (Mathieson, 1979). Various plant types occur (Fig. 7.8); canopy-forming species (*Agarum cribrosum* and *Laminaria* spp.) dominate the upper or mid-sublittoral zone, with an understorey of foliose, turf-forming and crustose species as well (Neushul, 1971; Foster, 1975a, b; Denley and Dayton, 1985). Within the deepest sublittoral zone, crustose or foliose forms dominate (Wilce, 1971; Mathieson, 1979; Norall et al., 1981), as they are adapted to low light regimes (Adey, 1970; Mathieson and Norall, 1975a, b). Many fleshy crustose red algae are restricted to the sublittoral zone (Wilce, 1971; Sears and Cooper, 1978), while crustose corallines show a broader distribution (Adey, 1966a).

Several perennial red algae have unequal proportions of their isomorphic generations within the sublittoral zone. Most often a dominance of tetrasporophytes has been recorded in deeper (or outer) portions of the sublittoral zone (Jones, 1956b; Mathieson and Burns, 1975; Craigie and Pringle, 1978; Norall et al., 1981; Kain, 1982, 1984, 1986; Whittick, 1984; Greenwell et al., 1985), while gametophyte dominance has been reported less frequently within the sublittoral zone (Dyck et al., 1985; May, 1986). Recent photosynthetic studies of *Chondrus crispus* and *Ptilota serrata*, both of which show patterns of diploid dominance, demonstrate higher photosynthetic rates for sporophytic than gametophytic plants (Mathieson and Norall, 1975a, b). Perennation and differential survival of tetraspores or juvenile gametophytes may maintain domination by tetrasporophytes (Rao, 1970; Barilotti and Silverthorne, 1972; Edwards, 1973; Dawes et al., 1974; Hansen and Doyle, 1976; Lee and Mathieson, 1983; May, 1986). For example, *P. serrata* exhibits extensive polar regeneration of "rhizoids" from fragments (Lee and Mathieson, 1983), which enhances vegetative propagation and colonization of rocky substrata (Mathieson, un-

TABLE 7.5

"Extinction depths" or lower limits of vegetation in the northwest Atlantic

Location	Extinction depth (m)	Collection method	Dominant organisms	References
Eastern Newfoundland and Labrador	−50 to −75	Dredging	Crustose coralline algae	Adey (1966a)
Southern Gulf of St. Lawrence	−20 to −30	SCUBA	Crustose coralline algae	Adey (1966a)
Southern Gulf of St. Lawrence (near Prince Edward Island)	−25	SCUBA	Algal cover (red), including crustose corallines, exceedingly sparse at −25 m	Bird et al. (1983), McLachlan et al. (1987)
Halifax, Nova Scotia	−40	SCUBA	Most species decrease markedly at −30 to −40 m; crustose corallines dominate; deep water *Ptilota serrata* and *Phycodrys rubens* etiolated delicate; *Agarum cribrosum* stunted)	Edelstein et al. (1969)
St. Margaret's Bay, Nova Scotia	−20 to −30	SCUBA	*Ptilota* association	Mann (1972a)
Nova Scotia	−50 to −70	Dredging	Crustose coralline algae	Adey (1966b)
Gulf of Maine	−20 to −30	SCUBA	Crustose coralline algae	Adey (1964, 1965, 1966b, c)
Southern Maine and New Hampshire (12 sites)	−13 to −32	SCUBA	Turf, crustose fleshy and coralline algae dominate; *Callophyllis cristata*, *Ptilota serrata* and etiolated *Fimbrifolium dichotomum*	Mathieson (1979), Mathieson et al. (1981)
Cape Ann, Massachusetts	−20	SCUBA	Below −16 m vegetation appreciably thinner; rocks mostly covered with fine gray silt; scattered *Agarum cribrosum*, dwarf *Phyllophora truncata*, thin *Rhodomela confervoides*, and many crustose coralline algae.	Lamb and Zimmerman (1964)
Boston Harbor (offshore), Massachusetts	−30	SCUBA	Crustose coralline algae	Harris and Mathieson (1976)
Jeffreys Ledge, Gulf of Maine, Massachusetts (offshore)	−45	SCUBA (Helgoland habitat)	Fleshy macroscopic red algae (particularly *Ptilota serrata*) dominate from −29 to −32 m, encrusting corallines from −38 m to −45 m; below this, endozoic and epizoic diatoms associated with sponges; *Ptilota* an important substratum for North Atlantic herring eggs.	Sears and Cooper (1978)
Central Gulf of Maine (offshore)	−63	Submersible	Crustose coralline algae	Steneck et al. (1986), Vadas and Steneck (1988)

publ. observ.). *Chondrus crispus* also exhibits extensive perennation and regeneration of its perennial holdfast (Taylor et al., 1975, 1981; see p. 143 below). Varying ecophysiological responses of isomorphic reproductive phases and/or susceptibility to grazing may also be significant in causing these distributional patterns (Hannach and Santelices, 1985; Luxoro and Santelices, 1989).

As shown in Table 7.5, the extinction depth (*sensu* Sears and Cooper, 1978) is extremely variable, ranging from approximately 20 to 30 m in the Gulf of Maine and the southern Gulf of St. Lawrence, to 50 to 75 m off eastern portions of Newfoundland and Nova Scotia (Adey, 1964, 1965, 1966a–c). Shallow and truncated vegetational limits occur in turbid and/or silty locations, and in sites with major discontinuities between rock and sand; deeper populations occur in clear-water rocky sites. Species richness and composition vary considerably between specific areas within the Northwest Atlantic. Several morphological differences are apparent for deep-water populations; some species are conspicuously etiolated (for example, *Ptilota serrata*), while others (for example, *Agarum cribrosum* and *Rhodomela confervoides*) are diminutive and/or delicate.

Sublittoral zone (isolated offshore communities)

Offshore rocky pinnacles in the Gulf of Maine exhibit several unique plant and animal populations (Tables 7.5 and 7.6). Pinnacles such as Pigeon Hill on Jeffreys Ledge and Ammen Rock on Cashes Ledge are small offshore sea mounts, which are distinct from nearshore ecosystems. Sears and Cooper (1978) described strong interrelationships between animals and seaweeds in two deep-water communities with twenty seaweed species on Jeffreys Ledge, approximately 5 km due east of Cape Ann, Massachusetts. For example, *Ptilota serrata* is an important substratum for eggs of the North Atlantic herring (*Clupea harengus*) and may reduce post-hatch mortality (Cooper et al., 1975). A shallower (−29 to −37 m) *P. serrata* association consists almost exclusively of foliose red algae, while the deeper *Lithothamnion* spp. association (−38 to −44 m) is dominated by encrusting coralline algae which extend to the lower limits of macroscopic plants. Besides *P. serrata, Phycodrys rubens* and *Phyllophora truncata* were the most conspicuous red algae, while only two green algae of minor importance (*Derbesia marina* and *Uronema curvata*), one coccoid blue-green alga, and no brown algae occurred. Crustose coralline algae covered 10 to 75% of available rocky substrata from −39.5 to −42.5 m and disappeared below −45 m. At greater depths only endozoic and epizoic diatoms (*Navicula* and *Nitzschia* spp.) occur in association with the sponge *Polymastia* sp. Sears and Cooper (1978) stated that variations of species composition between deep inshore and offshore regions occur (Tables 7.3 and 7.6), lending support to the view that multiple species associations exist within the sublittoral zone (Sears and Wilce, 1975).

Vadas and Steneck (1988) described three depth zones of algal dominance for a deep water rocky pinnacle (Ammen Rock) in the central Gulf of Maine: (1) leathery macrophytes (to −40 m), (2) foliose red algae (to −50 m), and (3) crustose algae (fleshy crusts to −55 m and coralline crusts to −63 m). *Metridium senile* dominated some vertical faces to −24 m and locally appeared to set the upper vertical limits of kelp and possibly foliose red algae. From −50 to −90 m, and below the shallow band of *M. senile*, patchy assemblages of sponges (*Haliclona, Phakellia, Polymastia* and *Subertechinus*), ascidians (*Aplidium, Ascidia* and *Botrylloides*), and two other anemones (*Bolocera* and *Tealia*) occurred. Also, invertebrates such as *Modiolus* and *Myxicola* were abundant in scattered locations at this depth range. Most conspicuous by their absence within the zones of algae were sea urchins, while other herbivores (for example, limpets and chitons) were also rare. They speculate that these record depths for seaweeds in the Gulf of Maine and for cold water environments may result from an absence of large herbivores and the high productivity potential of the benthos in these relatively clear offshore waters.

Hulbert et al. (1982) described two major benthic communities (−33 and −42 m) at Pigeon Hill on Jeffreys Ledge (Table 7.6, Fig. 7.1) and emphasized the significance of substratum orientation (Harris et al., 1979; Harris and Irons, 1982). Horizontal surfaces are dominated by an algal–polychaete community, while a sponge–tunicate community occurs on vertical surfaces (Lundalv, 1971; Gulliksen, 1978). The foliose red alga *Ptilota serrata*, the sabellid polychaete *Chone infundibuliformis*, and the terebellid polychaete *Thelepus cincinnatus* dominate the former community,

TABLE 7.6

Patterns of sublittoral zonation (offshore)

Sears and Cooper (1978) Jeffreys Ledge, Gulf of Maine	Hulbert et al. (1982) Pigeon Hill (Jeffreys Ledge), Gulf of Maine (−33 and −42 m stations)	Steneck et al. (1986) Vadas and Steneck (1988) Ammen Rock, Cashes Ledge, central Gulf of Maine	Sebens and Witman (1986) Cashes Ledge, Gulf of Maine
Ptilota association	Algal-polychaete community	Articulated calcified assemblage	Kelp and anemone zone
(fleshy reds)	(horizontal surfaces)	(−24 to −33 m)	(−22 to −36 m)
(−29 to −37 m)	*Chone infundibuliformis*	*Corallina officinalis*	*Agarum cribrosum*
Callophyllis cristata	*Corophium crassicorne*		*Laminaria saccharina*
Ceratocolax hartzii	*Haploops tubicola*	Filamentous assemblage	*Metridium senile*
Derbesia marina	*Ischyrocerus anguipes*	(−24 to −33 m)	
Gloeocapsa sp.	*Ophiopholis aculeata*	*Audouinella purpurea*	Sponge/ascidian/brachiopod and
Iophon nigricans	*Ophiura robusta*	*Uronema curvata*	anemone zone (−37 to −60 m)
Leptophytum foecundum	*Ptilota serrata* (absent below		*Bolocera tuediae*
Leptophytum laeve	−38 m)	Leathery macrophyte assemblage	*Tealia crassicornis*
Lithophyllum orbiculatum	*Thelepus cincinnatus*	(−24 to −40 m)	
Lithothamnion glaciale	+36 infaunal species (bivalves,	*Agarum cribrosum*	Sponge/polychaete/various invertebrates
Membranoptera alata	gastropods, ophiuroids and	*Laminaria longicruris*	(> −60 m)
Petrocelis	polychaetes)	*Laminaria* sp. (? new species)	
Phycodrys rubens	Sponge-tunicate community	*Metridium senile*	*Myxicola* sp.
Phyllophora truncata	(vertical surfaces)		
Phymatolithon laevigatum	abundant crustose coralline algae	Foliose-corticated assemblage	
Phymatolithon rugulosum	*Halichondria panicea*	(−24 to −50 m)	
Ptilota serrata	*Haliclona oculata*	*Callophyllis cristata*	
Scagelia pylaisaei	*Haliclona palmata*	*Membranoptera alata*	
Terebratulina septentrionalis	*Hymedesia* sp.	*Phycodrys rubens*	
Uronema curvata	*Iophon pattersoni*	*Phyllophora truncata*	
2 unidentified red crusts	*Myxilla fimbriata*	*Ptilota serrata*	
	Plocamionida ambigua		
Lithothamnion association	*Terebratulina septentrionalis*	Crustose assemblage,	
(encrusting corallines)		non-calcified forms	
(−38 to −44(45) m)		(−30 to −55 m)	
Leptophytum foecundum		*Cruoriopsis*	
Leptophytum laeve		*Petrocelis*	
Lithophyllum orbiculatum		*Peyssonnelia dubyi*	
Lithothamnion glaciale			
Phymatolithon laevigatum			
Phymatolithon rugulosum			

	Crustose assemblage, calcified forms
	(−24 to −63 m)
Endozoic and epizoic diatom association growing on sponges (*Polymastia* sp.)	*Leptophytum laeve*
(−45 m)	*Lithophyllum orbiculatum*
Navicula	*Lithothamnion lemoineae*
Nitzschia	
	Animal-dominated zone
	(−50 to −90 m)
	Aplidium
	Ascidia
	Bolocera
	Botrylloides
	Haliclona
	Modiolus
	Myxicola
	Phakellia
	Polymastia
	Subertechinus
	Tealia

which provides important three-dimensional habitat structure and secondary substrata for numerous infaunal species. The tubicolous amphipods *Corophium crassicorne, Haploops tubicola* and *Ischyrocerus anguipes* contribute to this polychaete matrix, while a variety of amphipods, caprellids, small asteroids, ophiuroids, brachiopods and ectoproct colonies inhabit the secondary substratum layer. Tunicates and brachiopods attach to the polychaete tubes. Approximately 36 infaunal species occur amongst the sediments trapped by polychaetes, that is, bivalves, gastropods, ophiuroids and polychaetes. Ophiuroids are the numerically dominant group, *Ophiura robusta* and *Ophiopholis aculeata* being the most abundant species. With the disappearance of *P. serrata* below -37 m (Sears and Cooper, 1978), polychaetes remain the dominant macrobenthic component of the community on upper horizontal surfaces at -42 m, while there is a simultaneous increase in densities of brachiopods, bivalves and gastropods. The most obvious differences between horizontal and vertical surfaces are that the primary substratum in the latter is relatively free of sediments, and *P. serrata* is virtually absent. At -33 m crustose coralline algae cover much of the primary substrata on vertical rock walls (Sebens, 1986a, b), surviving overgrowth by sponges and tunicates. Brachiopods (*Terebratulina septentrionalis*) show a patchy distribution on vertical and undercut rocky walls, as well as on secondary substrata (Noble et al., 1976; Witman, 1982; Witman and Cooper, 1983). At least nine species of sponges occur on vertical surfaces, comprising three major growth forms: thin sheet-like encrustations (*Halichondria panicea* and *Hymedesia* sp.), rounded globose forms (*Iophon pattersoni, Myxilla fimbriata*, and *Plocamionida ambigua*) and upright branching forms (*Haliclona oculata* and *H. palmata*). Thus, vertical communities are dominated by colonial growth forms (sponges and tunicates) which trap little sediment and have few secondary encrustations.

ORGANIZATION OF OPEN COASTAL ROCKY ECOSYSTEMS: LITTORAL AND SUBLITTORAL ZONES

General overview

Zonation patterns are controlled by a combination of physical and biological factors (Connell, 1972, 1975; Menge, 1976; Robles and Cubit, 1981; Dring, 1982; Hulbert et al., 1982; Levinton, 1982; Robles, 1982; Denley and Dayton, 1985; Lubchenco, 1986; Norton, 1986; Underwood, 1986). At the upper shoreline where abiotic stress (*sensu* Grime, 1979) is most extreme, few organisms occur (Table 7.2) and interspecific competition is reduced, while at lower elevations physical environmental conditions are more benign and competition and predation more intense (Chapman, 1979; Underwood and Denley, 1984; Denley and Dayton, 1985; Peckol et al., 1988).

In describing the upper limits of plants within the littoral zone Dring (1982) emphasized that the rate of water loss is largely dependent on the surface area/volume of the tissue, and bears no consistent relationship to the natural habitat of a species (Kristensen, 1968). The extent of photosynthetic recovery after drying is critical, being much greater for populations at upper than at lower elevations, including intraspecific differences (Bewley, 1979; Schonbeck and Norton, 1979b, c). According to Norton (1986) desiccation and overheating are the major factors involved in aerial emersion at the upper shoreline, whereas high irradiation, rainfall and frost are less significant. Repeated exposure to sub-critical desiccation drought-hardens plants, this tolerance being maximal in summer and minimal in winter (Schonbeck and Norton, 1979c). Even so, the uppermost populations of *Ascophyllum nodosum* and *Fucus spiralis* are periodically "pruned back" during summer, particularly after periods of drying conditions and neap tides (Schonbeck and Norton, 1978). Several intertidal snails compensate for temperature stress by evaporative water loss, concomitant with circulation of body tissue fluids; other invertebrates with hard body parts seal off the environment and resist desiccation more effectively than soft-bodied organisms (Levinton, 1982). Seaweeds growing on the lower shoreline exhibit a variety of drought-avoidance strategies (Schonbeck and Norton, 1979b, c; Norton, 1986), including low surface area/biomass ratios, mutual protection by overlapping fronds, and growth in dense mats in shaded sites and tide pools. Nutrient stress may be a significant limiting factor. Some plants at the upper shoreline are adept at utilizing the limited nutrient supply available during brief and intermittent periods of submergence (Schonbeck and Norton, 1979a).

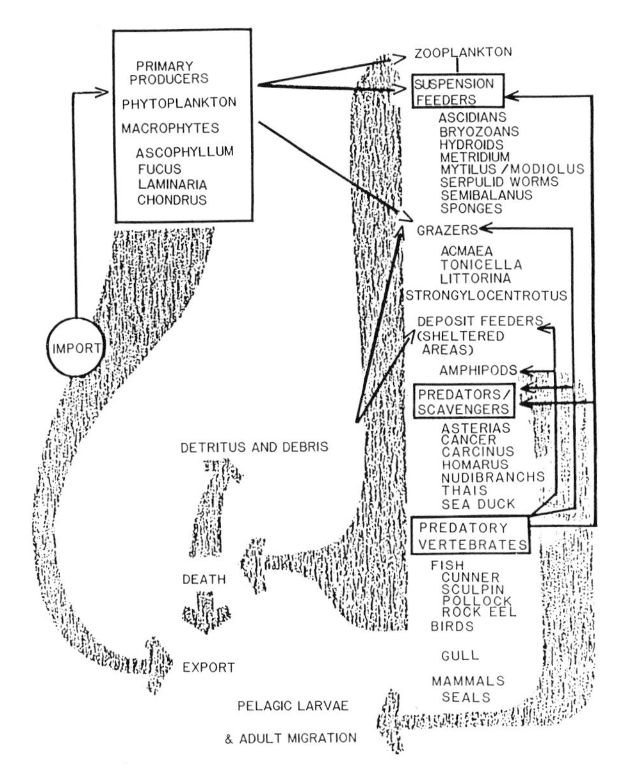

Fig. 7.10. Food web on a typical rocky shore within the northwest Atlantic. (Modified from Anonymous, 1974b.)

Several structural features are in common throughout the eulittoral zone, particularly within the mid and lower shoreline zones (Lubchenco and Menge, 1978; Underwood and Denley, 1984). Predator–prey interactions (Fig. 7.10) in relatively benign habitats control the abundance of functionally dominant competitors such as *Mytilus edulis*, allowing algae to colonize and persist (Newcombe, 1935; Menge, 1976; Menge and Sutherland, 1976; Suchanek, 1981, 1986). On exposed headlands most predators are ineffective in controlling competitively dominant species (such as mussels), because of increased risks associated with foraging (Menge, 1978a, b). Spatial and temporal escapes from predators and herbivores provide considerable structural diversity on rocky shorelines (Menge, 1976, 1978a, b; Lubchenco, 1978, 1980; Lubchenco and Menge, 1978; Lubchenco and Cubit, 1980; Menge and Lubchenco, 1981; Hughes, 1986). Such escapes are easily attained since predation and herbivore pressure are highly variable, and there are relatively few types of consumers (predators, herbivores and omnivores) in the system (Menge and Lubchenco, 1981; Menge, 1982). These escapes include both coexistence and non-coexistence refugia. For example, *M. edulis* can escape control due to spatially heterogeneous predation, allowing mixed populations of mussels and Irish moss with intermediate wave exposure. Similar escapes of plants from grazers are particularly significant in determining community structure (Menge, 1975). Thus, preferred feeding of the dominant herbivore *Littorina littorea* on several ephemeral algae allows persistence of *Chondrus crispus* and *Fucus* spp., as well as providing an antifouling function for the latter perennial seaweeds (Lubchenco, 1978, 1986). Seaweeds exhibit varying patterns of competitive interactions within the eulittoral zone. Ephemeral algae only compete successfully when herbivores are ineffective and wave action is benign, while perennial seaweeds with coexistence escapes from herbivores frequently compete with one another even in the presence of herbivores (Lubchenco, 1986).

In the absence of established predator–prey interactions or major competitors, introduced species may show explosive expansion. For example, the red alga *Dumontia contorta* has colonized the entire coastline of New England since 1910 (Bird et al., 1983). Similar patterns of introduction and rapid expansion have been reported for *Codium fragile* ssp. *tomentosoides* (Fralick and Mathieson, 1973; Carlton and Scanlon, 1985), *Bonnemaisonia hamifera* (Taylor, 1957; Dixon and Irvine, 1977), *Fucus serratus* (Dale, 1979), the common periwinkle *Littorina littorea* (Carlton et al., 1982), and several other species (Druehl, 1973; Carlton, 1979; Farnham, 1980).

Stress and disturbance (Dayton, 1975; Grime, 1979; Sousa, 1984) are often reduced within the lower eulittoral and sublittoral zones, causing intense competition for light and space between species occupying a deeper zone on the shore (Connell, 1961, 1972, 1975; Dayton, 1971; Paine, 1974; Dring, 1982; Kastendiek, 1982; P.D. Moore, 1982; P.G. Moore, 1983; Underwood and Denley, 1984; Denley and Dayton, 1985; Suchanek, 1986; Underwood, 1986; Peckol et al., 1988). For example, dense canopies of shallow kelp populations may impose a lower limit on *Fucus* spp. Competition with other seaweeds cannot account for the lower limits of kelps, since associated plants at greater depths are slow-growing, delicate forms, or custose and often calcified species (see pp. 128, 131). Similarly, the maximum depth of deep-water algae cannot be determined by competition with other seaweeds, since only bare rock and/or animals occur below them (Chapman, 1979). Dring (1982) emphasized that light requirements, or tolerance to low irradiance, is an ultimate determinant of the lower limits of most seaweeds, regardless of whether imposed by the physical environment or by competition with another species (Dayton, 1975; Menge and Sutherland, 1976; Cinelli et al., 1978; Schonbeck and Norton, 1980; Seip, 1980; Drew, 1983; Kastendiek, 1982). Other important abiotic factors structuring benthic sublittoral communities are temperature (Golikov and Scarlato, 1973), substratum characteristics (Thorson, 1946, 1957, 1966; Sanders, 1958, 1968; Davis, 1960; Scheltema, 1961; Holme and McIntyre, 1971; Rhoads and Young, 1971; Young and Rhoads, 1971; Meadows and Campbell, 1972; Gray, 1974; Harris and Mathieson, 1976; Hulbert et al., 1982; Hartnoll, 1983; Bowen et al., 1986), water movement (Sanders, 1960; Noble et al., 1976; Gulliksen, 1978; Hiscock, 1983; Sebens and Witman, 1986) and depth (Thorson, 1957; Sanders, 1968; Lie and Kelley, 1970; Golikov and Scarlato, 1973; Dayton et al., 1974; Bowen et al., 1986). In addition, predation (Paine, 1966, 1974, 1976a, b) and competition (Connell, 1961, 1972; Jackson, 1977) are key biological factors. Community complexity grows with increasing species diversity (Paine and Vadas, 1969; Sanders, 1969; Fager, 1971; Dayton et al., 1974).

In some cases the lower limits of sublittoral algae are controlled by herbivorous animals such as sea urchins (Himmelman, 1980; Vadas, 1985). Although grazing by limpets and littorinid snails within the littoral zone is important (Jones, 1948; Lodge, 1948; Burrows and Lodge, 1950; Southward, 1955; Underwood, 1979; Branch, 1981, 1986; Lubchenco and Gaines, 1981; Gaines and Lubchenco, 1982; Hawkins and Hartnoll, 1983), it usually affects species composition over the whole area, rather than vertical limits of a particular species. Recolonization after the *Torrey Canyon* oil spill in southeast England demonstrated that the upper limits of some sublittoral algae may be affected by grazing (Southward and Southward, 1978), as seaweeds (for example, *Laminaria digitata*) colonized areas up to 2 m higher after the death of limpets. Gradual recovery of limpet populations 5 to 10 years after the original disturbance was followed by re-establishment of the eulittoral/sublittoral boundary at its original height. Thus, the temporary absence of herbivores resulted in a rise of the upper limits of sublittoral algae to a new level imposed by abiotic stresses associated with emersion.

An understanding of recruitment patterns can help to interpret distribution and abundance of benthic organisms (Osman, 1977; Lewis, 1978; Suchanek, 1981, 1986; Birkeland, 1982; Yoshioka, 1982; Ebert, 1983; Chapman, 1984; Keser and Larson, 1984a, b; Watanabe, 1984; Caffey, 1985; Connell, 1985, 1986; Gaines and Roughgarden, 1985; Roughgarden et al., 1985; Bowman, 1986; Deysher and Dean, 1986; Norton, 1986; Wethey, 1986). Underwood and Denley (1984) suggested that variations in recruitment are important in interpreting the roles which competition and predation play in structuring intertidal communities. In describing recruitment patterns, several investigators have observed patchiness and differential persistence (see Barnes, 1956; Hruby and Norton, 1979; Hoffman and Ugarte, 1985; Zechman and Mathieson, 1985), both of which may be of critical importance. Dispersal ranges and settlement patterns of plant and animal propagules, including chemotactic settlement guides (cues) to settling larvae, are equally significant in establishing and maintaining zonation patterns (Doyle, 1974, 1975; Strathmann, 1974; Scheltema, 1977; Mackay and Doyle, 1978; Strathmann and Branscomb, 1979; Strathmann et al., 1981; Levinton, 1982).

Surface texture of rocky substrata can also regulate recruitment patterns of benthic communities (Harlin, 1974; Luther, 1976; Harlin and Lindbergh, 1977; Hartnoll, 1983; Denley and Dayton, 1985). For example, Harlin and Lindbergh (1977) found that *Corallina officinalis* reaches maximum abundance on artificial substrata with small particles of silica (0.1 to 0.5 mm), whereas *Ceramium* spp., *Chondrus crispus, Polysiphonia* spp., *Spongomorpha* spp. and *Ulva lactuca* show maximum biomass on coarser particles (0.5 to 2.0 mm). Thus, *C. crispus* shows significantly greater colonization and biomass on particles of 0.5 to 2.0 mm, while only stunted individuals occur on finer particles (0.1 to 0.5 mm) and very few on smooth surfaces.

Several investigators have described varying reproductive phenologies of individual seaweeds in different habitats (Dixon, 1965; Nienhuis, 1974; Norall et al., 1981; Mathieson, 1989), some of which may have adaptive significance (Sideman and Mathieson, 1983a, b). For example, Nienhuis (1974) found reduced sexual reproduction for some populations of *Rhizoclonium riparium* growing in stressful habitats (high vs. low shores), while Sideman and Mathieson (1983a, b) found differences in reproductive phenologies between upper and lower eulittoral populations of *Fucus distichus* subspecies, the former (ssp. *anceps*) having unimodal and the latter (ssp. *edentatus*) bimodal reproductive patterns. Among others, Suchanek (1981) has emphasized the adaptive significance of the length of initial reproductive maturation. Early reproductive maturation is often attributable to a high degree of habitat disturbance (Harper, 1977; Grime, 1979; Suchanek, 1981, 1986).

Upper eulittoral zone

The major factors structuring communities on emergent rocks within the upper eulittoral zone (exposed and sheltered) are those affecting settlement, growth and mortality of *Semibalanus balanoides*, including such density-dependent factors as wave shock and desiccation (Hatton, 1938; Menge, 1976). Predation by *Nucella lapillus* at sheltered sites and interspecific competition with *Mytilus edulis* play significant roles, but only near the lower distributional limits of *S. balanoides* (Menge, 1976; Menge and Sutherland, 1976; Menge and Lubchenco, 1981). On protected upper shorelines dense canopies of fucoid algae provide protection for increased foraging by *N. lapillus* and other predators, as well as inhibiting the recruitment of cyprid stages of barnacles (Lubchenco, 1986). Substratum heterogeneity (e.g. angle of inclination) causes an uplifting of *M. edulis* and *N. lapillus*, producing a patchy mosaic of barnacles, mussels and open space (Grant, 1977).

On some exposed upper shorelines the dwarf fucoid *Fucus distichus* ssp. *anceps* grows abundantly on *Semibalanus balanoides* (Sideman and Mathieson, 1983a), causing their mutual dislodgement due to wave disturbance and denudation, particularly during winter (Burrows and Lodge, 1950; Barnes and Topinka, 1969). By contrast on adjacent and less exposed habitats, *Fucus spiralis* replaces *F. distichus* ssp. *anceps* (Niemeck and Mathieson, 1976). In discussing other competitive interactions between *F. distichus* ssp. *anceps* and *S. balanoides*, Sideman and Mathieson (1983a) state that the alga does not have to compete as strongly for light on the upper shore as its relative *F. distichus* ssp. *edentatus* does within the lower eulittoral zone. That is, *S. balanoides* is the tallest competitor in a habitat dominated by dwarf forms of many species (Norton et al., 1981).

Within sheltered upper-shore tide pools, *Littorina littorea* controls the abundance and type of algae (Lubchenco, 1978), with the highest seaweed diversity occurring at intermediate snail densities (Connell, 1979). This unimodal relationship between algal species diversity and herbivore density occurs because the snail's preferred food (*Enteromorpha* spp.) is competitively dominant in tide pools. Moderate grazing allows inferior algal species (e.g. *Chondrus crispus*) to persist, whereas intense grazing eliminates most individuals and species. Lubchenco (1986) suggested that *Enteromorpha*-dominated pools harbor many crabs (*Carcinus maenas*) which prey on young *L. littorina*, while *Chondrus*-dominated pools lack this crab, probably because of intense predation by gulls (*Lárus argentatus* and *L. marinus*) (cf. Marsh, 1986).

On some semi-protected upper shorelines (e.g. in Maine), growth and colonization by the dominant fucoid *Fucus vesiculosus* is retarded by *Littorina littorea* (Keser and Larson, 1984a). However, grazing within such habitats is modest as compared with sheltered mid and lower shorelines (Lubchenco, 1980). Micrograzers such as the spring-tail *Anurida maritima* (Table 7.2), and various dipteran larvae (Lewis, 1964; Robles and Cubit, 1981; Robles, 1982) occur in high-tide pools and crevices. As noted by Vadas (1985), these and other micrograzers may play an important role in structuring benthic communities by acting as a larval filter, inhibiting recruitment of seaweed propagules.

Mid-eulittoral zone

On exposed mid-shore locations, interspecific competition between *Mytilus edulis* and *Semibalanus balanoides* is the dominant biological interaction (J.L. Menge, 1975; B.A. Menge, 1976; Steele,

1983). Although *M. edulis* is the competitive dominant of the two, it is a fugitive species having a distinct seasonal cycle of reproduction (Seed, 1975; Suchanek, 1981, 1986). Replacement of *S. balanoides* by *M. edulis* takes longer on vertical than on shelving surfaces (18–24, vs. 5 months), causing a patchy mosaic. Typically, a larger number of small mussels occur on exposed than on sheltered sites, as some predation by *Nucella lapillus* occurs on the former sites, as well as dislodgement of whole clusters of large populations by waves (Dayton, 1971, 1973; Harger and Landenberger, 1971; Levin and Paine, 1974; Paine and Levin, 1981; Steele, 1983; Suchanek, 1986). On many northern shores (e.g. Newfoundland), *M. edulis* is often swept away by ice long before it reaches its size potential.

Predation by sea-birds on mussel populations has been known for many years (Forbush and May, 1955; Feare and Summers, 1986), but its ecological significance has only recently been recognized. Menge (1976) attributed the disappearance of *Mytilus edulis* populations to winter storms, which can certainly happen. However, sea ducks (old squaw, common eider and black scoter) are major predators of mussels in the Gulf of Maine during winter months (Forbush and May, 1955; Stott, 1972; Stott and Olson, 1973; Harris, unpubl. observ.) as well as in Newfoundland (Goudie and Ankney, 1986; Figs. 7.10 and 7.11). Herring gulls (*Lárus argentatus*) are also important intertidal predators (Menge, 1976). During winter months, most fecal droppings of gulls within the supralittoral zone at the Isles of Shoals (Gulf of Maine) contain large numbers of mussel shells (Harris, unpubl. observ.). Marsh (1986) describes

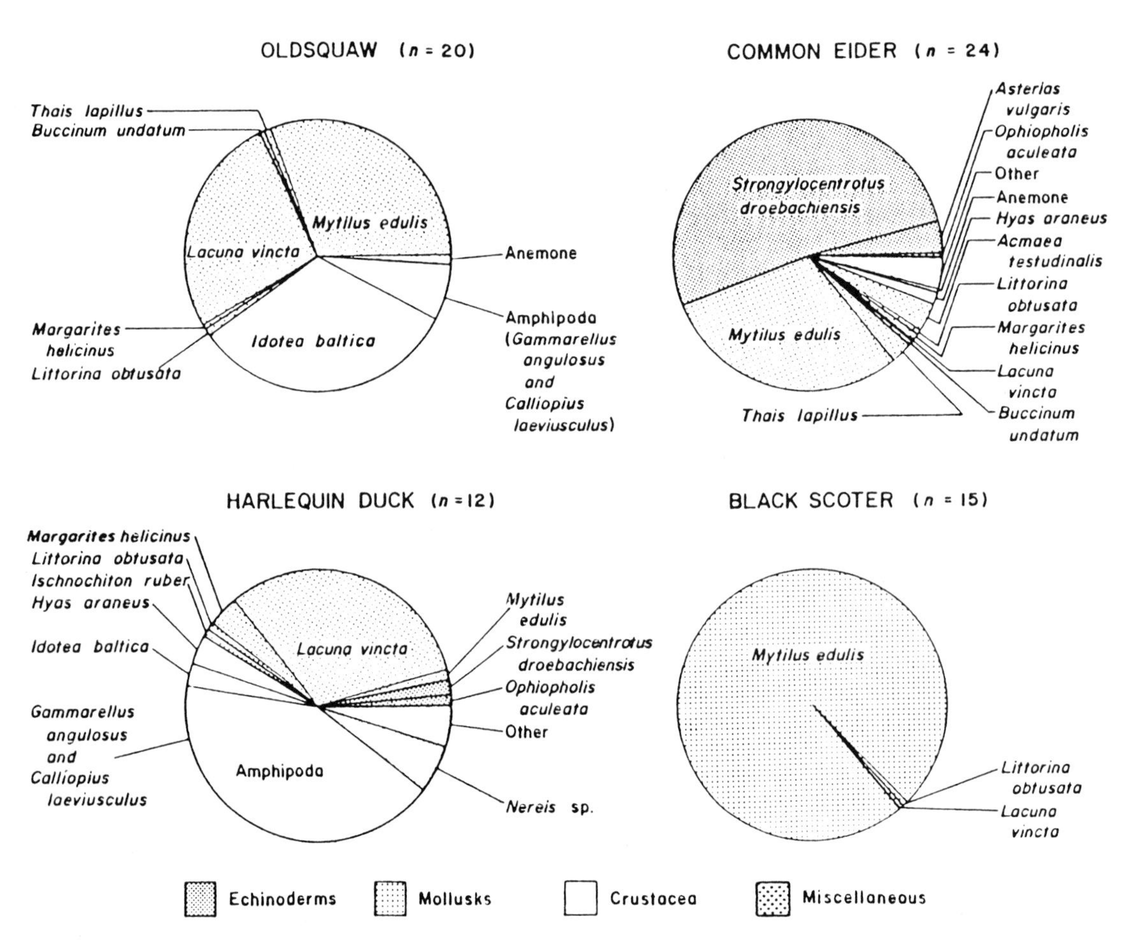

Fig. 7.11. Distributions of prey items, by mass, in diets of species of sea ducks collected at Cape St. Mary's, Newfoundland, based on examination of gullet contents. See previous comments regarding *Nucella* (= *Thais*) *lapillus*. (From Goudie and Ankney, 1986.)

gull predation on *Mytilus* spp. along the Oregon coast.

At protected mid-shore sites, predation and herbivory are major factors affecting space utilization (Menge and Sutherland, 1976; Menge, 1978a, b; Lubchenco, 1983, 1986; Steele, 1983). By clearing space and alleviating interspecific competition, *Nucella lapillus* and other predators of *Mytilus edulis* (Newcombe, 1935; Edwards et al., 1982; Suchanek, 1986) and *Semibalanus balanoides* allow the persistence of *Fucus vesiculosus* and *Ascophyllum nodosum* on semi-protected and protected sites, respectively (Keser and Larson, 1984a, b). Both fucoids are competitively inferior to many ephemeral algae, as well as to *M. edulis* and *S. balanoides* (Lubchenco, 1986). Because of irregular predation, *M. edulis* and *S. balanoides* occasionally escape elimination, causing patches and unpredictable patterns of species richness (Menge and Lubchenco, 1981; Steele, 1983). Several interactions between *M. edulis*, *S. balanoides* and *Fucus* spp. occur (Menge, 1976). For example, initial settlement and survival of *S. balanoides* is often reduced under an extensive *Fucus* canopy due to whiplash effects of the plants' fronds, while abrasive effects of *M. edulis* around *Fucus* spp. cause plant mortality (Menge, 1976, 1978a; Grant, 1977; Denley and Dayton, 1985). The periwinkle *Littorina littorea*, which migrates upward in spring and downward in fall (Williams and Ellis, 1975), affects algal recruitment, development, and zonation over a large portion of the shore (Lubchenco, 1980; Vadas, 1985; Petraitis, 1979, 1987). For example, *L. littorea*, as well as its congener *L. obtusata*, grazes heavily and preferentially on early successional algae like *Enteromorpha* spp., *Porphyra* spp. and *Ulva lactuca*, speeding up secondary succession to *Fucus vesiculosus* and subsequently *Ascophyllum nodosum* (Lubchenco, 1978, 1983, 1986). Germlings of *F. vesiculosus* may be so heavily grazed on smooth rock surfaces that their growth is prevented; more often they obtain refuge in small crevices or pits, among barnacles, or by reaching a size escape (5 cm long). A combination of grazing by *L. littorea* and competition with other algae excludes *F. vesiculosus* from mid-shore tide pools (Lubchenco, 1982).

Variability of life history and shell morphological features influence the functional role of predatory and herbivorous animals like *Nucella lapillus* and *Littorina saxatilis* (Kitching, 1986). On sheltered shores *N. lapillus* grows faster, produces fewer, larger offspring, attains a larger size at maturity, and experiences lower adult mortality than on exposed shores (Etter, 1986). Thick shells of *L. saxatilis* and other gastropods are associated with the presence of mobile boulders and shore crabs (Johannesson, 1986), as well as other predators (Hawkins and Hartnoll, 1983).

On semi-exposed sites like Jaffrey Point, New Hampshire, *Fucus vesiculosus* and *Semibalanus balanoides* show pronounced temporal variations in recruitment and persistence (Hardwick-Witman and Mathieson, 1983). Settlement of *S. balanoides* is greatest during spring (Barnes, 1956). *Fucus vesiculosus* exhibits a period of low settlement during winter followed by high recruitment in spring, which corresponds to its *in situ* reproductive periodicity (Mathieson et al., 1976; Mathieson, 1989).

On sheltered mid-shore locations *Fucus vesiculosus* recruits more rapidly and in greater numbers than *Ascophyllum nodosum* (Knight and Parke, 1950; Printz, 1956, 1959; Baardseth, 1970; Sundene, 1973; Keser and Larson, 1984a, b). Even so, *A. nodosum* eventually supplants *F. vesiculosus* and represents the "climax" perennial fucoid vegetation in sheltered sites (Lewis, 1978), because of its survivorship, growth characteristics and longevity (Keser and Larson, 1984b). For example, *A. nodosum* grows slower than *F. vesiculosus*, but its adult plants live longer (12 to 15 vs. 2 to 4 years).

Contrasting patterns of colonization and growth may influence community development. *Ascophyllum nodosum* settles earlier and more densely in the lower than in the mid-eulittoral zone; however, it recruits most successfully near its uppermost zone where few *Littorina littorea* occur (Busse, 1983). Maximum growth has been variously described at elevations ranging from the mid to lower eulittoral zone (Busse, 1983; Keser and Larson, 1984a, b; Peckol and Harlin, 1985). In contrast, *Fucus vesiculosus* colonizes earliest and grows fastest at the lower shoreline. Once a canopy of *A. nodosum* is established few understorey algae occur, including juvenile fucoids. Lubchenco (1986) suggested that low light, the physical barrier presented by a thick canopy, whiplash or other factors may exclude many seaweeds (Dayton, 1975; Sousa, 1979; Velimirov and Griffiths, 1979; Miller, 1984; Cousens, 1985; Denley and Dayton, 1985). Variable levels of grazing by *Littorina littorea* on understorey species have been described, ranging from minimal to substantial (Miller, 1984; Lubchenco, 1986). Overall, competitive interactions between adult fucoid algae (including *F. vesiculosus* on semi-protected sites) and potential colonists are probably the dominant biological interactions associated with these perennial macrophytes.

Physical disturbance due to sand scouring or burial can influence the occurrence of mid-shore fucoid algal populations (Chapman, 1943; Daly and Mathieson, 1977; Shaughnessy, 1983, 1986a, b). Typically, fucoid algae are absent on sand-covered rocks and are replaced by a *Cladophora/Enteromorpha/Sphacelaria* association (Chapman, 1943). Marginal populations of *Fucus vesiculosus* grow on sand-impacted sites like Bound Rock, New Hampshire, where irregular fluctuations in sand cover occur, but *Ascophyllum nodosum* is conspicuously absent (Lewis, 1964; Daly and Mathieson, 1977). Varying patterns of recruitment during sand disturbance (Keser and Larson, 1984a, b), vegetative regeneration from adult plants, and burial tolerance may all be significant in allowing the occurrence of *F. vesiculosus* but not *A. nodosum* in sand-disturbed habitats (Shaughnessy, 1983, 1986a, b).

Lower eulittoral and sublittoral zones (nearshore)

On exposed headlands most predators are excluded from the lower eulittoral zone due to strong wave action (Steele, 1983), allowing *Mytilus edulis* to achieve dominance over *Chondrus crispus* and *Semibalanus balanoides* (J.L. Menge, 1975; B.A. Menge, 1976, 1983; Lubchenco and Menge, 1978; Lubchenco, 1980). Conversely, on sheltered sites, heavy predation of mussels by *Asterias* spp. (Hancock, 1955, 1958; Menge, 1979; Hulbert, 1980), *Cancer* spp. (Kitching et al., 1959; Ebling et al., 1964; Muntz et al., 1965; Seed, 1969; Walne and Dean, 1972; Hulbert et al., 1976; Harris and Irons, 1982), *Carcinus maenas* (Elner, 1978; Elner and Hughes, 1978; Harris and Irons, 1982), *Homarus americanus* (Squires, 1970), *Tautogolabrus adspersus* (Chao, 1973; Shumway and Stickney, 1975; Harris and Irons, 1982), and sea ducks occurs (Graham, 1975; MacLeod, 1975; Incze and Lutz, 1980). Experimental exclusion studies on sheltered habitats confirm that persistence of *C. crispus* is a by-product of predation, as Irish moss may be replaced by mussels and barnacles under reduced predation (Lubchenco and Menge, 1978; Menge, 1983). In evaluating the relative contribution of predators, Menge (1983) suggested that, if one species becomes scarce, others may increase their effects and reduce variation in the total predation intensity exerted by the guild (Peterson, 1979). Increased diversity of the types of foraging characteristics in such guilds is an important component of enhanced predation intensity along gradients of decreased environmental rigor. In addition, substratum characteristics are important in determining habitat selection and potential predatory activity of crabs such as *Cancer borealis* (Richards, 1986).

Spatial heterogeneity and desiccation influence herbivory patterns on *Mytilus edulis*-dominated lower shorelines. Limpets, as well as littorinids, are often confined at low tides to moist cracks and mussel beds, causing limited foraging ranges and grazing halos on ephemeral algae (Seed, 1969). Thus, refuges for algae are created when herbivore access is restricted, as with microtopographic barriers (Petraitis, 1979; 1987).

Although *Fucus* spp. and fugitive ephemeral algae rapidly colonize open space within sheltered lower shorelines (Keser and Larson, 1984a; Sideman and Mathieson, 1983b), *Chondrus crispus* and/or *Mastocarpus stellatus* gradually dominate (Burns and Mathieson, 1972b; Lubchenco, 1978, 1980; Darley, 1982; Pringle and Mathieson , 1987). Thus, although *Fucus* spp. have higher rates of recruitment and growth than *C. crispus* and *M. stellatus*, they are inferior competitors, *Fucus* spp. being r-selected and both red algae K-selected (Pianka, 1970). Lubchenco (1980) demonstrated that, when fronds of *C. crispus* were experimentally removed, *Fucus* spp. germlings did not become established on residual holdfasts. The outcome of competitive interactions between *C. crispus*, *Fucus* spp. and *M. stellatus* is frequently modified by predation and disturbance (Lubchenco, 1978, 1980; Levinton, 1982). For example, *F. vesiculosus* is usually excluded by competition with *C. crispus* (see above), whereas Irish moss does not extend upwards due to its sensitivity to desiccation (Pringle and Mathieson, 1987). On the other hand, differences in mature sizes and growth rates cause varying patterns of competition and colonization. Thus, most plants of *F. distichus* ssp. *edentatus* become reproductive and die after two years, while *C. crispus* eventually dominates due to its longevity (Pringle and Mathieson, 1987; Sideman and Mathieson, 1983a, b).

During its recruitment, *Chondrus crispus* exhibits a high frequency of sporeling (holdfast) coalescence (Ring, 1970; Tveter and Mathieson, 1976; Tveter-Gallagher and Mathieson, 1980), resulting from discharge of spores in a sticky, mucilaginous plug (Chen and Taylor, 1976). Coalescence is associated with an earlier initiation and growth of fronds, which has obvious ecological significance if rocky substrata and other resources are limiting (Jones, 1956a). Once established, the perennial holdfast of Irish moss is a persistent (Connell, 1986) and stable source of new fronds (Taylor et al., 1975, 1981), which can endure substantial damage by storms, grazing, etc. and still regenerate successfully. Fronds of Irish moss exhibit a similar pattern of wound healing and regeneration from any part of their surface (Chen and McLachlan, 1972). Shoots of different ages are present on a single discoid holdfast, and, if left undamaged, the holdfast continues to produce shoot primordia for a number of years. Typically, fronds live 2 to 6 years, while holdfasts persist considerably longer (Taylor and Chen, 1973; Taylor et al., 1975, 1981). *Mastocarpus stellatus* exhibits an analogous pattern of sporeling coalescence (Chen et al., 1974; Tveter and Mathieson, 1976), enhanced growth, and retention of space by its perennial holdfast.

Substratum heterogeneity (i.e. angle of inclination), in combination with varying degrees of disturbance, results in a patchy mosaic of *Chondrus crispus* and *Mastocarpus stellatus* in the lower shoreline (p. 122). Differential exposure to wave action (Lewis, 1964), and greater tolerances to desiccation and freezing by *M. stellatus* may be operational in establishing its dominance on vertical surfaces, while most of the *C. crispus* is found on shelving and horizontal surfaces (Ring, 1970; Burns and Mathieson, 1972a, b; Mathieson and Burns, 1975; Green and Cheney, 1981; Green, 1983; Dudgeon, 1988; Dudgeon et al., 1989). As wave action determines the density of the periwinkle *Littorina littorea*, it also influences competitive interactions among seaweeds (Lubchenco and Menge, 1978; Lubchenco, 1980). Thus, at exposed sites with reduced numbers of snails, fugitive ephemeral algae flourish and out-compete *C. crispus*, particularly within tide pools (Lubchenco, 1978), while Irish moss dominates when high snail densities reduce green algal populations. Experimental removal of *L. littorea* from *Chondrus*-dominated pools allows takeover by *Enteromorpha*/*Ulva* (Lubchenco, 1978). Studies by Geiselman (1980) and Geiselman and McConnell (1981) showed restricted or inhibited feeding on adult Irish moss by *L. littorea*; however, Cheney (1982) found that periwinkles readily eat Irish moss sporelings less than a week old.

Physical disturbance due to ice and sand-scouring influences the dominant populations of *Fucus* and *Mastocarpus* on the lower shoreline. Mild short-term or highly localized ice action has little effect, except on fragile epiphytes, and communities quickly return to their original state (Hooper, 1981). Severe scouring in the Gulf of St. Lawrence and in Newfoundland (Taylor, 1975; Hooper, 1981; Bird et al., 1983; Lobban and Hanic, 1984; McLachlan et al., 1987) removes most perennial seaweeds and associated herbivores (e.g. *Littorina littorea, L. obtusata*, and *Acmaea testudinalis*), allowing a greater concentration of biomass among fugitive species (O'Clair et al., 1979) and vertical range extensions (Hooper, 1981). Thus, severe ice scouring suppresses temporal succession of dominant perennial taxa, restricts their presence to crevices or the sublittoral zone, and may alter zonation patterns (South, 1983). Loss of canopy protection indirectly affects viability of surviving subcanopy species by increasing exposure to desiccation and freezing (Hooper, 1981).

Sand-abraded rocky surfaces exhibit analogous groupings, including truncated distributional limits, erratic patterns of species richness, and reduced numbers of perennial taxa (Daly and Mathieson, 1977; Devinny and Volse, 1978; McLachlan et al., 1987). Typically, sand-impacted rock outcrops are dominated by opportunistic annuals (*Enteromorpha* spp., *Ulva lactuca*, etc.) and perennial psammophytic or stress-tolerant species (such as *Ahnfeltia plicata* and *Sphacelaria radicans*). The latter have a variety of adaptations to sandy habitats, including tough and wiry morphologies, thalli consisting of turfs, crusts, and perennial holdfasts that produce ephemeral blades, have extensive regeneration of upright shoots from their bases and incomplete life histories (Mathieson, 1968, 1982; Markham, 1973; Hay, 1981; Littler et al., 1983; Taylor and Hay, 1984). Although *Fucus vesiculosus* exhibits some of these features (for example, extensive frond regeneration), it is not a true psammophytic species (Shaughnessy, 1983, 1986a, b).

Disturbance as a result of substratum mobility is important in determining species diversity and community development within the lower eulittoral/sublittoral zone (Hartnoll, 1983; Davis, 1984; Davis and Wilce, 1984). Small, frequently disturbed cobbles are dominated by early-successional species, and large immobile rocks by late-successional perennials (Mann, 1972a; Sears and Wilce, 1975; Davis and Wilce, 1984; Pringle and Semple, 1984; Scheibling, 1986). Intermediate-sized cobbles have approximately equal numbers of annuals and perennials and patterns of maximum species diversity (Connell, 1979), caused by a nonequilibrium state (Sousa, 1979, 1984).

In New England, the structural dominance of *Chondrus crispus* on sheltered lower shorelines differs from that in the Bay of Fundy and in Europe, where mixed stands of Irish moss and *Mastocarpus stellatus* occur (Lubchenco, 1978; Lubchenco and Menge, 1978). Urchins (*Strongylocentrotus droebachiensis*) occur primarily within the sublittoral zone in New England, while in the Bay of Fundy they are conspicuous in both the lower eulittoral and sublittoral zones. Lubchenco (1978) speculated that urchin abundance in the former zone within the Bay of Fundy may be due to reduced gull predation, as well as diminished desiccation and wave action (Appendices I and II). When urchins are exceedingly abundant, algal species diversity and abundance are very low (Vadas, 1968), a pattern comparable to that described for *Littorina littorea*. However, the effects of these two herbivores differ, as urchins remove adult *Chondrus crispus*, while periwinkles only graze juvenile Irish moss (Cheney, 1982) or its epiphytes.

Competitive interactions (Seip, 1980; Dring, 1982; Denley and Dayton, 1985) between sublittoral seaweeds influence their distribution and productivity. For example, crustose coralline algae occupy considerable primary space (p. 125) and show strong competitive influences upon one another. By evaluating pair-wise interactions (i.e. overgrowth) Garwood et al. (1985) demonstrated that *Clathromorphum circumscriptum* dominates shallow rocks, while *Lithothamnion glaciale* is most conspicuous in deeper habitats (Adey, 1965, 1966a, b). Dense populations of shallow kelps cause intense competition for light and space, limiting subcanopy species (Dayton, 1975; Pearse and Hines, 1979; Duggins, 1980; Harkin, 1981; Moore, 1983; Keats et al., 1985). *Laminaria longicruris* exhibits a variety of competitive, opportunistic, and stress-tolerant traits (Grime, 1979), including rapid growth and recruitment (Gerard and Mann, 1979; Chapman and Gagné, 1980; Smith, 1985, 1986), high reproductive output (Chapman, 1984), ecotypic variations with respect to light optima and seasonal growth patterns (Gagné et al., 1982; Gerard, 1986), biomechanical properties (Babb, 1985), and high phenolic content (Johnson, 1984; Johnson and Mann, 1986b). By these multiple strategies *L. longicruris* is able to maintain its dominance in the face of disturbance (for example, by herbivores or storms) and varying levels of stress (Johnson and Mann, 1986b, c). The shorter-canopied and slower-growing *L. digitata* is more competitive and does better in hydrographically active situations (Smith, 1986; cf. also Dayton, 1985; Ojeda and Santelices, 1984). As noted previously, the upper limits of *Agarum cribrosum* and other seaweeds are often depressed where *Laminaria* spp. are abundant (Mann, 1972a, 1973; Chapman, 1973).

Weeding experiments conducted by Marshall et al. (1949) demonstrated that, when *Fucus* spp. were removed, *Chondrus crispus* increased substantially in size. In the Canadian Maritime Provinces, after extensive harvesting of Irish moss *Corallina officinalis* may invade *C. crispus* beds (MacFarlane, 1968). By contrast, unharvested populations of *C. crispus* along the coast of Maine may be overgrown by *Chaetomorpha* spp., prohibiting commercial exploitation (Foster, 1955, 1956). *Phyllophora pseudoceranoides* and *P. truncata* often compete with Irish moss below −5 m, or in the absence of *Furcellaria lumbricalis* (Prince, 1971; Mathieson and Burns, 1975; Pringle and Sharp, 1980; Holmsgaard et al., 1981; Pringle and Mathieson, 1987; see also Fig. 7.12). On sedimented rocky surfaces *Phyllophora* spp. usually replace *C. crispus*, presumably because of greater ability to withstand

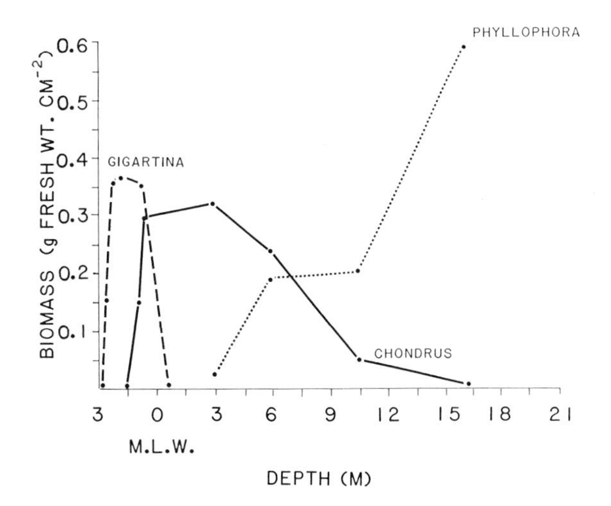

Fig. 7.12. Changes in biomass of *Chondrus crispus, Mastocarpus stellatus* (= *Gigartina stellata*) and *Phyllophora* spp. with position on the shore at Jaffrey Point, New Hampshire. (Modified from Mathieson and Burns, 1975.)

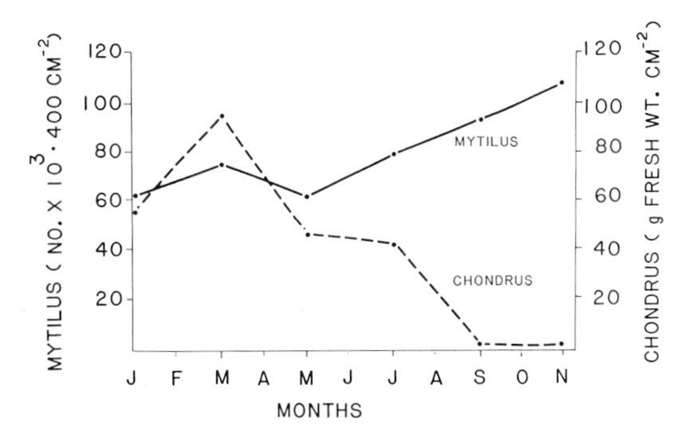

Fig. 7.13. Changes in biomass of *Chondrus crispus* and *Mytilus edulis* at Dover Point, New Hampshire. (Modified from Mathieson et al., 1983.)

holdfast burial (Newroth, 1970; Mathieson and Prince, 1973). In clear waters *Phyllophora* spp. are generally restricted to deeper habitats, whereas in more turbid areas they occur in abundance at much shallower depths, allowing variable competitive interactions with *C. crispus* and *F. lumbricalis* (Holmsgaard et al., 1981). Initial regrowth of *C. crispus* on denuded quadrats is reduced by algal and animal competition, particularly during spring (MacFarlane, 1952; Ring, 1970; Prince, 1971; Mathieson and Burns, 1975). Irish moss eventually dominates after re-establishment of its perennial holdfast and dense cluster of fronds (Pringle and Mathieson, 1987). Rapid spread of the introduced species *Fucus serratus* throughout the southern Gulf of St. Lawrence is probably due to the absence of *F. distichus* ssp. *edentatus* (Bird et al., 1983), as well as the inability of the two *Fucus* species to coexist (Dale, 1979).

Chondrus crispus and *Mytilus edulis* exhibit strong competition, particularly within the shallow sublittoral/lower eulittoral zone (MacFarlane, 1956, 1966; Lilly, 1968; Reynolds, 1971; Mathieson et al., 1983; see also pp. 122, 142). Gross fluctuations of mussel populations may occur depending upon abundance and viability of spat (Hardwick-Witman and Mathieson, 1983; Suchanek, 1986). Reynolds (1971) quantified the demise of Irish moss populations when settled upon by thousands of spat. As mussels grew, sand and silt were deposited amongst their byssal threads, reducing light and space, and the plants ultimately died (Fig. 7.13).

According to Keats et al. (1984a, 1985), the dominance of *Alaria esculenta* within the shallow sublittoral zone in eastern Newfoundland is mediated by an interaction between the periodicity and severity of ice-scouring episodes, together with destruction by *Strongylocentrotus droebachiensis*. Typically, urchin biomass within this zone is low and grazing has no detectable effects; however, see Himmelman (1980). In the absence of ice-scour, *A. esculenta* grows rapidly, forms a dense shade-producing canopy (Sundene, 1962) which inhibits understorey species and reduces species diversity (Fig. 7.14). Spring ice-scour, which is temporally and spatially variable, removes much of the *A. esculenta*, opening well-lighted rocky substrata for colonization. Scoured patches are colonized by ephemeral annuals (for example, *Chordaria flagelliformis* and *Dictyosiphon foeniculaceus*), resulting in increased species diversity. Total biomass returns to pre-scour levels approximately two months after ice-scouring, but diversity remains high due to coexistence of several annual seaweeds and unscoured patches of *A. esculenta*. A return to domination by *A. esculenta* takes less than a year. On the south coast of Newfoundland, where ice-scouring is rare, such extreme domination does not occur; rather, a rich and diverse community is present, including *A. esculenta* (Hooper et al., 1980; Hooper, 1981; South, 1983). In areas where ice-scouring does not occur, other types of disturbance (such as grazing and wave action) must be more important (Himmelman, 1980). Nutrient

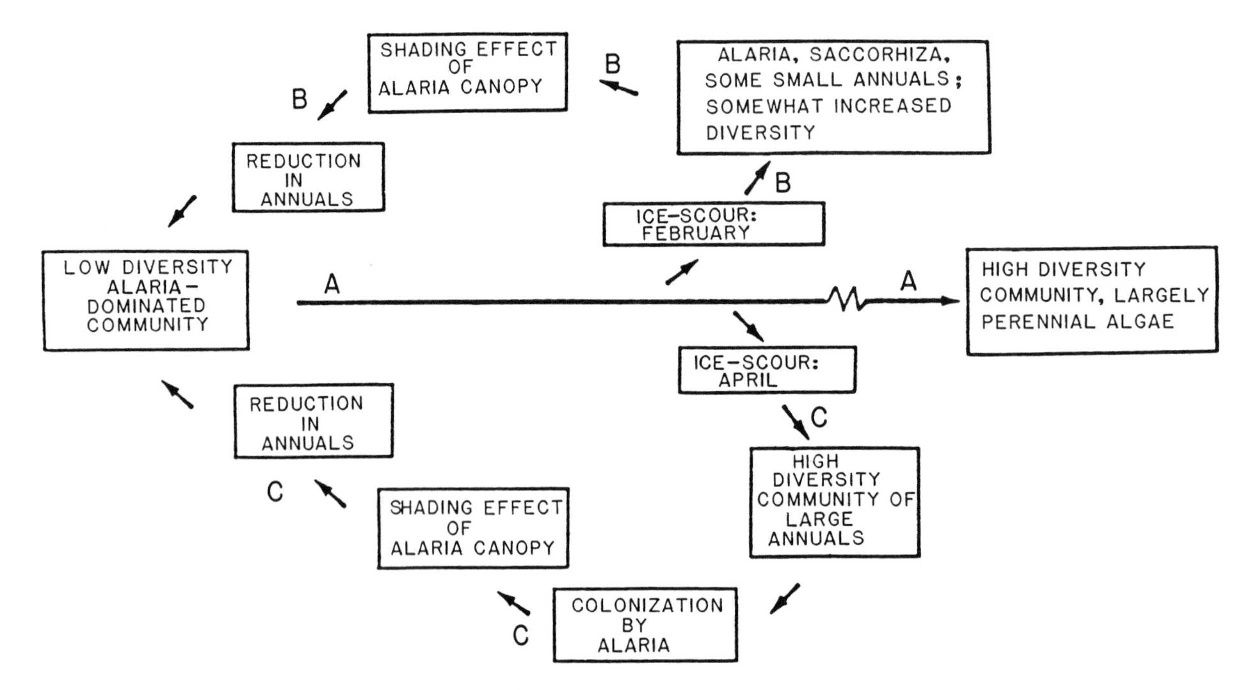

Fig. 7.14. Conceptual representation of the effect of ice-scour on an open coastal community in eastern Newfoundland. Pathway "A" is interrupted by periodic ice-scour and "B" or "C" predominates depending on the time of scouring. (Modified from Keats et al., 1985.)

kinetics of ephemeral species may determine primary succession of denuded surfaces. For example, *C. flagelliformis* colonizes patches scoured in early spring (April), taking advantage of short-term pulses of dissolved nutrients, and scavenging dissolved nitrates at low concentrations (Probyn and Chapman, 1982, 1983).

Free-living populations of crustose coralline algae (rhodoliths) are formed when ice removes them from their substrata (Himmelman, 1980; Hooper, 1980, 1981; Norton and Mathieson, 1983). Within the shallow sublittoral zone of Newfoundland many crustose coralline algae (*Clathromorphum circumscriptum, C. compactum, Lithothamnion glaciale* and *Phymatolithon laevigatum*) occur as free-living balls, which are formed by ice damage which fractures *in situ* plants, increasing numbers of individuals, and destroying their competitors.

Extensive populations of *Strongylocentrotus droebachiensis* have recently developed within the shallow sublittoral zone in the northwest Atlantic, causing "barren grounds" (*sensu* Pearse et al., 1970) of low productivity with only residual encrusting coralline algae, diatoms, and scattered foliose red species (Miller et al., 1971; Breen, 1974; Lang and Mann, 1976; Mann, 1977; Himmelman, 1980; Pringle et al., 1980, 1982; Wharton, 1980; Steneck, 1985, 1986; Pringle, 1986; Vadas et al., 1986). Formation of these barrens by urchin overgrazing has been documented throughout most of eastern Nova Scotia (Breen and Mann, 1976a, b; Stasko et al., 1980), Labrador and Newfoundland (Himmelman, 1980; Hooper, 1980), Prince Edward Island, the Bay of Fundy and the Gulf of Maine (Pringle et al., 1982; South, 1983; Witman, 1984). In some areas (northern New England and Newfoundland, for example), urchin populations have dominated for 15 to 20 years (Hooper, 1980; Pringle et al., 1980; Himmelman et al., 1983a; Keats, Steele and South, 1984; Harris, unpubl. observ.), while in others (for example, Atlantic coast of Nova Scotia and New Brunswick) recolonization of seaweeds has occurred after the death of urchins (Arnold, 1976; Miller and Colodey, 1983; Moore and Miller, 1983; Scheibling, 1984, 1986; Scheibling and Stephenson, 1984; Scheibling and Raymond, 1986). With reduced grazing pressure, ephemeral algae regrow quickly, followed by kelps. *Acmaea testudinalis, Margarites helicinus* and *Mytilus edulis* show a dramatic increase within a year (Breen and Mann,

1976a, b; Chapman, 1981; Wharton and Mann, 1981; Keats et al., 1982a; Himmelman et al., 1983a, b; Moore and Miller, 1983; Johnson, 1984; Johnson and Mann, 1986a; Novaczek and McLachlan, 1986; Pringle, 1986). Subsequently, limpet and mussel populations decrease dramatically, due to increased predation and competition with foliose algae. For example, *Laminaria* spp. overgrow and dislodge *Modiolus modiolus* due to drag during storms (Witman, 1984; Witman and Suchanek, 1984). A similar mechanism occurs with *Acmaea testudinalis* in the absence of urchins; limpets are also a characteristic component of urchin-dominated communities (Simenstad et al., 1978). *Chondrus crispus* takes approximately two years to grow from a spore to a reproductive plant (Pringle et al., 1980, 1982; Pringle and Mathieson, 1987), while other deep-water perennials like *Phyllophora* spp. take even longer, perhaps because of limited competitive abilities during recolonization and reduced ambient light (Holmsgaard et al., 1981; Novaczek and McLachlan, 1986).

Initial theories (Mann and Breen, 1972; Breen, 1974, 1980; Breen and Mann, 1976a, b; Mann, 1977; Wharton, 1980) postulated that overfishing of a "keystone" predator (Paine, 1966, 1969), the lobster, released urchins from predator control, allowing overgrazing of kelps. Subsequent studies have not substantiated the fact that urchins are preferred by lobsters (Elner and Campbell, 1987), nor that they are readily included in their natural diets (Dow, 1949; Squires, 1970; Himmelman and Steele, 1971; Ennis, 1973; McLeese, 1973; Reddin, 1973; Evans and Mann, 1977; Hirtle and Mann, 1978; Elner, 1980; Scarratt, 1980; Carter and Steele, 1982; Vadas et al., 1986). The initial lobster predation model was revised to incorporate a suite of predators, including fish, crabs and ducks (Mann, 1977, 1982; Elner, 1980; Pringle et al., 1980, 1982; Bernstein et al., 1981, 1983; Wharton and Mann, 1981; Bernstein and Mann, 1982). Pringle et al. (1982) suggested that commercial fishing of several fin-fish, including wolffish (*Anarhichas lúpus*), cod (*Gadus morhua*), cunner (*Tautogolabrus adspersus*), haddock (*Melanogrammus aeglefinus*), ocean pout (*Macrozoarces americanus*), sea raven (*Hemitripterus americanus*), and short horn sculpin (*Myoxocephalus scorpius*), reduced predation to a level that permitted urchin densities to explode. In a further characterization of urchin mortality, Hooper (1981) emphasized that predation can occur at all stages of urchin growth. Larvae are eaten in large numbers by the scyphomedusa *Cyanea capillata*, and also by anemones (*Metridium senile*), sponges, tunicates and larval fish. If plankton predation is high, low recruitment may occur. Small post-recruitment juveniles are still susceptible to many predators, including anemones (*Tealia crassicornis*), flounders (*Hippoglossoides platessoides*), gunnels (*Pholis gunnellus*), shannies (*Leptoclinus maculatus, Stichaeus punctatus, Ulvaria subbifurcata*), as well as juvenile crabs (*Cancer* and *Hyas* spp.), lobsters (*Homarus americanus*), sea stars (*Leptasterias tenera, Ophiopholis aculeata, Solaster endeca*), and cod (*G. morhua*). Various fish (*A. lúpus, G. morhua* and *H. platessoides*), sea birds, crabs (*Cancer* spp.), lobsters (*H. americanus*), anemones (*T. crassicornis*), and sea stars (*S. endeca*) are predators of larger-size urchins.

Recently, urchin populations in eastern Nova Scotia have declined dramatically. Miller (1985) suggested that a virulent protistan disease (tentatively identified as *Labyrinthomyxa* sp. by Li et al., 1982) caused this reduction, which has been observed in several locales (Hooper, 1980; Himmelman et al., 1983a; Miller and Colodey, 1983; Moore and Miller, 1983; Scheibling, 1984; Scheibling and Stephenson, 1984)[1]. In reviewing urchin/macroalgal associations, Pringle (1986) concluded that within eastern Canada only the coastline of southeast Nova Scotia alternates between urchin barrens and kelp beds, while the other major hydrographic areas (Appendices II–V) have either consistent urchin barrens or algal communities. Interestingly, the southern boundary of this mode of variability (i.e. southeast Nova Scotia) coincides with major discontinuities in commercial lobster landings, commercial fin-fish associations, and the southerly extension of the Nova Scotian Current which, in part, originates from discharge from the southern Gulf of St. Lawrence. Although this discharge varies annually, it is typically warmer and more nutrient-rich than adjacent water masses and has a higher organic load than the Scotian Shelf water (Appendices III and IV). Pringle (1986)

[1] *Labyrinthomyxa* is a questionable taxon and possibly synonymous with *Labyrinthula* (see Moss, 1986a, b).

concludes that warm-water temperatures, nutrient concentrations and organic loading of southeast Nova Scotian waters significantly influence, either directly or indirectly, sea urchin densities and ultimately community structure.

Since the demise of *Strongylocentrotus droebachiensis* in eastern Nova Scotia (see above), the numerically dominant herbivores in the shallow to mid-sublittoral zone are *Acmaea testudinalis*, *Lacuna vincta*, *Littorina littorea* and *Tonicella ruber* (Hooper, 1980). Manipulative field experiments indicate that, in the absence of sea urchins, colonization and growth of foliose macroalgae are limited by high densities of chitons ($\sim 920\ m^{-2}$), limpets ($\sim 300\ m^{-2}$) and periwinkles ($\sim 60\ m^{-2}$), as well as physical disturbance by wave action (Scheibling, 1986; Scheibling and Raymond, 1986). At sites where *Laminaria longicruris* has re-established, limpets and chitons are unable to graze its macroscopic sporophytes (Johnson and Mann, 1986b), while *Lacuna vincta* grazes directly upon adult kelps (Fralick et al., 1974). Despite this potential, direct consumption of kelp laminae by *L. vincta* only amounts to approximately 0.05% of total biomass available. Since grazing is concentrated on the frilled margins of laminae, a significant reduction in canopy area results without increased kelp mortality. Ultimately the canopy is able to close in during early spring when *Laminaria* spp. growth rates are high (Chapman and Craigie, 1977; Gerard and Mann, 1979; Anderson et al., 1981; Gagné et al., 1982). Thus, there is little opportunity for understorey species to benefit from temporarily high light levels (Harkin, 1981; Lubchenco, 1986). Understorey development is further impacted by *L. vincta* grazing, as it often occurs even more abundantly beneath kelp canopies (Harris, unpubl. observ.).

If severe ice damage occurs within the upper sublittoral zone in normally unscoured locations (e.g. Newfoundland) it causes seaweed destruction as well as high mortality of *Strongylocentrotus droebachiensis* (Hooper, 1981). Subsequent recolonization takes place deeper and more luxuriantly than usual. Unusually rapid spring thaws of ice within Newfoundland fjords, plus heavy rainfall and winds, can combine to produce a surface low-salinity layer which is fatal to sea urchins, but not to kelps or many other marine species (Hooper, 1980). In discussing long- and short-term effects of urchins on the composition of Newfoundland's sublittoral flora, South (1983) states that mass mortalities, as a result of ice-scouring and disease, are important in fostering colonization and enhanced diversity. It is unclear if "coralline barrens" described from Nova Scotia are the same as those in Newfoundland, as the latter may be formed by physical and biological factors (Hooper, 1980; Pringle et al., 1980; Pringle, 1986).

As noted previously (see p. 129), sublittoral populations of *Modiolus modiolus* form extensive beds and refuges from predation and grazing (Witman, 1984, 1985). Heavy predation of epibenthos outside mussel beds by a guild of generalist predators, including *Asterias vulgaris, Buccinum undatum, Cancer borealis, C. irroratus, Homarus americanus, Pseudopleuronectes americanus* and *Tautogolabrus adspersus* (Hulbert, 1980; Wheeler, 1980; Bernstein et al., 1981; Edwards et al., 1982; Suchanek, 1986), plays an important role in determining distribution and abundance patterns. For example, the densities of *Hiatella arctica, Ophiopholis aculeata* and *Strongylocentrotus droebachiensis* are significantly higher inside than outside these beds. Witman (1985) conducted predation experiments at the Isles of Shoals (Gulf of Maine) to test whether artificial mussel beds increased survival of invertebrates as compared to individuals in the open. Predation was highest during the day on *Hiatella arctica* but not for *Strongylocentrotus droebachiensis*. At night, crabs (*Cancer borealis* and *C. irroratus*) and lobsters (*Homarus americanus*) accounted for all attacks observed, while cunner (*Tautogolabrus adspersus*) and winter flounder (*Pseudopleuronectes americanus*) accounted for 71% of the total prey consumed during the day (Bernstein et al., 1981; Edwards et al., 1982). Quantitative sampling of infaunal communities inside and outside beds of *Modiolus modiolus* immediately after an episodic urchin grazing event showed that mussel-bed communities showed smaller changes in species composition, dominance, and diversity than adjacent epibenthic populations. Low mortality and prolonged persistence of *M. modiolus* (36 to 65 years; Wiborg, 1946; Rowell, 1967) allows long-term refuges under diverse biological disturbances (Connell, 1986). Persistence of mussel beds at shallow depths (−8 m) depends upon the ability of *M. modiolus* to escape predation by *Asterias vulgaris*, and an enhanced dislodgement by attached kelps (Merz, 1984; Witman, 1987). Frequency of kelp-mediated dislodgement depends upon the level of grazing by resident urchins (that is, the amount of canopy present) and the drag imparted by the kelps (Witman, 1984; Witman and Suchanek, 1984). Kelp-induced mortality before and after an episodic urchin grazing event was estimated at 31.2% and 5.2%, respectively.

Other transient refuges occur within the sublittoral zone in New England (Witman, 1984, 1985). Kelp holdfasts and canopies, understorey foliose and turf-forming red algae, as well as microhabitats beneath crustose coralline algae (Coull and Wells, 1983; Watanabe, 1984), provide transient spatial refuges from fish and invertebrate consumers. Cracks and crevices on vertical rock walls provide major refuges (Sebens, 1986a, b).

Sea ducks are major predators of benthic invertebrates and small fish within the upper sublittoral zone in New England (Stott, 1972; Stott and Olson, 1973), with varying patterns of habitat use and food preferences (Fig. 7.11). Red-breasted mergansers (*Mérgus serrátor*) and common goldeneyes (*Bucephala clangula*) are characteristic of rocky habitats, while oldsquaw (*Clangula hyemális*) occur in diverse coastal areas. Common goldeneyes feed primarily on amphipods (*Ampithoe rubricata*), isopods (*Idotea balthica* and *I. phosphorea*), crabs (*Cancer irroratus*) and gastropod populations (*Lacuna vincta* and *Littorina littorea*) within *Chondrus crispus* beds (Dawson, 1909; Madsen, 1954; Olney and Mills, 1963; Feare, 1967; Feare and Summers, 1986). Bivalves (*Mytilus edulis*), polychaetes and other invertebrates are eaten by *Bucephala clangula* in coastal areas, while in adjacent harbors eelgrass seeds and sand shrimp are preferred (Jacobs et al., 1981). Red-breasted mergansers primarily eat small fish (*Fundulus* spp. and *Menidia menidia*) found within Irish moss beds and rocky crevices (Kortright, 1953; Bent, 1962; Cronan and Halla, 1968). Gastropods (*Lacuna vincta* and *Littorina littorea*) are the major food of oldsquaw in rocky habitats, while bivalves, sand shrimp and isopods are consumed in adjacent sandy habitats and harbors (Lagler and Wienert, 1948; Mackay, 1982). Common eiders (*Somatéria mollissima*) and several species of scoter consume considerable numbers of *Mytilus edulis* (Graham, 1975; Incze and Lutz, 1980; Lutz, 1980). In discussing the impact of avian predation on *Mytilus* spp. and other invertebrates, Marsh (1986) stated that the effects of birds are long-term when they exploit patches of prey effectively and recruitment of prey is infrequent. Conversely, the effects are shorter when prey are hard to digest and have short life cycles, and birds are present for only part of the year (Schneider and Harrington, 1981).

Several other sessile and mobile animals characterize sublittoral communities (Nesis, 1962, 1965; Harris and Mathieson, 1976; Noble et al., 1976; Steele, 1983; Sebens, 1986a, b; see also Table 7.3). For example, in Newfoundland, capelin (*Mallotus villosus*) is the most important forage species (Steele, 1983). Typically, it lives in schools offshore, but during summer moves inshore to spawn on sand or gravel beaches. The spawning migration to shallow water is important, since both adults and eggs are sources of food for many species at a time when predators are reproducing, growing and/or putting on reserves for later reproduction. Cod (*Gadus morhua*) follow capelin inshore in a feeding migration, while Atlantic salmon (*Salmo salar*), long-finned squid (*Loligo pealei*), and resident shore fish, such as sculpin (*Hemitripterus americanus, Myoxocephalus aeneus, M. octodecimspinosus, M. scorpius, Triglops ommatistius*), feed heavily on capelin when they approach nearshore. Fish with small mouths, such as winter flounder (*Pseudopleuronectes americanus*) and haddock (*Melanogrammus aeglefinus*), feed on capelin eggs. Sea birds such as the herring gull (*Lárus argentatus*), great black-backed gull (*Lárus marinus*), black-legged kittiwake (*Rissa tridáctyla*), Atlantic puffin (*Fratércula árctica*), common murre (*Úria áalge*), thick-billed murre (*Úria lómvia*), and razorbills (*Alca tórda*) may feed on capelin, which is one of the reasons for the abundance of these birds in Newfoundland waters. Both Minke (*Balaenoptera acutorostrata*), and humpback whales (*Megaptera novaengliae*) feed extensively on capelin during their inshore feeding migrations. Thus, capelin play a major role in Newfoundland's food chain.

As outlined by Keats et al. (1982b), the distribution of deep-water *Agarum cribrosum* has been the subject of considerable interest and research. Vadas (1968), working on the Pacific coast of Washington, presented experimental evidence indicating that the presence of abundant populations depended on preferential feeding of urchins on other Laminariales, which reduced interspecific competition. Mann (1973) speculated that the presence of *A. cribrosum* within the mid-lower sublittoral zone in Nova Scotia might result from preferential grazing on its competitors (*Laminaria* spp.) by *Strongylocentrotus droebachiensis*. Experimental evidence for these grazing preferences has been documented (Himmelman, 1969, 1980; Vadas, 1977; Larson et al., 1980). For example, Himmelman (1980) found that survival of *A. cribrosum*, as well as *Phycodrys rubens* and *Ptilota serrata*, at depths where urchins predominate, is probably due to their marked unattractiveness to urchins. After studying the distribution of *A. cribrosum, Laminaria* spp. and *S. droebachiensis* at several coastal sites in Nova Scotia, Tremblay and Chapman (1980) concluded that the occurrence of *Agarum cribrosum* depended upon the presence of urchins on the one hand, and of *Laminaria* spp. on the other. In the absence of urchins, *Laminaria* spp. outcompete *A. cribrosum* for space or light. During conversion of a kelp forest to a barren ground, *A. cribrosum* can briefly exist as the only kelp with urchins, but only during a period of disequilibrium. Based upon a series of long-term field studies, Keats et al. (1982b) suggested that distribution and abundance of *A. cribrosum* in Newfoundland is determined by other factors. In Fortune Bay *A. cribrosum* and *Laminaria* spp. are abundant in the absence of urchins, while *A. cribrosum* occurs with urchins in Conception Bay. In the Gulf of Maine *A. cribrosum* is the dominant kelp below −12 m, even in areas where there are low densities of herbivores (Harris and Mathieson, 1976; Hulbert et al., 1976; Mathieson, 1979; Hulbert, 1980; Mathieson et al., 1981; Witman, 1984; Martin et al., 1988). Within these same areas *Laminaria* spp. may grow in deeper habitats where *A. cribrosum* dominates (Table 7.3), but densities of *Laminaria* spp. are very low and their distribution patchy.

Vertical or fouling communities

Vertical or slightly undercut rock walls below −3 m in the Gulf of Maine are primarily dominated by encrusting invertebrates, whereas a greater diversity of plants and animals occur on horizontal surfaces (Sebens, 1986a, b). Based upon a detailed analysis of percent coverage of encrusting organisms on two vertical rock walls in Massachusetts, Sebens (1986a, b) found that competition for space was intense and often mediated by predation. *Alcyonium siderium, Aplidium pallidum*, and *Metridium senile* can each dominate patches of wall indefinitely when few sea urchins are present, while crustose coralline algae (*Lithothamnion glaciale* and *Phymatolithon rugulosum*) dominate when urchins are abundant (Fig. 7.15). A series of complex hierarchical interactions occur between *Alcyonium siderium, Aplidium pallidum* and *Metridium senile*, determining space retention and acquisition. A thick and massive growth form like *M. senile* is at the top of the hierarchy, because of its stature and longevity, its infliction of extensive damage on *A. siderium* by its nematocysts, and its ability to hold off encroachment by *A. pallidum*. On the other hand, small individuals of *A. siderium* (Sebens, 1982) and *M. senile* are frequently overgrown by *A. pallidum*. Thus, once a patch of rock wall is dominated by any of these three species it may persist indefinitely. Under constant urchin grazing, *Lithothamnion glaciale* and *Phymatolithon rugulosum* dominate and only an occasional erect or encrusting sponge (*Halichondria panicea*) or ascidian occurs. At other sites *Metridium senile* is the most conspicuous sessile invertebrate on urchin-grazed rock walls (Harris, 1986). Although *L. glaciale* is clearly the competitive dominant among encrusting algae (Steneck, 1982b), it is susceptible to attack from at least two sources. Urchins break off its growing edges, and boring invertebrates (e.g. the polychaete *Dodecaceria coralii*) weaken the thallus, causing exfoliation (Adey, 1966b). The thinner and smoother surface of *P. rugulosum* does not suffer from polychaete boring and appears to be less easily damaged by continuous urchin scraping. An equilibrium coexistence between these crustose coralline algae occurs because of disproportionate damage to the competitive dominant, *L. glaciale*.

By comparing several indices of spatial interactions for these encrusting populations (growth, overgrowth, resistance, overgrowth weighted by abundance), Sebens (1986a, b) demonstrated varying strategies of growth and competition. Species resistant to overgrowth were generally large, slow-growing and not the most active overgrowers of other species. An evaluation of these indices

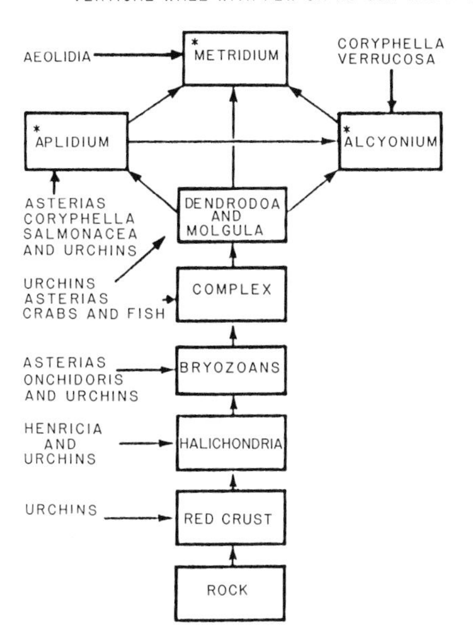

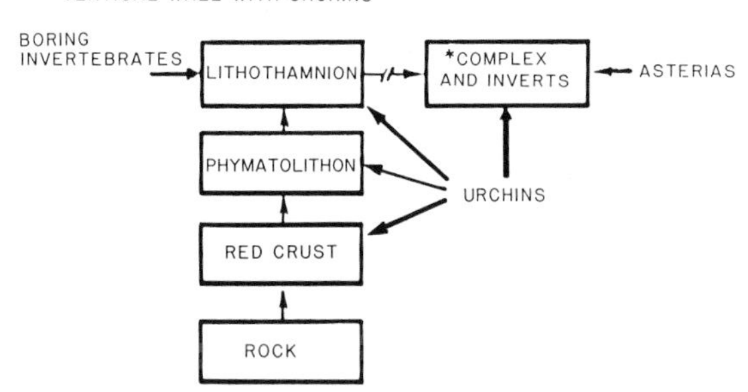

Fig. 7.15. Block diagram of succession, competition, and predation on sublittoral rock walls at sites in northern Massachusetts. Succession begins with cleared rock and moves upward to one or more alternate "climax" stages which would persist indefinitely without physical disturbance or predation on that species. Competitive abilities of sessile species are indicated by arrows pointing from poorer to better competitors. Disturbance (physical scraping or predation) can set succession back to cleared rock at any stage. Important predators are indicated by arrows impinging laterally on each stage (thicker arrows indicating more intense predation). A high diversity of encrusting species is maintained on vertical walls with few or no sea urchins, but with some sea stars and other predators, by frequent small-scale disturbances (predation). Heavy sea urchin predation on vertical walls results in a simple community of encrusting coralline algae which is prevented from reaching its climax. (Modified from Sebens, 1986b.)

showed significant correlations between exposed and sheltered sites but not between warm and cold seasons. Many encrusting species have the same competitive roles in slightly different communities (exposed vs. sheltered), but their importance varies as growth rates change seasonally. Crustose red algae like *Phymatolithon rugulosum* and *Peyssonnelia* sp. were often encountered alive under invertebrates that had been in position for several months. Harris and Irons (1982) proposed that the surface layer of sessile invertebrates was an important factor in competitive encounters, because organisms such as *Alcyonium siderium*, *Aplidium pallidum*, *Metridium senile* and various sponges cannot be grown upon due to their naked epidermal layer. Thus, they may be smothered by larger colonies or individuals, damaged by nematocysts, dislodged, or crawl aside (*M. senile*), but actual overgrowth (*sensu* Jackson, 1977) will not occur. As a result, any one of several species may dominate space depending upon which one initially recruits and attains a size large enough to escape predation.

Predation also plays an important role in structuring rock-wall communities. All of the competitive dominants listed by Sebens (1986b) are influenced by predators. *Alcyonium siderium* is preyed upon by *Coryphella verrucosa* (Sebens, 1986a, b), *Aplidium pallidum* by *Coryphella salmonacea* and sea urchins, *Metridium senile* by *Aeolidia papillosa* (Harris, 1986). Nudibranchs and urchins are most common and active during winter months, and at least urchins and *A. papillosa* are capable of altering the population structure of their prey. *Aeolidia papillosa* preferentially attacks small individuals of *M. senile*, leading to populations dominated by large individuals (Schick et al., 1979; Harris, 1986). Harris (1986) experimentally demonstrated size-selective predation on the nudibranch *Aeolidia papillosa* by cunners or wrasses (*Tautogolabrus adspersus*), partially reducing nudibranch predation pressure on *M. senile* when these fish were common.

The major competitive dominant in most fouling and rock-wall communities is *Mytilus edulis*, but it is so heavily preyed upon by asteroids (Menge, 1979), crabs (Harris and Irons, 1982), cunner (Chao, 1973; Harris and Irons, 1982), and urchins (Briscoe and Sebens, 1986) that it is uncommon except as spat. Sebens (1986a, b) described *Strongylocentrotus droebachiensis* as a major grazer on rock-walls, as well as its more obvious role as a grazer in algal-dominated systems.

Although most investigators have assumed that

urchins are herbivores (Lawrence, 1975), recent evidence suggests that *Strongylocentrotus droebachiensis* is capable of a predatory role in some communities and should be considered an omnivore (Scott, 1902; Thorson, 1966; Harris and Mathieson, 1976; Himmelman, 1980; Briscoe and Sebens, 1986; Sebens, 1986a, b). In discussing the functional role of *S. droebachiensis*, Briscoe and Sebens (1986) stated that urchins from kelp beds consume fewer mussels than those from sublittoral mussel beds, while Thorson (1966) suggested that *Laminaria* spp. covered with bryozoans (*Membranipora* spp.) were preferentially consumed as against algae alone (cf. Elmhirst, 1922; Ayling, 1978; Vance, 1979). It should be recalled that *S. droebachiensis* is a boreal species, which is most active during winter months; at this time foliose algae are heavily encrusted with sessile and motile fauna. [However, see Himmelman (1980) regarding contrasting patterns of urchin feeding in Newfoundland due to heavy wave conditions.] According to Miller and Mann (1973), *S. droebachiensis* fed on a diet of *Laminaria* spp. contained large concentrations of nitrogen-fixing bacteria within the gut, but they also leaked 40% of the organic matter ingested, incorporating only 4% as somatic tissue. In contrast, urchins taken from barren grounds did not contain nitrogen-fixing bacteria, presumably because they obtained enough nitrogen from diatoms. It is even more likely that urchins receive enough nitrogen in barren grounds, as well as in kelp beds, from amphipod tube mats and numerous animal recruits upon which they graze.

Little information is available on factors influencing community structure in sessile, suspension-feeding invertebrate assemblages dominating rock surfaces below −30 m. Sediment accumulation caused by suspension feeders favors infaunal motile forms, but inhibits growth of the few algal species growing at these depths. Predation by sea stars and fish is intense within deepwater-fouling communities. Hulbert (1980) has described a high diversity of sea stars below −30 m, several of which feed primarily on sessile invertebrates (*Porania* spp. and *Henricia* spp. on sponges, *Hippasteria* sp. on soft corals and anemones, and *Solaster* sp. on holothurians). Hulbert et al. (1976) recorded active foraging and high densities (200 per 25 m^2) of cunner on horizontal surfaces, while Witman and Cooper (1983) found that predation by cod kept the brachiopod *Terebratulina septentrionalis* from growing to a large size in this habitat.

Isolated offshore communities

As noted previously (p. 133 and Table 7.6), offshore rocky habitats exhibit unique zonation patterns and species assemblages not found in nearshore sites. On a deep-water pinnacle in the central Gulf of Maine, Vadas and Steneck (1988) found low levels of herbivory, whereas several other unique interactions occurred, including extensive competition between kelps (*Laminaria* sp. and *Agarum cribrosum*) and *Metridium senile* (p. 134). Based upon similar studies in the outer Gulf of Maine, Sebens and Witman (1986) concluded that wave surge and predators such as sea stars (*Ophiopholis aculeata* and *Ophiura robusta*), wolffish (*Anarhichas lúpus*), and cod (*Gadus morhua*), were the dominant forces structuring deep-water communities (−22 to −120 m). At Pigeon Hill (−33 to −42 m) off Cape Ann, Massachusetts, Hulbert et al. (1982) found that the relative effect and type of predation varied between upper horizontal communities (algae and polychaetes) and those of vertical rock walls (sponges and tunicates). Invertebrate predators such as urchins, prosobranch gastropods, crabs, and lobsters were more common on horizontal than vertical surfaces. Polychaetes and ophiuroids in upper horizontal communities were heavily preyed upon by demersal fish such as haddock (*Melanogrammus aeglefinus*) and yellowtail flounder (*Limanda ferruginea*). Stomach contents of adult haddock indicated that most (83%) were feeding on polychaetes and ophiuroids. Although there are predators upon vertical walls that feed on sponges and tunicates (*Euphrosine borealis* and *Henricia sanguinolenta*), their impact is modest. Sediments, which accumulate more extensively on horizontal than vertical surfaces, may foul suspension-feeding organisms (e.g. sponges and tunicates), and reduce their numbers (Harris and Mathieson, 1976).

Hulbert et al. (1982) described trophic classifications of benthic assemblages at Pigeon Hill according to the definitions of Fedra (1976). The community consists of 42% suspension feeders (57 species), 5% deposit feeders (7 species), 21% of species that are both deposit feeders and herbivores (30 species), 8% that feed on two trophic levels (deposit feeders and carnivores, 12 species), and 18% motile carnivores (25 species). Suspension feeders are distributed on both horizontal and vertical surfaces, but are more abundant on the latter (Harris and Mathieson, 1976). Motile invertebrate carnivores consist of asteroids, nudibranchs, polychaetes and small crustaceans, which are widely distributed. Primary consumers comprised 69% of the species, while 18% were secondary consumers.

Other biological interactions

Many other biological interactions occur within the littoral and sublittoral zones. Various fungi and bacteria cause extensive infections of seaweeds (Barton, 1901; Petersen, 1905; Lind, 1913; Chemin, 1927, 1931; Johnson and Sparrow, 1961; Boney, 1965; Ring, 1970; Andrews, 1976; Andrews and Goff, 1985; Kingham and Evans, 1986; Molina, 1986; Moss, 1986a, b; Garbary and Gautam, 1989). Epifaunal and epiphytic populations enhance frond breakage, and reduce growth, survivorship and reproduction of host plants (Lilly, 1968; Prince, 1971; Enright, 1977, 1979; Hartnoll, 1983; Witman and Suchanek, 1984; D'Antonio, 1985; Peckol and Harlin, 1985; Lubchenco, 1986; Seed, 1986). Lilly (1968) recorded approximately 50 species of invertebrates (adults and juveniles) growing on or amongst *Chondrus crispus*, the most abundant of which were polychaetes, amphipods, crustaceans, gastropods and pelecypods (Scarratt, 1972; Hicks, 1986). Similar patterns exist for epifaunal populations on *Corallina officinalis* (Keats et al., 1984), while crustose corallines have abundant epifaunas of molluscs and echinoderms, plus an infauna of boring pelecypods and ophiuroids (Adey, 1966b). Intertidal fucoid algae (*Ascophyllum nodosum* and *Fucus* spp.) as well as fleshy subtidal seaweeds (*Chondrus crispus, Desmarestia* spp., etc.) create habitats for numerous invertebrates, fish and encrusting algae which live under/amongst their canopies (Boaden et al., 1975; Doyle, 1975; Lubchenco, 1980, 1983; Johns and Mann, 1987; Keats et al., 1987). Many seaweeds are covered with extensive bryozoans (Rogick and Croasdale, 1949; Yonge, 1949; MacFarlane, 1952, 1968), causing reduced growth.

Herbivorous amphipods (*Ampithoe rubricata, Gammarus oceanicus, Hyale nilssoni*, and *Talorchestia longicornis*) occur on several seaweeds (Skutch, 1926; Lilly, 1968; Brenner et al., 1976; Nicotri, 1980; McBane, 1981; Shacklock and Croft, 1981; McBane and Croker, 1983; see also Fig. 7.10), and may cause extensive damage. Nematodes (e.g. *Halenchus* spp.) and isopods (e.g. *Idotea balthica*) can cause similar effects (Nicotri, 1980; Shacklock and Croft, 1981; Moore, 1986). For example, *Gammarus oceanicus* consumes fronds of Irish moss at a daily rate of 5% of its body weight; for *I. balthica* the rate is 20% per day (Shacklock and Croft, 1981). Talitrid amphipods (*Orchestia platensis*) eat fresh beach wrack dominated by *Ascophyllum nodosum* and *Chondrus crispus*, and also live oligochaetes, *Limulus polyphemus* eggs, and benthic diatoms (Behbehani and Croker, 1982).

In discussing the effects of epizoans and epiphytes (including *Laminaria* spp.) on underlying invertebrates, Witman and Suchanek (1984) stated that they can be positive, neutral or negative. A common benefit is protection from predation (Ross, 1971; Bloom, 1975; Vance, 1978; Hartnoll, 1983). Frequently, tubicolous polychaetes, bryozoans, and other colonial invertebrate epizoans have no appreciable effect on their hosts (Seed and O'Connor, 1981). Epizoans can affect the host adversely by smothering it (Burrows and Lodge, 1950; Dayton, 1973), by eroding the host's shell (Korringa, 1951), or causing the host to be torn loose from the substratum (Burrows and Lodge, 1950; Barnes and Topinka, 1969; Dayton, 1973; Paine, 1979; Witman, 1984; Witman and Suchanek, 1984). Epizoans may also influence survival of individual organisms (Rutzler, 1970) and the persistence of entire communities (Vance, 1978).

As outlined by Vadas (1985), marine herbivores have a wide range of morphological and behavioral adaptations for exploiting and increasing utilization of seaweeds. For example, radular morphology and function in molluscs seem to have undergone considerable evolution (Steneck and Watling, 1982; Steneck, 1983a, b), which permits grazing on a wide range of algal forms. In general discussions of algal community structure and herbivory, Steneck (1983b, 1985) suggested that herbivores can be placed in three major groups according to their feeding capabilities: (1) those that feed on algae but do not denude the primary substratum (polychaetes, malacostracans); (2) those that denude primary substratum but cannot excavate calcium carbonate (some molluscs, urchins and fishes); (3) those that excavate calcium carbonate (chitons, limpets, some urchins and parrotfishes). Some herbivores cannot consume certain seaweeds because of limitations in their feeding apparatus and body plan relative to structural and morphological characteristics of the algae. Only the third group of herbivores has an ecological impact on crustose coralline algae; interestingly, a "diffuse coevolution" (Futuyma and Slatkin, 1983) appears to have occurred between crustose coralline algae and coralline-grazing herbivores.

In describing adaptations of crustose coralline algae to herbivory, Steneck (1985, 1986) commented on the adaptive value of several anatomical, morphological and reproductive characteristics. Thick crustose species (e.g. *Clathromorphum circumscriptum*) dominate in zones of intense herbivory, while thinner corallines (e.g. *Phymatolithon* spp.) are less competitive under these conditions. The former species, which grows more slowly than the latter, maintains living tissue beneath its photosynthetic zone (upper 200 μm). Some thick crustose corallines (e.g. *C. circumscriptum*) depend upon grazing to prevent overgrowth by foliose seaweeds or epiphytic diatoms (Steneck, 1977, 1978, 1982a, b, 1985, 1986). By contrast, others such as *Phymatolithon* spp. have developed an innate antifouling mechanism by shedding their epithallial surface cells (Adey, 1964, 1966b, 1973; Johnson and Mann, 1986a), which is analogous to the situation in *Ascophyllum nodosum* and *Chondrus crispus* (Filion-Myklebust and Norton, 1981; Sieburth and Tootle, 1981; Moss, 1982; Russell and Veltkamp, 1984). Branched corallines with non-articulate protuberances (e.g. *Lithothamnion glaciale*) have several ecological advantages over unbranched corallines (Steneck, 1982c, 1986). Branches are an effective defense against some excavating herbivores (e.g. *Strongylocentrotus droebachiensis*), as only the branch tips are grazed (Steneck and Milliken, 1981). Branches also increase the photosynthetic and reproductive ability of crusts. Most thin, branched corallines (e.g. *L. glaciale*) have raised reproductive conceptacles, while thicker, unbranched crusts (e.g. *C. circumscriptum*) have sunken conceptacles, providing protection against invertebrate herbivores. Deeply sunken meristems are correlated with heavy grazing by molluscs. For example, the epithallus of *C. circumscriptum* protects it from grazing by the limpet *Acmaea testudinalis*, this adaptation being so specific that when limpets are excluded it dies because of excessive tissue (surficial) build-up (Adey, 1973; Steneck, 1977, 1982a, 1983a, 1986). In summarizing these anatomical and morphological features Steneck (1985, 1986) emphasized that thallus differentiation into several tissues (perithallus, hypothallus and epithallus), intercellular conduits ("vegetative fusion cells"), and conceptacles paved the way for successful evolution of corallines, as well as their coevolution with several herbivores.

Many foliose seaweeds have a tough texture that resists herbivores (Watson and Norton, 1985), while others harbor a variety of toxins that may discourage invertebrates (Levinton, 1982), fungi and other microbes (Sieburth, 1964, 1969; Burkholder and Sharma, 1969; Ragan, 1984; Harlin, 1987). Ephemeral seaweeds that dominate early successional stages have no major defenses against grazers (Lubchenco and Gaines, 1981; Lubchenco, 1986), whereas mechanically and chemically resistant seaweeds tend to dominate late successional stages (Littler and Littler, 1980; Steneck and Watling, 1982). [However, see Vadas (1979), Norton et al. (1982), Padilla (1985) and Steinberg (1986), regarding such strategies and functional groups based on thallus form.]

Active liberation of polyphenols (e.g. tannins), halogenated compounds and other biodynamic substances from seaweeds may serve as a chemical defense against herbivores (Augier and Hoffman, 1952; Peguy and Heim, 1961; Conover and Sieburth, 1964, 1966; Craigie and McLachlan, 1964; Sieburth and Conover, 1965; McLachlan and Craigie, 1966; Doty and Aguilar-Santos, 1970; Glombitza and Stoffelen, 1972; Stoffelen et al., 1972; Scheuer, 1973; Bhakuni and Silva, 1974; Faulkner and Anderson, 1974; Glombitza et al., 1974; Stallard and Faulkner, 1974; Fenical, 1975, 1980; Baker and Murphy, 1976; Ragan, 1976; Ragan and Craigie, 1976, 1978; Vadas, 1977; Sun and Fenical, 1979; Geiselman, 1980; Larson et al., 1980; Geiselman and McConnell, 1981; Anderson and Velimirov, 1982; Smithson and Cheney, 1984;

Steinberg, 1984, 1985; Bell and Cheney, 1985; Espinosa, 1985; Johnson and Mann, 1986b; Littler et al., 1986; Harlin, 1987). For example, tannin exudates from *Ralfsia verrucosa* inhibit the settlement of *Semibalanus balanoides* cyprids in tide pools (Conover and Sieburth, 1966; also cf. Magre, 1974). Studies by Geiselman (1980) and by Geiselman and McConnell (1981) have shown that polyphenols in *Ascophyllum nodosum* and *Fucus vesiculosus* are an effective chemical defense against the herbivorous snail *Littorina littorea*. A diet with as little as 1% polyphenols (dry weight) from these fucoids causes a significant reduction in *L. littorea* feeding, while a diet with 10% polyphenols almost totally inhibits feeding. The phenol and polyphenol contents within these fucoids and other New England seaweeds (such as *Codium fragile* ssp. *tomentosoides*, *Chondrus crispus*, *Petalonia fascia* and *Scytosiphon lomentaria* var. *lomentaria*) show significant differences, with only those of the fucoids being sufficiently high to inhibit snail feeding throughout the year. Spatial patterns of grazing by *Lacuna vincta* on *Laminaria longicruris* seem to be governed by the localized distribution of polyphenols, as well as variations in toughness and nutritional quality (Johnson and Mann, 1986b). Other kelps (e.g. *Agarum cribrosum*) with elevated and more uniformly distributed polyphenol levels are not preferred by urchins (Vadas, 1977; Himmelman, 1980; Larson et al., 1980).

Chondrus crispus inhibits growth of unicellular green algae (e.g. *Chlorella pyrenoidosa*) by producing hydrogen peroxide by the action of the enzyme hexose oxidase (Ikawa et al., 1969), which chemically resembles fungal oxidase, except for its high copper content (Sullivan and Ikawa, 1973). *Chondrus crispus* also contains compounds with diverse antifungal, antibacterial (Hornsey and Hide, 1974, 1976a, b; Biard et al., 1980), and antibiotic effects on many benthic diatoms (Khfaji and Boney, 1979; Huang and Boney, 1983, 1984, 1985). By contrast, the closely related red alga *Mastocarpus stellatus* does not cause reduced growth nor other suppressive effects in littoral diatoms (Hornsey and Hide, 1974). According to Tuchman and Blinn (1979), diatom–macrophyte interactions are closely related to metabolic competition, including possible allelopathy. Even though many algal compounds are suspected of providing protection against epiphytes and grazers (halogenated compounds of red algae), these roles are largely speculative (Lobban et al., 1985; Harlin, 1987).

Several investigators (Lubchenco and Cubit, 1980; Slocum 1980; Levinton, 1982; Steneck, 1983b) have discussed the adaptive significance of heteromorphic life histories of seaweeds as escapes from herbivory. Steneck (1983b) suggested that heteromorphic alternation of generations and ontogenetic changes diversify functional characteristics of species, and their vulnerability to herbivory changes accordingly. Slocum (1980) suggested that there is a differential susceptibility to grazers in gigartinalean red algae with crustose sporophytes (Setchell and Gardner, 1933); crusts are grazer-independent but are susceptible to overgrowth, while productive blades are more vulnerable to grazing. "Bet-hedging" may be important in maintaining two different phases in one life history (for example, *Mastocarpus papillatus* and *M. stellatus*). Lubchenco and Cubit (1980) hypothesize that herbivory patterns may explain the seasonal occurrence of *Petalonia fascia* and *Porphyra* spp. in New England, and that some heteromorphic life histories may have been selected as a means of avoiding variable but predictable grazing pressure. By contrast, Shannon et al. (1988) found that patterns of seasonal occurrence and abundance of the crustose and upright stages of *Petalonia fascia* and *Scytosiphon lomentaria* were not partitioned seasonally in relation to the dominant herbivore *Littorina littorea* (Dethier, 1981). Levington (1982) stated that the hypothesis presented by Lubchenco and Cubit does not explain the occurrence of isomorphic species prone to predation (e.g. *Enteromorpha* spp. and *Ulva lactuca*). It is difficult to prove that heteromorphy initially evolved as a response to grazing.

According to Littler and Littler (1983) the selective role of grazing in heteromorphic life histories may have been over-emphasized, as crustose stages of many seaweeds are adapted to withstand physical forces such as wave shearing, sand abrasion, extremes of temperatures and limited irradiance (Taylor and Littler, 1982; South, 1983). Thus, a crustose morphology sacrifices productivity and rapid growth for toughness and a large proportion of structural tissue, which enables it to be grazer-resistant and withstand physical disturbance (Littler and Littler, 1980). Incorporation of different morphological stages within one life history may reflect adaptations to different sources of mortality (Shannon, 1985; Shannon et al., 1988). For example, the two life-history stages of *Petalonia fascia* and *Scytosiphon lomentaria* have different tolerances to temperature and light (Wynne, 1969; Lüning, 1980; Correa et al., 1986), grazing (Lubchenco, 1978; Lubchenco and Cubit, 1980; Dethier, 1981), and physical disturbance (Littler and Littler, 1983).

In discussing life-history strategies of Newfoundland seaweeds, South (1983) stated that, while heteromorphic alternations are employed by a number of species as a potential means of surviving contrasting seasons, and of dispersal, it is not always evident what the advantages may be (for example, *Turnerella pennyi–Cruoria arctica/rosea* alternation: South et al., 1972). At the northern or southern extremes of their distribution, certain species may have truncated or incomplete life histories, or use vegetative propagation as the principal means of replication (Chen et al., 1969; South, 1972, 1975; Hooper and South, 1977; Whittick, 1978). It is conceivable that the evolution of alternation of generations has contributed significantly to the dispersal potential of many marine algae. The advantages of the heteromorphic life history to dispersal are evident in species such as *Bonnemaisonia hamifera*, where the tetrasporophyte occurs much farther north than the gametophyte (Chen et al. 1969; Dixon, 1973). Newfoundland populations of *Plumaria elegans* bear only paraspores, indicative of a triploid condition, as in northern Europe (Hehre and Mathieson, 1970). A typical triphasic life history is restricted to more southerly sites (Whittick, 1973).

Jackson and Hughes (1985) described life-history and growth strategies of sessile animals on coral reefs, which are equally relevant to those in algal-dominated and fouling communities in the northwest Atlantic. Growth and reproductive strategies of colonial and clonal forms provide a means by which species can occupy and hold expanding areas without sexual reproduction. At the same time these forms may increase the reproductive potential of a successful genotype, and maximize persistence and competitive ability by growing into newly-opened adjacent space, rather than waiting for recruitment of larval stages of the species. Clones and colonies with sheet-like growth-forms may also avoid or reduce competitive and predatory impacts by differential growth and regeneration after partial predation. The complex life histories and colonial growth-forms of many sessile invertebrates have strong parallels to the advantages proposed for seaweeds with heteromorphic life histories. Todd (1986) emphasized the prevalence of perennial life cycles in rocky-shore invertebrates, particularly with respect to acquisition and retention of primary space in the face of intense inter- and intra-specific competition.

PRIMARY PRODUCTIVITY OF BENTHIC COMMUNITIES

Mid-eulittoral zone

Most rocky intertidal shores in the northwest Atlantic are dominated by fucoid algae, and to a lesser extent by the red algae *Chondrus crispus* and *Mastocarpus stellatus* (pp. 118, 122). *Ascophyllum nodosum* contributes most of the carbon fixed by intertidal seaweeds, the maximum annual production in several Nova Scotian beds of *A. nodosum* being 2.82 kg dry weight $m^{-2} yr^{-1}$ (Cousens, 1984). However, both standing crop and production were extremely variable, depending upon wave action. MacFarlane (1952) estimated that the maximum standing crop of *A. nodosum* in southwestern Nova Scotia was 8 kg dry weight m^{-2}. Based upon MacFarlane's estimates of seasonal changes in standing crop, Westlake (1963) calculated that annual production by *A. nodosum* in southwestern Nova Scotia was 2.0 to 2.6 kg dry weight $m^{-2} yr^{-1}$. Cousens (1981) documented *A. nodosum* production and standing crop values for seven Nova Scotian locations along a gradient of wave exposure. Production increased from 0.15 to 1.72 kg dry weight $m^{-2} yr^{-1}$ with decreasing exposure. Maximum production values corresponded to 630 g carbon fixed $m^{-2} yr^{-1}$, assuming 36% carbon in the dry matter (Cousens, 1981). Standing crop, though displaying a somewhat similar trend to production, was slightly reduced under extreme as against moderate shelter. Topinka et al. (1981) examined fucoid biomass (both *A. nodosum* and *Fucus vesiculosus*) at 46 open coastal and estuarine sites in central Maine. Maximum standing crop was found at moderately exposed sites, that is, 3 to 6 on Ballantine's (1961) exposure scale (Fig. 7.7).

Reductions of standing crop similar to those documented by Cousens (1981) were observed under extreme shelter and exposure. In the former, winter ice "rafting" (that is, *in situ* freezing, followed by breakage and floating away) or ice scouring (Cousens, 1981; Topinka et al., 1981; Mathieson et al., 1982). At sheltered estuarine locations in New England, ice "rafting" may remove up to 50% of the winter fucoid standing crop (Mathieson et al., 1982). Pringle and Semple (1980) described the distribution of *A. nodosum* biomass along a vertical transect at West Pubnico, Nova Scotia (Fig. 7.5). Maximum biomass (~7.2 kg fresh weight m^{-2}) was recorded at +2.0 m.

Not only is the overall production of *Ascophyllum nodosum* affected by wave action, but biomass and production within its canopy varies with exposure. In exposed sites, the canopy is composed of slow-growing, short individuals (Cousens, 1985). With increasing shelter, the canopy is longer, with most biomass 30 to 45 cm above the substratum. Under sheltered conditions most production is contributed by a relatively few large fronds forming the canopy. Dominant plants shade and limit many shorter fronds (Cousens, 1981).

A large part of the annual fucoid production enters the near-shore detrital cycle (Fig. 7.10) by leaching of dissolved organic carbon and tissue fragmentation. In addition, *Ascophyllum nodosum* produces numerous lateral branches (receptacles) containing gametes. When mature, receptacles are shed over a relatively discrete time period (Mathieson et al., 1976; Bacon, 1983; Mathieson, 1989), providing an important pulse of organic carbon to near-shore and estuarine systems (Josselyn and Mathieson, 1978). According to Cousens (1986), reproductive biomass may comprise 10 to 29% of the standing crop of *A. nodosum*, while receptacles represent 41 to 70% of the plant's annual production. Josselyn and Mathieson (1978) cited even higher reproductive biomass allocation in *A. nodosum* (50%), while Robertson (1987) found maximum reproductive allocation patterns of 30% for *Fucus spiralis*. Russell (1979) stated that such patterns in the related taxon *Fucus vesiculosus* vary, depending upon the magnitude of competition for light and space (Abrahamson and Gadgil, 1973; Spitters, 1983; McCourt, 1985). Josselyn and Mathieson (1978) and Topinka et al. (1981) reported the levels of organic material released into near-shore communities by dehiscent *A. nodosum* receptacles. Typically the loss of total fucoid material is also maximal during reproductive periods (i.e. whole plants are lost), due to the weight and drag of their receptacles (Knight and Parke, 1950; Niemeck and Mathieson, 1976; Sideman and Mathieson, 1983b; Robertson, 1987).

Lower eulittoral-sublittoral (near-shore) zone

MacFarlane (1952) estimated that the average standing crop of *Chondrus crispus* in southwestern Nova Scotia was 0.45 to 1.25 kg fresh weight m^{-2}, while Foster (1953, 1954, 1955, 1956) recorded standing crops in Maine of 0.6 to 7.5 kg fresh weight m^{-2}. The largest biomass was measured in eastern Maine near the Bay of Fundy. On Prince Edward Island within the Gulf of St. Lawrence, Taylor (1973) recorded biomass values of sublittoral *C. crispus* between 210 to 231 g dry weight m^{-2} over a three-year period (1969 to 1971). In the same habitat he recorded 583 g dry weight m^{-2} of *Furcellaria lumbricalis*. See McLachlan et al. (1987) for further details of standing stocks and annual harvests (1967–1981) for the same locale. The vertical distribution of *C. crispus* biomass at West Pubnico, Nova Scotia, showed a maximum between +0.5 and −7.0 m (Fig. 7.5), where the biomass was 1058 g fresh weight m^{-2} (Pringle and Semple, 1980). Prince (1971) described seasonal patterns for the standing crop of *C. crispus* at Plymouth, Massachusetts: at −2 to −3 m annual biomass varied between 0.8 to 1.3 kg dry weight m^{-2}. Burns and Mathieson (1972b) reported monthly biomass measurements for New Hampshire populations of *Mastocarpus stellatus*. A simple estimate of annual production from their data (as the difference between maximum and minimum standing crop) is 6.5 kg fresh weight m^{-2} yr^{-1}. Seasonally, the standing crop of *M. stellatus* varied from 3.5 kg m^{-2} (February) to 10 kg m^{-2} (August).

Plant biomass in the mid-sublittoral zone is primarily dominated by kelps. MacFarlane (1952) reported a seasonal biomass maximum of 3.8 kg dry weight m^{-2} for *Laminaria longicruris* in southwestern Nova Scotia. Westlake (1963) used this figure to calculate an annual mean production of

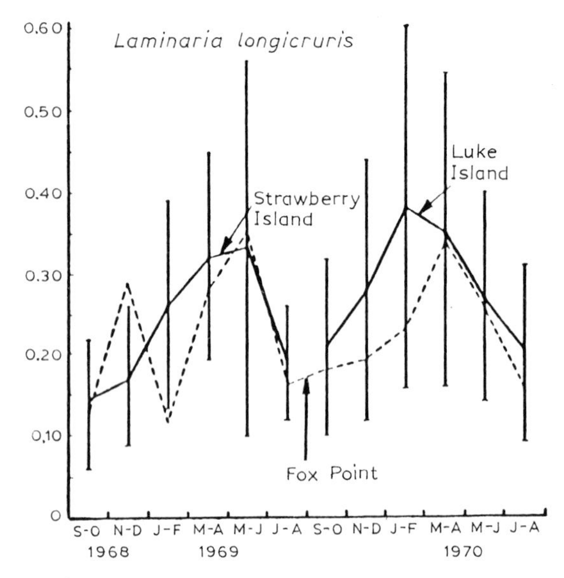

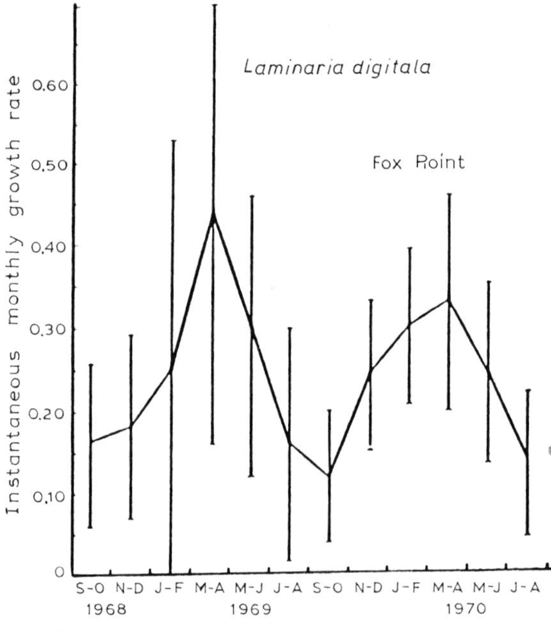

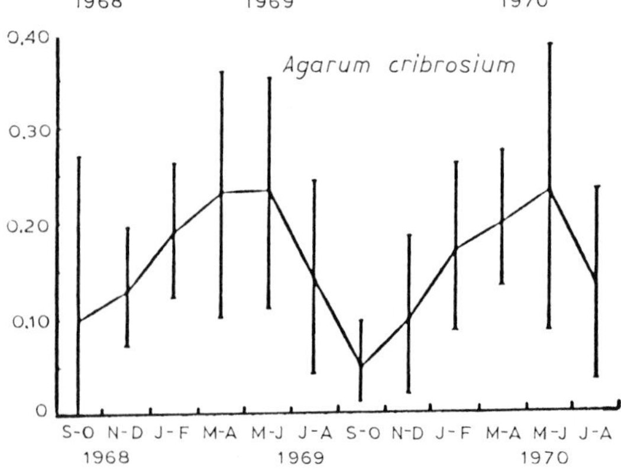

4.8 kg dry weight m^{-2} yr^{-1}. In St. Margaret's Bay, Nova Scotia, Mann (1972a, b) measured standing crops and annual production of littoral and sublittoral seaweeds. Within the kelp community, maximum biomass (16.01 kg fresh weight m^{-2}) occurred in a zone of mixed *L. digitata* and *L. longicruris* (Mann, 1972a). At greater depths, the mixed standing crop of *Agarum cribrosum* and *L. longicruris* totaled 4.88 kg dry weight m^{-2} (Table 7.4). For *A. cribrosum*, *L. digitata* and *L. longicruris*, maximum growth, as elongation, occurs during late winter and spring (Fig. 7.16). Average annual production for the seaweed zone (including littoral species) was calculated as 1750 g carbon-m^{-2} yr^{-1} (Mann, 1972b). Mann's (1972a, b) data were among the first to show the large organic production potential of temperate seaweed communities (Table 7.4). At West Pubnico, Nova Scotia, Pringle and Semple (1980) measured kelp biomass along a vertical transect. The mean standing crop of *L. longicruris* between +2.0 and −8.5 m was 944 g fresh weight m^{-2}, while that for *L. digitata* was 416 g fresh weight m^{-2} between +1.0 and −7.5 m (Fig. 7.5). Boden (1979) reported vertical differences in biomass and production of *Laminaria saccharina* at Appledore Island, Isles of Shoals, Gulf of Maine. Growth (as blade elongation) was maximal at −8 to −12 m (0.8–1.2 cm day^{-1}), decreasing both at shallower and deeper locations.

Several investigators have evaluated the influence of various environmental parameters on seasonal growth of kelp species (Chapman and Craigie, 1977, 1978; Hatcher et al., 1977; Boden, 1979; Gerard and Mann, 1979; Chapman and Gagné, 1980; Gagné and Mann, 1981; Gagné et al., 1982). In St. Margaret's Bay, Nova Scotia, growth of *Laminaria longicruris* is slowest from September to December (Mann, 1972b; Gagné and Mann, 1981; see also Fig. 7.17). During this period, the availability of external dissolved nitrogen and internal stored nitrogen may limit growth (Chapman and Gagné, 1980; Gagné et al., 1982). In addition, carbohydrate reserves become depleted (Hatcher et al., 1977; Chapman and Gagné, 1980; Gagné and Mann, 1981; Gagné et al., 1982).

Fig. 7.16. Seasonal growth patterns (cm $month^{-1}$) of seaweeds in eastern Canada. Vertical lines are standard deviations. (From Mann, 1972b.)

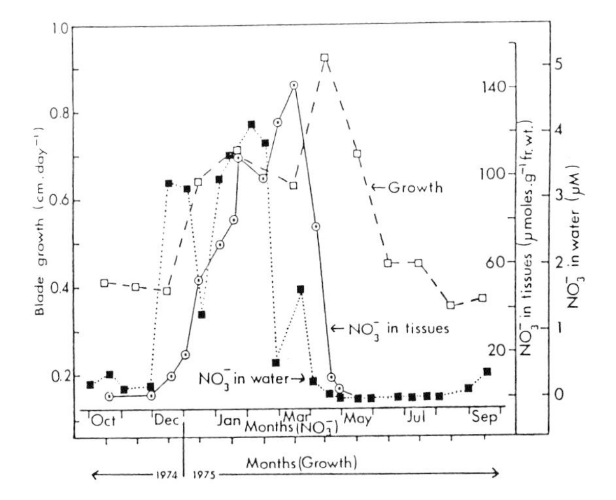

Fig. 7.17. Seasonal changes in growth rate, ambient nitrate and tissue nitrate at Fox Point, Nova Scotia. Growth rates are integrated over months; nutrient levels are shown for single points in time. (From Chapman and Gagné, 1980.)

Growth increases from December to March as dissolved inorganic nitrogen concentrations increase (Gagné and Mann, 1981). With increased nutrient availability, *L. longicruris* stores both inorganic and organic nitrogen reserves sufficient to support growth for several months, when irradiance is greatest, after ambient nitrogen concentrations decline (Chapman and Gagné, 1980; Gagné and Mann, 1981). However, in habitats without severe seasonal nutrient limitations, annual variations in growth and tissue chemistry are markedly different from this pattern. Under enhanced nitrogen availability (e.g. from upwelling), *L. longicruris* does not accumulate carbon reserves to support winter growth. Rather, maximum elongation takes place during the summer corresponding with seasonal irradiance patterns. Transplantation of kelp plants between these contrasting nutrient habitats indicates that there may be some degree of genetic differentiation with respect to seasonal growth and storage patterns (Chapman and Gagné, 1980).

APPENDIX I. GULF OF MAINE

Geography. The Gulf is a partially enclosed sea extending from Nantucket Shoals and Cape Cod, Massachusetts, on the west (70°W 42°N) to Cape Sable, Nova Scotia, on the east (65°W 43.5°N). Grand Manan, which is in the northeast corner of the Gulf, is the boundary of the Bay of Fundy (Fig. 7.1). The Gulf is approximately 968 km long, excluding coastal irregularities; it has an area of 93 240 km^2 and an average depth of 150 m. Approximately 30% of its length is exposed to the open Atlantic, that is, the shorelines of northern Massachusetts, New Hampshire, Maine and parts of New Brunswick and Nova Scotia. It extends seaward approximately 323 km, where it is separated from the North Atlantic by a rim of submarine shelves and banks, as well as by a major discontinuity of temperature and salinity associated with the Gulf Stream. Three major regions can be distinguished: north (Cape Elizabeth, Maine, to Grand Manan, New Brunswick), central (Cape Elizabeth to Cape Ann, Massachusetts), and southern (Cape Ann to Cape Cod Bay, Massachusetts), depending upon shore profiles, geology and tidal river discharges (cf. Geology). [Bigelow (1927, 1928); Bigelow and Schroeder (1953); Colton (1964); Anonymous (1974a, c); Apollonio (1979); Fefer and Schettig (1980a, b).]

Geology. The geology consists of a complex association of igneous and metamorphic rocks, most strata being of late Precambrian and Paleozoic ages. The central and southern portions tend to have lower shore profiles and a predominance of barrier beaches, as against the granitic headlands farther north. Islands are abundant in the northern (particularly) and central regions, while many fine harbors occur in southern portions. The central region is primarily composed of sandy barrier beaches, behind which extensive tidal marshes occur; the latter habitats are rare farther north. The majority of the largest estuarine waters (i.e. tidal rivers) entering the Gulf are in the north; only seven rivers of significant size occur in the central portions and none in the south. [Johnson (1925); Bloom (1960, 1963); Uchupi (1960, 1966); Borns (1963); Emery and Garrison (1967); Grant (1968); Schlee and Pratt (1970); Emery and Uchupi (1972); Anonymous (1974a, b, c); Ballard and Uchupi (1974); Schnitker (1974); Stuvier and Borns (1975); Fader et al. (1977); Apollonio (1979); Fefer and Schettig (1980a, b); Larsen and Doggett (1981).]

Circulation. The major circulation pattern is counterclockwise, but around Georges Bank and other shoals (Fig. 7.1), it is clockwise, with seasonal and annual variations. The average yearly drainage of fresh water is approximately 9.46×10^8 m^3, and primarily enters the Gulf of Maine north of Cape Elizabeth, Maine (i.e. 90%). This discharge provides the energy that generates the counterclockwise circulation in the Gulf, which is maximal in spring but continues year-round. During spring (May), one large cyclonic gyre encompasses the entire Gulf. It begins to slow down in June and by autumn and winter the southern side breaks down into a drift across Georges Bank. There is an inflow on the eastern side of the gyre from the Scotian Shelf and Brown Bank, particularly during winter. In southwestern Nova Scotia (the Lurcher Shoals) the drift may continue northward into the Bay of Fundy, or turn westward towards the coast of southern Maine, continue southward across Massachusetts Bay where it may divert into Cape Cod Bay through Great South Bay, or turn east, north of Georges Bank. South of Cape Cod, there is a sluggish southward movement of water. The Cape also serves as a dividing line between the Labrador Current and the Gulf Stream. [Bigelow (1914, 1917, 1927, 1928); Huntsman (1923); Redfield (1941); Haight (1942); Hachey et al. (1954); Day (1958); Chevrier (1959); Chevrier and Trites (1960); Colton (1964); Bumpus and Lauzier (1965); Lauzier (1967); Graham (1970a); Garrett (1972); Bumpus (1973); Anonymous (1974a, b, c, 1981b); Loucks et al. (1974); Sutcliffe et al. (1976); Apollonio (1979); Hopkins and Garfield (1979); Fefer and Schettig (1980b); DeWolfe (1981).]

Tides. Amplitudes of the semidiurnal tides are extremely variable spatially, due to resonance of the tidal wave in the constricted eastern Gulf and within the Bay of Fundy. South of Cape Cod, tidal ranges are rarely over 1.2 m; just to the north of the Cape, they rise to 2.1 to 3.0 m and continue to increase in a northerly and easterly direction. Mean and spring tidal ranges vary from 2.5 to 5.6 m and 2.9 to 6.4 m, respectively. Just to the north of the Gulf (within the Passamaquoddy Bay), tidal range is approximately 8.5 m, and even greater amplitudes occur upstream within the Bay of Fundy. [Rich (1929); Anonymous (1974a, b, c, 1984a); Apollonio (1979); Fefer and Schettig (1980b); Redfield (1980).]

Wave action and upwelling. Wave action ranges from high-energy granitic headlands (in the north) to low-energy shores with low profiles (in central and southern regions). Strong gradients of wave action occur within short distances, because of highly indented shorelines. Upwelling is common along the northeastern section, while localized upwelling occurs seasonally in central regions and more consistently within channel areas in the same regions. [Hulburt (1968, 1970); Hulburt and Corwin (1970); Anonymous (1974a, b, c); Yentsch et al. (1976); Apollonio (1979); Larsen and Doggett (1981).]

Temperatures. Because of extensive coastal upwelling (Garrett and Loucks, 1976) and tidal mixing, surface waters off southwestern Nova Scotia and eastern Maine are the coolest (Fig. 7.2), as well as the most saline. This turbulence results in small surface-to-bottom temperature differences year-round. Surface waters in the southwestern quarter of the Gulf (from Cape Ann and Jeffreys Ledge northward to Cape Elizabeth) are the warmest in summer and the least salty, while pronounced stratification occurs because of limited vertical mixing. Overall, surface water temperatures during summer show a maximum temperature variation of 10°C. In winter, surface temperatures tend to be cooler on the western than on the northern and eastern edges, as the former area is primarily exposed to cold westerly winds. The mean monthly temperature patterns for surface waters at Woods Hole, Massachusetts, which is southwest of the Gulf, are conspicuously warmer and more variable (0.5 to 21.6°C) than those on the east side of Cape Cod Canal (0.4 to 16.2°C), in central or northern Maine (0.4 to 14.2°C and 1.7 to 11.0°C respectively) or within southwestern Nova Scotia (2.9 to 11.2°C). [Bigelow (1927, 1928); Anonymous (1960, 1965, 1974a, b, c); Colton (1964, 1968, 1972); Colton et al. (1968); Hulburt (1968); Graham (1970b); Colton and Stoddard (1972); Hopkins and Garfield (1979); Fefer and Schettig (1980b); Mathieson and Penniman (1986a); Garrison and Brown (1987, 1989).]

Ice. Ice only forms in sheltered habitats, depending upon the severity of winter conditions (Fig. 7.3). [Anonymous (1974b); Fefer and Schettig (1980a).]

Salinity. The light fresh water that enters the Gulf (primarily in the northwest; cf. Circulation) flows out over and on top of the colder, denser sea water, resulting in unstable inshore waters, which are literally higher than offshore waters. Hence, the lowest salinities occur along the northern and western coastal areas, and the highest spring salinities are found on the eastern side of the Gulf. Surface salinities typically range from 28 to 33‰ in the open Gulf, with much greater ranges occurring in sites adjacent to estuaries (e.g. Great Bay, New Hampshire: 0 to 29‰). Typically, three water masses are distinguishable: (1) surface (28 to 33‰ and 0.4 to 16.2°C), (2) intermediate (32 to 33‰ and 1 to 6°C), and (3) bottom (31 to 33‰ and 1 to 17°C). [Bigelow (1927, 1928); Anonymous (1960, 1965, 1974a, b, c); Colton (1964); Colton et al. (1968); Graham (1970a, b); Apollonio (1979); Hopkins and Garfield (1979); Fefer and Schettig (1980b); Mathieson and Penniman (1986a); Garrison and Brown (1987, 1989).]

Nutrients. The pattern of nutrients varies geographically. Nutrient levels are higher, particularly that of inorganic nitrogen, in the northeastern section of the Gulf, due to extensive upwelling, while localized and scattered enriched areas exist because of upwelling in the central Gulf (cf. Wave Action and Upwelling). Off the coastline of southern Maine and New Hampshire where limited upwelling occurs, nutrients are generally highest from December to March (10 to 15 $\mu mol\,l^{-1}$); thereafter, a sharp decline occurs during spring (1.5 to 3 $\mu mol\,l^{-1}$). Intermediate levels are usually found during summer ($\sim$1 to 3 $\mu mol\,l^{-1}$) and they increase in the fall (3 to 8 $\mu mol\,l^{-1}$). [Rakestraw (1933); Redfield et al. (1937); Redfield and Keys (1938); Ketchum et al. (1958); Vaccaro (1963); Colton et al. (1968); Hulburt (1970); Hulburt and Corwin (1970); Anonymous (1974a, b, c); Daly et al. (1979); Fefer and Schettig (1980b); Norall et al. (1982).]

APPENDIX II. BAY OF FUNDY

Geography. The Bay of Fundy is located in the northeast corner of the Gulf of Maine (66°W 48°N to 64°W 46°N) (Fig. 7.1). It is bounded on the north and east by Nova Scotia, on the west by the coasts of Maine and New Brunswick, and to the south by Grand Manan. It consists of a large mass of water more than 161 km long from the mouth to the point at which its inner end forks into two arms. The average width of the bay is approximately 48 km. The margins and inner sections of the Bay are less than 75 m deep. [Bigelow (1927, 1928); Bell and MacFarlane (1933); Bigelow and Schroeder (1953); Stephenson and Stephenson (1954a, b, 1972); Colton (1964); Anonymous (1974a, c, 1979, 1981a, 1983); Apollonio (1979); Wilson et al. (1979); Cousens (1986).]

Geology. The coastline is steeply to moderately cliffed and primarily composed of hard granitic substrata, friable shale-like materials, or hard gray basalt. Shoreward from these cliffs massive and uneven rocky reefs occur, descending to low water. Irregular rocky platforms and masses alternate with fields of large and small boulders. Below low spring water level the shores often slope gradually, with a floor of rocks, pebbles and coarse gravel. In many areas the higher shoreline is dominated by angular-shaped rocks, while the lower has more rounded shapes due to erosion by pebbles and ice. [Johnson (1925); Stephenson and Stephenson (1954a, b, 1972); Klein (1963); MacFarlane (1964); Colinvaux (1966, 1970); Edelstein et al. (1970); Stone et al. (1970); Sherwin (1973); Anonymous (1974a, 1979, 1981a, 1983); Ballard and Uchupi (1974); Mott (1975); Railton (1975); MacLean and King (1976); Welsted (1976); Fader et al. (1977).]

Circulation. The northern branch of the counterclockwise current within the Gulf of Maine enters the Bay of Fundy southeast of Cape Mary, Nova Scotia. This inflow is most conspicuous at −50 m, but is recognizable from the surface to −100 m. A cyclonic circulation pattern (i.e. counterclockwise) is established within the northwest side of the basin, with the current flowing southwesterly from the Bay. Inflow into the Bay is maximal in summer and fall and minimal during winter. Outflow to the northern Gulf of Maine exhibits a seasonal variation, being minimal in winter and maximal during spring and early summer. Seven major drainage areas contribute a voluminous discharge of fresh water into the Bay, which, as in the Gulf of Maine, contributes to the counterclockwise circulation. Strong tidal currents (mean maximum ebb

3.9 knots or 201 cm s^{-1} near Gannett Rock, New Brunswick) are a characteristic feature of the Bay. [Mavor (1922); Huntsman (1923); Bigelow (1927, 1928); Watson (1936); MacGregor and McLellan (1952); Ketchum and Keen (1953); Hachey et al. (1954); Stephenson and Stephenson (1954a, b, 1972); Bumpus (1959, 1960, 1973); Chevrier (1959); Forrester (1959, 1960); Chevrier and Trites (1960); Hachey (1961); Colton (1964); MacFarlane (1964); Bumpus and Lauzier (1965); Lauzier (1967); Graham (1970a, b); Garrett (1972, 1984); Anonymous (1974c, 1981a, b, 1983, 1984b); Loucks et al. (1974); Metcalf et al. (1976); Sutcliffe et al. (1976); Moyse (1978); Apollonio (1979); DeWolfe (1981); Holloway (1981); Smith et al. (1984).]

Tides. Semidiurnal tides increase from the mouth (mean 2.7 m, spring 3.1 m) inland, with the world's largest tidal amplitude (mean 11.7 m, spring 13.3 m) occurring in the Minas Basin — the easternmost arm of the Bay. [Hind (1875); Hachey (1952); Stephenson and Stephenson (1954a, b, 1972); Anonymous (1974c, 1979, 1981a, 1983, 1984a); Moyse (1978); Apollonio (1979); Wilson et al. (1979); Fefer and Schettig (1980b); Redfield (1980); Cousens (1986).]

Wave action and upwelling. As the Bay faces 97 to 113 km of open water to the west and southwest, it experiences a range of conditions varying from relatively exposed to sheltered, as well as strong tidal currents (cf. Circulation). Extensive coastal upwelling occurs off the southwestern shore of Nova Scotia, primarily because of strong currents flowing parallel to the coast. [Watson (1936); Stephenson and Stephenson (1954a, b, 1972); Colinvaux (1966, 1970); Edelstein et al. (1970); Stone et al. (1970); Anonymous (1974c, 1979, 1981a, 1983); Garrett and Loucks (1976); Moyse (1978); Apollonio (1979); Hsueh and O'Brien (1981); Cousens (1982, 1986); Gagné et al. (1982).]

Temperatures. Mean surface water temperatures (Fig. 7.2), based upon long-term monthly means recorded at St. Andrews and Grand Manan, New Brunswick, near the southern boundary of the Bay of Fundy, range from 0.6 to 13.0°C (mean 7.0°C) and 1.7 to 11.4°C (mean 6.7°C), respectively. Thus, they are cooler in summer than on the Atlantic coast of Nova Scotia and within the Gulf of St. Lawrence, primarily because of extensive coastal upwelling as in the northern Gulf of Maine. [Bigelow (1927, 1928); Bailey et al. (1954); Stephenson and Stephenson (1954a, b, 1972); Bailey (1955); Forgeron (1959); Colton (1964, 1968, 1972); Lauzier and Hull (1969); Colton and Stoddard (1972); Anonymous (1974b, c, 1979, 1981a, 1983); Metcalf et al. (1976); Moyse (1978); Smith et al. (1984).]

Ice. Ice only forms in sheltered habitats, depending upon the severity of winter conditions (Fig. 7.3). [Hind (1875); Barnes (1913); Huntsman (1930); Anonymous (1946, 1979, 1981a, 1983); Stephenson and Stephenson (1954a, b, 1972); Colinvaux (1966, 1970); Moyse (1978); Gordon and Desplanque (1983).]

Salinity. The voluminous discharge of fresh water in spring (April to May) causes the Bay to be fresher than the Gulf of Maine during summer (cf. Circulation). An inflow of bottom water from the Gulf compensates for dilution from vertical mixing. Typically, surface water salinities are 30 to 33‰, except near sources of fresh water. [Bigelow (1927, 1928); Ketchum and Keen (1953); Stephenson and Stephenson (1954a, b, 1972); Forgeron (1959); Colton (1964); Colinvaux (1966, 1970); Graham (1970a, b); Sharaf El Din et al. (1970); Anonymous (1974c, 1979, 1981a, 1983); Metcalf et al. (1976); Moyse (1978); Apollonio (1979).]

Nutrients. At the mouth, nutrient-rich waters occur year round (e.g. >3.0 μmol l^{-1} dissolved inorganic nitrogen) due to upwelling (cf. Wave Action and Upwelling). [Anonymous (1974c, 1979, 1981a, 1983); Gagne et al. (1982); Keizer and Gordon (1985).]

APPENDIX III. ATLANTIC COAST OF NOVA SCOTIA

Geography. The eastern coastline of Nova Scotia is long and highly indented, primarily facing southeast (65°W 43.5°N to 59.7°W 47°N) (Fig. 7.1). Its coastal waters overlie a wide submarine platform, the Scotian Shelf, which is 91 to 183 m deep. An inshore strip of banks rises to depths of less than 91 m. [Bell and MacFarlane (1933); Stephenson and Stephenson (1954a, b, 1972); Anonymous (1974c); Cousens (1986).]

Geology. The coastline consists of massive granitic outcrops sloping into the sea at various angles, causing considerable variations in wave exposure. Within sheltered inlets the shores are more gentle, with some large scattered boulders. Gravel, stones and mud occur in the lower shoreline; mud dominates the offshore submarine platform, the Scotian Shelf. [Stephenson and Stephenson (1954a, b, 1972); MacFarlane (1964); Grant (1968); Kranck (1972); Sherwin (1973); Anonymous (1974a); Railton (1975); MacLean and King (1976); Owens and Bowen (1977).]

Circulation. The southwesterly flowing Scotian Current is chiefly driven by fresh-water outflow from the Gulf of St. Lawrence, which reaches the Shelf in late summer. Minimum and maximum transport rates (mean values) vary from ~ 100 to 300×10^3 m^3 s^{-1}. During periods of low fresh-water outflow from the Gulf of St. Lawrence, a clockwise gyre may occur south of Cape Breton Island. Only a small part of the easternmost southwesterly drift rounds Cape Sable and enters the northern Gulf of Maine and the Bay of Fundy. This drift ceases by June, and a series of cyclonic and anticyclonic eddies are established about the basins and banks off the Scotian Shelf. The area between Cape Sable and Shelburne (which is to the east) becomes a "dead area". [Hachey (1938, 1942, 1947); Hachey et al. (1954); Stephenson and Stephenson (1954a, b, 1972); MacFarlane (1964); Bumpus and Lauzier (1965); Lauzier (1967); Ingram (1972); Bumpus (1973); Anonymous (1974c); Loucks et al. (1974); Sutcliffe et al. (1976); Apollonio (1979); Lawrence (1979); Trites (1979a, b).]

Tides. Semidiurnal tides have mean and spring ranges of 0.7 to 2.2 m and 0.9 to 2.6 m, respectively. A clear gradient of increasing tidal amplitude occurs from the Atlantic coast into the Bay of Fundy, with a marked change both in amplitude and time of low water at Cape Sable. [Hachey (1938, 1942); Stephenson and Stephenson (1954a, b, 1972); Drinkwater (1979); Trites (1979a); Wilson et al. (1979); Anonymous (1984a); Cousens (1986).]

Wave action and upwelling. A pronounced gradient of diminishing wave action occurs from the open Atlantic into the Bay of Fundy. The highly indented shoreline (cf. Geology) allows a wide range of wave exposure over very short distances. Thus, vertical promontories are exposed to the direct forces of the open Atlantic, while adjacent bays and inlets of various types are extremely sheltered. Extensive upwelling is unknown. [Stephenson and Stephenson (1954a, b, 1972); MacFarlane (1964); Cousens (1982, 1986); Gagné et al. (1982).]

Temperature. The mean surface water temperatures (Fig. 7.2), based upon long-term monthly values, at Halifax Harbor and at the Sambro Lightship, Nova Scotia, range from 0.9 to 15.0°C (mean 7.6°C) and 1.2 to 15.9°C (mean 8.7°C) respectively. Thus, during summer, they are warmer than in the Bay of Fundy and the northern Gulf of Maine but cooler than within the Gulf of St. Lawrence. On the Scotian Shelf, three layers are recognizable (cf. Salinity); the bottom one is warmer (>5°C) and more saline than the middle (<5°C), while the surface layer is most variable. [Stephenson and Stephenson (1954a, b, 1972); Bailey (1955); Colton (1968, 1972); Colton et al. (1968); Lauzier and Hull (1969); Sharaf El Din et al. (1970); Colton and Stoddard (1972); Heath (1973); Anonymous (1974b).]

Ice. Ice forms in sheltered inlets and harbors (Fig. 7.3). During early spring, ice accumulated in the Gulf of St. Lawrence is discharged into the open Atlantic, and can spread more than 121 km from Cape Breton Island. East and south winds bring this sometimes as far west as Halifax. [Barnes (1913); Huntsman (1930); Anonymous (1946); Stephenson and Stephenson (1954a, b, 1972); Dinsmore (1972).]

Salinity. Surface-water salinities are usually <32.0‰. On the Scotian Shelf three layers of variable thickness occur, depending upon weather, season, and the behavior of adjacent Atlantic water. Surface-water salinities may be as low as 30.6‰; the intermediate layer, which is cooler than the bottom one (cf. Temperature), varies from 32.0 to 33.5‰, and the bottom layer has salinities >33.5‰. [Stephenson and Stephenson (1954a, b, 1972); Colton et al. (1968); Heath (1973); Anonymous (1974c); Apollonio (1979).]

Nutrients. A clear gradient of increasing nutrients (inorganic nitrogen) occurs from the Atlantic Ocean into the Bay of Fundy. For example, within St. Margaret's Bay abundant nitrogen (1.0 to 5.2 $\mu mol\ l^{-1}$) is present from December to March, while at the southwestern tip of Nova Scotia near Pubnico intermediate patterns occur (> 1.0 $\mu mol\ l^{-1}$ from October to March), whereas nitrogen is abundant year-round (> 3.0 $\mu mol\ l^{-1}$) near the mouth of the Bay of Fundy, due to local upwelling. [Platt and Irwin (1968, 1970, 1971); Colton et al. (1968); Mann (1972b); Chapman and Craigie (1977); Denman et al. (1977); Bugden et al. (1982); Gagné et al. (1982).]

APPENDIX IV. GULF OF ST. LAWRENCE

Geography. The Gulf is a large, shallow, enclosed sea connected to the Atlantic by three straits: (1) to the north by the narrow Strait of Belle Isle (56°W 54°N); (2) in the south by the still narrower Strait of Canso between Nova Scotia and Cape Breton Island (61.5°W 45.5°N); and (3) to the east by the widest strait, Cabot Strait, between Cape Breton Island and Newfoundland (60°W 46°N to 59°W 47.5°N) (see Fig. 7.1). Runoff from the St. Lawrence River (estuary) enters the northwestern corner of the Gulf. Two large islands (Anticosti and Prince Edward) lie within the Gulf, as well as several smaller islands (e.g. the Magdalens). The floor has a coastal strip < 91 m in depth and variable in width; it is widest in the southern part where it contains Magdalen Island, Magdalen Shallows, the coast of Prince Edward Island, and the coasts of New Brunswick, Nova Scotia and Cape Breton Island, which border on the Northumberland Strait and the Gulf of St. Lawrence. In addition, a deeper trench (the Laurentian Channel) occurs with depths > 183 m. The complicated topography of the region causes profound complications in its tides (cf. Tides). [Bell and MacFarlane (1933); Stephenson and Stephenson (1954a, b, 1972); MacFarlane (1966); Bird et al. (1983); Dickie and Trites (1983).]

Geology. The coastline, which faces either northward or westward, is basically composed of a soft red sandstone. Some areas have harder stone consisting of a mixture of boulders and pebbles, which are of Ordovician to middle Carboniferous age. Most shorelines are low-profiled cliffs, rapidly eroding due to the effects of waves and ice. At the foot of cliffs, rocky stretches (platforms) and sandy beaches may alternate with some rocky headlands. Platforms, which are either broken or continuous, have numerous crevices, boulders and pot-holes. Much of the surface between these irregularities has an extremely smooth, clean appearance. [Milligan (1949); Stephenson and Stephenson (1954a, b, 1972); MacFarlane (1966); Grant (1968); Loring and Nota (1973); Owens and Bowen (1977); Dadswell (1979); Bird et al. (1983); Dickie and Trites (1983); Pringle and Semple (1983, 1984); Lobban and Hanic (1984).]

Circulation. A southwest-flowing branch of the Labrador Current enters the Gulf of St. Lawrence on the Anticosti side of the Gaspé Passage. It then moves out to the open Atlantic via Cabot Strait with a mean outflow transport of $505 \times 10^3\ m^3\ s^{-1}$. An inward movement of sea water (mean $492 \times 10^3\ m^3\ s^{-1}$) also occurs on the Newfoundland side of Cabot Strait because of the clockwise circulation pattern of coastal waters around Newfoundland. A major inflow of fresh water (mean $19 \times 10^3\ m^3\ s^{-1}$) enters from the St. Lawrence River. This fresher water flows out through Cabot Strait, but on the Cape Breton side. Tidal currents are a characteristic feature of Northumberland Strait near Prince Edward Island. However, maximum velocities (2.3 knots or 118.3 $cm\ s^{-1}$, East Narrow) are weak compared to those within the Bay of Fundy. [Hachey et al. (1954); Stephenson and Stephenson (1954a, b, 1972); MacGregor (1956); Bumpus and Lauzier (1965); Lauzier (1965); MacFarlane (1966); Prest and Grant (1969); Prest (1970); Keyte and Trites (1971); Trites (1972, 1979a, b); Bumpus (1973); El Sabh (1976, 1977); Apollonio (1979); Dadswell (1979); Drinkwater (1979); Dickie and Trites (1983).]

Tides. The semidiurnal tides have a mean and spring tidal range of 0.4 to 2.0 m and 0.6 to 2.6 m, respectively. The complicated topography of the region (cf. Geography) causes pronounced complications, especially in southern parts of the Gulf. For example, on the north coast of Prince Edward Island there may be several consecutive days, twice in each month, where only one distinct high and low tide occur per day, while at other times there are the usual two high and two low tides. In addition, tides may sometimes remain at the same level for hours. [Stephenson and Stephenson (1954a, b, 1972); MacFarlane (1966); Drinkwater (1979); Bird et al. (1983); Dickie and Trites (1983); Anonymous (1984a); Lobban and Hanic (1984).]

Wave action and upwelling. Although the Gulf is an enclosed sea (cf. Geography), many areas (e.g. the north shore of Prince Edward Island) have strong wave action, as there is no land for nearly 322 km northward. Within Northumberland Strait there is a reduced fetch, as well as strong tidal currents. Pronounced variability of shore profiles and geographic features (i.e. embayments vs. promontories and steep vs. sloping shorelines) causes a wide range of wave action over a short distance. Sudden storms can cause extensive stirring and turbidity in the shallow waters. Wind-driven upwelling occurs in some locations. [Stephenson and Stephenson (1954a, b, 1972); MacFarlane (1964, 1966); Buckley et al. (1974); Dadswell (1979); Dickie and Trites (1983); McLachlan et al. (1987).]

Temperature. Long-term monthly values for mean surface water temperatures (Fig. 7.2) at North Point and Port Borden, Prince Edward Island, in the southern Gulf of St. Lawrence, range from 1.5 to 19.7°C (mean 7.2°C), and −1.1 to 18.5°C, (mean 7.6°C), respectively, with maxima >22°C. Thus, they have the warmest open coastal temperatures north of Cape Cod. However, during winter, inshore areas of the southern Gulf of St. Lawrence are cooler (icebound between December and mid-April) than most open coastal areas in the Gulf of Maine, the Bay of Fundy and on the Atlantic coast of Nova Scotia. In the northern Gulf of St. Lawrence (e.g. St. Mary's Island, Quebec), a narrower seasonal range and cooler summer temperatures occur (−1.6 to 10.8°C, mean 3.8°C). Thus, thermal patterns are cooler than in many areas to the south, as well as at several sites in Newfoundland. [Lauzier et al. (1951); Stephenson and Stephenson (1954a, b, 1972); Bailey (1955); Wilce (1959); MacFarlane (1966); Lauzier and Hull (1969); Hanic and Lobban (1971); Taylor (1975); Caddy et al. (1977); Prouse and Hargrave (1977); Lakshminarayana and Bourque (1979); Trites (1979a); Weiler and Keeley (1980); Bird et al. (1983); Dickie and Trites (1983); Lobban and Hanic (1984).]

Ice. The Gulf is icebound from early December to mid-April (Fig. 7.3), with both fast and pack ice occurring. Fast ice is produced by freezing of surface waters, while pack ice occurs because of its intrusion through the Strait of Belle Isle (a small amount), and due to *in situ* formation within the northern and western parts of the Gulf. The mass of ice accumulated in the Gulf is discharged into the Atlantic (via Cabot Strait) during early spring. In winter, the thickness of shore ice varies from approximately 0.46 to 1.02 m; if it is rafted and subjected to greater pressure, it may be 1.22 to 3.05 m or more in thickness. [Barnes (1913); Huntsman (1930); Anonymous (1946); Forward (1954); Stephenson and Stephenson (1954a, b, 1972); Wilce (1959); MacFarlane (1966); Dinsmore (1972); Dadswell (1979); O'Neill (1979); Bird et al. (1983); Dickie and Trites (1983); Lobban and Hanic (1984); McLachlan et al. (1987).]

Salinity. Within more "coastal" southern and central portions of the Gulf, a surface layer of some stability occurs above the thermocline, with salinities of 25.2 to 28.0‰ and temperatures >15°C, particularly during summer (cf. Temperature). Below the thermocline a body of water with higher salinities (30 to 32.4‰) and low temperatures (<5°C) occurs. In more sheltered areas like Northumberland Strait, surface salinities vary from 18.0 to 29.5‰. [Lauzier et al. (1951); Stephenson and Stephenson (1954a, b, 1972); MacFarlane (1966); Loring and Nota (1973); Taylor (1975); Lakshminarayana and Bourque (1979); Trites (1979a, b); Dickie and Trites (1983); Lobban and Hanic (1984).]

Nutrients. Nutrient levels are approximately three times higher than on the Scotian Shelf. [Steven (1971); Lakshminarayana and Bourque (1979); Bugden et al. (1982); Bird et al. (1983).]

APPENDIX V. NEWFOUNDLAND

Geography. Newfoundland (Fig. 7.1) occupies approximately 106 000 km^2 and is the world's sixteenth largest island, comparable to Cuba or Iceland. Its shoreline is highly indented and more than 8000 km long, and spans latitudes 45 to 52°N and longitudes 53 to 59°W. [Wilce (1959); Lee (1968); Mathieson et al. (1969); Hooper et al. (1980); South (1983); Steele (1983).]

Geology. The coastline consists chiefly of hard rocks of Precambrian (eastern shore) to Carboniferous age (western shore) with a southwestern to northeastern strike. Differential submergence has produced a characteristic linearity, the island being highest in the west due to postglacial rebound (since the Wisconsin Period) and tilting towards the east. Overall, the coastline is highly indented, rocky and very precipitous. Because of hardness of the rocks and youth of the coastline (10 000 years or less), few extensively eroded platforms exist except along the western shoreline, which is little dissected and consists of flat shelving rocks with low relief. Even so, the shoreline is typically mantled by a thin layer of unconsolidated Pleistocene sediments (till), which easily erode, forming rounded particles of various sizes. Much of the latter material is deposited in embayments to form pocket beaches, spits, barachois bars and tombolas. Sand beaches are uncommon; they are rare and small on the eastern shoreline, but larger and more common in central and western portions. Fine substrata (mud and silt) are almost entirely confined to the shores and bottoms of barachois ponds and the mouths of estuaries; hence, such deposits are small and scattered. [Schubert and Dunbar (1934); Wilce (1959); Mathieson et al. (1969); Hooper et al. (1980); Rogerson (1983); South (1983); Steele (1983).]

Circulation. The southwest-flowing Labrador Current moves along the island's eastern shore. The main flow passes around the southeast corner or the tail of the Grand Banks and mixes in eddies with warmer waters of the northeast-flowing North Atlantic Drift, a continuation of the Gulf Stream. A small branch of the Labrador Current enters the Strait of Belle Isle north of Newfoundland and flows into the Gulf of St. Lawrence. [Hachey et al. (1954); Bumpus and Lauzier (1965); Bumpus (1973); Sutcliffe et al. (1976); Apollonio (1979); Hooper et al. (1980); South (1983); Steele (1983).]

Tides. Mixed semidiurnal tides occur on the southwest coast, and semidiurnal on the west and north coasts. Tidal range is small (mean = 0.6 to 1.5 m, spring = 0.9 to 1.9 m), with maximum amplitudes occurring on middle portions of the eastern shore and on the south coast, particularly at the head of Placentia Bay (1.5 m). Elsewhere, the average tidal range is 1.0 m or less. [Wilce (1959); Dohler (1963); Lee (1968); Hooper et al. (1980); South (1983); Steele (1983); Anonymous (1984a).]

Wave action and upwelling. Most shores are semi- to highly exposed, the latter occurring on headlands and islands of the eastern side (e.g. the Avalon Peninsula). Sheltered sites are small, scattered and mainly restricted to heads of embayments, except where the shore is protected by offshore islands (Notre Dame and Bonavista Bays). Wave height increases from west to east. The Avalon Peninsula receives the largest waves and is often exposed to upwellings of cold intermediate waters. [Mathieson et al. (1969); Hooper et al. (1980); Banfield (1983); South (1983); Steele (1983).]

Temperature. The south-flowing Labrador Current reduces water temperatures and differentially affects vertical stratification. For example, surface water temperatures (Fig. 7.2) on the central west coast (Bonne Bay) and the southeastern shoreline (Avalon Peninsula) have approximately the same seasonal ranges (1 to 16°C as against 0 to 15°C). However, during summer and fall, surface water temperatures at Bonne Bay are somewhat higher than those on the Avalon Peninsula, while deeper waters in the former are considerably colder (approximately 1°C as against 5°C, at −50 m). Exposed open coastal sites like those on the Avalon Peninsula tend to have unstable thermoclines due to mixing, while deep embayments on the south and west coasts have stable thermoclines in summer and early fall, with warmer waters overlying the essentially Arctic water. [Bailey and Hachey (1951); Lauzier (1952, 1957); Bailey (1955); Leim (1957); Mathieson et al. (1969); Steele (1974, 1975, 1983); Hooper et al. (1980); South (1983).]

Ice. Fast ice occurs mostly in sheltered bays on the south, west and especially the northern coasts (Fig. 7.3). Pack ice, which is formed from breakage of land fast ice, drifts at the mercy of wind and current. When blown landward, it compresses into a solid mass developing great lateral pressures. It may also thicken as it piles up. With offshore winds pack ice blows out to sea and forms a loose mixture of ice and water; it may be formed around Newfoundland, but to a greater extent develops in the Labrador Sea, from which it drifts south in late winter and early spring, reaching its peak in March to April. At this time it usually reaches the Avalon Peninsula, but in a light ice year it may only reach much farther north. In a heavy year it may extend as far south as Placentia Bay. Pack ice also enters and blocks the Strait of Belle Isle, but the narrowness of the Strait prevents much ice from entering the Gulf of St. Lawrence. The south coast is typically free of pack ice. Icebergs, which are common and primarily carved from glaciers in Greenland, drift southward with the Labrador Current. Most travel offshore and circulate around the edge of Grand Banks, but some enter the Gulf of St. Lawrence, and many others reach the north and east coasts. [Barnes (1913); Huntsman (1930); Anonymous (1946); Wilce (1959); Dinsmore (1972); Hooper et al. (1980); Bolton (1983); South (1983); Steele (1983); Keats et al. (1985).]

Salinity. Surface-water salinities vary from 30 to 32‰ near-shore to 35‰ in the warm North Atlantic Drift south of the island. Near river mouths salinities are reduced. Thus, within estuaries and lagoons a thin layer of fresh water may occur which is $>18°C$ in summer; this surface layer overlies a thicker layer of cooler and denser sea water. [Lauzier (1957); Mathieson et al. (1969); Steele (1974, 1983); Hooper et al. (1980); South (1983).]

Nutrients. Inorganic nitrogen levels are relatively high in coastal waters, $>9\ \mu mol\ l^{-1}$ during winter. In April and May there is a dramatic decrease from winter maxima to summer minima ($2.2\ \mu mol\ l^{-1}$). In comparison to other sites on the Atlantic coast of Nova Scotia and in southern New England, these summer concentrations are relatively high. [Buggeln (1978).]

ACKNOWLEDGEMENT

Scientific Contribution No. 1503 from the New Hampshire Agricultural Experiment Station, also published as Jackson Estuarine Laboratory Contribution Number 204.

REFERENCES

Abbott, R.T., 1968. *Seashells of North America*. Golden Press, New York, 280 pp.

Abrahamson, W.E. and Gadgil, M., 1973. Growth form and reproductive effort in goldenrods (*Solidago*, Compositae). *Am. Nat.*, 107: 651–661.

Adey, W.H., 1964. The genus *Phymatolithon* in the Gulf of Maine. *Hydrobiologia*, 103: 377–420.

Adey, W.H., 1965. The genus *Clathromorphum* (Corallinaceae) in the Gulf of Maine. *Hydrobiologia*, 26: 539–573.

Adey, W.H., 1966a. Distribution of saxicolous crustose corallines in the northwestern North Atlantic. *J. Phycol.*, 2: 49–54.

Adey, W.H., 1966b. The genera *Lithothamnion, Leptophytum* (nov. gen.) and *Phymatolithon* in the Gulf of Maine. *Hydrobiologia*, 28: 321–370.

Adey, W.H., 1966c. The genus *Pseudolithophyllum* (Corallinaceae) in the Gulf of Maine. *Hydrobiologia*, 27: 479–497.

Adey, W.H., 1970. The effects of light and temperature on growth rates in boreal–subarctic crustose corallines. *J. Phycol.*, 6: 269–276.

Adey, W.H., 1971. The sublittoral distribution of crustose corallines on the Norwegian coast. *Sarsia*, 46: 41–58.

Adey, W.H., 1973. Temperature control of reproduction and productivity in a subarctic coralline alga. *Phycologia*, 12: 111–118.

Anderson, M.R., Cardinal, A. and Larochelle, J., 1981. An alternate growth pattern for *Laminaria longicruris*. *J. Phycol.*, 17: 405–411.

Anderson, R.J. and Velimirov, B., 1982. An experimental investigation of the palatability of kelp bed algae to the sea urchin *Parechinus angulosus*. *Mar. Ecol. Prog. Ser.*, 3: 357–373.

Andrews, J.H., 1976. Pathology of marine algae. *Biol. Rev.*, 51: 211–253.

Andrews, J.H. and Goff, L.J., 1985. Pathology. In: M.M. Littler and D.S. Littler (Editors), *Handbook of Phycological Methods, Ecological Field Methods: Macroalgae*. Cambridge Univ. Press, Cambridge, pp. 573–591.

Anonymous, 1946. *Ice Atlas of the Northern Hemisphere*. U.S. Hydrographic Office, Washington, D.C.

Anonymous, 1960. *Surface water temperature and salinity, Atlantic Coast, North and South America*. U.S. Dept. Comm., Coast and Geod. Serv. Publ. 31-1, 76 pp.

Anonymous, 1965. *Surface water temperature and salinity, Atlantic Coast, North and South America*. U.S. Dept. Comm., Coast and Geod. Serv. Publ. 31-1, 2nd rev., 88 pp.

Anonymous, 1974a. *A Socio-Economic and Environmental Inventory of the North Atlantic Region Including the Outer Continental Shelf and Adjacent Waters from Sandy Hook, New Jersey, to Bay of Fundy, Vol. 1. Bk. 1.* The Research Inst. Gulf of Maine, So. Portland, Maine, 519 pp.

Anonymous, 1974b. *A Socio-Economic and Environmental Inventory of the North Atlantic Region Including the Outer*

Continental Shelf and Adjacent Waters from Sandy Hook, New Jersey, to Bay of Fundy, Vol. 1. Bk. 2. The Research Inst. Gulf of Maine, So. Portland, Maine, 803 pp.

Anonymous, 1974c. *A Socio-Economic and Environmental Inventory of the North Atlantic Region Including the Outer Continental Shelf and Adjacent Waters from Sandy Hook, New Jersey, to Bay of Fundy, Vol. 1. Bk. 3.* The Research Inst. Gulf of Maine, So. Portland, Maine, 746 pp.

Anonymous, 1979. Bay of Fundy bibliography: supplement I. *Proc. N.S. Inst. Sci.*, 29: 313–314.

Anonymous, 1981a. Bay of Fundy bibliography: supplement II. *Proc. N.S. Inst. Sci.*, 31: 181–185.

Anonymous, 1981b. *Atlas of Tide Currents 1981: Bay of Fundy and Gulf of Maine.* Can. Hydrog. Serv., Dept. Fish. and Oceans, Ottawa, 36 pp.

Anonymous, 1983. Bay of Fundy bibliography: supplement III. *Proc. N.S. Inst. Sci.*, 33: 101–105.

Anonymous, 1984a. *Tide tables 1985, high and low water predictions, east coast of North and South America, including Greenland.* U.S. Dept. Comm. National Oceanic and Atmospheric Administration, Washington, D.C., 285 pp.

Anonymous, 1984b. *Tidal current tables 1985, Atlantic Coast of North America.* U.S. Dept. of Comm. National Oceanic and Atmospheric Administration, Washington, D.C., 241 pp.

Apollonio, S., 1979. *The Gulf of Maine.* Courier of Maine Books, Rockland, Maine, 59 pp.

Ardré, F., 1970. Contribution à l'étude des algues marines du Portugal. I. La flore. *Portug. Acta Biol. Ser. B Biogeogr.*, 10(1/4): 137–555.

Arnold, D.C., 1976. Local denudation of the sublittoral fringe by the green sea urchin, *Strongylocentrotus droebachiensis* (O.F. Muller). *Can. Field Nat.*, 90: 186–187.

Augier, J. and Hoffman, G., 1952. Sucre et brome dans le *Polysiphonia fastigiata. Bull. Soc. Bot. Fr.*, 99: 80–82.

Ayling, A.L., 1978. The relation of food availability and food preferences to the field diet of an echinoid *Evechinus chloroticus* (Valenciennes). *J. Exp. Mar. Biol. Ecol.*, 33: 223–235.

Baardseth, E., 1970. *Synopsis of biological data on knobbed wrack, Ascophyllum nodosum (Linnaeus) Le Jolis.* FAO Fish. Synop., No. 38 Rev. 1, 41 pp.

Babb, I., 1985. The biomechanics of kelps: their distribution, morphology, tensile strength and form efficiency. *Abstr. Twenty-fourth Northeast Algal Symp., Woods Hole, Mass.*

Bacon, L.C., 1983. Environmental variables influencing gamete release in *Ascophyllum nodosum*: a multivariate analysis. *Program and Abstr. Twenty-second Northeast Algal Symp., Woods Hole, Mass.*

Bailey, W.B., 1955. Summer surface temperatures in the Canadian Atlantic. *Fish. Res. Board Can. Prog. Rep.*, 63: 16–18.

Bailey, W.B. and Hachey, H.B., 1951. *Hydrographic features of the Strait of Belle Isle.* Joint Committee on Oceanography, Atlantic Group, St. Andrews, New Brunswick, 19 pp.

Bailey, W.B., MacGregor, D.G. and Hachey, H.B., 1954. Annual variations of temperature and salinity in the Bay of Fundy. *J. Fish. Res. Board Can.*, 11: 32–47.

Baker, J.T. and Murphy, V., 1976. *Compounds from Marine Organisms, Vol. 1.* CRC Press, Cleveland, Ohio, 226 pp.

Ballantine, W.J., 1961. A biologically defined exposure scale for the comparative description of rocky shores. *Field Studies*, 1: 1–19.

Ballard, R.D. and Uchupi, E., 1974. Geology of the Gulf of Maine. *Bull. Am. Assoc. Petroleum Geol.*, 58: 1156–1158.

Banfield, C.E., 1983. Climate. In: G.R. South (Editor), *Biogeography and Ecology of the Island of Newfoundland.* Junk, The Hague, pp. 37–106.

Barilotti, C.D. and Silverthorne, W., 1972. A resource management study of *Gelidium robustum.* In: K. Nisizawa (Editor), *Proc. Seventh Int. Seaweed Symp., Sapporo, Japan.* Univ. Tokyo Press, Tokyo, pp. 255–261.

Barnes, H., 1956. *Balanus balanoides* (L.) in the Firth of Clyde: the development and annual variation of the larval population, and the causative factors. *J. Animal Ecol.*, 25: 72–84.

Barnes, H. and Topinka, J.A., 1969. Effects of the nature of the substratum on the force required to detach a common littoral alga. *Am. Zool.*, 9: 753–758.

Barnes, H.T., 1913. *Report on the influence of icebergs and land on the temperature of the seas.* Ann. Rep. Dept. Mar. Fish., Ottawa, Sessional Paper 21c, 37 pp.

Barton, E.S., 1901. On certain galls in *Furcellaria* and *Chondrus. J. Bot.*, 39: 49–51.

Behbehani, M.I. and Croker, R.A., 1982. Ecology of beach wrack in northern New England with special reference to *Orchestia platensis. Estuarine Coastal Shelf Sci.*, 15: 611–620.

Bell, E.M. and Cheney, D.P., 1985. Comparison of food preference and phenolic content of intertidal New England seaweeds. *Abstr. Twenty-fourth Northeast Algal Symp., Woods Hole, Mass.*

Bell, H.P. and MacFarlane, C., 1933. The marine algae of the Maritime Provinces of Canada. II. A study of their ecology. *Can. J. Res.*, 9: 280–293.

Bent, A.C., 1962. *Life Histories of North American Wild Fowl, Vols. I and II.* Dover, New York, 685 pp.

Bernstein, B.B. and Mann, K.H., 1982. Changes in the nearshore ecosystem of the Atlantic Coast of Nova Scotia, 1968–81. *NATO Sci. Coun. Studies*, 5: 101–105.

Bernstein, B.B., Williams, B.E. and Mann, K.H., 1981. The role of behavioral responses to predators in modifying urchins' (*Strongylocentrotus droebachiensis*) destructive grazing and seasonal foraging patterns. *Mar. Biol.*, 63: 39–49.

Bernstein, B.B., Schroeter, S.C. and Mann, K.H., 1983. Sea urchin (*Strongylocentrotus droebachiensis*) aggregating behavior investigated by a subtidal multifactorial experiment. *Can. J. Fish. Aquat. Sci.*, 40: 1975–1986.

Bewley, J.D., 1979. Physiological aspects of desiccation tolerance. *Ann. Rev. Plant Physiol.*, 30: 195–238.

Bhakuni, D.S. and Silva, M., 1974. Biodynamic substances from marine flora. *Bot. Mar.*, 17: 40–51.

Biard, J.F., Verbist, J.F., Boterff, J. Le, Ragas, G. and Lecocq, M., 1980. Algues fixées de la côte Atlantique Française contenant des substances antibactériennes et antifungiques. *Planta Med. Suppl.*, pp. 136–151.

Bigelow, H.B., 1914. Explorations in the Gulf of Maine, July and August, 1912, by the U.S. Fisheries schooner *Grampus.*

Oceanography and notes on the plankton. *Bull. Mus. Comp. Zool.*, 58: 31–134.

Bigelow, H.B., 1917. Explorations of the coast water between Cape Cod and Halifax in 1914 and 1915, by the Fisheries schooner *Grampus*. Oceanography and plankton. *Bull. Mus. Comp. Zool.*, 61: 161–357.

Bigelow, H.B., 1927. Physical oceanography of the Gulf of Maine. *Bull. U.S. Bur. Fish.*, 40: 511–1027.

Bigelow, H.B., 1928. Exploration of the waters of the Gulf of Maine. *Geogr. Rev.*, 18: 232–260.

Bigelow, H.B. and Schroeder, W.C., 1953. Fishes of the Gulf of Maine. *Fish. Bull. U.S. Fish and Wildl. Serv.*, 53: 1–577.

Bird, C.J., Edelstein, T. and McLachlan, J., 1976. Investigations of the marine algae of Nova Scotia. XII. The flora of Pomquet Harbour. *Can. J. Bot.*, 54: 2726–2737.

Bird, C.J., Greenwell, M. and McLachlan, J., 1983. Benthic marine algal flora of the north shore of Prince Edward Island (Gulf of St. Lawrence), Canada. *Aquat. Bot.*, 16: 315–335.

Birkeland, C., 1982. Terrestrial runoff as a cause of outbreaks of *Acanthaster planci* (Echinodermata:Asteroidea). *Mar. Biol.*, 69: 175–185.

Bloom, A.L., 1960. *Late Pleistocene changes of sea level in southwestern Maine*. Maine Geol. Survey, 143 pp.

Bloom, A.L., 1963. Late Pleistocene fluctuations of sea level and post-glacial rebound in coastal Maine. *Am. J. Sci.*, 261: 862–879.

Bloom, S.A., 1975. The motile escape response of a sessile prey: a sponge-scallop mutualism. *J. Exp. Mar. Biol. Ecol.*, 17: 311–321.

Boaden, P.J.S., O'Connor, R.J. and Seed, R., 1975. The composition and zonation of a *Fucus serratus* community in Strangford Lough, Co. Down. *J. Exp. Mar. Biol. Ecol.*, 17: 111–136.

Boden, G.T., 1979. The effect of depth on summer growth of *Laminaria saccharina* (Phaeophyta, Laminariales). *Phycologia*, 18: 405–408.

Bolton, J.J., 1981. Community analysis of vertical zonation patterns on a Newfoundland rocky shore. *Aquat. Bot.*, 10: 299–316.

Bolton, J.J., 1983. Effects of short-term ice scouring on a Newfoundland rocky shore community. *Astarte*, 12: 39–48.

Boney, A.D., 1965. Aspects of the biology of seaweeds of economic importance. *Adv. Mar. Biol.*, 3: 105–253.

Borns, H.W., 1963. Preliminary report on the age and distribution of the late Pleistocene ice in north-central Maine. *Am. J. Sci.*, 261: 738–740.

Bousfield, E.L. and Laubitz, D.R., 1972. Station lists and new distribution records of littoral marine invertebrates of the Canadian Atlantic and New England regions. *Nat. Mus. Can. Publ. Biol. Oceanogr.*, 5: 1–51.

Bousfield, E.L. and Thomas, M.L.H., 1975. Postglacial changes in distribution of littoral marine invertebrates in the Canadian Atlantic region. *Proc. N.S. Inst. Sci. Supp.*, 3: 47–60.

Bowen, M., Shipman, J. and Kinner, P., 1986. Assessing variability in a New Hampshire coastal hard-bottom benthic community: a matter of scale. *Abstr. 1986 Benthic Ecol. Meet., Boston, Mass.*

Bowman, R.S., 1986. The biology of the limpet *Patella vulgata* L. in the British Isles: spawning time as a factor determining recruitment success. In: P.G. Moore and R. Seed (Editors), *The Ecology of Rocky Coasts*. Columbia Univ. Press, New York, pp. 178–193.

Branch, G.M., 1981. The biology of limpets: physical factors, energy flow and ecological interactions. *Oceanogr. Mar. Biol. Ann. Rev.*, 19: 235–380.

Branch, G.M., 1986. Limpets: their role in littoral and sublittoral community dynamics. In: P.G. Moore and R. Seed (Editors), *The Ecology of Rocky Coasts*. Columbia Univ. Press, New York, pp. 97–116.

Breen, P.A., 1974. *Relations among lobsters, sea urchins and kelp in Nova Scotia*. Ph.D. Thesis, Dalhousie Univ., Halifax, N.S., 190 pp.

Breen, P.A., 1980. Relations among lobster, sea urchins, and kelp in Nova Scotia. In: J.D. Pringle, G.J. Sharp and J.F. Caddy (Editors), *Proc. Workshop on the relationship between sea urchin grazing and commercial plant/animal harvesting. Can. Tech. Rep. Fish. Aquat. Sci.*, 954: 24–32.

Breen, P.A. and Mann, K.H., 1976a. Changing lobster abundance and the destruction of kelp beds by sea urchins. *Mar. Biol.*, 34: 137–142.

Breen, P.A. and Mann, K.H., 1976b. Destructive grazing of kelp by sea urchins in eastern Canada. *J. Fish. Res. Board Can.*, 33: 1278–1283.

Brenner, D., Valiela, I. and Van Raalte, C.D., 1976. Grazing by *Talorchestia longicornis* on an algal mat in a New England salt marsh. *J. Exp. Mar. Biol. Ecol.*, 22: 161–169.

Briggs, J.C., 1974. *Marine Zoogeography*. McGraw Hill, New York, 475 pp.

Briscoe, C.S. and Sebens, K.P., 1986. Omnivory in the green sea urchin *Strongylocentrotus droebachiensis* (Muller). *Abstr. 1986 Benthic Ecol. Meet., Boston, Mass.*

Buckley, D.E., Owens, E.H., Schafer, C.T., Vilks, G., Cranston, R.E., Rashid, M.A., Wagner, F.J.E. and Walker, D.A., 1974. Canso Strait and Chedabucto Bay: A multidisciplinary study of the impact of man on the marine environment. *Geol. Surv. Can. Pap.*, 74–30: 133–160.

Bugden, G.L., Hargrave, B.T., Sinclair, M.M., Tang, C.L., Theriault, J.-C. and Yeats, P.A., 1982. Freshwater runoff effects in the marine environment: the Gulf of St. Lawrence example. *Can. Tech. Rep. Fish. Aquat. Sci.*, 1078: 1–89.

Buggeln, R.G., 1978. Physiological investigations on *Alaria esculenta* (Laminariales, Phaeophyceae). IV. Inorganic and organic nitrogen in the blade. *J. Phycol.*, 14: 156–160.

Bumpus, D.F., 1959. *Sources of water in the Bay of Fundy contributed by surface circulation*. Rep. of the Int. Passamaquoddy Fish. Board to the Int. Joint Comm., Chap. 6, 10 pp.

Bumpus, D.F., 1960. Sources of water contributed to the Bay of Fundy by surface circulation. *J. Fish. Res. Board Can.*, 17: 181–197.

Bumpus, D.F., 1973. A description of the circulation on the continental shelf of the east coast of the United States. In: B.A. Warren (Editor), *Progress in Oceanography. Vol. 6*. Woods Hole Oceanographic Institute, Woods Hole, Mass., pp. 111–157.

Bumpus, D.F. and Lauzier, L.M., 1965. *Surface circulation on the continental shelf of eastern North America between*

Newfoundland and Florida. Serial Atlas of the Marine Environment, Folio F., Am. Geog. Soc., New York, N.Y.

Burkholder, P.R. and Sharma, G.M., 1969. Antimicrobial agents from the sea. *Lloydia*, 32: 466–483.

Burns, R.L. and Mathieson, A.C., 1972a. Ecological studies of economic red algae. II. Culture studies of *Chondrus crispus* Stackhouse and *Gigartina stellata* (Stackhouse) Batters. *J. Exp. Mar. Biol. Ecol.*, 8: 1–6.

Burns, R.L. and Mathieson, A.C., 1972b. Ecological studies of economic red algae III. Growth and reproduction of natural and harvested populations of *Gigartina stellata* (Stackhouse) Batters in New Hampshire. *J. Exp. Mar. Biol. Ecol.*, 9: 77–95.

Burrows, E.M. and Lodge, S.M., 1950. A note on the interrelationships of *Patella, Balanus* and *Fucus* on a semi-exposed coast. *Ann. Rep. 1949, Mar. Biol. Sta. Port Erin*, 62: 30–34.

Busse, P.K., 1983. Significance of small plants occurring high in the *Ascophyllum* zone. *Program and Abstr. Twenty-second Northeast Algal Symp., Woods Hole, Mass.*

Caddy, J.F., Amaratunga, T., Dadswell, M.J., Edelstein, T., Linkletter, L.E., McMullin, B.R., Stasko, A.B. and Van de Poll, H.W., 1977. *1975 Northumberland Strait project, Part I: Records of benthic fauna, flora, demersal fish and sedimentary data. Fish. Mar. Serv. MS Rep. 1431*, Fish. and Environ. Can., 46 pp.

Caffey, H.M., 1985. Spatial and temporal variation in settlement and recruitment of intertidal barnacles. *Ecol. Monogr.*, 55: 313–332.

Cambridge, M.L., Breeman, A.M., Van Oosterwijk, R. and Van den Hoek, C., 1984. Temperature responses of some North Atlantic *Cladophora* species (Chlorophyceae) in relation to their geographic distribution. *Helgol. Wiss. Meeresunters.*, 38: 349–363.

Caram, B. and Jónsson, S., 1972. Nouvel inventaire des algues marines de l'Islande. *Acta Bot Isl.*, 1: 5–31.

Carlton, J.T., 1979. *Caveat indigenae: the unintentional introduction of exotic species with target species in mariculture operations, and the potential ecological impact on the adjacent environment.* Int. Council Exploration Sea, 69th Plenary Session, Woods Hole, Mass, U.S.A., Mariculture Comm., C.M. 1981/F, 5 pp.

Carlton, J.T. and Scanlon, J.A., 1985. Progression and dispersal of an introduced alga: *Codium fragile* ssp. *tomentosoides* (Chlorophyta) on the Atlantic Coast of North America. *Bot Mar.*, 28: 155–165.

Carlton, J.T., Cheney, D.P. and Vermeij, G.J., 1982. Ecological effects and biogeography of an introduced marine species: the periwinkle, *Littorina littorea*. *Malacol. Rev.*, 15: 143–150.

Carter, J.A. and Steele, D.H., 1982. Stomach contents of immature lobsters (*Homarus americanus*) from Placentia Bay, Newfoundland. *Can. J. Zool.*, 60: 337–347.

Chao, L.N., 1973. Digestive system and feeding habits of the cunner, *Tautogolabrus adspersus*, a stomachless fish. *Fish. Bull.*, 71: 565–586.

Chapman, A.R.O., 1973. A critique of prevailing attitudes towards the control of seaweed zonation on the shore. *Bot. Mar.*, 16: 80–82.

Chapman, A.R.O., 1979. *Biology of Seaweeds*. Univ. Park Press, Baltimore, Md., 134 pp.

Chapman, A.R.O., 1981. Stability of sea urchin dominated barren grounds following destructive grazing of kelp in St. Margaret's Bay, eastern Canada. *Mar. Biol.*, 62: 307–311.

Chapman, A.R.O., 1984. Reproduction, recruitment and mortality in two species of *Laminaria* in southwest Nova Scotia. *J. Exp. Mar. Biol. Ecol.*, 78: 99–109.

Chapman, A.R.O. and Craigie, J.S., 1977. Seasonal growth in *Laminaria longicruris*: relations with dissolved inorganic nutrients and internal reserves of nitrogen. *Mar. Biol.*, 40: 197–205.

Chapman, A.R.O. and Craigie, J.S., 1978. Seasonal growth in *Laminaria longicruris*: relations with reserve carbohydrate storage and production. *Mar. Biol.*, 46: 209–213.

Chapman, A.R.O. and Gagné, J., 1980. Environmental control of kelp growth in St. Margaret's Bay and on the southwest shore of Nova Scotia. In: J.D. Pringle, G.L. Sharp and J.F. Caddy (Editors), *Proc. Workshop on the relationship between sea urchin grazing and commercial plant/animal harvesting. Can. Tech. Rep. Fish. Aquat. Sci.*, 954: 194–207.

Chapman, V.J., 1943. Zonation of marine algae on the sea shore. *Proc. Linn. Soc. London*, 154: 239–253.

Chemin, E., 1927. Action des bactéries sur quelques algues rouges. *Bull. Soc. Bot. Fr.*, 74: 441–451.

Chemin, E., 1931. Sur la présence de galles chez quelques Floridées. *Rev. Algol.*, 5: 315–325.

Chen, L.C.-M. and McLachlan, J., 1972. The life history of *Chondrus crispus* in culture. *Can. J. Bot.*, 50: 1055–1060.

Chen, L.C.-M. and Taylor, A.R.A., 1976. Scanning electron microscopy of early sporeling ontogeny of *Chondrus crispus*. *Can. J. Bot.*, 56: 672–678.

Chen, L.C.-M., Edelstein, T. and McLachlan, J., 1969. *Bonnemaisonia hamifera* Hariot in nature and in culture. *J. Phycol.*, 5: 211–220.

Chen, L.C.-M., Edelstein, T. and McLachlan, J., 1974. The life history of *Gigartina stellata* (Stackh.) Batt. (Rhodophyceae, Gigartinales) in culture. *Phycologia*, 13: 287–294.

Cheney, D.P., 1977. R & C/P — a new and improved ratio for comparing seaweed floras. *J. Phycol.*, 13 (Suppl.): 12.

Cheney, D. P., 1982. The determining effects of snail herbivore density on intertidal algal recruitment and composition. *J. Phycol.*, 18 (Suppl.): a8.

Chevrier, J.R., 1959. *Drift bottle measurements in the Quoddy region*. Rep. of the Int. Passamaquoddy Fish. Board to the Int. Joint Comm., Ch. 2, 13 pp.

Chevrier, J.R. and Trites, R.W., 1960. Drift bottle experiments in the Quoddy region, Bay of Fundy. *J. Fish Res. Board Can.*, 17: 743–762.

Cinelli, F., Fresi, E., Mazzella, L., Pansini, M., Pronzato, R. and Svoboda, A., 1978. Distribution of benthic phyto- and zoocoenoses along a light gradient in a superficial marine cave. In: B.F. Keegan, P.Ó. Céidigh and P.J.S. Boaden (Editors), *Biology of Benthic Organisms*. Pergamon Press, Oxford, pp. 173–183.

Coleman, D.C. and Mathieson, A.C., 1974. Investigation of New England marine algae. VI: Distribution of marine algae near Cape Cod, Massachusetts. *Rhodora*, 76: 537–563.

Coleman, D.C. and Mathieson, A.C., 1975. Investigations of

New England marine algae. VII: Seasonal occurrence and reproduction of marine algae near Cape Cod, Massachusetts. *Rhodora*, 77: 76–104.

Colinvaux, L.H., 1966. Distribution of marine algae in the Bay of Fundy, New Brunswick, Canada. In: E.G. Young and J.L. McLachlan (Editors), *Proc. Fifth Int. Seaweed Symp., Halifax, Canada*. Pergamon Press, Oxford, pp. 91–98.

Colinvaux, L.H., 1970. Marine algae of eastern Canada: a seasonal study in the Bay of Fundy. *Nova Hedwigia*, 19: 139–157.

Colton Jr., J.B., 1964. History of oceanography in the offshore waters of the Gulf of Maine. *U.S. Fish Wildl. Serv. Spec. Sci. Rep.*, 496: 1–18.

Colton Jr., J.B., 1968. A comparison of current and long-term temperatures of continental shelf waters, Nova Scotia to Long Island. *Int. Comm. Northwest Atlantic Fish. Res. Bull.*, 5: 111–129.

Colton Jr., J.B., 1972. Temperature trends and the distribution of groundfish in continental shelf waters, Nova Scotia to Long Island. *U.S. Nat. Mar. Fish. Serv. Fish. Bull.*, 70: 637–657.

Colton Jr., J.B. and Stoddard, R.R., 1972. *Average monthly seawater temperatures, Nova Scotia to Long Island, 1940–1959*. Am. Geogr. Soc. Serial Atlas of the Marine Environment Folio, 21: 2 pp.

Colton Jr., J.B., Marak, R.R., Nickerson, S. and Stoddard, R.R., 1968. *Physical, chemical, and biological observations on the continental shelf, Nova Scotia to Long Island, 1964–1966*. U.S. Dept. Interior, FWS Data Rept. 23, Washington, D.C., 190 pp.

Connell, J.H., 1961. The influence of interspecific competition and other factors on the distribution of the barnacle *Chthamalus stellatus*. *Ecology*, 42: 710–723.

Connell, J.H., 1972. Community interactions on marine rocky intertidal shores. *Ann. Rev. Ecol. Syst.*, 3: 169–192.

Connell, J.H., 1975. Some mechanisms producing structure in natural communities: a model and evidence from field experiments. In: M.L. Cody and J.M. Diamond (Editors), *Ecology and Evolution of Communities*. Belknap Press, Cambridge, Mass., pp. 460–490.

Connell, J.H., 1979. Tropical rain forests and coral reefs as open non-equilibrium systems. In: R.M. Anderson, B.D. Turner and L.R. Taylor (Editors), *Population Dynamics*. Blackwell Sci. Publ., Oxford, pp. 141–163.

Connell, J.H., 1985. The consequences of variation in initial settlement vs. post-settlement mortality in rocky intertidal communities. *J. Exp. Mar. Biol. Ecol.*, 93: 11–45.

Connell, J.H., 1986. Variation and persistence of rocky shore populations. In: P.G. Moore and R. Seed (Editors), *The Ecology of Rocky Coasts*. Columbia Univ. Press, New York, pp. 57–69.

Conover, J.T. and Sieburth, J.McN., 1964. Effect of *Sargassum* distribution on its epibiota and antibacterial activity. *Bot. Mar.*, 6: 147–157.

Conover, J.T. and Sieburth, J.McN., 1966. Effect of tannins excreted from Phaeophyta on plankton and animal survival in tidepools. In: E.G. Young and J.L. McLachlan (Editors), *Proc. Fifth Int. Seaweed Symp., Halifax, Canada*. Pergamon Press, Oxford, pp. 99–100.

Coomans, H.E., 1962. The marine mollusk fauna of the Virginian area as a basis for defining zoogeographical provinces. *Beaufortia*, 98: 83–104.

Cooper, R.A., Uzmann, J.R., Clifford, R.A. and Pecci, K.J., 1975. Direct observations of herring (*Clupea harengus harengus* L.) egg beds on Jeffreys Ledge, Gulf of Maine in 1974. *Int. Comm. North Atl. Fish. Res. Doc.*, pp. 75–93.

Coote, A.R. and Yeats, P.A., 1979. Distribution of nutrients in the Gulf of St. Lawrence. *J. Fish. Res. Board Can.*, 36: 122–131.

Correa, J., Novaczek, I. and McLachlan, J., 1986. Effect of temperature and daylength on morphogenesis of *Scytosiphon lomentaria* (Scytosiphonales, Phaeophyta) from eastern Canada. *Phycologia*, 25: 469–475.

Coull, B.C. and Wells, J.B.J., 1983. Refuges from fish predation: experiments with phytal meiofauna from the New Zealand rocky intertidal. *Ecology*, 64: 1599–1609.

Cousens, R., 1981. Variation in annual production by *Ascophyllum nodosum* (L.) Le Jolis with degree of wave exposure. In: T. Levring (Editor), *Proc. Tenth Int. Seaweed Symp., Göteborg, Sweden*. Walter de Gruyter, Berlin, pp. 253–258.

Cousens, R., 1982. The effect of exposure to wave action on the morphology and pigmentation of *Ascophyllum nodosum* (L.) Le Jolis in southeastern Canada. *Bot. Mar.*, 25: 191–195.

Cousens, R., 1984. Estimation of annual production by the intertidal brown alga *Ascophyllum nodosum* (L.) Le Jolis. *Bot. Mar.*, 27: 217–227.

Cousens, R., 1985. Frond size distributions and the effects of the algal canopy on the behavior of *Ascophyllum nodosum* (L.) Le Jolis. *J. Exp. Mar. Biol. Ecol.*, 92: 231–249.

Cousens, R., 1986. Quantitative reproduction and reproductive effort by stands of the brown alga *Ascophyllum nodosum* (L.) Le Jolis in southeastern Canada. *Estuarine Coastal Shelf Sci.*, 22: 495–507.

Craigie, J.S. and McLachlan, J., 1964. Excretion of colored ultraviolet-absorbing substances by marine algae. *Can. J. Bot.*, 42: 23–33.

Craigie, J.S. and Pringle, J.D., 1978. Spatial distribution of tetrasporophytes and gametophytes in four Maritime populations of *Chondrus crispus*. *Can. J. Bot.*, 56: 2910–2914.

Cronan, J.M. and Halla, B.F., 1968. *Fall and winter foods of Rhode Island waterfowl*. Rhode Island Dept. Natural Resources, Wildlife Pamphlet No. 7, 40 pp.

Dadswell, M.J., 1979. The Canso Causeway and its possible effects on regional inshore fisheries — an overview. *Fish. Mar. Serv. Tech. Rep.*, 834 (Part II): 1–13.

Dale, M.R.T., 1979. *The analysis of patterns of zonation and phytosociological structure of seaweed communities*. Ph.D. Thesis, Dalhousie Univ., Halifax, N.S., 331 pp.

Daly, M.A. and Mathieson, A.C., 1977. The effects of sand movement on intertidal seaweeds and selected invertebrates at Bound Rock, New Hampshire, USA. *Mar. Biol.*, 43: 45–55.

Daly, M.A., Mathieson, A.C. and Norall, T.L., 1979. *Temperature, salinity, turbidity and light attenuation in the Great Bay Estuary System, 1974–1978*. Jackson Estuarine Laboratory Publ. No. 85. Univ. New Hampshire, Durham, N.H., 72 pp.

Dana, J.D., 1852. Crustacea. In: *United States Exploring*

Expedition during the years 1838, 1839, 1840, 1841, 1842 under the command of Charles Wilkes, U.S.N. C. Sherman. Phila., 13: 692–1618.

Dana, J.D., 1853. On an isothermal oceanic chart illustrating the geographical distribution of marine animals. *Am. J. Sci.*, 16: 314–327.

D'Antonio, C., 1985. Epiphytes on the rocky intertidal red alga *Rhodomela larix* (Turner) C. Agardh: negative effects on the host and food for herbivores. *J. Exp. Mar. Biol. Ecol.*, 86: 197–218.

Darley, W.M., 1982. *Algal Biology: A Physiological Approach.* Blackwell, Oxford, 168 pp.

Davis, A.N., 1984. The role of physical disturbance in maintaining high algal diversity on a sublittoral cobble bottom (Plum Cove, Cape Ann, Massachusetts). *Program and Abstr. Twenty-third Northeast Algal Symp., Woods Hole, Mass.*

Davis, A.N. and Wilce, R.T., 1984. Ecology of sublittoral benthic algae from a cobble bottom: new evidence for the non-equilibrium view of community structure. *Br. Phycol. J.*, 19: 192.

Davis, H.C., 1960. Effects of turbidity producing materials in seawater on eggs and larvae of the clam *Venus mercenaria mercenaria. Biol. Bull.*, 118: 48–54.

Dawes, C.J., 1974. *Marine Algae of the West Coast of Florida.* Univ. Miami Press, Coral Gables, Fla., 201 pp.

Dawes, C.J., Mathieson, A.C. and Cheney, D.P., 1974. Ecological studies of Floridian *Eucheuma* (Rhodophyta, Gigartinales) I. Seasonal growth and reproduction. *Bull. Mar. Sci.*, 24: 235–273.

Dawson, W.L., 1909. *The Birds of Washington.* Occidental Publ., Seattle, Wash., 997 pp.

Day, C.G., 1958. Surface circulation in the Gulf of Maine deduced from drift bottles. *Fish. Bull. Fish Wildl. Serv.*, 58: 443–472.

Dayton, P.K., 1971. Competition, disturbance, and community organization: the provision and subsequent utilization of space in a rocky intertidal community. *Ecol. Monogr.*, 41: 351–389.

Dayton, P.K., 1973. Dispersion, dispersal, and persistence of the annual intertidal alga, *Postelsia palmaeformis* Ruprecht. *Ecology*, 54: 433–438.

Dayton, P.K., 1975. Experimental evaluation of ecological dominance in a rocky intertidal algal community. *Ecol. Monogr.*, 45: 137–159.

Dayton, P.K., 1985. Ecology of kelp communities. *Ann. Rev. Ecol. Syst.*, 16: 215–245.

Dayton, P.K., Robilliard, G.A., Paine, R.T. and Dayton, L.B., 1974. Biological accommodation in the benthic community at McMurdo Sound, Antarctica. *Ecol. Monogr.*, 44: 105–128.

De Sève, M.A., Cardinal, A. and Goldstein, M., 1979. Les algues marines benthiques des Iles-de-la-Madeleine (Québec). *Proc. N.S. Inst. Sci.*, 29: 223–233.

Den Hartog, C., 1959. The epilithic algal communities occurring along the coast of the Netherlands. *Wentia*, 1: 1–241.

Denley, E.J. and Dayton, P.K., 1985. Competition among macroalgae. In: M.M. Littler and D.S. Littler (Editors), *Handbook of Phycological Methods, Ecological Field Methods: Macroalgae.* Cambridge Univ. Press, Cambridge, pp. 511–531.

Denman, K., Irwin, B. and Platt, T., 1977. Phytoplankton productivity experiments and nutrient measurements at the edge of the Nova Scotia continental shelf between June 28 and July 12, 1976. *Fish. Mar. Serv. Res. Dev. Tech. Rep.*, 708: 1–205.

Dethier, M.N., 1981. Heteromorphic algal life histories: the seasonal pattern and response to herbivory of the brown crust *Ralfsia californa. Oecologia*, 49: 333–339.

Devinny, J.S. and Volse, L.A., 1978. The effects of sediments on the development of *Macrocystis pyrifera* gametophytes. *Mar. Biol.*, 48: 343–348.

DeWolfe, D.L., 1981. *Atlas of tidal currents; Bay of Fundy and Gulf of Maine.* Can. Hydrogr. Serv., Ottawa, 36 pp.

Deysher, L.E. and Dean, T.A., 1986. *In situ* recruitment of sporophytes of the giant kelp, *Macrocystis pyrifera* (L.) C.A. Agardh: effects of physical factors. *J. Exp. Mar. Biol. Ecol.*, 103: 41–63.

Dickie, L.M. and Trites, R.W., 1983. The Gulf of St. Lawrence. In: B.W. Ketchum (Editor), *Estuaries and Enclosed Seas. Ecosystems of the World, Vol. 26.* Elsevier, Amsterdam, Ch. 16, pp. 403–425.

Dinsmore, R.P., 1972. Ice and its drift into the Atlantic Ocean. *Int. Comm. Northwest Atl. Fish. Spec. Publ.*, 8: 89–128.

Dixon, P.S., 1965. Perennation, vegetative propagation and algal life histories with special references to *Asparagopsis* and other Rhodophyta. *Bot. Goth. Acta Univ. Goth.*, 3: 67–74.

Dixon, P.S., 1973. *Biology of the Rhodophyta.* Oliver and Boyd, Edinburgh, 285 pp.

Dixon, P.S. and Irvine, L.M., 1977. *Seaweeds of the British Isles, Vol. 1, Rhodophyta, Part 1: Introduction, Nemaliales, Gigartinales.* British Mus. Nat. History, London, 252 pp.

Dohler, G., 1963. *Tides in Canadian Waters.* Can. Hydrogr. Serv., Dept. Energy, Mines and Resources, 14 pp.

Doty, M.S. and Aguilar-Santos, G., 1970. Transfer of toxic algal substances in marine food chains. *Pac. Sci.*, 24: 351–355.

Dow, R.L., 1949. *The story of the Maine lobster* (*Homarus americanus*). Dept. Sea Shore Fish. Bull., 26 pp.

Doyle, R.W., 1974. Choosing between darkness and light: the ecological genetics of photobehavior in the planktonic larva of *Spirorbis borealis. Mar. Biol.*, 25: 311–317.

Doyle, R.W., 1975. Settlement of planktonic larvae: a theory of habitat selection in varying environments. *Am. Nat.*, 109: 113–126.

Drew, E.A., 1983. Light. In: R. Earll and D.G. Erwin (Editors), *Sublittoral Ecology.* Clarendon Press, Oxford, pp. 10–57.

Dring, M.J., 1982. *The Biology of Marine Plants.* Edward Arnold, London, 199 pp.

Drinkwater, K.F., 1979. Flow in the Strait of Canso and St. Georges Bay, Nova Scotia. *Fish. Mar. Serv. Tech. Rep.*, 834 (Pt. IV): 11–18.

Druehl, L.D., 1973. Marine transplantations. *Science*, 179: 12.

Dudgeon, S.R., 1988. Physiological consequences of freezing to photosynthetic metabolism in the intertidal macroalgae *Chondrus* and *Mastocarpus. Program and Abstr. Twenty-seventh Northeast Algal Symposium, Woods Hole, Mass.*

Dudgeon, S.R., Davison, I.R. and Vadas, R.L., 1989. Effect of

freezing on photosynthesis of intertidal macroalgae: relative tolerance of *Chondrus crispus* and *Mastocarpus stellatus* (Rhodophyta). *Mar. Biol.*, 10: 107–114.
Duggins, D.O., 1980. Kelp beds and sea otters: an experimental approach. *Ecology*, 61: 447–453.
Dyck, L., DeWreede, R.E. and Garbary, D., 1985. Life history phases in *Iridaea cordata* (Gigartinaceae): relative abundance and distribution from British Columbia to California. *Jpn. J. Phycol.*, 33: 225–232.
Earll, R. and Erwin, D.G., 1979. The species recording scheme — results of the 1977 season. In: J.C. Gamble and J.D. George (Editors), *Progress in Underwater Science. Vol. 4.* Pentech Press, London, pp. 105–120.
Earll, R. and Farnham, W., 1983. Biogeography. In: R. Earll and D.G. Erwin (Editors), *Sublittoral Ecology.* Clarendon Press, Oxford, pp. 165–208.
Ebert, T.A., 1983. Recruitment in echinoderms. In: M. Jangoux and J.M. Lawrence (Editors), *Echinoderm Studies, Vol. 1.* Balkema, Rotterdam, pp. 169–203.
Ebling, F.J., Kitching, J.A., Muntz, L. and Taylor, C.M., 1964. The ecology of Lough Ine. XII. Experimental observations of the destruction of *Mytilus edulis* and *Nucella lapillus* by crabs. *J. Animal Ecol.*, 33: 73–82.
Edelstein, T., Craigie, J.S. and McLachlan, J., 1969. Preliminary survey of the sublittoral flora of Halifax County. *J. Fish. Res. Board Can.*, 26: 2703–2713.
Edelstein, T., Chen, L. and McLachlan, J., 1970. Investigations of the marine algae of Nova Scotia. VIII. The flora of Digby Neck Peninsula, Bay of Fundy. *Can. J. Bot.*, 48: 621–629.
Edwards, D.C., Conover, D.O. and Sutter, III, F., 1982. Mobile predators and the structure of marine intertidal communities. *Ecology*, 64: 1175–1180.
Edwards, P., 1973. Life history studies of selected British *Ceramium* species. *J. Phycol.*, 9: 181–184.
Ekman, S., 1953. *Zoogeography of the Sea.* Sidgwick and Jackson, London, 417 pp.
El Sabh, M.I., 1976. Surface circulation pattern in the Gulf of St. Lawrence. *J. Fish. Res. Board Can.*, 33: 124–138.
El Sabh, M.I., 1977. Oceanographic features, currents and transport in Cabot Strait. *J. Fish. Res. Board Can.*, 34: 516–528.
Ellis, D.V. and Wilce, R.T., 1961. Arctic and subarctic examples of intertidal zonation. *Arctic*, 14: 224–235.
Elmhirst, R., 1922. Habits of *Echinus esculentus. Nature*, 110: 667.
Elner, R.W., 1978. The mechanics of predation by the shore crab, *Carcinus maenas* (L.), on the edible mussel, *Mytilus edulis* L. *Oecologia*, 36: 333–344.
Elner, R.W., 1980. Predation on the sea urchin (*Strongylocentrotus droebachiensis*) by the American lobster (*Homarus americanus*) and the rock crab (*Cancer irroratus*). In: J.D. Pringle, G.J. Sharp and J.F. Caddy (Editors), *Proc. workshop on the relationship between sea urchin grazing and commercial plant/animal harvesting. Can. Tech. Rep. Fish. Aquat. Sci.*, 954: 48–65.
Elner, R.W. and Campbell, A., 1987. Natural diets of lobster *Homarus americanus* from barren ground and macroalgal habitats off southwestern Nova Scotia, Canada. *Mar. Ecol. Prog. Ser.*, 37: 131–140.
Elner, R.W. and Hughes, R.N., 1978. Energy maximization in the diet of the shore crab, *Carcinus maenas. J. Anim. Ecol.*, 47: 103–116.
Emery, K.O. and Garrison, L.E., 1967. Sea level 7000 to 20 000 years ago. *Science*, 157: 684–687.
Emery, K.O. and Uchupi, E., 1972. *Western North Atlantic Ocean: Topography, rocks, structure, water, life and sediments.* Am. Assoc. Petrol. Geologists, Mem. 17, 532 pp.
Ennis, G.P., 1973. Food, feeding, and condition of lobster, *Homarus americanus*, throughout the seasonal cycle in Bonavista Bay, Newfoundland. *J. Fish. Res. Board Can.*, 30: 1905–1909.
Enright, C.T., 1977. *Competitive interaction between Chondrus crispus (Rhodophyceae) and the weed species Ulva lactuca (Chlorophyceae) in Chondrus aquaculture.* M.Sc. Thesis, Dalhousie Univ., Halifax, N.S., 79 pp.
Enright, C.T., 1979. Competitive interaction between *Chondrus crispus* (Florideophyceae) and *Ulva lactuca* in *Chondrus* aquaculture. In: A. Jensen and J.R. Stein (Editors), *Proc. Ninth Int. Seaweed Symp., Santa Barbara, California.* Science Press, Princeton, pp. 209–218.
Ernst, J., 1968. The life forms of some perennial marine algae of Roscoff and their vertical distribution. *Bot. Mar.*, 11: 36–39.
Erwin, D.G., 1983. The community concept. In: R. Earll and D.G. Erwin (Editors), *Sublittoral Ecology.* Clarendon Press, Oxford, pp. 144–164.
Espinosa, R., 1985. Differential defense mechanisms to snail grazing by two species of the brown algal *Fucus* (Phaeophyceae). *Abstr. Twenty-fourth Northeast Algal Symp., Woods Hole, Mass.*
Etter, R.J., 1986. Interpopulation variation in the life history characteristics of an intertidal snail. *Abstr. 1986 Benthic Ecol. Meet., Boston, Mass.*
Evans, P. and Mann, K.H., 1977. Selection of prey by American lobsters (*Homarus americanus*) when offered a choice between sea urchins and crabs. *J. Fish. Res. Board Can.*, 34: 2203–2207.
Fader, G.B., King, L.H. and MacLean, B., 1977. Surficial geology of the eastern Gulf of Maine and Bay of Fundy. *Geol. Surv. Can. Pap.*, 76–17: 1–22 pp.
Fager, E.W., 1971. Patterns in the development of a marine community. *Limnol. Oceanogr.*, 16: 241–253.
Farlow, W.G., 1881. *The marine algae of New England.* Rep. U.S. Commiss. Fish and Fisheries for 1879, pp. 1–210.
Farnham, W.F., 1980. Studies on aliens in the marine flora of southern England. In. J.H. Price, D.E.G. Irvine and W.F. Farnham (Editors), *The Shore Environment, Vol. 2: Ecosystems.* Systematics Assoc. Spec. Vol. 17(b). Academic Press, London, pp. 875–914.
Faulkner, D.J. and Anderson, R.J., 1974. Natural products chemistry of the marine environment. In: E. Goldberg (Editor), *The Sea. Vol. 5.* Wiley, New York, pp. 679–714.
Feare, C.J., 1967. The effect of predation by shorebirds on a population of dogwhelks *Thais lapillus. Ibis*, 109: 474.
Feare, C.J. and Summers, R.W., 1986. Birds as predators on rocky shores. In: P.G. Moore and R. Seed (Editors), *The Ecology of Rocky Coasts.* Columbia Univ. Press, New York, pp. 249–264.

Fedra, K., 1976. On the ecology of a North Adriatic benthic community: distribution, standing crop, and composition of the macrobenthos. *Mar. Biol.*, 38: 129–145.

Fefer, S.I. and Schettig, P.A., 1980a. *An ecological characterization of coastal Maine (north and east of Cape Elizabeth).* U.S. Fish Wildl. Serv., U.S. Dept. Interior, Biol. Serv. Prog., Vol. 1, 158 pp.

Fefer, S.I. and Schettig, P.A., 1980b. *An ecological characterization of coastal Maine (north and east of Cape Elizabeth).* U.S. Fish Wildl. Serv., U.S. Dept. Interior, Biol. Serv. Prog., Vol. 2, 396 pp.

Feldmann, J., 1937. Recherches sur la végétation marine de la Méditerranée. La côte des Albères. *Rev. Algol.*, 10: 1–339.

Feldmann, J., 1954. Inventaire de la flora marine de Roscoff. Algues, champignons, lichens et spermatophytes. *Trav. Stn. Biol. Roscoff. Suppl.*, 6: 1–152.

Feldmann, J. and Magne, F., 1964. Additions à l'inventaire de la flore marine de Roscoff. Algues, champignons, lichens. *Trav. Stn. Biol. Roscoff, N.S.*, 15: 1–28.

Femino, R.J. and Mathieson, A.C., 1980. Investigations of New England marine algae IV. The ecology and seasonal succession of tide pool algae at Bald Head Cliff, York, Maine, USA. *Bot. Mar.*, 23: 319–332.

Fenical, W., 1975. Halogenation in the Rhodophyta: a review. *J. Phycol.*, 11: 245–259.

Fenical, W., 1980. Distributional and taxonomic features of toxin-producing marine algae. In: I.A. Abbott, M.S. Foster and L.F. Eklund (Editors), *Pacific Seaweed Aquaculture.* Calif. Sea Grant College Program, Inst. Mar. Resources, Univ. Calif., La Jolla, U.S.A., pp. 144–151.

Filion-Myklebust, C.C. and Norton, T.A., 1981. Epidermis shedding in the brown seaweed *Ascophyllum nodosum* (L.) Le Jolis. *Mar. Biol. Lett.*, 2: 45–51.

Forbush, E.H. and May, J.B., 1955. *A Natural History of American Birds of Eastern and Central North America.* Bramhall House, New York, 552 pp.

Forgeron, F.D., 1959. *Temperature and salinity in the Quoddy Region.* Rep. of the Int. Passamaquoddy Fish. Board to the Int. Joint Comm., Chap. 1, 23 pp.

Forrester, W.D., 1959. *Current measurements in Passamaquoddy Bay and the Bay of Fundy, 1957 and 1958.* Rep. of the Int. Passamaquoddy Fish. Board to the Int. Joint Comm. Appendix I (Oceanography) Chap. 3, 73 pp.

Forrester, W.D., 1960. Current measurements in the Passamaquoddy Bay and the Bay of Fundy, 1957 and 1958. *J. Fish. Res. Board Can.*, 17: 727–729.

Forward, C.N., 1954. Ice distribution in the Gulf of St. Lawrence during the break-up season. Dept. Mines and Techn. Surveys, Geographical Branch. *Geograph. Bull.*, 6: 45–84.

Foster, M.S., 1975a. Algal succession in a *Macrocystis pyrifera* forest. *Mar. Biol.*, 32: 313–329.

Foster, M.S., 1975b. Regulation of algal community development in a *Macrocystis pyrifera* forest. *Mar. Biol.*, 32: 331–342.

Foster, W.S., 1953. *Sea moss (Chondrus crispus), survey Washington County.* Gen. Bull. No. 3, Maine Dept. Sea Shore Fish., Augusta, Me., 12 pp.

Foster, W.S., 1954. *Sea moss, (Chondrus crispus), survey West Point to Pemaquid Neck.* Gen. Bull. No. 4, Maine Dept. Sea Shore Fish., Augusta, Me., 13 pp.

Foster, W.S., 1955. *Sea moss, (Chondrus crispus), survey Pemaquid Point to Owls Head.* Gen. Bull. No. 5, Maine Dept. Sea Shore Fish., Augusta, Me., 8 pp.

Foster, W.S., 1956. *Sea moss (Chondrus crispus), survey Isleboro, Long Island, Maine.* Gen. Bull. No. 6, Maine Dept. Sea Shore Fish., Augusta, Me., 3 pp.

Fralick, R.A. and Mathieson, A.C., 1973. Ecological studies of *Codium fragile* ssp. *tomentosoides* in New England. *Mar. Biol.*, 19: 127–132.

Fralick, R.A., Turgeon, K.W. and Mathieson, A.C., 1974. Destruction of kelp populations by *Lacuna vincta* (Montagu). *Nautilus*, 88: 112–114.

Futuyma, D.J. and Slatkin, M., 1983. *Coevolution.* Sinauer, Sunderland, Mass., 555 pp.

Gagné, J.A. and Mann, K.H., 1981. Comparison of growth strategy in *Laminaria* populations living under differing seasonal patterns of nutrient availability. In: T. Levring (Editor), *Proc. Tenth Int. Seaweed Symp., Göteborg, Sweden.* Walter de Gruyter, Berlin, pp. 297–302.

Gagné, J.A., Mann, K.H. and Chapman, A.R.O., 1982. Seasonal patterns of growth and storage in *Laminaria longicruris* in relation to differing patterns of availability of nitrogen in the water. *Mar. Biol.*, 69: 91–101.

Gaines, S.D. and Lubchenco, J., 1982. A unified approach to marine plant-herbivore interactions. 2. Biogeography. *Ann. Rev. Ecol. Syst.*, 13: 111–138.

Gaines, S. and Roughgarden, J., 1985. Larval settlement rate: a leading determinant of structure in an ecological community of the marine intertidal zone. *Proc. Nat. Acad. Sci.*, 82: 3707–3711.

Ganong, W.F., 1891. Southern invertebrates on the shores of Acadia. *Proc. Trans. R. Soc. Can.*, 8: 167–185.

Garbary, D.J. and Gautam, A., 1989. The *Ascophyllum, Polysiphonia, Mycosphaerella* symbiosis I. Population ecology of *Mycosphaerella* from Nova Scotia. *Bot. Mar.*, 32: 181–196.

Garrett, C.J.R., 1972. Tidal resonance in the Bay of Fundy and Gulf of Maine. *Nature*, 238: 441–443.

Garrett, C.J.R., 1984. Tides and tidal power in the Bay of Fundy. *Endeavour, N.S.*, 8: 58–64.

Garrett, C.J.R. and Loucks, R.H., 1976. Upwelling along the Yarmouth shore of Nova Scotia. *J. Fish. Res. Board Can.*, 33: 116–117.

Garrison, K.M. and Brown, W.S., 1987. *Hydrographic survey in the Gulf of Maine, summer 1986.* Tech. Rep. No. UNH-MP-T/DR-SG-87-12, Univ. New Hampshire, Durham, N.H., 211 pp.

Garrison, K.M. and Brown, W.S., 1989. *Hydrographic survey in the Gulf of Maine, September 1987.* Tech. Rep. No. UNH-MP-T/DR-SG-89-6, Univ. New Hampshire, Durham, N.H., 158 pp.

Garwood, P.E., Vadas, R.L. and Ojeda, F.P., 1985. Competitive interactions among crustose coralline algae. *Abstr. Twenty-fourth Northeast Algal Symp., Woods Hole, Mass.*

Geiselman, J.A., 1980. *Ecology of chemical defenses of algae against the herbivorous snail, Littorina littorea, in the New England rocky intertidal community.* Ph.D. Thesis, Mass.

Inst. Techn./Woods Hole Oceanogr. Inst., Woods Hole, Mass., 206 pp.
Geiselman, J.A. and McConnell, O.J., 1981. Polyphenols in brown algae *Fucus vesiculosus* and *Ascophyllum*: chemical defenses against the marine herbivorous snail, *Littorina littorea*. *J. Chem. Ecol.*, 7: 1115–1133.
Gerard, V.A., 1986. Searching for light ecotypes of *Laminaria*. *Abstr. Twenty-fifth Northeast Algal Symp., Woods Hole, Mass.*
Gerard, V.A. and Mann, K.H., 1979. Growth and production of *Laminaria longicruris* (Phaeophyta) populations exposed to different intensities of water movement. *J. Phycol.*, 15: 33–41.
Gislèn, T., 1930. *Epibiosis of the Gullmar Fjord. I. A study in sociology.* Skr. Kristinebergs Zool. Sta. 1877–1937, 3–4, 503 pp.
Glombitza, K.W., Murawski, U., Bielaczek, J. and Egge, H., 1974. Bromophenole aus Rhodomelaceen (Antibiotica aus Algen 9. mitt.). *Planta Med.*, 25: 105–114.
Glombitza, K.W. and Stoffelen, H., 1972. 2,3-Dibromo-5-hydroxylbenzyl-1, 4-disulfat (dikalium salz) aus Rhodomelaceen (Antibiotica aus Algen 6. mitt.). *Planta Med.*, 22: 391–395.
Golikov, A.N. and Scarlato, O.A., 1973. Comparative characteristics of some ecosystems of the upper regions of the shelf in tropical, temperate, and arctic waters. *Helgol. Wiss. Meeresunters.*, 24: 219–234.
Gordon Jr., D.C. and Desplanque, C., 1983. Dynamics and environmental effects of ice in the Cumberland Basin of the Bay of Fundy. *Can. J. Fish. Aquat. Sci.*, 40: 1331–1342.
Goudie, R.I. and Ankney, C.D., 1986. Body size, activity budgets, and diets of sea ducks wintering in Newfoundland. *Ecology*, 67: 1475–1482.
Graham Jr., F., 1975. *Gulls, a Social History.* Random House, New York, 179 pp.
Graham, J.J., 1970a. Coastal currents of the western Gulf of Maine. *Int. Comm. Northwest Atl. Fish. Res. Bull.*, 7: 19–31.
Graham, J.J., 1970b. Temperature, salinity and transparency observation coastal Gulf of Maine, 1962–65. *U.S. Fish. Wildlife Serv. Data Rep.*, 42: 1.
Grant, D.R., 1968. Recent submergence in Nova Scotia and Prince Edward Island, Canada. *Geol. Surv. Can. Pap.*, 68: 162–164.
Grant, W.S., 1977. High intertidal community organization on a rocky headland in Maine, USA. *Mar. Biol.*, 44: 15–25.
Gray, J.S., 1974. Animal-sediment relationships. In: H. Barnes (Editor), *Oceanography and Marine Biology, Annual Review, Vol. 12*, pp. 223–262.
Green, J.E., 1983. *Factors controlling the vertical zonation of two intertidal seaweeds: Chondrus crispus Stackhouse and Gigartina stellata (Stackhouse) Batters.* M.Sc. Thesis, Northeastern Univ., Boston, Mass., 134 pp.
Green, J.E. and Cheney, D., 1981. Factors controlling the vertical zonation of *Gigartina stellata* and *Chondrus crispus*. *Abstr. Twentieth Northeast Algal Symp., Woods Hole, Mass.*
Greenwell, M., Lazo, M.L. and McLachlan, J., 1985. The haploid:diploid ratio of Irish moss (*Chondrus crispus* Stackh; Gigartinaceae) in Prince Edward Island, Canada. In: *Programme and Book of Abstract Twelfth Int. Seaweed Symp., Sao Paulo, Brazil.*
Grime, J.P., 1979. *Plant Strategies and Vegetation Processes.* Wiley, Chichester, 223 pp.
Guiry, M.D., 1978. A consensus and bibliography of Irish seaweeds. *Bibl. Phycologia*, 44: 1–287.
Guiry, M.D., 1984. Photoperiodic and temperature responses in the growth and tetrasporogenesis of *Gigartina acicularis* (Rhodophyta) from Ireland. *Helgol. Wiss. Meeresunters.*, 38: 335–347.
Guiry, M.D., West, J.A., Kim, D.-H. and Masuda, M., 1984. Reinstatement of the genus *Mastocarpus* Kutzing (Rhodophyta). *Taxon*, 33: 53–63.
Gulliksen, B., 1978. Rocky bottom fauna in a submarine gully at Lappkalven, Finmark, northern Norway. *Estuarine Coastal Mar. Sci.*, 7: 361–372.
Hachey, H.B., 1938. The origin of the cold water layer of the Scotian Shelf. *Trans. R. Soc. Can.*, 32: 29–42.
Hachey, H.B., 1942. The waters of the Scotian Shelf. *J. Fish. Res. Board Can.*, 5: 377–397.
Hachey, H.B., 1947. Water transports and current patterns for the Scotian Shelf. *J. Fish. Res. Board Can.*, 7: 1–16.
Hachey, H.B., 1952. The general hydrography of the waters of the Bay of Fundy. *Fish. Res. Board Can. MS. Rep. Biol. Sta.*, 455: 1–100.
Hachey, H.B., 1961. Oceanography and Canadian Atlantic Waters. *Fish. Res. Board Can. Bull.*, 134: 1–120.
Hachey, H.B., Hermann, F. and Bailey, W.N., 1954. The waters of the ICNAF Convention area. *Int. Comm. Northwest Atl. Fish. Ann. Proc.*, 4: 67–102.
Haight, F.J., 1942. *Coastal currents along the Atlantic coast of the United States.* U.S. Coast and Geodetic Surv. Spec. Publ. No. 230, 73 pp.
Hall Jr., C.F., 1964. Shallow-water marine climates and molluscan provinces. *Ecology*, 45: 226–234.
Hancock, D.A., 1955. The feeding behaviour of starfish on Essex oyster beds. *J. Mar. Biol. Assoc. U.K.*, 34: 313–331.
Hancock, D.A., 1958. Notes on starfish on an Essex oyster bed. *J. Mar. Biol. Assoc. U.K.*, 37: 565–589.
Hanic, L.A. and Lobban, C., 1971. *Marine zonation research, Prince Edward Island National Park.* Project 05/1–14, Dep. Indian Affairs and Northern Development, National and Historic Parks Branch, Ottawa, 151 pp.
Hannach, G. and Santelices, B., 1985. Ecological differences between the isomorphic reproductive phases of two species of *Iridaea* (Rhodophyta: Gigartinales). *Mar. Ecol. Prog. Ser.*, 22: 291–303.
Hansen, J.E. and Doyle, W.T., 1976. Ecology and natural history of *Iridaea cordata* (Rhodophyta; Gigartinaceae): population structure. *J. Phycol.*, 12: 273–278.
Hardwick-Witman, M.N. and Mathieson, A.C., 1983. Intertidal macroalgae and macroinvertebrates: seasonal and spatial abundance patterns along an estuarine gradient. *Estuarine Coastal Shelf Sci.*, 16: 113–129.
Harger, J.R.E. and Landenberger, D.E., 1971. The effects of storms as a density dependent mortality factor on populations of sea mussels. *Veliger*, 14: 195–201.
Harkin, E., 1981. Fluctuations in epiphyte biomass following *Laminaria hyperborea* canopy removal. In: T. Levring

(Editor), *Proc. Tenth Int. Seaweed Symp., Göteborg, Sweden.* Walter de Gruyter, Berlin, pp. 303–308.
Harlin, M.M., 1974. The surfaces seaweeds grow on may be a clue to their control. *Maritimes*, pp. 7, 8.
Harlin, M.M., 1987. Allelochemistry in marine macroalgae. *CRC Crit. Rev. Plant Sci.*, 5: 237–249.
Harlin, M.M. and Lindbergh, H.J., 1977. Selection of substrata by seaweeds: optimal surface relief. *Mar. Biol.*, 40: 33–40.
Harper, J.L., 1977. *Population Biology of Plants.* Academic Press, New York, 892 pp.
Harris, L.G., 1973. Nudibranch associations. In: T.C. Cheng (Editor), *Current Topics In Comparative Pathobiology, Vol. II.* Academic Press, Baltimore, Md., pp. 213–315.
Harris, L.G., 1986. Size-selective predation in a sea anemone, nudibranch and fish food chain. *Veliger*, 29: 38–47.
Harris, L.G. and Irons, K.P., 1982. Substrate angle and predation as determinants in fouling community succession. In: J. Cairns (Editor), *Artificial Substrates.* Ann Arbor Sci. Publ., Ann Arbor, Mich., pp. 131–174.
Harris, L.G. and Mathieson, A.C., 1976. *Field studies on benthic communities in the New England offshore mining environmental study (NOMES).* Final Report, Environmental Res. Lab. National Oceanic and Atmospheric Admin., Miami, 130 pp.
Harris, L.G., Hulbert, A.W., Witman, J.D. and Pecci, K.J., 1979. A comparison by depth and substrate angle of the subtidal benthic communities at Pigeon Hill in the Gulf of Maine. *Am. Zool.*, 19: 792.
Hartnoll, R.G., 1983. Substratum. In: R. Earll and D.G. Erwin (Editors), *Sublittoral Ecology.* Clarendon Press, Oxford, pp. 97–124.
Harvey, W.H., 1852. Nereis Boreali-Americana, Part I, Melanospermae. *Smithson. Contrib. Knowl.*, 3: 1–150.
Hatcher, B.G., Chapman, A.R.O. and Mann, K.H., 1977. An annual carbon budget for the kelp *Laminaria longicruris. Mar. Biol.*, 44: 85–86.
Hatton, H., 1938. Essais de bionomie explicative sur quelques espèces intercotidales d'algues et d'animaux. *Ann. Inst. Oceanogr. Monaco*, 17: 241–348.
Hay, M.E., 1981. The functional morphology of turf-forming seaweeds: persistence in stressful marine habitats. *Ecology*, 62: 739–750.
Hawkins, S.J. and Hartnoll, R.G., 1983. Grazing of intertidal algae by marine invertebrates. *Oceanogr. Mar. Biol. Ann. Rev.*, 21: 195–282.
Heath, R.A., 1973. Variability of water properties and circulation of St. Margaret's Bay, Nova Scotia. *Fish. Res. Board Can. Tech. Rep.*, 404: 1–56.
Hecht, A.D., 1969. Miocene distribution of Molluskan provinces along the east coast of the United States. *Geol. Soc. Am. Bull.*, 80: 1617–1620.
Hedgpeth, J.W., 1957. Marine biogeography. In: J.W. Hedgpeth (Editor), Treatise on Marine Ecology and Paleoecology. *Geol. Soc. Am. Mem.*, 67: 359–382.
Hehre, E.J. and Mathieson, A.C., 1970. Investigations of New England marine algae III. Composition, seasonal occurrence and reproductive periodicity of the marine Rhodophyceae in New Hampshire. *Rhodora*, 72: 194–239.
Hicks, G.R.F., 1986. Meiofauna associated with shore algae. In: P.G. Moore and R. Seed (Editors), *The Ecology of Rocky Coasts.* Columbia Univ. Press, New York, pp. 36–56.
Himmelman, J.H., 1969. *Some aspects of the ecology of Strongylocentrotus droebachiensis in eastern Newfoundland.* M.Sc. Thesis, Memorial Univ. Newfoundland, St. John's, Newfoundland, 159 pp.
Himmelman, J.H., 1980. The role of the green sea urchin, *Strongylocentrotus droebachiensis*, in the rocky subtidal region of Newfoundland. In: J.D. Pringle, G.J. Sharp and J.F. Caddy (Editors), *Proc. Workshop on the relationship between sea urchin grazing and commercial plant/animal harvesting.* Can. Tech. Rep. Fish. Aquat. Sci., 954: 92–119.
Himmelman, J.H., 1985. Urchin feeding and macroalgal distribution in Newfoundland, eastern Canada. *Nat. Can.*, 111: 337–348.
Himmelman, J.H. and Steele, D.H., 1971. Foods and predators of the green sea urchin *Strongylocentrotus droebachiensis* in Newfoundland waters. Mar. Biol., 9: 315–322.
Himmelman, J.H., Cardinal, A. and Bourget, E., 1983a. Community development following removal of urchins, *Strongylocentrotus droebachiensis*, from the rocky subtidal zone of the St. Lawrence Estuary, eastern Canada. *Oecologia*, 59: 27–39.
Himmelman, J.H., Lavergne, Y., Axelsen, F., Cardinal, A. and Bourget, E., 1983b. Sea urchins in the Saint Lawrence Estuary: their abundance, site-structure, and suitability for commercial exploitation. *Can. J. Fish. Aquat. Sci.*, 40: 476–486.
Hind, H.Y., 1875. The ice phenomena and the tides of the Bay of Fundy. *Can. Month. Nat. Rev.*, 8: 189–203.
Hirtle, R.W.M. and Mann, K.H., 1978. Distance chemoreception and vision in the selection of prey by American lobsters (*Homarus americanus*). *J. Fish. Res. Board Can.*, 35: 1006–1008.
Hiscock, K., 1983. Water movement. In: R. Earll and D.G. Erwin (Editors), *Sublittoral Ecology.* Clarendon Press, Oxford, pp. 58–96.
Hiscock, K., 1986. Aspects of the ecology of rocky sublittoral areas. In: P.G. Moore and R. Seed (Editors), *The Ecology of Rocky Coasts.* Columbia Univ. Press, New York, pp. 290–328.
Hoffman, A.J. and Ugarte, R., 1985. The arrival of propagules of marine macroalgae in the intertidal zone. *J. Exp. Mar. Biol. Ecol.*, 92: 83–95.
Holloway, P.E., 1981. Longitudinal mixing in the upper reaches of the Bay of Fundy. *Estuarine Coastal Shelf Sci.*, 13: 495–515.
Holme, N.A. and McIntyre, A.D., 1971. *Methods for the study of Marine Benthos.* Int. Biol. Prog. Handbook 16. Blackwell, Oxford, 334 pp.
Holmsgaard, J.E., Greenwell, M. and McLachlan, J., 1981. Biomass and vertical distribution of *Furcellaria lumbricalis* and associated algae. In: T. Levring (Editor), *Proc. Tenth Int. Seaweed Symp. Göteborg, Sweden.* Walter de Gruyter, Berlin, pp. 309–314.
Hooper, R.G., 1975. *Bonne Bay Marine Resources: An ecological and biological assessment.* Parks Canada Atlantic Regional Off. Contract No. AR074–83, Vol. 1, 295 pp., Vol. 2 (maps), 109 pp.

Hooper, R.G, 1980. Observations on algal-grazer interactions in Newfoundland and Labrador. In: J.D. Pringle, G.J. Sharp and J.F. Caddy (Editors), *Proc. Workshop on the relationship between sea urchin grazing and commercial plant/animal harvesting. Can. Tech. Rep. Fish. Aquat. Sci.*, 954: 120–124.

Hooper, R.G., 1981. Recovery of Newfoundland benthic marine communities from sea ice. In: G.E. Fogg and W.E. Jones (Editors), *Proc. Eighth Int. Seaweed Symp., Bangor, North Wales.* Univ. College North Wales Mar. Sci. Lab. Publ., Menai Bridge, pp. 360–366.

Hooper, R.G., 1982. The Straits of Belle Isle marine benthic biogeographic boundary. *Br. Phycol. J.*, 17: 233.

Hooper, R.G. and South, G.R., 1977. Additions to the benthic marine algal flora of Newfoundland III, with observations on species new to eastern Canada and North America. *Nat. Can.*, 104: 383–394.

Hooper, R.G., South, G.R. and Whittick, A., 1980. Ecological and phenological aspects of the marine phytobenthos of the island of Newfoundland. In: J.H. Price, D.E.G. Irvine and W.F. Farnham (Editors), *The Shore Environment. Vol. 2: Ecosystems* Systematics Assoc. Spec. Vol. 17(b). Academic Press, London, pp. 395–423.

Hopkins, T.S. and Garfield, N., 1979. Gulf of Maine intermediate water. *J. Mar. Res.*, 37: 103–139.

Hornsey, I.S. and Hide, D., 1974. The production of antimicrobial compounds by British marine algae. I. Antibiotic producing algae. *Br. Phycol. J.*, 9: 353–361.

Hornsey, I.S. and Hide, D., 1976a. The production of antimicrobial compounds by British marine algae. II. Seasonal variation in production of antibiotics. *Br. Phycol. J.*, 11: 63–67.

Hornsey, I.S. and Hide, D., 1976b. The production of antimicrobial compounds by British marine algae. III. Distribution of antimicrobial activity within the algal thallus. *Br. Phycol. J.*, 11: 175–181.

Hruby, T. and Norton, T.A., 1979. Algal colonization on rocky shores in the Firth of Clyde. *J.Ecol.*, 67: 65–77.

Hsueh, Y. and O'Brien, J.J., 1981. Steady coastal upwelling induced by an along shore current. *J. Phys. Oceanogr.*, 1: 180–186.

Huang, R. and Boney, A.D., 1983. Effects of diatom mucilage on the growth and morphology of marine algae. *J. Exp. Mar. Biol. Ecol.*, 67: 79–89.

Huang, R. and Boney, A.D., 1984. Growth interactions between littoral diatoms and juvenile marine algae. *J. Exp. Mar. Biol. Ecol.*, 81: 21–45.

Huang, R. and Boney, A.D., 1985. Individual and combined interactions between littoral diatoms and sporelings of red algae. *J. Exp. Mar. Biol. Ecol.*, 85: 101–111.

Hughes, R.N., 1986. Rocky shore communities: catalysts to understanding predation. In: P.G. Moore and R. Seed (Editors), *The Ecology of Rocky Coasts.* Columbia Univ. Press, New York, pp. 223–233.

Hulbert, A.W., 1980. *The functional role of Asterias vulgaris Verrill (1866) in three subtidal communities.* Ph.D. Thesis, Univ. New Hampshire, Durham, New Hampshire, 174 pp.

Hulbert, A.W., McEdward, L., Powers, J., Black, L., Perez, J., Richardson, E., Runyon, F. and Tacy, K., 1976. *Role of predation in subtidal community zonation.* 1976–77 Sea Grant Program Rep., Univ. New Hampshire, Durham, N.H., 51 pp.

Hulbert, A.W., Harris, L.G. and Witman, J.D., 1981. Subtidal community zones on hard substrate in the Gulf of Maine. *Am. Zool.*, 20: 883.

Hulbert, A.W., Pecci, K.J., Witman, J.D., Harris, L.G., Sears, J.R. and Cooper, R.A., 1982. *Ecosystem definition and community structure of the macrobenthos of the NEMP monitoring station at Pigeon Hill in the Gulf of Maine.* NOAA Tech. Memor. NMFS-F/NEC-14, U.S. Dept. Comm., Woods Hole, Mass., 143 pp.

Hulburt, E.M., 1968. Stratification and mixing in coastal waters of the western Gulf of Maine during summer. *J. Fish. Res. Board Can.*, 25: 2609–2621.

Hulburt, E.M., 1970. Competition for nutrients by marine phytoplankton in oceanic, coastal and estuarine regions. *Ecology*, 51: 475–484.

Hulburt, E.M. and Corwin, J., 1970. Relation of the phytoplankton to turbulence and nutrient renewal in Casco Bay, Maine. *J. Fish. Res. Board Can.*, 27: 2081–2090.

Humm, H.J., 1969. Distribution of marine algae along the Atlantic coast of North America. *Phycologia*, 7: 43–53.

Huntsman, A.G., 1923. The importance of tidal and other oscillations in oceanic circulation. *Proc. Trans. R. Soc. Can.*, 17: 15–20.

Huntsman, A.G., 1930. Arctic ice on our eastern coast. *Bull. Biol. Board Can.*, 13: 1–12.

Hutchins, L.W., 1947. The bases for temperature zonation in geographic distribution. *Ecol. Monogr.*, 17: 25–35.

Ikawa, M., Ma, D.S., Meeker, G.B. and Davis, R.P., 1969. Use of *Chlorella* in mycotoxin and phycotoxin research. *J. Agric. Food Chem.*, 17: 425–429.

Incze, L.W. and Lutz, R.A., 1980. Mussel culture: an east coast perspective. In: R.A. Lutz (Editor), *Mussel Culture and Harvest.* Elsevier, Amsterdam, pp. 99–140.

Ingram, R.G., 1972. Winter surface currents around Cape Breton Island. *J. Fish. Res. Board Can.*, 30: 121–123.

Jackson, J.B.C., 1977. Competition on marine hard substrata: the adaptive significance of solitary and colonial strategies. *Am. Nat.*, 111: 743–767.

Jackson, J.B.C. and Hughes, T.B., 1985. Adaptive strategies of coral-reef invertebrates. *Am. Sci.*, 73: 265–274.

Jacobs, R.P.W.M., Den Hartog, C., Braster, B.F. and Carriere, F.C., 1981. Grazing of the seagrass *Zostera noltii* by birds at Terschelling (Dutch Wadden Sea). *Aquat. Bot.*, 10: 241–259.

Johannesson, B., 1986. Shell morphology of *Littorina saxatilis* Olivi: the relative importance of physical factors and predation. *J. Exp. Mar. Biol. Ecol.*, 102: 183–195.

Johns, P.M. and Mann, K.H., 1987. An experimental investigation of juvenile lobster habitat preference and mortality among habitats of varying structural complexity. *J. Exp. Mar. Biol. Ecol.*, 109: 275–285.

Johnson, C.R., 1984. *Ecology of the kelp Laminaria longicruris and its principal grazers in the rocky subtidal of Nova Scotia.* Ph.D. Thesis, Dalhousie Univ., Halifax, Nova Scotia, 280 pp.

Johnson, C.R. and Mann, K.H., 1986a. The crustose coralline alga, *Phymatolithon* Foslie, inhibits the overgrowth of seaweeds without relying on herbivores. *J. Exp. Mar. Biol. Ecol.*, 96: 127–146.

Johnson, C.R. and Mann, K.H., 1986b. The importance of plant defence abilities to the structure of subtidal seaweed communities: the kelp *Laminaria longicruris* de la Pylaie survives grazing by the snail *Lacuna vincta* (Montagu) at high population densities. *J. Exp. Mar. Biol. Ecol.*, 97: 231–267.

Johnson, C.R. and Mann, K.H., 1986c. Multiple strategies in the kelp *Laminaria longicruris* in the rocky subtidal of Nova Scotia. *Abstr. Twenty-fifth Northeast Algal Symp., Woods Hole, Mass.*

Johnson, D.S., 1925. *The New England-Acadian shoreline.* New York, Hafner, 608 pp.

Johnson, D.S. and Skutch, A.F., 1928a. Littoral vegetation on a headland of Mt. Desert Island, Maine. I. Submersible or strictly littoral vegetation. *Ecology*, 9: 188–215.

Johnson, D.S. and Skutch, A.F., 1928b. Littoral vegetation on a headland of Mt. Desert Island, Maine. II. Tide-pools and the environment and classification of submersible plant communities. *Ecology*, 9: 307–338.

Johnson, D.S. and Skutch, A.F., 1928c. Littoral vegetation on a headland of Mt. Desert Island, Maine. III. Adlittoral or nonsubmersible region. *Ecology*, 9: 429–448.

Johnson Jr., T.W. and Sparrow Jr., F.K., 1961. *Fungi in Oceans and Estuaries.* J. Cramer, New York, 668 pp.

Jones, N.S., 1948. Observations and experiments on the biology of *Patella vulgata* at Port St. Mary, Isle of Man. *Proc. Trans. Liverpool Biol. Soc.*, 56: 60–77.

Jones, W.E., 1956a. Effect of spore coalescence on the early development of *Gracilaria verruscosa* (Hudson) Papenfuss. *Nature*, 178: 426–427.

Jones, W.E., 1956b. *The littoral and sublittoral marine algae of Bardsey.* Bardsey Observatory 1955 Rep., pp. 40–51.

Jónsson, S. and Gunnarsson, K., 1978. Botnpórungar i sjó vid Island. *Greiningalykill. Hafrannsoknir*, 15: 1–94.

Joosten, A.M.T. and Van den Hoek, C., 1986. World-wide relationships between red seaweed floras: a multivariate approach. *Bot. Mar.*, 29: 195–214.

Josselyn, M.N. and Mathieson, A.C., 1978. Contribution of receptacles from the fucoid *Ascophyllum nodosum* to the detrital pool of a north temperate estuary. *Estuaries*, 1: 258–261.

Kain, J.M., 1982. The reproductive phenology of nine species of Rhodophyta in the subtidal region of the Isle of Man. *Br. Phycol. J.*, 17: 321–331.

Kain, J.M., 1984. *Plocamium cartilagineum* in the Isle of Man: why are there so many tetrasporophytes? *Br. Phycol. J.*, 19: 195.

Kain, J.M., 1986. Plant size and reproductive phenology of six species of Rhodophyta in subtidal Isle of Man. *Br. Phycol. J.*, 21: 129–138.

Kapraun, D.F., 1980. Floristic affinities of North Carolina inshore benthic marine algae. *Phycologia*, 19: 245–252.

Kastendiek, J., 1982. Competitor-mediated coexistence: interactions among three species of benthic macroalgae. *J. Exp. Mar. Biol. Ecol.*, 62: 201–210.

Keats, D.W., 1985. *The effects on benthic marine macroalgae and herbivorous invertebrates of the experimental removal of sea urchins at a periodically ice-scoured sublittoral site in eastern Newfoundland.* Ph.D. Thesis, Memorial Univ., Newfoundland, St. John's, Newfoundland.

Keats, D.W., South, G.R. and Steele, D.H., 1982a. Experimental assessment of the effect of *Strongylocentrotus droebachiensis* on subtidal algal communities in Newfoundland, Canada. *Br. Phycol. J.*, 17: 234–235.

Keats, D.W., South, G.R. and Steele, D.H., 1982b. The occurrence of *Agarum cribrosum* (Mert.) Bory (Phaeophyta, Laminariales) in relation to some of its competitors and predators in Newfoundland. *Phycologia*, 21: 189–191.

Keats, D.W., South, G.R. and Steele, D.H., 1984a. Dominance of the upper sublittoral by *Alaria esculenta*: modification by ice scour. *Program and Abstr. Twenty-third Northeast Algal Symp., Woods Hole, Mass.*

Keats, D.W., Steele, D.H. and South, G.R., 1984b. Depth-dependent reproductive output of the green sea urchin, *Strongylocentrotus droebachiensis* (O.F. Muller), in relation to the nature and availability of food. *J. Exp. Mar. Biol. Ecol.*, 80: 77–91.

Keats, D.W., South, G.R. and Steele, D.H., 1985. Algal biomass and diversity in the upper subtidal at a pack-ice disturbed site in eastern Newfoundland. *Mar. Ecol. Prog. Ser.*, 25: 151–158.

Keats, D.W., Steele, D.H. and South, G.R., 1987. The role of fleshy macroalgae in the ecology of juvenile cod (*Gadus morhua* L.) in inshore waters off eastern Newfoundland. *Can. J. Zool.*, 65: 49–53.

Keizer, P.D. and Gordon, D.C., 1985. Nutrient dynamics in Cumberland Basin-Chignecto Bay, a turbid macrotidal estuary in the Bay of Fundy, Canada. *Neth. J. Sea Res.*, 19: 193–205.

Keser, M. and Larson, B.R., 1984a. Colonization and growth dynamics of three species of *Fucus*. *Mar. Ecol. Prog. Ser.*, 15: 125–134.

Keser, M. and Larson, B.R., 1984b. Colonization and growth of *Ascophyllum nodosum* (Phaeophyta) in Maine. *J, Phycol.*, 20: 83–87.

Ketchum, B.H. and Keen, D.J., 1953. The exchanges of fresh and salt waters in the Bay of Fundy and in Passamaquoddy Bay. *J. Fish. Res. Board Can.*, 10: 97–124.

Ketchum, B.H., Vaccaro, R.F. and Corwin, N., 1958. The annual cycle of phosphorus and nitrogen in New England coastal waters. *J. Mar. Res.*, 17: 282–301.

Keyte, F.K. and Trites, R.W., 1971. Circulation patterns in the Gulf of St. Lawrence: June, 1968 and September, 1969. *Fish. Res. Board Can. Tech. Rep.*, 271: 1–108.

Khfaji, A.K. and Boney, A.D., 1979. Antibiotic effects of crustose germlings of the red alga *Chondrus crispus* Stackh. on benthic diatoms. *Ann. Bot.*, 43: 231–232.

Kingham, D.L. and Evans, L.V., 1986. The *Pelvetia–Mycosphaerella* interrelationship. In: S.T. Moss (Editor), *The Biology of Marine Fungi.* Cambridge Univ. Press, Cambridge, pp. 177–187.

Kingsbury, J.M., 1976. *Transect study of the intertidal biota of Star Island, Isles of Shoals.* Shoals Mar. Lab. Publ., Cornell Univ., Ithaca, N.Y., 66 pp.

Kitching, J.A., 1986. The ecological significance and control of shell variability in dogwhelks from temperate rocky shores. In: P.G. Moore and R. Seed (Editors), *The Ecology of*

Rocky Coasts. Columbia Univ. Press, New York, pp. 234–248.

Kitching, J.A., Sloane, J.F. and Ebling, F.J., 1959. The ecology of Lough Ine. VII. Mussels and their predators. *J. Anim. Ecol.*, 28: 331–341.

Klein, G. de V., 1963. Bay of Fundy intertidal zone sediments. *J. Sed. Petrol.*, 33: 844–854.

Knight, M. and Parke, M., 1950. A biological study of *Fucus vesiculosus* L. and *F. serratus* L. *J. Mar. Biol. Assoc. U.K.*, 29: 439–514.

Kornmann, P. and Sahling, P.H., 1977. Meeresalgen von Helgoland. *Helgol. Wiss. Meeresunters.*, 29: 1–289.

Korringa, P., 1951. The shell of *Ostrea edulis* as a habitat. *Neth. J. Zool.*, 10: 32–152.

Kortright, F.H., 1953. *The Ducks, Geese and Swans of North America*. Wildlife Management Inst., Washington, D.C. and the Stackpole Co., Harrisburg, Pa., 476 pp.

Kranck, K., 1972. Geomorphological development and post-Pleistocene sea-level changes, Northumberland Strait, Maritime Provinces. *Can. J. Earth Sci.*, 9: 835–844.

Kristensen, L., 1968. Surf influence on the thallus of fucoids and the rate of desiccation. *Sarsia*, 34: 69–82.

Lagler, K.F. and Wienert, C.C., 1948. Food of the oldsquaw in Lake Michigan. *Wilson Bull.*, 60: 118.

Lakshminarayana, J.S.S. and Bourque, H., 1979. Changes in the water quality of the Northumberland Strait, N.B. *Fish. Mar. Serv. Tech. Rep.*, 834 (Pt. IV): 67–80.

Lamb, M. and Zimmerman, M.H., 1964. Marine vegetation of Cape Ann, Essex County, Massachusetts. *Rhodora*, 66: 217–254.

Lang, C. and Mann, K.H., 1976. Changes in sea urchin populations after the destruction of kelp beds. *Mar. Biol.*, 36: 321–326.

Larsen, P.F. and Doggett, L.F., 1981. *The ecology of Maine's intertidal habitats*. Maine State Planning Office and Bigelow Lab for Ocean Sciences, 183 pp.

Larson, B.R., Vadas, R.L. and Keser, M., 1980. Feeding and nutritional ecology of the sea urchin *Strongylocentrotus droebachiensis* in Maine, U.S.A. *Mar. Biol.*, 59: 49–62.

Lauzier, L.M., 1952. Effect of storms on the water conditions in the Magdalen Shallows. *J. Fish. Res. Board Can.*, 8: 332–339.

Lauzier, L.M., 1957. Hydrographic features of the waters of the Bay of Islands, Newfoundland, in the autumn. *Fish. Res. Board Can. Bull.*, 111: 287–317.

Lauzier, L.M., 1965. Drift bottle observations in Northumberland Strait, Gulf of St. Lawrence. *J. Fish. Res. Board Can.*, 22: 353–368.

Lauzier, L.M., 1967. Bottom residual drift on the continental shelf area of the Canadian Atlantic Coast. *J. Fish. Res. Board Can.*, 24: 1845–1858.

Lauzier, L.M. and Hull, J.H., 1969. Coastal station data temperatures along the Canadian Atlantic Coast, 1921–1969. *Fish. Res. Board Can. Tech. Rep.*, 150: 1–25.

Lauzier, L.M., Trites, R.W. and Hachey, H.B., 1951. *Some features of the surface layer of the Gulf of St. Lawrence*. Joint Comm. Ocean., Atlantic Group, St. Andrews, New Brunswick, 19 pp.

Lawrence, D.J., 1979. *Flow patterns in Chedabucto Bay, Nova Scotia*. Fish. Mar. Serv. Tech. Rep. 834, pt. IV, pp. 19–30.

Lawrence, J.M., 1975. On the relationships between marine plants and sea urchins. In: H. Barnes (Editor), *Oceanography and Marine Biology Annual Review, Vol. 13*, pp. 213–286.

Lawson, G.W., 1978. The distribution of seaweed floras in the tropical and subtropical Atlantic Ocean: a quantitative approach. *Bot. J. Linn. Soc.*, 76: 177–193.

Lee, R.K.S., 1968. A collection of marine algae from Newfoundland. 1. Introduction and Phaeophyta. *Nat. Can.*, 95: 957–978.

Lee, R.K.S., 1980. A catalogue of the marine algae of the Canadian Arctic. *Nat. Mus. Canada Publ. Bot.*, 9: 1–32

Lee, T.F. and Mathieson, A.C., 1983. Polar regeneration in *Ptilota serrata* Kutzing. *Program and Abstr. Twenty-second Northeast Algal Symp., Woods Hole, Mass.*

Leim, A.H., 1956. Faunal zones in the seas adjacent to Newfoundland, the Maritimes and Quebec. *Rep. Biol. Sta. Fish. Res. Board Can.*, 617: 1–7.

Leim, A.H., 1957. Summary of results under the Atlantic Herring Investigation Committee. *Fish. Res. Board Can. Bull.*, 111: 1–16.

Levin, S.A. and Paine, R.T., 1974. Disturbance, patch formation, and community structure. *Proc. Nat. Acad. Sci.*, 71: 2744–2747.

Levinton, J.S., 1982. *Marine Ecology*. Prentice-Hall Inc., Englewood Cliffs, N.J., 526 pp.

Levring, T., 1960. Submarine Licht und die Algenvegetation. *Bot. Mar.*, 1: 67–73.

Lewis, J.R., 1964. *The Ecology of Rocky Shores*. English Univ. Press, London, 323 pp.

Lewis, J.R., 1978. The role of physical and biological factors in the distribution and stability of rocky shore communities. In: B.F. Keegan, P.Ó. Céidigh and P.J.S. Boaden (Editors), *Biology of Benthic Organisms*. Pergamon, Oxford, pp. 417–423.

Li, M.F., Cornick, J.W. and Miller, R.J., 1982. Studies of recent mortalities of the sea urchin (*Strongylocentrotus droebachiensis*) in Nova Scotia. *Int. Counc. Explor. Mer*, CM 1982/L: 46.

Lie, U. and Kelley, J.C., 1970. Benthic infauna communities off the coast of Washington and Puget Sound: identification and distribution of the communities. *J. Fish. Res. Board. Can.*, 27: 621–651.

Lilly, G.R., 1968. Some aspects of the ecology of Irish moss, *Chondrus crispus* (L.) Stackhouse in Newfoundland waters. *Fish. Res. Board Can. Tech. Rep.*, 43: 1–44.

Lind. J., 1913. *Danish Fungi as Represented in the Herbarium of E. Rostrup*. Gyldendalske Boghandel, Copenhagen, 650 pp.

Littler, D.S. and Littler, M.M., 1986. *The unnoticed reef builders*. Ocean Realm, Fall 1986, pp. 80–83.

Littler, M.M. and Littler, D.S., 1980. The evolution of thallus form and survival strategies in benthic marine macroalgae: field and laboratory tests of a functional form model. *Am. Nat.*, 116: 25–44.

Littler, M.M. and Littler, D.S., 1983. Heteromorphic life-history strategies in the brown alga *Scytosiphon lomentaria* (Lyngb.) Link. *J. Phycol.*, 19: 425–431.

Littler, M.M., Martz, D.R. and Littler, D.S., 1983. Effects of recurrent sand deposition on rocky intertidal organisms:

importance of substrate heterogeneity in a fluctuating environment. *Mar. Ecol. Prog. Ser.*, 11: 129–139.

Littler, M.M., Taylor, P.R. and Littler, D.S., 1986. Plant defense associations in the marine environment. *Coral Reefs*, 5: 63–71.

Lobban, C.S. and Hanic, L.A., 1984. Rocky shore zonation at North Rustico and Prim Point, Prince Edward Island. *Proc. N.S. Inst. Sci.*, 34: 25–40.

Lobban, C.S., Harrison, P.J. and Duncan, M.J., 1985. *The Physiological Ecology of Seaweeds*. Cambridge Univ. Press, Cambridge, 242 pp.

Lodge, S.M., 1948. Algal growth in the absence of *Patella* on an experimental strip of foreshore, Port St. Mary, Isle of Man. *Proc. Trans. Liverpool Biol. Soc.*, 56: 78–83.

Loring, D.H. and Nota, D.J.G., 1973. *Morphology and sediments of the Gulf of St. Lawrence*. Bull. 182, Fish. Mar. Serv., Ottawa, 147 pp.

Loucks, R.H., Trites, R.W., Drinkwater, K.F. and Lawrence, D.J., 1974. Summary of physical, biological, socio-economic and other factors relevant to potential oil spills of the Passamaquoddy Region of the Bay of Fundy. Section 1, Physical oceanographic characteristics. *Fish. Res. Board Can. Tech. Rep.*, 428: 1–60.

Lubchenco, J., 1978. Plant species diversity in a marine intertidal community: importance of herbivore food preference and algal competitive abilities. *Am. Nat.*, 112: 23–39.

Lubchenco, J., 1980. Algal zonation in a New England rocky intertidal community: an experimental analysis. *Ecology*, 61: 333–344.

Lubchenco, J., 1982. Effects of grazers and algal competitors on fucoid colonization in tide pools. *J. Phycol.*, 18: 544–550.

Lubchenco, J., 1983. *Littorina* and *Fucus*: effects of herbivores, substratum heterogeneity, and plant escapes during succession. *Ecology*, 64: 1116–1123.

Lubchenco, J., 1986. Relative importance of competition and predation: early colonization by seaweeds in New England. In: J. Diamond and T.J. Case (Editors), *Community Ecology*. Harper and Row, New York, pp. 537–555.

Lubchenco, J. and Cubit, J., 1980. Heteromorphic life histories of certain marine algae as adaptations to variations in herbivory. *Ecology*, 61: 676–687.

Lubchenco, J. and Gaines, S.D., 1981. A unified approach to marine plant-herbivore interactions. I. Populations and communities. *Ann. Rev. Ecol. Syst.*, 12: 405–437.

Lubchenco, J. and Menge, B.A., 1978. Community development and persistence in a low rocky intertidal zone. *Ecol. Monogr.*, 48: 67–94.

Lubinsky, I., 1980. Marine bivalve molluscs of the Canadian central and eastern Arctic: faunal composition and zoogeography. *Can. Bull. Fish. Aquat. Sci.*, 207: 1–111.

Lund, S., 1959. The marine algae of east Greenland. I. Taxonomical part. *Medd. Groenl.*, 156: 1–247.

Lundalv, T., 1971. Quantitative studies on rocky-bottom biocoenoses by underwater photogrammetry: a methodological study. *Thalassia Jugosl.*, 71: 201–208.

Lüning, K., 1970. Tauchuntersuchungen zur Vertikalverteilung der sublitoralen Helgolander Algenvegetation. *Helgol. Wiss. Meeresunters.*, 21: 271–291.

Lüning, K., 1980. Control of algal life-history by daylength and temperature. In: J.H. Price, D.E.G. Irvine and W.F. Farnham (Editors), *The Shore Environment, Vol. 2: Ecosytems*. Systematics Association Special Vol. 17(6). Academic Press, London, pp. 915–945.

Lüning, K., 1984. Temperature tolerance and biogeography of seaweeds: the marine algal flora of Helgoland (North Sea) as an example. *Helgol. Wiss. Meeresunters.*, 38: 305–317.

Luther, G., 1976. Bewuchsuntersuchungen auf Natursteinsubstraten im Gezeitenbereich des nordsylter Wattenmeeres: Algen. *Helgol. Wiss. Meeresunters.*, 28: 318–351.

Lutz, R.A., 1980. Introduction: mussel culture and harvest in North America. In: R.A. Lutz (Editor), *Mussel Culture and Harvest*. Elsevier, Amsterdam, pp. 1–17.

Luxoro, C. and Santelices, B., 1989. Additional evidence for ecological differences among isomorphic reproductive phases of *Iridaea laminarioides* (Rhodophyta: Gigartinales). *J. Phycol.*, 25: 206–212.

MacFarlane, C.I., 1952. A survey of certain seaweeds of commercial importance in southwest Nova Scotia. *Can. J. Bot.*, 30: 78–97.

MacFarlane, C.I., 1956. *Irish moss in the Maritime Provinces*. N.S. Res. Found., Halifax, Nova Scotia, 20 pp.

MacFarlane, C.I., 1964. A comparison of two marine drumlin regions in Nova Scotia. In: A. Davy de Virville and J. Feldmann (Editors), *Proc. Fourth Int. Seaweed Symp., Biarritz, France*. MacMillan, New York, pp. 240–247.

MacFarlane, C.I., 1966. Sublittoral surveying for commercial seaweeds in Northumberland Strait. In: E.G. Young and J.L. McLachlan (Editors), *Proc. Fifth Int. Seaweed Symp., Halifax, Can*. Pergamon, Oxford, pp. 169–171.

MacFarlane, C.I., 1968. *Chondrus crispus Stackhouse — a synopsis*. N.S. Res. Found., Halifax, Nova Scotia, 47 pp.

MacGregor, D.G., 1956. Currents and transport in Cabot Strait. *J. Fish. Res. Board. Can.*, 13: 435–448.

MacGregor, D.G. and McLellan, J.J., 1952. Current measurements in the Grand Manan Channel. *J. Fish. Res. Board Can.*, 9: 213–222.

Mackay, G.H., 1982. Habits of the oldsquaw (*Clangula hyemalis*) in New England. *Auk*, 9: 330–337.

Mackay, T.F.C. and Doyle, R.W., 1978. An ecological and genetic analysis of the settling behaviour of a marine polychaete. I. Probability of settlement and gregarious behaviour. *Heredity*, 40: 1–12.

MacLean, B. and King, L.H., 1976. Geology of the Scotian Shelf. *Geol. Surv. Can. Pap.*, 74–31, 1–31.

MacLeod, L.L., 1975. *Experimental blue mussel (Mytilus edulis) culture in Nova Scotia waters*. Interim Rep. 1. Fish. Mar. Sci., Pictou, Nova Scotia, 38 pp.

Madsen, F.J., 1954. On the food habits of the diving ducks in Denmark. *Dan. Rev. Game Biol.*, 2: 157–266.

Magre, E.J., 1974. *Ulva lactuca* L. negatively affects *Balanus balanoides* (L.) (Cirripedia: Thoracica) in tidepools. *Crustaceana*, 27: 231–234.

Mann, K.H., 1972a. Ecological energetics of the seaweed zone in a marine bay on the Atlantic coast of Canada. I. Zonation and biomass of seaweeds. *Mar. Biol.*, 12: 1–10.

Mann, K.H., 1972b. Ecological energetics of the seaweed zone in a marine bay on the Atlantic coast of Canada. II. Productivity of the seaweeds. *Mar. Biol.*, 14: 199–209.

Mann, K.H., 1973. Seaweeds: their productivity and strategy for growth. *Science*, 182: 975–981.
Mann, K.H., 1977. Destruction of kelp-beds by sea urchins: a cyclical phenomenon or irreversible degradation? *Helgol. Wiss. Meeresunters.*, 30: 455–467.
Mann, K.H., 1982. Kelp, sea urchins and predators, a review of strong interactions in rocky subtidal systems of eastern Canada, 1970–1980. *Neth. J. Sea Res.*, 16: 414–423.
Mann, K.H. and Breen, P.A., 1972. The relation between lobster abundance, sea urchins, and kelp beds. *J. Fish. Res. Board Can.*, 29: 603–605.
Markham, J.W., 1973. Observations on the ecology of *Laminaria sinclairii* on three northern Oregon beaches. *J. Phycol.*, 9: 336–341.
Marsh, C.P., 1986. Rocky intertidal community organization: the impact of avian predators on mussel recruitment. *Ecology*, 67: 771–786.
Marshall, S.M., Newton, L. and Orr, A.P., 1949. *A study of certain British seaweeds and their utilization in the preparation of agar.* London, H.M.S.O., 184 pp.
Martin, P.D., Truchon, S.P. and Harris, L.G., 1988. *Strongylocentrotus droebachiensis* populations and community dynamics at two depth-related zones over an 11-year period. In: R.D. Burke, P.V. Mladenov, P. Lambert and R.L. Parsley (Editors), *Echinoderm Biology*. Balkema, Rotterdam, pp. 475–482.
Mathieson, A.C., 1968. Morphology and life history of *Phaeostrophion irregulare* Setchell *et* Gardner. *Nova Hedwigia*, 13: 293–318.
Mathieson, A.C., 1979. Vertical distribution and longevity of subtidal seaweeds in northern New England, U.S.A. *Bot. Mar.*, 30: 511–520.
Mathieson, A.C., 1982. Field ecology of the brown alga *Phaeostrophion irregulare* Setchell *et* Gardner. *Bot. Mar.*, 25: 67–85.
Mathieson, A.C., 1989. Phenological patterns of northern New England seaweeds. *Bot. Mar.*, 32, in press.
Mathieson, A.C. and Burns, R.L., 1975. Ecological studies of economic red algae V. Growth and reproduction of natural and harvested populations of *Chondrus crispus* Stackhouse in New Hampshire. *J. Exp. Mar. Biol. Ecol.*, 17: 137–156.
Mathieson, A.C. and Dawes, C.J., 1975. Seasonal studies of Florida sublittoral marine algae. *Bull. Mar. Sci.*, 25: 46–65.
Mathieson, A.C. and Hehre, E.J., 1986. A synopsis of New Hampshire seaweeds. *Rhodora*, 88: 1–139.
Mathieson, A.C. and Norall, T.L., 1975a. Physiological studies of subtidal red algae. *J. Exp. Mar. Biol. Ecol.*, 20: 237–247.
Mathieson, A.C. and Norall, T., 1975b. Photosynthetic studies of *Chondrus crispus*. *Mar. Biol.*, 33: 207–213.
Mathieson, A.C. and Penniman, C.A., 1986a. A phytogeographic interpretation of the marine flora from the Isles of Shoals. U.S.A. *Bot. Mar.*, 29: 413–434.
Mathieson, A.C. and Penniman, C.A., 1986b. Species composition and seasonality of New England seaweeds along an open coastal-estuarine gradient. *Bot. Mar.*, 29: 161–176.
Mathieson, A.C. and Prince, J.S., 1973. Ecology of *Chondrus crispus* Stackhouse. In: M.J. Harvey and J. McLachlan (Editors), *Chondrus crispus*. N.S. Inst. Sci. Publ., Halifax, Nova Scotia, Can., pp. 53–79.
Mathieson, A.C., Dawes, C.J. and Humm, H.J., 1969. Contributions to the marine algae of Newfoundland. *Rhodora*, 71: 110–159.
Mathieson, A.C., Shipman, J.W., O'Shea, J.R. and Hasevlat, R.C., 1976. Seasonal growth and reproduction of estuarine fucoid algae in New England. *J. Exp. Mar. Biol. Ecol.*, 25: 273–284.
Mathieson, A.C., Hehre, E.J. and Reynolds, N.B., 1981. Investigations of New England marine algae I: A floristic and descriptive ecological study of the marine algae at Jaffrey Point, New Hampshire. *Bot. Mar.*, 24: 521–532.
Mathieson, A.C., Penniman, C.A., Busse, P.K. and Tveter-Gallagher, E., 1982. Effects of ice on *Ascophyllum nodosum* within the Great Bay Estuary System of New Hampshire–Maine. *J. Phycol.*, 18: 331–336.
Mathieson, A.C., Neefus, C. and Emerich-Penniman, C., 1983. Benthic ecology in an estuarine tidal rapid. *Bot. Mar.*, 26: 213–230.
Mavor, J.W., 1922. The circulation of the water in the Bay of Fundy. I. Introduction and drift bottle experiments. *Contrib. Can. Biol.*, 1: 101–124.
May, G., 1986. Life history variations in a predominantly gametophytic population of *Iridaea cordata* (Gigartinaceae, Rhodophyta). *J. Phycol.*, 22: 448–455.
McAlice, B.J., 1981. On the post-glacial history of *Acartia tonsa* (Copepoda: Calanoida) in the Gulf of Maine and the Gulf of St. Lawrence. *Mar. Biol.*, 64: 266–272.
McBane, C.D., 1981. *Studies on the ecology of Hyale nilssoni (Rathke) 1843, an algal-inhabiting amphipod crustacean.* Ph.D. Thesis, Univ. New Hampshire, Durham, N.H., 156 pp.
McBane, C.D. and Croker, R., 1983. Animal-algal relationships of the amphipod *Hyale nilssoni* (Rathke) in the rocky intertidal. *J. Crustacean Biol.*, 3: 592–601.
McCourt, R.M., 1985. Reproductive biomass allocation in three *Sargassum* species. *Oecologia*, 67: 113–117.
McLachlan, J. and Bird, C.J., 1984. Geographical and experimental assessment of the distribution of *Gracilaria* species (Rhodophyta, Gigartinales) in relation to temperature. *Helgol. Wiss. Meeresunters.*, 38: 319–334.
McLachlan, J. and Craigie, J.S., 1966. Antialgal activity of some simple phenols. *J. Phycol.*, 2: 133–135.
McLachlan, J., Greenwell, M., Bird, C.J. and Holmsgaard, J.E., 1987. Standing stocks of seaweeds of commercial importance on the north shore of Prince Edward Island, Canada. *Bot. Mar.*, 30: 277–289.
McLeese, D.W., 1973. Olfactory responses of lobsters (*Homarus americanus*) to solutions from prey species and to seawater extracts and chemical fractions of fish muscle and effects of antennule ablation. *Mar. Behav. Physiol.*, 2: 237–249.
Meadows, P.S. and Campbell, J.I., 1972. Habitat selection by aquatic invertebrates. *Adv. Mar. Biol.*, 10: 271–382.
Menge, B.A., 1976. Organization of the New England rocky intertidal community: role of predation, competition, and environmental heterogeneity. *Ecol. Monogr.*, 46: 355–393.
Menge, B.A., 1978a. Predation intensity in a rocky intertidal community: effect of an algal canopy, wave action and desiccation on predator feeding rates. *Oecologia*, 34: 17–35.
Menge, B.A., 1978b. Predation intensity in a rocky intertidal

community: relation between predator foraging activity and environmental harshness. *Oecologia*, 34: 1–16.

Menge, B.A., 1979. Coexistence between the seastars *Asterias vulgaris* and *A. forbesi* in a heterogeneous environment: a non-equilibrium explanation. *Oecologia*, 41: 245–272.

Menge, B.A., 1982. Effects of feeding on the environment: Asteroidea. In: M. Jangoux and J.M. Lawrence (Editors), *Echinoderm Nutrition*. Balkema, Rotterdam, pp. 521–551.

Menge, B.A., 1983. Components of predation intensity in the low zone of the New England rocky intertidal region. *Oecologia*, 58: 141–155.

Menge, B.A. and Lubchenco, J., 1981. Community organization in temperate and tropical rocky intertidal habitats: prey refuges in relation to consumer pressure gradients. *Ecol. Monogr.*, 51: 429–450.

Menge, B. and Sutherland, J.P., 1976. Species diversity gradients: synthesis of the roles of predation, competition and temporal heterogeneity. *Am. Nat.*, 110: 351–369.

Menge, J.L., 1975. *Effects of herbivores on community structure of the New England rocky intertidal region: distribution, abundance and diversity of algae*. Ph.D. Thesis, Harvard University, Cambridge, Mass., 165 pp.

Merz, A.A., 1984. Benthic organisms and hydromechanical adaptations. *Program and Abstr. Twenty-third Northeast Algal Symp., Woods Hole, Mass.*

Metcalf, C.D., Dadswell, M.J., Gillis, G.F. and Thomas, M.L.H., 1976. Physical, chemical and biological parameters of the Saint John River Estuary, New Brunswick, Canada. *Fish. Mar. Ser. Res. Dev. Tech. Rep.*, 686: 1–42.

Michanek, G., 1979. Phytogeographic provinces and seaweed distribution. *Bot. Mar.*, 22: 375–391.

Miller, R.J., 1985. Succession in sea urchin and seaweed abundance in Nova Scotia, Canada. *Mar. Biol.*, 84: 275–286.

Miller, R.J. and Colodey, A.G., 1983. Widespread mass mortalities of the green sea urchin in Nova Scotia, Canada. *Mar. Biol.*, 73: 263–267.

Miller, R.J. and Mann, K.H., 1973. Ecological energetics of the seaweed zone in a marine bay on the Atlantic Coast of Canada. III. Energy transformation by sea urchins. *Mar. Biol.*, 18: 99–114.

Miller, R.J., Mann, K.H. and Scarratt, D.J., 1971. Production potential of a seaweed-lobster community in eastern Canada. *J. Fish. Res. Board Can.*, 28: 1733–1738.

Miller, S.L., 1984. The population biology of *Ascophyllum nodosum:* biological and physical factors influencing the survivorship of germlings. *Program and Abstr. Twenty-third Northeast Algal Symp., Woods Hole, Mass.*

Milligan, G.C., 1949. *Geological survey of Prince Edward Island*. Dept. Indust. Nat. Res., Charlottetown, P.E.I., 83 pp.

Molina, F.I., 1986. *Petersenia pollagaster* (Oomycetes): An invasive fungal pathogen of *Chondrus crispus* (Rhodophyceae). In: S.T. Moss (Editor), *The Biology of Marine Fungi*. Cambridge Univ. Press, Cambridge, pp. 165–175.

Moore, D.S. and Miller, R.J., 1983. Recovery of macroalgae following widespread sea urchin mortality with a description of the nearshore hard-bottom habitat on the Atlantic Coast of Nova Scotia. *Can. Tech. Rep. Fish. Aquat. Sci.*, 1230: 1–94.

Moore, P.D., 1982. Measuring competition in plant communities. *Nature*, 298: 514–515.

Moore, P.G., 1983. Biological interactions. In: R. Earll and D.G. Erwin (Editors), *Sublittoral Ecology*. Clarendon Press, Oxford, pp. 125–143.

Moore, P.G., 1986. Levels of heterogeneity and the amphipod fauna of kelp holdfasts. In: P.G. Moore and R. Seed (Editors), *The Ecology of Rocky Coasts*. Columbia Univ. Press, New York, pp. 274–289.

Moss, B.L., 1982. The control of epiphytes by *Halidrys siliquosa* (L.) Lyngb. (Phaeophyta, Cystoseiraceae). *Phycologia*, 21: 185–188.

Moss, S.T. (Editor), 1986a. *The Biology of Marine Fungi*. Cambridge Univ. Press, Cambridge, 382 pp.

Moss, S.T., 1986b. Biology and phylogeny of the Labyrinthulales and Thraustochytriales. In: S.T. Moss (Editor), *The Biology of Marine Fungi*. Cambridge Univ. Press, Cambridge, pp. 105–129.

Mott, R.J., 1975. Post-glacial history and environments in southwestern New Brunswick. In: J.G. Ogden III and M.J. Harvey (Editors), *Environmental Change in the Maritimes*. N.S. Inst. Sci. Publ., pp. 67–82.

Moyse, C.M., 1978. Bay of Fundy environmental and tidal power bibliography. *Fish. Mar. Ser. Tech. Rep.*, 822: 1–125.

Muntz, L., Ebling, F.J. and Kitching, J.A., 1965. The ecology of Lough Ine. XIV. Predatory activities of large crabs. *J. Anim. Ecol.*, 34: 315–329.

Nesis, K.N., 1962. Soviet investigations of the benthos of the Newfoundland–Labrador fishing area. In: Y.Y. Marti (Editor), *Soviet Fisheries Investigations in the North West Atlantic*, pp. 214–220.

Nesis, K.N., 1965. Biocenosi i biomassa benthosa Newfaundlendsko–Lavradorskovo raiona. *Trudy Vses Nauchno-issled. Inst. Morsk. Rib. Chiz i Okean*, 57: 453–489.

Neushul, M., 1971. The kelp community of seaweeds. *Nova Hedwigia*, 32: 265–267.

Newcombe, C.L., 1935. A study of the community relationships of the sea mussel, *Mytilus* L. *Ecology*, 16: 234–243.

Newroth, P.R., 1970. *A study of the genus Phyllophora Greville*. Ph.D. Thesis, Univ. New Brunswick, Fredericton, N.B., 311 pp.

Nicotri, M.E., 1980. Factors involved in herbivore food preference. *J. Exp. Mar. Biol. Ecol.*, 42: 13–26.

Niemeck, R.A. and Mathieson, A.C., 1976. An ecological study of *Fucus spiralis*. *J. Exp. Mar. Biol. Ecol.*, 24: 33–48.

Nienhuis, P.H., 1974. Variability in the life cycle of *Rhizoclonium riparium* (Roth) Harv. (Chlorophyceae: Cladophorales) under Dutch estuarine conditions. *Hydrobiol. Bull.*, 8: 172–178.

Noble, J.P., Logan, A. and Webb, R.G., 1976. The recent *Terebratulina* community in the rocky subtidal zone of the Bay of Fundy, Canada. *Lethaia*, 9: 1–17.

Norall, T.L., Mathieson, A.C. and Emerich-Penniman, C., 1982. *Nutrient and hydrographic data for the Great Bay Estuarine System, New Hampshire–Maine, Part I, September, 1973–December, 1975*. Jackson Estuarine Laboratory Contribution No. 150, Univ. New Hampshire, Durham, N.H., 102 pp.

Norall, T.L., Mathieson, A.C. and Kilar, J.A., 1981. Reproduc-

tive ecology of four subtidal red algae. *J. Exp. Mar. Biol. Ecol.*, 54: 119–136.
Norton, T.A., 1986. The zonation of seaweeds on rocky shores. In: P.G. Moore and R. Seed (Editors), *The Ecology of Rocky Coasts*. Columbia Univ. Press, New York, pp. 7–21.
Norton, T.A. and Mathieson, A.C., 1983. Biology of unattached seaweeds. In: F. Round and D. Chapman (Editors), *Progress in Phycological Research, Vol. 2*. Elsevier, Amsterdam, pp. 331–386.
Norton, T.A. and Milburn, J.A., 1972. Direct observation on the sublittoral marine algae of Argyll, Scotland. *Hydrobiologia*, 40: 55–68.
Norton, T.A., Mathieson, A.C. and Neushul, M., 1981. Morphology and environment. In: C.S. Lobban and M. Wynne (Editors), *Biology of Seaweeds*. Blackwell, Oxford, pp. 421–451.
Norton, T.A., Mathieson, A.C. and Neushul, M., 1982. A review of some aspects of form and function in seaweeds. *Bot. Mar.*, 25: 501–510.
Novaczek, I. and McLachlan, J., 1986. Recolonization by algae of the sublittoral habitat of Halifax County, Nova Scotia, following the demise of sea urchins. *Bot. Mar.*, 29: 69–73.
Novaczek, I., Bird, C.J. and McLachlan, J., 1987. Phenology and temperature tolerance of the red algae *Chondria baileyana, Lomentaria baileyana, Griffithsia globulifera*, and *Dasya baillouviana* in Nova Scotia. *Can. J. Bot.*, 65: 57–62.
O'Clair, C.E., Hanson, J.L., Myren, R.T., Gharrett, J.A., Merrell Jr., T.R. and MacKinnon, J.S., 1979. *Reconnaissance of intertidal communities in the eastern Bering Sea and the effects of ice-scour on community structure. Environmental Assessment of the Alaskan Continental Shelf, Final Rep. of the Principal Investigators, Vol. 10: Biological studies*, U.S. Dept. Interior, Bureau of Land Management, pp. 1–419.
Ojeda, F.P. and Santelices, B., 1984. Ecological dominance of *Lessonia nigrescens* (Phaeophyta) in central Chile. *Mar. Ecol. Prog. Ser.*, 19: 83–91.
Olney, P.J.S. and Mills, D.H., 1963. The food and feeding habits of goldeneye (*Bucephala clangula*) in Great Britain. *Ibis*, 105: 293–300.
O'Neill, A.D.J., 1979. Climate and ice in the Strait of Canso region. *Fish. Mar. Ser. Tech. Rep.*, 834 (Pt. IV): 1–10.
Orton, J.H., 1920. Sea-temperature, breeding and distribution in marine animals. *J. Mar. Biol. Assoc. U.K.*, 12: 339–366.
Osman, R.W., 1977. The establishment and development of a marine epifaunal community. *Ecol. Monogr.*, 47: 37–63.
Ott, F.D., 1973. The marine algae of Virginia and Maryland including the Chesapeake Bay area. *Rhodora*, 75: 258–296.
Owens, E.H. and Bowen, A.J., 1977. Coastal environments of the Maritime Provinces. *Maritime Sediments*, 13: 1–31.
Padilla, D.K., 1985. Structural resistance of algae to herbivores, a biomechanical approach. *Mar. Biol.*, 90: 102–109.
Paine, R.T., 1966. Food web complexity and species diversity. *Am. Nat.*, 100: 65–75.
Paine, R.T., 1969. A note on trophic complexity and community stability. *Am. Nat.*, 103: 91–93.
Paine, R.T., 1974. Intertidal community structure. *Oecologia*, 15: 93–120.
Paine, R.T., 1976a. Biological observations on a subtidal *Mytilus californianus* bed. *Veliger*, 19: 125–129.
Paine, R.T., 1976b. Size-limited predation: an observational and experimental approach with *Mytilus* and *Pisaster* interaction. *Ecology*, 57: 858–873.
Paine, R.T., 1979. Disaster, catastrophe, and local persistence of the sea palm *Postelsia palmaeformis*. *Science*, 205: 685–687.
Paine, R.T. and Levin, S.A., 1981. Intertidal landscapes: disturbance and the dynamics of pattern. *Ecol. Monogr.*, 51: 145–178.
Paine, R.T. and Vadas, R.L., 1969. The effects of grazing by sea urchins *Strongylocentrotus* spp. on benthic algal populations. *Limnol. Oceanogr.*, 14: 710–719.
Pankow, H., 1971. *Algenflora der Ostsee — I. Benthos. (Blau-, Gruen-, Braun- und Rotalgen)*. Gustav Fischer, Stuttgart, 419 pp.
Parke, M. and Dixon, P.S., 1976. Checklist of British marine algae, third rev. *J. Mar. Biol. Assoc. U.K.*, 56: 527–594.
Pearse, J.S. and Hines, A.H., 1979. Expansion of a central California kelp forest following the mass mortality of sea urchins. *Mar. Biol.*, 51: 83–91.
Pearse, J.S., Clark, M.E., Leighton, D.L., Mitchell, C.T. and North, W.J., 1970. Marine waste-disposal and sea urchin ecology. In: W.J. North (Editor), *Kelp habitat improvement project, annual report 1969–1970*, Calif. Inst. Tech., Pasadena, Calif., appendix, pp. 1–93.
Peckol, P. and Harlin, M.M., 1985. Growth, survivorship, and photosynthetic performance of intertidal and subtidal Rhode Island populations of *Ascophyllum nodosum* (L.) Le Jol. *Abstr. Twenty-fourth Northeast Algal Symp., Woods Hole, Mass.*
Peckol, P., Harlin, M.M. and Krumscheid, P., 1988. Physiological and population ecology of intertidal and subtidal *Ascophyllum nodosum* (Phaeophyta). *J. Phycol.*, 24: 192–198.
Peguy, M. and Heim, R., 1961. A propos de l'activité antibiotique présumeé d'une solution aqueuse du corps phénolique brome et sulphone extrait du *Polysiphonia lanosa* (Linné) Tandy (Floridée, Rhodomelacée). *Comp. Rend. Acad. Sci. Paris, Ser. D.*, 252: 2134–2135.
Petersen, H.E., 1905. Contributions à la connaissance des Phycomycètes marins (Chytridinae Fisher). *Oversigt K. Dan. Vidensk. Selsk. Forh.*, 1904: 439–488.
Peterson, C.H., 1979. The importance of predation and competition in organizing the intertidal epifaunal communities of Barnegat Inlet, New Jersey. *Oecologia*, 39: 1–24.
Petraitis, P.S., 1979. *Behavior of Littorina littorea and its role in maintaining refuges for sessile organisms*. Ph.D. Thesis, State Univ. New York, Stony Brook.
Petraitis, P.S., 1987. Factors organizing rocky intertidal communities of New England: herbivory and predation in sheltered bays. *J. Exp. Mar. Biol. Ecol.*, 109: 117–136.
Pianka, F.R., 1970. On r- and K-selection. *Am. Nat.*, 104: 592–597.
Pielou, E.C., 1977. The latitudinal spans of seaweed species and their patterns of overlap. *J. Biogeography*, 4: 299–311.
Pielou, E.C., 1978. Latitudinal overlap of seaweed species: evidence for quasi-sympatric speciation. *J. Biogeography*, 5: 227–238.

Pielou, E.C., 1979. *Biogeography*. Wiley-Interscience, New York, 351 pp.

Platt, T. and Irwin, B., 1968. Primary productivity measurements in St. Margaret's Bay, 1967. *Fish. Res. Board. Can. Tech. Rep.*, 77: 1–123.

Platt, T. and Irwin, B., 1970. Primary productivity measurements in the St. Margaret's Bay, 1968–70. *Fish. Res. Board Can. Tech. Rep.*, 203: 1–68.

Platt, T. and Irwin, B., 1971. Phytoplankton production and nutrients in Bedford Basin, 1969–70. *Fish. Res. Board Can. Tech. Rep.*, 247: 1–172.

Prentice, S.A. and Kain, J.M., 1976. Numerical analysis of subtidal communities on rocky shores. *Estuarine Coastal Mar. Sci.*, 4: 65–70.

Prest, V.K., 1970. Quaternary geology of Canada. In: R.W.J. Douglas (Editor), *Geology and economic minerals of Canada. 5th ed. Geol. Surv. Can. Econ. Rep.*, 1: 676–764.

Prest, V.K. and Grant, D.R., 1969. Retreat of the last ice sheet from the Maritime Provinces–Gulf of St. Lawrence region. *Geol. Surv. Can. Pap.*, 69–33: 1–15.

Prince, J.S., 1971. *An ecological study of the marine red alga Chondrus crispus in the waters off Plymouth, Massachusetts.* Ph.D. Thesis, Cornell Univ., Ithaca, N.Y., 193 pp.

Pringle, J.D., 1986. A review of urchin/macro-algal associations with a new synthesis for nearshore, eastern Canadian waters. *Monogr. Biol.*, 4: 191–218.

Pringle, J.D. and Mathieson, A.C., 1987. *Chondrus crispus* Stackhouse. In: M.S. Doty, J.F. Caddy and B. Santelices (Editors), *Case Studies of Seven Commercial Seaweed Resources.* FAO Fisheries Techn. Pap., 281, Food and Agriculture Organization, United Nations, Rome, pp. 49–122.

Pringle, J.D. and Semple, R.E., 1980. The benthic algal biomass, commercial harvesting, and *Chondrus* growth and colonization off southwestern Nova Scotia. In: J.D. Pringle, G.J. Sharp and J.F. Caddy (Editors), *Proc. Workshop on the relationship between sea urchin grazing and commercial plant/animal harvesting. Can. Tech. Rep. Fish. Aquat. Sci.*, 954: 144–169.

Pringle, J.D. and Semple, R.E., 1983. A description of the major commercial Irish moss (*Chondrus crispus* Stackh.) beds off western Prince Edward Island. *Can. MS Rep. Fish. Aquat. Sci.*, 1725: 1–117.

Pringle, J.D. and Semple, R.E., 1984. Dragrake harvesting intensity in Irish moss (*Chondrus crispus* Stackh.) beds in the southern Gulf of St. Lawrence. In: C.J. Bird and M.A. Ragan (Editors), *Proc. Eleventh Int. Seaweed Symp., Qingdao, China.* Junk, Dordrecht, pp. 342–346.

Pringle, J.D. and Sharp, G.J., 1980. Multispecies resource management of economically important marine plant communities of eastern Canada. *Helgol. Wiss. Meeresunters.*, 33: 711–720.

Pringle, J.D., Sharp, G.J. and Caddy, J.F., 1980. Proceedings of the workshop on the relationship between sea urchin grazing and commercial plant/animal harvesting. *Can. Tech. Rep. Fish. Aquat. Sci.*, 954: 1–273.

Pringle, J.D., Sharp, G.J. and Caddy, J.F., 1982. Interactions in kelp bed ecosystems in the northwest Atlantic: review of a workshop. In: M.C. Mercer (Editor), *Multispecies approaches of fisheries management advice*, Can. Spec. Publ. Fish. Aquat. Sci., pp. 108–115.

Printz, H., 1956. Recuperation and recolonization in *Ascophyllum.* In: T. Braarud and N.A. Sorensen (Editors), *Proc. Second Int. Seaweed Symp., Trondheim, Norway.* Pergamon Press, New York, pp. 194–197.

Printz, H., 1959. Investigations of the failure of recuperation and repopulation in cropped *Ascophyllum nodosum. Skr. Nor. Vidensk. Akad. Oslo Mat.-Natvidensk. Kl.*, 3: 1–15.

Probyn, T.A. and Chapman, A.R.O., 1982. Nitrogen uptake characteristics of *Chordaria flagelliformis* (Phaeophyta) in batch mode and continuous mode experiments. *Mar. Biol.*, 71: 129–133.

Probyn, T.A. and Chapman, A.R.O., 1983. Summer growth of *Chordaria flagelliformis* (O.F. Muell.) C. Ag.: physiological strategies in a nutrient stressed environment. *J. Exp. Mar. Biol. Ecol.*, 73: 243–271.

Prouse, N.J. and Hargrave, B.T., 1977. Chlorophyll, carbon and nitrogen in suspended and sedimented particulate matter in St. Georges Bay, Nova Scotia. *Fish. Mar. Ser. Tech. Rep.*, 721: 1–75.

Ragan, M.A., 1976. Physodes and the phenolic compounds of brown algae: composition and significance of physodes *in vivo. Bot. Mar.*, 19: 145–154.

Ragan, M.A., 1984. Bioactivities in marine genera of Atlantic Canada: the unexplored potential. *Proc. N.S. Inst. Sci.*, 34: 83–132.

Ragan, M.A. and Craigie, J.S., 1976. Physodes and the phenolic compounds of brown algae: isolation and characterization of phloroglucinol polymers from *Fucus vesiculosus* (L.). *Can. J. Biochem.*, 54: 66–73.

Ragan, M.A. and Craigie, J.S., 1978. Phenolic compounds in brown and red algae. In: J.A. Hellebust and J.S. Craigie (Editors), *Handbook of Phycological Methods, Vol. II.* Cambridge Univ. Press, Cambridge, pp. 157–179.

Railton, J.B., 1975. Post-glacial history of Nova Scotia. In: J.G. Ogden III and M.J. Harvey (Editors), *Environmental Change in the Maritimes.* N.S. Inst. Sci., pp. 37–42.

Rakestraw, N.W., 1933. Studies on the biology and chemistry of the waters of the Gulf of Maine I. Chemistry of the waters of the Gulf of Maine in August, 1932. *Biol. Bull.*, 64: 149–158.

Rao, K.R., 1970. Studies on growth cycle and phycocolloid content in *Hypnea musciformis* (Wulf) Lamour. *Bot. Mar.*, 13: 163–165.

Reddin, D., 1973. *The ecology and feeding habits of the American lobster (Homarus americanus Milne-Edwards, 1837) in Newfoundland.* M.Sc. Thesis, Memorial Univ., Newfoundland, St. John's, Newfoundland, 101 pp.

Redfield, A.C., 1941. The effect of the circulation of water on the distribution of the calanoid community in the Gulf of Maine. *Biol. Bull.*, 80: 86–110.

Redfield, A.C., 1980. *The tides of waters of New England and New York.* Woods Hole Oceanographic Institution, Woods Hole, Mass., 108 pp.

Redfield, A.C. and Keys, A.B., 1938. The distribution of ammonia in the waters of the Gulf of Maine. *Biol. Bull.*, 74: 83–92.

Redfield, A.C., Smith, H.P. and Ketchum, B.H., 1937. The

cycle of organic phosphorus in the Gulf of Maine. *Biol. Bull.*, 73: 421–443.

Reynolds, N.B., 1971. *The ecology of a New Hampshire tidal rapid.* Ph.D. Thesis, Univ. New Hampshire, Durham, N.H., 101 pp.

Rhoads, D.C. and Young, D.K., 1971. Animal-sediment relations in Cape Cod Bay, Massachusetts. II. Reworking by *Molpadio oolitica* (Holothuroidea). *Mar. Biol.*, 11: 255–261.

Rich, W.H., 1929. *Fishing grounds of the Gulf of Maine.* Rep. U.S. Comm. Fish. for 1929. U.S. Dept. Commerce, Bureau Fish., Washington, D.C., pp. 51–117.

Richards, A., 1986. Habitat selection and predator avoidance: ontogenetic shifts in habitat use by Jonah crabs. *Abstr. 1986 Benthic Ecol. Meet., Boston, Mass.*

Rietema, H. and Van den Hoek, C., 1984. Search for possible latitudinal ecotypes in *Dumontia contorta* (Rhodophyta). *Helgol. Wiss. Meeresunters*, 38: 389–399.

Ring, P.D., 1970. *Development and ecophysiological studies of Chondrus crispus (L.) Stackh.* M.Sc. Thesis, Univ. Maine, Orono, Maine, 73 pp.

Robertson, B.L., 1987. Reproductive ecology and canopy structure of *Fucus spiralis* L. *Bot. Mar.*, 30: 475–482.

Robles, C., 1982. Disturbance and predation in an assemblage of herbivorous Diptera and algae on rocky shores. *Oecologia*, 54: 23–31.

Robles, C.D. and Cubit, J., 1981. Influence of biotic factors in an upper intertidal community; dipteran larvae grazing on algae. *Ecology*, 62: 1536–1547.

Rogerson, R.J., 1983. Geological evolution. In: G.R. South (Editor), *Biogeography and Ecology of the Island of Newfoundland.* Junk, The Hague, pp. 5–35.

Rogick, M.D. and Croasdale, H., 1949. Studies on marine bryozoa. III. Woods Hole region Bryozoa associated with algae. *Biol. Bull.*, 96: 32–69.

Ross, D.M., 1971. Protection of the hermit crab (*Dardamus* spp.) from octopus by the commensal sea anemone (*Caliactis* spp.). *Nature*, 230: 401, 402.

Roughgarden, J., Iwasa, Y. and Baxter, C., 1985. Theory of population processes for marine organisms. I. Demography of an open population with space-limited recruitment. *Ecology*, 66: 54–67.

Rowell, T.W., 1967. *Some aspects of the ecology, growth, and reproduction of the horse mussel Modiolus modiolus.* M.Sc. Thesis, Queens Univ., Kingston, Ontario, Canada, 138 pp.

Rueness, J., 1977. *Norsk Algeflora.* Universitetsforlaget, Oslo, 266 pp.

Runnstrom, S., 1928. Über die Thermopathie der Fortpflanzung und Entwicklung mariner Tiere in Beziehung zu ihrer geographischen Verbreitung: eine experimentell ökologische Studie. *Bergens Mus. Arbok. Naturvitensk Rekke*, 1927 (2): 1–67.

Runnstrom, S., 1930. Weitere Studien über die Temperaturanpassung der Fortpflanzung und Entwicklung mariner Tiere. *Bergens Mus. Arbok. Naturvitensk Rekke*, 1929 (19): 1–46.

Runnstrom, S., 1936. Die Anpassung der Fortpflanzung und Entwicklung mariner Tiere und die Temperaturverhaltnisse verschiedener Verbreitungsgebiete. *Bergens Mus. Arbok. Naturvitensk Rekke*, 1936 (3): 1–46.

Russell, G., 1973. The litus line: a re-assessment. *Oikos*, 24: 158–161.

Russell, G., 1979. Heavy receptacles in estuarine *Fucus vesiculosus* L. *Estuarine Coastal Mar. Sci.*, 9: 659–661.

Russell, G. and Veltkamp, C.J., 1984. Epiphyte survival on skin-shedding macrophytes. *Mar. Ecol. Prog. Ser.*, 18: 149–153.

Rutzler, K., 1970. Spatial competition among Porifera: solution by epizoism. *Oecologia*, 5: 85–95.

Sanders, H.L., 1958. Benthic studies in Buzzards Bay. I. Animal sediment relationships. *Limnol. Oceanogr.*, 111: 245–258.

Sanders, H.L., 1960. Benthic studies in Buzzards Bay. III. The structure of the soft bottom community. *Limnol. Oceanogr.*, 5: 138–153.

Sanders, H.L., 1968. Marine benthic diversity: a comparative study. *Am. Nat.*, 102: 243–282.

Sanders, H.L., 1969. Benthic marine diversity and the Stability–Time Hypothesis. *Brookhaven Symp. Biol.*, 22: 71–81.

Scarratt, D.J., 1972. Investigations into the effects on lobsters of raking Irish moss. *Fish. Res. Board Can. Tech. Rep.*, 329: 1–20.

Scarratt, D.J., 1980. The food of lobsters. In: J.D. Pringle, G.J. Sharp and J.F. Caddy (Editors), Proceedings of the workshop on the relationship between sea urchin grazing and commercial plant/animal harvesting. *Can. Tech. Rep. Fish. Aquat. Sci.*, 954: 66–91.

Scheibling, R.E., 1984. Echinoids, epizootics and ecological stability in the rocky subtidal off Nova Scotia, Canada. *Helgol. Wiss. Meeresunters.*, 37: 233–242.

Schiebling, R.E., 1986. *Macroalgal succession following mass mortalities of sea urchins off Nova Scotia.* Abstr. Twenty-fifth Northeast Algal Symp., Woods Hole, Mass.

Scheibling, R.E. and Raymond, B.G., 1986. Macroalgal succession on a cobble bottom following mass mortalities of sea urchins. *Abstr. 1986. Benthic Ecol. Meet., Boston, Mass.*

Scheibling, R.E. and Stephenson, R.L., 1984. Mass mortality of *Strongylocentrotus droebachiensis* (Echinodermata: Echinoidea) off Nova Scotia, Canada. *Mar. Biol.*, 78. 153–164.

Scheltema, R.S., 1961. Metamorphosis of the veliger larvae of *Nassarius obsoletus* (Gastropoda) in response to bottom sediment. *Biol. Bull.*, 120: 92–109.

Scheltema, R.S., 1977. Dispersal of marine invertebrate organisms: paleobiogeographic and biostratigraphic implications. In: E.G. Kauffman and J.E. Hazel (Editors), *Concepts and Methods of Biostratigraphy.* Hutchinson and Ross, Stroudsburg, Pa., pp. 73–108.

Scheuer, P.J., 1973. *Chemistry of Marine Natural Products.* Academic Press, New York, 201 pp.

Schick, J.M., Hoffman, R.J. and Lamb, A.N., 1979. Asexual reproduction, population structure and genotype-environment interactions in sea anemones. *Am. Zool.*, 19: 699–713.

Schlee, J. and Pratt, R.M., 1970. *Atlantic continental shelf and slope of the United States — Gravels of the northeastern part.* U.S. Geol. Survey Professional Paper 529-H, 39 pp.

Schneider, C.W., Suyemoto, M.M. and Yarish, C., 1979. An annotated check list of Connecticut seaweeds. *Conn. Geol. Nat. Hist. Surv. Bull.*, 108: 1–20.

Schneider, D.C. and Harrington, B.A., 1981. Timing of

shorebird migration in relation to prey depletion. *Auk*, 98: 801–811.

Schnitker, D., 1974. Postglacial emergence of the Gulf of Maine. *Geol. Soc. Am. Bull.*, 85: 491–494.

Schonbeck, M.W. and Norton, T.A., 1978. Factors controlling the upper limits of fucoid algae on the shore. *J. Exp. Mar. Biol. Ecol.*, 31: 303–313.

Schonbeck, M.W. and Norton, T.A., 1979a. The effects of brief periodic submergence on intertidal fucoid algae. *Estuarine Coastal Mar. Sci.*, 8: 205–211.

Schonbeck, M.W. and Norton, T.A., 1979b. An investigation of drought avoidance in intertidal fucoid algae. *Bot. Mar.*, 22: 133–144.

Schonbeck, M.W. and Norton, T.A., 1979c. Drought-hardening in the upper-shore seaweeds *Fucus spiralis* and *Pelvetia canaliculata*. *J. Ecol.*, 67: 687–696.

Schonbeck, M. and Norton, T., 1980. Factors controlling the lower limits of fucoid algae on the shore. *J. Exp. Mar. Biol. Ecol.*, 43: 131–150.

Schubert, C. and Dunbar, C.O., 1934. Stratigraphy of western Newfoundland. *Geol. Soc. Am. Mem.*, 1: 1–123.

Scott, F.H., 1902. Food of the sea urchin (*Strongylocentrotus droebachiensis*). *Contr. Can. Biol.*, (1902): 49–54.

Searles, R.B., 1984. Seaweed biogeography of the mid-Atlantic coast of the United States. *Helgol. Wiss. Meeresunters.*, 38: 259–271.

Searles, R.B. and Schneider, C.W., 1978. A checklist and bibliography of North Carolina seaweeds. *Bot. Mar.*, 21: 99–108.

Sears, J.R. and Cooper, R.A., 1978. Descriptive ecology of offshore, deep-water, benthic algae in the temperate western North Atlantic Ocean. *Mar. Biol.*, 44: 309–314.

Sears, J.R. and Wilce, R.T., 1975. Sublittoral, benthic marine algae of southern Cape Cod and adjacent islands: seasonal periodicity, associations, diversity and floristic composition. *Ecol. Monogr.*, 45: 337–365.

Sebens, K.P., 1982. Competition for space: growth rate, reproductive output and escape in size. *Am. Nat.*, 120: 189–197.

Sebens, K.P., 1986a. Community ecology of vertical walls in the Gulf of Maine, USA: small scale processes and alternative community states. In: P.G. Moore and R. Seed (Editors), *The Ecology of Rocky Coasts*. Columbia Univ. Press, New York, pp. 346–371.

Sebens, K.P., 1986b. Spatial relationships among encrusting marine organisms in the New England subtidal zone. *Ecol. Monogr.*, 56: 73–96.

Sebens, K. and Witman, J., 1986. Community organization of deep rocky habitats in the outer Gulf of Maine. *Abstr. 1986 Benthic Ecol. Meet., Boston, Mass.*

Seed, R., 1969. The ecology of *Mytilus edulis* L. (Lamellibranchiata) on exposed shores. II. Growth and mortality. *Oecologia*, 3: 317–350.

Seed, R., 1975. Reproduction in *Mytilus* (Mollusca: Bivalvia) in European waters. *Publ. Staz. Zool. Napoli*, 39 (Suppl.): 317–334.

Seed, R., 1986. Ecological pattern in the epifaunal communities of coastal macroalgae. In: P.G. Moore and R. Seed (Editors), *The Ecology of Rocky Coasts*. Columbia Univ. Press, New York, pp. 23–35.

Seed, R. and O'Connor, R.J., 1981. Community organization in marine algal epifaunas. *Ann. Rev. Ecol. Syst.*, 12: 49–74.

Seip, K.L., 1980. A mathematical model of competition and colonization in a community of marine benthic algae. *Ecol. Model.*, 10: 77–104.

Setchell, W.A., 1915. The law of temperature connected with distribution of the marine algae. *Ann. Mo. Bot. Gard.*, 2: 287–305.

Setchell, W.A., 1917. Geographical distribution of the marine algae. *Science*, 45: 197–204.

Setchell, W.A., 1920. The temperature interval in the geographical distribution of marine algae. *Science*, 52: 187–190.

Setchell, W.A., 1922. Cape Cod in its relation to the marine flora of New England. *Rhodora*, 24: 1–11.

Setchell, W.A. and Gardner, N.L., 1933. A preliminary survey of *Gigartina*, with special reference to its Pacific North American species. *Univ. Calif. Publ. Bot.*, 17: 255–340.

Shacklock, P.F. and Croft, G.B., 1981. Effect of grazers on *Chondrus crispus* in culture. *Aquaculture*, 22: 331–342.

Shannon, R.K., 1985. *Phenology and life history of Petalonia fascia and Scytosiphon lomentaria (Scytosiphonales: Phaeophyta) in New Hampshire.* M.S. Thesis, Univ. New Hampshire, Durham, N.H., 80 pp.

Shannon, R.K., Crow, G.E. and Mathieson, A.C., 1988. Seasonal abundance and recruitment patterns of *Petalonia fascia* (O.F. Muller) Kuntze and *Scytosiphon lomentaria* (Lyngbye) Link var. *lomentaria*. *Bot. Mar.*, 31: 207–214.

Sharaf El Din, S.H., Hassan, E.M. and Trites, R.W., 1970. The physical oceanography of St. Margaret's Bay. *Fish. Res. Board Can. Tech. Rep.*, 219: 1–242.

Shaughnessy, F.J., 1983. Physiological ecology of sand-buried *Fucus vesiculosus* (Phaeophyceae). *Program and Abstr. Twenty-second Northeast Algal Symp., Woods Hole, Mass.*

Shaughnessy, F.J., 1986a. *Effects of sand on the density, growth, morphology and photosynthate of Fucus vesiculosus L. at Seabrook, New Hampshire.* M.Sc. Thesis, Univ. New Hampshire, Durham, N.H., 186 pp.

Shaughnessy, F.J., 1986b. Effects of sand on the distribution of *Fucus vesiculosus* L. *Abstr. Twenty-fifth Northeast Algal Symp., Woods Hole, Mass.*

Sherwin, D.F., 1973. Scotian shelf and Grand Banks. In: R.G. McCrossan (Editor), *The future petroleum provinces of Canada — their geology and potential, Calgary.* Can. Soc. Petrol. Geol., pp. 519–559.

Shumway, S.E. and Stickney, R.P., 1975. Notes on the biology and food habits of the cunner. *N.Y. Fish Game J.*, 22: 71–79.

Sideman, E.J. and Mathieson, A.C., 1983a. Ecological and genecological distinctions of a high intertidal dwarf form of *Fucus distichus* (L.) Powell in New England. *J. Exp. Mar. Biol. Ecol.*, 72: 171–188.

Sideman, E.J. and Mathieson, A.C., 1983b. The growth, reproductive phenology, and longevity of non-tide pool *Fucus distichus* (L.) Powell in New England. *J. Exp. Mar. Biol. Ecol.*, 68: 111–127.

Sieburth, J.McN., 1964. Antibacterial substances produced by marine algae. *Dev. Ind. Microbiol.*, 5: 124–134.

Sieburth, J.McN., 1969. Studies on algal substances in the sea. III. The production of extracellular organic matter by littoral marine algae. *J. Exp. Mar. Biol. Ecol.*, 3: 290–309.

Sieburth, J.McN. and Conover, J.T., 1965. *Sargassum* tannin, an antibiotic which retards fouling. *Nature*, 208: 52–53.

Sieburth, J.McN. and Tootle, J.L., 1981. Seasonality of microbial fouling on *Ascophyllum nodosum* (L.) Le Jolis, *Fucus vesiculosus* L., *Polysiphonia lanosa* (L.) Tandy and *Chondrus crispus* Stackh. *J. Phycol.*, 17: 57–64.

Simenstad, C.A., Estes, J.A. and Kenyon, K.W., 1978. Aleuts, sea otters, and alternate stable-state communities. *Science*, 200: 403–411.

Skutch, A.F., 1926. On the habits and ecology of the tube building amphipod *Ampithoe rubricata* Montagn. *Ecology*, 7: 481–502.

Slocum, C.J., 1980. Differential susceptibility to grazers in two phases of an intertidal alga: advantages of heteromorphic generations. *J. Exp. Mar. Biol. Ecol.*, 46: 99–110.

Smith., B.D., 1985. Recovery following experimental harvesting of *Laminaria longicruris* and *L. digitata* in southwestern Nova Scotia. *Helgol. Wiss. Meeresunters.*, 39: 83–101.

Smith, B.D., 1986. Implications of population dynamics and interspecific competition for harvest management of the seaweed *Laminaria*. *Mar. Ecol. Prog. Ser.*, 33: 7–18.

Smith, G.J.D., Jovellanos, C.L. and Gaskin, D.E., 1984. Near-surface bio-oceanographic phenomena in the Quoddy Region, Bay of Fundy. *Can. Tech. Rep. Fish. Aquat. Sci.*, 1280: 1–124.

Smith, R.M., 1967. Sublittoral ecology of marine algae on the North Wales coast. *Helgol. Wiss. Meeresunters.*, 15: 467–479.

Smithson, B.A. and Cheney, D., 1984. An ecological study of phenolics in the intertidal brown alga *Fucus*. *Program and Abstr. Twenty-third Northeast Algal Symp., Woods Hole, Mass.*

Sousa, W.P., 1979. Experimental investigations of disturbance and ecological succession in a rocky intertidal algal community. *Ecol. Monogr.*, 49: 227–254.

Sousa, W.P., 1984. The role of disturbance in natural communities. *Ann. Rev. Ecol. Syst.*, 15: 353–391.

South, G.R., 1972. On the life history of *Tilopteris mertensii* (Turn. in Sm.) Kuetz. In: K. Nisizawa (Editor), *Proc. Seventh Int. Seaweed Symp., Sapporo, Japan*. Univ. Tokyo Press, Tokyo, pp. 73–79.

South, G.R., 1975. Contributions to the flora of marine algae of eastern Canada. III. Order Tilopteridales. *Nat. Can.*, 103: 693–702.

South, G.R., 1976a. *Checklist of marine algae from Newfoundland, Labrador, and the French Islands of St. Miquelon . . . first revision*. Memorial Univ. Newfoundland Mar. Sci. Res. Lab., St. John's, Newfoundland. Tech. Rep. 19, 35 pp.

South, G.R., 1976b. A checklist of marine algae of eastern Canada — first revision. *J. Mar. Biol. Assoc. U.K.*, 56: 817–843.

South, G.R., 1983. Benthic marine algae. In: G.R. South (Editor), *Biogeography and Ecology of the Island of Newfoundland*. Junk, The Hague, pp. 385–420.

South, G.R. and Hill, R.D., 1970. Studies on marine algae of Newfoundland. I. Occurrence and distribution of free-living *Ascophyllum nodosum* in Newfoundland. *Can. J. Bot.*, 48: 1697–1701.

South, G.R., Hooper, R.G. and Irvine, L.M., 1972. The life history of *Turnerella pennyi* Schmitz. *Br. Phycol. J.*, 7: 221–233.

Southward, A.J., 1955. The population balance between limpets and seaweeds on wave-beaten rocky shores. *Ann. Rep. Mar. Biol. Stat. Port Erin*, 68: 20–29.

Southward, A.J. and Southward, E.C., 1978. Recolonization of rocky shores in Cornwall after use of toxic dispersants to clean up the Torrey Canyon spill. *J. Fish. Res. Board Can.*, 35: 682–706.

Spitters, C.J.T., 1983. An alternative approach to the analysis of mixed cropping experiments. 2. Marketable yield. *Neth. J. Agric. Sci.*, 31: 143–155.

Squires, H.J., 1970. Lobster (*Homarus americanus*) fishery and ecology in Port au Port Bay, Newfoundland, 1960–65. *Proc. Natl. Shellfish. Assoc.*, 60: 22–39.

Stallard, M.O. and Faulkner, D.J., 1974. Chemical constituents of the digestive gland of the sea hare *Aplysia californica*. I. Importance of diet. *Comp. Biochem. Physiol.*, 49B: 25–36.

Stasko, A.B., Campbell, A. and Graham, D.E., 1980. Sea urchin (*Strongylocentrotus droebachiensis*) distribution around western Nova Scotia. In: J.D. Pringle, G.J. Sharp and J.F. Caddy (Editors), *Proc. Workshop on the Relationship between Sea Urchin Grazing and Commercial Plant/Animal Harvesting*. Can. Tech. Rep. Fish. Aquat. Sci., 954: 225–236.

Steele, D.H., 1974. *Temperature and salinity cycles at the Marine Sciences Research Laboratory, Logy Bay, Newfoundland.* Mem. Univ. Newfoundland Mar. Sci. Res. Lab. Tech. Rep. 12, 21 pp.

Steele, D.H., 1975. Marine climate and the biogeography of the surface waters in the northwest Atlantic. *Nat. Can.*, 102: 189–198.

Steele, D.H., 1983. Marine ecology and zoogeography. In: G.R. South (Editor), *Biogeography and Ecology of the Island of Newfoundland.* Junk, The Hague, pp. 421–465.

Steinberg, P.D., 1984. Algal chemical defense against herbivores: allocation of phenolic compounds in the kelp *Alaria marginata*. *Science*, 223: 405–407.

Steinberg, P.D., 1985. Feeding preferences of *Tegula funebralis* and chemical defenses of marine brown algae. *Ecol. Monogr.*, 55: 333–349.

Steinberg, P.D., 1986. Chemical defenses and the susceptibility of tropical marine brown algae to herbivores. *Oecologia*, 69: 628–630.

Steneck, R.S., 1977. Crustose coralline-limpet interaction in the Gulf of Maine. *J. Phycol.*, 13: 65.

Steneck, R.S., 1978. *Factors influencing the distribution of crustose coralline algae (Rhodophyta, Corallinaceae) in the Damariscotta River, Maine.* M.Sc. Thesis, Univ. Maine, Orono, Me., 58 pp.

Steneck, R.S., 1982a. A limpet-coralline algal association: adaptations and defenses between a selective herbivore and its prey. *Ecology*, 63: 507–522.

Steneck, R.S., 1982b. *Adaptive trends in the ecology and evolution of crustose coralline algae (Rhodophyta, Corallinaceae)*. Ph.D. Thesis, John Hopkins Univ., Baltimore, Md., 253 pp.

Steneck, R.S., 1982c. Adaptive trends in branching crustose coralline algae: patterns in space and time. *Geol. Soc. Am. Bull.*, 14: 86.

Steneck, R.S., 1983a. Escalating herbivory and resulting adaptive trends in calcareous algal crusts. *Paleobiology*, 9: 44–61.

Steneck, R.S., 1983b. Algal community structure and herbivory: a general model based on functional groups. *Program and Abstr. Twenty-second Northeast Algal Symp., Woods Hole, Mass.*

Steneck, R.S., 1985. Adaptations of crustose coralline algae to herbivory: patterns in space and time. In: D.F. Toomey and M.H. Nitecki (Editors), *Paleoalgology*. Springer, Berlin, pp. 352–366.

Steneck, R.S., 1986. Ecology of coralline algal crusts: convergent patterns and adaptive strategies. *Ann. Rev. Ecol. Syst.*, 17: 273–303.

Steneck, R.S. and Milliken, B., 1981. The branching morphology of crustose corallines as a structural defense against herbivores and a refuge for filamentous algae. *Abstr. Proc. Twentieth Northeast Algal Symp., Woods Hole, Mass.*

Steneck, R.S. and Watling, L., 1982. Feeding capabilities and limitations of herbivorous molluscs: a functional group approach. *Mar. Biol.*, 68: 299–319.

Steneck, R.S., Vadas, R.L. and Babb, I., 1986. *The structure and zonation of deep benthic algal communities in Maine*. Abstr. 1986 Benthic Ecol. Meet., Boston, Mass.

Stephenson, T.A. and Stephenson, A., 1954a. Life between tidemarks in North America, IIIA. Nova Scotia and Prince Edward Island: description of the region. *J. Ecol.*, 42: 14–45.

Stephenson, T.A. and Stephenson, A., 1954b. Life between tidemarks in North America, IIIB. Nova Scotia and Prince Edward Island: the geographical features of the region. *J. Ecol.*, 42: 46–70.

Stephenson, T.A. and Stephenson, A., 1972. *Life Between Tidemarks on Rocky Shores*. W.H. Freeman, San Francisco, 425 pp.

Steven, D.M., 1971. International Biological Programme study of the Gulf of St. Lawrence. In: *Proc. Second Gulf of St. Lawrence Workshop*. Bedford Institute of Oceanography, Dartmouth, N.S., pp. 146–159.

Stewart, J.G., 1984. Algal distributions and temperature: test of an hypothesis based on vegetative growth rates. *Bull. S. Calif. Acad. Sci.*, 83: 57–68.

Stoffelen, H., Glombitza, K.W., Murawski, U., Bielaczek, J. and Egge, H., 1972. Bromophenole aus *Polysiphonia lanosa* (L.) Tandy. (antibiotica aux algen 7. mitt.). *Planta Med.*, 22: 396–401.

Stone, R.A., Hehre, E.J., Conway, J.M. and Mathieson, A.C., 1970. A preliminary checklist of the marine algae of Campobello Island, New Brunswick, Canada. *Rhodora*, 72: 313–338.

Stott, R.S., 1972. *Habitat usage and populations of sea ducks on the New Hampshire coastline*. M.Sc. Thesis, Univ. New Hampshire, Durham, N.H., 120 pp.

Stott, R.S. and Olson, D.P., 1973. Food-habitat relationship of sea ducks on the New Hampshire coastline. *Ecology*, 54: 996–1007.

Strathmann, R.R., 1974. The spread of sibling larvae of sedentary marine invertebrates. *Am. Nat.*, 108: 29–44.

Strathmann, R.R. and Branscomb, E.S., 1979. Adequacy of cue to favorable sites used by settling larvae of two intertidal barnacles. In: S.E. Stancyk (Editor), *Reproductive Ecology of Marine Invertebrates*. Univ. South Carolina Press, Columbia, S.C., pp. 77–89.

Strathmann, R.R., Branscomb, E.S. and Vedder, K., 1981. Fatal errors in set as a cost of dispersal and the influence of intertidal flora on set of barnacles. *Oecologia*, 48: 13–18.

Stuvier, M. and Borns Jr., H.W., 1975. Late Quaternary marine invasion in Maine; its chronology and associated crustal movement. *Bull. Geol. Soc. Am.*, 86: 99–104.

Suchanek, T.H., 1981. The role of disturbance in the evolution of life history strategies in the intertidal mussels *Mytilus edulis* and *Mytilus californianus*. *Oecologia*, 50: 143–152.

Suchanek, T.H., 1986. Mussels and their role in structuring rocky shore communities. In: P.G. Moore and R. Seed (Editors), *The Ecology of Rocky Coasts*. Columbia Univ. Press, New York, pp. 70–96.

Sullivan Jr., J.D. and Ikawa, M., 1973. Purification and characterization of hexose oxidase from the red alga *Chondrus crispus*. *Biochim. Biophys. Acta*, 309: 11–22.

Sun, H.H. and Fenical, W., 1979. Rhipocephalin and rhipocephenal; toxic feeding deterrents from the tropical marine alga *Rhipocephalus phoenix*. *Tetrahedron Lett.*, 8: 685–688.

Sundene, O., 1953. The algal vegetation of Oslofjord. *Skr. Norske Vidensk. Ak. Oslo*, 2: 1–244.

Sundene, O., 1962. The implications of transplants and culture experiments on the growth and distribution of *Alaria esculenta*. *Nytt Mag. Bot.*, 9: 1–6.

Sundene, O., 1973. Growth and reproduction in *Ascophyllum nodosum* (Phaeophyceae). *Nor. J. Bot.*, 20: 249–255.

Sutcliffe Jr., W.H., Loucks, R.H. and Drinkwater, K.F., 1976. Coastal circulation and physical oceanography of the Scotian Shelf and the Gulf of Maine. *J. Fish. Res. Board Can.*, 33: 98–115.

Sutherland, J.P., 1974. Multiple stable points in natural communities. *Am. Nat.*, 108: 859–873.

Sverdrup, H.U., Johnson, M.W. and Fleming, R.H., 1942. *The Oceans*. Prentice-Hall, Englewood Cliffs, N.J., 1087 pp.

Taylor, A.R.A., 1973. *Studies of populations of Chondrus crispus (Irish moss) and Furcellaria fastigiata in Prince Edward Island waters during 1971*. Tech. Rep. Indust. Develop. Branch, Fish. Mar. Ser. Can., 109 pp.

Taylor, A.R.A., 1975. *The Chondrus crispus-Furcellaria fastigiata community at Campbell's Cove, Prince Edward Island*. Tech. Rep. 88, Indust. Dev. Branch, Fish. and Mar. Ser., Environ. Can., 72 pp.

Taylor, A.R.A. and Chen, L.C.-M., 1973. The biology of *Chondrus crispus* Stackhouse: systematics, morphology and life history. In: M.J. Harvey and J. McLachlan (Editors), *Chondrus crispus*. N.S. Inst. Sci. Publ., Halifax, Nova Scotia, pp. 1–21.

Taylor, A.R.A., Chen, L.C.-M., Smith, B.D. and Staples, L.S., 1975. *Chondrus holdfasts in natural populations and in culture*. Indust. Dev. Branch, Fish. and Mar. Ser., Environ. Can., 8 pp.

Taylor, A.R.A., Chen, L.C.-M., Smith, B.D. and Staples, L.S., 1981. *Chondrus* holdfasts in natural populations and in culture. In: G.E. Fogg and W.E. Jones (Editors), *Proc. Eighth Int. Seaweed Symp., Bangor, North Wales*. Univ. College North Wales Mar. Sci. Lab. Publ., Menai Bridge, pp. 140–145.

Taylor, P.R. and Hay, M.E., 1984. Functional morphology of intertidal seaweeds: adaptive significance of aggregate vs. solitary forms. *Mar. Ecol. Prog. Ser.*, 18: 295–302.

Taylor, P.R. and Littler, M.M., 1982. The roles of compensatory mortality, physical disturbance and substrate retention in the development and organization of a sand-influenced rocky intertidal community. *Ecology*, 63: 135–146.

Taylor, W.R., 1957. *Marine Algae of the Northeastern Coast of North America*. Univ. Michigan Press, Ann Arbor, 509 pp.

Taylor, W.R. and Bernatowicz, A.J., 1969. Distribution of marine algae about Bermuda. *Bermuda Biol. Res. Sta. Spec. Publ.*, 1: 1–42 pp.

Thorson, G., 1946. Reproduction and larval development of Danish marine bottom invertebrates with special reference to the planktonic larvae in the Sound (Oresund). *Meddel. Komm. Danmarks Fisk. Havund.*, 4: 1–523.

Thorson, G., 1957. Bottom communities (sublittoral or shallow shelf). In: J.W. Hedgepeth (Editor), Treatise on Marine Ecology and Paleoecology. *Geol. Soc. Am. Memoir.*, 67: 461–534.

Thorson, G., 1966. Some factors influencing the recruitment and establishment of marine benthic communities. *Neth. J. Sea Res.*, 3: 267–293.

Todd, C.D., 1986. Reproductive strategies of north temperate rocky shore invertebrates. In: P.G. Moore and R. Seed (Editors), *The Ecology of Rocky Coasts*. Columbia Univ. Press, New York, pp. 203–219.

Topinka, J., Tucker, L. and Korjeff, W., 1981. The distribution of fucoid biomass along central coastal Maine. *Bot. Mar.*, 24: 311–319.

Tremblay, C. and Chapman, A.R.O., 1980. The local occurrence of *Agarum cribrosum* in relation to the presence or absence of its competitors and predators. *Proc. N.S. Inst. Sci.*, 30: 165–170.

Trites, R.W., 1972. The Gulf of St. Lawrence from a pollution viewpoint. In: *Marine Pollution and Sea Life*. FAO Publ., Rome, Italy, pp. 59–72.

Trites, R.W., 1979a. Some physical oceanographic features in relation to the Canso Causeway — an overview. *Fish. Mar. Ser. Tech. Rep.*, 834 (pt. II): 14–29.

Trites, R.W., 1979b. Comments on residual current patterns in the inshore area south of Cape Breton Island. *Fish. Mar. Ser. Tech. Rep.*, 834 (pt IV): 41–50.

Tuchman, M. and Blinn, D.W., 1979. Comparison of attached algal communities on natural and artificial substrata along a thermal gradient. *Br. Phycol. J.*, 14: 243–254.

Tveter, E. and Mathieson, A.C., 1976. Sporeling coalescence in *Chondrus crispus* (Rhodophyceae). *J. Phycol.*, 12: 110–118.

Tveter-Gallagher, E. and Mathieson, A.C., 1980. An electron microscopic study of sporeling coalescence in the red alga *Chondrus crispus*. *Scanning Electron Micros.*, III: 570–580.

Uchupi, E., 1960. Topography and structure of Northeast Channel, Gulf of Maine. *Bull. Am. Assoc. Petroleum Geol.*, 50: 105–167.

Uchupi, E., 1966. Structural framework — the Gulf of Maine. *J. Geophys. Res.*, 71: 3014–3028.

Underwood, A.J., 1979. Ecology of intertidal gastropods. *Adv. Mar. Biol.*, 16: 111–210.

Underwood, A.J., 1986. Physical factors and biological interactions: the necessity and nature of ecological experiments. In: P.G. Moore and R. Seed (Editors), *The Ecology of Rocky Coasts*. Columbia Univ. Press, New York, pp. 372–390.

Underwood, A.J. and Denley, E.J., 1984. Paradigms, explanations and generalizations in models for the structure of intertidal communities on rocky shores. In: D.R. Strong, D. Simberloff, L.G. Abele and A.B. Thistle (Editors), *Ecological Communities: Conceptual Issues and the Evidence*. Princeton Univ. Press, Princeton, N.J., pp. 151–180.

Vaccaro, R.F., 1963. Available nitrogen and phosphorus and the biochemical cycle in the Atlantic off New England. *J. Mar. Res.*, 21: 284–301.

Vadas, R.L., 1968. *The ecology of Agarum and the kelp bed community*. Ph.D. Thesis, Univ. Washington, Seattle, 280 pp.

Vadas, R.L., 1977. Preferential feeding: an optimization strategy in sea urchins. *Ecol. Monogr.*, 47: 337–371.

Vadas, R.L., 1979. Seaweeds: an overview; ecological and economic importance. *Experientia*, 35: 429–432.

Vadas, R.L., 1985. Herbivory. In: M.M. Littler and D.S. Littler (Editors), *Handbook of Phycological Methods, Ecological Field Methods: Macroalgae*. Cambridge Univ. Press, Cambridge, pp. 531–572.

Vadas, R.L. and Steneck, R.S., 1988. Zonation of deep water benthic algae in the Gulf of Maine. *J. Phycol.*, 24: 338–346.

Vadas, R.L., Elner, R.W., Garwood, P.E. and Babb, L.G., 1986. Experimental evaluation of aggregation behavior in the sea urchin *Strongylocentrotus droebachiensis* — a reinterpretation. *Mar. Biol.*, 90: 433–448.

Van den Hoek, C., 1975. Phytogeographic provinces along the coasts of the northern Atlantic Ocean. *Phycologia*, 14: 317–330.

Van den Hoek, C., 1979. The phytogeography of *Cladophora* (Chlorophyceae) in the northern Atlantic Ocean, in comparison to that of other benthic algal species. *Helgol. Wiss. Meeresunters.*, 32: 374–393.

Van den Hoek, C., 1982a. The distribution of benthic marine algae in relation to the temperature regulation of their life histories. *Biol. J. Linn. Soc.*, 18: 81–144.

Van den Hoek, C., 1982b. Phytogeographic distribution groups of benthic marine algae in the North Atlantic Ocean. A review of experimental evidence from life history studies. *Helgol. Wiss. Meeresunters.*, 35: 153–214.

Van den Hoek, C., 1982c. A taxonomic revision of the American species of *Cladophora* (Chlorophyceae) in the North Atlantic Ocean and their geographic distribution. *Verh. K. Ned. Akad. Wet.*, 236 pp.

Van den Hoek, C., 1984. World-wide latitudinal and longitudinal seaweed distribution patterns and their possible causes, as illustrated by the distribution of Rhodophytan genera. *Helgol. Wiss. Meeresunters.*, 38: 227–257.

Van den Hoek, C. and Donze, M., 1967. Algal phytogeography of the European Atlantic coasts. *Blumea*, 15: 63–89.

Van Dongen, A., 1955. The preference of *Littorina obtusata* for Fucaceae. *Arch. Neerl. Zool.*, 2: 373–386.

Van Hofsten, N., 1915. Die Echinodermen des Eisfjords. *Kongl. Svenska Vetens.-Akad. Handl.*, 54: 1–282.

Vance, R.R., 1978. A mutualistic interaction between a sessile marine clam and its epibionts. *Ecology*, 59: 679–685.

Vance, R.R., 1979. Effects of grazing by the sea urchin, *Centrostephanus coronatus* on prey community composition. *Ecology*, 60: 537–546.

Velimirov, B. and Griffiths, C.L., 1979. Wave-induced kelp movement and its importance for community structure. *Bot. Mar.*, 22: 169–172.

Vernberg, F.J. and Vernberg, W.B., 1970. Lethal limits and the zoogeography of the faunal assemblages of coastal Carolina waters. *Mar. Biol.*, 6: 26–32.

Walne, P.R. and Dean, G.J., 1972. Experiments on predation by the shore crab, *Carcinus maenas* (L.), on *Mytilus* and *Mercenaria*. *J. Cons.*, 34: 190–199.

Watanabe, J.M., 1984. The influence of recruitment, competition, and benthic predation on the spatial distributions of three species of kelp forest gastropods (Trochidae: *Tegula*). *Ecology*, 65: 920–936.

Watling, L., 1979. Zoogeographic affinities of northeastern North American gammaridean Amphipoda. *Bull. Biol. Soc. Wash.*, 3: 256–282.

Watson, D.C. and Norton, T.A., 1985. The physical characteristics of seaweed thalli as deterrents to littorine grazers. *Bot. Mar.*, 28: 383–387.

Watson, E.E., 1936. Mixing and residual currents in tidal waters as illustrated in the Bay of Fundy. *J. Biol. Board Can.*, 2: 141–208.

Weiler, J.D.M. and Keeley, J.R., 1980. *Monthly sea surface temperatures for the Gulf of St. Lawrence*. Mar. Environ. Data Ser. Tech. Rep. 7, Dept. Fish. and Oceans, Ottawa, 43 pp.

Welsted, J.W., 1976. Post-glacial emergence of the Bay of Fundy coast: an analysis of the evidence. *Can. Geogr.*, 20: 367–383.

Westlake, D.F., 1963. Comparisons of plant productivity. *Biol. Rev.*, 38: 384–425.

Wethey, D.S., 1986. Local and regional variation in settlement and survival in the littoral barnacle *Semibalanus balanoides* (L.): patterns and consequences. In: P.G. Moore and R. Seed (Editors), *The Ecology of Rocky Coasts*. Columbia Univ. Press, New York, pp. 194–202.

Wharton, W.G., 1980. The distribution of sea urchin-dominated barren grounds along the south shore of Nova Scotia. In: J.D. Pringle, G.J. Sharp and J.F. Caddy (Editors), *Proc. Workshop on the relationship between sea urchin grazing and commercial plant/animal harvesting. Can. Tech. Rep. Fish. Aquat. Sci.*, 945: 33–47.

Wharton, W.G. and Mann, K.H., 1981. Relationship between destructive grazing by the sea urchin, *Strongylocentrotus droebachiensis*, and the abundance of American lobster, *Homarus americanus*, on the Atlantic Coast of Nova Scotia. *Can. J. Fish. Aquat. Sci.*, 38: 1339–1349.

Wheeler, A., 1980. Fish algal relations in temperate waters. In: J.H. Price, D.E.G. Irvine and W.F. Farnham (Editors), *The Shore Environment. Vol. 2: Ecosystems*. Systematics Assoc. Spec. Vol. 17(b). Academic Press, London, pp. 677–698.

Whittick, A., 1973. *The taxonomy, life history and ecology of some species of the Ceramiaceae (Rhodophyta) in the northwest Atlantic*. Ph.D. Thesis, Memorial Univ. Newfoundland, St. John's, Newfoundland, 368 pp.

Whittick, A., 1978. The life history and phenology of *Callithamnion corymbosum* (Rhodophyta: Ceramiaceae) in Newfoundland. *Can. J. Bot.*, 56: 2497–2499.

Whittick, A., 1984. The Newfoundland Ceramiaceae, why are there so many tetrasporophytes? *Br. Phycol. J.*, 19: 201.

Wiborg, K.F., 1946. Undersokeiser over oskjellet (*Modiola modiolus*) (L.). *Fiskeridir. Skr. Ser. Havunders.*, 8: 1–85.

Wilce, R.T., 1959. The marine algae of the Labrador Peninsula and northwest Newfoundland (ecology and distribution). *Bull. Nat. Mus. Can.*, 158: 1–103.

Wilce, R.T., 1971. Some remarks on chrysophytes and fleshy red and brown crusts. In: N.W. Riser and G.A. Carlson (Editors), *Cold Water Inshore Marine Biology*. Northeastern Univ. Sci. Inst. Publ., Boston, Mass., pp. 17–25.

Williams, I.C. and Ellis, C., 1975. Movements of the common periwinkle, *Littorina littorea* (L.) on the Yorkshire coast in winter and the influence of infection with larval *Digenea*. *J. Exp. Mar. Biol. Ecol.*, 17: 45–58.

Wilson, J.S., Bird, C.J., McLachlan, J. and Taylor, A.R.A., 1979. *An annotated checklist and distribution of benthic marine algae of the Bay of Fundy*. Memorial Univ. Newfoundland Occas. Pap. Biol. 2, 65 pp.

Witman, J.D., 1982. *Disturbance and contrasting patterns of population structure in the brachiopod Terebratulina septentrionalis from two subtidal habitats in the Gulf of Maine*. M.Sc. Thesis, Univ. New Hampshire, Durham, N.H., 61 pp.

Witman, J.D., 1984. *Ecology of rocky subtidal communities: the role of Modiolus modiolus and the influence of disturbance, competition, and mutualism*. Ph.D. Thesis, Univ. New Hampshire, Durham, N.H., 199 pp.

Witman, J.D., 1985. Refuges, biological disturbance, and rocky subtidal community structure in New England. *Ecol. Monogr.*, 55: 421–445.

Witman, J.D., 1987. Subtidal coexistence: storms, grazing, mutualism, and the zonation of kelps and mussels. *Ecol. Monogr.*, 57: 167–187.

Witman, J.D. and Cooper, R.A., 1983. Disturbance and contrasting patterns of population structure in the brachiopod *Terebratulina septentrionalis* (Couthouy) from two subtidal habitats. *J. Exp. Mar. Biol. Ecol.*, 73: 57–59.

Witman, J.D. and Suchanek, T.H., 1984. Mussels in flow: drag and dislodgement by epizoans. *Mar. Ecol. Prog. Ser.*, 16: 259–268.

Wood, R.D. and Villalard-Bohnsack, M., 1974. Marine algae of Rhode Island. *Rhodora*, 76: 399–421.

Woodin, S.A. and Jackson, J.B.C., 1979. Interphyletic competition among marine benthos. *Am. Zool.*, 19: 1029–1043.

Wright, W.R. and Worthington, L.V., 1970. *The water masses of the North Atlantic Ocean; a volumetric census of temperature and salinity*. Ser. Atlas Mar. Environ., 19, 8 pp.

Wynne, M.J., 1969. Life history and systematic studies of some Pacific North American Phaeophyceae (brown algae). *Univ. Calif. Publ. Bot.*, 50: 1–88.

Yarish, C., Breeman, A.M. and Van den Hoek, C., 1984. Temperature, light, and photoperiod responses of some Northeast American and west European endemic rhodophytes in relation to their geographic distribution. *Helgol. Wiss. Meeresunters.*, 38: 273–304.

Yarish, C., Breeman, A.M. and Van den Hoek, C., 1986.

Survival strategies and temperature responses of seaweeds belonging to different biogeographic distribution groups. *Bot. Mar.*, 29: 215–230.

Yentsch, C.S., Skea, W., Laird, J.C. and Hopkins, T.S., 1976. *A report on ocean color-observations during the Apollo–Soyoz Mission.* Tech. Rep. 6–76, Bigelow Lab. Ocean Sci., West Boothbay Harbor, Me.

Yonge, C.M., 1949. *The Sea Shore.* Collins Ltd., London, 311 pp.

Yoshioka, P.M., 1982. Role of plankton and benthic factors in the population dynamics of the bryozoan *Membranipora membranacea. Ecology*, 63: 457–468.

Young, D.K. and Rhoads, D.C., 1971. Animal-sediment relations in Cape Cod Bay, Massachusetts. I. A transect study. *Mar. Biol.*, 11: 242–254.

Zechman, F.W. and Mathieson, A.C., 1985. The distribution of seaweed propagules in estuarine, coastal and offshore waters of New Hampshire, U.S.A. *Bot. Mar.*, 27: 283–294.

Zottoli, R., 1978. *Introduction to Marine Environments.* Second ed. Mosby, St. Louis, 252 pp.

Chapter 8

LITTORAL AND INTERTIDAL SYSTEMS IN THE MID-ATLANTIC COAST OF THE UNITED STATES

ROBERT J. ORTH, KENNETH L. HECK JR. and ROBERT J. DIAZ

INTRODUCTION

The littoral and intertidal habitats of the mid-Atlantic region of the U.S.A., delineated here for the purpose of the review as the area south of Cape Cod, Massachusetts (~41°30′N) to and including Georgia (~30°40′N) (Fig. 8.1), can best be characterized by the notable lack of natural hard rocky substrates (with the predominant substrate being quartz sand) and the extreme temperature ranges (up to 40°C) that organisms are subjected to over an annual cycle. Hard substrates in this region are either man-made (rock jetties and wooden pilings) or biologically generated. These biologically structured habitats include oyster and worm reefs as well as seagrass beds and drift algae. The latter two habitats are restricted to low-energy environments, that is, in coastal lagoons and bays behind barrier islands, and estuaries, and do not occur along the open, outer coastline. Seagrass beds as well as species of freshwater plants that have invaded estuaries are present in the shoal areas from Massachusetts to North Carolina. They have not been reported from South Carolina and Georgia (Fig. 8.1). A prominent feature of the intertidal area along the mid-Atlantic region is the expansive system of salt marshes, dominated by *Spartina alterniflora*. An estimated 537 846 ha can be found from Rhode Island to Georgia with South Carolina and Georgia containing 69% of the total (Reimold, 1977). We are concentrating our review on soft substrates and seagrass beds because, at present, little information is available on drift algal beds and intertidal and shallow subtidal oyster reefs, and salt marshes were the subject of an earlier volume in this series (Chapman, 1977).

PHYSICAL SETTING AND SEDIMENTS

A variety of habitat settings can be found along the east coast of the United States. These range from high-salinity coastal ponds, lagoons and bays behind barrier islands and spits to brackish-water habitats found in estuaries, with the Chesapeake Bay being the largest estuary in North America. Sand beaches facing the Atlantic Ocean are a common habitat along the entire east coast.

The predominant sediment type in this region in the shallows is sand, with quartz being the most abundant constituent, comprising greater than 90% of the sand fraction in most areas. Feldspar is present in far lesser quantities but can be locally important (Folger, 1972). The specific sediment structure in any locality will depend on water depth, sediment source, exposure to wave action, fetch, and faunal and floral composition.

Although numerous benthic and sediment studies have been conducted in this region, each one describing the particular sediment parameters unique to that area, only one study showed an overall picture of the textural variation of beach sands along the entire mid-Atlantic coast (Emery and Uchupi, 1972). Samples across the shore zone were taken every 20 km from the Gulf of Maine to Texas (Fig. 8.2), and analysis of these data (Fig. 8.3) showed that the beach sands had regional characteristics:

"Beaches of Cape Cod, the offshore islands, and the coast as far west as the entrance to Long Island Sound (samples 14 to 32) are somewhat similar in their variety but are coarser than north of Cape Cod, probably in response to the greater number of eroding cliffs of glacial till. The only fine-grained beaches in this segment were in and near the protected waters of Narragansett

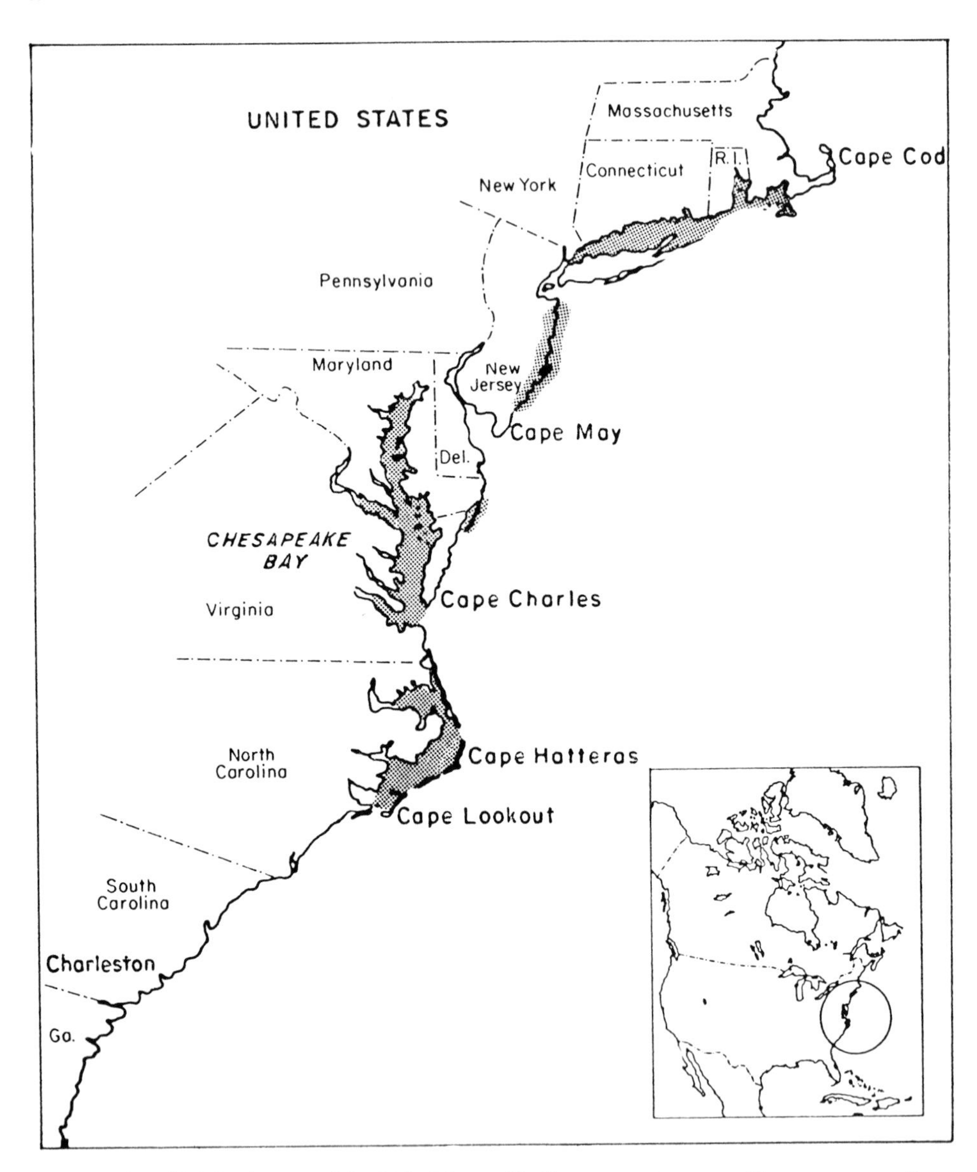

Fig. 8.1. The mid-Atlantic coast of the United States. Shading represents areas with seagrasses or other species of submerged aquatic vegetation tolerant of low salinities.

Bay. Beaches along the south side of Long Island (samples 33 to 42) exhibited a smooth westward decrease in grain size in response to the absence of cliffs and major streams. A less smooth decrease in grain size continues southward from Long Island to the coastal indentation of Georgia (samples 43 to 100); variations are caused by local stream supplies and by seafloor erosion (Emery and Uchupi, 1972)."

The predominant source of sediment reaching the ocean is from rivers and streams and from erosion of existing beaches and mainland bordering the ocean (Emery and Uchupi, 1972). Biogenic deposition of calcium carbonate occurs in warmer waters and is not a factor in the mid-Atlantic. Estuaries and lagoons can also receive large quantities of beach sand from the continental shelf because of a net landward flow of bottom water that can transport bottom sediments. A significant seasonal interaction occurs between beaches and the nearshore environment where beaches are narrower in the spring after winter storms and

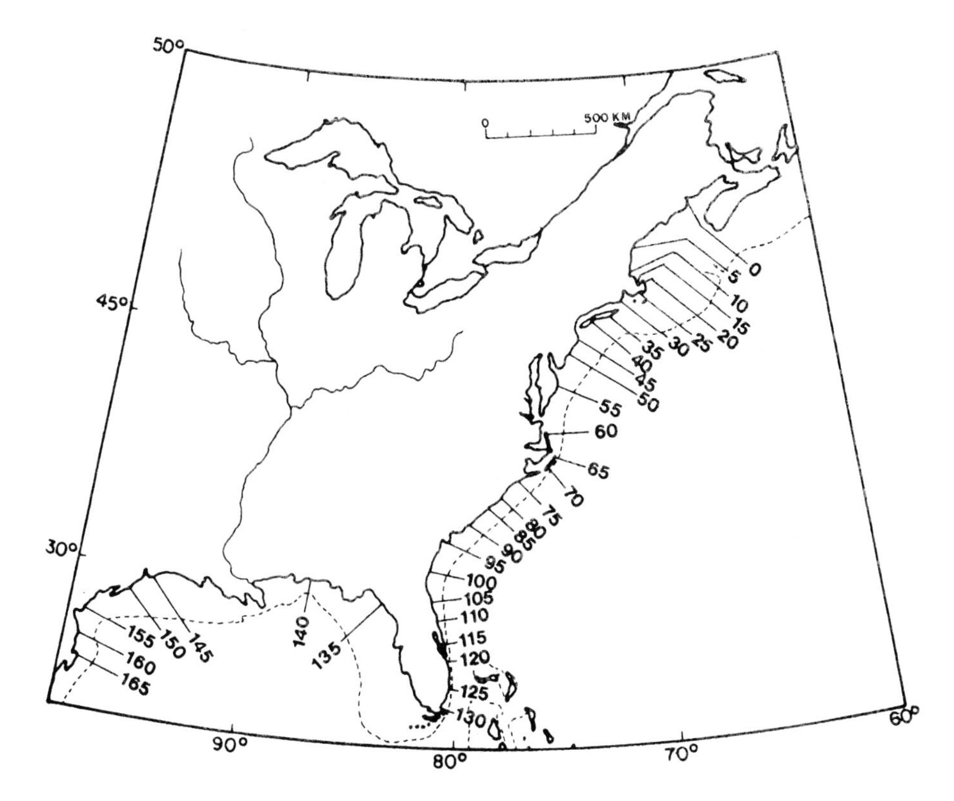

Fig. 8.2. Location where sand samples were collected from foreshores of beaches along the Atlantic and Gulf coast of the United States. Textural variations are presented in Fig. 8.3. (From Emery and Uchupi, 1972; reprinted by permission of American Association of Petroleum Geologists.)

widest in autumn with movement of sand back onshore after milder summer conditions.

Biological structures (such as worm tubes and seagrasses) have the ability to modify sediment characteristics either through the binding of the sediments, the baffling of currents and wave action or from the stabilizing/destabilizing influences of floral and faunal components found associated with the tubes or grasses in the adjacent sediment (Orth, 1977a; Eckman et al., 1981; Fonseca et al., 1982; Ward et al., 1984; Luckenbach, 1986). For example, sediment characteristics were significantly different along a transect from bare sand through a *Zostera marina* dominated seagrass meadow in the Chesapeake Bay (Fig. 8.4). Particle size and degree of sorting were greatest and organic carbon content least in bare sand, while particle size and degree of sorting decreased and organic carbon increased from the edges of the bed to the center where *Z. marina* was most dense. Modal class shift was also evident along this transect.

ENVIRONMENTAL VARIABLES

Climatic variations, reflected in the large range of air and water temperatures observed in most locations along the mid-Atlantic coast, are considerable. This is dissimilar to the Pacific coast of North America where the severity is reduced by the maritime influence.

Latitudinal gradients in surface water temperature for several selected stations are available from work by Orth (1977b) and show not only the range of minimum and maximum water temperatures but also the annual variation at these sites (Fig. 8.5). All sites along the east coast exhibit surface water temperature ranges of at least 20 to 25°C (Massachusetts, −1 to 23°C; New York–New Jersey, 1 to 25°C; Virginia–Maryland, 2 to 28°C; North Carolina, 3 to 30°C; and Georgia, 9 to 33°C).

Salinities along the open, coastal areas range between 33 and 37‰, although this can be much

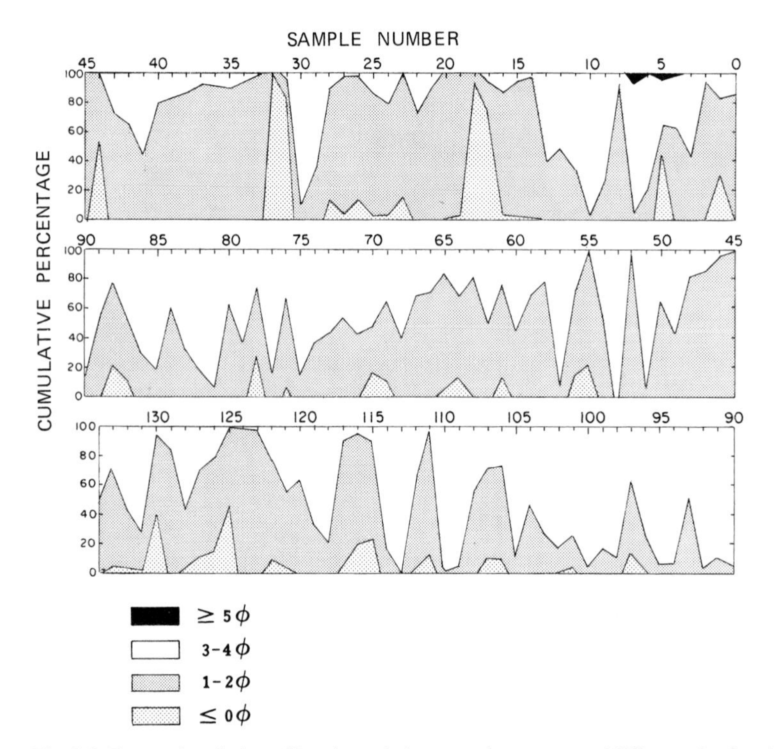

Fig. 8.3. Textural variation of beach sands (presented as percent of different size fractions of sediment in ϕ units, where $\phi = -\log_2$ mm) from the different station locations (sample number is station location) identified in Fig. 8.2. (Revised from Emery and Uchupi, 1972; reprinted by permission of American Association of Petroleum Geologists.)

less where the coast is close to an estuary with substantial flow (for example, Chesapeake or Delaware Bay). Salinity gradients from 0 to 33‰ are found in the east coast estuaries, although the steepness of the gradients will be a function of the size and depth of the estuary and freshwater inflow.

Interstitial salinities can be affected by the proximity of the freshwater table. Lucy (1976) found that interstitial salinities at sediment depths of 15 and 30 cm at a Chesapeake Bay shallow water area where the water table was close to the surface were consistently 9 to 13‰ below that of overlying water (14.5–20.6‰). At a site further upriver where the water table was much deeper, interstitial salinities were only reduced 1.3‰ from the overlying water.

Dominant winds are generally northerly in autumn and winter (September–March) and southerly in spring and summer (April–August). Winds are strongest in the winter and weakest during summer, resulting in calmest seas in summer and roughest in autumn and winter.

Tides along the Atlantic coast are of the semi-diurnal type (12 h., 25 min period). Tidal heights range from 0.2 m in North Carolina sounds to 2.5 m in Georgia. Spring tides can raise tidal heights from 0.05 m to approximately 0.5 m. Tidal heights are generally greatest along the outer, open coast and decrease in magnitude in estuaries, bays and sounds (NOAA, 1986). Water levels in bays and lagoons with restricted entrances to the sea and where lunar tide influence is minimal are very sensitive to wind. Tidal height changes due to winds can be greater than the lunar tides.

Environmental conditions tend to be more extreme in the intertidal and shallow subtidal areas than adjacent, offshore deeper water. During winter, ice formation is a common occurrence in intertidal areas from Virginia northward, which can cause severe ice scour as well as the removal of sediment and fauna when these become embedded in the ice. Diurnal fluctuations in water temperature can be extreme. Johnson (1965) found that water temperature on a shallow sand flat in California could fluctuate as much as 8°C on a daily basis. Although these data are from the west

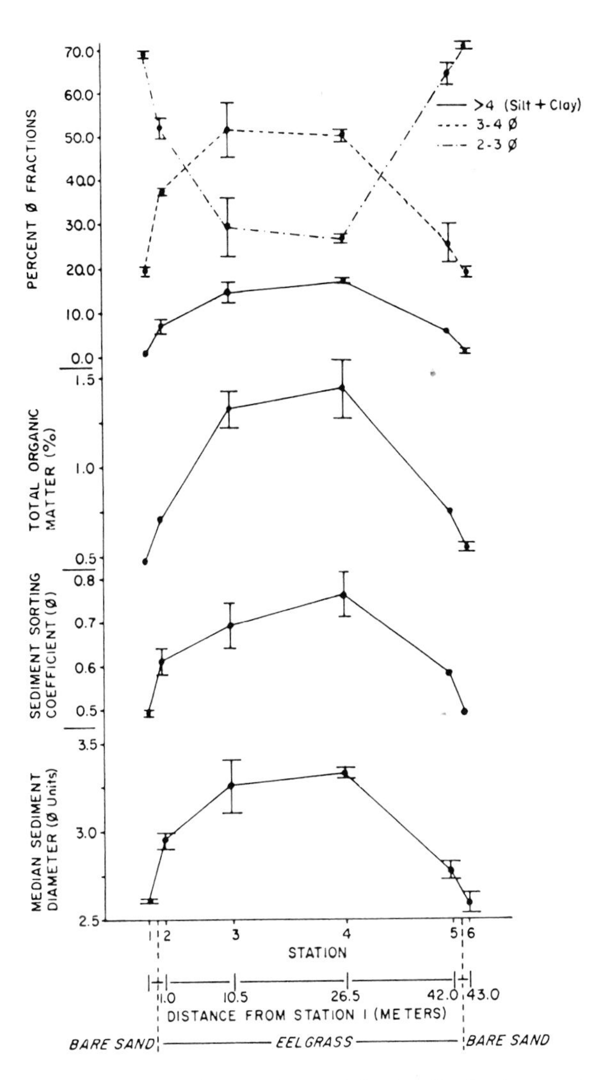

Fig. 8.4. Mean and range for particle size distribution, total organic matter, sediment sorting, and median sediment diameter for six stations located along a transect across a *Zostera marina* meadow in the Chesapeake Bay. (From Orth, 1977a; reprinted by permission of South Carolina Press.)

coast of the United States, one would expect temperature changes to be even greater for east coast areas where air temperatures can exceed 35°C during the summer period.

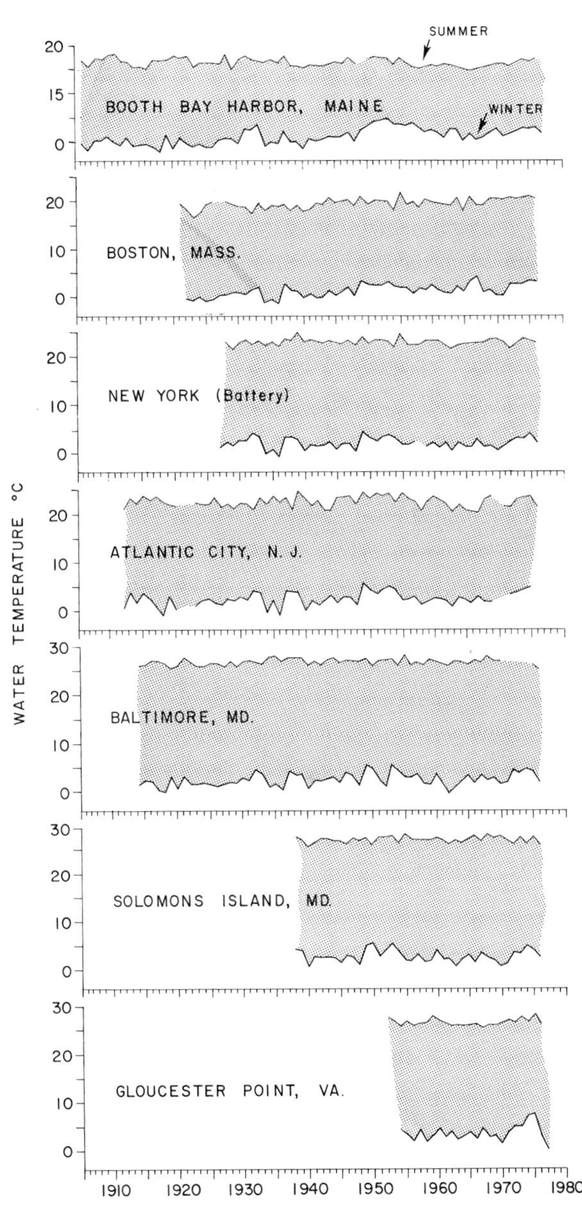

Fig. 8.5. Annual range and year-to-year variation in coldest and warmest surface water temperatures at selected sites along the Atlantic coast of the United States. (From Orth, 1977b.)

FAUNAL AND FLORAL CHARACTERISTICS

Soft substrates

Sediment composition is the dominant physical feature by which benthic assemblages may be easily classified (see reviews by Gray, 1974; Rhoads, 1974). In the following discussion assemblages are related to their position along gradients from fine to coarse sediments in different localities along the eastern coast of the United States. The effect of other variables (such as wave exposure, temperature) on benthic species composition and abundance are discussed within the sediment classification scheme.

Sand

Interstitial organisms are found in sandy substrates with large median grain sizes whereas a greater percentage of burrowers are found on finer substrates (Coull and Fleeger, 1977). These interstitial taxa, primarily meiobenthos (63 to 500 µm), are dominated by ciliates, although harpacticoid copepods, turbellarians and gastrotrichs also occur when grain size is between 100 and 200 µm. In sand with grain sizes exceeding 200 µm many other taxa occur, including tardigrades, ostracods, and archiannelids (Fenchel, 1978).

Little information is available on meiofaunal populations in Atlantic coast sediments, other than that reported by Coull and co-workers from South Carolina (Coull and Fleeger, 1977; Coull et al., 1979; Bell, 1980) and scattered studies elsewhere (see, for example, Tietjen, 1969). Meiofaunal communities are dominated by nematodes, copepods, ostracods, polychaetes, amphipods and the juveniles and larvae of bivalves.

The abundance of dominant meiofaunal taxa may be on the order of 10^6 m^{-2}, which is roughly 3 orders of magnitude larger than macrofaunal ($\geq$500 µm) abundances, and meiofaunal production may rival that of macrofauna (Coull et al., 1979). In addition, epibenthic and burrowing species are frequently fed upon by juvenile fish (Alheit and Scheibel, 1982) and larger invertebrates (Bell and Coull, 1980), although their contribution to higher trophic levels has been difficult to quantify.

Filter-feeding species of crustaceans, bivalve molluscs and polychaetes dominate the macroinfauna of sandy substrates of the Atlantic coast of the United States. In low-energy intertidal environments haustoriid amphipods are dominant taxa (Dexter, 1967; Croker, 1970; Holland and Polgar, 1976; Dörjes, 1977; Holland and Dean, 1977; Maurer and Aprill, 1979; Leber, 1982; McDermott, 1983). In latitudes below about 35°N, mole crabs (*Emerita talpoida*) and coquina clams (*Donax* sp.) also dominate high-energy intertidal substrates (Dexter, 1969; Matta, 1977; Diaz and DeAlteris, 1982), composing up to 98% of the macrofauna on a North Carolina high energy beach (Leber, 1982). Subtidally, filter-feeding bivalves (*Solen viridis*, *Tellina* sp.), polychaetes (*Onuphis* sp., *Scolelepis squamata*), echinoderms (*Mellita quinquesperforata*, *Moira atropos*), and crustaceans (*Callianassa major*, *Ogyrides* sp.) may be locally abundant. Species occurrences for the macrofauna of sand substrates are shown in Fig. 8.6 according to their distribution along the tidal gradient (Leber, 1982). Abundance values range from about 1 to 12×10^3 ind. m^{-2} (Table 8.1), being considerably less than those reported from muddy substrates.

Larger predatory macroinvertebrates of sandy substrates include portunid crabs (*Arenaeus cribarius*, *Callinectes sapidus*, *Ovalipes ocellatus*) and horseshoe crabs (*Limulus polyphemus*) in the subtidal and ghost crabs (*Ocypode quadrata*) in the intertidal (Leber, 1982; McDermott, 1983). Fishes commonly occurring over sandy substrates include *Menidia menidia* (silversides), *Menticirrhus* sp. (kingfish), *Paralichthys* sp. (flounders), *Pomatomus saltatrix* (bluefish) and *Trachinotus falcatus* (pompano), among others (Leber, 1982; McDermott, 1983).

Shorebirds may be very abundant on Atlantic coast beaches and intertidal flats, especially during stopovers at Atlantic coast staging areas during annual migration flights. For example, during spring migrations over one million shorebirds occur within the confines of the Delaware Bay as they stop over on their flight from South America to the Arctic (Wander and Dunne, 1981; Myers, 1986). Dominant species include *Arenária intérpres* (ruddy turnstone), *Calidris alba* (sanderling), *C. canútus* (red knot), *C. pusilla*

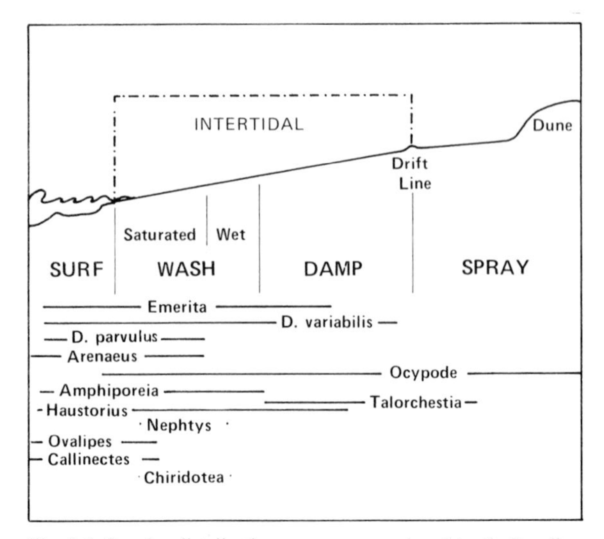

Fig. 8.6. Species distribution among zones in a North Carolina high-wave-energy sandy beach. (From Leber, 1982; reprinted by permission of *Bulletin of Marine Science*.)

TABLE 8.1

Macrofauna of sandy beaches from New Jersey to Georgia. (Modified from Maurer and Aprill, 1979.)

Study area	New Jersey	Delaware	North Carolina	South Carolina	Georgia
Refs.[a]	[1]	[2]	[3]	[4]	[5]
Type of beach		Protected	Semi-protected	Protected	Semi-protected
Sieve size (mm)	0.5	1.0	0.5	1.0	0.8
Density (m^{-2})	–	881	963	1440	12 565
Biomass ($g\ m^{-2}$)	1.6	2.65[b]	–	–	–
Dominant species	*Scolelepis squamata* *Donax variabilis* *Amphiporeia virginiana*, *Haustorius canadensis*	*Scoloplos fragilis* *Heteromastus filiformis* *Neohaustorius biarticulatus*	*Neohaustorius schmitzi* *Acantho-haustorius millsi* *Donax variabilis*	*Acantho-haustorius millsi* *Pseudo-haustorius caroliniensis*	*Acantho-haustorius* sp. *Bathyporeia* sp. *Neohaustorius schmitzi* *Donax variabilis*

[a] References:
[1] McDermott (1983).
[2] Maurer and Aprill (1979).
[3] Dexter (1969).
[4] Holland and Polgar (1976).
[5] Howard and Dörjes (1972).
[b]Ash-free dry matter.

(semipalmated sandpiper), and *Limnódromus griseus* (short-billed dowitcher). In addition, laughing gulls (*Lárus atricilla*) are also conspicuous components of the Delaware Bay shore-zone assemblage.

Shorebirds may be significant predators of infaunal species of sandy substrates, although relatively little quantitative information is available to determine the magnitude of the impact of shorebird foraging. In the British Isles, oystercatchers (*Haemátopus ostralegus*) have been reported to remove up to 40% of the populations of cockles (*Cerastoderma edule*) in a winter's foraging (Sutherland, 1982). Along the west coast of the United States, shore birds foraging in sand substrates did not significantly reduce infaunal densities, although significant reductions occurred in experiments on muddy substrates (Quammen, 1982, 1984). Along the mid-Atlantic coast, two studies have addressed the question of shorebird feeding on infauna. Schneider and Harrington (1981) used exclosures and found that some shorebirds were important predators of polychaetes and amphipods. However, substrate types were not specified for each set of experiments although experiments were carried out in both mud and sand, and, in addition, other predators had access to the exclosures. Botton (1984a) found no significant impact of shorebird foraging on sand substrates and concluded that his caging studies support the notion of little impact on infaunal taxa in sandy substrates.

Mud

Muddy substrates along the mid-Atlantic coast of the United States often support large standing stocks of benthic invertebrates whose average body sizes are small compared to those in sandy substrates (Rhoads, 1974). For example, abundance of muddy infauna may be in the range of 10^4 to 10^5 ind m^{-2} (McCall, 1977; Woodin, 1978; Holland et al., 1980; Summerson and Peterson, 1984), although densities of macro-infauna seem to be lower on Atlantic than Pacific coasts of the United States (Woodin, 1976).

Along the Atlantic coast, muddy substrates are populated by assemblages of meiofaunal burrowers, such as nematodes, harpacticoid copepods (Coull and Fleeger, 1977), annelids (Bell, 1980), ostracods and foraminiferans (Matera and Lee, 1972) as they are in other parts of the world

(Fenchel, 1978). As is true for sandy substrates along the mid-Atlantic coasts, relatively little is known of these meiofaunal communities in muddy substrates.

Much more is known of the macrofaunal organisms in muddy substrates where deposit-feeding polychaetes and bivalves are dominant taxa from New England to Georgia (cf. Dörjes, 1977; McCall, 1977; Mountford et al., 1977; Summerson and Peterson, 1984). Common polychaetes include *Capitella* cf. *capitata, Nephtys incisa, Owenia fusiformis* and *Streblospio benedicti* (especially in disturbed habitats), the tube-dwelling *Diopatra cuprea* and several species of chaetopterid (parchment) worms. Bivalves include *Abra aequalis, Ensis directus, Macoma* spp. and *Tellina* spp. Tube-building amphipods (*Ampelisca abdita*), burrowing decapods (*Upogebia* sp.) and ophiuroids (*Hemipholis* sp.), as well as representatives of other phyla, may be locally abundant.

Important predators on muddy substrates include mobile macroinvertebrates such as blue crabs (*Callinectes sapidus*) and horseshoe crabs (*Limulus polyphemus*) (Virnstein, 1977; Woodin, 1981; Botton, 1984b) and bottom feeding fishes such as croaker (*Micropogonias undulatus*) and spot (*Leiostomus xanthurus*) (Holland et al., 1980). As discussed previously, the impact of shorebird foraging may be a significant source of mortality for the benthos, especially polychaetes, in muddy sediments (Schneider and Harrington, 1981; Botton, 1984a).

Seagrass and algal habitats

Three species of seagrass are present in the Mid-Atlantic region: *Zostera marina, Halodule wrightii*, and *Ruppia maritima. Zostera marina* is the dominant species of submerged vegetation and occurs along the east coast of North America from North Carolina to Nova Scotia. *Halodule wrightii*, with a pantropical distribution (Den Hartog, 1970), is found only in North Carolina. *Ruppia maritima* is not considered a true seagrass species, but because of its pronounced tolerance for salinity and temperature has been able to grow and compete successfully in marine and brackish-water habitats.

Under more brackish-water conditions and advancing to freshwater areas, a diverse array of freshwater submerged aquatic species have extended their distribution into these areas. These include *Elodea* spp., *Myriophyllum spicatum, Najas* spp., *Potamogeton pectinatus, P. perfoliatus, Vallisneria americana* and *Zannichelia palustris*.

Most seagrass communities in this region are normally composed of *Z. marina* and *R. maritima*, either in mixed or pure stands. Only in North Carolina all three species are found (Thayer et al., 1984). Where *Z. marina* is found to co-occur with either or both species, *H. wrightii* and *R. maritima* are usually present in the shallowest locations (in some cases, exposed at low tide) and *Z. marina* in the deeper sections of the grass bed, with a mixture of the species at intermediate depths (Orth et al., 1979; Wetzel and Penhale, 1983). Description of the more brackish-water grass communities has been concentrated in the Chesapeake Bay (Kemp et al., 1983; Orth and Moore, 1984) and North Carolina (Davis and Brinson, 1976), where communities can consist of as many as six species (Anderson and Macomber, 1980). Grass beds may be continuous stands or may be a heterogeneous array of sandy and muddy areas interspersed among the vegetation.

Benthic microalgae are important constituents of intertidal and shallow water ecosystems due to their high year-long productivity, fast turnover, and utilization as food by primary consumers. These communities are composed primarily of motile, pennate diatoms, though flagellates and blue-green algal mats do occur and are sometimes very abundant (Leach, 1970; Marshall et al., 1971; Gallagher and Daiber, 1974).

Benthic macroalgae can be extremely productive in systems that afford suitable habitats for growth (Mann, 1973). They are very conspicuous components of any rocky shore zone. The absence of significant hard substrata along the east coast has relegated these components to pilings, oyster shell, rock jetties and seagrasses (Taylor, 1957). Dominant species include *Ceramium* spp., *Enteromorpha* spp., *Fucus vesiculosus, Gracilaria* spp., *Lyngbya* spp., *Polysiphonia* spp. and *Ulva lactuca*. Humm (1969) recognized seven distributional groups of species along the North Atlantic coast (Fig. 8.7). All of the benthic species that occur between Cape Cod and Florida are eurythermal and are elements of either the tropical Atlantic or the cooler, upper North Atlantic waters.

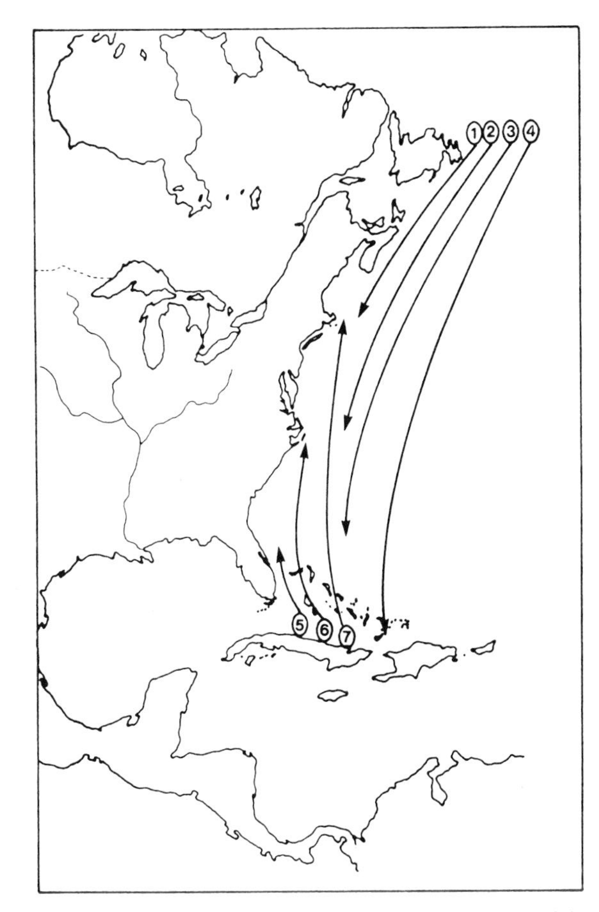

Fig. 8.7. Diagrammatic representation of the distribution of the inshore marine algae of the Atlantic coast of North America in which Humm recognized seven distributional groups. The algae of the Atlantic coast of North America can be divided into two floras, one centered in tropical waters, the other in cold waters north of Cape Cod. Numbers represent species extending from Arctic waters south to Cape Cod (1), Arctic to Beaufort, North Carolina (2), Arctic to Cape Kennedy, Florida (3), Arctic to the Tropics (4), Tropics to Cape Kennedy (5), Tropics to Beaufort (6), Tropics to Cape Cod (7). (From Humm, 1969; reprinted by permission of *Phycologia*.)

Detached macroalgae sometimes form large drifts which may, at times, be considerable in shallow bays and lagoons. Little is known about the ecological significance of these drift algal beds, but in systems that lack any form of habitat structure, drift algae may offer significant cover for smaller animals.

Because seagrass beds provide a firm substrate for faunal attachment, provide protection from predators, stabilize sediments, affect current and wave regime and are a food resource, the density and diversity of faunal and floral components in east coast grass beds are usually much greater than the fauna and flora of adjacent unvegetated areas (Thayer et al., 1975; Orth, 1977a; Heck and Orth, 1980; Homziak et al., 1982; Peterson, 1982; Kemp et al., 1984; Summerson and Peterson, 1984).

Macro- and micro-algal epiphytes are significant components of seagrass beds (Harlin, 1980) and can serve as major sources of food for herbivorous fauna. Brauner (1975) recorded 79 species of macroalgae occurring on seagrasses in North Carolina, while Marsh (1970) recorded 28 species of macroalgae on *Zostera marina* in the York River, Virginia. Phaeophyta and Rhodophyta were dominant in North Carolina, making up 71% of all algal species, while in the York River 86% of the species were in the Chlorophyta and Rhodophyta. Sieburth and Thomas (1973) recorded the temporal sequence of colonization of *Z. marina* by microflora in Rhode Island. Initially, *Cocconeis scutellum* formed a unialgal mat followed by other diatoms such as *Amphora, Navicula, Nitzschia* and *Pleurosigma*.

Infaunal densities of invertebrates in grass beds have ranged from 300 to 15 000 ind m^{-2} with polychaetes, oligochaetes, bivalves and peracarid crustaceans being important constituents (Orth, 1973; Thayer et al., 1975). Dominant polychaetes include *Heteromastus filiformis, Nereis* spp., *Polydora ligni* and *Streblospio benedicti*. Bivalves include *Gemma gemma, Mya arenaria* and *Tellina versicolor*. Peracarid crustaceans include *Ampelisca* spp. and *Edotea triloba*. Infaunal densities have been shown to be a function of the blade density of the bed and the location within the bed. In a study of infauna in a *Z. marina* bed in Chesapeake Bay, Orth (1977a) demonstrated an increase in both density and diversity along a transect from bare sand across the edge to the center (Fig. 8.8), with the most significant changes occurring from the bare sand to just inside the grass-bed boundary. Although this increase was attributed to sediment stabilization by *Z. marina*, the ability of the roots and rhizomes to prevent predators from gaining access to prey contributes to this difference (Blundon and Kennedy, 1982). Peterson (1982) found densities of two bivalve species, *Mercenaria mercenaria* and *Chione cancellata*, to be higher in seagrass than on bare sand (11.3 versus 0.4 and 10.3 versus 0.4 m^{-2}, respectively) in North Carolina. He attributed the higher densities to the

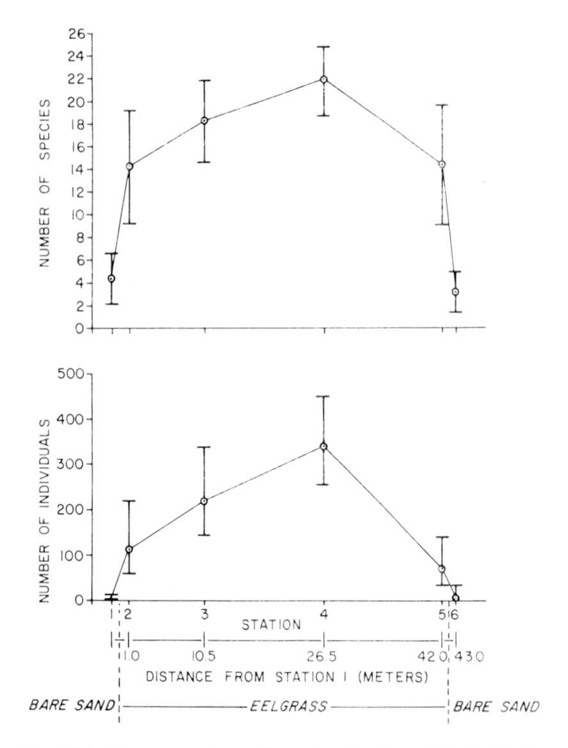

Fig. 8.8. Mean number of species (top) and individuals of invertebrates per core ($0.07\ m^2$) (bottom) for six stations located along a transect across a *Zostera marina* meadow in the Chesapeake Bay (vertical bars are 95% confidence intervals). (From Orth, 1977a; reprinted by permission of South Carolina Press.)

function of the roots and rhizomes serving not only as a refuge but also binding sediments and increasing sediment compaction. Peterson also noted that the shallow-dwelling species, *C. cancellata*, was more susceptible to predators than the deeper-dwelling form, *M. mercenaria*. Thus, not only is the vegetation important in creating prey refugia, but the survival of a prey species also depends on its particular lifestyle (Orth et al., 1984). Summerson and Peterson (1984) found that in North Carolina, the abundance of epibenthos in seagrass beds was fifty-two times that on sand flats, while the infauna was three times as abundant. Although epibenthic predators were more numerous in the grass bed, it was suggested that seagrass beds served as refuges, similar to coral reefs where species use vegetation as a shelter by day and forage over sand under protection of night.

Epifaunal densities can range from 80 to 20 000 ind m^{-2} (Marsh, 1973; Thayer et al., 1975). Gastropods, amphipods and isopods dominate the seagrass epifauna. *Ampithoe* spp., *Bittium varium*, *Crepidula convexa*, *Cymadusa compta*, *Erichsonella attenuata*, *Gammarus mucronatus*, *Idotea balthica*, *Melita appendiculata*, *Mitrella lunata* and *Paracerceis caudata* are dominant members of the less vagile fauna (Nagle, 1968; Marsh, 1973, 1976) while vagile forms include *Callinectes sapidus*, *Crangon septemspinosa*, *Hippolyte* spp., *Neopanope sayi* and *Palaeomonetes* spp.

Hydrodynamics appears to be an important factor in the population ecology of seagrass epifauna. Eckman (1987) found that hydrodynamics in a New York eelgrass meadow exerted a stronger influence on bivalve recruitment than interblade abrasion or possible differences among sites in the nature of the seagrass blade substratum.

Experimental work with epifaunal amphipods in North Carolina as prey and *Lagodon rhomboides* and *Palaeomonetes vulgaris* as predators showed that predation rates decreased with increasing seagrass blade density but not in a linear function (Nelson, 1979a, b, 1980). In addition, the susceptibility to predation depended on the lifestyle of the amphipod, infaunal and tube-building amphipods being less susceptible than epifaunal and free-living forms.

Field experiments with tethered crabs also showed lower rates of predation in vegetated than in unvegetated areas, although the greatest amount of protection provided for young crabs occurred at intermediate *Zostera marina* densities (Heck and Wilson, 1987; Wilson et al., 1987).

Studies of fish community assemblages have been concentrated in the Chesapeake Bay (Orth and Heck, 1980; Weinstein and Brooks, 1983; Heck and Thoman, 1984) and North Carolina (Thayer et al., 1975; Adams, 1976a, b). In general, nekton communities are more diverse in vegetated areas compared to unvegetated areas and were shown to have distinct diel, tidal and seasonal fluctuations in abundance and composition. Common species include *Anchoa* spp. (anchovies), *Apeltes quadracus* (sticklebacks), *Bairdiella chrysoura* (perch), *Lagodon rhomboides* (pinfish), *Leiostomus xanthurus* (spot) and *Sygnathus fuscus* (pipefish).

Although waterfowl studies are sparse for this region, seagrasses can have a dramatic effect on waterfowl populations. The decline of *Z. marina* in the 1930s due to the wasting disease resulted in the drastic reduction in brant populations along the East Coast (Cottam, 1934). More current work in the Chesapeake Bay indicated that the Canada goose (*Bránta canadénsis*) consumed about 21% of the standing crop of seagrasses (Wilkins, 1982).

FUNCTIONAL RELATIONS

Biological interactions

Mutualistic relationships and the aggregation of species around biogenic structures, such as worm tubes, are important in littoral habitats. On the mid-Atlantic coast some well-known examples are the occurrence of pinnotherid crabs in chaetopterid worm tubes (Williams, 1984), and the large increases in infaunal density that occur near the tube clusters of the polychaete *Diopatra cuprea* (Woodin, 1978, 1981). These infaunal increases are presumably due to decreased foraging success by crabs and fish predators, as well as increased settlement of larvae among tube dwellers. Similarly, bacterial and meiofaunal populations may be greater in areas surrounding well-oxygenated burrows (Yingst and Rhoads, 1980). However, recent experimental work suggests that biotically mediated effects on sediment stability by these *Diopatra* tubes rather than the tube itself are responsible for lowering erosion thresholds in these areas and thus allowing for increased macrofaunal abundance (Luckenbach, 1986). In addition, adult aggregations of polychaete species around *Diopatra* tubes can inhibit recruitment of some taxa, such as, for example, the bivalve *Mulinia lateralis* (Luckenbach, 1987).

In soft-bottom communities the effects of competition seem much less important than on hard substrates, and predation seems to be the dominant biological interaction (Peterson, 1979; Dayton, 1984). For example, caging studies in Maryland, Virginia and North and South Carolina have shown that predation on both macrofaunal and meiofaunal populations inside cages is greatly reduced from that seen in unprotected areas (Virnstein, 1977; Bell and Coull, 1980; Holland et al., 1980; Woodin, 1981; Kneib and Stiven, 1982; Summerson and Peterson, 1984). Feeding activities may also produce significant changes in populations of organisms that are not consumed. For example, sediment disturbance during the excavation for food by blue crabs (*Callinectes sapidus*) and horseshoe crabs (*Limulus polyphemus*) results in reductions in the abundance and species number of non-food items such as small polychaetes and bivalves (Woodin, 1978; Botton, 1984b). Such "disturbance effects" may be important, depending on the intensity and frequency of the disturbance and the size of the disturbing agent; however, recent experimental evidence has shown that reductions of benthic invertebrate densities in a South Carolina shallow mud bottom by a caridean shrimp, *Palaemonetes pugio*, were due to predation rather than disturbance effects by the foraging activities of the shrimp (Kneib, 1985).

Predation levels may also be related to substrate type, that is, sand versus mud. Smith and Coull (1987) experimentally determined that predation on meiofauna by juvenile spot (*Leiostomus xanthurus*) in South Carolina is greater in muddy than in sandy substrata. They attributed these differential predation effects to the higher densities of meiofauna in mud than in sand and the inability of spot to pass sand grains through its gill rakers.

Just as predators may control animal abundance, grazers may control bacteria and algal abundance. For example, Pace et al. (1979) showed that removal of the grazing gastropod *Ilyanassa obsoleta* from mud flats in Georgia resulted in increased microbial and algal biomass.

Competitive interactions in soft substrates do not often result in competitive exclusion such as observed on hard substrates. On the west coast of the United States there is evidence that exclusion may occur between sediment-stabilizing and sediment-destabilizing species (Brenchley, 1981; Wilson, 1981). Whether similar relationships will be found on the Atlantic coast remains to be seen. In addition, it has been suggested that infaunal invertebrates on the east coast of the United States exist at densities below those required for significant competitive interactions to occur (Woodin, 1976; Peterson, 1979). Also, competitive effects are usually sublethal when they occur in soft-bottom habitats, resulting in reduced growth (Peterson, 1979) or vertical separation (Levinton, 1977).

Biomass, production, decomposition

Fauna

Standing stocks of invertebrates in the littoral and intertidal habitats are variable through time, salinity, and quality of the habitat (sand/mud, vegetated/unvegetated). From a broad study of these habitats in the Maryland–Virginia–North Carolina area, Diaz et al. (1982) found standing stocks to be highest in the spring or fall. Summer was found to be the time of minimum standing stock. Superimposed on the temporal variation in standing stock are spatial patterns. Organisms tend to be patchy in distribution, producing areas of high and low standing stock. If temporal and spatial variability are averaged, then the trends in quality of littoral habitats can be seen (Table 8.2). Overall, vegetated areas have highest standing stocks of all groups of organisms except molluscs, which tend to be higher in sand or mud. Abundance tends to follow this pattern except for the species that attain large size such as the clams and crabs, where largest individuals are found in unvegetated habitats.

Although descriptive studies of high-energy sand-beach habitats exist, there are no good estimates of biomass; however, the abundance of organisms reported by Dexter (1969), Matta (1977), Diaz and DeAlteris (1982) and Leber (1982) suggest that there is a substantial standing stock present. This standing stock is also very mobile, with distinct horizontal beach migrations related to tides and season.

Few estimates of secondary production have been made in the littoral habitats of the mid-Atlantic coast. While standing stock provides estimates of existing biomass, the flow of energy through littoral and intertidal habitats provides a better estimate of the value of any particular habitat. Productivity of these areas must be substantial, judging by the number of bird and fish species that utilize these habitats as feeding grounds.

The most detailed estimates of productivity come from vegetated habitats, very little work having been done in sand or mud. Diaz and Fredette (1982) estimated the productivity of the nine most important species of higher-level consumers in a polyhaline mixed *Zostera–Ruppia* grass bed. They found the yearly production of these nine species to be 40.7 g dry weight m^{-2} (Table 8.3), which is a high value when compared to other studies (Sanders, 1956; Warwick and Price, 1975; Warwick et al., 1978; Robertson, 1979). They estimated that 53 metric tonnes of dry tissue was then produced and available for consumption by other higher or lower trophic levels for the entire grass bed of 140 hectares. This represented approximately 6×10^{10} individuals who were born, grew, survived, and died in the grass bed each year (Diaz and Fredette, 1982).

Secondary production from bare mud and sand flats is much less than that recorded from vegetated flats. Data from sand flats indicate they are the least productive.

While much of the productivity of these habitats goes to supporting higher level consumers and standing stock, the exact amount is unknown, as is the proportion of the productivity that goes to decomposers. Research into the pathways and flow rates that distribute the energy produced is needed.

Seagrasses and algae

Peak measurements for *Zostera marina* growth along the east coast generally occur progressively later with increasing latitude reflecting the differential heating patterns evident at the different geographical locations (Burkholder and Doheny, 1968; Wetzel and Penhale, 1983; Thayer et al., 1984). Silberhorn et al. (1983) showed that all stages of flowering occurred at approximately the same temperature, but the latitudinal differences in temperature regimes resulted in a shift in the developmental stages later in the year at more northerly locations.

Peak above-ground biomass for *Z. marina* along the mid-Atlantic coast ranges from 106 to 2040 g dry weight m^{-2}, while peak below-ground biomass ranges from 50 to 960 g dry weight m^{-2} (Burkholder and Doheny, 1968; Penhale, 1977; Thorne-Miller et al., 1983; Thayer et al., 1984). Leaf productivity measurements range from 0.59 to 1.23 g C m^{-2} d^{-1}, while root-rhizome productivity (though not as frequently measured as leaf productivity) ranges from 0.15 to 0.28 g C m^{-2} d^{-1} (Thayer et al., 1984). These values are well within ranges reported worldwide for *Z. marina* (Jacobs, 1984).

Epiphytic algae can be a significant component of the total biomass of a *Z. marina* bed. Penhale

TABLE 8.2

Average abundance and standing stock of fauna from soft-bottom littoral and intertidal habitats of Maryland, Virginia, and North Carolina (from Diaz et al., 1982)

Habitat/type		Annelids		Molluscs		Isopods and amphipods		Crabs		Shrimp		Fish
		ind m^{-2}	g m^{-2}	ind m^{-2}	g m^{-2}	ind m^{-2}	g m^{-2}	ind $(100\ m^2)^{-1}$	g $(100\ m^2)^{-1}$	ind $(100\ m^2)^{-1}$	g $(100\ m^2)^{-1}$	ind $(100\ m^2)^{-1}$
Dense grass												
Estuarine	P	31 490	–	2670	–	12 680	–	6.6	13.8	147.2	35.3	101
Estuarine	M–P	17 290	49.4	890	34.3	3650	16.6	3.9	96.1	30.0	12.7	103
Ocean sound	M–P	8230	15.5	90	0.4	910	4.0	6.2	9.2	183.5	40.7	77
Estuarine	LM	10 200	63.2	490	38.7	1590	4.5	0.2	11.2	16.5	4.2	411
Ocean sound	LM	1570	35.2	30	11.3	200	0.4	4.6	27.4	5.7	1.7	113
Patchy grass												
Estuarine	M–P	4240	32.7	460	70.0	290	1.1	2.4	35.5	15.6	4.9	27
Ocean sound	M–P	30	0.9	0	0.0	30	0.1	–	–	–	–	–
Estuarine	LM	19 290	185.0	140	0.7	21 710	30.0	0.0	0.0	2.7	0.7	255
Tidal	F	1490	2.1	570	2.7	60	0.3	0.0	0.0	0.0	0.0	4
Unvegetated mud												
Estuarine	M–P	6510	67.0	470	165.0	390	3.7	2.3	37.1	60.4	12.3	146
Estuarine	LM	4490	13.4	430	524.0	1340	12.0	0.1	0.1	0.2	0.1	23
Unvegetated sand												
Estuarine	P	1560	–	520	–	240	–	0.9	2.1	2.5	0.8	39
Estuarine	M–P	700	10.1	430	174.0	220	0.9	0.4	3.6	0.1	0.1	2
Ocean sound	M–P	1790	12.9	60	0.3	390	0.7	0.5	0.8	3.3	0.7	50
Estuarine	LM	2270	22.7	110	0.3	1800	3.4	0.0	0.0	0.2	0.1	3

P = Polyhaline; M = Mesohaline; LM = Lower mesohaline; F = Freshwater.

TABLE 8.3

Production estimates for species from mid-Atlantic littoral and intertidal habitats

Taxa	Production g dry wt m^{-2} yr^{-1}	P/B ratio	Habitat	Refs.[a]
Polychaetes				
Nephtys incisa	9.3	2.2	mud flat	[4]
Pectinaria gouldii	1.7	1.9	mud flat	[4]
Molluscs				
Bittium varium	0.2	3.2	grass bed	[1]
Gemma gemma	0.7	5.9	grass bed	[1]
Pandora gouldiana	6.1	6.1	mud flat	[4]
Yoldia limatula	3.0	3.2	mud flat	[4]
Amphipods				
Gammarus mucronatus	8.0	24.5	grass bed	[1]
G. mucronatus	11.4	76.8	macroalgae	[2]
G. mucronatus	5.8	36.8	grass bed	[2]
Isopods				
Edotea triloba	2.0	11.3	grass bed	[1]
Erichsonella attenuata	17.6	18.9	grass bed	[1]
Idotea balthica	1.0	8.5	grass bed	[1]
Decapods				
Callinectes sapidus	8.9	4.7	grass bed	[1]
C. sapidus	2.7[b]	0.5	grass bed	[3]
C. sapidus	0.07[b]	0.15	sand flat	[3]
Crangon septemspinosa	0.5	4.4	grass bed	[1]
C. septemspinosa	0.4[b]	0.3	grass bed	[3]
C. septemspinosa	0.04[b]	0.07	sand flat	[3]
Palaemonetes vulgaris	1.8	3.4	grass bed	[1]
P. vulgaris	0.7[b]	0.9	grass bed	[3]
Xanthidae	0.9[b]	0.62	grass bed	[3]
Xanthidae	0.002[b]	1.00	sand flat	[3]

[a] References:
[1] Diaz and Fredette (1982).
[2] Fredette and Diaz (1986).
[3] Penry (1982).
[4] Sanders (1956).
[b] Measurements over 207 days only.

(1977) reported that 17 to 52% of the total dry weight of *Z. marina* beds was due to epiphytes. Productivity of this component averaged 0.2 g C m^{-2} d^{-1}.

In areas where two or more species are present, peak biomass values can vary depending on whether the species occur monospecifically or mixed. Orth and Moore (1982) found that, in an area where *Zostera marina* and *Ruppia maritima* occur, the above-ground biomass of each species was usually greater when the species occurred in a monospecific stand than when both species grew together, suggesting some competitive interference between the two species. In the mixed zone, *Zostera marina* appeared to have the competitive advantage. In North Carolina, similar differences for the *Z. marina–Halodule wrightii* system may also occur, since leaf biomass values for *Z. marina* in a mixed bed were higher than those for *H. wrightii* (Thayer et al., 1984).

Production estimates given above indicate that seagrasses (and their attached epiphytes) produce significant amounts of organic matter during the year. A relatively small proportion of this material is directly consumed, as the major herbivore group is wintering waterfowl. A significant amount of production is transported from the bed and is deposited on the shoreline, in deeper, offshore sediments, or adjacent salt marsh creek sediments (Wilson et al., 1985). This material becomes available to other consumers during decomposition, when microbial processes enhance the nutritional value of the seagrass (Thayer et al., 1977).

Decomposition rates can vary depending on the physical setting where this process occurs. Thayer et al. (1984) showed a more rapid decay of *Z. marina* in North Carolina under continuously submerged conditions compared to alternating wet and dry conditions.

Standing stocks of macroalgae can fluctuate and may be highly variable between sites. Thorne-Miller et al. (1983) reported biomass of macroalgae to be 13 to 46% of the total submerged macrophytes in Rhode Island ponds and in one pond even 100%. Standing crop estimates in some species equaled that of adjacent seagrasses (for example, *Ulva lactuca* – 370 g dry weight m^{-2}; *Enteromorpha plumosa* – 400 g dry weight m^{-2}).

Benthic microalgal communities can be very productive and frequently exceed phytoplankton production in shallow and turbid coastal waters (Leach, 1970; Marshall et al., 1971). Annual production estimates for east coast intertidal mud flats and sand flats have ranged from 113 to 184 g C m^{-2} (Baillie and Welsh, 1980; Rizzo and Wetzel, 1985), while on submerged mud and sand annual production estimates have ranged from 107 to 280 g C m^{-2} (Nowicki and Nixon, 1985; Rizzo and Wetzel, 1985).

Food webs

In an early attempt to develop an annual energy budget of a *Zostera marina* system in North Carolina, Thayer et al. (1975) delineated several of the major interactions in these energy pathways. They estimated that 55% of the net production of *Z. marina*, phytoplankton and benthic microalgae in the bed was consumed by the macrofauna, which was considered high since epiphytes and benthic microautotrophs were not included. More recent evidence, both from stable carbon isotope ratios and experimental manipulative work in the Chesapeake Bay and North Carolina, indicates that the epiphytes may be a key source of food for many of the smaller invertebrates, which in turn support the larger carnivores (Thayer et al., 1978; Van Montfrans et al., 1982, 1984). Translocation experiments in Rhode Island and North Carolina with nitrogen and phosphorus have shown that these nutrients move from the sediments to the leaves, and are then excreted and can be taken up by the epiphytes (Harlin, 1973; Penhale and Smith, 1977; Penhale and Thayer, 1980; Thursby and Harlin, 1984). These epiphytes are an important food source; they appear to have been underestimated both in their importance as food and as contributors to the carbon dynamics in estuaries (Smith and Penhale, 1980). Elimination of epiphytic grazers can affect leaf production since epiphytes have been shown to affect plant growth negatively (Orth and Van Montfrans, 1984; Van Montfrans et al., 1984).

Numerous species have been reported to contain living *Z. marina* in their guts (Thayer et al., 1984), but this consumption is probably accidental as these animals are actually consuming associated plants and animals. The only animals that actually consume the grasses for their nutritional needs are the wintering waterfowl, for example, brant, widgeon, redheads, geese, etc. Although data are still sparse, Thayer et al. (1977) suggested that the major pathway for *Z. marina* utilization is via the detrital food chain.

Benthic microalgae are important not only because of their substantial productivity but because they have a rapid turnover, can be consumed directly by herbivorous animals, and are productive in winter months when other autotrophic production is low. Most studies of grazing of microalgae along the east coast have been done in salt marshes, and here these algae serve as an important food source for a diverse array of organisms, including crabs, benthic copepods, gastropods, amphipods, and polychaetes (Nixon and Oviatt, 1973). Given the extensive sand and mud flats in this region, benthic microalgae are a primary food source for many of the organisms occurring in these habitats.

Nutrient cycling

Tidal-flat sediments are important sites for converting complex organic detrital material into a simpler, more utilizable form (Gosselink et al., 1974; Delaune et al., 1976; Nixon, 1981; Boynton and Kemp, 1985). Large detrital particles are mechanically broken down by a number of infaunal and epifaunal organisms (Frankenburg and Smith, 1967; Fenchel, 1970; Hargrave,

1970a, b; Nixon and Oviatt, 1973; Tenore and Rice, 1980). These smaller particles are more easily colonized by micro-organisms (Odum, 1968; Pomeroy et al., 1977; Christian and Wetzel, 1978). In addition, these smaller particles with their associated microbial flora are both more easily ingestible and more nutritious for detritivores than the original detrital particles (Burkholder and Bornside, 1957; Teal, 1962; Odum, 1968; Schultz and Quinn, 1973; De la Cruz and Gabriel, 1974; Gosselink and Kirby, 1974; Squiers and Good, 1974; Pomeroy et al., 1977; Wetzel, 1977).

Sediments continuously replenish the phosphate concentration of overlying waters (Pomeroy, 1961; Oppenheimer and Ward, 1963; Pomeroy et al., 1965). Sediments in Doboy Sound, Georgia, contain enough exchangeable phosphate to replace water concentrations 25 times (Pomeroy et al., 1965). The rate of exchange for phosphate, as well as for other nutrients, is governed both by physical factors such as absorption and desorption to and from sediments and tidal flushing, and by biological factors such as biochemical transformations, bioturbation, biodeposition, and translocation by macrophytic plants (Gessner, 1960; Kuenzler, 1961; Pomeroy, 1961; Reimold, 1972; Day et al., 1973; Zeitzschel, 1980; Aller, 1982; Kaplan, 1983; Smetacek, 1984; Aller and Aller, 1986). The abundance of phosphate in tidal-flat sediments and its rapid exchange is the main reason that phosphate is seldom limiting to macrophytes or benthic autotrophic communities.

In contrast to the elements sulfur and phosphorus, nitrogen is frequently described as a limiting nutrient for primary producers, such as benthic micro-algae, marsh grasses, and submerged macrophytes (Williams, 1972; Day et al., 1973; Nixon and Oviatt, 1973; Estrada et al., 1974). In intertidal sediments nitrogen may be supplied by nitrogen fixation, runoff, rain, and remineralization of organic matter (Whitney et al., 1975; Haines et al., 1977; Carpenter et al., 1978; Harrison, 1978; Martens et al., 1978; Valiela et al., 1978). Recent work indicates that remineralization is the most important source of nitrogen in sediments (Haines et al., 1977; Nixon, 1981; Blackburn and Henriksen, 1983; Twilley and Kemp, 1986; Carpenter and Capone, 1983; Andersen et al., 1984). Remineralization rates have been found to be greatest at the sediment–water interface with up to 70% of the total nitrogen activity occurring in the top centimeter of sediments (Haines et al., 1977; Andersen et al., 1984).

Nutrient cycling in seagrass beds is affected by (1) inputs of organic material from the plants and epiphytes, (2) the roots of rhizomes which are not only vehicles for translocation of nutrients and oxygen but bind sediments and restrict water flow, thus reducing potential exchange with the overlying medium, and (3) the large numbers of faunal inhabitants (both in and above the sediment surface) that ingest and excrete particulate organic matter and aerate subsurface sediments. All of the above processes result in nutrient concentrations and budgets significantly different from those of unvegetated sediments. This is evident for a *Zostera marina* bed in North Carolina where organic matter, ammonium and total nitrogen concentrations are greater in vegetated sediments, even at depth (Kenworthy et al., 1982) (Fig. 8.9). In addition, Kenworthy et al. (1982) found the largest pools of nitrogen, the finest sediment texture and the greatest organic matter content in sediments associated with the center of seagrass meadows, intermediate at the edges of the bed and small isolated patches of grass, and least in unvegetated substrate. Similar patterns were found for sediment grain size and faunal distribution in the Chesapeake Bay (Orth, 1977a) (Figs. 8.4 and 8.8).

Nitrogen fixation appears to be important in grass beds and has been detected both in anaerobic intact sediments of the rhizosphere of *Z. marina* (Capone, 1983) and aerobically on the surfaces of roots and rhizomes (Capone and Budin, 1982). Capone (1983) estimates that 28% of the nitrogen required by plants in a Long Island *Z. marina* bed may be supplied by nitrogen fixation.

The effect of submerged macrophyte communities on material budgets can vary. Kemp et al. (1983) estimated the participation of plants in the upper Chesapeake Bay in three budgets (Table 8.4) ranging from moderate (carbon and nitrogen) to very large (sediments). The budgets were estimated for the 1960s when submerged macrophytes dominated the shallows of the upper bay. With the recent changes in these plant communities in the bay (Orth and Moore, 1983), many of the important plant interactions have diminished or been lost.

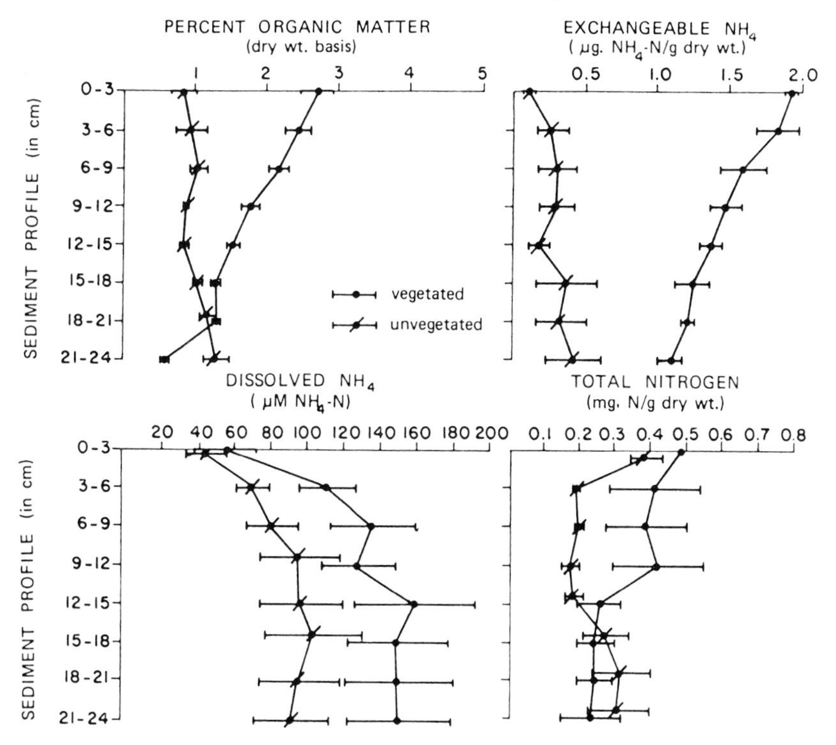

Fig. 8.9. Sediment profiles for organic matter and three nitrogen cycle intermediates in vegetated and unvegetated sediments. (From Kenworthy et al., 1982; reprinted by permission of *Oecologia*.)

TABLE 8.4

Estimated influence of submersed vascular plant communities on selected materials budgets for upper Chesapeake Bay in 1960. (From Kemp et al., 1984; reprinted by permission of Academic Press.)

Sources (or sinks)[a]	Material fluxes[a]	
	Mass flux ($Mt\ y^{-1}$)	Perc. of inputs
Organic carbon		
Rivers	0.8×10^5	11
Phytoplankton	3.8×10^5	56
Macrophytes	2.2×10^5	33
Sediments		
Rivers	2.2×10^6	65
Shore erosion	0.6×10^6	35
(Macrophytes)	(3.6×10^6)	(210)
Nitrogen		
Rivers	50×10^3	92
Sewage	4.2×10^3	8
(Macrophytes)	(3.6×10^3)	(7)

[a] Estimates modified from Boynton (1982), including entire area of upper Bay plus tributaries above Potomac River mouth ($1.5 \times 10^9\ m^2$). Values in parentheses indicate that submersed vascular plant (macrophyte) communities act as material sinks.

ACKNOWLEDGEMENT

Contribution No. 1386 for the Virginia Institute of Marine Science, School of Marine Science, College of William and Mary, Gloucester Point, Va.

REFERENCES

Adams, S.M., 1976a. The ecology of eelgrass, *Zostera marina* (L.), fish communities. I. Structural analysis. *J. Exp. Mar. Biol. Ecol.*, 22: 269–291.

Adams, S.M., 1976b. The ecology of eelgrass, *Zostera marina* (L.), fish communities. II. Functional analysis. *J. Exp. Mar. Biol. Ecol.*, 22: 293–311.

Alheit, J. and Scheibel, W., 1982. Benthic harpacticoids as a source of food for fish. *Mar. Biol.*, 70: 141–147.

Aller, J.Y. and Aller, R.C., 1986. Evidence for localized enhancement of biological activity associated with tube and burrow structures in deep-sea sediments at the HEBBLE site, western North Atlantic. *Deep-Sea Res.*, 33: 755–790.

Aller, R.C., 1982. The effects of macrobenthos on chemical properties of marine sediment and overlying water. In: P.C. McCall and J.S. Tevesz (Editors), *Animal–Sediment Relations*. Plenum Press, New York, pp. 53–102.

Andersen, T.K., Jensen, M.H. and Sprensen, J., 1984. Diurnal variation of nitrogen cycling in coastal, marine sediments. I. Denitrification. *Mar. Biol.*, 83: 171–176.

Anderson, R.R. and Macomber, R.T., 1980. *Distribution of submersed vascular plants, Chesapeake Bay, Maryland.* U.S. EPA. Final Rep., Chesapeake Bay Program, Grant No. R805970, 126 pp.

Baillie, P. and Welsh, B.L., 1980. The effect of tidal resuspension on the distribution of intertidal epibenthic algae in an estuary. *Estuarine Coastal Shelf Sci.*, 10: 165–180.

Bell, S.S., 1980. Meiofauna–macrofauna interactions in a high salt marsh habitat. *Ecol. Monogr.*, 50: 487–505.

Bell, S.S. and Coull, B.C., 1980. Experimental evidence for a model of juvenile macrofauna–meiofauna interactions. In: K.R. Tenore and B.C. Coull (Editors), *Marine Benthic Dynamics*. Univ. of S. Carolina Press, Columbia, S.C., pp. 179–192.

Blackburn, T.H. and Henriksen, K., 1983. Nitrogen cycling in different types of sediment from Danish waters. *Limnol. Oceanogr.*, 28: 477–493.

Blundon, J.A. and Kennedy, V.S., 1982. Refuges for infaunal bivalves from blue crab, *Callinectes sapidus* (Rathbun), predation in Chesapeake Bay. *J. Exp. Mar. Biol. Ecol.*, 65: 67–81.

Botton, M.L., 1984a. Effects of laughing gull and shorebird predation on the intertidal fauna at Cape May, N.J. *Estuarine Coastal Shelf. Sci.*, 18: 209–220.

Botton, M.L., 1984b. The importance of predation by horseshoe crabs, *Limulus polyphemus*, to an intertidal sand flat community. *J. Mar. Res.*, 42: 139–161.

Boynton, W.R., 1982. Ecological role and value of submerged macrophyte communities: a scientific summary. In: E.G. Macalister, D.A. Barker and M. Kasper (Editors), *Chesapeake Bay Program Technical Studies: A Synthesis.* U.S.E.P.A. NTIS, Springfield, Va, pp. 428–502.

Boynton, W.R. and Kemp, W.M., 1985. Nutrient regeneration and oxygen consumption by sediment along an estuarine salinity gradient. *Mar. Ecol. Prog. Ser.*, 2: 45–55.

Brauner, J.F., 1975. Seasonality of epiphytic algae on *Zostera marina* at Beaufort, North Carolina. *Nova Hedwigia*, 26: 125–133.

Brenchley, G.A., 1981. Disturbance and community structure: an experimental study of bioturbation in marine soft-bottom environments. *J. Mar. Res.*, 39: 767–790.

Burkholder, P.R. and Bornside, G.H., 1957. Decomposition of marsh grass by aerobic marine bacteria. *Bull. Torrey Bot. Club*, 84: 366–383.

Burkholder, P.R. and Doheny, T.E., 1968. *The biology of eelgrass.* Contrib. No. 1227, Lamont Geol. Obs., 120 pp.

Capone, D.G., 1983. N_2 fixation in seagrass communities. *Mar. Tech. Soc. J.*, 17: 32–37.

Capone, D.G. and Budin, J.M., 1982. Nitrogen fixation associated with rinsed roots and rhizomes of the eelgrass *Zostera marina. Plant Physiol.*, 70: 1601–1604.

Carpenter, E.J. and Capone, D.G., 1983. *Nitrogen in the Marine Environment.* Academic Press, New York, 900 pp.

Carpenter, E.J., Van Raalte, C.D. and Valiela, I., 1978. Nitrogen fixation in a Massachusetts salt marsh. *Limnol. Oceanogr.*, 23: 318–327.

Chapman, V.J. (Editor), 1977. *Wet Coastal Ecosystems, Ecosystems of the World, Vol. 1.* Elsevier, Amsterdam, 428 pp.

Christian, R.R. and Wetzel, R.L., 1978. Interaction between substrate, microbes, and consumers of *Spartina* detritus in estuaries. In: M. Wiley (Editor), *Estuarine Interactions.* Academic Press, New York, pp. 93–113.

Cottam, C., 1934. The eelgrass shortage in relation to waterfowl. *Am. Game Conf. Trans.*, 20: 272–279.

Coull, B.C. and Fleeger, J.W., 1977. Long-term variation and community dynamics of meiobenthic copepods. *Ecology*, 58: 1136–1143.

Coull, B.C., Bell, S.S., Savory, A.M. and Dudley, B.W., 1979. Zonation of meiobenthic copepods in a southeastern United States salt marsh. *Estuarine Coastal Mar. Sci.*, 9: 181–188.

Croker, R.A., 1970. Intertidal sand macrofauna from Long Island, New York. *Chesapeake Sci.*, 11: 134–137.

Davis, G.J. and Brinson, M.M., 1976. *The submersed macrophytes of the Pamlico River estuary, North Carolina.* Rep. No. 112, Univ. North Carolina Water Resources Res. Inst., 202 pp.

Day, J.W., Smith, W.G., Wagner, P. and Stowe, W.C., 1973. *Community structure and carbon budget of a salt marsh and shallow bay estuarine system in Louisiana.* Centr. Wetlands Resour. Publ. No. LSU-SG-72–04, Louisiana State Univ., Baton Rouge, La.

Dayton, P.K., 1984. Processes structuring some marine communities: are they general? In: D.R. Strong, Jr., D. Simberloff, L.G. Abele and A.B. Thistle (Editors), *Ecological Communities.* Princeton Univ. Press, Princeton, N.J., pp. 181–200.

De la Cruz, A.A. and Gabriel, B.C., 1974. Caloric, elemental and nutritive changes in decomposing *Juncus roemerianus* leaves. *Ecology*, 55: 882–886.

Delaune, R.D., Patrick Jr., W.H. and Brannon, J.M., 1976. *Nutrient transformation in Louisiana salt marsh soils.* Central Wetlands Resour. Publ. No. LSU-T-76–009, Louisiana State Univ., Baton Rouge, La., 37 pp.

Den Hartog, C., 1970. *The Seagrasses of the World.* North-Holland, Amsterdam, 275 pp.

Dexter, D.M., 1967. Distribution and niche diversity of haustoriid amphipods in North Carolina. *Chesapeake Sci.*, 8: 187–192.

Dexter, D.M., 1969. Structure of an intertidal sandy beach community in North Carolina. *Chesapeake Sci.*, 10: 93–98.

Diaz, R.J. and DeAlteris, J.T., 1982. *Long-term changes in beach fauna at Duck, North Carolina.* Misc. Rep. 82–12, U.S. Army Corps of Engineers, Coastal Engineering Research Center, Fort Belvoir, Va., 48 pp.

Diaz, R.J. and Fredette, T.J., 1982. Secondary production of some dominant macroinvertebrate species inhabiting a bed of submerged vegetation in the lower Chesapeake Bay. In: R.L. Wetzel, K.L. Webb, P.A. Penhale, R.J. Orth, D.F. Boesch, G.W. Boehlert and J.V. Merriner (Editors), *Structural and functional aspects of the biology of submerged aquatic macrophyte communities in the lower Chesapeake Bay.* Final Rep. Environmental Protection Agency/Chesapeake Bay Program, Grant No. R805974, pp. 68–79.

Diaz, R.J., Markwith, G., Orth, R.J., Rizzo, W. and Wetzel, R., 1982. *Examination of tidal flats, Vol. 1.* Research Rep.

FHWA/RD-80/181. Fed. Highway Admin. Washington, D.C., 76 pp.

Dörjes, J., 1977. Marine macrobenthic communities of the Sapelo Island, Georgia, region. In: B.C. Coull (Editor), *Ecology of Marine Benthos*. Univ. of S. Carolina Press, Columbia, S.C., pp. 399–421.

Eckman, J.E., 1987. The role of hydrodynamics in recruitment, growth, and survival of *Argopecten irradians* (L.) and *Anomia simplex* (D'Orbegny) within eelgrass meadows. *J. Exp. Mar. Biol. Ecol.*, 106: 165–191.

Eckman, J.E., Nowell, A.R.M. and Jumars, P.A., 1981. Sediment destabilization by animal tubes. *J. Mar. Res.*, 39: 361–374.

Emery, K.O. and Uchupi, E., 1972. *Western North Atlantic Ocean: topography, rocks, structure, water, life and sediments.* Memoir 17, Am. Ass. Petroleum Geologists, Tulsa, Okla., 532 pp.

Estrada, M., Valiela, I. and Teal, J.M., 1974. Concentration and distribution of chlorophyll in fertilized plots in a Massachusetts salt marsh. *J. Exp. Mar. Biol. Ecol.*, 14: 47–56.

Fenchel, T., 1970. Studies on the decomposition of organic detritus derived from the turtle grass *Thalassia testudinum*. *Limnol. Oceanogr.*, 15: 15–20.

Fenchel, T., 1978. The ecology of micro- and meiobenthos. *Annu. Rev. Ecol. System.*, 9: 99–121.

Folger, D.W., 1972. *Characteristics of estuarine sediments of the United States.* Prof. Paper No. 742, U.S. Geol. Survey, 94 pp.

Fonseca, M.S., Fisher, J.S., Zieman, J.C. and Thayer, G.W., 1982. Influence of the seagrass *Zostera marina* L. on current flow. *Estuarine Coastal Shelf Sci.*, 15: 351–364.

Frankenburg, D. and Smith Jr., K.L., 1967. Coprophagy in marine animals. *Limnol. Oceanogr.*, 12: 443–449.

Fredette, T.J. and Diaz, R.J., 1986. Secondary production of the crustacean *Gammarus mucronatus* Say (Amphipoda: Gammaridae) in warm temperate estuarine habitats, York River, Virginia. *J. Crustacean Biol.*, 6: 729–741.

Gallagher, J.L. and Daiber, F.C., 1974. Primary production of edaphic algal communities in a Delaware salt marsh. *Limnol. Oceanogr.*, 19: 390–395.

Gessner, F., 1960. Untersuchungen uber den phosphathaushalt des Amazonas. *Int. Rev. Gesamten Hydrobiol. Hydrogr.*, 45: 339–345.

Gosselink, J.G. and Kirby, C.J., 1974. Decomposition of salt-marsh grass, *Spartina alterniflora*. *Limnol. Oceanogr.*, 19: 825–832.

Gosselink, J.G., Odum, E.P. and Pope, R.M., 1974. *The value of a tidal marsh.* Urban and Regional Development Center, Univ. of Florida, Gainesville, Fla.

Gray, J.S., 1974. Animal–sediment relationships. *Oceanogr. Mar. Biol. Ann. Rev.*, 12: 223–261.

Haines, E., Chalmers, A., Hanson, R. and Sherr, B., 1977. Nitrogen pools and fluxes in a Georgia salt marsh. In: M. Wiley (Editor), *Estuarine Processes. II. Circulation, Sediments and Transfer of Material in the Estuary.* Academic Press, New York, pp. 241–254.

Hargrave, B.T., 1970a. The utilization of benthic microflora by *Hyalella azteca* (Amphipoda). *J. Animal Ecol.*, 39: 427–437.

Hargrave, B.T., 1970b. The effect of a deposit feeding amphipod on the metabolism of benthic microflora. *Limnol. Oceanogr.*, 15: 21–30.

Harlin, M.M., 1973. Transfer of products between epiphytic marine algae and host plants. *J. Phycol.*, 9: 243–248.

Harlin, M.M., 1980. Seagrass epiphytes. In: R.C. Phillips and C.P. McRoy (Editors), *Handbook of Seagrass Biology, An Ecosystem Perspective.* Garland STPM Press, New York, pp. 117–151.

Harrison, S.C., 1978. Experimental measurements of nitrogen remineralization in coastal waters. *Limnol. Oceanogr.*, 23: 684–694.

Heck Jr., K.L. and Orth, R.J., 1980. Structural components of eelgrass (*Zostera marina*) meadows in the lower Chesapeake Bay — Decapod crustacea. *Estuaries*, 3: 289–295.

Heck Jr., K.L. and Thoman, T.A., 1984. The nursery role of seagrass meadows in the upper and lower reaches of the Chesapeake Bay. *Estuaries*, 7: 70–92.

Heck Jr., K.L. and Wilson, K.A., 1987. Predation rates on decapod crustaceans in latitudinally separated seagrass communities: a study of spatial and temporal variation using tethering techniques. *J. Exp. Mar. Biol. Ecol.*, 107: 87–100.

Holland, A.F. and Dean, J.M., 1977. The community biology of intertidal macrofauna inhabiting sand bars in the North Inlet area, South Carolina. In: B.C. Coull (Editor), *Ecology of Marine Benthos*. Univ. of S. Carolina Press, Columbia, S.C., pp. 423–438.

Holland, A.F. and Polgar, T.T., 1976. Seasonal changes in the structure of an intertidal community. *Mar. Biol.*, 37: 341–348.

Holland, A.F., Mountford, N.K., Hiegel, M.H., Kaumeyer, K.R. and Mihursky, J.A., 1980. Influence of predation on infaunal abundance in upper Chesapeake Bay (USA). *Mar. Biol.*, 57: 221–235.

Homziak, J., Fonseca, M.S. and Kenworthy, W.J., 1982. Macrobenthic community structure in a transplanted eelgrass (*Zostera marina*) meadow. *Mar. Ecol. Prog. Ser.*, 9: 211–221.

Howard, J.W. and Dörjes, J., 1972. Animal–sediment relationships in two beach-related tidal flats; Sapelo Island, Georgia. *J. Sediment. Petrol.*, 42(3): 608–623.

Humm, H.J., 1969. Distribution of marine algae along the Atlantic coast of North America. *Phycol*, 7: 43–53.

Jacobs, R.P.W.M., 1984. Biomass potential of eelgrass (*Zostera marina* L.). *Crit. Rev. Plant Sci.*, 2: 49–80.

Johnson, R.G., 1965. Temperature variation in the infaunal environment of a sand flat. *Limnol. Oceanogr.*, 10: 114–120.

Kaplan, W.A., 1983. Nitrification. In: E.J. Carpenter and D.G. Capone (Editors), *Nitrogen in the Marine Environment.* Academic Press, New York, pp. 139–190.

Kemp, W.M., Boynton, W.R., Twilley, R.R., Stevenson, J.C. and Means, J.C., 1983. The decline of submerged vascular plants in the upper Chesapeake Bay: Summary of results concerning possible causes. *Mar. Tech. Soc. J.*, 17: 78–89.

Kemp, W.M., Boynton, W.R., Twilley, R.R., Stevenson, J.C. and Ward, L.G., 1984. Influences of submersed vascular plants on ecological processes in upper Chesapeake Bay.

In: V.S. Kennedy (Editor), *The Estuary as a Filter*. Academic Press, New York, pp. 367–394.

Kenworthy, W.J., Zieman, J.C. and Thayer, G.W., 1982. Evidence for the influence of seagrasses on the benthic nitrogen cycle in a coastal plain estuary near Beaufort, North Carolina (USA). *Oecologia*, 54: 152–158.

Kneib, R.T., 1985. Predation and disturbance by grass shrimp, *Palaemonetes pugio* Holthuis, in soft-substratum benthic invertebrate assemblages. *J. Exp. Mar. Biol. Ecol.*, 93: 91–102.

Kneib, R.T. and Stiven, A.E., 1982. Benthic invertebrate responses to size and density manipulations of the common mummichog, *Fundulus heteroclitus*, in an intertidal salt marsh. *Ecology*, 63: 1518–1532.

Kuenzler, E.J., 1961. Structure and energy flow of a mussel population in a Georgia salt marsh. *Limnol. Oceanogr.*, 11: 191–197.

Leach, J.H., 1970. Epibenthic algal production in an intertidal mud flat. *Limnol. Oceanogr.*, 15: 514–521.

Leber, K.M., 1982. Seasonality of macroinvertebrates on a temperate, high wave energy sandy beach. *Bull. Mar. Sci.*, 32: 86–98.

Levinton, J.S., 1977. Ecology of shallow water deposit-feeding communities, Quisset Harbor, Massachusetts. In: B.C. Coull (Editor), *Ecology of Marine Benthos*. Univ. of S. Carolina Press, Columbia, S.C., pp. 191–228.

Luckenbach, M.W., 1986. Sediment stability around animal tubes: the roles of hydrodynamic processes and biotic activity. *Limnol. Oceanogr.*, 31: 779–787.

Luckenbach, M.W., 1987. Effects of adult infauna on new recruits: implications for the role of biogenic structures. *J. Exp. Mar. Biol. Ecol.*, 105: 197–206.

Lucy, J.A., 1976. *The reproductive cycle of Mya arenaria L. and distribution of juvenile clams in the upper portion of the nearshore zone of the York River, Virginia*. M.A. Thesis, College of William and Mary, Williamsburg, Va., 131 pp.

Mann, K.H., 1973. Seaweeds: Their productivity and strategy for growth. *Science*, 182: 975–981.

Marsh, G.A., 1970. *A seasonal study of Zostera epibiota in the York River*, Virginia. Ph.D. Dissertation. College of William and Mary, Williamsburg, Va., 155 pp.

Marsh, G.A., 1973. The *Zostera* epifaunal community in the York River, Virginia. *Chesapeake Sci.*, 14: 87–97.

Marsh, G.A., 1976. Ecology of the gastropod epifauna of eelgrass in a Virginia estuary. *Chesapeake Sci.*, 17: 182–187.

Marshall, N., Oviatt, C.A. and Skaven, D.M., 1971. Productivity of the benthic microflora of shoal estuarine environments in southern New England. *Int. Rev. Gesamten Hydrobiol. Hydrogr.*, 56: 947–956.

Martens, C.S., Berner, R.A. and Rosenfeld, J.K., 1978. Interstitial water chemistry of anoxic Long Island Sound sediments. 2. Nutrient regeneration and phosphate removal. *Limnol. Oceanogr.*, 23: 605–617.

Matera, N.J. and Lee, J.J., 1972. Environmental factors affecting the standing crop of Foraminifera in sublittoral and psammolittoral communities of a Long Island salt marsh. *Mar. Biol.*, 14: 89–103.

Matta, J.F., 1977. *Beach faunal study of the CERC field research facility, Duck, North Carolina*. Misc. Rep. 77–6. U.S. Corps of Engineers, Coastal Engr. Res. Cntr., Fort Belvoir, Va., 48 pp.

Maurer, D. and Aprill, G., 1979. Intertidal benthic invertebrates and sediment stability at the mouth of Delaware Bay. *Int. Rev. Gesamten Hydrobiol. Hydrogr.*, 64: 379–403.

McCall, P., 1977. Community patterns and adaptive strategies of the infaunal benthos of Long Island Sound. *J. Mar. Res.*, 35: 221–266.

McDermott, J.J., 1983. Food web in the surf zone of an exposed sandy beach along the mid-Atlantic coast of the United States. In: A. McLachlan and T. Erasmus (Editors), *Sandy Beaches as Ecosystems*. Junk, The Hague, pp. 529–538.

Mountford, N.K., Holland, A.F. and Mihursky, J.A., 1977. Identification and description of macrobenthic communities in the Calvert Cliffs region of Chesapeake Bay. *Chesapeake Sci.*, 18: 360–369.

Myers, J.P., 1986. Sex and gluttony on Delaware Bay. *Nat. Hist.*, 5/86: 69–78.

Nagle, J.S., 1968. Distribution of the epibiota of macrobenthic plants. *Contr. Mar. Sci.* (Univ. Texas), 13: 105–144.

Nelson, W.G., 1979a. An analysis of structural pattern in an eelgrass (*Zostera marina* L.) amphipod community. *J. Exp. Mar. Biol. Ecol.*, 39: 231–264.

Nelson, W.G., 1979b. Experimental studies of selective predation on amphipods: consequences for amphipod distribution and abundance. *J. Exp. Mar. Biol. Ecol.*, 38: 225–245.

Nelson, W.G., 1980. The biology of eelgrass (*Zostera marina* L.) amphipods. *Crustaceana*, 39: 59–89.

Nixon, S.W., 1981. Remineralization and nutrient cycling in coastal marine ecosystems. In: B.J. Neilson and L.E. Cronin (Editors), *Estuaries and Nutrients*. Humana Press, Clifton, N.J., pp. 111–138.

Nixon, S.W. and Oviatt, C.A., 1973. Ecology of a New England salt marsh. *Ecol. Monogr.*, 43: 463–498.

NOAA, 1986. *Tide Tables, East Coast of North and South America*. U.S. Dept. of Commerce, National Ocean Service, Washington, D.C., 288 pp.

Nowicki, B.L. and Nixon, S.W., 1985. Benthic community metabolism in a coastal lagoon ecosystem. *Mar. Ecol. Prog. Ser.*, 22: 21–30.

Odum, W.E., 1968. The ecological significance of fine particle selection by the striped mullet, *Mugil cephalus*. *Limnol. Oceanogr.*, 13: 92–98.

Oppenheimer, C.H. and Ward, R.A., 1963. Release and capillary movement of phosphorus in exposed tidal sediments. In: C.H. Oppenheimer (Editor), *Symposium on Marine Microbiology*. Charles C. Thomas, Springfield, Ill., pp. 664–673.

Orth, R.J., 1973. Benthic infauna of eelgrass, *Zostera marina*, beds. *Chesapeake Sci.*, 14: 258–269.

Orth, R.J., 1977a. The importance of sediment stability in seagrass communities. In: B.C. Coull (Editor), *Ecology of Marine Benthos*. Univ. of South Carolina Press, Columbia, S.C., pp. 281–300.

Orth, R.J., 1977b. *Zostera marina*. In: J.C. Stevenson and N. Confer (Editors), *Summary of available information on Chesapeake Bay submerged vegetation*, Final Rep. U.S. Fish and Wildlife Ser. 14–16–0008–1255, pp. 58–106.

Orth, R.J. and Heck Jr., K.L., 1980. Structural components of eelgrass (*Zostera marina*) meadows in the lower Chesapeake Bay — fishes. *Estuaries*, 3: 278–288.

Orth, R.J. and Moore, K.A., 1982. *The biology and propagation of Zostera marina, eelgrass, in the Chesapeake Bay, Virginia*, Final Rep. U.S. Environmental Protection Agency, Grant. No. R805953, 187 pp.

Orth, R.J. and Moore, K.A., 1983. Chesapeake Bay: an unprecedented decline in submerged aquatic vegetation. *Science*, 222: 51–53.

Orth, R.J. and Moore, K.A., 1984. Distribution and abundance of submerged aquatic vegetation in Chesapeake Bay: an historical perspective. *Estuarine Bull.*, 7: 531–540.

Orth, R.J. and Van Montfrans, J., 1984. Epiphyte-seagrass relationships with an emphasis on the role of micrograzing: A review. *Aquat. Bot.*, 18: 43–69.

Orth, R.J., Moore, K.A. and Gordon, H.H., 1979. *Distribution and abundance of submerged aquatic vegetation in the lower Chesapeake Bay, Virginia.* U.S. EPA, Fin. Rep., Chesapeake Bay Program. EPA-600/8–79–029/SAV1.

Orth, R.J., Heck Jr., K.L. and Van Montfrans, J., 1984. Faunal communities in seagrass communities: A review of the influence of plant structure and prey characteristics on predator–prey relationships. *Estuarine Bull.*, 7: 339–350.

Pace, M.L., Shimmel, S. and Darley, W.M., 1979. The effect of grazing by a gastropod, *Nassarius obsoletus*, on the benthic microbial community of a salt marsh mudflat. *Estuarine Coastal Mar. Sci.*, 9: 121–134.

Penhale, P.A., 1977. Macrophyte-epiphyte biomass and productivity in an eelgrass (*Zostera marina* L.) community. *J. Exp. Mar. Biol. Ecol.*, 26: 211–224.

Penhale, P.A. and Smith Jr., W.O., 1977. Excretion of dissolved organic carbon by eelgrass (*Zostera marina*) and its epiphytes. *Limnol. Oceanogr.*, 22: 400–407.

Penhale, P.A. and Thayer, G.W., 1980. Uptake and transfer of carbon and phosphorus by eelgrass (*Zostera marina* L.) and its epiphytes. *J. Exp. Mar. Biol. Ecol.*, 42: 113–123.

Penry, D.L., 1982. *Utilization of a Zostera marina and Ruppia maritima habitat by four decapods with emphasis on Callinectes sapidus.* M.A. Thesis, College of William and Mary, Williamsburg, VA, 101 pp.

Peterson, C.H., 1979. Predation, competitive exclusion, and diversity in the soft-bottom benthic communities of estuaries and lagoons. In: R.J. Livingston (Editor), *Ecological Processes in Coastal and Marine Systems.* Plenum, New York, pp. 233–264.

Peterson, C.H., 1982. Clam predation by whelks (*Busucon* spp.): Experimental tests of the importance of prey size, prey density, and seagrass cover. *Mar. Biol.*, 66: 159–170.

Pomeroy, L.R., 1961. Experimental studies of the turnover of phosphate in marine environments. In: V. Schultz and A.W. Klement (Editors), *Radioecology. Proc. 1st Symp. on Radioecology, Washington, D.C.* U.S. Atomic Energy Commission, pp. 163–169.

Pomeroy, L.R., Smith, E.E. and Grant, C.M., 1965. The exchange of phosphate between estuarine waters and sediments. *Limnol. Oceanogr.*, 10: 167–175.

Pomeroy, L.R., Bancroft, K., Breed, J., Christian, R.R., Frankenburg, D., Hall, J.R., Maurer, L.G., Wiebe, W.J., Wiegert, R.G. and Wetzel, R.L., 1977. Flux of organic matter through a salt marsh. In: M. Wiley (Editor), *Estuarine Processes. II. Circulation, sediments and transfer of materials in the estuary.* Academic Press, New York, pp. 270–279.

Quammen, M.L., 1982. Influence of subtle substrate differences on feeding by shorebirds on intertidal mudflats. *Mar. Biol.*, 71: 339–343.

Quamman, M.L., 1984. Predation by shorebirds, fish and crabs on invertebrates in intertidal mudflats: an experimental test. *Ecology*, 65: 529–537.

Reimold, R.J., 1972. The movement of phosphorus through the salt marsh cordgrass, *Spartina alterniflora* (Loisel.). *Limnol. Oceanogr.*, 17: 606–611.

Reimold, R.J., 1977. Mangals and salt marshes of eastern United States. In: V.J. Chapman (Editor), *Wet Coastal Ecosystems, Ecosystems of the World, Vol. 1.* Elsevier, Amsterdam, pp. 157–166.

Rhoads, D.C., 1974. Organism–sediment relations on the muddy sea floor. *Oceanogr. Mar. Biol. Ann. Rev.*, 12: 263–300.

Rizzo, W.M. and Wetzel, R.L., 1985. Intertidal and shoal benthic community metabolism in a temperate estuary: Studies of spatial and temporal scales of variability. *Estuaries*, 8: 342–351.

Robertson, A.I., 1979. The relationship between annual production: biomass ratios and life spans for marine macrobenthos. *Oecologia*, 38: 193–202.

Sanders, H.L., 1956. Oceanography of Long Island Sound 1952–1954. X. The biology of marine bottom communities. *Bull. Bingham Oceanogr. Collect.*, 15: 345–414.

Schneider, D. and Harrington, B.A., 1981. Timing of shorebird migration in relation to prey depletion. *Auk*, 98: 801–811.

Schultz, D.M. and Quinn, J.G., 1973. Fatty acid composition of organic detritus from *Spartina alterniflora. Estuarine Coastal Mar. Sci.*, 1: 177–190.

Sieburth, J.Mc. and Thomas, D., 1973. Fouling on eelgrass (*Zostera marina* L.). *J. Phycol.*, 9: 46–50.

Silberhorn, G.M., Orth, R.J. and Moore, K.A., 1983. Anthesis and seed production in *Zostera marina* L. (eelgrass) from the Chesapeake Bay. *Aquat. Bot.*, 15: 133–144.

Smetacek, V., 1984. The supply of food to the benthos. In: M.J. Fasham (Editor), *Flows of Energy and Materials in Marine Ecosystems: Theory and Practice.* NATO Adv. Res. Inst., Washington, D.C., pp. 517–547.

Smith, L.D. and Coull, B.C., 1987. Juvenile spot (Pisces) and grass shrimp predation on meiobenthos in muddy and sandy substrata. *J. Exp. Mar. Biol. Ecol.*, 105: 123–136.

Smith Jr., W.O. and Penhale, P.A., 1980. The heterotrophic uptake of dissolved organic carbon by eelgrass (*Zostera marina* L.) and its epiphytes. *J. Exp. Mar. Biol. Ecol.*, 48: 233–242.

Squires, E.R. and Good, R.E., 1974. Seasonal changes in the productivity, caloric content, and chemical composition of a population of salt marsh cord grass (*Spartina alterniflora*). *Chesapeake Sci.*, 15: 63–71.

Summerson, H.C. and Peterson, C.H., 1984. Role of predation in organizing benthic communities of a temperate-zone seagrass bed. *Mar. Ecol. Prog. Ser.*, 15: 63–77.

Sutherland, W.J., 1982. Spatial variation in the predation of cockles by oystercatchers at Treath Melynog, Anglesey. II. The pattern of mortality. *J. Anim. Ecol.*, 51: 491–500.

Taylor, W.R., 1957. *Marine Algae of the Northeastern Coast of*

North America. The Univ. of Michigan Press, Ann Arbor, Mich., 509 pp.
Teal, J.M., 1962. Energy flow in the salt marsh ecosystem of Georgia. *Ecology*, 43: 614–624.
Tenore, K.R. and Rice, D.L., 1980. A review of trophic factors affecting secondary production of deposit feeders. In: K.R. Tenore and B.C. Coull (Editors), *Marine Benthic Dynamics*. University of South Carolina Press, Columbia, S.C., pp. 325–340.
Thayer, G.W., Adams, S.M. and LaCroix, M.W., 1975. Structural and functional aspects of a recently established *Zostera marina* community. *Estuarine Res.*, 1: 518–540.
Thayer, G.W., Engel, D.W. and LaCroix, M.W., 1977. Seasonal distribution and changes in the nutritive quality of living, dead and detrital fractions of *Zostera marina* L. *J. Exp. Mar. Biol. Ecol.*, 30: 109–127.
Thayer, G.W., Parker, P.L., LaCroix, M.W. and Fry, B., 1978. The stable carbon isotope ratio of some components of an eelgrass *Zostera marina* bed. *Oecologia*, 35: 1–12.
Thayer, G.W., Kenworthy, W.J. and Fonseca, M.S., 1984. *The ecology of eelgrass meadows of the Atlantic Coast: A community profile*. U.S. Fish and Wildlife Ser. FWS/OBS-84/02, 147 pp.
Thorne-Miller, B., Harlin, M.M., Thursby, G.B., Brady-Campbell, M.M. and Dworetzky, B.A., 1983. Variations in the distribution and biomass of submerged macrophytes in five coastal lagoons in Rhode Island, U.S.A. *Bot. Mar.*, 26: 231–242.
Thursby, G.B. and Harlin, M.M., 1984. Interaction of leaves and roots of *Ruppia maritima* in the uptake of phosphate, ammonia and nitrate. *Mar. Biol.*, 83: 61–67.
Tietjen, J.H., 1969. The ecology of shallow water meiofauna in two New England estuaries. *Oecologia*, 2: 251–291.
Twilley, R.R. and Kemp, W.M., 1986. The relation of denitrification potentials to selected physical and chemical factors in sediments of Chesapeake Bay. In: D. Wolfe (Editor), *Estuarine Variability*. Academic Press, New York, pp. 277–293.
Valiela, I., Teal, J.M., Volkman, S., Shafer, D. and Carpenter, E.J., 1978. Nutrient and particulate fluxes in a salt marsh ecosystem: Tidal exchanges and inputs by precipitation and groundwater. *Limnol. Oceanogr.*, 23: 798–812.
Van Montfrans, J., Orth, R.J. and Vay, S., 1982. Preliminary studies of grazing by *Bittium varium* on eelgrass periphyton. *Aquat. Bot.*, 14: 75–89.
Van Montfrans, J., Wetzel, R.L. and Orth, R.J., 1984. Epiphyte–grazer relationships in seagrass meadows: consequences for seagrass growth and production. *Estuaries*, 7: 289–309.
Virnstein, R.W., 1977. The importance of predation by crabs and fishes on benthic fauna in Chesapeake Bay. *Ecology*, 58: 1199–1217.
Wander, W. and Dunne, P., 1981. *Species and numbers of shorebirds on the Delaware bayshore of New Jersey*. Occ. Paper 140, N.J. Audubon Soc., Records of New Jersey birds, 7: 59–64.
Ward, L.G., Kemp, W.M. and Boynton, W.R., 1984. The influence of waves and seagrass communities on suspended sediment dynamics in an estuarine embayment. *Mar. Geol.*, 59: 85–103.
Warwick, R.M. and Price, J.R., 1975. Macrofauna production in an estuarine mud-flat. *J. Mar. Biol. Ass. U.K.*, 55: 1–18.
Warwick, R.M., George, C.L. and Davies, J.R., 1978. Annual macrofauna production in a *Venus* community. *Estuarine Coastal Mar. Sci.*, 7: 215–241.
Weinstein, M.P. and Brooks, H.A., 1983. Comparative ecology of nekton residing in a tidal creek and adjacent seagrass meadow: community composition and structure. *Mar. Ecol. Prog. Ser.*, 12: 15–27.
Wetzel, R.L., 1977. Carbon resources of a benthic salt marsh invertebrate *Nassarius obsoletus* Say (Mollusca: Nassariidae). In: M. Wiley (Editor), *Estuarine Processes, Vol. II. Circulation, Sediments and Transfer of Material in the Estuary*. Academic Press, New York, pp. 293–308.
Wetzel, R.L. and Penhale, P.A., 1983. Production ecology of seagrass communities in the lower Chesapeake Bay. *Mar. Tech. Soc. J.*, 17: 22–31.
Whitney, D.E., Woodwell, G.M. and Howarth, R.W., 1975. Nitrogen fixation in Flax Pond: A Long Island salt marsh. *Limnol. Oceanogr.*, 20: 640–643.
Wilkins, E.W., 1982. *Waterfowl utilization of a submerged vegetation (Zostera marina and Ruppia maritima) bed in the lower Chesapeake Bay*. M.A. Thesis, College of William and Mary, Williamsburg, Va., 83 pp.
Williams, A.B., 1984. *Shrimps, lobsters and crabs of the Atlantic coast*. Smithsonian Institution Press, Washington, D.C., 550 pp.
Williams, R.B., 1972. *Nutrient levels and phytoplankton productivity in the estuary, Coastal Marsh and Estuary Symposium*. Louisiana State University, Baton Rouge, La., July 17–18, 1972.
Wilson, J.O., Valiela, I. and Swain, T., 1985. Sources and concentrations of vascular plant material in sediments of Buzzards Bay, Massachusetts, U.S.A. *Mar. Biol.*, 90: 129–137.
Wilson, K.A., Heck Jr., K.L. and Able, K.W., 1987. Juvenile blue crab, *Callinectes sapidus* survival: an evaluation of eelgrass, *Zostera marina*, as refuge. *Fish. Bull.*, 85: 53–58.
Wilson, W.H., 1981. Sediment-mediated interactions in a densely populated infaunal assemblage: the effects of the polychaete *Abarenicola pacifica*. *J. Mar. Res.*, 39: 735–748.
Woodin, S.A., 1976. Adult-larval interactions in dense infaunal assemblages: patterns of abundance. *J. Mar. Res.*, 34: 25–41.
Woodin, S.A., 1978. Refuges, disturbance and community structure: a marine soft-bottom example. *Ecology*, 59: 274–284.
Woodin, S.A., 1981. Disturbance and community structure in a shallow water sand flat. *Ecology*, 62: 1052–1066.
Yingst, I.Y. and Rhoads, D.C., 1980. The role of bioturbation in the enhancement of bacterial growth rates in marine sediments. In: K.R. Tenore and B.C. Coull (Editors), *Marine Benthic Dynamics*. Univ. of S. Carolina Press, Columbia, S.C., pp. 407–421.
Zeitzschel, B., 1980. Sediment-water interactions in nutrient dynamics. In: K.R. Tenore and B.C. Coull (Editors), *Marine Benthic Dynamics*. Univ. of S. Carolina Press, Columbia, S.C., pp. 195–218.

Chapter 9

THE TROPICAL WESTERN ATLANTIC INCLUDING THE CARIBBEAN SEA

CLINTON J. DAWES, EARL D. McCOY and KENNETH L. HECK Jr.

INTRODUCTION

The Caribbean region is the geographical area encompassing the coasts of Colombia and Venezuela in South America, the east coast of Central America from Panama north to the Yucatán Peninsula, the Caribbean Sea and the Caribbean island arc. The Florida Keys and Bermuda, situated in the Florida Current, are also included in this review because the western tropical Atlantic region, as defined by the 20°C isotherm in winter (Van den Hoek, 1984), includes not only the area which lies within the Tropic of Cancer, including the Caribbean Sea, but also areas under the influence of the Gulf Stream and the Florida Current.

PHYSICAL CHARACTERISTICS

The Caribbean Sea is a small sea or irregular channel lying between latitudes 23° and 10°N and bounded virtually on all sides by land, with the eastern end (60°W longitude) opening toward the tropical part of the North Atlantic (western Sargasso Sea) and the western end (85°W longitude) opening into the Gulf of Mexico between the Yucatán Peninsula and Cuba. The northern and southern margins of the Caribbean are formed by belts of strongly folded sedimentary rocks. The western margin consists of a complex of folded sediments and volcanic chains of Central America, while the eastern margin is the predominantly volcanic chain of the Lesser Antilles (West Indian) island arc. The types of hard substrata available in the littoral region thus range from black igneous rock to a variety of light-colored sedimentary and coral limestones. The Caribbean topography is rugged with deep (to 7200 m) trenches adjacent to steep-sided ridges, many of which form islands. Because seawater flowing into or out of the Caribbean basin must pass over a sill about 250 m below sea level, the sea is fairly isolated from the influence of deep water of the tropical North and South Atlantic.

The tropical western Atlantic experiences heavy summer rainfall and a relatively dry winter. Open-ocean salinities range from 32 to 37‰ and water transparency is high in the central Caribbean. Surface water temperatures range from 20 to 25°C in winter and from 28 to 30°C in summer in central and southern Caribbean waters (Vroman, 1968). Northern waters have lower winter (15°C) and higher summer (31–33°C) temperatures over shallow shelves and reefs (Dawes et al., 1974).

Upwelling occurs in various portions of the tropical western Atlantic. An upwelling off the north coast of Isla Margarita (Venezuela) results in lower, stable surface-water temperatures of 19 to 23°C, and an upward shift of seaweeds and invertebrates into the littoral zone (Diaz-Piferrer, 1969).

FLORISTIC STUDIES

The first record of marine algae from the Caribbean was published by Sloane (1707–1725) based on collections made in Jamaica in 1687. Species lists published since this time have been reviewed by Taylor (1960) and by Papenfuss (1972). Floras published since the work by Taylor (1960) are listed in Table 9.1. Diaz-Piferrer (1969) published a list of algae for the entire Caribbean including 223 genera and 673 species, comprising

TABLE 9.1

Floristic studies (species numbers) of algae of the western tropical Atlantic since the work by Taylor (1960)

Location	Green algae	Brown algae	Red algae	Total
General Caribbean				
Diaz-Piferrer (1969)	221	103	496	820
South America				
Colombia				
Brattström (1980)	12	10	38	60
Kapraun (1972)	14	6	13	33
Venezuela				
Diaz-Piferrer (1970)	36	20	71	127
De Rios (1972)	42	26	97	165
Central America				
Yucatán Peninsula				
Taylor (1972)	33	11	35	79
Belize				
Norris and Bucher (1982)	52	23	90	165
Tsuda and Dawes (1974)	39	19	34	92
Guatemala				
Bird and McIntosh (1979)	6	1	20	27
Nicaragua				
Phillips et al. (1982)	34	17	46	97
Panama				
Earle (1972)	38	20	60	118
Island Arc				
Cuba				
Diaz-Piferrer (1964a)	22	11	52	317[a]
Jamaica				
Chapman (1961, 1963)	123	47	161	331
Dominican Republic				
Diaz-Piferrer (1964b)	63	39	121	223[a]
Williams et al. (1983)	28	10	46	84
Puerto Rico				
Diaz-Piferrer (1963, 1970)	27	15	61	311
Almodovar and Ballantine (1983)	113	53	218	384
Netherlands Antilles				
Diaz-Piferrer (1964c)	2	7	13	239[a]
Vroman (1968)	73	34	111	218
Northern sites				
Bermuda				
Taylor and Bernatowicz (1969)	55	34	91	180
Thomas (1985)				
Content Key, Florida				
Croley and Dawes (1970)	79	29	150	258

[a]Totals include previous collections by other investigators.

81% of the total algal flora of the tropical western Atlantic as listed by Taylor (1960), excluding members of the Cyanophyta. Of the 673 species, 15.7% are limited to the Caribbean, 24.8% range northward, 6.7% range southward, and 34.1% show wide geographical distribution patterns to the north and south.

Phytogeographic studies indicate that the Caribbean and tropical western Atlantic have the richest algal flora of the entire Atlantic Ocean (Van den Hoek, 1975, 1984; Michanek, 1979). Lawson (1978) noted high endemism (55%) and a much richer algal flora in the tropical western Atlantic than in the tropical eastern Atlantic (see also

Ch. 12). He suggested that the rise of the Isthmus of Panama in the late Pliocene allowed three million years for separation and evolution of a distinct algal flora and fish fauna (1% in common with Pacific fish) and a widely divergent invertebrate fauna (cf. Briggs, 1974).

Taxonomic relationships among the algae of the Caribbean and other areas were examined by comparing the actual numbers of species held in common among several locations with the numbers expected by chance (see Raup and Crick, 1979). Species lists were gathered for Bermuda (Taylor and Bernatowicz, 1969), western Florida (Dawes, 1974), Content Key, Florida (Croley and Dawes, 1970), Puerto Rico (Almodovar and Ballantine, 1983), Carrie Bow Cay, Belize (Norris and Bucher, 1982), the Atlantic coast of Panama (Earle, 1972), the Pacific coast of Panama (Earle, 1972), and Venezuela (Taylor, 1960; Diaz-Piferrer, 1970). Expected commonality between two locations was calculated by drawing a sample from the total species pool equivalent in size to the actual number of species at the first location, replacing the species, drawing a sample equivalent in size to the actual number of species at the second location, and tabulating how many species were selected twice.

The results presented in Table 9.2 show that the flora from the Pacific coast in Panama is distinct from the others, and that the remaining seven locations are more similar to one another than would be "expected". A relationship exists among the algal floras from Bermuda through the Caribbean and supports the phytogeographic hypothesis of Van den Hoek (1984) and Lawson (1978).

FAUNISTIC STUDIES

Owing to the remarkable richness of animal life, complete faunal lists for Caribbean locations are exceedingly rare. A comprehensive volume on the biota of the Bermuda islands (Sterrer, 1986) is available, as well as general handbooks for several Caribbean locations. More commonly, selected taxa, usually macro-invertebrates or fishes, are studied. Many Caribbean taxa appear to be closely related to elements of the eastern Pacific fauna, often as twin species, owing to the relatively short time they have been separated by the Isthmus of Panama (Glynn, 1972; Briggs, 1974). The western Atlantic is more diverse than the tropical eastern Atlantic and Pacific Oceans, but is usually less diverse than the Indo-Pacific Oceans, as discussed by Abele (1972) for crustacean taxa and by McCoy and Heck (1976) for plants, invertebrates, and fish.

COMMUNITIES

The major littoral communities can be identified by their relationship to substratum types and salinity regimes. Seagrass communities are dealt with in a separate section of this review (p. 223); mangrove and coral-reef communities are dealt

TABLE 9.2

Actual numbers of species of algae held in common (first entry) and expected numbers of species held in common (second entry) among eight locations

Location	1	2	3	4	5	6	7	8
Bermuda (1)	–	102/81	106/80	143/120	76/49	67/36	14/22	99/81
West Florida (2)		–	137/109	184/163	78/67	74/49	17/30	124/110
Content Key (3)			–	187/162	85/66	76/49	15/30	116/109
Puerto Rico (4)				–	122/99	110/74	19/45	189/164
Belize (5)					–	65/30	12/18	71/67
Atlantic coast, Panama (6)						–	15/13	65/50
Pacific coast, Panama (7)							–	12/30
Venezuela (8)								–

with in other volumes of the series (Chapman, 1977; Dubinsky, 1990).

Hard substrata and zonation

The tidal range is small in much of the Caribbean region: 29 cm in Colombia (Brattström, 1980), 30 cm in Puerto Rico (Vroman, 1968), 65 cm on the Bahama Bank (Newell et al., 1959), 70 cm in Bermuda (Thomas, 1985), and 45 cm in the Florida Keys (Dawes et al., 1974). On the other hand, the tidal range is over 300 cm on the east coast of Florida and the outer Bahama Islands, and 70 to 110 cm in Barbados (Lewis, 1960). Studies of the zonation pattern on hard substrata on the coasts of southern Florida include Soldier's Key in Biscayne Bay (Voss and Voss, 1954) and some of the Florida Keys (Croley and Dawes, 1970; Stephenson and Stephenson, 1950). Zonation studies (Table 9.3) have also been published for the Bahama Bank (Newell et. al., 1959), the northern Lesser Antilles (Vroman, 1968), Barbados (Lewis, 1960), Curaçao (Van den Hoek et al., 1975, 1978), Isla Margarita (Rodriguez, 1959), and Colombia (Brattström, 1980). For the purposes of this review, the intertidal and subtidal zones will be divided into the supralittoral, eulittoral, and sublittoral.

A study of rocky-shore zonation carried out at nine sites on the Colombian coast showed that the eight zones (including maritime) shifted vertically among sites in response to the amount of wave action (Brattström, 1980). Biological boundaries of the supralittoral and eulittoral zones were not identified precisely because of seasonal variation in wave activity, rainfall, and insolation. Brattström (1980) concluded that a consistent pattern existed between sites within the Caribbean and western tropical Atlantic similar to that described by Stephenson and Stephenson (1950).

The zonation patterns on hard substrata described by Croley and Dawes (1970) for Content Key, Florida, by Newell et al. (1959) for the Great Bahama Bank, and by Brattström (1980) for Santa Maria, Colombia were compared. We tested the proposition that the distribution of organisms coincided with the zones identified by these workers (that is, that the overlap in distributions of organisms among zones was less than expected by chance), using the technique of McCoy et al. (1986). Although a few zones overlapped most did not, indicating that the overall pattern of presences and absences of species was not dictated by zonation pattern. We also tested the proposition that the distributions of organisms along gradients were not independent (that is, distributions overlapped more than expected by chance) for the Great Bahama Bank, using the technique of Pielou (1977). The actual overlap could not be distinguished from the expected overlap under the conditional hypothesis [see Pielou (1977) for details], indicating that distributions of species are largely independent of one another.

It appears that, in many cases, zonation "patterns" are described by physical characteristics, presence of "indicator species", and other qualitative means. It is sometimes difficult even to be sure of the exact criteria used for identification of zones. Quantitative approaches, which should be standard practice, do not often seem to be employed, although conspicuous exceptions exist (see for instance, Van den Hoek et al., 1975, 1978).

Supralittoral. The supralittoral or spray community of the tropical western Atlantic is subject to wave action, and thus extensive development is limited to exposed coasts [the "littoral fringe" of the littoral zone (Croley and Dawes, 1970); the "upper eulittoral" (Van den Hoek, 1975; Van den Hoek et al., 1978); the "littoral margin" (Brattström, 1980)]. The grey and black regions described by Stephenson and Stephenson (1950) for this zone result from the presence of endo- and epilithic blue-green algae. Macroscopic algae of the supralittoral zone can include filamentous species of *Bostrychia*, *Catenella*, *Murrayella* and *Polysiphonia*. Littorinoid gastropods (*Littorina* spp., *Nerita* spp., *Nodilittorina tuberculata*, *Tectarius muricatus*) and isopods (*Ligia baudiniana*) are characteristic of the supralittoral or black zone (Stephenson and Stephenson, 1950; Newell et al., 1959; Van den Hoek, 1975; Brattström, 1980; Thomas, 1985).

Eulittoral. The eulittoral zone is usually divided into an upper, more desiccated region, and a lower area which is usually wetted by wave action at low tide. The upper eulittoral area ["upper yellow zone" (Stephenson and Stephenson, 1950); "green zone" (Lewis, 1960); "barnacle and vermetid and white zones" (Brattström, (1980)] is distinguished

TABLE 9.3

Zonation diagrams for sites in the tropical western Atlantic. (Modified from Brattström, 1980.)

Sublittoral zone		Littoral zone					Supralittoral zone/ Maritime zone	Region
		Eulittoral zone		Littoral fringe				
Infralittoral zone	Infralittoral fringe	Midlittoral zone		Supralittoral fringe				Florida Keys Stephenson and Stephenson (1950)
		Lower yellow zone	Upper yellow zone	Black zone	Grey zone	White zone		
Sublittoral zone Upper sublittoral region		Littoral zone					Maritime zone	Content Key, Florida Croley and Dawes (1970)
		Lower eulittoral zone	Upper eulittoral zone	Littoral fringe				
Millepora coralline community lip zone				Littorine community				Great Bahama Bank Newell et al. (1959)
		Lower yellow zone	Upper yellow zone	Black zone	Grey zone	White zone		
Sublittoral region		Eulittoral region		Supralittoral margin				Northern Lesser Antilles Vroman (1968)
Surf zone		Pink zone	Green zone	Black zone	Yellow zone	Weather zone		Barbados Lewis (1960)
Algal belt		Splash or *Balanus* zone		Spray or *Littorina* zone				Isla Margarita, Venezuela Rodriguez (1959)
		Lower zone	Upper zone					
Sublittoral (region)		Littoral (region)		Littoral margin			Supralittoral region	Colombia Brattström (1980)
Zone of algal mat and *Palythoa* zone		Zone of mixed algae	Barn— vermetid zone	White zone	Spray zone		Maritime zone	

by an increase in macroscopic algae and invertebrates. Species of the saccate green alga *Valonia*, together with filamentous red and green algae, can be found in fissures of the limestone (Croley and Dawes, 1970). Fuzz or turf algae are common in this region (Stephenson and Stephenson, 1950), together with small barnacles (*Chthamalus*, *Tetraclita*) and vermetids (*Petaloconchus*, *Spiroglyphus*). Brattström (1980) divided the upper eulittoral region on the Colombian coast into two zones: an upper white or naked zone (yellowish-grey) containing blue-green and bleached calcareous algae (lithothamnia), and a lower barnacle–vermetid zone containing the crustose brown algal genus *Ralfsia* and the turf-forming red algal genus *Laurencia*.

The lower eulittoral region ["mixed algal zone" (Brattström, 1980)] shows an increased variety of filamentous (*Centroceras clavulatum*, *Giffordia* spp., *Polysiphonia* spp., *Sphacelaria* spp.) and fleshy (*Acanthophora spicifera*, *Laurencia* spp., *Lobophora variegata*, *Sargassum* spp.) algae (Croley and Dawes, 1970; Van den Hoek et al., 1975). Lithothamnia may be common in this zone or become overgrown with turf-type red (*Laurencia*) and brown (*Colpomenia*, *Lobophora*, *Ralfsia*) algae. Barnacles (*Chthamalus* spp., *Tetraclita* spp.) and mussels (*Mytilus exustus*) are common in this yellow-colored region (Stephenson and Stephenson, 1950; Brattström, 1980).

The lower eulittoral region of the tropical western Atlantic can support a complex algal "fuzz" (turf, mat) (Bernatowicz, 1952; Taylor, 1971), which Brattström (1980) refers to as the "mixed algal zone". In more protected areas, the algal turf can reach 0.5 to 1 cm in height. In Bermuda, the fuzz consists primarily of *Bostrychia tenella* and *Polysiphonia howei* (Bernatowicz, 1952), while in Colombia it includes red algae [*Centroceras*, *Ceramium*, *Hypnea*, *Laurencia* (Brattström, 1980)]. In the lower eulittoral region, the algal turf may include green (*Cladophora fuliginosa*) and red (*Centroceras clavulatum*) filamentous algae, as well as more wiry or fleshy members of the red algal order Gelidiales (Taylor, 1971).

The turf algae in the lower eulittoral region at Galeta Point on the coast of Panama occur in actively accreting reef habitats stressed by surf or sand scouring, or habitats which are extensively grazed (Hay, 1981c). Distribution of the algal turf community with depth indicates that these species are most common on the reef flat and shallow reef slopes, where desiccation or high levels of herbivory are common. The algal turf community is gradually replaced by larger individual plants in deeper portions of the reef. Hay (1981c) noted that the dominant turf-forming seaweeds were *Dictyota* spp., *Halimeda opuntia*, and *Laurencia papillosa*. He found that, although turf growth resulted in lower (33 to 61%) productivity compared with free individuals, turfs suffered less physiological damage during desiccating low tides, and less biomass was lost to herbivores. Further, he showed that spatial partitioning of photosynthetic activity caused the upper portions of the turf algae to be more productive. The growing tips were damaged regularly, resulting in increased branching and compaction of the turf.

Sublittoral. Most studies of zonation in the tropical western Atlantic have been limited to the supralittoral and eulittoral zones down to the surf region (sublittoral fringe) of the sublittoral zone (Table 9.4). The sublittoral zone usually is divided into upper (shallow, to about −5 m, including a sublittoral fringe), lower (−5 to −35 m), and deep (below −35 m) regions. In areas lacking extensive coral growth, the upper sublittoral is usually characterized by perennial red algae, including calcareous (*Jania*, *Lithothamnion*), fleshy (*Eucheuma*, *Hypnea*, *Laurencia*), and ephemeral filamentous forms (*Centroceras*). Brown algae (*Dictyota*, *Padina*, *Sargassum*) also may occur there, as described for Content Key (Croley and Dawes, 1970), and the brown seaweeds contribute to the algal mat (Brattström, 1980). Filamentous algae also occur as fuzzes on the rocks (Bernatowicz, 1952) or as algal epiphytes (Navarro and Almadovar, 1974; Ballantine, 1979). The diversity and abundance of algal species is higher in this region than in the lower portion of the eulittoral zone; collections made on a shallow (0 to −4 m) limestone shelf of the Florida Keys revealed 112 species off Molasses Key and 98 species at Bahia Honda Key (Mathieson and Dawes, 1975). Species composition at these two sites was similar to that reported for Content Key (Croley and Dawes, 1970) and the Dry Tortugas (Taylor, 1928).

The upper sublittoral region is also rich in

TABLE 9.4

Sublittoral studies of benthic marine algae

Area	Ref.[a]	Depth (m)	Seasonality	Method	Number of species			
					Green	Brown	Red	Total
Southwest coast of Florida	[1]	19 stations, −5 to −7	Single set of collections	Dredge and dive	50	28	70	148
Bahia Honda Key, Florida	[2]	0 to −1	1 year, monthly	Dive, transects	31	11	56	98
Molasses Key, Florida	[2]	0 to −4	1 year, monthly	Dive, transects	32	9	71	112
Content Key, Florida	[3]	0 to −10	2.5 years	Dive, transects	79	29	150	258
Curaçao	[4]	0 to −20	Jan., Feb.	Dive, band transects	22	5	23	50
	[5]	0 to −40	Apr., May	Dive, band transects	11	4	26	41
Colombia	[6]	0 to −1	Apr., May	Dive, line transects	12	10	38	60

[a]References:
[1] Dawes et al. (1967).
[2] Mathieson and Dawes (1975).
[3] Croley and Dawes (1970).
[4] Van den Hoek et al. (1975).
[5] Van den Hoek et al. (1978).
[6] Brattström (1980).

animals, including fixed or sessile hydrozoa (*Halocordyle*, *Millepora*), zooanthids (*Palythoa, Zoanthus*), sea anemone (*Spirobranchus*), and numerous barnacles and chitons (Stephenson and Stephenson, 1950; Newell et al., 1959; Rodriguez, 1959; Brattström, 1980).

In the lower sublittoral region, shell hash, gravel and sand support a more limited flora than continuous limestone outcroppings (Dawes et al., 1967). Taylor (1971) noted a distinct drop in species diversity when moving from −10 m (189 spp.) to −25 m (138 spp.) to −50 m (63 spp.) in the tropical western Atlantic (Table 9.5). Few species of algae (green: 2 species; brown: 1 species; red: 6 species) have upper distributional limits deeper than −10 m; thus, separation of the upper and lower sublittoral regions can be made only with regard to species abundance and not presence (Taylor, 1971). In contrast, a distinct set of sublittoral species is evident on coral reefs such as Glover's Reef atoll off Belize [four zones (Dawes, 1981)] and the fringing reef off Curaçao [six zones (Van den Hoek et al., 1975, 1978)]. These distinctive zones appear to reflect coral distribution patterns, at least to about −30 m in depth. Grazing is known to be an important determinant in algal distribution on coral reefs (Van den Hoek et al., 1975, 1978), but little information is available (Dawes et al., 1967).

Deep water. The benthic algal flora of deep-water shelves and banks (below −35 m) in the sublittoral zone is poorly known (Table 9.5). Recent studies using submersibles have indicated that algae extend to great depths in clear tropical waters. A crustose coralline red alga was reported from −268 m on a sea-mount off San Salvador in the Bahamas, where photon flux density was only 0.007 to 0.009 $\mu E\ m^{-2}\ s^{-1}$ (Littler et al., 1985). This study documented depth limits for *Lobophora*

TABLE 9.5

Studies on numbers of species of deep-water algae in the tropical western Atlantic

Location	Reference	To			
		30 m	50 m	70 m	90 m
West coast of Florida	Dawes and Van Breedveld (1969)				
Green algae		22	12	6	–
Brown algae		16	14	10	–
Red algae		43	27	22	–
Total		81	53	38	–
Caribbean region	Taylor (1971)				
Green algae		–	36	–	14
Brown algae		–	10	–	1
Red algae		–	17	–	8
Total		–	63	–	23
East coast of Florida	Hanisak and Blair (unpubl. data)				
Green algae		25	25	1	1
Brown algae		6	12	4	2
Red algae		74	104	44	25
Total		105	141	49	28

(−88 m), *Halimeda* (−130 m), and *Peyssonnelia* (−189 m); all three genera are common in the tropical western Atlantic.

A deep-water algal community, dominated by filamentous and fleshy forms, similar to the lower eulittoral region, has been reported for a fringing reef off Curaçao at −55 to −65 m (Van den Hoek et al., 1978). In their study, Van den Hoek et al. showed that a tall algal turf (to 5 cm high) and a lack of coral cover were characteristic of this deep-water flora. The authors suggested that the large algal biomass in the deeper portion of the fringing reef may reflect low grazing pressure, as absolute growth rates must be very low under limited light (<1% of surface radiation). The brown alga *Lobophora variegata* and 18 other algae showed a discontinuous distribution; they occurred in the upper sublittoral region and in deep water, but not in between. Grazing pressure was cited as the basis for the discontinuous distribution, because of the large populations of herbivorous fish in intermediate depths.

Information about the biomass of deep-water algae is limited, but estimates of percentage cover have been made for the San Salvador sea-mount (Littler et al., 1985) and the fringing reef off Curaçao (Van den Hoek et al., 1978). In the latter study, the percentage cover of crustose coralline, fleshy, and filamentous algae ranged from 20 to 50% at depths from −55 to −65 m. The San Salvador study recorded 10% cover of a crustose coralline at −240 m. Video and dredge data of deep-water (−61 to −75 m) algal communities off the Marquesas Keys (in the Florida Keys) have indicated that the green alga *Anadyomene menziesii* forms large (to 0.5 m tall) beds that are continuous over a number of kilometers (Dawes, unpubl. data). Such biomass indicates high productivity, although no direct measurements are available.

Algae characteristic of the tropical western Atlantic extend into more subtropical or temperate areas in deeper waters, such as the west coast of Florida (Dawes and Van Breedveld, 1969) and the coast of North Carolina (Searles and Schneider, 1980). In a study of deep-water species from the east coast of Florida, Hanisak and Blair (1988) identified a total of 208 species of algae – 42 Chlorophyta, 19 Phaeophyta and 147 Rhodophyta.

Soft substrata communities

Sandy substrata. There is an almost complete absence of information on both interstitial and meiofaunal taxa from Caribbean sandy substrata. Even macrofaunal species of sandy substrata are known only from a few Caribbean sites.

The intertidal sandy-beach fauna in Panama is dominated numerically by isopods (such as *Ancinus brasilensis*, *Excirolana brazilensis*), polychaetes (*Scolelepsis agilis*) and cumacean crustaceans (*Cyclaspis* sp.), while biomass dominants include the bivalve *Donax denticulatus* and the gastropod *Terebra cinerea* (Dexter, 1972, 1979). In addition, ghost crabs (*Ocypode quadrata*) forage over the entire intertidal zone, and several other decapod crustaceans, including mole crabs (*Emerita portoricensis*, *Hippa testudinaria*, *Lepidopa dexterae*, *L. richmondae*), commonly occur intertidally along Caribbean sand beaches (Abele, 1972, 1976; Dexter, 1979).

Subtidally, filter-feeding bivalves (*Divaricella quadrisulcata*, *Laevicardium laevigatum*, *Tellina martinicensis*), polychaetes (*Clymenella torquata*, *Haploscoloplos fragilis*, *Scoloplos riseri*), echinoderms (*Amphioplus* sp., *Mellita quinquesperforata*), and crustaceans (*Callianassa* sp.) occur as infauna (O'Gower and Wacasey, 1967; Jackson, 1972; Murina et al., 1974; Rosenberg, 1975); predaceous gastropods (*Conus*, *Murex* spp.,) and echinoderms occur on the substratum surface (Heck, 1979).

Quantitative estimates of infaunal biomass suggest that standing crop is relatively low compared to temperate (Wade, 1972) and other tropical regions (Dexter, 1979), while diversity is quite similar to that reported from temperate regions (Dexter, 1972, 1979). For example, an average density of 190 ind. m^{-2} was reported for Panamanian intertidal sand beaches (Dexter, 1979), one to two orders of magnitude less than abundances reported for North Carolina (963 m^{-2}: Dexter, 1979) or Georgia (12 566 m^{-2}: Howard and Dörjes, 1972).

Predators on sandy substrata include portunid and calappid crabs (for instance *Arenaeus cribrarius*: Abele, 1976; Heck, 1979), fishes and shorebirds. Published work on either the occurrence or abundance of sandy-bottom fishes or shorebirds is almost completely lacking, despite their abundance and potential importance as predators. Several species of surf-zone fishes recorded from the Gulf of Mexico (*Arius felis*, *Chloroscombrus chrysurus*, *Trachinotus falcatus*) by Yanez-Arancibia and Day (1982) are probably also present in the Caribbean.

Muddy substrata. Intertidal mud flats are limited in extent owing to the rather small tidal range in the Caribbean. When present, they are often associated with mangroves. As is true for sandy substrata, virtually nothing is known of either the micro- or meio-benthos of Caribbean muddy substrata. There are few accounts of the infauna of such habitats. Warner (1969) and Abele (1972, 1976) have described the crabs (such as *Eurytium limosum*, *Panopeus herbstii*, *Sesarma* spp., *Uca* spp.) that occur intertidally in the subsurface tunnels known as "crab runs". Subtidally, muddy areas are dominated by deposit feeders, including polychaetes of the families Spionidae and Magelonidae, bivalves (e.g. *Diplodonata punctata*, *Tellina martinicensis*), and echinoderms (*Amphioplus coniortodes*, *Moira atropos*) (Wade, 1972; Murina et al., 1974; Singletary and Moore, 1974; Rosenberg, 1975). Epifaunal species may include gastropods (*Aplysia dactylomela*, *Murex* spp.) and decapod crustaceans (*Pagurus* sp., *Periclimenes* spp.) (Wade, 1972).

Densities of infauna are often greater in muddy than in sandy substrates (Gray, 1974), but tropical infaunal densities in the Caribbean are low when compared to temperate areas. Wade (1972), for example, reported densities of 240 ind. m^{-2} from Kingston harbor, Jamaica, which are two to three orders of magnitude lower than values reported from North Carolina (1960–5035 ind. m^{-2}: Grassle, 1967) or Massachusetts (8985 ind. m^{-2}: Sanders, 1960). There are no studies that have specifically addressed the nektonic and bird predators associated with muddy Caribbean substrata.

Seagrass communities

Species and distribution. Five genera of seagrasses occur in the tropical western Atlantic (Den Hartog, 1970). Species found are *Halodule wrightii*, *Halophila decipiens*, *H. engelmannii*, *H. johnsonii*, *Ruppia maritima*, *Syringodium filiforme*, and *Thalassia testudinum*. Low temperatures appear to be a limiting factor for northern or southern extensions of the three dominant Caribbean seagrasses, *T. testudinum* (turtle grass), *S. filiforme* (manatee grass), and *H. wrightii* (shoal grass) (McMillan, 1979, 1984; Barber and Behrens, 1985).

Distribution of particular seagrass species within the tropical western Atlantic appears to be limited horizontally by salinity (Phillips, 1960a) and vertically by water transparency (Vicente and Rivera, 1982). Extensive and dense beds of most seagrass species usually occur in waters less than 10 m deep (Humm, 1973). Two of the three dominant species, *Syringodium filiforme*, and *Thalassia testudinum*, are commonest in open coastal waters with stable salinities of 20 to 36‰ and subtropical to tropical temperatures (Phillips, 1960a–c, 1962; Humm, 1973; Dawes, 1974; Zieman, 1975). *Halodule wrightii* is often found in estuarine waters with salinities of 10 to 25‰, but it will form dense stands in open coastal, high-salinity waters (Dawes, 1974) and in shoals or tidal flat regions of water movement and exposure.

Halophila decipiens (= *H. baillonis*) and *H. engelmanni* are smaller understorey species occurring in communities dominated by the larger seagrasses. They appear to be limited to areas of high (> 25‰) salinities. *Halophila decipiens* can occur in shallow water (Eiseman, 1980), but more commonly it is found in deep waters (−20 to −100 m), forming meadows in the sand (Dawes and Van Breedveld, 1969; Dawes, 1974; Dawes et al., 1989). *Halophila engelmanni* is common in shallow waters, mixed with *Syringodium filiforme* and *Thalassia testudinum* (Dawes, 1974; Dawes et al., 1986). Eiseman (1980) has described a third species of *Halophila*, *H. johnsonii*, from the east coast of Florida in the Indian River.

Ruppia maritima is a subtidal angiosperm, not always included in lists of the seagrasses because it also occurs in estuarine to fresh waters (1–5‰), with meadows extending into the mouths of rivers (Phillips, 1960a; Humm, 1973). The plant also occurs in areas of high salinities such as fringing mangrove swamps in the Florida Keys. It forms extensive meadows at the sublittoral fringe, where it may be exposed at low tide.

Seagrass epiphytes. The study of organisms living on seagrass blades has suffered from an "observational approach" (Harlin, 1980). Such topics as competition, production, and trophic status among these organisms have only recently been considered. Published lists for Florida coasts indicate that a variety of algae grow epiphytically on seagrass blades (Table 9.6). Information regarding the epifauna of tropical western Atlantic seagrasses is limited. A study of *Thalassia testudinum* blades in Barbados (Lewis and Hollingworth, 1982) recorded 90 metazoan species from 11 phyla. The most important group was the nematodes (> 70%), the majority of which were detrital feeders. Copepods were the second most important group. Almost 90% of the epifauna of seagrass

TABLE 9.6

Species numbers of algal epiphytes of seagrasses in the tropical western Atlantic

Location	Total no. of species	Seasonality	Algal groups[a]				Reference
			1	2	3	4	
West coast of Florida							
Crystal Bay	46	Non seasonal	5	7	8	26	Phillips (1960c)
Boca Ciega Bay, Tampa Bay, and Anclote Anchorage	68	Non seasonal	13	11	4	40	Phillips (1960b)
Anclote Anchorage	66	Seasonal	14	13	8	31	Ballantine and Humm (1975)
East coast of Florida							
Indian River	41	Seasonal	4	10	10	17	Hall and Eiseman (1981)
St. Lucia	49	Seasonal	5	9	7	28	Phillips (1961)
All Florida	113	Non seasonal	10	15	19	69	Humm (1964)

[a]1 — Cyanophyceae (Cyanobacteria); 2 — Chlorophyceae; 3 — Phaeophyceae; 4 — Rhodophyceae.

TABLE 9.7

Species numbers of macroalgae within seagrass communities from the tropical western Atlantic

Site	Ref.[a]	Total	Cyanophyceae	Chlorophyceae	Phaeophyceae	Rhodophyceae
West coast of Florida						
Anclote Anchorage	[1]	122	18	39	17	50
Seven Sites	[2]	30	–	11	2	17
East coast of Florida						
Indian River	[3]	120	8	30	19	63
Indian River	[4]	63	3	12	9	39

[a]References:
[1] Hamm and Humm (1976).
[2] Dawes (1986) (dominant species only).
[3] Phillips (1961).
[4] Benz et al. (1979) (drift species only).

blades were grazers on the epiflora or the detrital feeders. A faunal "succession" occurred during community aging in a manner similar to that reported for the temperate seagrass *Zostera marina* (Lewis and Hollingworth, 1982).

Seagrass macroalgae. Both drift and attached macroalgae are important components of seagrass communities (Table 9.7). Drift seaweeds often show distinct seasonal variation in biomass, with large spring blooms (Phillips, 1961; Benz et al., 1979; Dawes et al., 1985). The most common drift red alga in seagrass communities is *Laurencia poitei* (Josselyn, 1977; Dawes et al., 1985). Other drift algae common to Florida seagrass beds include the red algal species *Acanthophora spicifera*, *Gracilaria verrucosa*, *Hypnea cervicornis*, *H. musciformis*, and *Spyridia filamentosa* (Benz et al., 1979; Dawes et al., 1985). Attached macroalgae common in seagrass communities include members of the green algal order Caulerpales (South, 1983; Dawes, 1986), of which rhizophytic species of *Caulerpa*, *Halimeda*, *Penicillus* and *Udotea* are usually dominant. Although seaweeds are common in the tropical western Atlantic, little is known regarding their production in seagrass beds, and the most detailed studies have been carried out on the west coast of Florida (Hamm and Humm, 1976; Dawes et al., 1979, 1985).

Seagrass macrofauna. Brook (1975) published a list of invertebrates and fish present in *Thalassia testudinum* beds off Card Sound, Florida. The fauna of seagrass beds has been reviewed by Kikuchi and Pérès (1977) and by Ogden (1980), and caging experiments separating the predatory impact of fishes from that of decapods in seagrass beds were carried out in the Indian River, Florida, by Young et al. (1976).

The macrofauna of seagrass beds is exceedingly rich, as is the meiofauna (Bell et al., 1984). Qualitatively, infaunal species differ very little from those found on unvegetated sandy or muddy substrata, although infaunal abundances are typically much greater than in unvegetated areas (O'Gower and Wacasey, 1967; Orth, 1977), presumably because of reduced predator effectiveness in seagrass roots and rhizomes (cf. Peterson, 1982; Orth et al., 1984). Mobile epifaunal species are usually not found outside seagrass habitats, and consist of crustaceans, molluscs, echinoderms and fishes (Moore et al. 1963a, b; Heck, 1974, 1979; Thorhaug and Gessner, 1977; Ogden and Lobel, 1978; Weinstein and Heck, 1979; Martin and Cooper, 1981; Yanez-Arancibia et al., 1982; Virnstein et al., 1984; Bauer, 1985). Characteristic species include caridean and penaeid shrimps (for example, *Hippolyte* spp., *Penaeus* spp., *Periclimenes* spp.), grazing snails (*Cerithium* sp., *Modulus modulus*), herbivorous urchins (*Lytechinus variegatus*, *Tripneustes ventricosus*), starfish (*Oreaster reticulatus*) and fishes (pipefishes, juvenile snappers, grunts and groupers, parrotfishes and wrasses). In addition, green turtles (*Chelonia mydas*) are characteristic and important grazers of seagrass (Thayer et al., 1984).

PHOTOSYNTHESIS AND PRODUCTIVITY

Algal productivity

Photosynthetic rates have been measured for drift *Cladophora* in Bermuda (Bach and Josselyn, 1979), in shallow reefs (Wanders, 1976a, b) and deep reefs (Vooren, 1981) off Curaçao; for 43 species of reef algae in Belize (Littler et al., 1983); on the reef flats of the Florida Keys for *Eucheuma* (Mathieson and Dawes, 1974), *Batophora* (Morrison, 1984), and *Sargassum* (Prince, 1980); for 54 species from Cuba (Buesa, 1977); for four deep-water species of *Halimeda* from the Bahamas (Jensen et al., 1985); and for 14 species of macroalgae found on mangrove roots in Puerto Rico (Burkholder and Almodovar, 1973).

Buesa (1977) studied 54 Cuban species and found a wide range of photosynthetic rates among species of green, red, and brown algae and among seagrasses, probably due to the variety of functional-form types that he measured (Littler et al., 1983). The 54 species had a mean production of 6.6 mg O_2 (g dry wt)$^{-1}$ h^{-1}, which was not significantly different from that reported for Hawaiian species [6.9 mg O_2 (g dry wt)$^{-1}$ h^{-1}; Doty, 1971]. In a study of algae in Belize, Littler et al. (1983) reported that filamentous and sheet algae had photosynthetic rates of 5.65 and 5.06 mg C (g dry wt)$^{-1}$ h^{-1}, respectively, which was 2 to 6 times higher than coarsely branched and leathery species [1.06 and 0.88 mg C (g dry wt)$^{-1}$ h^{-1} respectively] and 15 times higher than crustose species [0.18 mg C (g dry wt)$^{-1}$ h^{-1}]. The two free-floating species of *Sargassum* in the Sargasso Sea had average photosynthetic rates of 1.2 mg C (g dry wt)$^{-1}$ h^{-1} north of 30°N latitude and 0.42 mg C (g dry wt)$^{-1}$ h^{-1} south of 30°N latitude (Carpenter and Cox, 1974).

Few studies have calculated productivity (g C m^{-2} day^{-1}). Carpenter and Cox (1974) calculated an average yearly productivity of the free-floating *Sargassum–Dicothrix* complex as 1.0 mg C m^{-2} day^{-1}, with a seasonal high of 4.2 mg C m^{-2} day^{-1} in May. The productivity is only 0.5 to 2.0% of phytoplankton productivity in the northwestern Sargasso Sea. The drift form of *Cladophora prolifera* was found to have a carbon fixation rate of 3.1 g C m^{-2} day^{-1} in Bermuda (Bach and Josselyn, 1979).

Productivity studies of Caribbean algae have been carried out in shallow and deep reefs off Curaçao. Wanders (1976a) found that net productivity of a shallow (-0.5 to -3 m) reef algal–coral community was 17 g C m^{-2} day^{-1}, which included the productivity of encrusting corallines (2.7 g C), fleshy and filamentous algae (3.3 g C) and corals (11.0 g C). Wanders (1976b) showed that a *Sargassum* bed in the same region containing *S. platycarpum* (1.32 g C m^{-2} day^{-1}), *Dictyopteris justii* (0.46 g C m^{-2} day^{-1}) and *Dictyota dentata* (0.6 g C m^{-2} day^{-1}) had a net productivity of 2.38 g C m^{-2} day^{-1}.

Productivity of benthic turf-forming seaweeds from deeper reefs off Curaçao has been estimated to be 1.8 g C m^{-2} day^{-1} at -10 m and 1.4 g C m^{-2} day^{-1} at -25 m (Vooren, 1981). Crustose coralline algae were the dominant plant group at -25 m, with an average net fixation rate of 0.21 g C m^{-2} day^{-1} per species component of the vegetation. The productivities of these deep-water communities are about 1/2 to 1/3 those reported for shallow-water communities in the same region (Wanders, 1976a). The low productivity of deep-water algae (Jensen et al., 1985; Littler et al., 1985) appears to be related to decreasing light levels.

Seagrass communities

Previous studies of seagrass productivity have measured rates of (1) blade production (length, dry weight: Table 9.8); (2) oxygen production; and (3) ^{14}C uptake. Using leaf-marking methods, Ziemán (1974) found that the productivity of *Thalassia testudinum* was very high, with blade turnover times as short as 10 to 12 days (Table 9.8).

Using light/dark chambers and an oxygen probe, Jones (1968) found a maximum net photosynthetic rate of 700 mg O_2 m^{-2} h^{-1} in blades of *Thalassia testudinum* from Biscayne Bay, Florida. However, about 1/3 to 1/4 of that productivity was by seagrass epiphytes. Using ^{14}C uptake, the photosynthetic carbon pathways in the three dominant seagrasses of the Caribbean have been shown to be of the C_3 type, although *T. testudinum* and *Halodule wrightii* had high initial incorporation rates into organic acids, like C_4 plants (Beer and Wetzel, 1982). Productivity, as measured by ^{14}C uptake, for the three dominant Caribbean species reached 17.3 mg C (g dry wt)$^{-1}$ h^{-1} in leaves of

TABLE 9.8

Seagrass productivity in the tropical western Atlantic

Species/location	Blade length (mm day^{-1})	Blade weight (g dry wt day^{-1})	Reference
Thalassia testudinum			
Biscayne Bay, Florida	2–5	2.3–5.0	Zieman (1975)
Biscayne Bay, Florida	2–4	2.2–10.0	Wood et al. (1960)
Boca Ciega Bay, Florida	3.4	0.4–0.7	Phillips (1960b)
Anclote Key, Florida	3.5	–	Dawes and Lawrence (1980)
Cuba	3.0	4.9	Buesa (1977)
Jamaica	1.4	4.6	Greenway (1974)
Barbados	5.3–7.1	3.1	Patriquin (1973)
Bermuda	6.8	1.2–14.1	Patriquin (1973)
Syringodium filiforme			
Indian River, Florida	9–31	–	Fry (1983)
Halodule wrightii			
Indian River, Florida	up to 8.5	3.0	Virnstein (1982)

Halodule wrightii (Williams and McRoy, 1982), which is 10 times higher than values for temperate seagrasses.

Secondary production

No attempts have been made to estimate total secondary production for any Caribbean habitat, and there are few data for secondary production even by single species or taxonomic assemblages. Generalizing from standing-stock estimates on soft substrata, secondary production is lower than in corresponding temperate habitats. However, the small size of tropical species (cf. Bauer, 1985) and their more continuous reproduction may actually result in higher productivity. Estimates of fish productivity ranging from 7.4 to 8.5 g wet wt. m^{-2} yr^{-1} have been made for turtle grass and lagoonal habitats in Mexico (Amezcua Linares and Yanez-Arancibia, 1980; Yanez-Arancibia and Day, 1982), but these estimates are complicated by difficulties in ageing fish and estimating growth rates.

GRAZING AND FOOD WEBS

Algal communities

The presence of an algal turf at Carrie Bow Key in Belize and at Galeta Point in Panama is in part attributable to fish grazing (Hay, 1981a, b). Also, the grazing effect of *Diadema antillarium* in Discovery Bay, Jamaica, reduced the algal biomass (Sammarco, 1982), especially for algal turf species (Carpenter, 1981).

Grazing by fish has been well documented in the Caribbean. Hay (1981b) found that sand-plain species of macroalgae (occurring in water depths from −11 to −14 m on scattered hard substrata) off Galeta Point in Panama were excluded from reefs by intense grazing. Caging experiments in a fringing reef off Curaçao demonstrated that, without grazing by fish, fleshy and filamentous species of macroalgae overgrew and shaded out the reef-building and consolidating crustose corallines (Wanders, 1977). Lobel (1980) found that damselfishes (Pomacentridae) in Panama selectively ate the epiphytic layer of the algal mat growing inside their territories, and that this selective grazing resulted in a thick mat of tough *Gelidium* turf which prevented coralline development (see also Hinds and Ballantine, 1987). Lobel (1980) concluded that such effects could result in the eventual collapse of the calcareous reef system. The parrotfish *Sparisoma radians* shows a distinct grazing preference at St. Croix for *Thalassia* blades with epiphytes (Lobel and Ogden, 1981).

Micrograzers also may play an important role in maintaining coral reefs in the Caribbean. Amphipods at densities of one individual per square centimeter of water surface can control filamentous algal species in a coral reef sustained in an artificial reef microcosm (Brawley and Adey, 1981).

Grazing in seagrass communities

Grazers in the tropical western Atlantic seagrass communities include gammarid amphipods (Zimmerman et al., 1979), gastropods (Randall, 1964), echinoids (Greenway, 1976; Lawrence, 1976), fish (Tribble, 1981; Fry et al., 1982), and the green turtle, *Chelonia mydas* (Björndal, 1980; Thayer et al., 1982; Zieman et al., 1984). Grazing may be intense. For example, selective grazing by fish in seagrass meadows near coral reefs in the San Blas Islands off Panama removes *Syringodium filiforme*, resulting in monospecific beds of the slower growing seagrass *Thalassia testudinum* (Tribble, 1981).

The grazing pattern of the green turtle reflects utilization of seagrass blade energetics. Björndal (1980) found that the turtles first graze the old blades of *Thalassia testudinum*, and then return after a few weeks to regraze the younger, non-epiphytized blades. Young blades contain a higher percentage of protein, a lower level of ash (Dawes and Lawrence, 1980), and a higher caloric value. Green turtles also feed on the leaf bases of *T. testudinum*, where the blade meristems occur. The leaf bases have a higher proportion of nitrogen and lower lignin content than leaves (Zieman et al., 1984). Thayer et al. (1982) found that an average green turtle excretes 2.9 g nitrogen daily in its fecal material, while only 0.04 g nitrogen would be released daily from decomposition of the same amount of *T. testudinum* leaf material; thus, the time for decomposition and subsequent enrichment of growing seagrasses by the detrital food chain is reduced.

Algal epiphytes of seagrass blades may play a larger role as primary producers in food webs of seagrass meadows than the actual seagrass blades. Recent studies in Florida using the ratio of carbon isotopes ($^{13}C/^{14}C$) have shown that the algal epiphytes often are the primary food of grazers in seagrass meadows on both the west (Kitting et al., 1984) and east coasts (Fry, 1984). Using carbon isotope methods, Fry et al. (1982) found that, in seagrass meadows of Nicaragua and the Virgin Islands, benthic algae and seagrasses together contribute at least 48 to 76% of the carbon found in fish.

Decomposition, detritus, and export

Detritus derived from benthic seagrasses and macroalgae is an important link between primary and secondary production in shallow-water areas (Fenchel, 1970, 1977). The decomposition of organic detritus derived from *Thalassia testudinum* involves a complex microbial assemblage consisting of bacteria, ciliates, and diatoms, which collectively can consume oxygen at a rate of 0.7 to 1.4 mg O_2 (g dry wt)$^{-1}$ h^{-1} (Fenchel, 1970). Leaves of *T. testudinum* have a protein:carbohydrate ratio of 1:15, and in litter bags decompose to a fine-particulate and dissolved state within six months (Newell et al., 1984). Animals may use only a small amount of the particulate detrital matter derived from seagrass blades because of the large amount of indigestible cellulose and lignin (up to 40%: Dawes, 1986). Instead, they graze the bacteria and fungi which colonize the blade particles (Fenchel, 1977). Newell et al. (1984) recorded a decomposition time of 52 days for *T. testudinum* leaves and showed that C:N ratios decreased slightly for the first 31 days, increasing thereafter.

Export of seagrass blades and macroalgae has been studied in the Virgin Islands (Zieman et al., 1979; Josselyn et al., 1983; 1986) and the deep sea of the Caribbean (Wolff, 1976). Leaves of *Thalassia testudinum* usually sink after detachment and move along the bottom toward deeper waters (Zieman et al., 1979). The leaves may provide a habitat for many deep-sea sessile animals, such as protozoans, actinians, and gastropods (Wolff, 1976). *Syringodium filiforme* blades, which possess a much larger lacunal system, tend to float and can be exported offshore for much longer distances (Zieman et al., 1979). Extremely high productivities and times of export to submarine canyons were demonstrated for seagrass blades (4 to 18 days) and macroalgae (0.1 to 4 days) in the Virgin Islands (Josselyn et al., 1983; 1986).

REFERENCES

Abele, L.G., 1972. Comparative habitat diversity and faunal relationships between the Pacific and Caribbean

Panamanian decapod Crustacea: a preliminary report with some remarks on the crustacean fauna of Panama. *Bull. Biol. Soc. Wash.*, 2: 125–138.

Abele, L.G., 1976. Comparative species composition and relative abundance of decapod crustaceans in marine habitats of Panama. *Mar. Biol.*, 38: 263–278.

Almodovar, L.R. and Ballantine, D.L., 1983. Checklist of benthic marine macroalgae plus additional records from Puerto Rico. *Caribb. J. Sci.*, 198: 7–20.

Amezcua Linares, F. and Yanez-Arancibia, A., 1980. Ecologia de los sistemas fluvio-lazunares asociados a Laguna de Terminos. El habitat y estructura de las comunidades de peces. *Anal. Centro Ci. Mar Limnol., Univ. Nal. Auton Mexico*, 7: 69–118.

Bach, S.D. and Josselyn, M.N., 1979. Production and biomass of *Cladophora prolifera* (Chlorophyta, Cladophorales) in Bermuda. *Bot. Mar.*, 22: 163–169.

Ballantine, D.L., 1979. The distribution of algal epiphytes on macrophyte hosts offshore from La Parguera, Puerto Rico. *Bot. Mar.*, 22: 107–111.

Ballantine, D. and Humm, H.J., 1975. Benthic algae of the Anclote estuary. I. Epiphytes of seagrass leaves. *Fla. Sci.*, 38: 150–162.

Barber, B.J. and Behrens, P.J., 1985. Effects of elevated temperature on seasonal *in situ* leaf productivity of *Thalassia testudinum* Banks ex Konig and *Syringodium filiforme* Kutzing. *Aqua. Bot.*, 22: 61–69.

Bauer, R.T., 1985. Diel and seasonal variation in species composition and abundance of caridean shrimps (Crustacea, Decapoda) from seagrass meadows of the north coast of Puerto Rico. *Bull. Mar. Sci.*, 36: 150–162.

Beer, S. and Wetzel, R.G., 1982. Photosynthetic carbon fixation pathways in *Zostera marina* and three Florida seagrasses. *Aquat. Bot.*, 13: 141–146.

Bell, S.S., Walters, K. and Kem, J.C., 1984. Meiofauna from seagrass habitats: A review and prospectus for future research. *Estuaries*, 7: 331–338.

Benz, M.C., Eiseman, N.J. and Gallagher, E.E., 1979. Seasonal occurrence and variation in standing crop of a drift algal community in the Indian River, Florida. *Bot. Mar.*, 22: 413–420.

Bernatowicz, A.J., 1952. Seasonal aspects of the Bermuda algal flora. *Pap. Mich. Acad. Sci. Arts Lett.*, 36: 3–8.

Bird, K.T. and McIntosh, R.P., 1979. Notes on the marine algae of Guatemala. *Rev. Biol. Trop.*, 27: 163–169.

Björndal, K.A., 1980. Nutrition and grazing behavior of the green turtle *Chelonia mydas*. *Mar. Biol.*, 56: 147–154.

Brattström, H., 1980. Rocky-shore zonation in the Santa Marta area, Colombia. *Sarsia*, 65: 163–226.

Brawley, S.H. and Adey, W.H., 1981. The effect of micrograzers on algal community structure in a coral reef microcosm. *Mar. Biol.*, 61: 167–177.

Briggs, J.C., 1974. *Marine Zoogeography*. McGraw-Hill, New York, 475 pp.

Brook, I.M., 1975. *Some aspects of the trophic relationships among the higher consumers in a seagrass community (Thalassia testudinum) in Card Sound, Florida.* Ph.D. dissertation, Univ. of Miami, Miami, Fla., 133 pp.

Buesa, R.J., 1977. Photosynthesis and respiration of some tropical marine plants. *Aquat. Bot.*, 3: 203–216.

Burkholder, P.R. and Almodovar, L.R., 1973. Studies on mangrove algal communities in Puerto Rico. *Fl. Sci.*, 36: 66–74.

Carpenter, E.J. and Cox, J.L., 1974. Production of pelagic *Sargassum* and a blue-green epiphyte in the western Sargasso Sea. *Limnol. Oceanogr.*, 19: 429–441.

Carpenter, R.C., 1981. Grazing by *Diadema antillarum* (Philippi) and its effects on the benthic algal community. *J. Mar. Res.*, 39: 749–765.

Chapman, V.J., 1961. *The Marine Algae of Jamaica. Part I. Myxophyceae and Chlorophyceae.* Institute of Jamaica, Kingston, 159 pp.

Chapman, V.J., 1963. *The Marine Algae of Jamaica. Part II. Phaeophyceae and Rhodophyceae.* Institute of Jamaica, Kingston, 201 pp.

Chapman, V.J. (Editor), 1977. *Wet Coastal Ecosystems, Ecosystems of the World, Vol. 1.* Elsevier, Amsterdam, 428 pp.

Croley, F.C. and Dawes, C.J., 1970. Ecology of the algae of a Florida Key. I. A preliminary checklist, zonation and seasonality. *Bull. Mar. Sci.*, 20: 165–185.

Dawes, C.J., 1974. *Marine Algae of the West Coast of Florida.* University of Miami Press, Coral Gables, Fla., 201 pp.

Dawes, C.J., 1981. *Marine Botany.* Wiley, New York, 628 pp.

Dawes, C.J., 1986. Seasonal proximate constituents and caloric values in seagrasses and algae on the west coast of Florida. *J. Coast. Res.*, 2: 25–32.

Dawes, C.J. and Lawrence, J.M., 1980. Seasonal changes in the proximate constituents of the seagrasses *Thalassia testudinum, Halodule wrightii*, and *Syringodium filiforme. Aquat. Bot.*, 8: 371–380.

Dawes, C.J. and Van Breedveld, J.F., 1969. Benthic marine algae. Marine Research Laboratory, Department of Natural Resources, State of Florida, St. Petersburg, Florida. *Memoirs of the Hourglass Cruises*, 1(2): 1–47.

Dawes, C.J., Earle, S.A. and Croley, F.C., 1967. The offshore benthic flora of the southwest coast of Florida. *Bull. Mar. Sci.*, 17: 211–231.

Dawes, C.J., Mathieson, A.C. and Cheney, D.P., 1974. Ecological studies of Floridian *Eucheuma* (Rhodophyta, Gigartinales). I. Seasonal growth and reproduction. *Bull. Mar. Sci.*, 24: 235–273.

Dawes, C.J., Bird, K., Durako, M., Goddard, R., Hoffman, W. and McIntosh, R., 1979. Chemical fluctuations due to seasonal and cropping effects on an algal-seagrass community. *Aquat. Bot.*, 6: 79–86.

Dawes, C.J., Hall, M.O. and Reichert, R., 1985. Seasonal biomass and energy content in seagrass communities on the west coast of Florida. *J. Coastal Res.*, 1: 255–262.

Dawes, C.J., Chan, M., Chinn, R., Koch, E.W., Lazar, A. and Tomasko, D.A., 1986. Proximate composition, photosynthetic and respiratory responses of the seagrass *Halophila engelmanii* from Florida. *Aquat Bot.*, 27: 195–201.

Dawes, C.J., Lobban, C.S. and Tomasko, D.A., 1989. A comparison of the physiological ecology of the seagrasses *Halophila decipiens* Ostenfeld and *H. johnsonii* Eiseman from Florida. *Aquat. Bot.*, 33: 149–154.

Den Hartog, C., 1970. *The Sea-grasses of the World.* North-Holland, Amsterdam, 275 pp.

De Rios, N.R., 1972. Contribucion al estudio sistematico de las

algas macroscopicas de las costas de Venezuela. *Acta Bot. Venez.*, 7: 219–324.
Dexter, D.M., 1972. Comparative community structure of the sandy beaches of Panama. *Bull. Mar. Sci.*, 22: 449–462.
Dexter, D.M., 1979. Community structure and seasonal variation in intertidal Panamanian beaches. *Estuarine Coastal Mar. Sci.*, 9: 543–558.
Diaz-Piferrer, M., 1963. Adiciones a la flora marina de Puerto Rico. *Caribb. J. Sci.*, 3: 215–235.
Diaz-Piferrer, M., 1964a. Adiciones a la flora marin de Cuba. *Caribb. J. Sci.*, 4: 353–371.
Diaz-Piferrer, M., 1964b. Las investigaciones ficologicas en el Caribe. La flora marina de la Republica Dominicana. *Moscosoa*, 1: 1–9.
Diaz-Piferrer, M., 1964c. Adiciones a la flora marina de las Antillas Holandesas Curazao y Bonaire. *Caribb. J. Sci.*, 4: 513–543.
Diaz-Piferrer, M., 1969. Distribution of the marine benthic flora of the Caribbean Sea. *Caribb. J. Sci.*, 9: 151–178.
Diaz-Piferrer, M., 1970. Adiciones a la flora marina de Venezuela. *Caribb. J. Sci.*, 10: 159–198.
Doty, M.S., 1971. The productivity of benthic frondose algae at Waikiki Beach, 1967–1968. *Univ. Hawaii Bot. Sci. Pap.*, 22: 1–119.
Dubinsky, Z. (Editor), 1990. *Coral Reefs. Ecosystems of the World, Vol. 25.* Elsevier, Amsterdam, 550 pp.
Earle, S.A., 1972. A review of the marine plants of Panama. *Bull. Biol. Soc. Wash.*, 2: 69–88.
Eiseman, N.J., 1980. A new species of seagrass, *Halophila johnsonii* from the Atlantic coast of Florida. *Aquat. Bot.*, 9: 15–19.
Fenchel, T., 1970. Studies on the decomposition of organic detritus derived from the turtle grass *Thalassia testudinum. Limnol. Oceanogr.*, 15: 14–20.
Fenchel, T., 1977. Aspects of the decomposition of seagrasses. In: C.P. McRoy and C. Helfferich (Editors), *Seagrass Ecosystems. A Scientific Perspective.* Marcel Dekker, New York, pp. 123–146.
Fry, B., 1983. Leaf growth in the seagrass *Syringodium filiforme* Kutz. *Aquat. Bot.*, 16: 361–368.
Fry, B., 1984. $^{13}C/^{12}C$ ratios and the trophic importance of algae in Florida *Syringodium filiforme* seagrass meadows. *Mar. Biol.*, 79: 11–19.
Fry, B., Lutex, R., Northam, M. and Parker, P.L., 1982. A $^{13}C/^{12}C$ comparison of food webs in Caribbean seagrass meadows and coral reefs. *Aquat. Bot.*, 14: 389–398.
Glynn, P., 1972. Ecology of the shallow waters of Panama. *Bull. Biol. Soc. Wash.*, 2: 13–30.
Grassle, J.F., 1967. *Influence of environmental variation on species diversity in benthic communities on the continental slope and shelf.* Ph.D. Thesis, Duke University, Durham, N.C., 194 pp.
Gray, J.S., 1974. Animal–sediment relationships. *Oceanogr. Mar. Biol. Annu. Rev.*, 12: 223–261.
Greenway, M., 1974. The effects of cropping on the growth of *Thalassia testudinum* (Konig) in Jamaica. *Aquaculture*, 4: 199–206.
Greenway, M., 1976. The grazing of *Thalassia testudinum* in Kingston Harbour, Jamaica. *Aquat. Bot.*, 2: 117–126.
Hall, M.O. and Eiseman, N.J., 1981. The seagrass epiphytes of the Indian River, Florida. I. Species list with descriptions and seasonal occurrences. *Bot. Mar.*, 24: 139–146.
Hamm, D. and Humm, H.J., 1976. Benthic algae of the Anclote estuary. II. Bottom-dwelling species. *Flor. Sci.*, 39: 209–229.
Hanisak, M.D. and Blair, S.M., 1988. The deep-water macroalgal community of the east Florida continental shelf (USA). *Helgol. Wiss. Meeresunters.*, 42: 133–163.
Harlin, M.M., 1980. Seagrass epiphytes. In: R. Phillips and C.P. McRoy (Editors), *Handbook of Seagrass Biology. An Ecosystem Perspective.* Garland STPM Press, New York, pp. 117–152.
Hay, M.E., 1981a. Spatial patterns of grazing intensity on a Caribbean barrier reef: Herbivory and algal distribution. *Aquat. Bot.*, 11: 97–109.
Hay, M.E., 1981b. Herbivory, algal distribution, and the maintenance of between-habitat diversity on a tropical fringing reef. *Am. Nat.*, 118: 520–540.
Hay, M.E., 1981c. The functional morphology of turf-forming seaweeds: Persistence in stressful marine habitats. *Ecology*, 62: 739–750.
Heck Jr., K.L., 1974. Comparative species richness, composition and abundance of invertebrates in Caribbean seagrass (*Thalassia*) meadows. *Mar. Biol.*, 41: 335–348.
Heck Jr., K.L., 1979. Some determinants of the composition and abundance of motile macroinvertebrate species in tropical and temperate turtlegrass (*Thalassia testudinum*) meadow. *J. Biogeogr.*, 6: 183–200.
Hinds, P.A. and Ballantine, D.L., 1987. Effects of the Caribbean threespot damselfish, *Stegastes planifrons* (Cuvier), on algal lawn composition. *Aquat. Bot.*, 21: 299–308.
Howard, J.D. and Dörjes, J., 1972. Animal–sediment relationships in two beach-related tidal flats; Sapelo Island, Georgia. *J. Sediment. Petrol.*, 42: 608–623.
Humm, H.J., 1964. Epiphytes of the seagrass, *Thalassia testudinum*, in Florida. *Bull. Mar. Sci. Gulf Caribb.*, 14: 306–341.
Humm, H.J., 1973. Seagrasses. In: J.I. Jones, R.E. Ring, M.O. Rinkel and R.E. Smith (Editors), *A Summary of Knowledge of the Eastern Gulf of Mexico.* Florida Institute of Oceanography, State University System of Florida, St. Petersburg, pp. IIIC: 1–10.
Jackson, J.B.C., 1972. The ecology of the molluscs of *Thalassia* communities, Jamaica, W.I. II. Molluscan population variability along a stress gradient. *Mar. Biol.*, 14: 304–337.
Jensen, P.R., Gibson, R.A., Littler, M.M. and Littler, D.S., 1985. Photosynthesis and calcification in four deep-water *Halimeda* species (Chlorophyceae, Caulerpales). *Deep Sea Res.*, 32: 451–464.
Jones, J.A., 1968. *Primary productivity by the tropical marine turtle grass, Thalassia testudinum Konig, and its epiphytes.* Ph.D. Dissertation. Univ. of Miami, Coral Gables, Flor., 196 pp.
Josselyn, M.N., 1977. Seasonal changes in the distribution and growth of *Laurencia poitei* (Rhodophyceae, Ceramiales) in a subtropical lagoon. *Aquat. Bot.*, 3: 217–229.
Josselyn, M.N., Cailliet, G.M., Niesen, T.M., Cowen, R., Hurley, A.C., Connor, J. and Hawes, S., 1983. Composition, export and faunal utilization of drift vegetation in the Salt River submarine canyon. *Estuarine Coastal Shelf Sci.*, 17: 447–465.

Josselyn, M.N., Fonseca, M., Niesen, T. and Larson, R., 1986. Biomass, production and decomposition of a deep-water seagrass, *Halophila decipiens* Ostenf. *Aquat. Bot.*, 25: 47–61.
Kapraun, D.F., 1972. Notes on the benthic marine algae of San Andres, Colombia. *Caribb. J. Sci.*, 12: 199–203.
Kikuchi, T. and Pérès, J.M., 1977. Consumer ecology of seagrass beds. In: C.P. McRoy and C. Helfferich (Editors), *Seagrass Ecosystems. A Scientific Perspective.* Marcel Dekker, New York, pp. 147–193.
Kitting, C.L., Fry, B. and Morgan, M.D., 1984. Detection of inconspicuous epiphytic algae supporting food webs in seagrass meadows. *Oecologia*, 62: 145–149.
Lawrence, J.M., 1976. Absorption efficiencies of four species of tropical echinoids fed *Thalassia testudinum. Thalassia Jugosl.*, 12: 201–205.
Lawson, G.W., 1978. The distribution of seaweed floras in the tropical and subtropical Atlantic Ocean: a quantitative approach. *Bot. J. Linnean Soc.*, 76: 177–193.
Lewis, J.B., 1960. The fauna of rocky shores of Barbados, West Indies. *Can. J. Zool.*, 38: 391–435.
Lewis, J.B. and Hollingworth, C.E., 1982. Leaf epifauna of the seagrass *Thalassia testudinum. Mar. Biol.*, 71: 41–49.
Littler, M.M., Littler, D.S. and Taylor, P.R., 1983. Evolutional strategies in a tropical barrier reef system: Functional-form groups of marine macroalgae. *J. Phycol.*, 19: 229–237.
Littler, M.M., Littler, D.S., Blair, S.M. and Norris, J.N., 1985. Deepest known plant life discovered on an uncharted seamount. *Science*, 227: 57–59.
Lobel, P.S., 1980. Herbivory by damselfishes and their role in coral reef community ecology. *Bull. Mar. Sci.*, 30: 273–289.
Lobel, P.S. and Ogden, J.C., 1981. Foraging by the herbivorous parrotfish, *Sparisoma radians. Mar. Biol.*, 64: 173–183.
Martin, F.D. and Cooper, M., 1981. A comparison of fish faunas found in pure stands of two tropical seagrasses of *Thalassia testudinum* and *Syringodium filiforme. North East Gulf Sci.*, 5: 31–38.
Mathieson, A.C. and Dawes, C.J., 1974. Ecological studies of Floridian *Eucheuma* (Rhodophyta, Gigartinales). II. Photosynthesis and respiration. *Bull. Mar. Sci.*, 24: 274–288.
Mathieson, A.C. and Dawes, C.J., 1975. Seasonal studies of Florida sublittoral marine algae. *Bull. Mar. Sci.*, 25: 46–65.
McCoy, E.D. and Heck Jr., K.L., 1976. Biogeography of corals, seagrasses, and mangroves: an alternative to the center of origin concept. *Syst. Zool.*, 25: 201–210.
McCoy, E.D., Bell, S.S. and Walters, K., 1986. A technique for placing biotic boundaries along environmental gradients. *Ecology*, 67: 749–759.
McMillan, C., 1979. Differentiation in response to chilling temperatures among populations of three marine spermatophytes, *Thalassia testudinum*, *Syringodium filiforme*, and *Halodule wrightii. Am. J. Bot.*, 66: 810–819.
McMillan, C., 1984. The distribution of tropical seagrasses with relation to their tolerance of high temperatures. *Aquat. Bot.*, 19: 217–220.
Michanek, G., 1979. Phytogeographic provinces and seaweed distribution. *Bot. Mar.*, 22: 375–391.
Moore, H.B., Jutare, T., Jones, J.A., McPherson, B.F. and Roper, C., 1963a. A contribution to the biology of *Tripneustes ventricosus. Bull. Mar. Sci.*, 18: 261–279.
Moore, H.B., Jutare, T., Bauer, J.C. and Jones, J.A., 1963b. The biology of *Lytechinus variegatus. Bull. Mar. Sci.*, 13: 23–35.
Morrison, D., 1984. Seasonality of *Batophora oerstedi* (Chlorophyta), a tropical macroalga. *Mar. Ecol. Prog. Ser.*, 14: 235–244.
Murina, V.V., Chu Khchin, V.D., Gomez, O. and Suarez, G., 1974. Quantitative distribution of bottom macrofauna in the upper sublittoral zone of the northwestern part of Cuba. In: *Investigations of the Central American Seas.* Indian National Science Documentation Centre, New Delhi, pp. 242–259.
Navarro, J.N. and Almodovar, L.R.,1974. Epiphytism in the marine benthic algae in Puerto Rico. *Flor. Sci.*, 36: 128–132.
Newell, N.D., Imbrie, J., Purdy, E.G. and Thurber, D.L., 1959. Organism communities and bottom facies, Great Bahama Bank. *Bull. Am. Mus. Nat. Hist.*, 117: 177–228.
Newell, S.Y., Fell, J.W., Statzell-Tallman, A., Miller, C. and Cefalu, R., 1984. Carbon and nitrogen dynamics in decomposing leaves of three coastal marine vascular plants of the subtropics. *Aquat. Bot.*. 19: 183–192.
Norris, J.N. and Bucher, K.E., 1982. Marine algae and seagrasses from Carrie Bow Cay, Belize. In: K. Rutzler and I.G. Macintyre (Editors), *The Atlantic Barrier Reef Ecosystem at Carrie Bow Cay, Belize 1: Structure and Communities. Smithsonian Contrib. Mar. Sci.*, 12: 167–223.
Ogden, J.C., 1980. Faunal relationships in Caribbean seagrass beds. In: R.C. Phillips and C.P. McRoy (Editors), *Handbook of Seagrass Ecology: An Ecosystem Perspective.* Garland STPM Press, New York, pp. 173–198.
Ogden, J.C. and Lobel, P.S., 1978. The role of herbivorous fishes and urchins in coral reef communities. *Environm. Biol. Fish.*, 3: 49–63.
O'Gower, A.K. and Wacasey, J.W., 1967. Animal communities associated with *Thalassia*, *Diplanthera*, and sand beds in Biscayne Bay. I. Analysis of communities in relation to water movement. *Bull. Mar. Sci.*, 17: 175–210.
Orth, R.J., 1977. The importance of sediment stability in seagrass communities. In: B.C. Coull (Editor), *Ecology of Marine Benthos.* University of South Carolina Press, Columbia, S.C., pp. 281–300.
Orth, R.J., Heck Jr., K.L. and Van Montfrans, J., 1984. Faunal communities in seagrass beds: a review of the influence of plant structure and prey characteristics on predator–prey relationships. *Estuaries*, 7: 339–350.
Papenfuss, G.F., 1972. On the geographical distribution of some tropical marine algae. *Proc. Int. Seaweed Symp.*, 7: 45–51.
Patriquin, D., 1973. Estimation of growth rate, production, and age of the marine angiosperm *Thalassia testudinum* Konig. *Caribb. J. Sci.*, 13: 111–123.
Peterson, C.H., 1982. Clam predation by whelks (*Busycon* spp.): Experimental tests of the importance of prey size, prey density and seagrass cover. *Mar. Biol.*, 66: 159–170.
Phillips, R.C., 1960a. *Observations on the ecology and distribution of the Florida Seagrasses.* Florida State Board of Conservation, Marine Laboratory, St. Petersburg, Professional Pap. No. 2, 72 pp.
Phillips, R.C., 1960b. Ecology and distribution of marine algae

found in Tampa Bay, Boca Ciega Bay, and at Tarpon Springs, Florida. *Q. J. Flor. Acad. Sci.*, 23: 222–259.

Phillips, R.C., 1960c. The ecology of marine plants of Crystal Bay, Florida. *Q. J. Flor. Acad. Sci.*, 23: 328–337.

Phillips, R.C., 1961. Seasonal aspect of the marine algal flora of St. Lucie Inlet and adjacent Indian River, Florida. *J. Flor. Acad. Sci.*, 24: 135–147.

Phillips, R.C., 1962. *Distribution of seagrasses in Tampa Bay, Florida.* Florida State Board of Conservation, Marine laboratory, St. Petersburg, Special Scientific Rep. No. 6, 12 pp.

Phillips, R.C., Vadas, R.L. and Ogden, N., 1982. The marine algae and seagrasses of the Miskito Bank, Nicaragua. *Aquat. Bot.*, 13: 187–195.

Pielou, E.C., 1977. The latitudinal spans of seaweed species and their patterns of overlap. *J. Biogeogr.*, 4: 299–311.

Prince, J.S., 1980. The ecology of *Sargassum pteropleuron* Grunow (Phaeophyceae, Fucales) in the waters off south Florida. II. Seasonal photosynthesis and respiration of *S. pteropleuron* and comparison of its phenology with that of *S. polyceratium* Montagne. *Phycologia*, 19: 190–193.

Randall, J.E., 1964. Notes on the biology of the queen conch, *Strombus gigas*. *Bull. Mar. Sci. Gulf Caribb.*, 14: 246–295.

Raup, D.M. and Crick, R.E., 1979. Measurement of faunal similarity in paleontology. *J. Paleontol.*, 53: 1213–1227.

Rodriguez, G., 1959. The marine communities of Margarita Island, Venezuela. *Bull. Mar. Sci. Gulf Caribb.*, 9: 237–280.

Rosenberg, R., 1975. Stressed tropical benthic faunal communities off Miami, Florida. *Ophelia*, 14: 93–112.

Sammarco, P.W., 1982. Effects of grazing by *Diadema antillarum* Philippi (Echinodermata: Echinoidea) on algal diversity and community structure. *J. Exp. Mar. Biol. Ecol.*, 65: 83–105.

Sanders, H.L., 1960. Benthic studies in Buzzard's Bay. III. The structure of the soft bottom community. *Limnol. Oceanogr.*, 5: 138–153.

Searles, R.B. and Schneider, C.W., 1980. Biogeographic affinities of the shallow and deep water benthic marine algae of North Carolina. *Bull. Mar. Sci.*, 30: 732–736.

Singletary, R.L. and Moore H.B., 1974. A redescription of the *Amphioplus coniortodes — Ophionepthys limnicola* community of Biscayne Bay, Florida. *Bull. Mar. Sci.*, 24: 690–699.

Sloane, H., 1707–1725. Submarine plants. In: *A voyage to the Islands Madera, Barbados, Nieves, St. Christophers and Jamaica.* London, Vol. 1, pp. 49–64; Vol. 2; pp. 157–274.

South, G.R., 1983. A note on two communities of seagrasses and rhizophytic algae in Bermuda. *Bot. Mar.*, 26: 243–248.

Stephenson, T.A. and Stephenson, A., 1950. Life between tidemarks in North American I. The Florida Keys. *J. Ecol.*, 38: 354–364.

Sterrer, W., 1986. *Marine flora and fauna of Bermuda.* Wiley, New York, 900 pp.

Taylor, W.R., 1928. *The marine algae of Florida, with special reference to the Dry Tortugas.* Carnegie Institute of Washington, Publ. No. 379, 219 pp.

Taylor, W.R., 1960. *Marine Algae of the Eastern Tropical and Subtropical Coasts of the Americas.* The University of Michigan Press, Ann Arbor, 870 pp.

Taylor, W.R., 1971. Notes on algae from the tropical Atlantic Ocean IV. *Br. Phycol. J.*, 6: 145–156.

Taylor, W.R., 1972. Marine algae of the Smithsonian-Bredin Expedition to Yucatan — 1960. *Bull. Mar. Sci.*, 22: 34–44.

Taylor, W.R. and Bernatowicz, A.J., 1969. *Distribution of marine algae about Bermuda.* Bermuda Biological Station for Research, Spec. Publ. No. 1, St. George's West, 42 pp.

Thayer, G.W., Engel, D.W. and Bjorndal, K.A., 1982. Evidence for short-circuiting of the detritus cycle of seagrass beds by the green turtle, *Chelonia mydas* L. *J. Exp. Mar. Biol. Ecol.*, 62: 173–183.

Thayer, G.W., Björndal, K.A., Ogden, J.C., Williams, S.L. and Zieman, J.C., 1984. Role of larger herbivores in seagrass communities. *Estuaries*, 7: 351–376.

Thomas, M.L.H., 1985. Littoral community structure and zonation on the rocky shores of Bermuda. *Bull. Mar. Sci.*, 37: 857–870.

Thorhaug, A. and Gessner, M.A., 1977. Seagrass community dynamics in a subtropical estuarine lagoon. *Aquaculture*, 12: 253–277.

Tribble, G.W., 1981. Reef-based herbivores and the distribution of two seagrasses (*Syringodium filiforme* and *Thalassia testudinum*) in the San Blas Islands (Western Caribbean). *Mar. Biol.*, 65: 277–281.

Tsuda, R.T. and Dawes, C.J., 1974. Preliminary checklist of the marine benthic plants from Glover's Reef, British Honduras. *Atoll Res. Bull.* No. 173, 13 pp.

Van den Hoek, C., 1975. Phytogeographic provinces along the coasts of the northern Atlantic Ocean. *Phycologia*, 14: 317–330.

Van den Hoek, C., 1984. World-wide latitudinal and longitudinal seaweed distribution patterns and their possible causes, as illustrated by the distribution of the rhodophytan genera. *Helgol. Wiss. Meeresunters.*, 38: 227–257.

Van den Hoek, C., Cortel-Breeman, A.M. and Wanders, J.B.W., 1975. Algal zonation in the fringing coral reef of Curaçao, Netherlands Antilles, in relation to zonation of corals and gorgonians. *Aquat. Bot.*, 1: 269–308.

Van den Hoek, C., Breeman, A.M., Bak, R.P.M. and Van Buurt, G., 1978. The distribution of algae, corals and gorgonians in relation to depth, light attenuation, water movement and grazing pressure in the fringing coral reef of Curaçao, Netherlands Antilles. *Aquat. Bot.*, 5: 1–46.

Vicente, V.P. and Rivera, J.A., 1982. Depth limits of the seagrass *Thalassia testudinum* (Konig) in Jobos and Guayanilla Bays, Puerto Rico. *Caribb. J. Sci.*, 17: 73–80.

Virnstein, R.W., 1982. Leaf growth of the seagrass *Halodule wrightii* photographically measured *in situ*. *Aquat. Bot.*, 12: 209–218.

Virnstein, R.W., Nelson, W.G., Lewis III, F.G. and Howard, R.K., 1984. Latitudinal patterns in seagrass epifauna: do patterns exist and can they be explained? *Estuaries*, 7: 310–330.

Vooren, C.M., 1981. Photosynthetic rates of benthic algae from the deep coral reef of Curaçao. *Aquat. Bot.*, 10: 143–154.

Voss, G.L. and Voss, N.A., 1954. An ecological survey of Soldier Key, Biscayne Bay, Florida. *Bull. Mar. Sci. Gulf Caribb.*, 5: 203–224.

Vroman, M., 1968. The marine algal vegetation of St. Martin, St. Eustatius and Saba (Netherlands Antilles). *Stud. Flora Curaçao*, 2: 1–120.

Wade, B.A., 1972. A description of a highly diverse soft bottom community in Kingston Harbour, Jamaica. *Mar. Biol.*, 13: 57–69.

Wanders, J.B.W., 1976a. The role of benthic algae in the shallow reef of Curaçao (Netherlands Antilles). I. Primary productivity in the coral reef. *Aquat. Bot.*, 2: 235–270.

Wanders, J.B.W., 1976b. The role of benthic algae in the shallow reef of Curaçao (Netherlands Antilles) II: Primary productivity of the *Sargassum* beds on the north-east coast submarine plateau. *Aquat. Bot.*, 2: 327–336.

Wanders, J.B.W., 1977. The role of benthic algae in the shallow reef of Curaçao (Netherlands Antilles) III: The significance of grazing. *Aquat. Bot.*, 3: 357–390.

Warner, G.F., 1969. The occurrence and distribution of crabs in a Jamaican mangrove swamp. *J. Anim. Ecol.*, 38: 379–389.

Weinstein, M.P. and Heck Jr., K.L., 1979. Ichthyofauna of seagrass meadows along the Caribbean coast of Panama and in the Gulf of Mexico: composition, structure and community ecology. *Mar. Biol.*, 50: 97–107.

Williams Jr., E.H., Clavijo, I., Kimmel, J.J., Colin, P.L., Carela, C.D., Bardales, A.T., Armstrong, R.A., Williams, L.B., Boulon, R.H. and Garcia, J.R., 1983. A checklist of marine plants of the South Coast of the Dominican Republic. *Caribb. J. Sci.*, 19: 39–53.

Williams, S.L. and McRoy, C.P., 1982. Seagrass productivity: The effect of light on carbon uptake. *Aquat. Bot.*, 12: 321–344.

Wolff, T., 1976. Utilization of seagrass in the deep sea. *Aquat. Bot.*, 2: 161–174.

Wood, E.J.F., Odum, W.E. and Zieman, J.C., 1960. Influence of seagrasses on the productivity of coastal lagoons. In: *Lagunas Coasteras, un Simposio, Mem. Simp. Int. Lagunas Costeras*. UNAM-UNESCO, Mexico, D.F., pp. 495–502.

Yanez-Arancibia, A. and Day Jr., J.W., 1982. Ecological characterization of Terminos Lagoon, a tropical lagoon–estuarine system in the southern Gulf of Mexico. In: P. Laserre and H. Postma (Editors), *Coastal Lagoons. Oceanol. Acta Paris*, 5 (Suppl. 4): 431–440.

Yanez-Arancibia, A., Lara-Dominguez, A.L., Sanchez-Gil, P., Vargas Maldonado, I., Chavance, P., Amezcua Linares, F., Aguirre Leon, A. and Diaz Ruiz, S., 1982. Ecosystem dynamics and nichthemeral and seasonal programming of fish community structure in a tropical estuarine inlet, Mexico. In: P. Laserre and H. Postma (Editors), *Coastal Lagoons. Oceanol. Acta Paris*, 5 (Suppl. 4): 417–429.

Young, D.E., Buzas, M.A. and Young, M.W., 1976. Species densities of macrobenthos associated with seagrass: A field experimental study of predation. *J. Mar. Res.*, 34: 577–591.

Zieman, J.C., 1974. Methods for the study of the growth and production of turtle grass, *Thalassia testudinum* Konig. *Aquaculture*, 4: 139–143.

Zieman, J.C., 1975. Seasonal variation of turtle grass, *Thalassia testudinum* Konig, with reference to temperature and salinity effects. *Aquat. Bot.*, 1: 107–123.

Zieman, J.C., Thayer, G.W., Robblee, M.B. and Zieman, R.T., 1979. Production and export of sea grasses from a tropical bay. In: R.J. Livingston (Editor), *Ecological Processes in Coastal and Marine Systems*. Plenum Press, New York, pp. 21–34.

Zieman, J.C., Iverson, R.L. and Ogden, J.C., 1984. Herbivory effects on *Thalassia testudinum* leaf growth and nitrogen content. *Mar. Ecol. Prog. Ser.*, 15: 151–158.

Zimmerman, R., Gibson, R. and Harrington, J., 1979. Herbivory and detritivory among gammaridean amphipods from a Florida seagrass community. *Mar. Biol.*, 54: 41–47.

Chapter 10

OPEN COAST INTERTIDAL AND SHALLOW SUBTIDAL ECOSYSTEMS OF THE NORTHEAST PACIFIC

MICHAEL S. FOSTER, ANDREW P. DE VOGELAERE, JOHN S. OLIVER, JOHN S. PEARSE and CHRISTOPHER HARROLD

INTRODUCTION

The open coastal wave-exposed intertidal and shallow subtidal ecosystems in the northeast Pacific are among the most diverse and well known in the world. In this chapter, we summarize what is known about the ecology of these ecosystems with emphasis on present community structure and organization. We include the geographic area of the eastern Pacific from the Aleutian Islands, Alaska, U.S.A. (53°N, 178°E) to the tip of Baja California, Mexico (23°N, 110°W; Fig. 10.1). Ecosystems discussed are (1) the rocky intertidal zone; (2) rocky subtidal reefs down to the compensation depth (1% of surface irradiance); and (3) intertidal and subtidal beaches down to the compensation depth. More protected bay and estuarine ecosystems are discussed in other volumes in this series (Chapman, 1977; Ketchum, 1983).

This diversity of ecosystems, the size of the region, and the vast amount of information available make a detailed review impossible. Fortunately there are recent reviews of northeast Pacific intertidal shores in general (Carefoot, 1977; Ricketts et al., 1985), rocky shores (Moore and Seed, 1986; Foster et al., 1988), and shallow subtidal rocky reefs (Dayton, 1985; Foster and Schiel, 1985; Schiel and Foster, 1986). In addition, excellent descriptions of the marine flora and fauna are also available for many parts of the region [for example, California: Abbott and Hollenberg (1976) — macroalgae; Smith and Carlton (1975) and Morris et al. (1980) — invertebrates; Miller and Lea (1972) — fishes]. We have used these references extensively for this summary, and encourage the reader to consult them for more detailed information. Moreover, because of our own expertise and the location of most studies, this review is biased towards the coasts of California, Oregon, Washington, and British Columbia. Our approach is to summarize the general abiotic environment of the region, review its biogeography, discuss the community structure (species composition, distribution and abundance) and organization (causes of structure) of each ecosystem, and finally review what we know about the trophic structure and energetics of these systems. Unless otherwise qualified, all statements below refer only to these ecosystems in the northeast Pacific between Alaska and Baja California, Mexico.

ABIOTIC ENVIRONMENT

Geological setting

The northeast Pacific coastline stretches nearly 20 000 km from Alaska to the tip of Baja California, Mexico (Clarke, 1984; Lane, 1985). In regions like Alaska with numerous coastal crenulations, the coastline distance can be increased by a factor of 5.5 if offshore islands, bays and rivers are included to the head of tide water (Lane, 1985).

Rocks in this region are mainly sedimentary. However, small pockets of granite occur just south of San Francisco and around Monterey, and the Aleutian Islands are primarily volcanic (Kinney, 1966). The intertidal substrata range in morphology from massive rock benches to sandy beaches. The California coast consists of 41% rock, 38% sand, 11% marsh and 10% man-made structures (U.S. Army Engineer Division, 1971). The sand and rock substrata are generally intermixed, but sand beaches are dominant south of Santa Bar-

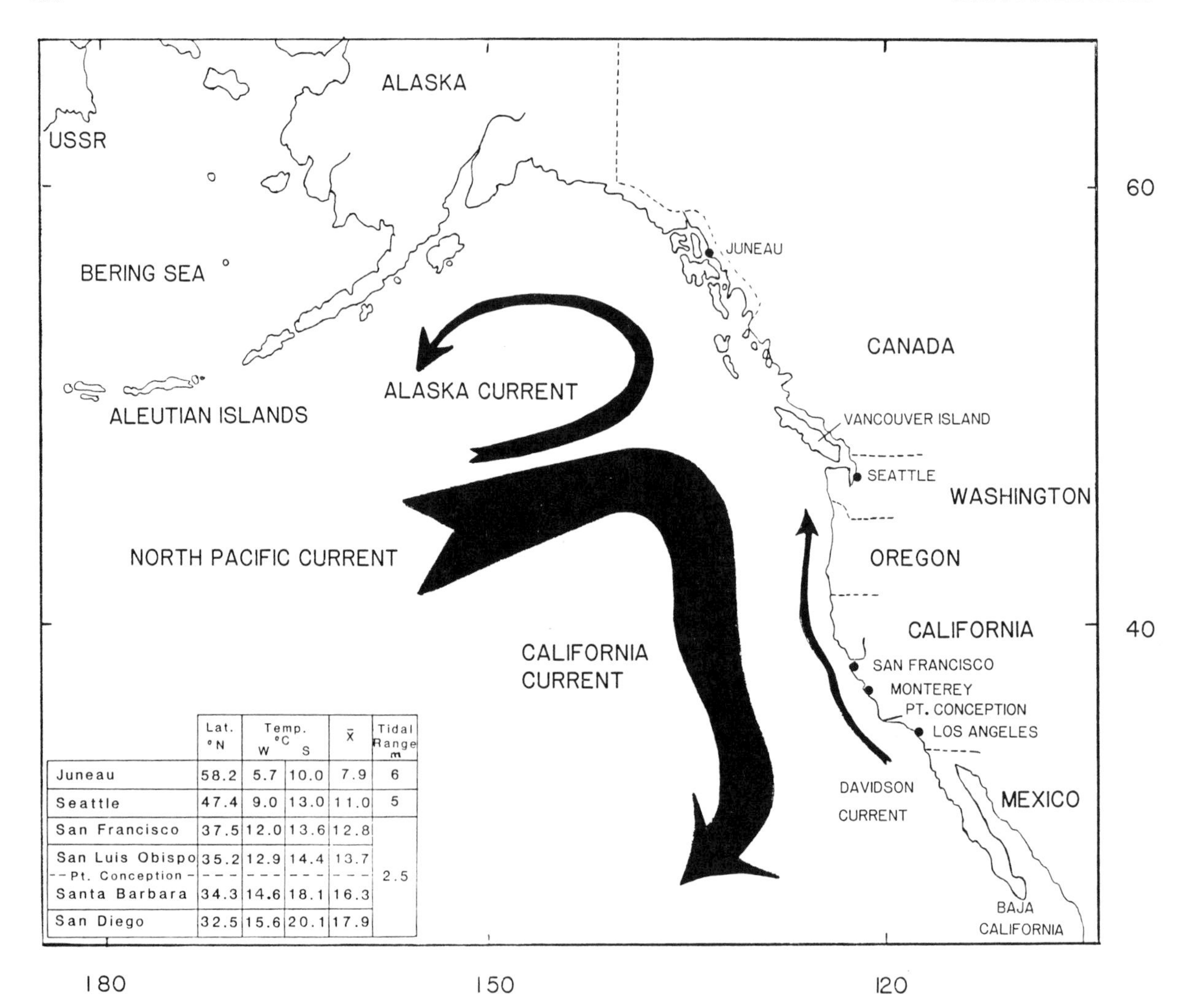

	Lat. °N	Temp. °C W	Temp. °C S	$\bar{X}$	Tidal Range m
Juneau	58.2	5.7	10.0	7.9	6
Seattle	47.4	9.0	13.0	11.0	5
San Francisco	37.5	12.0	13.6	12.8	2.5
San Luis Obispo	35.2	12.9	14.4	13.7	
-- Pt. Conception --	---	---	---	---	
Santa Barbara	34.3	14.6	18.1	16.3	
San Diego	32.5	15.6	20.1	17.9	

Fig. 10.1. Currents, temperatures and tidal ranges in the northeast Pacific. Temperatures calculated from monthly oceanographic data (1984–1985) for stations closest to the cities are indicated. Raw data from Oceanographic Monthly Summary, U.S. Department of Commerce, Washington, D.C. Tidal ranges from Gorshkov (1976). (*W*) Mean for November through April; (*S*) mean for May through October; ($\bar{X}$) yearly mean.

bara. Very conservative estimates of beach expanses in Oregon and Washington are 45% and 31% of the respective coasts [derived from Wiedemann's (1984) dune records]. The south-central region of Alaska (Yakutat Bay to the southern Kenai Peninsula) has been studied extensively (Sears and Zimmerman, 1977; Lippincott, 1980). Substratum types are mixed, with 56% rock or gravel and 44% sand or mud.

Currents

The characteristics of the northeast Pacific coastal waters are influenced by two main surface currents and smaller-scale local phenomena. The nearshore water masses move mainly by way of the Alaska Current and the California Current (Fig. 10.1). The Alaska Current is part of a small gyre surrounded on three sides by Canada, the Aleutian Islands, and mainland Alaska. The California Current is linked to larger-scale circulation patterns, as is demonstrated by Japanese fishing buoys sometimes floating to the California coast. These buoys are carried through the large North Pacific gyre by the Kuroshio Current, into the North Pacific Current, and finally into the California Current.

The California Current brings cold water from

the north down along the coast to Pt. Conception, where it flows offshore (Bolin and Abbott, 1963). This causes a relatively sharp temperature gradient with a rapid increase to the south (Fig. 10.1). However, during the winter months, southwesterly winds drive the Davidson Current (Fig. 10.1) northward from southern California to the U.S.–Canadian border. These currents may be altered during large-scale "El Niño" oceanographic events (Cowen, 1985).

Drift card studies, particularly in California, have revealed much about local movements of surface waters. Such studies in Monterey Bay reveal considerable variability in local current patterns (Griggs, 1974; Fig. 10.2A). Currents may vary on a yearly basis and on shorter time scales within one set of card releases. They may also vary on the scale of hours in both speed and direction (Fig. 10.2B). Thus, large-scale movements of water associated with the major currents determine the general oceanographic setting of nearshore communities, but there are many small-scale changes in the flow of water along the coast that may affect dispersal of, and food available to, nearshore populations.

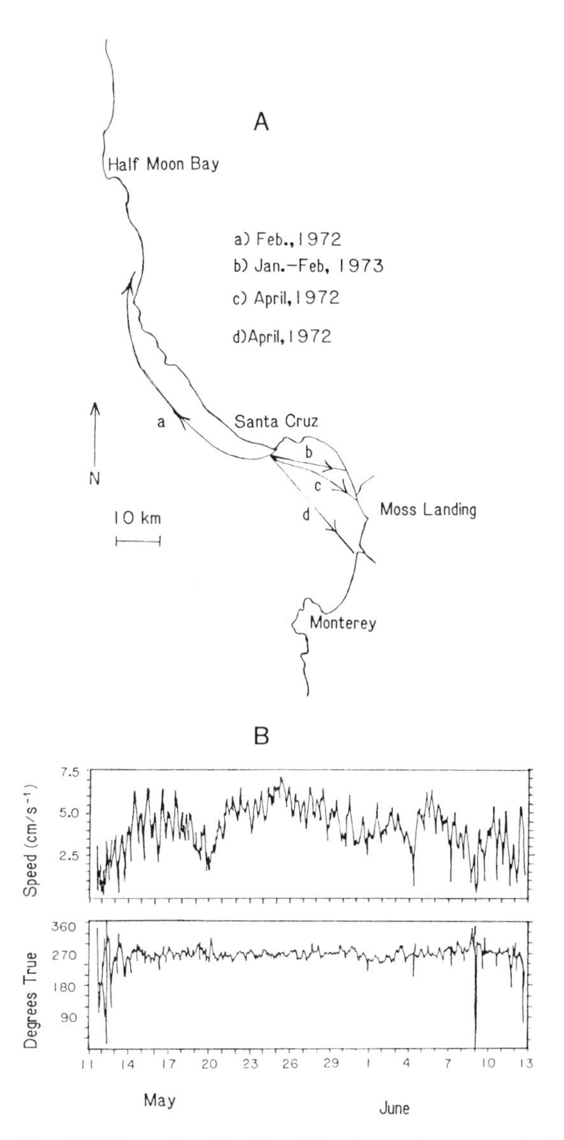

Fig. 10.2. Examples of local variation in nearshore currents. (A) Selected drift-card returns indicating yearly current differences and variation within one release (modified from Griggs, 1974). (B) Variation in local current speed and direction at one station near Santa Barbara, Ca. (Courtesy Kinetic Laboratories, Inc., Santa Cruz, Ca.)

Temperature

Ocean surface temperatures generally decrease with increasing latitude (Sverdrup et al., 1942; Fig. 10.1). In the northeast Pacific they range from 5°C near the Aleutian Islands to 20°C near Baja California (Miles et al., 1982). The temperature distributions are related to current patterns, as indicated by the cold water associated with the California Current, and the sharp temperature gradient at Pt. Conception (see section on Currents above). Temperatures along the coast will increase during El Niño. This is an anomalous oceanographic event related to the reversal of trade winds, during which warm water extends up the northeast Pacific coast (Vallis, 1986).

Surface temperatures vary seasonally with changes in incoming solar radiation, ocean currents, and the character of prevailing winds (Sverdrup et al., 1942). Mid-latitudes of the Pacific Ocean (around 40°N) have an annual temperature range of 5°C, while water at the pole remains consistently cold. A semiannual variation is present near the Equator which corresponds to cycles of incoming solar radiation (Sverdrup et al., 1942).

Temperature is also affected by coastal upwelling, a seasonal feature most prevalent off California and parts of Baja California, Mexico (Sverdrup et al., 1942; Dawson, 1951; Parrish et al., 1981). From the beginning of March until July (and occasionally at other times), colder, deep water (below 200 m) rises and replaces the surface layers.

In regions of intense upwelling (around 35° and 40°N) surface temperatures may be lower in the spring than in the winter. The upwelled bottom water also affects nearshore biotic communities because it is nutrient-rich (see section on Nutrients below).

Temperature usually varies with exposure to air in the intertidal zone and with depth in the subtidal zone. Intertidal temperatures during low tide can be similar to air temperatures. However, if the substratum remains moist (as on mud flats and porous rocks), temperature variation is reduced. Temperature decreases with depth in the shallow subtidal zone (for example, 3°C at 10 m: Jackson, 1977), but water in this habitat remains fairly well mixed by waves, surge and, in some areas, upwelling. However, south of Pt. Conception there is often a shallow thermocline within 10 m of the surface that separates wind-mixed surface water from cooler, bottom water (Zimmerman and Kremer, 1984; Zimmerman and Robertson, 1985).

Tides and waves

Tides along the northeast Pacific coast are mixed semidiurnal; two highs and two lows occur each lunar day, with successive high waters and successive low waters each having different heights. The differences between high and low tides range from 1.6 to 6.0 m, generally increasing with higher latitude (Gorshkov, 1976). In Alaska, some exceptionally high ranges are the result of particular coastal or bay geometry.

The rise and fall of tides create the intertidal zone habitat, an environment of extreme fluctuation of physical parameters. Yet the effects of tides are modified by latitudinal, seasonal and local variability. At high latitudes the daylength is very short during the winter months, so drying during this season is relatively mild. Desiccation of intertidal organisms is also ameliorated in areas such as Monterey with a high frequency of coastal fog during the spring and summer.

Much of the coast in this region is exposed to the full force of waves generated in the Pacific Ocean, particularly the coast from Pt. Conception to the Canadian border. Large waves from the northwest, west, and southwest occur seasonally associated with winter storms. The southern California mainland is generally protected from these waves by offshore islands. However, coasts in this area are exposed to waves from the southwest generated by southern storms and occasional hurricanes which originate off Mexico and Central America in summer (Emery, 1960). In addition to their mechanical effects on nearshore organisms, waves also modify the physical conditions of the intertidal zone. They may wash over the entire intertidal zone on exposed shores during low tides, while in bays and estuaries only boat wakes affect general tide patterns. So again, as will be shown throughout this review, local variability in general patterns is common.

Light

Primary production in the water column is dependent on water clarity, and the compensation depth in the region may vary from a few to over 30 m. It is locally affected by terrestrial run-off, sediments resuspended by wave surge, plankton abundance, particulate matter from large stands of algae or seagrasses, and shading from a variety of plant canopies. Generally, all but the latter contribute to more turbid water in mainland than in island sites (Ebeling et al., 1980a). Light rarely penetrates deeper than 5–15 mm into mud and sand beaches. Light also varies with season and latitude.

Light has other important ecological and biological effects in addition to driving primary production. Many kelps require a certain amount of blue light for the induction of fertility in gametophytes (Lüning and Neushul, 1978; Lüning, 1980). Seasonal reproduction of animals has also been related to light intensity and length of day (Pearse and Eernisse, 1982). Behavior patterns related to light include diurnal vertical migrations of plankton (Enright and Honegger, 1977), larval taxis towards and away from light depending on stage of development (Bousfield, 1955), and day–night changes in fish distribution and activity in kelp forests (Ebeling and Bray, 1976).

Salinity, oxygen, and nutrients

Salinity, oxygen, and nutrients can affect the distribution and abundance of organisms. These factors are much more variable between site and habitat than they are with latitude.

Salinity of the sea surface varies slightly with latitude, depending on the amounts of rain and evaporation. Coastal waters may have higher local variability depending on the proximity to river mouths. There also may be fresh-water seepage floating to the surface from ground water tables. In the intertidal zone, salinity can decrease rapidly when a low tide and heavy rain coincide. The highest ranges of salinity are, of course, in estuaries. Where rivers meet the ocean, salinity varies with river output and direction of tide to form a mobile gradient ranging from fresh to 35‰.

The distribution of oxygen in the ocean is mainly controlled through exchange with the atmosphere, by photosynthesis, and by respiration. In the habitats considered, oxygen concentrations may only become limiting in bays with little tidal movement, and in sediments. Oxygen concentrations in mud bottoms may decrease to an anoxic level at depths as little as 5 mm. Burrowing organisms play an important role in mixing sediment layers and allowing oxygen to penetrate deeper.

Nutrient concentrations change with upwelling and turnover of surface water by wind mixing, and these changes are often correlated with phytoplankton (Cushing, 1959; Bolin and Abbott, 1963) and macroalgal growth (summarized by Carefoot, 1977). Foster and Schiel (1985) have reviewed the nutrient needs of macroalgae and concluded that, although few algae have been examined in detail, nitrates appear to be the most important limiting nutrient in the region, particularly in southern California. Nitrate concentrations must be 1 to 2 μM for the typical growth rates of giant kelps (Gerard, 1982). Levels below this are found in southern California during the summer and fall when warm water masses move up from the south (Jackson, 1977; Wheeler and North, 1981; North et al., 1982). Nitrogen concentrations vary widely with upwelling and amount of terrestrial runoff (North et al., 1982). They are also influenced locally by tides and internal waves. Zimmerman and Kremer (1984) found that, within a kelp forest in southern California, there was frequently more variability in nutrient concentrations between depths on a given day than between seasons.

BIOGEOGRAPHY

The coastline from Alaska to Baja California includes three of the four major faunal zones: cold-temperate, warm-temperate, and tropical. Several faunal provinces have been suggested, the precise number depending on the investigator. The boundaries between the provinces are based on distributional patterns of several major taxonomic groups, and the precise location of boundaries is still a matter of debate and research. Over a dozen schemes for dividing the region into faunal provinces have been proposed over the past 120 years, and the scheme one employs is mostly a matter of personal choice. Hall (1964), Valentine (1966), and Briggs (1974) each reviewed previous schemes and each proposed their own. For this discussion, we have used the scheme (Fig. 10.3) proposed by Briggs (1974), which includes four geographic provinces.

A number of different animals (echinoderms, nemerteans, ascidians, hydroids, bryozoans, fishes, and molluscs) and algae have been used as the basis for establishing biogeographic provinces. Areas that include the range limits of many species are considered as boundaries, and areas with a high degree of endemism are considered provinces. However, the precise locations of these boundaries and provinces remain topics of debate, because it is difficult to establish objective criteria for defining them. Moreover, collection records used to define boundaries may be incomplete and affected by the current state of taxonomy for a particular species or group (Druehl, 1981). Valentine (1966), Hayden and Dolan (1976), and Seapy and Littler (1980) have utilized computer-assisted techniques, such as cluster analysis, to provide more objective criteria for describing faunal provinces. In general, their proposed biogeographic schemes agree with those of previous workers.

Biogeographic differences in algal floras have been discussed by Scagel (1963), and have recently been reviewed in detail by Murray et al. (1980) for California. Distributional end points for northern and southern species are particularly common along the California coast, producing a rich and diverse flora (Abbott and Hollenberg, 1976; Murray et al., 1980). With the exception of the region around Pt. Conception (Fig. 10.3), and perhaps

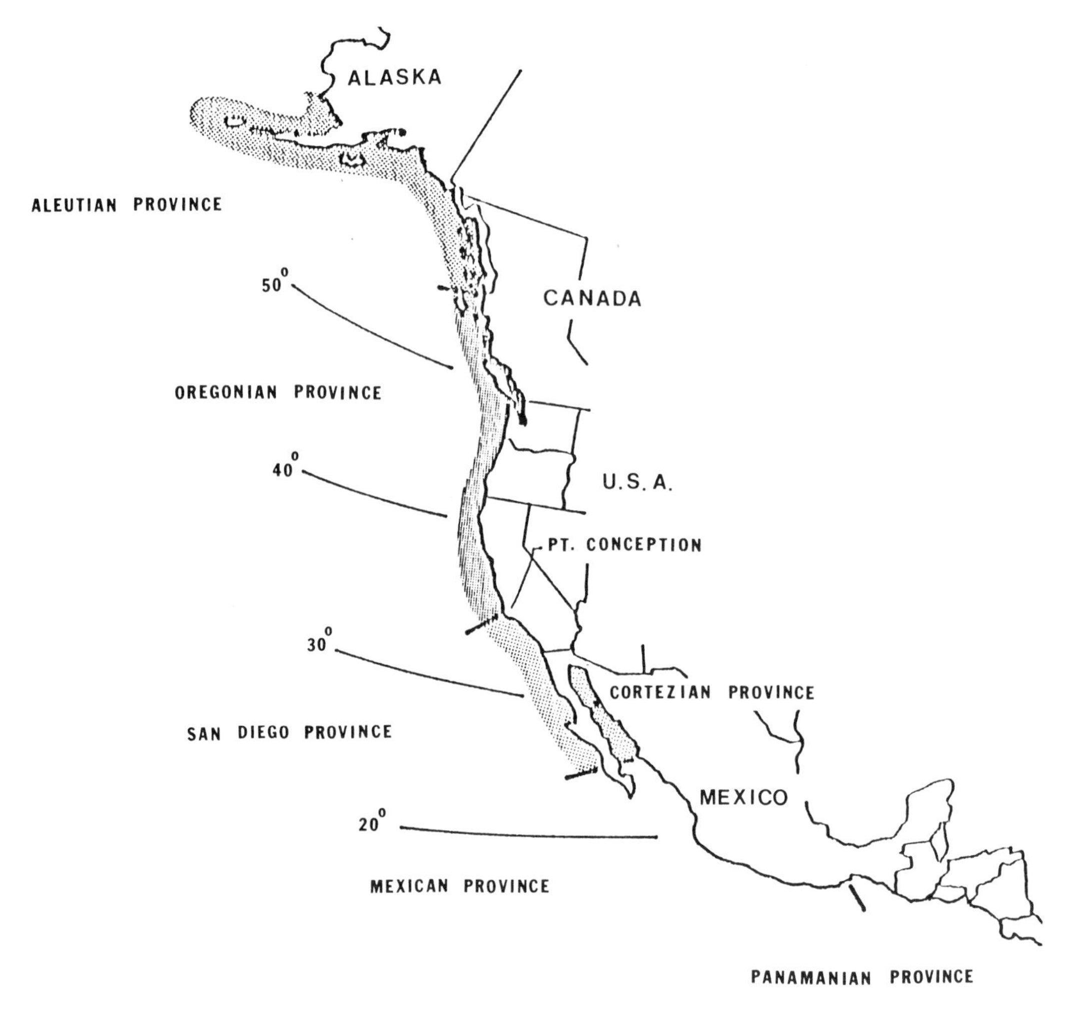

Fig. 10.3. Zoogeographical provinces of the Pacific coast of North America. (Modified from Briggs, 1974.)

Monterey, changes in the flora with latitude are gradual (Scagel, 1963; Murray et al., 1980).

What are the underlying factors responsible for biogeographic patterns in this region? Temperature is the causal agent most often cited (see, for instance, Hutchins, 1947), and the association of temperate algal species with areas of cold, upwelled water along the Baja California coast (Dawson, 1951) suggests that temperature can be important. The relatively rapid changes in algal species composition around Pt. Conception are also associated with changes in temperature (Murray et al., 1980), but other factors such as changes in nutrients (Jackson, 1977) and circulation patterns (Cowen, 1985) are probably important as well. To our knowledge, no one has suggested a correlation with oceanographic conditions for the observed floral break around Monterey; perhaps it is related to the diversity and extent of rocky habitats in the area and to the intensity of collection.

A direct correlation between faunal distributions and temperature has not yet been shown. Valentine (1966) attributed this difficulty to the fact that no single criterion for a "temperature factor" can be found, and concluded that overall provincial and sub-provincial patterns are controlled by thermal regimes, but not everywhere by the same thermal attributes. Fritchman (1962) concluded that the ranges of limpets studied in central California were determined by failure of reproduction at unsuitable temperatures, and that these temperatures were different for different species. Similarly, Hall

(1964) suggested that the number of consecutive days or months that shallow sea temperature is appropriate for reproduction and early growth may be the critical factor limiting molluscan distributions. Studies by Cowen (1985) on current patterns and the recruitment of the sheep-head (*Semicossyphus pulcher*) suggest that hydrographic constraints on dispersal are most important in producing the faunal break at Pt. Conception. Others have argued for the importance of integrating a variety of oceanographic features (Hayden and Dolan, 1976) or for the importance of other factors such as moistening of intertidal organisms (Mokyevsky, 1960), the effects of sea ice (Wethey, 1985), the effects of grazers on algal distribution (Gaines and Lubchenco, 1982), and geographic variation in disturbance (Sousa et al., 1981).

The provincial boundaries shown in Fig. 10.3 coincide with positions of changes in the flow of major current systems (Fig. 10.1). The northern boundary of the Oregonian Province at Dixon Entrance coincides with a transitional area where the eastward flowing West Wind Drift splits into a northward flowing current and the southward flowing California Current (Briggs, 1974). The boundary between the Oregonian and San Diego Provinces at Pt. Conception coincides with the departure of the California Current from the coastline. The southern California coastline and inner southern California islands are bathed by the northwesterly flow of the Southern California Eddy (Seapy and Littler, 1980). The temperate/tropical boundary between the San Diego and Mexican Provinces probably arises from the declining influence of the California Current as it turns westward away from the coastline just north of the boundary. Some workers have suggested a faunal boundary at Monterey Bay (Hall, 1964; Valentine, 1966; Hayden and Dolan, 1976), but Hartman and Zahary (1983) found no evidence supporting this view. As suggested for the algae above, the Monterey boundary may be an artifact of habitat diversity and sampling intensity in this area.

Although each province has characteristic assemblages of species, considerable species overlap occurs, even across such well-defined boundaries as Pt. Conception. The degree of overlap may also change with time. Many species of warm-water fishes are known to find their way as far north as Monterey Bay during periods of anomalous oceanographic conditions such as the 1982/83 El Niño. If such anomalies are sufficiently long-lasting, overlap will also show up in other groups of organisms as well.

OPEN COAST ROCKY INTERTIDAL ECOSYSTEMS

Patterns of distribution

Rocky shores are rich in species, especially in areas where fog reduces the risk of desiccation during summer low tides (Stephenson and Stephenson, 1972). Rocky shore communities also vary in composition both geographically and locally, especially relative to wave exposure and sand movement. With the exception of a few local investigations (for example, Dayton, 1971, 1975a; Seapy and Littler, 1978; Littler et al., 1983; D'Antonio, 1986) and general natural-history observations (such as, for instance, Stephenson and Stephenson, 1972; Carefoot, 1977; Ricketts et al., 1985), this variation has not been systematically described. Thus, it is difficult to summarize overall patterns and their causes, and the overview provided below should be considered as suggestive only.

The most obvious pattern of distribution on rocky shores is the vertical zonation of species. Several schemes of subdividing rocky shores into horizontal bands or zones have been proposed, and the major ones have been reviewed by Ricketts et al. (1985). Included in their discussion of intertidal zonation is Ricketts' own scheme and, because it was developed for the Northeast Pacific shores, we have adopted this scheme as a basis for comparison (Table 10.1).

The Ricketts scheme of intertidal zonation is based on the extent of tidal exposure. According to this scheme there are four well-defined horizontal bands, zones, or associations. These are discussed below using examples of organisms commonly found in central California. The highest or splash zone, Zone 1, is only infrequently wetted by storm waves and spray. It is mainly bare rock or covered with small green and blue-green algae. Larger green (*Enteromorpha* spp., *Ulva* spp.) or red (*Porphyra* spp., *Bangia vermicularis*) algae and masses of benthic diatoms may be present, especi-

TABLE 10.1

Patterns of zonation

Ricketts et al. (1985) Pacific Coast, N. America	Carefoot (1977) Vancouver Island, B.C.	Pearse (1980) Año Nuevo Island
LITTORINA KEENAE *Porphyra* *Cladophora* *Enteromorpha* *Ligia* *Balanus glandula*	*PORPHYRA*	SPLASH ZONE Blue-green algae Lichens *Porcellio scaber* *Ligia pallasii* *Littorina keenae*
	BARNACLE ZONE *Balanus glandula* *Chthamalus*	
BALANUS GLANDULA *Pelvetia* *Littorina scutulata* *Tegula funebralis*	MIXED BARNACLES/SEAWEEDS *Balanus glandula* *Mastocarpus* *Fucus/Pelvetiopsis*	HIGH ZONE *Porphyra* spp. *Mastocarpus papillatus* *Endocladia muricata* *Pelvetiopsis limitata* *Balanus glandula* *Chthamalus dalli* *Mytilus californianus* *Phragmatopoma californica* *Pollicipes polymerus* *Anthopleura elegantissima* *Collisella digitalis* *Collisella scabra* *Tegula funebralis* *Littorina scutulata/plena* *Nuttallina californica*
MYTILUS *Pollicipes* *Nucella emarginata* *Katharina* *Nuttallina*	MUSSELS/GOOSE BARNACLES *Mytilus californianus* *Pollicipes*	MID ZONE *Iridaea flaccida* *Anthropleura elegantissima* *Anthropleura xanthogrammica* *Dodecaceria fewkesi* *Collisella digitalis* *Collisella pelta* *Collisella scabra* *Notoacmaea scutum* *Tegula funebralis* *Tegula brunnea* *Pisaster ochraceus*
	BARNACLES/ALGAE *Balanus cariosus* *Ulva* *Halosaccion* Whelks Limpets	
PHYLLOSPADIX Laminarians	HEDOPHYLLUM Chitons Sea stars *Phyllospadix*	LOW ZONE *Phyllospadix* *Laminaria* spp. *Egregia menziesii* Sponges Bryozoans Tunicates

TABLE 10.1 (continued)

Stephenson and Stephenson (1972) Pacific Grove, Ca.	Ferguson (1984) Big Sur Coast, Ca.	Seapy and Littler (1978) Cayucos, Ca. (sea stack)	Caplan and Boolootian (1967) San Nicolas Island, Ca.
SUPRALITTORAL FRINGE *Littorina planaxis* *Ligia* *Pachygrapsus*	SPLASH ZONE *Littorina keenae* *Littorina scutulata* *Ligia*		LITTORAL FRINGE *Chthamalus fissus* *Littorina keenae*
UPPER MID-INTERTIDAL *Balanus glandula* *Tetraclita*	HIGH ZONE *Porphyra* *Pachygrapsus* *Collisella digitalis/ austrodigitalis* *Collisella scabra* *Lottia*	*BALANUS/ CHTHAMALUS*	UPPER EULITTORAL *Littorina scutulata* *Collisella digitalis*
LOWER MID-INTERTIDAL *Chthamalus dalli* *Tegula funebralis* *Thais* *Nuttallina*	MID ZONE *Pagurus* *Tegula funebralis* *Anthopleura elegantissima* *Haliotis* *Lottia* *Katharina* *Mytilus/ Pollicipes* (only on offshore rocks)	*MYTILUS/ POLLICIPES* *ENDOCLADIA* *EGREGIA/ LITHOPHYLLUM*	MIDDLE EULITTORAL *Mytilus californianus* *Lottia* *Collisella scabra*
INFRALITTORAL FRINGE *Alaria* *Lessoniopsis*	LOW ZONE Sponges Bryozoans Tunicates *Mopalia* *Tonicella* *Leptasterias* *Phyllospadix*	*CORALLINA/ PHYLLOSPADIX* MIXED REDS/ *PHYLLOSPADIX* MIXED REDS	LOWER EULITTORAL *Corallina* *Lithothamnion* *Nuttallina* *Strongylocentrotus*

ally in winter and spring (Castenholtz, 1961; Cubit, 1984). The few animals that occupy this zone include the limpet *Collisella digitalis*, other gastropods such as *Littorina keenae*, and isopods (*Ligia* spp.). Zone 2 is the high intertidal, and is usually exposed to air for a long period at least once a day. It is characterized by dense populations of barnacles (*Balanus glandula*) and frequently referred to as the barnacle zone. In addition to barnacles, the algae *Endocladia muricata*, *Mastocarpus papillatus* (= *Gigartina papillata*) and *Pelvetia fastigiata* are conspicuous and characteristic members of this zone. The tiny snail *Littorina scutulata* (= *L. plena*), the turban snail *Tegula funebralis*, and several species of limpets also occupy this zone. Zone 3, often called the mussel zone, is the middle intertidal. It is generally exposed to air for relatively short periods twice a day and, on moderately to fully exposed shores, its most conspicuous members are mussels (predominantly *Mytilus californianus*) and gooseneck barnacles (*Pollicipes polymerus*). Other characteristic species, especially on more protected shores, are the predatory snail *Nucella emarginata*, the chitons *Katharina tunicata* and *Nuttallina californica*, and the red alga *Iridaea flaccida*. Zone 4 is the low intertidal and is only uncovered by the lowest tides; it may be covered with water most days of each month. This zone is typically identified by carpets of surf-grass (*Phyllospadix* spp.), various kelps (particularly *Laminaria setchellii*) and a variety of red algae. Although these plants are most conspicuous, sponges, hydroids, and ascidians also occur in this zone.

Many studies of vertical zonation of intertidal organisms along the northeast Pacific coast have turned up a pattern similar to that of Ricketts. In Table 10.1 we also present other zonation patterns reported from British Columbia to southern California. Species assemblages or zones are shown in capitals, indicating what species are typical or characteristic of the assemblage. These species may or may not have been explicitly given by the authors. Within each column the assemblages are grouped vertically in their order of occurrence on the shore, without reference to tidal level since the studies are from several different locales with different exposure to surf. The order of species within a group does not necessarily reflect their vertical position in that group. Common to all seven studies are four distinct zones: a high splash or littorine zone [littorines were absent from this zone in the study by Seapy and Littler (1978)], a *Balanus* zone, a mussel zone, and a low *Phyllospadix* zone. Stephenson and Stephenson (1949) did not include a mussel zone, while Seapy and Littler (1978) and Carefoot (1977) distinguished additional assemblages between these common zones. The similarities among the seven sites could be misleading, however, because all the studies were done in similar habitats: steeply sloping rocky shores exposed to moderate to severe wave action. It is difficult to know whether such shores are "typical" because the relative abundance of the different kinds of shores (broad, flat benches,

Fig. 10.4. Vertical distribution of large intertidal organisms on a steeply sloping rocky shore near Monterey, Ca. The site is protected from direct wave exposure. (A) Supralittoral green alga, *Prasiola meridionalis*; (B) "bare zone" of limpets and littorines; (C) clumps of the goose barnacle *Pollicipes polymerus*; (D) clumps of the red alga *Endocladia muricata*; (E) a band of the barnacle *Tetraclita rubescens*; (F) *Iridaea flaccida* and other red algae; (G) various red algae including *Prionitis* spp. and *Gigartina* spp.; (H) surf-grass, *Phyllospadix* spp.

boulder fields, etc.) found between British Columbia and Baja California has not been determined.

Plants are frequently the most conspicuous organisms on intertidal shores in the region (Fig. 10.4). Large brown algae such as *Fucus distichus*, *Alaria* spp., *Hedophyllum sessile*, and *Laminaria* spp. tend to be most common between Alaska and Oregon, while red algae become more conspicuous as one proceeds south into California. Small red algae such as articulated corallines, various filamentous species, and encrusting coralline algae, are particularly common in southern California (Sousa et al., 1981; Stewart, 1982). Latitudinal trends in animals are less obvious; species composition changes gradually but similar groups (such as littorines, barnacles, limpets) are found at similar tidal levels throughout the region (Foster et al., 1988).

In addition to variation with tidal height, distinct changes in composition can be found locally along gradients of exposure to waves (Fig. 10.5) and sand inundation. High wave exposure results in a suite of species dominated by algae such as *Postelsia palmaeformis*, *Lessoniopsis littoralis*, and corallines, and animals such as mussels and goose barnacles, while a variety of more delicate plants and animals occur in more sheltered locations (Ricketts et al., 1985). Sand inundation selects for species capable of withstanding sand burial and abrasion (for instance, the algae *Laminaria sinclairii*, *Zonaria farlowii*, *Rhodomela larix*, and *Gymnogongrus* spp. and the anemone, *Anthopleura elegantissima*), and those that can recruit, grow, and reproduce during periods when sand is not present (for instance, the algae *Chaetomorpha linum*, *Ulva lobata*, and *Enteromorpha intestinalis*; Dahl, 1971; Markham and Newroth, 1972; Markham, 1973; Littler et al., 1983; D'Antonio, 1986).

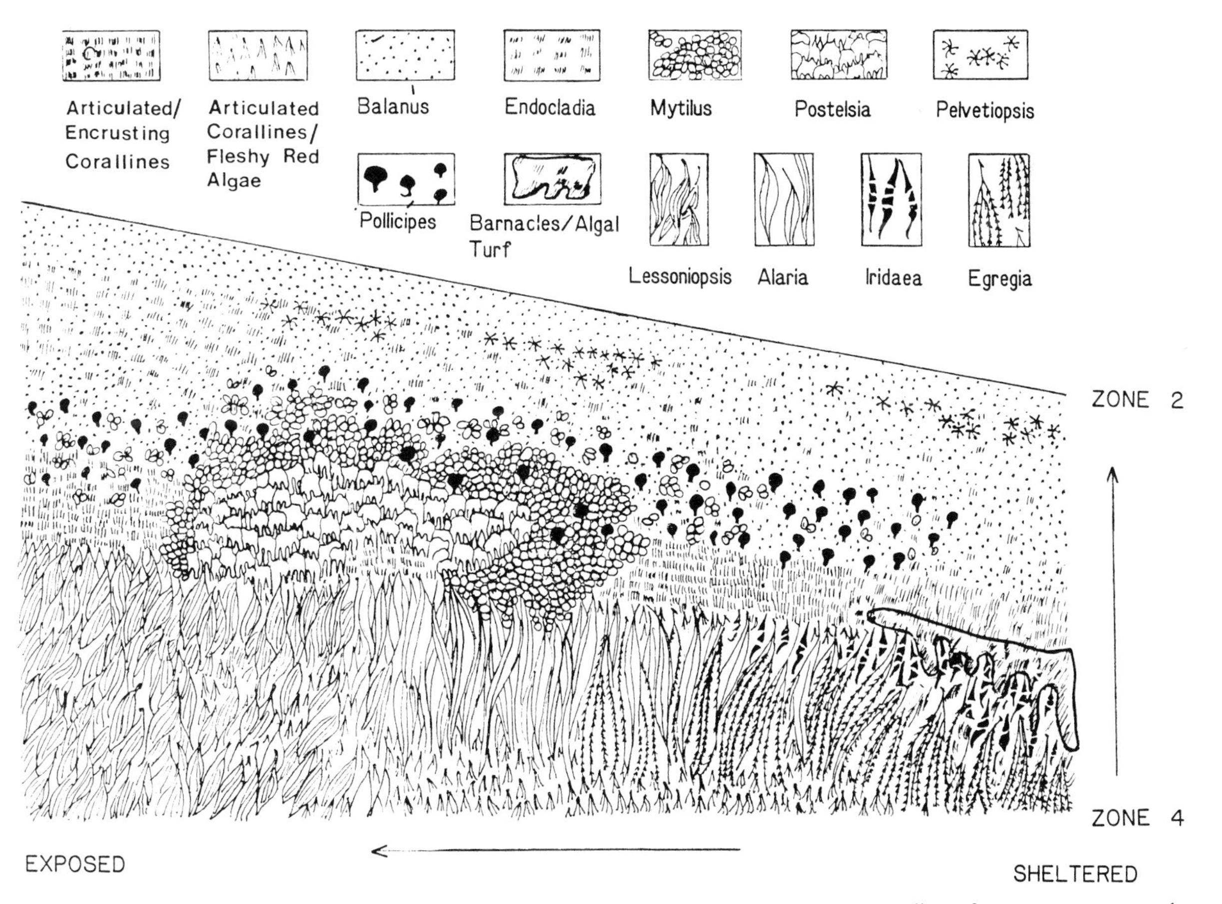

Fig. 10.5. Changes in the distribution and abundance of conspicuous plants and animals along a gradient of wave exposure on rocky shores near Monterey, Ca. (Redrawn from Stephenson and Stephenson, 1972.)

Species generally have a clumped distribution within a particular zone even when wave exposure is spatially constant (Foster et al., 1988), producing a mosaic with the more abundant, large species dominating the observed pattern. Such clumping is obvious on all shores, and numerous examples have been described in the literature (such as, for example, *Postelsia palmaeformis* and other algae — Dayton, 1973; Paine, 1979; Gunnill, 1980a, b; limpets — Frank, 1965, 1982; anemones — Francis, 1973; sea urchins — Paine and Vadas, 1969).

In addition to variation in distribution at various spatial scales, these ecosystems also vary in time. Temporal patterns are not obvious or well described; the necessary long-term studies are few. A general seasonal summer maximum and winter minimum cover has been found for many algae in one long-term study in central California (Fig. 10.6). Most of the algae in this study were perennial and it is likely that variation in cover was due to changes in the abundance of vegetative parts, not individual plants. Seasonality may be

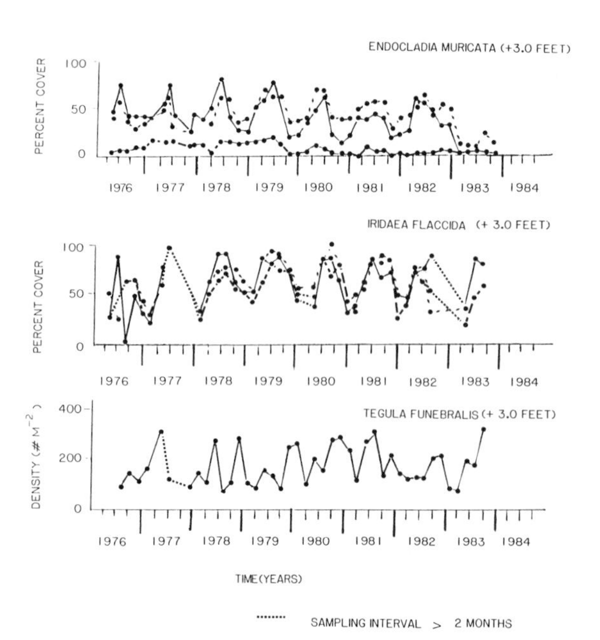

Fig. 10.6. Seasonal and year-to-year changes in the abundance of two perennial red algae (*Endocladia muricata* and *Iridaea flaccida*) and the black turban snail (*Tegula funebralis*) at sites near San Luis Obispo, Ca. Data are from three fixed 0.25 m^2 quadrats, approximately 1 m above mean sea level. Cover determined under 60 randomly placed points in each quadrat. Density is from sum of counts in three quadrats (from PG&E, 1984.)

more pronounced in Alaska where, at one site examined, the cover of the large, apparently annual brown alga *Alaria* sp. varied between approximately 20% (March, 1975) and 100% (August, 1975) (Dames and Moore, 1976). A similar large seasonal variation in Zone 1 algal species such as *Bangia vermicularis* and *Porphyra* spp. occurs at sites in Oregon and elsewere (Cubit, 1984). Seasonal trends in the abundance of adult animals are less apparent (Fig. 10.6), although seasonality in reproduction is common (Foster et al., 1988).

Little is known about long-term variation in rocky intertidal ecosystems, most information coming from opportunistic studies after some large-scale episodic disturbance. The results of long-term, detailed sampling studies, such as those of Gunnill's (1980a, b) on various intertidal algae near San Diego, suggest considerable year-to-year variation in recruitment, superimposed on seasonal trends. On the other hand, decade-long records for the chiton *Katharina tunicata* (Giese, 1969) and the sea-star *Pisaster ochraceus* (Halberg et al., 1969) indicate considerable stability in abundance at the sites examined. Populations of mussels and perhaps other intertidal prey of the sea otter, *Enhydra lutris*, have been altered as this mammal has expanded its range in Alaska and California (Simenstad et al., 1978; Estes and VanBlaricom, 1985). Catastrophic disease (Dungan et al., 1982) may also lead to long-term changes in intertidal systems. Other more localized disturbances such as earthquakes, severe storms, and pollution have also induced long-term changes in intertidal ecosystems in the region (see Foster et al., 1988).

Organization of rocky intertidal ecosystems

Of all the spatial patterns discussed above, the one that has historically received the most attention is vertical zonation. Until recently, the most widely accepted hypothesis concerning the cause of zonation was that of critical tidal levels. These levels are positions on the shore where some parameter associated with the tidal cycle, such as time of exposure to air, changes abruptly with a small change in tidal height. The fact that zonation patterns are directly caused by such changes is intuitively appealing, because limits to distribution of intertidal organisms are also abrupt. Colman

(1933) working on British shores, and Hewatt (1937) and Doty (1946) working on the northeast Pacific coast, correlated the vertical distributions of intertidal organisms with critical levels in the tidal cycles associated with abrupt changes in the hours of exposure to air. Shotwell (1950) correlated the upper limits of *Collisella* with critical tidal levels responsible for extremes in desiccation. The critical tidal level hypothesis has been criticized by Connell (1972) largely because the correlation between critical tidal levels and vertical limits of distribution of organisms is not objectively assessed and, looking at the published data, not very precise. Underwood (1978) refuted the hypothesis for British shores because a re-calculation of annual emersion times against height on the shore yielded a smooth, monotonic curve without sharp increases in emersion time, and because the boundaries of species distributions were randomly distributed, not clumped as expected if they were associated with critical tidal levels. In addition, much recent work has shown that biological interactions, behavior, and recruitment can be very important in regulating the vertical distribution of rocky intertidal organisms [see reviews by Connell (1972) and Foster et al. (1988)]. Thus, as a general hypothesis, critical tide levels has been rejected, although it may apply to a specific distribution in a particular place.

The hypothesis of Connell (1972) that upper limits of distribution are determined by physical factors, while lower ones are determined by biological factors, now appears to be more generally accepted. However, Underwood and Denley (1984) have provided a series of alternative hypotheses, mainly dealing with the settlement of faunal planktonic stages, which may modify this view. In addition, the upper limits of high intertidal algae can be regulated by grazers (Castenholz, 1961; Underwood, 1980; Cubit, 1984; Underwood and Jernakoff, 1984).

In sharp contrast to the voluminous literature on vertical zonation, there is little experimental evidence for the causes of distributional changes along exposure gradients; most available information consists of descriptions of exposed versus sheltered areas (for example, Ricketts et al., 1985). In a situation analogous to those described above for geographic distribution and early studies of vertical zonation, numerous correlations between horizontal distribution of species and gradients of wave exposure (Lewis, 1964; Stephenson and Stephenson, 1972; Carefoot, 1977) suggest that wave exposure is the direct cause of the distributions. However, the few experimental studies that have been done indicate that wave exposure may act indirectly by altering competitive interactions and predation (Connell, 1975; Dayton, 1975a; Menge, 1976; Peterson, 1979a). Differences in recruitment (Roughgarden et al., 1984; Connell, 1985), variations in growth rates, intraspecific competition, and substratum stability (Page, 1986), may also influence local population structure.

Dayton's (1971, 1973, 1975a) initial work in Washington State has stimulated many recent investigations of the causes of distributional patterns within intertidal zones. Variation in substratum stability [e.g. boulders *vs.* massive rock: Sousa (1979a, b)] and heterogeneity [e.g. surface texture: Harlin and Lindbergh (1977); presence of crevices: Menge (1978), Lubchenco (1983), Petraitis (1983); variations in slope: Haven (1971), Frank (1982), Gaines (1985); tide pools: Dethier (1984)] contribute to patterns directly through microhabitat diversity, and indirectly by altering biological interactions. In addition, physical disturbances such as waves, log battering, sand burial and abrasion, and biological disturbances caused by the activities of grazers and predators, provide new space for colonization [for instance, Duggins and Dethier (1985); for a review see Sousa (1985)]. These disturbances result in patches of different successional age whose composition is, in turn, affected by a variety of processes including recruitment, growth, competition, and grazing (Sousa, 1984; De Vogelaere, 1987). Finally, this patchiness has cascading effects on understorey algae and smaller invertebrates that use the microhabitats produced by larger plants and animals (see, for example, Glynn, 1965; Dayton, 1975a; Suchanek, 1979; Gunnill, 1983).

Seasonal variation in intertidal algal populations has been associated with a number of climatic and other factors. Hansen (1977) suggested that light may limit the growth of *Iridaea cordata* in winter, and that maturation is genetically controlled. Emerson and Zedler (1978) found that the perennial articulated coralline *Lithothrix aspergillum* declined in cover during the summer at a site near San Diego as a result of increased temperatures and desiccation during low tides. Occasional, widespread loss of the blades of perennial species occurs in Monterey when hot weather (no fog) coincides with summer low tides (Foster, pers. obs.). Gunnill (1980a) found that seasonal variation in the abundance of six intertidal algal species was largely due to variation in recruitment. Recruitment was commonest in spring/summer, and appeared to be related to changes in reproduction. He concluded that these population fluctuations were related to a complex of environmental factors including cloud cover, temperature, and wave action. Stewart (1983) has shown distinct seasonal patterns of abundance in several algal turf species. These patterns were correlated with seasonal changes in sand accumulation. Winter blooms of high intertidal algae appear to be related to more favorable abiotic conditions, which allow their growth to exceed the ability of grazers to remove them (Castenholz, 1961; Lubchenco and Cubit, 1980; Cubit, 1984).

Daily, tidal, and seasonal changes in climate influence intertidal animals in a variety of ways. Winter storms normally remove old or unstable sessile animals [such as mussels, barnacles, and certain tubeworms: Harger and Landenberger (1971), Grant (1977), Mayer et al. (1981)] and many plants with animal associates [such as blades of algae and surf-grass with limpets and snails: Black (1976), Gunnill (1983)]. Such seasonal removal of intertidal organisms provides cleared areas for new recruits, and is important in maintaining the mosaics of species (Paine and Levin, 1981). Episodic changes in climate may have severe impacts on resident organisms (Dungan et al., 1982). However, as pointed out by earlier workers (Hewatt, 1937; MacGinitie, 1938; Gislen, 1943, 1944; Glynn, 1965), seasonal fluctuations are generally moderate along the northeast Pacific coast, and overall winter/summer differences in rocky intertidal ecosystems are not marked.

Marked temporal changes or rhythms in the activities of many or most intertidal animals in the region and elsewhere do occur (see reviews in the work by DeCoursey, 1976; Naylor and Hartnoll, 1979). In particular, feeding, growth, and reproductive activities of intertidal animals often display distinct temporal patterns that are driven by daily, tidal and/or seasonal rhythms. These activity patterns may have major, but as yet largely unexplored, impacts on patterns of distribution and abundance.

ROCKY SUBTIDAL ECOSYSTEMS

Patterns of distribution

Nearshore rocky subtidal ecosystems in the northeast Pacific are, along with similar habitats in Australia and New Zealand, probably the most species rich and productive of all temperate ecosystems (Foster and Schiel, 1985). Adequate light for plant growth, the import of oceanic plankton, generally abundant nutrients from mixing and terrestrial runoff, and lack of desiccation have presumably encouraged the richness and productivity of this ecosystem. In addition, the common presence of surface canopy kelps adds further structural complexity to the system and, by exploiting the higher light levels at the surface, higher primary production.

Studies of this system were rare until the availability of simple diving apparatus in the 1950's. Since that time numerous subtidal studies have been done, especially in Alaska (Dayton, 1975b; Calvin and Ellis, 1978; Estes et al., 1978; Duggins, 1980), central and southern California (reviewed by North and Hubbs, 1968; North, 1971a; Foster and Schiel, 1985) and Canada and Washington State (Neushul, 1967; Foreman, 1977; Lindstrom and Foreman, 1978; Druehl and Wheeler, 1986).

Most shallow subtidal rocky reefs commonly have a marked three-dimensional structure provided by large stipitate and float-bearing kelps, and descriptions of the overall distribution of both plants and mobile animals are commonly made in the context of vertical layering or vertical position (Fig. 10.7). Surface-canopy kelps such as *Macrocystis* occupy the entire water column, producing a forest-like structure. Stipitate kelps such as *Pterygophora californica* and *Laminaria* spp. may form an additional vegetation layer within two meters of the bottom in kelp forests. A third layer of foliose red and brown algae, as well as articulated corallines, is common beneath the understorey kelps (Fig. 10.7, layer 2), with a final turf layer of filamentous and encrusting species on the bottom.

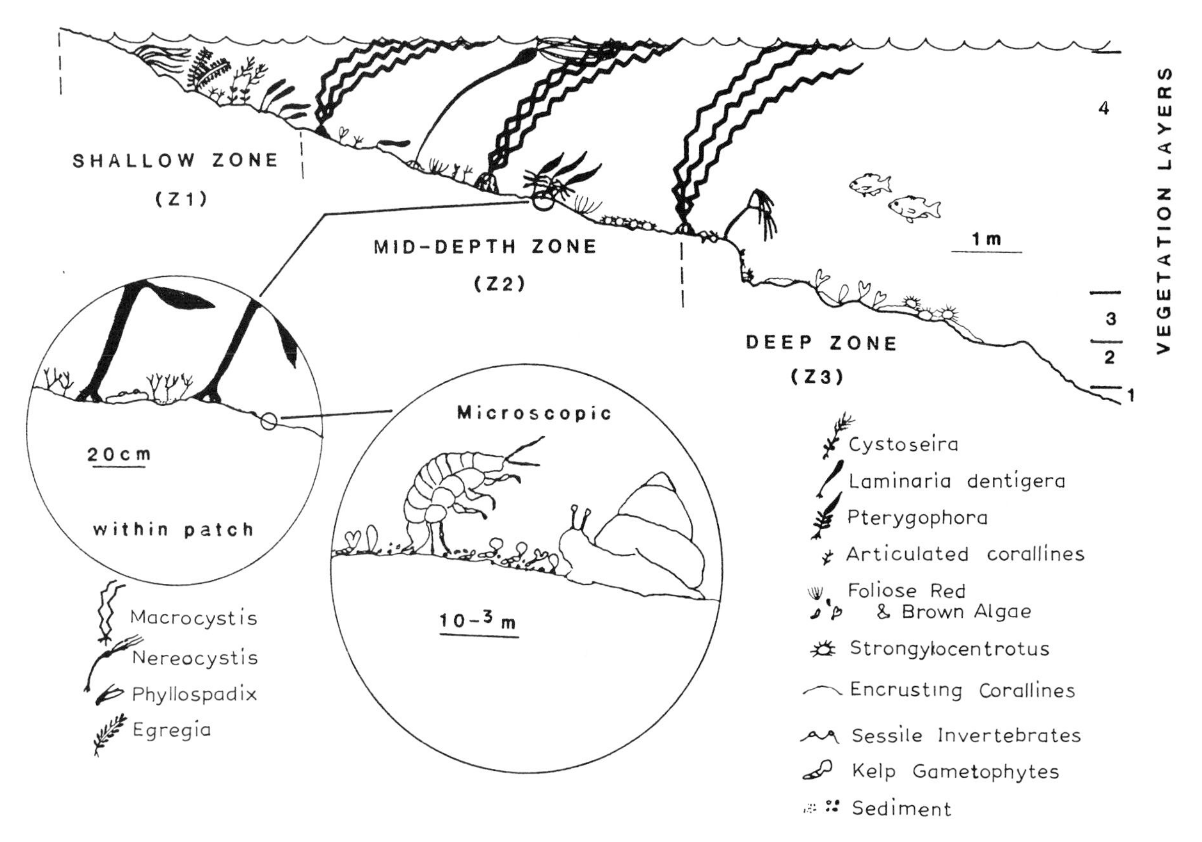

Fig. 10.7. Giant kelp-forest community structure at different depths and spatial scales. (From Foster and Schiel, 1988.)

Numerous crustaceans (Coyer, 1984) and molluscs (Lowry et al., 1974; Riedman et al., 1981) are associated with giant kelp fronds, some at particular depths along fronds. Over 300 species of invertebrates have been described on the bottom at a single site (Pequegnat, 1964). As many as 125 species of fish have been reported from rocky reefs and kelp forests in southern California (Feder et al., 1974). The fishes represent a variety of feeding types (Choat, 1982), and show an overall spatial segregation into surface–midwater species and bottom species (Ebeling et al., 1980a). Marine birds, especially cormorants, pelicans, and gulls, and marine mammals, including the sea otter (*Enhydra lutris*: Kenyon, 1969) and harbor seal (*Phoca vitulina*), commonly occur in kelp-forest habitats (Foster and Schiel, 1985).

There is considerable variation in species composition over the range of this ecosystem. Common surface-canopy algae are *Alaria fistulosa* in southwestern Alaska, *Macrocystis integrifolia* and *Nereocystis luetkeana* from eastern Alaska to Pt. Conception, California, and *Macrocystis pyrifera* from near Santa Cruz, California to Baja California, Mexico (Fig. 10.8). As in the intertidal zone, the limited data suggest that large understorey brown algae are more diverse and abundant in the northern part of the region, while red algae are commoner in the southern part (Dawson et al., 1960; Neushul, 1967; Devinny and Kirkwood, 1974; Dayton, 1975a; Calvin and Ellis, 1978; Lindstrom and Foreman, 1978; Breda and Foster, 1985). However, understorey composition can be highly variable on a local scale as well (Foster and Schiel, 1985). There are changes in the invertebrate fauna within the region, the commonest being the gradual replacement over latitude of one species by another in the same general taxonomic category [such as changes in species of the molluscan genus *Tegula*, and in different genera and species of sea urchins: Morris et al. (1980), Harrold and Pearse (1987)]. The fish assemblages can be quite different, especially north and south of Pt. Conception. Southern waters are generally less turbid, less

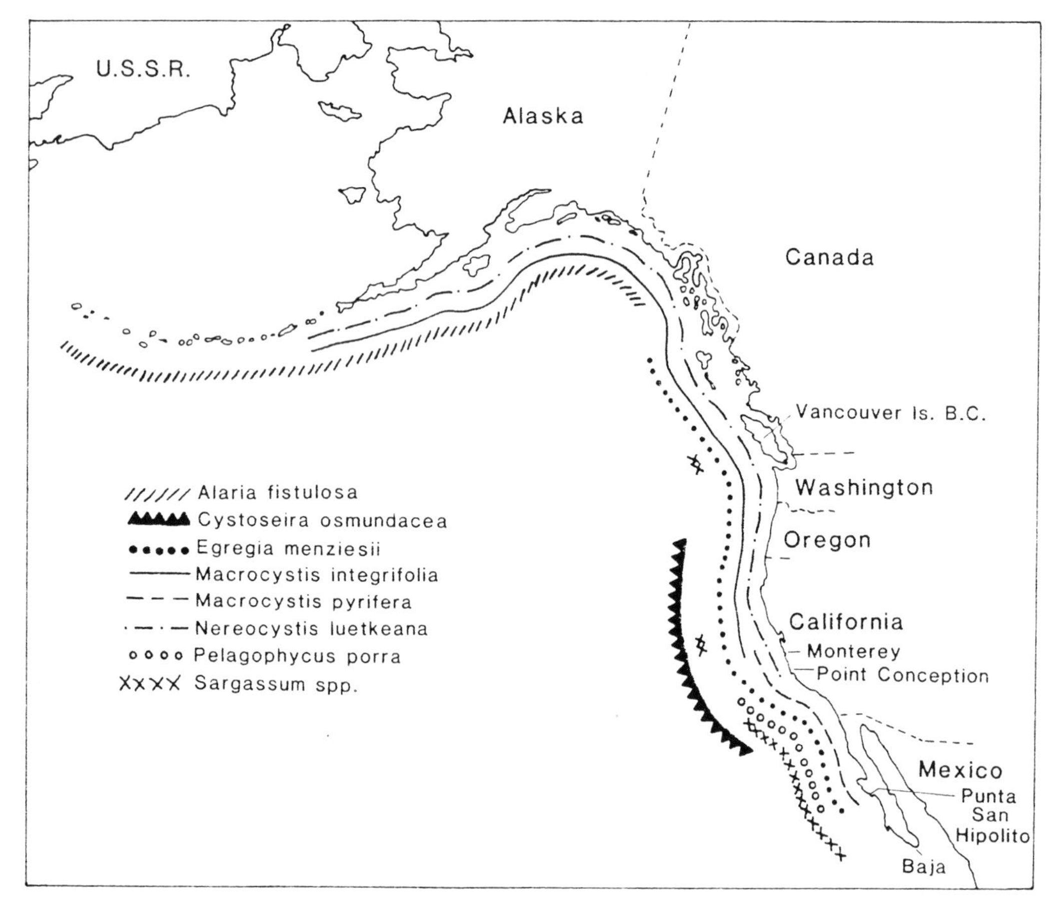

Fig. 10.8. Distribution of algae that form surface canopies in the northeast Pacific. (Redrawn from Druehl, 1970.)

turbulent, and warmer, and tropically derived species or families are much more prevalent there. These families include the Clinidae (clinids), Gobiidae (gobies), Pomacentridae (damsel-fishes), Labridae (wrasses), Serranidae (basses), and Kyphosidae (sea chubs) (Ebeling et al., 1980a, b). Temperate families such as the Embiotocidae (surf-perches), Scorpaenidae (rockfishes), Hexagrammidae (greenlings), and Cottidae (sculpins) are more diverse and abundant to the north (Ebeling et al., 1980a, b). These more temperate families generally also have more species per family, particularly the rockfishes.

As more descriptive surveys have become available, it is apparent that the structure of nearshore reef communities can be quite different at sites within a particular local area, often separated by distances of less than one kilometer (Foster et al., 1983). The depth distribution of the entire community may vary from site to site (Foster and Schiel, 1985). Within the community, some of the most obvious differences occur in the species composition of seaweeds, including canopy kelps (Devinny and Kirkwood, 1974; Foster, 1982a; Dayton et al., 1984). Differences in assemblages of invertebrates (Turner et al., 1968; Foster, 1982a) and fishes (Larson and DeMartini, 1984) have also been described.

These ecosystems also vary in composition with depth (Fig. 10.7). In his pioneering study of depth distribution, Neushul (1967) described three broad subtidal zones or associations of algae at sites in Washington State and, although species composition may change, this pattern is generally true at other locations within the region (Aleem, 1956; Neushul, 1965; Dayton, 1975b; Lindstrom and Foreman, 1978; Foster and Schiel, 1985). Robust plants such as *Laminaria setchellii*, *Phyllospadix* spp., and articulated coralline algae are common in shallow water. Surface-canopy plants are more abundant at intermediate depths, and red algae and sessile invertebrates tend to be the dominant species in deep water.

The zonation of invertebrates and fishes with depth is less obvious and not as well described. Sessile invertebrates are usually commonest on steeply sloping surfaces, vertical walls, and in deep water (Foster and Schiel, 1985). When these animals are very abundant on shallow reefs, they may show distinct zonation patterns along depth gradients and with changes in substratum relief (Pequegnat, 1964). In central California and perhaps elsewhere, these types of invertebrates are also more common on more oceanic sites such as offshore reefs and steeply sloping nearshore habitats (M.S. Foster, pers. observ.). Small mobile invertebrates (such as crustaceans, polychaetes, molluscs, brittle-stars) are abundant in kelp holdfasts (Andrews, 1945; Ghelardi, 1971), mats of coralline algae (Dearn, 1987), patches of sand (Hammer, 1981), and other cryptic subhabitats. These and other mobile animals may also associate with particular species or types of understorey algae (Lowry et al., 1974; Hammer and Zimmerman, 1979). The depth distribution of other animals may be associated with the abundance of macroalgae or the foraging abilities of predators [e.g. sea urchins: Mattison et al. (1977), Estes et al. (1978), and Dean et al. (1984)]. With exceptions such as some perches (Hixon, 1980) and rockfishes (Larson, 1980), fish distributions appear to be more strongly related to substratum relief (including that provided by plants) than depth (Quast, 1971; Ebeling et al., 1980a; Larson and DeMartini, 1984).

Perhaps the safest generalization that can be made about the distribution of plants and animals within a particular depth is that it is usually patchy. Discrete patches or stands of overstorey kelps (Rosenthal et al., 1974; Harrold and Reed, 1985), understorey kelps (Dayton et al., 1984; Reed and Foster, 1984; Schiel and Foster, 1986), large sea urchins and abalone (Lowry and Pearse, 1973; Foster and Schiel, 1988), and sessile invertebrates (Pequegnat, 1964) are notable examples.

All long-term studies of this type of ecosystem have found considerable temporal variation over a number of time scales. Daily changes in activity and distribution of demersal zooplankton (Hammer, 1981), sea urchins (Nelson and Vance, 1979), and fish (Ebeling and Bray, 1976) have been documented. Seasonal changes in surface-canopy kelp abundance have been found in British Columbia (Druehl and Wheeler, 1986), Washington State (Neushul, 1967) and central California (Miller and Geibel, 1973; Kimura and Foster, 1984). Seasonal cycles of kelp recruitment (Reed and Foster, 1984) and the abundance of understorey algae (Foster, 1982a; Breda and Foster, 1985), spider crabs (Hines, 1982), and juvenile rockfish (Miller and Geibel, 1973) are a regular feature of many kelp forests of central California. Adult fish show some variation with season at sites in southern California (Ebeling et al., 1980b). Seasonal variability appears to decrease from Alaska to Baja California, but there are exceptions [for example, giant kelp recruitment at two sites in British Columbia (Druehl and Wheeler, 1986)].

Long-term, interannual changes in populations and communities have also been documented. Rosenthal et al. (1974) noted the episodic loss of patches of giant kelp in southern California. The recent (1982/83) El Niño oceanographic event further emphasized the effects of such episodic events on shallow subtidal ecosystems, particularly in southern California. These effects include a reduction in plant growth (Zimmerman and Robertson, 1985), widespread plant mortality (Dayton and Tegner, 1984a) and, in some cases, invertebrate mortality [e.g. *Patiria miniata* and *Strongylocentrotus franciscanus*: Ebeling et al., (1985), T.A. Dean (pers. observ.), M.S. Foster (pers. observ.)], and changes in fish distribution (R.N. Lea, pers. commun.). Additional changes have resulted from the effects of sea-otter foraging as the range of these animals has increased (Estes et al., 1978; VanBlaricom and Estes, 1988). Finally, long-term changes in both intertidal and subtidal ecosystems have occurred near large sewer outfalls in southern California (Grigg and Kiwala, 1970; Bascom, 1983).

Organization of rocky subtidal ecosystems

Latitudinal changes in rocky subtidal ecosystems have been primarily correlated with temperature. Experimental confirmation of temperature effects, however, are rare. Moreover, temperature varies in complex ways in the subtidal zone and it is difficult to determine which attributes of the temperature regime (for example, highest, highest for some length of time, etc.) to test. Temperature may also be correlated with other environmental factors such as nutrients (Jackson, 1977; Zimmerman and Kremer, 1984). Sorting out these factors and other geographic changes, such as water motion (Foster and Schiel, 1985), grazing (Sousa et al., 1981; Gaines and Lubchenco, 1982), dispersal (Cowen, 1985), and geological history, remains a challenging problem for marine biogeographers.

Local variability between sites has been ascribed to a number of causes, with differences in substratum type (relative amounts of sand *vs.* rock) and composition (rock hardness and stability), exposure to wave action, turbidity, and grazing being of primary importance (Foster, 1982a; Dayton et al., 1984; Dayton, 1985; Harrold and Reed, 1985). Exposure to oceanic swells can vary over short distances depending on the orientation of the coast and the presence of offshore islands, and is probably the most common cause of mortality in adult *Macrocystis* spp. (Rosenthal et al., 1974; Dayton et al., 1984; Druehl and Wheeler, 1986). The overall species composition and abundance of the algal association in general can also be correlated with disturbance by wave action, especially when the substratum does not provide a firm attachment (Devinny and Kirkwood, 1974; Foster, 1982a; Breda and Foster, 1985). Between-site differences in sessile invertebrates have been attributed to differences in sedimentation and sand burial (Ostarello, 1973; Grigg, 1975).

Other potentially important factors affecting community structure at different sites include localized differences in recruitment, food availability, and predation. The latter has been of considerable interest as it affects the distribution of large sea urchins. The presence or absence of sea otters affects the abundance and distribution of large sea urchins, which may, in turn, greatly influence community structure (Leighton et al., 1966; Lowry and Pearse, 1973; Estes et al., 1978; Duggins, 1980; VanBlaricom, 1984). Other predators, disease, and storms may also affect the local and larger-scale distribution of sea urchins (Pearse et al., 1977; Tegner and Dayton, 1981; Cowen et al., 1982; Cowen, 1983; Ebeling et al., 1985).

Local differences in fish assemblages have generally been attributed to differences in substratum topography, including that provided by plants (Quast, 1971; Larson and DeMartini, 1984; Stephens et al., 1984; Ebeling and Laur, 1985). Water clarity and the presence of an algal turf that harbors food for some fishes can also be important (Ebeling et al., 1980a, b.).

Many of the same factors that affect structure at a particular site also affect structure at different depths. Sand may determine both the upper and lower depth limits of the community. Where rock is continuous, changes in algal species composition with depth are commonly correlated with changes in light availability. Kelps, for example, generally do not grow at depths where light intensity is less than 1% of that at the surface (Lüning, 1981). The plants themselves produce shade that varies with depth-associated changes in their abundance and species composition (Neushul, 1971; Dayton, 1975b; Pearse and Hines, 1979; Dayton et al., 1984; Gerard, 1984a; Reed and Foster, 1984; Dean, 1985). Changes in water motion (North, 1971b) and competition with other plants (Santelices and Ojeda, 1984) may determine the upper limit of distribution of *Macrocystis pyrifera*. Increased temperatures in shallow water may occasionally be responsible for depth-related mortalities of invertebrates (T.A. Dean, pers. commun.; V.A. Gerard, pers. commun.), and increased temperature and decreased nutrients may be associated with reduced growth of giant kelp (Gerard, 1984b; Zimmerman and Robertson, 1985; Dean and Jacobsen, 1986). Depth distributions can also vary in response to changes in predation [for example, sea urchins: Estes et al. (1978), Schroeter et al. (1983)], competitive interactions [(for example, fish: Hixon (1980), Larson (1980); algae: Kastendiek (1982a)] and food availability [for example, sea urchins: Mattison et al. (1977)].

The most important primary determinant of within-depth patchiness is probably variation in microhabitat. Variations in primary surfaces (rock, sand) and secondary substrate (organisms) over spatial scales of microscopic to tens of meters produces a mosaic of different micro-environments including crevices, areas with sand or cobble, rocks of different slopes, algal mats and holdfasts, etc. This substratum patchiness appears to affect the distribution of organisms in numerous direct and indirect ways. Additional primary contributors to patchiness include variability in dispersal (Anderson and North, 1966; Dayton et al., 1984; Reed et al., 1988) and disturbance (Connell and Keough, 1985). However, the importance of these factors in the distribution of organisms in the rocky subtidal zone is largely based on qualitative observations; the actual causes in terms of specific physical and/or biological processes are poorly known. Examples range from differences in mortality in the microscopic stages of kelp caused by sediment burial (Devinny and Volse, 1978) to behavioral and feeding associations of animals with particular microhabitats (see, for example, Ebeling et al., 1980a). Predation, competitive exclusion, and other secondary interactions amongst the organisms themselves contribute additional patchiness (for examples see Choat, 1982; Dayton, 1985; Schiel and Foster, 1986).

Temporal change at time scales of less than one year is affected by numerous factors. On the scale of hours, change is most commonly due to behavioral patterns associated with changes in light (Ebeling and Bray, 1976; Hammer, 1981). Seasonal patterns are often associated with changes in storm activity. *Macrocystis pyrifera* and other surface-canopy kelps are particularly susceptible to damage by water motion (ZoBell, 1971; Rosenthal et al., 1974; Foster, 1982a; Dayton et al., 1984). Changes in their canopy area and density produce additional changes in the plants growing beneath (Foster, 1982a; Reed and Foster, 1984). Mortality of spider crabs (Hines, 1982) and sea urchins (Cowen et al., 1982; Ebeling et al., 1985) is also associated with increased water motion. Moreover, storm-related changes in algal abundance can lead to changes in sea urchin foraging, causing areas to change from a forested to deforested condition (Ebeling et al., 1985; Harrold and Reed, 1985). Seasonal changes in oceanographic conditions, such as upwelling, are correlated with juvenile rockfish recruitment in central California (Miller and Geibel, 1973).

Superimposed upon these shorter-scale patterns are episodic events that can have significant short- and long-term effects on community structure (Dayton and Tegner, 1984b). Such events include year-to-year variations in storm intensity (Cowen et al., 1982; Foster, 1982a; Ebeling et al., 1985) and changes in hydrographic conditions that may alter temperature and nutrients (Dayton and Tegner, 1984a; Zimmerman and Robertson, 1985) and affect larval and spore dispersal (Cowen, 1985; Reed et al., 1988). The effects of the 1982/83 El Niño were particularly strong in southern California, as exceptionally severe swells occurred just prior to changes in nutrients and temperature. Expanding ranges of sea otter populations between Alaska and Baja California have had a dramatic impact on sites where adult sea urchins were abundant and grazing attached plants. In such cases, the removal of sea urchins by sea otters has resulted in increased algal abundance and perhaps changes at other trophic levels (Estes et al., 1978; VanBlaricom and Estes, 1988; Duggins et al., 1989). Localized impacts from sewer outfalls have occurred in southern California associated with increases in turbidity (Eppley et al., 1972), sludge accumulation on the bottom (Grigg and Kiwala, 1970), and the discharge of toxic chemicals (Burnett, 1971). As sewage treatment has improved, the affected areas have begun to return to a more natural state (reviewed by Bascom, 1983).

SANDY INTERTIDAL AND SUBTIDAL BEACHES

Patterns of distribution

Sandy beaches are highly dynamic habitats where the primary substrate is often moved by wave-generated bottom currents (Bascom, 1964). The entire surface deposit, where most infaunal animals live, can be transported onshore, offshore or along the beach during a single storm. The effects of wave disturbance are highly variable in space and time. Sandy beaches occur both on shorelines exposed directly to oceanic swell and in more protected coastal areas. Some beaches are narrow and steep. Others are broad with gradual slopes. In highly exposed regions, the sand is mixed or replaced with gravel or cobble. Compared to the sandy beach, the rocky shore presents a highly stable substratum. The animals of the sandy beach are well adapted to a mobile substratum. Along the Pacific coast, the best general description of the natural history of this fauna is that of Ricketts et al. (1985). For the outer Pacific coast, most of our observations of sandy beaches come from California (also see Hedgpeth, 1957).

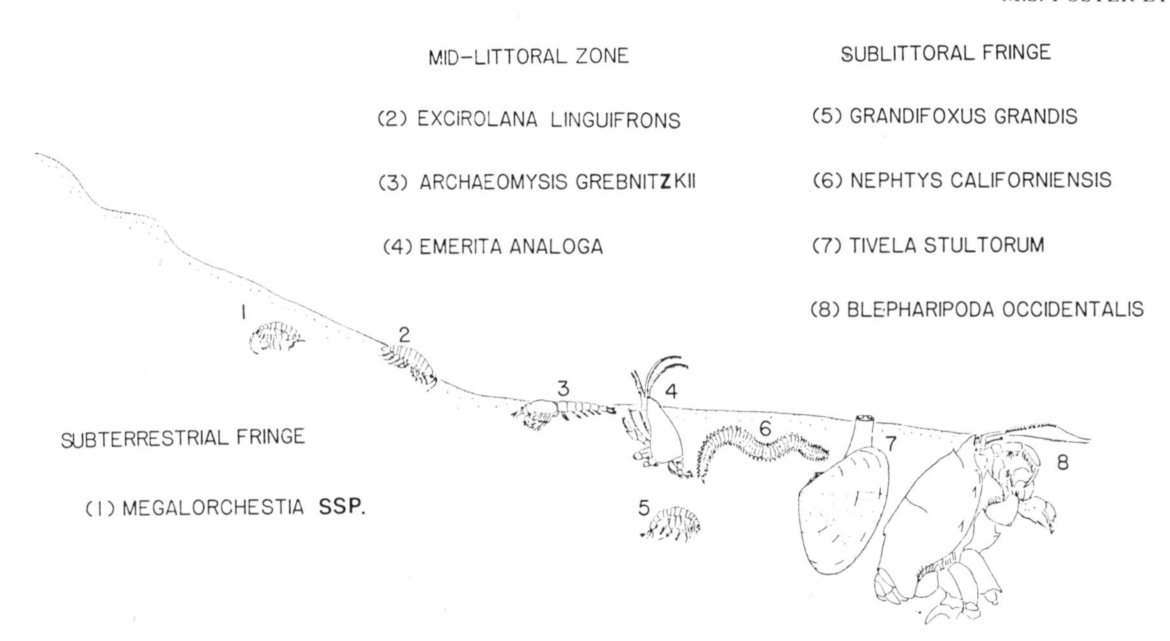

Fig. 10.9. Schematic zonation of animals along an intertidal beach in Monterey Bay, Ca., showing the most conspicuous or abundant animals in the three major zones.

Like the rocky shore, faunal zonation is the most widely explored pattern of distribution along sandy beaches. Although there is considerable movement of animals up and down the beach with changes in wave action and tide, the general zonation scheme proposed by Dahl (1952) also applies to the intertidal sandy beaches in the northeast Pacific (Ricketts et al., 1985). The three basic zones (Fig. 10.9) are (1) the uppermost subterrestrial fringe (talitrid belt), (2) a broad midlittoral zone (*Cirolana* belt), and (3) the sublittoral fringe.

Talitrid amphipods or beach hoppers are the most conspicuous animals in the upper zone. They are primarily scavengers with nocturnal feeding habits. Because beach hoppers are easy to observe and catch, their general movements, orientation and circadian rhythms are tractable to study and well explored (Bowers, 1964; Craig, 1971, 1973; Enright, 1961; Hartwick, 1975; Bousfield, 1981). Male beach hoppers fight for burrows with their antennas. In addition to beach hoppers, this upper zone harbors a number of terrestrial isopods and insects that are scavengers and predators (Hamner et al., 1968; Hamner, 1969; Craig, 1970; Hayes, 1974, 1977; Orth et al., 1977; Morris et al., 1980; Richards, 1982; Ricketts et al., 1985). In respect of predation by staphylinid beetles, Richards (1982) found it more profitable for them to feed on large beach hoppers, which are easy to detect but difficult to capture. Although the number of prey on the beach may be too low to support specialization on larger individuals, beach hoppers commonly occur in dense patches around drift deposits and carrion.

Drifting macroalgae, especially the giant kelp *Macrocystis pyrifera* and the bull kelp *Nereocystis luetkeana*, are commonly deposited in the upper zone (Fig. 10.10). Most of the upper beach residents are attracted to drift deposits for food and perhaps shelter. After one winter storm in Monterey Bay, we observed deposits of drift algae 1 m thick, 10 m wide and hundreds of meters long. Similar massive drift deposits occur in southern California (ZoBell, 1971). Few studies have examined the dispersal of drift deposits of algae, their residence time on the beach, how they are used by different beach animals, and the eventual sink for most of the organic matter. The age of drift patches affects their colonization by beach fauna and the subsequent interactions among these animals (Moore and Legner, 1974; Yaninek, 1980). Old patches harbor various insect larvae, flies, beetles, isopods, amphipods, nematodes and annelid worms (Kompfner, 1974; Morris et al., 1980; Yaninek, 1980). The movement and deposition

Fig. 10.10. Drift giant kelp on an intertidal beach near Santa Barbara, Ca., after a storm.

of drift algae create a highly dynamic mosaic of habitat patches on the beach (Zobell, 1971; Yaninek, 1980). While a number of important patterns have been documented, drift patches are amenable to complementary field and laboratory experiments, which could make this mosaic one of the best understood ecological systems on sandy beaches.

Several crustaceans are the most conspicuous animals in the midlittoral zone, particularly a cirolanid marine isopod and the mole crab *Emerita analoga*. These two groups often occur in sharp zones, with the isopod at slightly higher tidal elevations than the mole crab. They can number thousands per square meter (Johnson, 1976) and are probably the most important prey for shorebirds that forage on intertidal sandy beaches (Myers et al., 1980a; Connors et al., 1981). Sanderlings (Scolopacidae) apparently maintain feeding territories where these crustacean prey are most abundant (Myers et al., 1979, 1980b).

The mole crab is a suspension feeder, which moves up and down the beach with the tide and deep into the sediment during storms (Cubit, 1969). Research on this species (the best explored on the sandy beach) has included studies of distribution and dispersion (Barnes and Wenner, 1961; Cubit, 1969; Efford, 1965; Dillery and Knapp, 1970; Burton, 1979), feeding (Efford, 1966), reproduction (Efford, 1967, 1969; Cox and Dudley, 1968; Fusaro, 1980), and population dynamics (Efford, 1970; Fusaro, 1978). The cirolanid isopods are scavengers and predators. They eat dead animals and attack living mole crabs. They also eat human flesh and will open numerous wounds if allowed. Like the mole crabs, isopods move up and down the beach probably in response to the mechanical stimulus of wave action (Enright, 1965).

The lowest intertidal zone (the sublittoral fringe) harbors the greatest number of species, including amphipod, mysid, and anomuran crustaceans, polychaete worms, and clams. When clams such as the bean clam, *Donax gouldii*, the razor clam, *Siliqua patula*, and the Pismo clam, *Tivela stultorum*, are present, they are usually the most conspicuous animals on the beach. Each species has a different geographical range, as is found for most of the species we discuss. This rich, low intertidal zone grades into the broad subtidal breaker zone, which is the most poorly sampled portion of the sandy beach. Many of the lower intertidal animals are found in much deeper water. For example, the large spiny mole crab *Blepharipoda occidentalis* (Fig. 10.9) occurs in the lower intertidal, but is most abundant in the shallow subtidal beach. Its young often recruit in water depths of −10 to −15 m, many hundreds of meters from the breaker zone. In general, faunal zones are better defined and narrower along the upper and middle intertidal beach.

Historically the most conspicuous animal in the lower intertidal zone along many Californian beaches has been the large Pismo clam, *Tivela stultorum*. It was extensively exploited by a commercial fishery in the first half of this century and supported an active sport fishery until recently when the California sea otter expanded its range into the best clamming areas in Monterey Bay, Moro Bay and Pismo Beach. Pismo clam fisheries disappeared along every beach invaded by sea otters (Stephenson, 1977; Kvitek and Oliver, 1988). Like other relatively large and long-lived benthic invertebrates (Fukuyama and Oliver, 1985), Pismo clam populations are often dominated by one size or age class (Fitch, 1965). Since the recruitment of Pismo clams is highly variable in time and space (Fitch, 1950; Coe and Fitch, 1950), larval availability, settlement and early mortality may be the key factors limiting the re-establishment of intertidal populations.

The sedimentary habitat is highly dynamic and complex through the wash zone, the line of breakers and into the shallow offshore environment on sandy beaches. The bed forms vary from planar to large lunate mega-ripples, which often maintain their relative positions while migrating in response to changes in waves and tides (Clifton et al., 1971). As previously noted, the fauna in the breaker zone is very poorly known, but our observations indicate it is sparse relative to deeper water (see also Oliver et al., 1980). In contrast, there are a number of conspicuous epifaunal invertebrates seaward from the line of breakers that live on and near the surface of the mobile sea floor (Fig. 10.11).

The sand dollar, *Dendraster excentricus*, forms dense beds in this inner offshore zone (*sensu* Oliver et al., 1980) and plays a major role in structuring benthic communities directly inside and outside the bed (Morin et al., 1985). Generally the sand dollar bed has a distinct seaward edge and a much less distinct shoreward edge. The entire bed migrates seaward and shoreward in response to wave action (Oliver et al., 1980; Morin et al., 1985). The sand dollar bed apparently acts as a barrier to the movement of several other epifaunal species. In southern California, the Pismo clam, the purple olive snail (*Olivella biplicata*), and the sea pansy (*Renilla köllikeri*) live almost exclusively on the landward side of the bed, and the moon snail (*Polinices altus*), the sea pen (*Stylatula elongata*), and several sea stars (such as *Astropecten verrilli* and *Pisaster* spp.) live on the seaward edge (Morin et al., 1985). In much more wave-protected sand flats, there are also significant differences in the structure of infaunal assemblages inside and outside of sand dollar beds (Smith, 1981; Highsmith, 1982). Similar differences probably occur along the outer coast as well (Morin et al., 1985).

The offshore region is seaward of the influence of the sand dollar bed and is divided into a rich crustacean zone and a polychaete worm zone (Oliver et al., 1980). The rich crustacean zone is numerically dominated by infaunal amphipods, ostracods and cumaceans. While these groups can swim, the phoxocephalid and haustoriid amphipods are well developed sand burrowers and often the most abundant infauna (Slattery, 1985). Haustoriids are generally more abundant in shallower water and swim the least. There is a gradual increase in the number of tube-dwelling and relatively sedentary benthic species with increasing water depth in the offshore region (Oliver et al.,

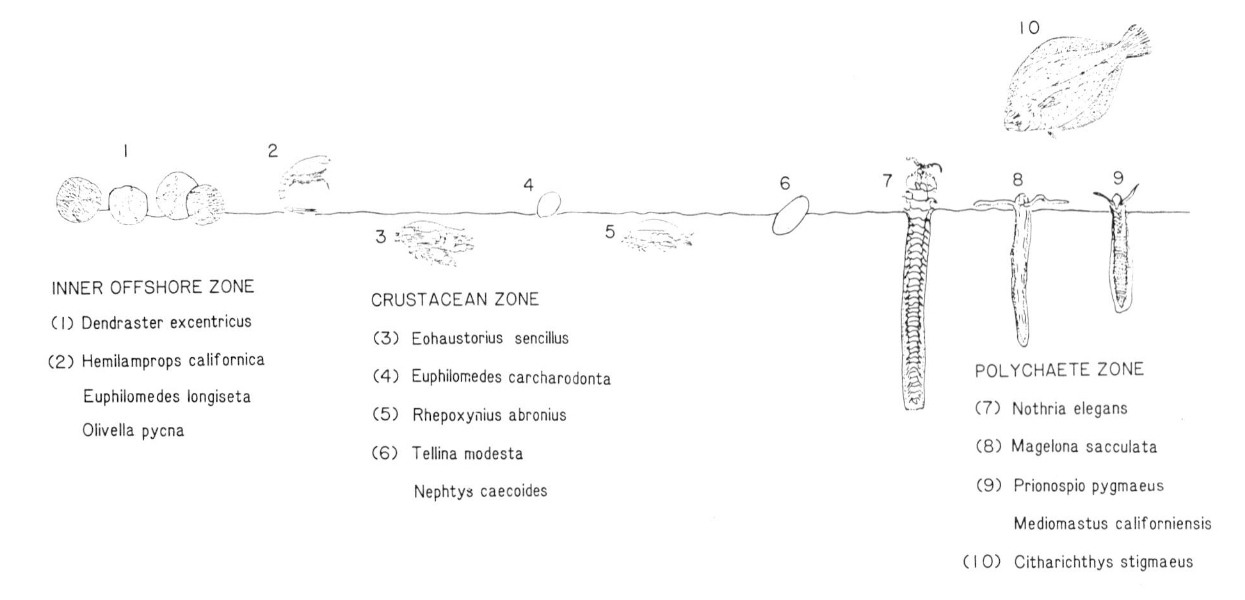

Fig. 10.11. Schematic zonations of animals along the subtidal, offshore beach in Monterey Bay, Ca., showing the most conspicuous or abundant animals in the three major zones just beyond the breaker zone.

1980). In Monterey Bay, a deeper polychaete worm zone occurs in water depths below −14 to −18 m. The most abundant and conspicuous sedentary polychaetes are the relatively large tube-dwelling onuphid *Nothria elegans* (Fig. 10.11), and several smaller spionids and magelonids. This gradual shift from a shallow zone of mobile peracarid crustaceans to a zone of more sedentary polychaete worms occurs in southern and northern California (Barnard, 1963; VanBlaricom, 1978; Oliver et al., 1980), Oregon (Carey, 1965, 1972) and Washington State (Lie, 1969; Lie and Kisker, 1970).

The abundance and trophic habitats of benthic fishes are best known seaward of the breaker zone, where there are more fishes and they are easier to sample. The intertidal and breaker zone are dominated by various species of perch, and the number of species and individuals of flatfishes increases with increasing depth in the offshore zone (Hulberg and Oliver, 1980). The distribution and abundance of meiofauna is poorly known along the sandy beaches of our region [see Wieser (1959) for more protected sites]. In central Monterey Bay, copepods and nematodes numerically dominate this fauna from the intertidal beach into the offshore crustacean and polychaete zones (Oakden, 1984).

Organization of sandy beach ecosystems

Wave-generated bottom currents move large quantities of sediment. Along the intertidal and shallow subtidal beach, the wash zone and line of breakers move up and down the beach with changes in the tide and wave action. Seasonal and episodic storms may move the largest quantities of sediment (Bascom, 1964) and may have the greatest influence on community structure. A variety of descriptive-correlative evidence suggests that wave disturbance has a strong influence on the offshore zonation of the mobile peracarid crustaceans and relatively sedentary polychaete worms (Oliver et al., 1980). The evidence includes: (1) a positive correlation between water depth and the numbers of tube-dwellers, burrow-dwellers, and commensal animals that apparently cannot establish or maintain populations in shifting sediments; (2) other natural-history patterns related to depth and thus to substrate disturbance; (3) a positive correlation between the strength of wave activity and the width and depth limits of the faunal zones (that is, when wave disturbance is more intense, the crustacean zone ends and the polychaete zone begins in deeper water); (4) a correspondence between the largest decrease in polychaete population size and the season and location of greatest wave activity (winter months at the shallowest station); and (5) a marked similarity between community zonation along a depth-dependent gradient of oscillatory substrate motion (gently sloping sand flats) and the zonation along a constant depth gradient of creeping substrate motion (submarine canyon ridge). Seasonal movements and burrowing of sand dollars are also related to changes in wave activity (Oliver et al., 1980; Morin et al., 1985). Within the intertidal zone, storm waves and sediment movement also have a dramatic effect on animal behavior and abundance, but most of the residents are highly mobile burrowers that can adjust to rapid substrate erosion and deposition (Ricketts et al., 1985). For example, mole crabs move up and down the beach with the tide, actively feed in the wash zone, and burrow to avoid storms (Cubit, 1969).

Infaunal animals probably respond to smaller-scale sedimentary features caused by wave action. Mole crabs sometimes aggregate in beach cusps formed by wave wash (Cubit, 1969). Different animals may live in the crests and troughs of sand ripples. Migrating ripples may harbor different animals from relatively immobile ripples. This scale of sampling has not been done on our beaches. In addition, there are no experimental studies that examine the role of wave disturbance in organizing population or community patterns. While this is probably the most important physical process in low intertidal and shallow subtidal beaches (Oliver et al., 1980), desiccation, interstitial water, beach slope, sediment size and related physical characteristics of the sediment may be more important along the upper beach (see, for instance, Edwards, 1969; Hamner et al., 1968).

The best-known biological interaction along the sandy beaches is predation. However, there are only a few studies that evaluate the population or community consequences of these predators. Nevertheless, the foraging habitats, diets and number of predators suggest they play major roles in organizing beach communities.

The largest benthic predator on sandy beaches is

the gray whale, *Eschrichtius robustus*. Gray whales forage just seaward of the breaker zone in spring, during their northern migration from California to British Columbia (Nerini, 1984), as well as during summer within feeding grounds, which are only well documented along the west coast of Vancouver Island (Murison et al., 1984; Oliver et al., 1984; Oliver and Kvitek, 1984; Oliver and Slattery, 1985; Kvitek and Oliver, 1986). Some benthic prey communities are highly resilient to the feeding disturbance caused by gray whales (Oliver and Slattery, 1985a), but others are not. Gray whales had apparently destroyed a dense bed of onuphid polychaetes along a wide sandy beach in British Columbia as a result of many years of summer feeding (Kvitek and Oliver, 1986). Since the gray whale population is apparently expanding its summer feeding range, feeding gray whales may become more common along many sandy beaches south of the Bering Sea. Even a few gray whales can have a dramatic effect on less resilient prey populations and communities. Another example of such a dramatic effect by other predators is Fager's (1964) observation of the destruction of a dense bed of oweniid polychaetes by skates and rays just behind the breaker zone at the Scripps beach. This dense tube mat had not recovered after many years (Davis and VanBlaricom, 1978).

The recovery of the gray whale and sea otter populations after commercial exploitation provides unique opportunities to explore the effects of these large predators on newly invaded prey communities. There are no comparable opportunities among the other potentially important vertebrate and invertebrate predators of the sandy beaches. The tremendous numbers of Pismo clams that were commercially harvested along the sandy beaches in California and Mexico may have been present because of the earlier and thorough commercial hunting of sea otters (Estes and VanBlaricom, 1985). In the presence of a sea otter population that was not hunted by humans, Pismo clams were probably rarely if ever more abundant than they are today, providing little or no potential for even a sport fishery (Kvitek and Oliver, 1988). While sea otters have a dramatic effect on the numbers of large Pismo clams, their foraging activities apparently have little effect on the associated benthic communities of the beach (Kvitek and Oliver, 1988).

Bird predation is most intense in the intertidal zone, where isopods and mole crabs are probably the major prey (Myers et al., 1980a; Connors et al., 1981). Existing field studies show how birds forage in relation to changes in these prey populations (Myers et al., 1979, 1980b), but do not evaluate how birds influence them (see the work by Quammen, 1982).

Among all the predators on the sandy beach, the community role of rays has been most thoroughly explored (VanBlaricom, 1982). Since rays dig pits to excavate their prey, their feeding disturbance can be mimicked in field experiments to explore their community effects. In southern California, ray pits are colonized by a number of benthic animals, especially amphipod crustaceans. The most abundant amphipods in this community apparently feed on trapped organic material. The feeding disturbance of rays produces a persistent mosaic of patches in various stages of infaunal recolonization (VanBlaricom, 1982). In addition to feeding rays, there are a large number of demersal flatfish and perch that feed on beach infauna. While these numerous predators are surely important in establishing and maintaining population and community patterns, their role is still poorly understood (Hulberg and Oliver, 1980).

Sea stars (starfish) are abundant predators along some beaches. In southern California, *Astropecten verrilli* is very abundant in the offshore zone and may be important in reducing the numbers of portunid crabs (VanBlaricom, 1982), which can have a major influence on the structure of infaunal communities (Virnstein, 1977; VanBlaricom, 1982). The numbers of other crabs (e.g. *Cancer* spp.: Gotshall, 1977) may be kept so low by predators — fishes and others — that they rarely play a major role along sandy beaches. Sea stars may influence the distribution of sand dollar beds as well. The sharp and dense outer edge of the bed apparently acts as a barrier to shoreward movement and prevents sea stars from foraging here, because they must leave the substrate and risk tumbling by wave surge (Birkeland and Chia, 1971; Kastendiek, 1982b; Morin et al., 1985).

Phoxocephalid amphipods are abundant and widespread predators among the smaller infauna (Oliver et al., 1982). In McMurdo Sound, Antarctica, it was found that they control the species composition and population structure of soft-bodied infauna (Oliver and Slattery, 1985b). Although phoxocephalids may play similar roles along our subtidal beaches, the many other predators present and the highly dynamic and disruptive physical conditions confound efforts to assess their relative importance (Oliver et al., 1982). The effects of predators among the smaller infauna (Commito, 1982; Ambrose, 1984), on sandy beach communities are unknown. The predacious glycerid polychaete, *Glycinde polygnatha*, specializes in feeding on tube-dwelling spionid polychaetes along the offshore beach, and like so many other infauna is itself a major prey item of demersal fish (Silberstein, 1987).

Thorson (1966) found it remarkable that any invertebrate can successfully recruit into the hostile and hungry neighborhoods of the benthos. The sand dollar is a good example of the potential importance of recruitment and predators. In central Monterey Bay, the sand dollar bed is dominated by one large size class, as are a number of other sand dollar beds along the coast (Fager, 1968; Merrill and Hobson, 1970; Morin et al., 1985). Individuals smaller than 2 to 3 cm are extremely rare. The smallest individuals (< 1 mm) are most abundant in water depths of −10 to −24 m, considerably seaward of the adult bed, where wave disturbance is reduced. None of these early recruits ever attained a size greater than 1 cm during the five-year study of Oliver et al. (1980). Since smaller individuals (< 1–2 cm) are only common in sand dollar beds protected from oceanic swell (Birkeland and Chia, 1971; Cameron and Rumrill, 1982; Highsmith, 1982) and in deeper water like the exposed Monterey Bay site, substrate disturbance may limit larval recruitment.

There are several voracious "filters" (*sensu* Dayton et al., 1974) that small sand dollar recruits must pass through on the beach. As noted earlier, nothing is known about the meiofaunal "filter". The next "filter" is the numerous, active and predacious phoxocephalid amphipods (Oliver et al., 1982). Sand dollars can escape this "filter" by growing larger (Oliver and Slattery, 1985b). The next "filter" may be small commensal crabs, which are abundant deep in the sediment (Oliver et al., 1980) and may emerge from burrows and tubes to feed at night. Unlike the surface-active *Cancer* and portunid crabs (VanBlaricom, 1982), pinnixids and other commensal crabs enjoy a deep-sediment refuge from predators. When sea stars are abundant they are also an effective "filter" over a range of prey size classes (VanBlaricom, 1982; Fukuyama and Oliver, 1985). Perhaps the most important and poorly known "filter" on the offshore beach consists of the many small juvenile flatfish. If a young sand dollar grows large enough to avoid the juvenile fish "filter", it is available to an abundant and rich group of adult demersal fishes on the beach (Hulberg and Oliver, 1980; VanBlaricom, 1982). How does a small sand dollar ever grow up and why are the beds dominated by one size class?

Our hypothesis is that the predacious "filters" prevent successful pulses of recruitment, and only a few individuals escape until the "filters" are destroyed or disrupted. Perhaps every decade or so a very large storm sweeps away most of the beach fauna, including the sand dollar bed, down to water depths of 10 m or more (Oliver et al., 1980). The next sand dollar cohort settles offshore, grows rapidly, and migrates into the zone where the adults usually live. This is the dominant cohort until the next disturbance event.

The sandy beach does not provide as much structure for spatial refuges from predators as the rocky intertidal and subtidal zones. Burrowing into the sediment is the major refuge for beach infauna. There are other refuges. The sand dollar bed prevents predators on the seaward edge from reaching prey on the shoreward side (Kastendiek, 1982b; Morin et al., 1985). However, few refuges are absolute. For example, feeding rays disrupt the refuge of burrowing amphipods, and make them available to flatfish foraging in association with the rays (VanBlaricom, 1982).

The best example of competition among sandy beach fauna involves the interactions of sand dollars and a number of other crawling epifauna (Morin et al., 1985). Sand dollars appear to exclude these species from their space. Like the rocky shore, the competition for space is much more easily demonstrated than competition for food. Despite problems with the cage exclusion of fish predators on the beach (Hulberg and Oliver, 1980), there is no evidence from any cage studies in soft-bottom habitats that predators reduce the numbers of a competitive dominant that can monopolize space, as mussels do on the rocky shore (Peterson, 1979b).

The importance of competition has been inferred from life-history and distribution patterns of several sand-burrowing amphipods along the east coast of North America (Croker, 1967a, b; Dexter, 1967; Sameoto, 1969a, b, c). In Monterey Bay, the phoxocephalid and haustoriid amphipods have distinct life-history patterns which help explain their depth zonation (Slattery, 1985). They reproduce at different times of the year and have different sex ratios, swimming behaviors and diets. In addition to different diets, the phoxocephalids show behavioral preferences for native sediment found at the water depth or zone where each species is most abundant (Oakden, 1984). While competition is often used to explain resource partitioning such as habitat selection, dietary shifts, and other life-history differences among related species, there is no direct evidence that competition is important in structuring the sand-burrowing amphipod assemblages along this coast (Oliver et al., 1980; Oakden, 1984; Slattery, 1985).

Life history differences and resource partitioning are often considered the result of past competitive interactions. While the phoxocephalid and haustoriid patterns can be interpreted this way, their history is quite different from that of the haustoriids on Atlantic coast beaches of North America. Phoxocephalids apparently originated in the Southern Hemisphere in the Pacific Ocean (Barnard and Drummond, 1978) and, though they have spread into other oceans, they are still most diverse and numerous in the Pacific (Barnard, 1960). They are not important members of beach communities of eastern North America. In contrast, haustoriids originated in the Atlantic Ocean and only recently invaded the Pacific, where only one genus lives on the North American beaches. *Eohaustorius* apparently came through an open Panama seaway (Slattery, 1985). This relatively new invader co-occurs with an older phoxocephalid fauna on the west-coast beaches. A more diverse haustoriid fauna with a very different history and essentially no phoxocephaliids inhabits east-coast beaches.

Biogeographic patterns also confound more local interpretations of community organization on the beach. The amphipod *Acuminodeutopus heteruropus* is the second most abundant infaunal animal along the offshore Scripps Beach. It is the best invader of pits made by feeding rays (VanBlaricom, 1982). With the exception of this amphipod, the other numerical dominants at Scripps are similar to those found in central Monterey Bay (Oliver et al., 1980). The absence of *Acuminodeutopus* may be related to the absence of important ray disturbance in Monterey Bay. Although ray feeding may help to maintain the high abundance of *Acuminodeutopus* at Scripps, this amphipod group is generally more diverse and abundant in warmer latitudes (Barnard, 1969) and might never become abundant in Monterey Bay even with high rates of ray disturbance.

The role of human disturbance is most pronounced along the intertidal beaches in southern California where the upper beach is often concrete, drift material is collected in beach sweepers, and the sources of beach sand are greatly reduced or gone. The greatest threats are the reduction in river flow and sediment inputs, and the effects of human structures (Bascom, 1964). The effects of human trampling on beach fauna are unknown. While there is much greater public appreciation for the sandy beach environment, human bureaucracy managed to construct the newest addition of the Moss Landing Marine Laboratories into the intertidal beach despite the cries of resident oceanographers. Still, the first scene of Jacob Bronowski's classic and optimistic film essay on the Ascent of Man begins on a warm night watching grunion spawn on the sandy beach.

ENERGETICS AND TROPHIC RELATIONSHIPS

Primary productivity

With few exceptions (such as, for instance, barnacles in intertidal zone 2, sites with extensive mussel beds, subtidal reefs dominated by invertebrates), rocky intertidal and subtidal ecosystems in the region are visually dominated by large algae and surf-grass. Even barnacle zones and mussel beds may have a high cover of plants that grow attached to these sessile animals (Foster, 1982b; Sousa, 1984) and, in the subtidal zone, plants may form a surface canopy above sessile animals (M.S. Foster, pers. observ.). Blinks (1955) was one of the first investigators to recognize that this high biomass of photosynthetic tissue in the northeast Pacific and other regions probably indicated high primary productivity, and he provided some of the first measurements showing that this was true. Blinks was also the first to relate this high productivity to the presence of surface canopies utilizing light before it is reduced by extinction in the water column, to the abundance of nutrients close to shore, and to the enhanced uptake of nutrients by attached plants over which the water flows. Additional primary production is added by phytoplankton in the overlying water, but this is substantially lower per unit area [generally less than 0.3 kg C m^{-2} yr^{-1}: reviewed by Mann (1982)] than macroalgal production in the region (Blinks, 1955).

Measurements by Blinks (1955) in central California and more recent studies in southern California and Baja California, Mexico, (Littler and Murray, 1974; Littler et al., 1979; Littler and Littler, 1981) have found net intertidal primary productivities of between 0.15 and 3.2 kg C m^{-2} yr^{-1} [multiplying daily rates summarized by Littler and Murray (1974) by 365]. The comprehensive study by Littler and Murray (1974) suggested 0.5 kg C m^{-2} yr^{-1} may be a rough mean value for the northeast Pacific where intertidal algal biomass is generally high; Mann (1982) suggested 0.1 as a mean for the world. Values tend to be higher where plants with high surface/volume ratios are abundant and where physiological stress, such as desiccation during prolonged exposure, is reduced (Littler and Murray, 1974; Littler and Littler, 1981). Littler et al. (1979) found that primary productivity varied with season, maximum productivity being associated with high temperatures and long day-lengths of summer and fall. These seasonal changes in productivity are probably greater at more northerly latitudes, but comparative measurements are unavailable.

Studies of primary productivity for subtidal reefs have focused on *Macrocystis pyrifera*. Estimates using a variety of physiological and field growth techniques vary from 0.5 to 3.5 kg C m^{-2} yr^{-1} (reviewed by Coon, 1982). Mann (1982) suggested a rough mean of around 1.0 kg C m^{-2} yr^{-1}. Wheeler and Druehl (1986) estimated the productivity of *Macrocystis integrifolia* at a site in British Columbia to be 1.3 kg C m^{-2} yr^{-1}. The yearly production of associated understorey vegetation has not been determined. Foster and Schiel (1985) suggest that, if this were included, then perhaps *M. pyrifera* forests would turn out to be the most productive of all marine ecosystems.

Seasonal variation in primary production on subtidal reefs has not been thoroughly investigated. Gerard (1976) found that *Macrocystis pyrifera* production at a site in central California was correlated with standing crop, the highest values of both generally occurring in summer and early fall. A similar seasonal cycle of productivity occurred for *M. integrifolia* at sites in British Columbia (Wheeler and Druehl, 1986). In both cases, seasonal variation in water motion, through its effects on standing stocks, was the most important regulating variable. Seasonal and episodic variation in nutrients may commonly regulate productivity in southern California (Jackson, 1977; Zimmerman and Kremer, 1984). In contrast to surface-canopy kelps, the productivity of understorey vegetation may be most closely linked to variations in overstorey canopy shading, so that maximum productivity or growth occurs whenever overstorey canopies are reduced (Johansen and Austin, 1970; Foster, 1982a; Heine, 1983).

The ranges in the above estimates of primary productivity are considerable, and presumably are due in part to variation in abiotic conditions and species composition and abundance at the sites examined. However, different measurement techniques also appear to produce different estimates (Foster and Schiel, 1985). Estimates based on physiological measurements (^{14}C uptake, O_2 production) tend to be around five times higher than those based on growth rates, and estimates from growth rates tend to be around ten times higher than those based on harvesting techniques. Given these difficulties, perhaps the only general conclusion to be made is qualitative: subtidal reefs in the region are more productive than the intertidal zone, and both are more productive than pelagic ecosystems.

All observations indicate that yearly primary production greatly exceeds standing stocks or biomass, indicating rapid biomass turnover. Blinks (1955) estimated turnover times for intertidal seaweeds ranging from 10 to 120 days (mean of 40 days), and Gerard (1976) estimated *Macrocystis pyrifera* turnover at 6.6 yr^{-1} for her fairly protected site in central California. In contrast to their terrestrial counterparts, high turnover rates appear typical amongst seaweeds (Mann, 1982).

Considerable organic matter is imported to the beach as drifting macroalgae (Fig. 10.10) and plankton. This production can be the most important source of energy for beach consumers. Except for the relationship between diatom production and the razor-clam (*Siliqua patula*) fishery along the Olympic Peninsula in Washington (Lewin and Mackas, 1972), and faunal associations with drift macroalgae in northern California (Yaninek, 1980), little is known about the amounts and fate of imported energy. There is also production from the benthic microflora and fauna (Steele and Baird, 1968). None of these small plants or animals has been studied in any detail on the wave-exposed sandy beaches in the northeast Pacific. There are no published data on primary production or even standing stocks of the rich diatom flora attached to sand grains and migrating through interstitial water (Hopkins, 1966).

Trophic relationships

Studies of energy flow have generally shown a diversity of energetic pathways from producers to consumers in these systems (reviewed by Mann, 1982). As foreshadowed by Blinks (1955) and summarized by Khailov and Burlakova (1969),

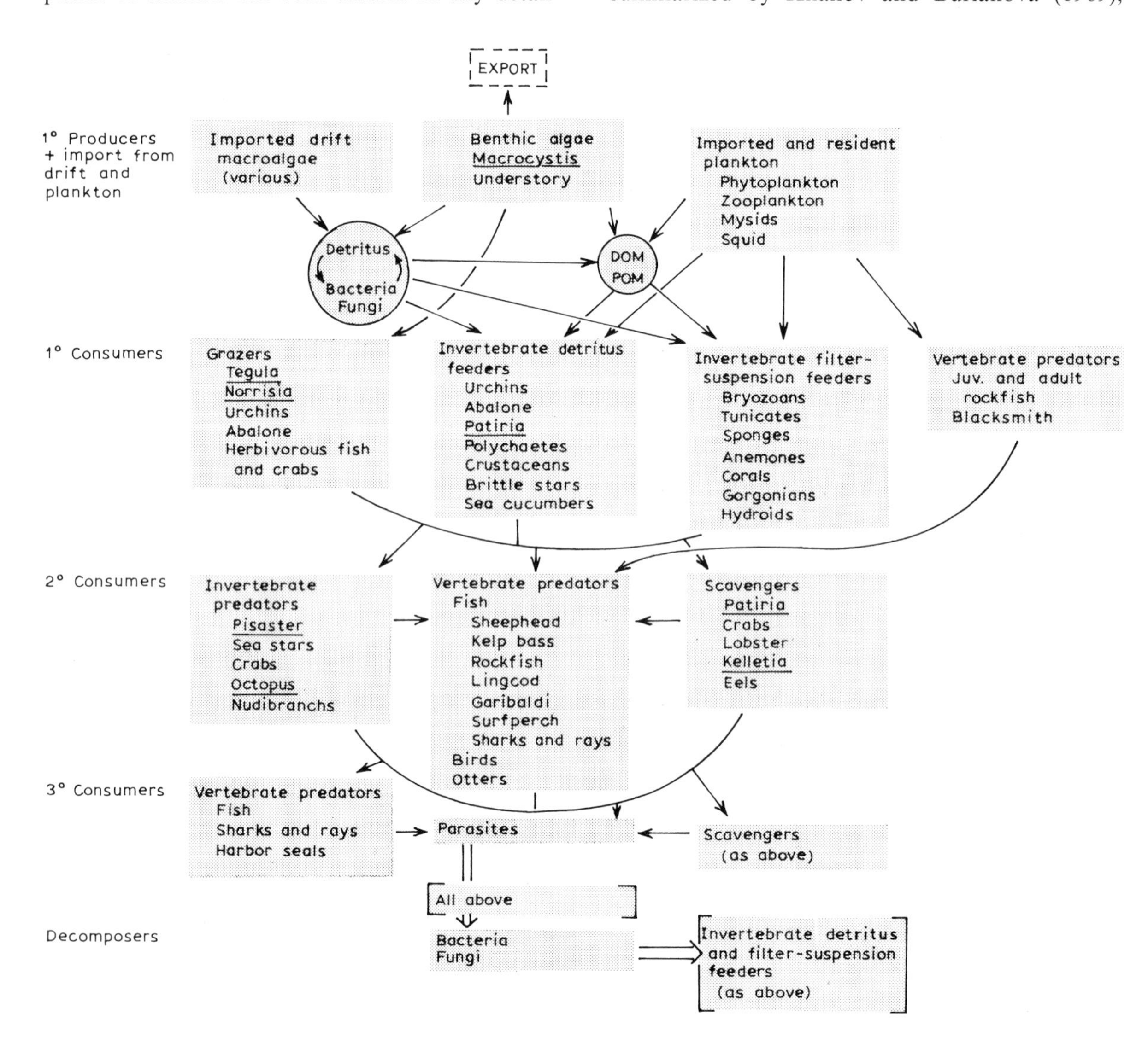

Fig. 10.12. Generalized food web for a kelp forest. Arrows indicate the direction of energy flow. (From Foster and Schiel, 1985.)

much of the macroalgal primary production is utilized as drift, and a considerable portion of this drift is exported from the rocky shores to other ecosystems (Gerard, 1976), including sandy beaches (ZoBell, 1971; Yaninek, 1980), nearshore oceanic areas (as surface drift: Mitchell and Hunter, 1970), and the benthos below the euphotic zone (North, 1971b; Cailliet and Lea, 1977). Figure 10.12 is a simplified food web for a hypothetical *Macrocystis pyrifera* forest based on numerous natural-history observations of diets (see, for example, Rosenthal et al., 1974; Foster and Schiel, 1985). Although the taxa may change (for instance, more barnacles than bryozoans, more limpets and fewer turban snails), similar general trophic relationships occur in the intertidal zone (Carefoot, 1977).

The most thorough energetic analysis of nearshore marine ecosystems in the northeast Pacific was done in a protected kelp forest in central California by Gerard (1976). She estimated that 10% of the primary production of *Macrocystis pyrifera* (excluding dissolved organic matter and detritus lost from attached plants) was consumed by grazers eating attached plants. The remaining 90% became drift algae, of which 50% was consumed within the forest. The rest was exported from the forest. The large proportion of production moving through this drift pathway is probably general in macroalgal systems (Khailov and Burlakova, 1969; Mann, 1982; Duggins et al., 1989), and the production and utilization of drifting plant material may thus be an important process determining differences in detritivore abundance in areas with and without giant kelp (Foster and Schiel, 1985). If, as indicated by Bedford and Moore (1984) for Scottish beds of *Laminaria saccharina*, direct consumption of detached plant biomass is by small detritivores (such as crustaceans, polychaetes, echinoderms) rather than decomposers (bacteria and fungi), then energetic pathways based on macroalgal drift are fundamentally different from those based on vascular-plant litter. The consequence may be more efficient energy transfer to higher trophic levels.

Also of interest is that the direct consumption of attached plants by grazers is apparently low. Gerard's (1976) figure of less than 10% may be typical (Miller and Mann, 1973). This seems curious, as seaweeds generally lack the unpalatable structural tissue found in vascular plants. Most seaweed biomass occurs above the substratum, so perhaps there is decreased efficiency and increased mortality associated with foraging on plants in the water column. Toxic or unpalatable chemicals may also play a role (Steinberg, 1985). On the other hand, because of the morphology of large seaweeds and their susceptibility to removal by water motion, these low rates of direct consumption can cause major structural alterations in kelp communities. In the case of sea urchins, grazing on the stipes of attached plants may cause them to be dislodged and drift out of the system (reviewed by Harrold and Pearse, 1987). Such phenomena illustrate the problem of trying to determine the causes of differences in primary production and community structure from measurements of material cycles and/or energy flow alone (Paine, 1980, 1984; Foster and Schiel, 1985).

The high intertidal zone may represent the one habitat where there is a consistent, energetically based grazer–algal interaction controlling structure. Herbivorous limpets and littorines can control algal biomass in this habitat (Castenholtz, 1961), and the interplay between grazing intensity and seasonal, abiotically induced changes in primary production accounts for much of the observed community structure (Cubit, 1984). This strong linkage (*sensu* Paine, 1980) may also be an important selective agent for the heteromorphic life histories observed in many of the algae living in this habitat (Lubchenco and Cubit, 1980).

ACKNOWLEDGEMENT

The preparation of the rocky intertidal portion of this review was supported in part by the U.S. Minerals Management Service, Contract No. 14–12–0001–30057.

REFERENCES

Abbott, I.A. and Hollenberg, G.J., 1976. *Marine Algae of California*. Stanford University Press, Stanford, Ca., 827 pp.

Aleem, A.A., 1956. Quantitative underwater study of benthic communities inhabiting kelp beds off California. *Science*, 123: 183.

Ambrose, W.G., 1984. Increased emigration of the amphipod

Rhepoxynius abronius (Barnard) and the polychaete *Nephtys caeca* (Fabricius) in the presence of invertebrate predators. *J. Exp. Mar. Biol. Ecol.*, 80: 67–75.

Anderson, E.K. and North, W.J., 1966. In situ studies of spore production and dispersal in the giant kelp, *Macrocystis pyrifera*. *Proc. Int. Seaweed Symp.*, 5: 73–86.

Andrews, H.L., 1945. The kelp beds of the Monterey region. *Ecology*, 26: 24–37.

Barnard, J.L., 1960. The amphipod family Phoxocephalidae in the eastern Pacific Ocean, with analysis of other species and notes for a revision of the family. *Allan Hancock Pacific Expeditions*, 18: 169–375.

Barnard, J.L., 1963. Relationship of benthic Amphipoda to invertebrate communities of inshore sublittoral sands of Southern California. *Pac. Natural.*, 3: 439–467.

Barnard, J.L., 1969. The families and genera of marine gammaridean Amphipoda. *Bull. U.S. Nat. Mus.*, 271: 1–535.

Barnard, J.L. and Drummond, M.M., 1978. Gammaridean Amphipoda of Australia, Part III. The Phoxocephalidae. *Smithson. Contrib. Zool.*, 245: 1–551.

Barnes, N.B. and Wenner, A.M., 1961. Seasonal variation in the sand crab *Emerita analoga* (Decapoda, Hippidae) in the Santa Barbara area of California. *Limnol. Oceanogr.*, 13: 465–475.

Bascom, W.N. (Editor), 1983. *The effects of waste disposal on kelp communities.* Southern California Coastal Water Research Project, Long Beach, 328 pp.

Bascom, W.N., 1964. *Waves and Beaches: The Dynamics of the Ocean Surface.* Doubleday, Garden City, N.Y., 267 pp.

Bedford, A.P. and Moore, P.G., 1984. Macrofaunal involvement in the sublittoral decay of kelp debris: the detritivore community and species interactions. *Estuarine Coastal Shelf Sci.*, 18: 97–111.

Birkeland, C. and Chia, F.S., 1971. Recruitment risk, growth, age and predation in two populations of *Dendraster excentricus* (Eschoscholtz). *J. Exp. Mar. Biol. Ecol.*, 6: 265–278.

Black, R., 1976. The effects of grazing by the limpet, *Acmaea insessa*, on the kelp, *Egregia laevigata*, in the intertidal zone. *Ecology*, 57: 267–277.

Blinks, L.R., 1955. Photosynthesis and productivity of littoral marine algae. *J. Mar. Res.*, 14: 363–373.

Bolin, R.L. and Abbott, D.P., 1963. Studies on the marine climate and phytoplankton of the central coast area of California, 1954–1960. *Calif. Coop. Oceanic Fish. Invest. Rep.*, 9: 23–45.

Bousfield, E.L., 1955. Ecological control of the occurrence of barnacles in the Miramichi estuary. *Nat. Mus. Can. Bull.*, 137: 1–69.

Bousfield, E.L., 1981. Evolution in North Pacific coastal marine amphipod crustacans. In: *Evolution Today*. G. Scudder and J. Reveal (Editors), Hunter Institute of Botanical Documentation, Pittsburg, Pa., pp. 69–89.

Bowers, D.E., 1964. Natural history of two beach-hoppers of the genus *Orchestoidea* (Crustacea: Amphipoda) with reference to their complemental distribution. *Ecology*, 45: 677–696.

Breda, V.A. and Foster, M.S., 1985. Composition, abundance, and phenology of foliose red algae associated with two central California kelp forests. *J. Exp. Mar. Biol. Ecol.*, 94: 115–130.

Briggs, J.C., 1974. *Marine Zoogeography*. McGraw-Hill, New York, 475 pp.

Burnett, R., 1971. DDT residues: distributions of concentrations in *Emerita analoga* (Stimpson) along coastal California. *Science*, 174: 606–608.

Burton, R.S., 1979. Depth regulatory behavior of the first stage zoea larvae of the sand crab *Emerita analoga* Stimpson (Decapoda: Hippidae). *J. Exp. Mar. Biol. Ecol.*, 37: 255–270.

Cailliet, G.M. and Lea, R.N., 1977. Abundance of the "rare" zoarcid, *Maynea californica* Gilbert, 1915, in the Monterey Canyon, Monterey Bay, California. *Calif. Fish Game*, 63: 253–261.

Calvin, N.J. and Ellis, R.J., 1978. Quantitative and qualitative observations on *Laminaria dentigera* and other subtidal kelps of southern Kodiak Island, Alaska. *Mar. Biol.*, 47: 331–336.

Cameron, R.A. and Rumrill, S.S., 1982. Larval abundance and recruitment of the sand dollar *Dendraster excentricus* in Monterey Bay, California, USA. *Mar. Biol.*, 71: 197–202.

Caplan, R.I. and Boolootian, R.A., 1967. Intertidal ecology of San Nicolas Island. In: R.N. Philbrick (Editor), *Proc. Symp. on the Biology of the California Islands.* Santa Barbara Botanic Garden, Santa Barbara, pp. 203–217.

Carefoot, T., 1977. *Pacific Seashores.* Univ. Washington Press, Seattle, Wash., 208 pp.

Carey, A.G., 1965. Preliminary studies of animal-sediment interrelationships off the central Oregon coast. In: *Ocean Science and Ocean Engineering, Vol. 1.* Marine Technology Society, Washington, D.C., pp. 100–110.

Carey, A.G., 1972. Ecological observations on the benthic invertebrates from the central Oregon continental shelf. In: A.T. Pruter and D.L. Alverson (Editors), *The Columbia River Estuary and Adjacent Ocean Waters, Bioenvironmental Studies.* University of Washington Press, Seattle, Wash., pp. 422–443.

Castenholz, R.W., 1961. The effect of grazing on marine littoral diatom populations. *Ecology*, 42: 783–794.

Chapman, V.J. (Editor), 1977. Wet Coastal Ecosystems, *Ecosystems of the World, Vol. 1.* Elsevier, Amsterdam, 428 pp.

Choat, J.H., 1982. Fish feeding and the structure of benthic communities in temperate waters. *Annu. Rev. Ecol. Syst.*, 13: 423–449.

Clarke, C.E. (Editor), 1984. *Corpus Almanac & Canadian Source Book.* Kirsh, Southam Communications, Don Mills, Ont., unpag.

Clifton, H.E., Hunter, R.E. and Phillips, R.L., 1971. Depositional structures and processes in the nonbarred high-energy nearshore. *J. Sediment. Petrol.*, 41: 651–670.

Coe, W.R. and Fitch, J.E., 1950. Population studies, local growth rates and reproduction of the Pismo clam (*Tivela stultorum*). *J. Mar. Res.*, 9: 188–210.

Colman, J., 1933. The nature of the intertidal zonation of plants and animals. *J. Mar. Biol. Ass. U.K.*, 18: 435–476.

Commito, J.A., 1982. Importance of predation by infaunal polychaetes in controlling the structure of a soft-bottom community in Maine, USA. *Mar. Biol.*, 68: 77–81.

Connell, J.H., 1972. Community interactions on marine rocky intertidal shores. *Annu. Rev. Ecol. Syst.*, 3: 169–192.
Connell, J.H., 1975. Some mechanisms producing structure in natural communities: a model and some evidence from field experiments. In: M.L. Cody and J. Diamond (Editors). *Ecology and Evolution of Communities*. Belknap Press, Cambridge, Mass., pp. 460–490.
Connell, J.H., 1985. The consequences of variation in initial settlement vs. post-settlement mortality in rocky intertidal communities. *J. Exp. Mar. Biol. Ecol.*, 93: 11–46.
Connell, J.H. and Keough, M.J., 1985. Disturbance and patch dynamics of subtidal marine animals on hard substrata. In: S.T.A. Pickett and P.S. White (Editors), *The Ecology of Natural Disturbance and Patch Dynamics*. Academic Press, Orlando, pp. 125–151.
Connors, P.G., Myers, J.P. and Connors, C.S., 1981. Interhabitat movements by sanderlings in relation to foraging profitability and the tidal cycle. *Auk*, 98: 49–64.
Coon, D., 1982. Primary productivity of macroalgae in North Pacific America. In: O.R. Zaborsky (Editor), *CRC Handbook of Biosolar Resources*. CRC Press, Boca Raton, Fla., pp. 447–454.
Cowen, R.K., 1983. The effect of sheephead (*Semicossyphus pulcher*) predation on red sea urchin (*Strongylocentrotus franciscanus*) populations: an experimental analysis. *Oecologia*, 58: 249–255.
Cowen, R.K., 1985. Large scale pattern of recruitment by the labrid, *Semicossyphus pulcher*: Causes and implications. *J. Mar. Res.*, 43: 719–742.
Cowen, R.K., Agegian, C.R. and Foster, M.S., 1982. The maintenance of community structure in a central California kelp forest. *J. Exp. Mar. Biol. and Ecol.*, 64: 189–201.
Cox, G.W. and Dudley, G.H., 1968. Seasonal pattern of reproduction of the sand crab, *Emerita analoga*, in southern California, *Ecology*, 49: 746–751.
Coyer, J.A., 1984. The invertebrate assemblage associated with the giant kelp, *Macrocystis pyrifera*, at Santa Catalina Island, California: a general description with emphasis on amphipods, copepods, mysids, and shrimps. *Fish. Bull. U.S.*, 82: 55–66.
Craig, P.C., 1970. The behavior and distribution of the intertidal sand beetle, *Thinopinus pictus* (Coleoptera: Staphylinidae). *Ecology*, 51: 1012–1017.
Craig, P.C., 1971. An analysis of the concept of lunar orientation in *Orchestoidea corniculata* (Amphipoda). *Anim. Behav.*, 19: 368–374.
Craig, P.C., 1973. Behavior and distribution of the sand-beach amphipod *Orchestoidea corniculata*. *Mar. Biol.*, 23: 101–109.
Croker, R.A., 1967a. Niche diversity in five sympatric species of intertidal amphipods (Crustacea: Haustoriidae). *Ecol. Monogr.*, 37: 173–200.
Croker, R.A., 1967b. Niche specificity of *Neohaustorius schmitze* and *Haustorius* sp. (Crustacea: Amphipoda) in North Carolina. *Ecology*, 48: 971–975.
Cubit, J.D., 1969. Behavior and physical factors causing migration and aggregation of the sand crab *Emerita analoga* (Stimpson). *Ecology*, 50: 118–123.
Cubit, J.D., 1984. Herbivory and the seasonal abundance of algae on a high intertidal rocky shore. *Ecology*, 65: 1904–1917.
Cushing, D.A., 1959. The seasonal variation in oceanic production as a problem in population dynamics. *J. Cons. Int. Explor. Mer.*, 24: 455–465.
Dahl, A.L., 1971. Development, form and environment in the brown alga *Zonaria farlowii* (Dictyotales). *Bot. Mar.*, 14: 76–112.
Dahl, E., 1952. Some aspects of the ecology and zonation of the fauna of sandy beaches. *Oikos*, 4: 1–27.
Dames and Moore, 1976. *Final report: marine plant community studies — Kachemak Bay, Alaska*. Dames and Moore, Anchorage, 288 pp.
D'Antonio, C.M., 1986. Role of sand in the domination of hard substrata by the intertidal alga *Rhodomela larix*. *Mar. Ecol. Prog. Ser.*, 27: 263–275.
Davis, N. and VanBlaricom, G.R., 1978. Spatial and temporal heterogeneity in a sand bottom epifaunal community of invertebrates in shallow water. *Limnol. Oceanogr.*, 23: 417–427.
Dawson, E.Y., 1951. A further study of upwelling and associated vegetation along Pacific Baja California, Mexico. *J. Mar. Res.*, 10: 39–58.
Dawson, E.Y., Neushul, M. and Wildman, R.D., 1960. Seaweeds associated with kelp beds along southern California and northwestern Mexico. *Pac. Nat.*, 1: 1–81.
Dayton, P.K., 1971. Competition, disturbance, and community organization: the provision and subsequent utilization of space in a rocky intertidal community. *Ecol. Monogr.*, 41: 351–389.
Dayton, P.K., 1973. Dispersion, dispersal and persistence of the annual intertidal alga, *Postelsia palmaeformis* Ruprecht. *Ecology*, 54: 433–438.
Dayton, P.K., 1975a. Experimental evaluation of ecological dominance in a rocky intertidal algal community. *Ecol. Monogr.*, 45: 137–159.
Dayton, P.K., 1975b. Experimental studies of algal canopy interactions in a sea otter dominated kelp community at Amchitka Island, Alaska. *Fish. Bull.*, 73: 230–237.
Dayton, P.K., 1985. Ecology of kelp communities. *Annu. Rev. Ecol. Syst.*, 16: 215–245.
Dayton, P.K. and Tegner, M.J., 1984a. Catastrophic storms, El Niño, and patch stability in a southern California kelp community. *Science*, 224: 283–285.
Dayton, P.K. and Tegner, M.J., 1984b. The importance of scale in community ecology: a kelp forest example with terrestrial analogs. In: P.W. Price, C.N. Slobodchikoff and W.S. Gaud (Editors), *A New Ecology: Novel Approaches to Interactive Systems*. Wiley, New York, pp. 457–481.
Dayton, P.K., Robilliard, G.A., Paine, R.T. and Dayton, L.B., 1974. Biological accommodation in the benthic community at McMurdo Sound, Antarctica. *Ecol. Monogr.*, 44: 105–128.
Dayton, P.K., Currie, V., Gerrodette, T., Keller, B.D., Rosenthal, R. and Ven Tresca, D., 1984. Patch dynamics and stability of some California kelp communities. *Ecol. Monogr.* 54: 253–289.
Dean, T.A., 1985. The temporal and spatial distribution of underwater quantum irradiation in a southern California kelp forest. *Estuarine Coastal Shelf Sci*, 21: 835–844.

Dean, T.A. and Jacobsen, F.R., 1986. Nutrient-limited growth of juvenile kelp, *Macrocystis pyrifera*, during the 1982–1984 "El Niño" in southern California. *Mar. Biol.*, 90: 597–601.

Dean, T.A., Schroeter, S.C. and Dixon, J.D., 1984. Effects of grazing by two species of sea urchins (*Strongylocentrotus franciscanus* and *Lytechinus anamesus*) on recruitment and survival of two species of kelp (*Macrocystis pyrifera* and *Pterygophora californica*). *Mar. Biol.*, 78: 301–313.

Dearn, S.L, 1987. *The fauna of subtidal articulated coralline mats: composition, dynamics, and effects of spatial heterogeneity*. M.Sc. Thesis, California State Univ., Stanislaus, 51 pp.

DeCoursey, P.J. (Editor), 1976. *Biological Rhythms in the Marine Environment*. Univ. South Carolina Press, Columbia, S.C., 283 pp.

Dethier, M.N., 1984. Disturbance and recovery in intertidal pools: maintenance of mosaic patterns. *Ecol. Monogr.*, 54: 99–118.

Devinny, J.S. and Kirkwood, P.D., 1974. Algae associated with kelp beds of the Monterey Peninsula, California. *Bot. Mar.*, 17: 100–106.

Devinny, J.S. and Volse, L.A., 1978. Effects of sediments on the development of *Macrocystis pyrifera* gametophytes. *Mar. Biol.*, 48: 343–348.

De Vogelaere, A.P., 1987. *Rocky intertidal patch succession after disturbance: effects of severity, size, and position within patch*. M.Sc. Thesis, California State Univ., San Francisco, Ca., 91 pp.

Dexter, D.M., 1967. Distribution and niche diversity of haustoriid amphipods in North Carolina. *Chesapeake Sci.*, 8: 187–192.

Dillery, D.G. and Knapp, L.V., 1970. Longshore movements of the sand crab, *Emerita analoga* (Decapoda, Hippidae). *Crustaceana*, 18: 233–240.

Doty, M.S., 1946. Critical tide factors that are correlated with the vertical distribution of marine algae and other organisms along the Pacific Coast. *Ecology*, 27: 315–328.

Druehl, L.D., 1970. The pattern of Laminariales distribution in the northeast Pacific. *Phycologia*, 9: 237–247.

Druehl, L.D., 1981. Geographical distribution. In: C.S. Lobban and M.J. Wynne (Editors), *The Biology of Seaweeds*. Univ. of California Press, Berkeley, Ca., pp. 306–325.

Druehl, L.D. and Wheeler, W.N., 1986. Population biology of *Macrocystis integrifolia* from British Columbia, Canada, *Mar. Biol.*, 90: 173–179.

Duggins, D.O., 1980. Kelp beds and sea otters: an experimental approach. *Ecology*, 61: 447–453.

Duggins, D.O. and Dethier, M.N., 1985. Experimental studies of herbivory and algal competition in a low intertidal habitat. *Oecologia*, 67: 183–191.

Duggins, D.O., Simenstad, C.A. and Estes, J.A., 1989. Magnification of secondary production by kelp detritus in coastal marine ecosystems. *Science*, 245: 170–173.

Dungan, M.L., Miller, T.E. and Thompson, D.A., 1982. Catastrophic decline of a top carnivore in the Gulf of California rocky intertidal zone. *Science*, 216: 989–991.

Ebeling, A.W. and Bray, R.N., 1976. Day versus night activity of reef fishes in a kelp forest of Santa Barbara, California. *Fish. Bull. U.S.*, 74: 703–717.

Ebeling, A.W. and Laur, D.R., 1985. The influence of plant cover on surfperch abundance at an offshore temperate reef. *Environ. Biol. Fish.*, 12: 169–179.

Ebeling, A.W., Larson, R.J. and Alevizon, W.S., 1980a. Habitat groups and island–mainland distribution of kelp-bed fishes off Santa Barbara, California. In: D.M. Power (Editor), *Multidisciplinary Symposium on the California Islands*. Santa Barbara Museum of Natural History, Santa Barbara, CA, pp. 404–431.

Ebeling, A.W., Larson, R.J., Alevizon, W.S. and Bray, R.N., 1980b. Annual variability of reef-fish assemblages in kelp forests off Santa Barbara, California. *Fish. Bull. U.S.*, 78: 361–377.

Ebeling, A.W., Laur, D.R. and Rowley, R.J., 1985. Severe storm disturbances and reversal of community structure in a southern California kelp forest. *Mar. Biol.*, 84: 287–294.

Edwards, D.C., 1969. Zonation by size as an adaptation for intertidal life in *Olivella biplicata*. *Am. Zool.* 9: 399–417.

Efford, I.E., 1965. Aggregation in the sand crab, *Emerita analoga* (Stimpson). *J. Animal Ecol.*, 34: 63–75.

Efford, I.E., 1966. Feeding in the sand crab, *Emerita analoga* (Stimpson). *Crustaceana*, 10: 167–182.

Efford, I.E., 1967. Neoteny in sand crabs of the genus *Emerita* (Anomura, Hippidae). *Crustaceana*, 13: 81–93.

Efford, I.E., 1969. Egg size in the sand crab *Emerita analoga* (Decapoda, Hippidae). *Crustaceana*, 16: 15–26.

Efford, I.E., 1970. Recruitment to sedentary marine populations as exemplified by the sand crab, *Emerita analoga* (Decapoda, Hippidae). *Crustaceana*, 18: 293–309.

Emerson, S.E. and Zedler, J.B., 1978. Recolonization of intertidal algae: an experimental study. *Mar. Biol.*, 44: 315–324.

Emery, K.O., 1960. *The Sea off Southern California*. Wiley, New York, 366 pp.

Enright, J.T., 1961. Lunar orientation of *Orchestoidea corniculata* Stout (Amphipoda). *Biol. Bull.*, 120: 148–156.

Enright, J.T., 1965. Entrainment of a tidal rhythm. *Science*, 147: 864–867.

Enright, J.T. and Honegger, H.W., 1977. Diurnal vertical migration: adaptive significance and timing. Pt. 2. Test of the model: details of timing. *Limnol. Oceanogr.*, 22: 873–886.

Eppley, R.W., Carlucci, A.F., Holm-Hansen, O., Kiefer, D., McCarthy, J.J. and Williams, P.M., 1972. Evidence for eutrophication in the sea near southern California sewage outfalls. *Cal. Coop. Ocean. Fish. Invest. Rep.*, 16: 74–83.

Estes, J.A. and VanBlaricom, G.R., 1985. Sea otters and shellfisheries. In: R.A. Beverton, D. Lavigne and J. Beddington (Editors), *Conflicts Between Marine Mammals and Fisheries*. Allen and Unwin, London, pp. 187–235.

Estes, J.A., Smith, N.S. and Palmisano, J.F., 1978. Sea otter predation and community organization in the western Aleutian Islands. *Ecology*, 59: 822–833.

Fager, E.W., 1964. Marine sediments: effects of a tube-building polychaete. *Science*, 143: 356–359.

Fager, E.W., 1968. A sand-bottom epifaunal community of invertebrates in shallow water. *Limnol. Oceanogr.*, 13: 448–464.

Feder, H.M., Turner, C.H. and Limbaugh, C., 1974. *Observations on fishes associated with kelp beds in southern*

California. California Department of Fish and Game, Fish Bulletin 160, Sacramento, 144 pp.

Ferguson, A. (Editor). 1984. *Intertidal plants and animals of the Landels-Hill Big Creek Reserve*. Center for Marine Studies, University of California, Santa Cruz, Ca., 106 pp.

Fitch, J.E., 1950. The Pismo clam. *Calif. Fish Game.*, 36: 285–312.

Fitch, J.E., 1965. A relatively unexploited population of Pismo clams, *Tivela stultorum* (Mawe 1823) (Veneridae). *Proc. Malacological Soc. London*, 36: 309–312.

Foreman, R.E., 1977. Benthic community modification and recovery following intensive grazing by *Strongylocentrotus droebachiensis*. *Helgol. Wiss. Meeresunter.*, 30: 468–484.

Foster, M.S., 1982a. The regulation of macroalgal associations in kelp forests. In: L. Srivastava (Editor), *Synthetic and Degradative Processes in Marine Macrophytes*. Walter de Gruyter, Berlin, pp. 185–205.

Foster, M.S., 1982b. Factors controlling the intertidal zonation of *Iridaea flaccida* (Rhodophyta). *J. Phycol.*, 18: 285–294.

Foster, M.S. and Schiel, D.R., 1985. *The ecology of giant kelp forests in California: a community profile*. U.S. Fish and Wildlife Service Biol. Rep. 85(7.2), Slidell, La., 152 pp.

Foster, M.S. and Schiel, D.R., 1988. Kelp communities and sea otters: keystone species or just another brick in the wall? In: G.R. VanBlaricom and J.A. Estes (Editors), *The Community Ecology of Sea Otters*. Springer, Berlin, pp. 92–115.

Foster, M.S., Carter, J.W. and Schiel, D.R., 1983. The ecology of kelp communities. In: W. Bascom (Editor), *The Effects of Waste Disposal on Kelp Communities*. Southern California Coastal Water Research Project, Long Beach, pp. 53–69.

Foster, M.S., De Vogelaere, A.P., Harrold, C., Pearse, J.S. and Thum, A.B., 1988. *Causes of spatial and temporal patterns in rocky intertidal communities of central and northern California*. Mem. Calif. Acad. Sci. No. 9, 45 pp.

Francis, L., 1973. Intraspecific aggression and its effect on the distribution of *Anthopleura elegantissima* and some related sea anemones. *Biol. Bull.*, 144: 73–92.

Frank, P.W., 1965. The biodemography of an intertidal snail population. *Ecology*, 46: 831–844.

Frank, P.W., 1982. Effects of winter feeding on limpets by black oystercatchers, *Haematopus bachmani*. *Ecology*, 63: 1352–1362.

Fritchman, H.K., 1962. A study of the reproductive cycle in California Acmaeidae (Gastropoda). Part IV. *Veliger*, 4: 134–140.

Fukuyama, A.K. and Oliver, J.S., 1985. Sea star and walrus predation on bivalves in Norton Sound, Bering Sea, Alaska. *Ophelia*, 24: 17–36.

Fusaro, C., 1978. Growth rate of the sand crab, *Emerita analoga* (Hippidae), in two different environments. *Fish. Bull. U.S.*, 76: 369–375.

Fusaro, C., 1980. Temperature and egg production by the sand crab, *Emerita analoga* (Stimpson) (Decapoda, Hippidae). *Crustaceana*, 38: 55–60.

Gaines, S.D., 1985. Herbivory and between-habitat diversity: the differential effectiveness of defenses in a marine plant. *Ecology*, 66: 473–485.

Gaines, S.D. and Lubchenco, J., 1982. A unified approach to marine plant–herbivore interactions. 2. biogeography. *Ann. Rev. Ecol. Syst.*, 13: 111–138.

Gerard, V.A., 1976. *Some aspects of material dynamics and energy flow in a kelp forest in Monterey Bay, California*. Ph.D. Thesis, University of California, Santa Cruz, Ca., 173 pp.

Gerard, V.A., 1982. In situ rates of nitrate uptake by giant kelp, *Macrocystis pyrifera* (L.) C. Agardh, tissue differences, environmental effects, and predictions of nitrogen-limited growth. *J. Exp. Mar. Biol. Ecol.*, 62: 211–224.

Gerard, V.A., 1984a. The light environment in a giant kelp forest: influence of *Macrocystis pyrifera* on spatial and temporal variability. *Mar. Biol.*, 84: 189–195.

Gerard, V.A., 1984b. Physiological effects of El Niño on giant kelp in southern California. *Mar. Biol. Lett.*, 5: 317–322.

Ghelardi, R.J., 1971. "Species" structure of the animal community that lives in *Macrocystis pyrifera* holdfasts. *Nova Hedwigia*, 32: 381–420.

Giese, A.C., 1969. A new approach to the biochemical composition of the mollusc body. *Oceanogr. Mar. Biol. Annu. Rev.*, 7: 175–229.

Gislèn, T., 1943. Physiographical and ecological investigations concerning the littoral of the northern Pacific. Section I. *Lunds Univ. Arrskrift. N.F. Avd. 2*, 39: 1–63.

Gislèn, T., 1944. Physiographical and ecological investigations concerning the littoral of the northern Pacific. Section II–IV. *Lunds Univ. Arsskr. Avd. 2*, 40: 1–91.

Glynn, P.W., 1965. Community composition, structure, and interrelationships in the marine intertidal *Endocladia muricata-Balanus glandula* association in Monterey Bay, California. *Beaufortia*, 12: 1–198.

Gorshkov, G. (Editor). 1976. *World Ocean Atlas, Vol. 1. Pacific Ocean*. Permagon Press, Elmsford, NY, 302 pp.

Gotshall, D.W., 1977. Stomach contents of northern California dungeness crabs, *Cancer magister*. *Calif. Fish Game*, 63: 43–51.

Grant, W.S., 1977. High intertidal community organization on a rocky headland in Maine, USA. *Mar. Biol.*, 44: 15–25.

Grigg, R.W., 1975. Age structure of a longevous coral: a relative index of habitat suitability and stability. *Am. Nat.*, 109: 647–657.

Grigg, R.W. and Kiwala, R.S., 1970. Some ecological effects of discharged waste on marine life. *Calif. Fish Game*, 56: 145–155.

Griggs, G.B., 1974. Nearshore current patterns along the central California coast. *Estuarine Coastal Mar. Sci.*, 2: 395–405.

Gunnill, F.C., 1980a. Recruitment and standing stocks in populations of one green alga and five brown algae in the intertidal zone near La Jolla, California during 1973–1977. *Mar. Ecol. Prog. Ser.*, 3: 231–243.

Gunnill, F.C., 1980b. Demography of the intertidal brown alga *Pelvetia fastigiata* in southern California, USA. *Mar. Biol.*, 59: 169–179.

Gunnill, F.C., 1983. Seasonal variations in the invertebrate faunas of *Pelvetia fastigiata* (Fucaceae): effects of plant size and distribution. *Mar. Biol.*, 73: 115–130.

Halberg, F., Halberg, F. and Giese, A.C., 1969. Estimation of objective parameters for circa annual rhythms in marine invertebrates. *Rass. Neur. Veg.*, 23: 173–186.

Hall, C.A., 1964. Shallow-water marine climates and molluscan provinces. *Ecology*, 45: 226–234.

Hammer, R.M., 1981. Day-night differences in the emergence of demersal zooplankton from a sand substrate in a kelp forest. *Mar. Biol.*, 62: 275–280.

Hammer, R.M. and Zimmerman, R.C., 1979. Species of demersal zooplankton inhabiting a kelp forest ecosystem off Santa Catalina Island, California. *Bull. S. Calif. Acad. Sci.*, 78: 199–206.

Hamner, W.M., 1969. The behavior and life history of a sand-beach isopod, *Tylos punctatus*. *Ecology*, 50: 442–453.

Hamner, W.M., Smyth, M. and Mulford, E.D., 1968. Orientation of the sand-beach isopod *Tylos punctatus*. *Anim. Behav.*, 16: 405–409.

Hansen, J.E., 1977. Ecology and natural history of *Iridaea cordata* (Gigartinales, Rhodophyta) growth. *J. Phycology*, 13: 395–402.

Harger, J.R.E. and Landenberger, D.E., 1971. The effect of storms as a density dependent mortality factor on populations of sea mussels. *Veliger*, 14: 195–201.

Harlin, M.M. and Lindbergh, J.M., 1977. Selection of substrata by seaweeds: optimal surface relief. *Mar. Biol.*, 40: 33–40.

Harrold, C. and Pearse, J.S., 1987. The ecological role of echinoderms in kelp forests. In: J.M. Lawrence (Editor), *Echinoderm Studies, Vol. II*. Balkema, Rotterdam, pp. 137–233.

Harrold, C. and Reed, D.C., 1985. Food availability, sea urchin grazing, and kelp forest community structure. *Ecology*, 66: 1160–1169.

Hartman, M.J. and Zahary, R.G., 1983. Biogeography of protected rocky intertidal communities of the northeastern Pacific. *Bull. Mar. Sci.*, 33: 729–735.

Hartwick, R.F., 1975. *Orientation behavior in beach hopper of the genus Orchestoidea: capacities and strategies*. Ph.D. Thesis, Univ. California, San Diego, Ca., 202 pp.

Haven, S.B., 1971. Niche differences in the intertidal limpets *Acmaea scabra* and *A. digitalis*. *Veliger*, 13: 231–248.

Hayden, B.P. and Dolan, R., 1976. Coastal marine fauna and marine climates of the Americas. *J. Biogeogr.*, 3: 71–81.

Hayes, W.B., 1974. Sand-beach energetics: importance of the isopod *Tylos punctatus*. *Ecology*, 55: 838–847.

Hayes, W.B., 1977. Factors affecting the distribution of *Tylos punctatus* (Isopoda, Oniscoidea) on beaches in southern California and northern Mexico. *Pacific Sci.*, 31: 165–186.

Hedgpeth, J.W., 1957. Sandy beaches. *Mem. Geol. Soc. Am. 67.*, 1: 587–608.

Heine, J.N., 1983. Seasonal productivity of two red algae in a central California kelp forest. *J. Phycol.*, 19: 146–152.

Hewatt, W.G., 1937. Ecological studies on selected marine intertidal communities of Monterey Bay, California. *Am. Midl. Nat.*, 18: 161–206.

Highsmith, R.C., 1982. Induced settlement and metamorphosis of sand dollar (*Dendraster excentricus*) larvae in predator-free sites: adult sand dollar beds. *Ecology*, 63: 329–337.

Hines, A.H., 1982. Coexistence in a kelp forest: size, population dynamics, and resource partitioning in a guild of spider crabs (Brachiura, Majidae). *Ecol. Monogr.*, 52: 179–198.

Hixon, M.A., 1980. Competitive interactions between California reef fishes of the genus *Embiotoca*. *Ecology*, 61: 918–932.

Hopkins, J.T., 1966. The role of water in the behaviour of an estuarine mud-flat diatom. *J. Mar. Biol. Assoc. U.K.*, 46: 617–626.

Hulberg, L.W. and Oliver, J.S., 1980. Caging manipulations in marine soft-bottom communities: the importance of animal interactions and sedimentary habitat modifications. *Can. J. Fish. Aquat. Sci.*, 37: 1130–1139.

Hutchins, L.W., 1947. The basis for temperature zonation in geographical distribution. *Ecol. Monogr.*, 17: 325–335.

Jackson, G.A., 1977. Nutrients and production of the giant kelp *Macrocystis pyrifera* off southern California. *Limnol. Oceanogr.*, 22: 979–995.

Johansen, H.W. and Austin, L.F., 1970. Growth rates in the articulated coralline *Calliarthron* (Rhodophyta). *Can. J. Bot.*, 48: 125–182.

Johnson, W.S., 1976. Biology and population dynamics of the intertidal isopod *Cirolana harfordi*. *Mar. Biol.*, 36: 343–350.

Kastendiek, J.E., 1982a. Competitor-mediated coexistence: interactions among three species of benthic macroalgae. *J. Exp. Mar. Biol. Ecol.*, 62: 201–210.

Kastendiek, J.E., 1982b. Factors determining the distribution of the sea pansy, *Renilla kollikeri*, in a subtidal sand-bottom habitat. *Oecologia*, 52: 340–347.

Kenyon, K.W., 1969. *The sea otter in the eastern Pacific Ocean*. U.S. Government Printing Office, U.S. Department of the Interior, North American Fauna 68, Washington, D.C., 352 pp.

Ketchum, B.H. (Editor), 1983. *Estuaries and Enclosed Seas, Ecosystems of the World, Vol. 26*. Elsevier, Amsterdam, 500 pp.

Khailov, K.M. and Burlakova, Z.P., 1969. Release of dissolved organic matter by marine seaweeds and distribution of their total organic production to inshore communities. *Limnol. Oceanogr.*, 14: 521–527.

Kimura, R.S. and Foster, M.S., 1984. The effects of harvesting *Macrocystis pyrifera* on the algal assemblage in a giant kelp forest. *Hydrobiol.*, 116/117: 425–428.

Kinney, D.M., 1966. *Geology in National Atlas (map 1: 7,500,000 scale)*. U.S. Geological Survey, Washington, D.C.

Kompfner, H., 1974. Larvae and pupae of some wrack dipterans on a California beach (Diptera: Coelopidae, Anthomyiidae, Sphaeroceridae). *Pan-Pac. Entomol.*, 50: 44–52.

Kvitek, R.G. and Oliver, J.S., 1986. Side-scan sonar impressions of gray whale feeding grounds along Vancouver Island, Canada. *Cont. Shelf Res.*, 6: 639–654.

Kvitek, R.G. and Oliver, J.S., 1988. Sea otter foraging habits in soft-bottom environments. In: G.R. VanBlaricom and J.A. Estes (Editors), *The Community Ecology of Sea Otters*. Springer, New York, pp. 22–47.

Lane, H.U. (Editor), 1985. *The World Almanac Book of Facts*. Newspaper Enterprise Assoc., New York, 928 pp.

Larson, R.J., 1980. Competition, habitat selection and the bathymetric segration of two rockfish (*Sebastes*) species. *Ecol. Monogr.*, 50: 221–239.

Larson, R.J. and DeMartini, E.E., 1984. Abundance and vertical distribution of fishes in a cobble-bottom kelp forest off San Onofre, California. *Fish. Bull. U.S.*, 82: 37–53.

Leighton, D.L., Jones, L.G. and North, W.J., 1966. Ecological relationships between kelp and sea urchins in southern California. *Proc. Int. Seaweed Symp.*, 5: 141–153.

Lewin, J. and Mackas, D., 1972. Blooms of surf-zone diatoms along the coast of the Olympic Peninsula, Washington. I. Physiological investigations of *Chaetoceros armatum* and *Asterionella socialis* in laboratory cultures. *Mar. Biol.*, 16: 171–181.

Lewis, J.R., 1964. *The Ecology of Rocky Shores*. The English Univ. Press, London, 323 pp.

Lie, U., 1969. Standing crop of benthic infauna in Puget Sound and off the coast of Washington. *J. Fish. Res. Board Can.*, 26: 55–62.

Lie, U. and Kisker, D.S., 1970. Species composition and structure of benthic infauna communities off the coast of Washington. *J. Fish. Res. Board Can.*, 27: 2273–2285.

Lindstrom, S.C. and Foreman, R.E., 1978. Seaweed associations of the Flat Top Islands, British Columbia: a comparison of community methods. *Syesis*, 11: 171–185.

Lippincott, W.H., 1980. Littoral Zone Biota. *Environmental assessment of the Alaskan continental shelf northwest gulf of Alaska interim synthesis report*. Science Applications, Inc 2760 29th Street, Boulder, Colorado 80301, U.S. Department of Commerce, U.S. Department of Interior, pp. 109–122.

Littler, M.M. and Littler, D.S., 1981. Intertidal macrophyte communities from Pacific Baja California and the upper Gulf of California: relatively constant vs. environmentally fluctuating systems. *Mar. Ecol. Prog. Ser.*, 4: 145–158.

Littler, M.M. and Murray, S.N., 1974. The primary productivity of marine macrophytes from a rocky intertidal community. *Mar. Biol.*, 27: 131–135.

Littler, M.M., Murray, S.N. and Arnold, K.E., 1979. Seasonal variations in net photosynthetic performance and cover of intertidal macrophytes. *Aquat. Bot.*, 7: 35–46.

Littler, M.M., Martz, D.R. and Littler, D.S., 1983. Effects of recurrent sand deposition on rocky intertidal organisms: importance of substrate heterogeneity in a fluctuating environment. *Mar. Ecol. Prog. Ser.*, 11: 129–139.

Lowry, L.F. and Pearse, J.S., 1973. Abalones and sea urchins in an area inhabited by sea otters. *Mar. Biol.*, 23: 213–219.

Lowry, L.F., McElroy, A.J. and Pearse, J.S., 1974. The distribution of six species of gastropod molluscs in a California kelp forest. *Biol. Bull.*, 147: 386–396.

Lubchenco, J., 1983. *Littorina* and *Fucus*: the effects of herbivores, substratum heterogeneity, and plant escapes during succession. *Ecology*, 64: 1116–1123.

Lubchenco, J. and Cubit, J.D., 1980. Heteromorphic life histories of certain marine algae as adaptations to variations in herbivory. *Ecology*, 61: 676–687.

Lüning, K., 1980. Critical levels of light and temperature regulating the gametogenesis of three *Laminaria* species (Phaeophyceae). *J. Phycol.*, 16: 1–16.

Lüning, K., 1981. Photobiology of seaweeds: ecophysical effects. *Proc. Int. Seaweed Symp.*, 10: 36–55.

Lüning, K. and Neushul, M., 1978. Light and temperature demands for growth and reproduction of laminarian gametophytes in southern and central California. *Mar. Biol. (Berl.)*, 45: 297–309.

MacGinitie, G.E., 1938. Littoral marine communities. *Am. Midl. Nat.*, 21: 28–55.

Mann, K.H., 1982. *Ecology of Coastal Waters: a Systems Approach*. Univ. California Press, Berkeley, CA, 322 pp.

Markham, J.W., 1973. Observations on the ecology of *Laminaria sinclairii* on three northern Oregon beaches. *J. Phycol.*, 9: 336–341.

Markham, J.W. and Newroth, P.R., 1972. Observations on the ecology of *Gymnogongrus linearis* and related species. *Proc. Int. Seaweed Symp.*, 7: 126–130.

Mattison, J.E., Trent, J.D., Shanks, A.L., Akin, T.B. and Pearse, J.S., 1977. Movement and feeding activity of red sea urchins (*Strongylocentrotus franciscanus*) adjacent to a kelp forest. *Mar. Biol.*, 39: 25–30.

Mayer, D.L., Lebednik, P.A. and Selak, P.J., 1981. Diablo Canyon Power Plant 316(a) demonstration annual report 1978. In: D.W. Behrens and E.A. Banuet-Hutton (Editors), *Environmental Investigations of Diablo Canyon 1978*. Pacific Gas and Electric Co., San Ramon, unpag.

Menge, B.A., 1976. Organization of the New England rocky intertidal community: role of predation, competition, and environmental heterogeneity. *Ecol. Monogr.*, 46: 355–393.

Menge, B.A., 1978. Predation intensity in a rocky intertidal community: effect of an algal canopy, wave action, and desiccation on predator foraging rates. *Oecologia*, 34: 17–35.

Merrill, R.J. and Hobson, E.S., 1970. Field observations of *Dendraster excentricus*, a sand dollar of western North America. *Am. Midl. Nat.*, 83: 585–624.

Miles, E., Sherman, J., Fluharty, D., Gibbs, S. and Tanaka, S. (Editors), 1982. *Atlas of Marine Use in the North Pacific Region*. Univ. California Press, Berkeley, Ca., 103 pp.

Miller, D.J. and Geibel, J.J., 1973. *Summary of blue rockfish and lingcod life histories; a reef ecology study; and giant kelp, Macrocystis pyrifera, experiments in Monterey Bay, California*. California Department of Fish and Game, Fish Bulletin 158, Sacramento, 137 pp.

Miller, D.J. and Lea, R.N., 1972. *Guide to coastal marine fishes of California*. California Department of Fish and Game, Fish Bulletin 157, Sacramento, 235 pp.

Miller, R.J. and Mann, K.H., 1973. Ecological energetics of the seaweed zone in a marine bay on the Atlantic coast of Canada III. Energy transformations by sea urchins. *Mar. Biol.*, 18: 99–114.

Mitchell, C.T. and Hunter, J.R., 1970. Fishes associated with drifting kelp, *Macrocystis pyrifera*, off the coast of southern California and northern Baja California. *Calif. Fish Game*, 56: 288–297.

Mokyevsky, O.B., 1960. Geographical zonation of marine littoral types. *Limnol. Oceanogr.*, 5: 389–396.

Moore, I. and Legner, E.F., 1974. Succession of the coleopterous fauna in wrack. *Wasmann J. Biol.*, 31: 289–290.

Moore, P.G. and Seed, R. (Editors), 1986. *The Ecology of Rocky Coasts*. Columbia Univ. Press, New York, 467 pp.

Morin, J.G., Kastendiek, J.E., Harrington, A. and Davis, N., 1985. Organization and patterns of interactions in a subtidal sand community of an exposed coast. *Mar. Ecol. Prog. Ser.*, 27: 163–185.

Morris, R.H., Abbott, D.P. and Haderlie, E.C., 1980. *Intertidal Invertebrates of California*. Stanford Univ. Press, Stanford, 690 pp.

Murison, L., Murie, D., Morin, K. and Curiel, J.S., 1984.

Aspects of feeding observed in the gray whale (*Eschrichtius robustus* Lillbeborg) along the west coast of Vancouver Island. In: M.L. Jones, S. Leatherwood and S. Swartz (Editors), *The Gray Whale*. Academic Press, New York, pp. 451–463.

Murray, S.N., Littler, M.M. and Abbott, I.A., 1980. Biogeography of the California marine algae with emphasis on the southern California islands. In: D.M. Power (Editor), *The California Islands: Proceedings of a Multidisciplinary Symposium*. Haagen Printing, Santa Barbara, Ca., pp. 325–339.

Myers, J.P., Connors, P.G. and Pitelka, F.A., 1979. Territory size in wintering sanderlings: the effects of prey abundance and intruder density. *Auk*, 96: 551–561.

Myers, J.P., Williams, S.L. and Pitelka, F.A., 1980a. An experimental analysis of prey availability for sanderlings (Aves: Scolopacidae) feeding on sandy beach crustaceans. *Can. J. Zool.*, 58: 1564–1574.

Myers, J.P., Connors, P.G. and Pitelka, F.A., 1980b. Optimal territory size in the sanderling: compromises in a variable environment. In: A.C. Kamil and T.D. Sargent (Editors), *Foraging Behaviour: Ecological, Ethological, and Psychological Approaches*. Garland Press, New York, pp. 135–158.

Naylor, E. and Hartnoll, R.G. (Editors), 1979. *Cyclic Phenomena in Marine Plants and Animals*. Pergamon Press, Oxford, 477 pp.

Nelson, B.V. and Vance, R.R., 1979. Diel foraging patterns of the sea urchin *Centrostephanus coronatus* as a predator avoidance strategy. *Mar. Biol.*, 51: 251–258.

Nerini, M., 1984. A review of gray whale feeding ecology. In: M.L. Jones, S.L. Swartz and S. Leatherwood (Editors), *The Gray Whale*. Academic Press, New York, pp. 423–450.

Neushul, M., 1965. SCUBA diving studies of the vertical distribution of benthic marine plants. In: *Proc. 5th Marine Biology Symp*. Acta Universitatis, Goteburg, pp. 161–176.

Neushul, M., 1967. Studies of subtidal marine vegetation in western Washington. *Ecology*, 48: 83–94.

Neushul, M., 1971. Submarine illumination in *Macrocystis* beds. *Nova Hedwigia*, 32: 241–254.

North, W.J. (Editor), 1971a. The biology of giant kelp beds (*Macrocystis*) in California. *Nova Hedwigia*, 32: 1–600.

North, W.J., 1971b. Introduction and background: the biology of giant kelp beds (*Macrocystis*) in California. *Nova Hedwigia*, 32: 1–97.

North, W.J. and Hubbs, C.L., 1968. *Utilization of kelp bed resources in southern California*. California Department of Fish and Game, Fish Bulletin 139, Sacramento, Ca., 264 pp.

North, W.J., Gerard, V. and Kuwabara, J., 1982. Farming *Macrocystis* at coastal and oceanic sites. In: L.M. Srivastava (Editor), *Synthetic and Degradative Processes in Marine Macrophytes*. Walter de Gruyter, New York, pp. 247–264.

Oakden, J.M., 1984. Feeding and substrate preference in five species of phoxocephalid amphipods from central California. *J. Crustacean Biol.*, 4: 233–247.

Oliver, J.S. and Kvitek, R.G., 1984. Side-scan sonar records and diver observations of gray whale (*Eschrichtius robustus*) feeding grounds. *Biol. Bull. (Woods Hole)*, 167: 264–269.

Oliver, J.S. and Slattery, P.N., 1985. Destruction and opportunity on the sea floor: effects of grey whale feeding. *Ecology*, 66: 1965–1975.

Oliver, J.S. and Slattery, P.N., 1985b. Effects of crustacean predators on species composition and population structure of soft-bodied infauna from McMurdo Sound, Antarctica. *Ophelia*, 24: 155–175.

Oliver, J.S., Slattery, P.N., Hulberg, L.W. and Nybakken, J.W., 1980. Relationships between wave disturbance and zonation of benthic invertebrate communities along a subtidal high-energy beach in Monterey Bay, California. *Fish. Bull. U.S.*, 78: 437–454.

Oliver, J.S., Oakden, J.M. and Slattery, P.N., 1982. Phoxocephalid amphipod crustaceans as predators on larvae and juveniles in marine soft-bottom communities. *Mar. Ecol. Prog. Ser.*, 7: 179–184.

Oliver, J.S., Slattery, P.N., Silberstein, M.A. and O'Connor, E.F., 1984. Gray whale feeding on dense ampeliscid amphipod communities near Bamfield, British Columbia. *Can. J. Zool.*, 62: 41–49.

Orth, R.E., Moore, I. and Fisher, T.W., 1977. Year-round survey of Stohylinidae of a sandy beach in southern California. *Wasmann. J. Biol.*, 35: 169–195.

Ostarello, G.L., 1973. Natural history of the hydrocoral *Allopora californica* Verrill (1866). *Biol. Bull.*, 45: 548–564.

Page, H.M., 1986. Differences in population structure and growth rate of the stalked barnacle *Pollicipes polymerus* between a rocky headland and an offshore oil platform. *Mar. Ecol. Prog. Ser.*, 29: 157–164.

Paine, R.T., 1979. Disaster, catastrophe, and local persistence of the sea palm *Postelsia palmaeformis*. *Science*, 205: 685–687.

Paine, R.T., 1980. Food webs: linkage, interaction strength and community infrastructure. *J. Anim. Ecol.*, 49: 667–685.

Paine, R.T., 1984. Some approaches to modeling multispecies systems. In: R.M. May (Editor), *Exploitation of Marine Communities*. Springer, Berlin, pp. 191–207.

Paine, R.T. and Levin, S.A., 1981. Intertidal landscapes: disturbance and the dynamics of pattern. *Ecol. Monogr.*, 51: 145–178.

Paine, R.T. and Vadas, R.L., 1969. The effects of grazing by sea urchins, *Strongylocentrotus* spp., on benthic algal populations. *Limnol. Oceanogr.*, 14: 710–719.

Parrish, R.H., Nelson, C.S. and Bakun, A., 1981. Transport mechanisms and reproductive success of fishes in the California Current. *Biol. Oceanogr.*, 1: 175–203.

Pearse, J.S., 1980. Intertidal animals. In: B.J. LeBouef and S. Kaza (Editors), *The Natural History of Año Nuevo*. Boxwood Press, Pacific Grove, pp. 205–236.

Pearse, J.S. and Eernisse, D.J., 1982. Photoperiodic regulation of gametogenesis and gonadal growth in the sea star *Pisaster ochraceus*. *Mar. Biol.*, 67: 121–125.

Pearse, J.S. and Hines, A.H., 1979. Expansion of a central California kelp forest following mass mortality of sea urchins. *Mar. Biol.*, 51: 83–91.

Pearse, J.S., Costa, D.P., Yellin, M.B. and Agegian, C.R., 1977. Localized mass mortality of red sea urchins, *Strongylocentrotus franciscanus*, near Santa Cruz, California. *Fish. Bull. U.S.*, 75: 645–648.

Pequegnat, W.E., 1964. The epifauna of a California siltstone reef. *Ecology*, 45: 272–283.

Peterson, C.H., 1979a. The importance of predation and competition in organizing the intertidal epifaunal communities of Barnegat Inlet, New Jersey. *Oecologia*, 39: 1–24.

Peterson, C.H., 1979b. Predation, competitive exclusion, and diversity in the soft-sediment benthic communities of estuaries and lagoons. In: R.J. Livingston (Editor), *Ecological Processes in Coastal and Marine Systems*. Plenum Press, New York, pp. 233–264.

Petraitis, P.S., 1983. Grazing patterns of the periwinkle and their effect on sessile intertidal organisms. *Ecology*, 64: 522–533.

PG&E (Pacific Gas and Electric Co.), 1984. *Thermal effects monitoring program, 1983 annual report, Diablo Canyon Nuclear Power Plant*. Pacific Gas and Electric Co., San Francisco, unpag.

Quammen, M.L., 1982. Influence of subtle substrate differences on feeding by shorebirds on intertidal mudflats. *Mar. Biol.*, 71: 339–343.

Quast, J.C., 1971. Fish fauna of the rocky inshore zone. *Nova Hedwigia*, 32: 481–507.

Reed, D.C. and Foster, M.S., 1984. The effects of canopy shading on algal recruitment and growth in a giant kelp forest. *Ecology*, 65: 937–948.

Reed, D.C., Laur, D.R. and Ebeling, A.W., 1988. Variation in algal disperal and recruitment: the importance of episodic recruitment. *Ecol. Monogr.*, 58: 321–335.

Richards, L.J., 1982. Prey selection by an intertidal beetle: field test of an optimal diet model. *Oecologia*, 55: 325–332.

Ricketts, E.F., Calvin, J., Hedgpeth, J.W. and Phillips, D.W., 1985. *Between Pacific Tides*. Stanford Univ. Press, Stanford, Ca., 652 pp.

Riedman, M.L., Hines, A.H. and Pearse, J.S., 1981. Spatial segregation of four species of turban snails (Gastropoda: *Tegula*) in central California. *Veliger*, 24: 97–102.

Rosenthal, R.J., Clarke, W.D. and Dayton, P.K., 1974. Ecology and natural history of a stand of giant kelp, *Macrocystis pyrifera*, off Del Mar, California. *Fish. Bull. U.S.*, 72: 670–684.

Roughgarden, J., Gaines, S. and Iwasa, Y., 1984. Dynamics and evolution of marine populations with pelagic larval dispersal. In: R.M. May (Editor), *Exploitation of Marine Communities*. Springer, Berlin, pp. 111–128.

Sameoto, D.D., 1969a. Comparative ecology, life histories, and behaviour of intertidal sand-burrowing amphipods (Crustacea: Haustoriidae) at Cape Cod. *J. Fish. Res. Board Can.*, 26: 361–388.

Sameoto, D.D., 1969b. Physiological tolerances and behaviour responses of five species of Haustoriidae (Amphipoda: Crustacea) to five environmental factors. *J. Fish. Res. Board Can.*, 26: 2283–2298.

Sameoto, D.D., 1969c. Some aspects of the ecology and life cycle of three species of subtidal sand-burrowing amphipods (Crustacea: Haustoriidae). *J. Fish. Res. Board Can.*, 26: 1321–1345.

Santelices, B. and Ojeda, F.P., 1984. Effects of canopy removal on the understory algal community structure of coastal forests of *Macrocystis pyrifera* from southern South America. *Mar. Ecol. Prog. Ser.*, 14: 165–173.

Scagel, R.F., 1963. Distribution of attached marine algae in relation to oceanographic conditions in the northeast Pacific. In: M.J. Dunbar (Editor), *Marine Distributions, Royal Society of Canada Spec. Publ. No. 5*. University of Toronto Press, Toronto, pp. 37–50.

Schiel, D.R. and Foster, M.S., 1986. The structure of subtidal algal stands in temperate waters. *Oceanogr. Mar. Biol. Ann. Rev.*, 24: 265–307.

Schroeter, S.C., Dixon, J. and Kastendiek, J., 1983. Effects of the starfish *Patiria miniata* on the distribution of the sea urchin *Lytechinus anamesus* in a southern California kelp forest. *Oecologia*, 56: 141–147.

Seapy, R.R. and Littler, M.M., 1978. The distribution, abundance, community structure, and primary productivity of macroorganisms from two central California rocky intertidal habitats. *Pac. Sci.*, 32: 293–314.

Seapy, R.R. and Littler, M.M., 1980. Biogeography of rocky intertidal macroinvertebrates of the southern California islands. In: D.M. Power (Editor), *The California Islands: Proc. Multidisciplinary Symp*. Santa Barbara Museum of Natural History, Santa Barbara, Ca., pp. 307–323.

Sears, H.S. and Zimmerman, S.T., 1977. *Alaska Intertidal Survey Atlas*. U.S. Department of Commerce, National Marine Fisheries Service, Auke Bay, Ala., unpag.

Shotwell, J.A., 1950. The vertical zonation of *Acmaea*, the limpet. *Ecology*, 31: 647–649.

Silberstein, M.A., 1987. *Feeding ecology of the polychaete worm Glycinde polygnatha Hartman 1950, an infaunal predator, with notes on life history*. M.Sc. Thesis, California State University, San Jose, CA, 75 pp.

Simenstad, C.A., Estes, J.A. and Kenyon, K.W., 1978. Aleuts, sea otters and alternate stable-state communities. *Science*, 200: 403–411.

Slattery, P.N., 1985. Life histories of infaunal amphipods from subtidal sands of Monterey Bay, California. *J. Crustacean Biol.*, 5: 635–649.

Smith, A.L., 1981. Comparison of macrofaunal invertebrates in sand dollar (*Dendraster excentricus*) beds and in adjacent areas free of sand dollars. *Mar. Biol.*, 65: 191–198.

Smith, R.I. and Carlton, J.T., 1975. *Light's Manual: Intertidal Invertebrates of the Central California Coast*. Univ. California Press, Berkeley, Ca., 716 pp.

Sousa, W.P., 1979a. Disturbance in marine intertidal boulder fields: the nonequilibrium maintenance of species diversity. *Ecology*, 60: 1225–1239.

Sousa, W.P., 1979b. Experimental investigations of disturbance and ecological succession in a rocky intertidal algal community. *Ecol. Monogr.*, 49: 227–254.

Sousa, W.P., 1984. Intertidal mosaics: the effects of patch size and a heterogeneous pool of propagules on algal succession. *Ecology*, 65: 1918–1935.

Sousa, W.P., 1985. Disturbance and patch dynamics on rocky intertidal shores. In: S.T. Pickett and P.S. White (Editors), *The Ecology of Natural Disturbance and Patch Dynamics*. Academic Press, New York, pp. 101–124.

Sousa, W.P., Schroeter, S.C. and Gaines, S.D., 1981. Latitudinal variation in intertidal algal community structure: the influence of grazing and vegetative propagation. *Oecologia*, 48: 297–307.

Steele, J.H. and Baird, I.E., 1968. Production ecology of a sandy beach. *Limnol. Oceanogr.*, 13: 14–25.

Steinberg, P.D., 1985. Feeding preferences of *Tegula funebralis* and chemical defenses of marine brown algae. *Ecol. Monogr.*, 55: 333–349.

Stephens, J.S., Morris, P.A., Zerba, K. and Love, M., 1984. Factors affecting fish diversity on a temperate reef: the fish assemblage off Palos Verdes Point, 1974–1981. *Environm. Biol. Fish.*, 11: 259–275.

Stephenson, M.D., 1977. Sea otter predation on Pismo clams in Monterey Bay. *Calif. Fish Game*, 63: 117–122.

Stephenson, T.A. and Stephenson, A., 1949. The universal features of zonation between tide-marks on rocky shores. *J. Ecol.*, 37: 289–305.

Stephenson, T.A. and Stephenson, A., 1972. *Life Between Tidemarks on Rocky Shores*. Freeman, San Francisco, 425 pp.

Stewart, J.G., 1982. Anchor species and epiphytes in intertidal algal turf. *Pac. Sci.*, 36: 45–60.

Stewart, J.G., 1983. Fluctuations in the quantity of sediments trapped among algal thalli on intertidal rock platforms in southern California. *J. Exp. Mar. Biol. Ecol.*, 73: 205–211.

Suchanek, T.H., 1979. *The Mytilus californianus community: studies on composition, structure, organization, and dynamics of a mussel bed*. Ph.D. Thesis, University of Washington, Seattle, Wash., 285 pp.

Sverdrup, H.U., Johnson, M.W. and Fleming, R.H., 1942. *The Oceans*. Prentice-Hall, Englewood Cliffs, NJ, 1087 pp.

Tegner, M.J. and Dayton, P.K., 1981. Population structure, recruitment and mortality of two sea urchins (*Strongylocentrotus franciscanus* and *S. purpuratus*) in a kelp forest. *Mar. Ecol. Prog. Ser.*, 5: 255–268.

Thorson, G., 1966. Some factors influencing the recruitment and establishment of marine benthic communities. *Neth. J. Sea Res.*, 3: 267–293.

Turner, C.H., Ebert, E.E. and Given, R.R., 1968. *The marine environment offshore from Point Loma, San Diego County*. California Dept. of Fish and Game, Fish Bulletin 140, Sacramento, Ca., 85 pp.

U.S. Army Engineer Division, 1971. *National Shoreline Study, California Regional Inventory*. Dames and Moore, San Francisco, Ca., 106 pp.

Underwood, A.J., 1978. A refutation of critical tidal levels as determinants of the structure of intertidal communities on British shores. *J. Exp. Mar. Biol. Ecol.*, 33: 261–276.

Underwood, A.J., 1980. The effects of grazing by gastropods and physical factors on the upper limits of distribution of intertidal macroalgae. *Oecologia*, 46: 201–213.

Underwood, A.J. and Denley, E.J., 1984. Paradigms, explanations, and generalizations in models for the structure of intertidal communities on rocky shores. In: D.R. Strong, D. Simberloff, L.G. Abele and A.B. Thistle (Editors), *Ecological Communities: Conceptual Issues and the Evidence*. Princeton Univ. Press, Princeton, pp. 151–180.

Underwood, A.J. and Jernakoff, P., 1984. The effects of tidal height, wave-exposure, seasonality and rock-pools on grazing and the distribution of intertidal macroalgae in New South Wales. *J. Exp. Mar. Biol. Ecol.*, 75: 71–96.

Valentine, J.W., 1966. Numerical analysis of marine molluscan ranges on the extratropical northeastern Pacific shelf. *Limnol. Oceanogr.*, 11: 198–211.

Vallis, G.K., 1986. El Niño: a chaotic dynamical system? *Science*, 232: 243–245.

VanBlaricom, G.R., 1978. *Disturbance, predation, and resource allocation in a high-energy sublittoral sand-bottom ecosystem: Experimental analyses of critical structuring processes for the infaunal community*. Ph.D. Thesis, Univ. California, San Diego, Ca., 328 pp.

VanBlaricom, G.R., 1982. Experimental analyses of structural regulation in a marine sand community exposed to oceanic swells. *Ecol. Monogr.*, 52: 283–305.

VanBlaricom, G.R., 1984. Relationships of sea otters to living marine resources in California: a new perspective. *Collection of papers presented at the Ocean Studies Symposium*. California Coastal Commission, Sacramento, pp. 361–381.

VanBlaricom, G.R. and Estes, J.A. (Editors), 1988. *The Community Ecology of Sea Otters*. Springer, Berlin, 247 pp.

Virnstein, R.W., 1977. The importance of predation by crabs and fishes on benthic infauna in Chesapeake Bay. *Ecology*, 58: 1199–1217.

Wethey, D.S., 1985. Catastrophe, extinction, and species diversity: a rocky intertidal example. *Ecology*, 66: 445–456.

Wheeler, P.A. and North, W.J., 1981. Nitrogen supply, tissue composition and frond growth rates for *Macrocystis pyrifera*. *Mar. Biol. (Berl.)*, 64: 59–69.

Wheeler, W.N. and Druehl, L.D., 1986. Seasonal growth and productivity of *Macrocystis integrifolia* in British Columbia, Canada. *Mar. Biol.*, 90: 181–186.

Wiedemann, A.M., 1984. *The ecology of pacific northwest coastal sand dunes: a community profile*. Biological Report FWS/OBS-84/04, U.S. Fish and Wildlife Service, Slidell, La., 130 pp.

Wieser, W., 1959. The effect of grain size on the distribution of small invertebrates inhabiting the beaches of Puget Sound. *Limnol. Oceanogr.*, 4: 181–194.

Yaninek, J.S., 1980. *Beach wrack: phenology of an imported resource and utilization by macroivertebrates of sandy beaches*. M.A. Thesis, Univ. California, Berkeley, Ca., 159 pp.

Zimmerman, R.C. and Kremer, J.N., 1984. Episodic nutrient supply to a kelp forest ecosystem in southern California. *J. Mar. Res.*, 42: 591–604.

Zimmerman, R.C. and Robertson, D.L., 1985. Effects of El Niño on local hydrography and growth of the giant kelp, *Macrocystis pyrifera*, at Santa Catalina Island, California. *Limnol. Oceanogr.*, 30: 1298–1302.

ZoBell, C.E., 1971. Drift seaweeds on San Diego County beaches. *Nova Hedwigia*, 32: 269–314.

Chapter 11

SOUTHERN CALIFORNIA ROCKY INTERTIDAL ECOSYSTEMS

MARK M. LITTLER, DIANE S. LITTLER, STEVEN N. MURRAY and ROGER R. SEAPY

INTRODUCTION

The Southern California Bight (Fig. 11.1) is defined as the triangular area with its apex at Pt. Conception and its southeastern and southwestern corners at the U.S./Mexico international border and the Tanner and Cortez submarine banks, respectively (Anonymous, 1973). This region is one of the most physiographically intriguing continental shelves of the world's oceans. Sheppard and Emery (1941) aptly termed this area the Southern California Borderland, due to its submarine relief as an expanded continental shelf. Topographically, the Borderland consists of about fourteen deep basins separated by a series of ridges, banks and offshore islands formed during the Miocene between 11 and 20 million years ago (Valentine and Lipps, 1967).

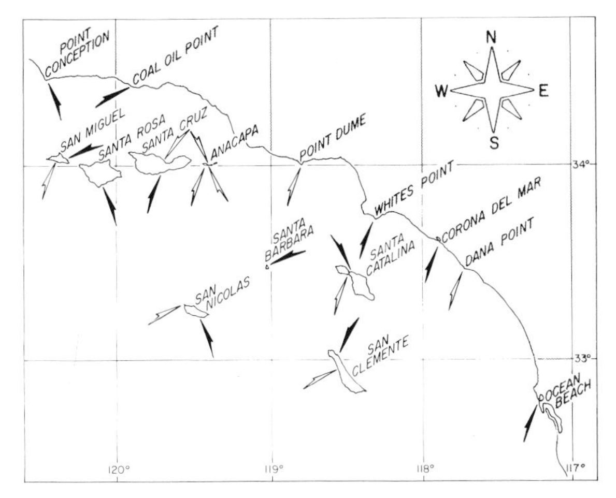

Fig. 11.1. Location of the 22 study sites; stations where biomass data were taken are shown by dark arrows (details given in Table 11.1).

Much is known about the climatology of the Bight (Kimura, 1974); however, published biological information still remains scarce. Cockerell (1939) was among the first to point out the biological importance of unusual mixing of cold and warm waters in the region. Changing climatic patterns results in a complex intermingling of physical conditions, which are reflected in the broad spectrum and variability of the biological systems within the Bight (Littler, 1980a, b) and the mixing of cold and warm temperate floras (Murray and Littler, 1981) and faunas (Seapy and Littler, 1980).

The predominant driving force of water circulation in the Southern California Bight is the California Current. This system, a portion of the eastern limb of the clockwise North Pacific Gyre, flows southward along the western coast of the United States. At Pt. Conception, the coastline turns toward the east, but the California Current continues to flow south to southeast. In a broad area to the south of Cortez Bank, surface flow is turned eastward and then northward through the Channel Islands, forming the Southern California Eddy or Countercurrent (Fig. 11.2). The development and strength of the Southern California Eddy are seasonal (Hickey, 1979), being strongest in the summer and weakest in the winter.

In addition to the southerly flowing California Current, the west coast of the United States is influenced by the deep California Undercurrent (core of flow at -200 to -300 m), which originates by the submergence of northward-flowing equatorial surface water beneath the California Current along a front extending southwestward off southern Baja California (Reid et al., 1958). During late fall and early winter, the

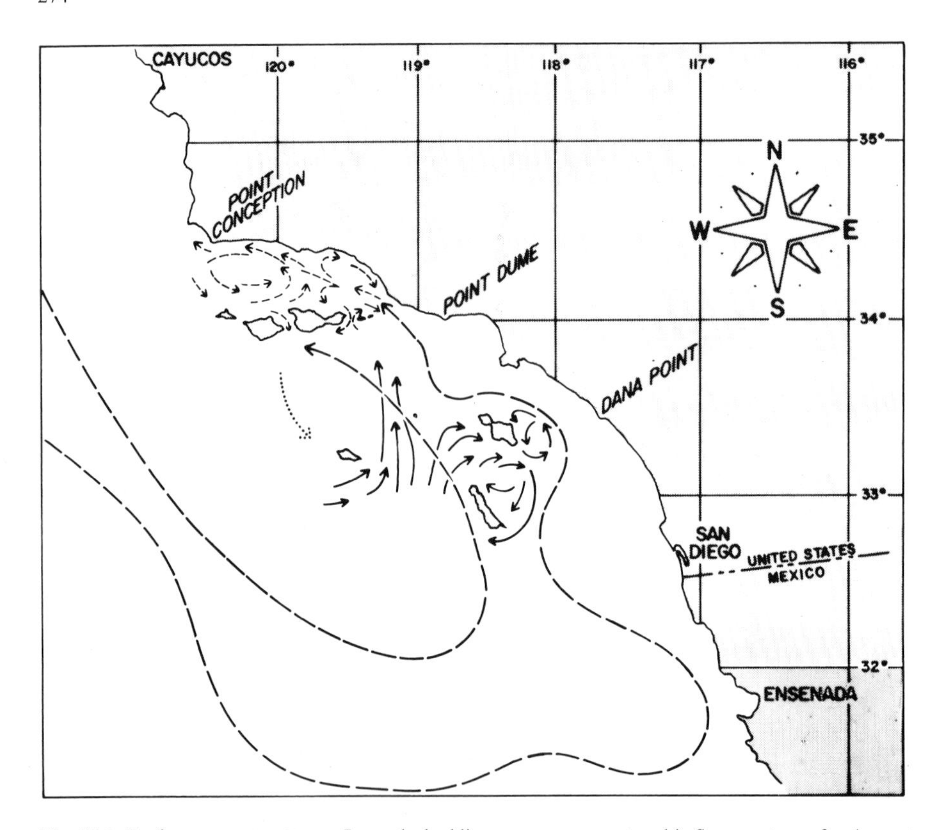

Fig. 11.2. Surface current patterns. Long dashed lines are mean geostrophic flow contours for August averaged for a 16-yr period; short dashed lines indicate surface current flow during August 1969 based on drift bottle studies; solid lines are surface currents derived from 10 m drogue releases during October 1958, while the single dotted line indicates a current system proposed by Neushul et al. (1967). (From Seapy and Littler, 1980.)

northerly winds decrease and are replaced by winds from the south to southwest to the north of Cape Mendocino and by northwesterly or southerly winds from Cape Mendocino to Pt. Conception (Hickey, 1979). North of Pt. Conception the California Undercurrent rises to the surface (Hickey, 1979) and flows northward along the coast as far as Vancouver Island, British Columbia (Schwartzlose, 1963). This northward surface flow is the Davidson Current, which is maximal (to 25 cm s^{-1}) in January (Reid and Schwartzlose, 1962). Whether the core of the California Undercurrent remains at a depth of −200 to −300 m to the south of Pt. Conception is unknown (Hickey, 1979). However, during winter months the northerly flow of the Southern California Eddy along the coast can be continuous with the Davidson Current at Pt. Conception (Hickey, 1979).

The mean monthly surface-water temperatures over the Southern California Borderland range from a low of 13°C in March and April to 20°C in August and September (Jones, 1971). Although the area over the continental shelf undergoes considerable mixing from differential currents and waves, the system becomes extensively stratified throughout the summer months, with the depth of the thermocline rarely exceeding 50 m (Jones, 1971).

Another hydrographic feature is the pattern of wind-driven upwelling. In southern California this process occurs in localized areas along both mainland and island shores, where strong steady winds displace surface water offshore to be replaced by cooler subsurface water containing high levels of nutrients. Upwelling is most intense in April, May and June, although wind conditions occasionally result in non-seasonal upwelling (Jones, 1971). Upwelling is most intense south of capes and points (for example, Pt. Conception) that extend into nearshore current streams (Reid et al., 1958). Such upwelling of deep nutrient-laden water may partially account for the high productivity and biotic richness within the Southern California Bight. Wind conditions are also important, in that major reversals occur predominantly throughout late fall and winter (Kimura, 1974). These result in strong, hot and dry "Santa Ana" winds arising from the inland desert regions. For southern California intertidal areas, daytime tidal emersion is pronounced during the fall and winter months (Seapy and Hoppe, 1973; Sousa, 1979a; Littler, 1980a, b; Seapy and Littler, 1982) when low tides occur during afternoon hours. Consequently, when "Santa Ana" wind conditions are severe during these periods of daytime aerial exposure, extreme desiccation stress can result (Gunnill, 1979, 1980a; Seapy and Littler, 1982).

An important ecological factor related to water movement in the Southern California Bight is the protection of certain mainland shores and the mainland sides of islands from open ocean swell and storm waves. This leads to higher wave-energy regimes on the unprotected outer island shores, with marked effects on their biological communities (Table 11.1). Nearly all of the southern California mainland coastline is protected to some degree by the outlying islands (Ricketts et al., 1968). The only mainland areas receiving direct westerly swell are those near Los Angeles and San Diego (Fig. 11.1). The combination of the prevailing Southern California Eddy and its secondary eddies (Fig. 11.2), the deflection of currents and swell by the islands and local, wind-driven coastal upwelling result in a complex hydrographic environment.

Rocky shorelines occur intermittently throughout the entire coastal area of the Southern California Bight, and are interspersed with sandy beaches and inlets to lagoons, estuaries and marinas. The intertidal habitats of the offshore Southern California islands consist of approximately 62.5% bedrock, 16.0% boulder beaches and 21.5% sandy beaches (Table 11.2). On the mainland, in the upper half of the intertidal zone, sandy beaches total 75.1% of the shoreline with the remainder comprised of 16.5% bedrock and 8.4% boulder beach. Sandy beaches on the mainland constitute 64.2% of the lower intertidal zone, bedrock 29.1% and boulders 6.7%.

A number of substratum types are represented among the rocky intertidal habitats (Table 11.1), ranging from hard, irregular breccia to smooth sandstone or siltstone. Some sites (especially Coal Oil Point and San Nicolas Island) are heavily inundated on a yearly basis by sand, which scours and removes organisms and often completely buries them. At some sites, sand consistently inundates the upper intertidal zone (Stewart, 1983), especially in many mainland habitats, thereby eliminating an entire zonal component from the biota. The widespread presence of extensive loose boulder fields (for example, at Whites Point, Corona del Mar and Dana Point) constitutes another form of environmental instability, curbing community development. The existence of such natural disturbances in rocky intertidal habitats (Sousa, 1979a; Littler and Littler, 1984) has important implications in interpreting changes associated with coastal exploitation and resource development.

The 22 study sites sampled during 1977–1978 (Littler, 1979) under the Bureau of Land Management (now Minerals Management Services) Outer Continental Shelf Program (Fig. 11.1, Table 11.1) form the basis for this synopsis. Fifteen of these sites were situated on the eight islands, while the remainder were located on the mainland (Fig. 11.1). The islands range in size from 2.5 to 249 km^2 and are positioned from 20 to 90 km from the mainland (Power, 1980). The northern Channel Islands (San Miguel, Santa Rosa, Santa Cruz and Anacapa) form a discrete geographic group with some affinity to the northern mainland sites of Government Point and Coal Oil Point. Similarly, the southern islands (San Nicolas, Santa Barbara, Santa Catalina, San Clemente) together with the four sites to the south of Santa Monica Bay (Whites Point, Corona del Mar, Dana Point and Ocean Beach) may be considered a discrete unit. A general trend of increasing water temperatures to the south is largely offset by exposure of the outermost islands (San Miguel and San Nicolas) to the eastern margin of the cold California Current system (Fig. 11.2). Thus, a division of island sites into northern and southern on the basis of geographic location related to the temperature regime is not accurate. This differential in surface temperatures mediated by water circulation patterns has also been substantiated by satellite thermal imagery (Hendricks, 1977).

Before 1970, relatively little information was available concerning the structure of the intertidal

TABLE 11.1

Physiographic attributes of the 22 rocky intertidal habitats studied

Study area	Latitude and longitude	Water temperature	Substrata	Tidal range (m)	Wave exposure	Disturbance source	Sand cover
Government Point	34°26′35″N 120°27′06″W	Cold	Monterey shale/ siltstone	−0.3 to +2.1	Exposed (heavy)	Oil seeps	Mid eulittoral
Coal Oil Point	34°24′27″N 119°52′40″W	Cold (moderate)	Monterey shale/ siltstone	−0.6 to +0.9	Exposed (moderate)	Oil seeps	Extensive
Malibu	34°00′42″N 118°47′30″W	Warm (moderate)	Monterey shale/ siltstone	−0.3 to +2.4	Moderate	None	Low eulittoral
Whites Point	33°34′11″N 118°19′39″W	Warm to intermediate	Diatomaceous Monterey shale and unstable boulders	−0.3 to +0.9	Exposed (moderate)	Domestic wastes	High eulittoral cobbles
Corona del Mar	33°35′14″N 117°51′54″W	Warm to intermediate	Unstable granitic boulders on sandstone/ siltstone	−0.3 to +0.9	Exposed (moderate)	Human usage (extensive)	High eulittoral
Dana Point	33°35′25″N 117°42′44″W	Warm (moderate)	Granitic boulders	0.0 to +2.1	Moderate	None	Low eulittoral
Ocean Beach, San Diego	32°44′35″N 117°15′15″W	Warm to intermediate	Poorly consolidated friable sandstone	+0.3 to +4.0	Exposed	None	None
Cuyler Harbor, San Miguel Island	34°02′55″N 120°20′08″W	Cold	Irregular volcanic flow breccia	−0.3 to +2.7	Exposed (moderate)	None	Low eulittoral
Crook Point, San Miguel Island	34°01′28″N 120°22′43″W	Cold	Sandstone	+0.3 to +2.7	Heavy	None	None
Santa Rosa Island	33°53′31″N 120°06′31″W	Cold (moderate)	Smooth sandstone	+0.3 to +3.4	Exposed (moderate)	None	Low eulittoral
Willows Anchorage, Santa Cruz Island	33°57′43″N 119°45′16″W	Intermediate	Irregular volcanic	+0.3 to +4.0	Surge	None	None

Prisoners Cove, Santa Cruz Island	34°01′14″N 119°41′14″W	Intermediate	Volcanic rock	−0.6 to +2.4	Low	None	None
Frenchys Cove, Anacapa Island	34°00′31″N 119°24′21″W	Intermediate	Volcanic boulders	−0.3 to +2.7	Low	Human usage (light)	None
Cat Rock, Anacapa Island	34°00′19″N 119°25′05″W	Intermediate	Volcanic rock	0.0 to +2.1	Moderate	None	None
Disturbed, Anacapa Island	34°00′24″N 119°24′38″W	Intermediate	Volcanic rock	−0.3 to +2.7	Moderate	Human usage (heavy)	High eulittoral
Dutch Harbor, San Nicolas Island	33°12′54″N 119°28′22″W	Cold	Sandstone	−0.3 to +1.5	Exposed (moderate)	None	Extensive
West Point, San Nicolas Island	33°16′43″N 119°34′41″W	Cold	Sandstone	0.0 to +1.5	Heavy	None	None
Santa Barbara Island	33°28′43″N 119°01′36″W	Intermediate	Vesicular volcanic rock	+0.3 to +3.7	Surge (heavy)	None	None
Fisherman Cove, Santa Catalina Island	33°26′47″N 118°29′04″W	Warm	Vesicular volcanic rock	−0.6 to +3.0	Protected	None	None
Catalina Harbor, Santa Catalina Island	33°25′42″N 118°30′42″W	Warm	Vesicular volcanic rock	−0.3 to +2.1	Protected	None	None
Wilson Cove, San Clemente Island	33°00′06″N 118°33′03″W	Warm	Stable granitic boulders	−0.3 to +2.1	Protected	None	None
Northwest Coast, San Clemente Island	33°58′06″N 118°34′18″W	Warm	Large volcanic boulders and rocks	−0.3 to +1.5	Moderate	None	None

TABLE 11.2

Relative amounts of substratum types observed in the rocky intertidal zone of the Southern California Bight. Based on aerial survey by helicopter during low-tide periods (From Littler and Littler, 1978, 1979.)

System	Bedrock	Boulder	Sand
Islands			
San Miguel Island	63.7	0.2	36.1
Santa Rosa Island	61.8	5.0	33.3
Santa Cruz Island	66.2	14.8	19.2
Anacapa Island	70.0	14.8	15.2
Santa Barbara Island	73.6	22.2	4.2
Santa Catalina Island	35.3	49.5	15.3
San Nicolas Island	60.7	4.6	34.7
San Clemente Island	68.6	17.3	14.0
Island mean	62.5	16.0	21.5
Mainland			
Upper intertidal	16.5	8.4	75.1
Lower intertidal	29.1	6.7	64.2
Mainland mean	22.8	7.5	69.6

ecosystems of the Southern California Bight. Reviews of the existing literature (Bright, 1974; Murray, 1974) pointed out the paucity of data (mostly biological), particularly for the offshore islands. For these eight islands, limited taxonomic lists were available only for a few isolated localities (Hewatt, 1946; Dawson, 1949; Dawson and Neushul, 1966; Neushul et al., 1967; Nicholson and Cimberg, 1971; Seapy, 1974; Sims, 1974). Quantitative ecological data could be found in only a few published papers (see, for instance, Hewatt, 1946; Caplan and Boolootian, 1967; Murray and Littler, 1974; Littler and Murray, 1974, 1975). Knowledge of biological communities inhabiting rocky intertidal habitats along the southern California mainland was also very limited (e.g., Dawson, 1959, 1965; Widdowson, 1971; Nicholson and Cimberg, 1971).

During the past ten years, however, there has been a dramatic increase in ecological information for rocky intertidal habitats of the Southern California Bight. A number of accounts of ecological research performed on island (Littler and Murray, 1977, 1978; Murray and Littler, 1978; Seapy and Littler, 1982; Taylor and Littler, 1982; Littler et al., 1983) and mainland (Emerson and Zedler, 1978; Sousa, 1979a, b, 1980; Gunnill, 1980a, b, 1983, 1985; Stewart and Myers, 1980; Sousa et al., 1981; Murray and Littler, 1984) rocky intertidal habitats have been published. A significant increase in the understanding of southern California intertidal communities resulted from an extensive three-year sampling program sponsored by the U.S. Department of Interior, Bureau of Land Management (now Minerals Management Service). The data obtained during 1975–1978 as part of this program represent quantitative ecological accounts of the distributions and abundances of the macrobiota for representative sites (Fig. 11.1), and provide the best single source of information on the structure of rocky intertidal communities within the Southern California Borderland. This is because identical field methodologies and analytical techniques were employed to obtain and compare populational and community data over the entire geographic extent of the Bight. For these reasons, this research, which to date has been available in its entirety only in governmental reports of limited distribution, forms the basis of our review. The reader is referred to more limited accounts of the results of earlier phases of this program (Littler, 1980a, b) for a detailed description of the sampling and analytical methods used consistently for all 22 sites.

COMMUNITY COMPOSITION AND STRUCTURE

Ecological assessments of the macro-epibiota inhabiting each site were performed throughout different seasons using a quadrat sampling technique (Littler, 1980a, b). These assessments provided data concerning the variable species compositions as well as distributions and abundances of the rocky intertidal communities for each of the stations. Both undisturbed (photogrammetric) and disturbed (harvest) samples were taken to provide data on population coverage (%) and biomass (wet and dry).

Species composition

A total of 539 taxa of macrophytes and macroinvertebrates was recorded during the three

TABLE 11.3

Numbers of taxa by major taxonomic groups

Major groups		Number of taxa collected
Macrophytes		
Bacillariophyta		1
Chlorophyta		23
Cyanophyta		2
Phaeophyta		47
Rhodophyta		149
Spermatophyta		2
Total		224
Macroinvertebrates		
Annelida	– Polychaeta	10
Arthropoda	– Crustacea	25
Cnidaria	– Anthozoa	11
Cnidaria	– Hydrozoa	19
Chordata	– Ascidiacea	20
Echinodermata	– Asteroidea	10
Echinodermata	– Echinoidea	2
Echinodermata	– Holothuroidea	3
Ectoprocta (Bryozoa)		4
Entoprocta		1
Mollusca	– Bivalvia	16
Mollusca	– Cephalopoda	1
Mollusca	– Gastropoda	131
Mollusca	– Polyplacophora	20
Porifera	– Calcarea	5
Porifera	– Demospongiae	37
Total		315

TABLE 11.4

Taxa common to all 22 study sites throughout 1975–1978

Blue-green algae	*Acmaea (Collisella) limatula*
Bossiella orbigniana spp. *dichotoma*	*Acmaea (Collisella) pelta*
Ceramium eatonianum	*Acmaea (Collisella) scabra*
Ceramium sinicola	*Anthopleura elegantissima*
Corallina officinalis var. *chilensis*	*Balanus glandula*
Corallina vancouveriensis	*Chthamalus dalli*
Cryptopleura spp. (4)	*Chthamalus fissus*
Crustose Corallinaceae (2)	*Cyanoplax hartwegii*
Egregia menziesii	*Littorina planaxis*
Gelidium coulteri	*Littorina scutulata*
Gelidium pusillum	*Nuttallina californica*
Gigartina canaliculata	*Nuttallina fluxa*
Polysiphonia spp. (6)	*Pachygrapsus crassipes*
Rhodoglossum affine	*Pagurus* spp. (2)
Ulva californica	*Phragmatopoma californica*
Ulva lobata	*Tetraclita rubescens*

years of research. The number of macrophyte taxa (224) was considerably less than the number of macroinvertebrates (315 taxa, Table 11.3). Over half of the macrophytes belonged to the Rhodophyta (149), followed by the Phaeophyta (47). Two species of seagrasses (Spermatophyta) and 23 Chlorophyta accounted for virtually all of the remaining taxa; efforts were made to quantify the abundances of encrusting blue-green algae, but these were all treated as a single entity for ecological analyses. Of the macroinvertebrates (Table 11.3), Mollusca (168) and particularly Gastropoda (131) contributed the most taxa, followed by Porifera (37), Cnidaria (30) and Arthropoda (25). Thirteen macrophyte and 14 macroinvertebrate taxa were found at all 22 study sites (Table 11.4). Ubiquitous macrophytes were mostly coralline red algae, or small filamentous and sheet-like species, while the macroinvertebrate taxa common to all sites were mostly species of limpets, littorine gastropods and barnacles. Relatively more taxa had restricted distributions (that is, found at fewer than four sites) compared to those that were broadly distributed (that is, present at more than four different localities, Fig. 11.3). This may reflect a high degree of environmental variation between habitats throughout the Bight.

Abundances of macrophyte and macroinvertebrate populations

Because space and light are believed (Connell, 1972) to be the primary limiting resources in rocky intertidal habitats, cover measurements probably represent the most meaningful method of quantifying abundances of macrophytes and sessile macroinvertebrates. Nearly all of the sites investigated were characterized by an upper intertidal zone of encrusting blue-green algae; thus, this taxonomic category averaged the greatest cover (41.4%). Blue-green algae were followed by the red alga *Gigartina canaliculata* (mean of 7.2% cover), the surf grasses *Phyllospadix scouleri* and *P. torreyi* (5.5%), the articulated coralline alga *Corallina officinalis* var. *chilensis* (5.3%) and the multi-

Fig. 11.3. Distribution of numbers of taxa as a function of the numbers of study sites in which they occurred.

species crustose Corallinaceae (5.0%). Sessile macroinvertebrates with the greatest mean cover were the sand-castle worm *Phragmatopoma californica* (6.3%), the small acorn barnacles *Chthamalus dalli* and *C. fissus* (4.5% combined) and the mussel *Mytilus californianus* (3.8%).

For the mobile macroinvertebrates, abundances are probably best represented by estimates of density (numbers of individual animals). Mobile macroinvertebrates with the greatest mean densities were the periwinkle *Littorina planaxis* (311 m^{-2}) and the limpets *Acmaea (Collisella) conus* and *A. scabra* (79 m^{-2}). Sessile macroinvertebrates were, however, more numerous with *Chthamalus dalli* and *C. fissus* having a combined average density of 3642 m^{-2}; other numerous sessile forms included *Phragmatopoma californica* (625 m^{-2}), the barnacles *Balanus glandula* (205 m^{-2}) and *Tetraclita squamosa rubescens* (204 m^{-2}), *Mytilus californianus* (162 m^{-2}) and the anemone *Anthopleura elegantissima* (87 m^{-2}).

Harvested samples taken at 12 of the sites provide important biomass information. Organic dry weight (biomass in g m^{-2}, after drying at 60°C, less inorganic parts and ash) constitutes an ecologically meaningful estimate of abundance, as it represents the quantity of bound food energy available to higher trophic levels. *Egregia menziesii*, a large brown kelp, had by far the greatest average organic dry biomass (119 g m^{-2}), followed by *Phyllospadix scouleri* (86 g m^{-2}), the rockweed *Pelvetia fastigiata* (64 g m^{-2}), *Gigartina canaliculata* (50 g m^{-2}), *Phyllospadix torreyi* (46 g m^{-2}) and *Eisenia arborea* (45 g m^{-2}). The combined figures for *Phyllospadix* make this seagrass genus the greatest contributor of biomass in the southern California rocky intertidal. Of the macroinvertebrates, *Mytilus californianus* averaged the greatest organic dry biomass (51 g m^{-2}), followed by *Anthopleura elegantissima* (26 g m^{-2}), the bivalve *Pseudochama exogyra* and the related *Chama arcana* (6 g m^{-2}), the tube worm *Dodecaceria fewkesi* (5 g m^{-2}), the purple sea urchin *Strongylocentrotus purpuratus* (4 g m^{-2}) and *Tetraclita rubescens* (4 g m^{-2}).

Species assemblages and intertidal zonation

The population data obtained for the individual quadrats were subjected to classification analysis to determine species assemblages and zonation patterns throughout the Bight. The Bray–Curtis similarity index (Bray and Curtis, 1957) was used to establish affinities between samples (after Smith, 1976). Classification analysis of the cover data revealed a total of 46 discrete biotic assemblages, which were used to characterize the zonation patterns for each of the sites (Table 11.5). This revealed 11 groupings characteristic of the supralittoral and upper intertidal zones, 23 found in the mid-intertidal and 12 in the lower intertidal zone.

The dendrogram generated from classification of the cover data for all stations (Fig. 11.4) separates the sites into two subgroups related largely to the influence of sand. Most of the stations assigned to

TABLE 11.5.

Distributional patterns of dominant species assemblages (as determined by Bray–Curtis cluster analysis based on cover) in relation to tidal height for each of the 22 study sites (abbreviations as in Table 11.1)

High intertidal

Site	Blue-green algae	*Chthamalus fissus/dalli*	*Peyssonneliaceae/Hildenbrandiaceae*	*Endocladia muricata*	*Chlorophycean crust*	*Pseudolithoderma nigra*	*Enteromorpha* spp.	*Chaetomorpha linum*	*Acmaea* spp.	Bare rock	*Littorina* spp.
Government Point	●	●								●	
Coal Oil Point											
Cuyler Harbor, SMI	●	●		●							●
Crook Point, SMI	●			●							
Prisoners Cove, SCR	●	●					●				
Willows Anchorage, SCR	●										
Malibu	●	●		●							
Frenchys Cove, ANA	●										●
Cat Rock, ANA	●	●									
Disturbed, ANA	●	●									
Santa Rosa Island	●			●							
Whites Point	●								●		
Corona Del Mar	●										
Santa Barbara Island	●										●
Dana Point	●	●									
Fisherman Cove, SCA	●	●				●					
Catalina Harbor, SCA	●	●									
Dutch Harbor, SNI	●	●						●			
West Point, SNI	●		●		●						
Wilson Cove, SCL	●	●				●					
Northwest Coast, SCL	●	●		●		●					
Ocean Beach, San Diego	●						●				

High eulittoral filamentous epiphytes; turfs

Site	*Polysiphonia* spp.	*Herposiphonia* spp.	Rhodophycean turf	*Gelidium* turf	*Centroceras clavulatum*	*Ceramium* spp.
Government Point						
Coal Oil Point			●			
Cuyler Harbor, SMI						
Crook Point, SMI						
Prisoners Cove, SCR						
Willows Anchorage, SCR						
Malibu						
Frenchys Cove, ANA						
Cat Rock, ANA						
Disturbed, ANA		●				
Santa Rosa Island						
Whites Point		●				●
Corona Del Mar		●		●	●	
Santa Barbara Island						
Dana Point					●	
Fisherman Cove, SCA						
Catalina Harbor, SCA	●					
Dutch Harbor, SNI						
West Point, SNI						
Wilson Cove, SCL						
Northwest Coast, SCL						
Ocean Beach, San Diego						

Mid intertidal

Site	*Anthopleura elegantissima*	*Gigartina leptorhynchos*	*Pelvetia fastigiata*	*Corallina* spp.	Crustose coralline	*Hesperophycus harveyanus*	*Mytilus californianus*	*Tetraclita squamosa rubescens*	*Phragmatopoma californica*	Sand	*Gigartina canaliculata*	*Rhodoglossum affine*	Articulated coralline crust	*Balanus glandula*	*Chondria californica*	Ralfsiaceae	*Ulva* spp.
Government Point	●		●						●		●						
Coal Oil Point	●										●						
Cuyler Harbor, SMI			●	●			●		●		●						
Crook Point, SMI			●				●		●		●						●
Prisoners Cove, SCR							●				●			●			
Willows Anchorage, SCR				●	●			●			●						
Malibu				●			●		●		●						
Frenchys Cove, ANA				●		●			●		●	●				●	
Cat Rock, ANA			●						●		●						
Disturbed, ANA			●				●		●			●					
Santa Rosa Island							●		●								
Whites Point		●		●						●	●		●		●	●	
Corona Del Mar				●	●												
Santa Barbara Island				●	●			●			●						
Dana Point			●	●	●						●					●	
Fisherman Cove, SCA			●	●	●	●		●			●						
Catalina Harbor, SCA			●	●	●			●									
Dutch Harbor, SNI	●			●							●						
West Point, SNI	●			●	●												
Wilson Cove, SCL				●				●			●	●					
Northwest Coast, SCL				●	●				●		●						
Ocean Beach, San Diego				●	●				●								

Low intertidal

Site	*Lithothrix aspergillum*	*Cryptopleura* spp.	*Phyllospadix torreyi/scouleri*	*Halidrys dioica*	*Laminaria* spp.	*Egregia menziesii*	*Eisenia arborea*	*Bossiella* spp.	*Gelidium purpurascens*	*Pterocladia capillaceae*	*Strongylocentrotus purpuratus*	*Dodecaceria fewkesi*
Government Point	●		●		●							
Coal Oil Point		●	●									
Cuyler Harbor, SMI			●									●
Crook Point, SMI												
Prisoners Cove, SCR			●									
Willows Anchorage, SCR			●									
Malibu			●									●
Frenchys Cove, ANA			●									
Cat Rock, ANA												
Disturbed, ANA			●									
Santa Rosa Island			●									
Whites Point	●										●	
Corona Del Mar						●						
Santa Barbara Island				●				●				
Dana Point		●								●		
Fisherman Cove, SCA				●		●	●					
Catalina Harbor, SCA							●			●		
Dutch Harbor, SNI			●									●
West Point, SNI												
Wilson Cove, SCL				●		●						
Northwest Coast, SCL			●				●		●			
Ocean Beach, San Diego			●									

subgroup A (Fig. 11.4) were exposed to seasonal sand inundation and scouring, whereas those belonging to subgroup B were characteristically associated with more consistently sand-free habitats. The stations in subgroup A all contained relatively large populations of the sand-castle worm *Phragmatopoma californica*, along with abundant *Mytilus californianus*, *Chthamalus dalli* and *C. fissus*; in contrast, subgroup B sites were dominated by cover of the articulated coralline algae *Corallina officinalis* var. *chilensis* and *C. vancouveriensis*.

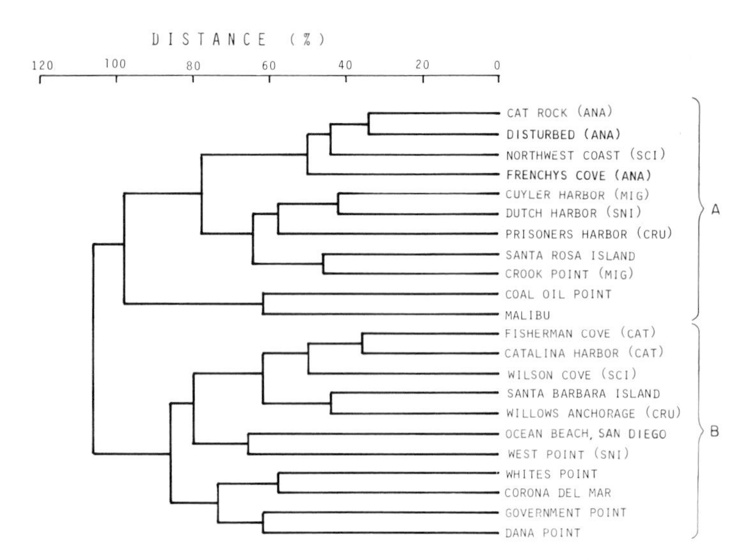

Fig. 11.4. Dendrogram of differential clustering of the 22 study sites based on mean cover values of individual macrophytes and macroinvertebrates (level of similarity indicated by Bray–Curtis percent distance). ANA: Anacapa Island; SCI: San Clemente Island; MIG: San Miguel Island; SNI: San Nicolas Island; CRU: Santa Cruz Island; CAT: Santa Catalina Island.

Biogeography

The Southern California Borderland has historically been regarded as part of the warm-temperate biogeographic province of the eastern North Pacific (Fig 11.5), a region commonly reported (see, for instance, Briggs, 1974; Hayden and Dolan, 1976; Brusca and Wallerstein, 1979) to extend from Pt. Conception to Punta Eugenio, Mexico. Previous biogeographic studies (Newell, 1948; Newman, 1979), however, have emphasized the transitional nature of the biotic assemblages in the Bight (Fig. 11.5), because species indicative not only of the warm-temperate zone but also of the cold-temperate province to the north of Point Conception co-occur in waters off southern California. Distributional analyses of binary (presence/absence) data for macrophytes (Murray et al., 1980; Murray and Littler, 1981) and macroinvertebrates (Seapy and Littler, 1980) sampled from mainland and island sites during our three-year program support this viewpoint and confirm that the southern California rocky intertidal zone contains a mixture of cold- and warm-temperate biotic associations. Island (San Miguel, San Nicolas, Santa Rosa) and mainland (Government Point) sites closest to the cold California Current (Fig. 11.2) generally had intertidal communities comparable to those of the cold-temperate coastline north of Pt. Conception (see Seapy and Littler, 1979; Horn et al., 1983). In contrast, southern mainland (Dana Point, Corona Del Mar, Whites Point) and island (San Clemente, Santa Catalina) sites most remote from California Current waters contained warm-water biotic associations (Murray et al., 1980; Murray and Littler, 1981; Seapy and Littler, 1980). The communities of island sites (Santa Cruz, Santa Barbara, Anacapa) and mainland localities (Malibu, Ocean Beach, Coal Oil Point) receiving variable and mixed exposure to the cold- and warm-water currents had intermediate biological affinities.

The dendrogram exhibiting the relationship between the 12 sites for which organic dry biomass data were obtained (Fig. 11.6) showed remarkable comparability with the biogeographic patterns described above derived solely from binary data. The more southeasterly sites exposed to the warmer waters of the Southern California Countercurrent (Fig. 11.2; Corona Del Mar, Whites Point, San Clemente Island, Santa Catalina Island) grouped together. The more northerly sites (Government Point, Coal Oil Point) and those islands with greatest proximity to California Current influence (San Miguel, San Nicolas, Santa Rosa) sorted as a unit along with the stations at Ocean

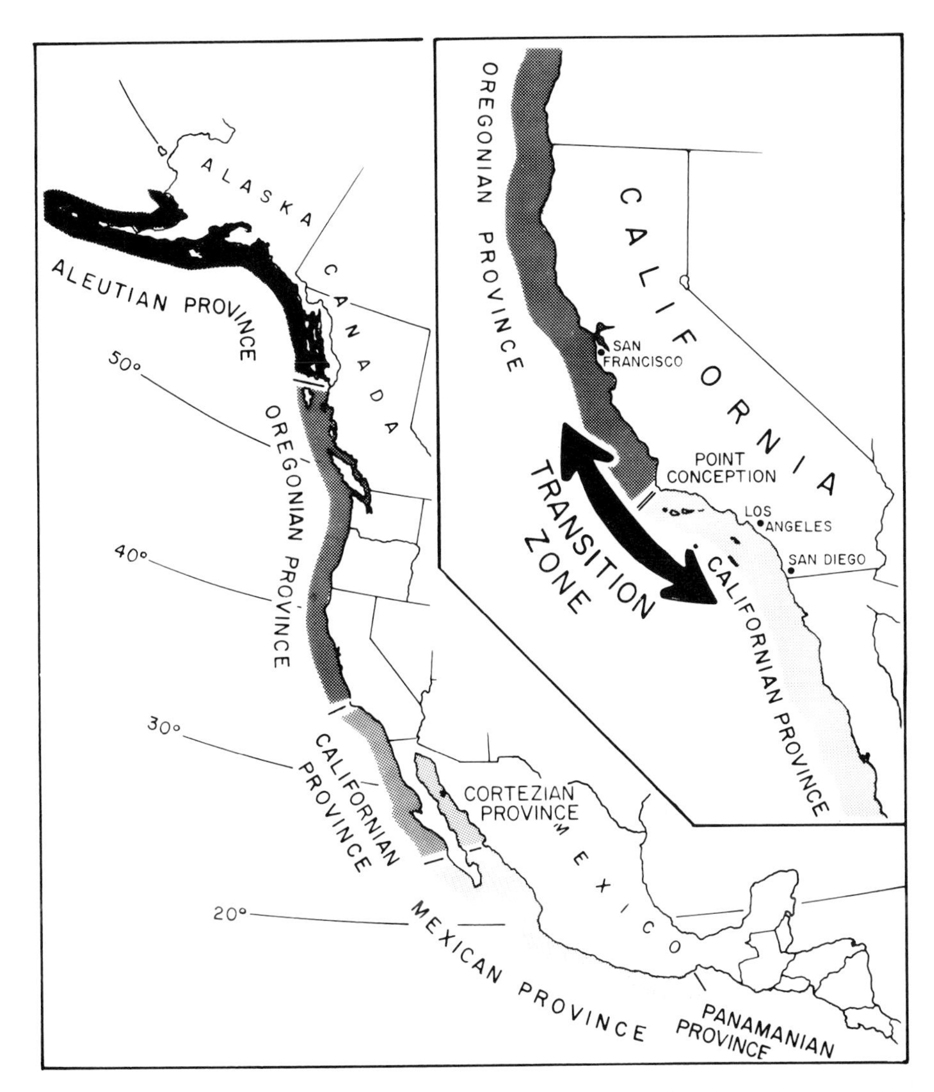

Fig. 11.5. Zoogeographic provinces of the Pacific coast of North America. (From Seapy and Littler, 1980.)

Beach, Santa Barbara Island and Santa Cruz Island, all of which receive mixed exposure to warm- and cold-water currents.

Cover and biomass

The overall abundances of the macro-epibiota varied concurrent with differences in the species composition and zonation patterns. Since the data were averaged by tidal interval over the entire intertidal range (see Littler, 1980a, b), individual abundance means were affected by the vertical extent of the shoreline. For example, biota were not present within the upper intertidal and supra-

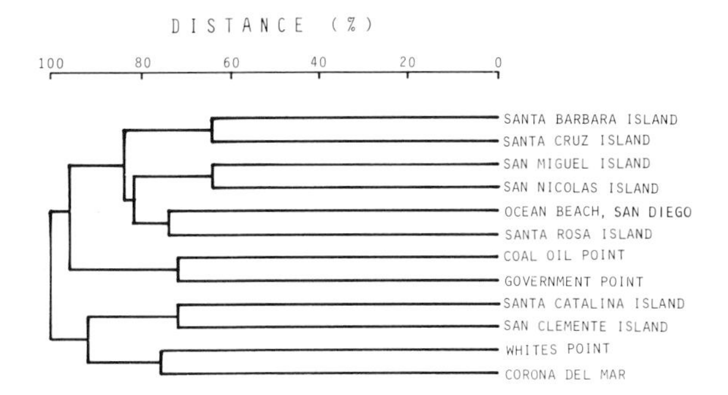

Fig. 11.6. Dendrogram of differential clustering of the 12 study sites based on mean dry organic biomass values of individual macrophytes and macroinvertebrates (Bray–Curtis percent distance).

littoral at one island (San Nicolas) and three mainland (Coal Oil Point, Whites Point, Corona Del Mar) sites because of persistent sand inundation or boulder cover of the uppermost rocky substrata. The absence of fixed and uncovered rocky substrata on the upper shore is of frequent occurrence in southern California, particularly for mainland localities (Table 11.2; Littler, 1978a, 1979). Consequently, many habitats are characterized by poor development or lack of upper shoreline biota. The grand means for each of the mainland and island sites reflect these differences. For example, the island biotas averaged 814 g m^{-2} compared to 614 g m^{-2} for the mainland sites in terms of organic dry biomass; cover data revealed a similar trend, with the island sites averaging 130% (due to stratification) compared to 119% for the mainland stations. These differences, although influenced by the lack of well-populated supralittoral and upper intertidal habitats for several of the mainland sites, were even more marked when comparisons were restricted to lower shore communities.

For the mid through lower intertidal regions (Table 11.6), island sites averaged 1496 g m^{-2}, a value nearly twice that for mainland stations (788

TABLE 11.6

Mean yearly macrophyte (M) and macroinvertebrate (I) dry organic biomass (g m^{-2}) and cover (%) comparisons between sites averaged for the three lower 0.3 m wide eulittoral intervals

Sites	Biomass (g m^{-2})			Cover (%)		
	M	I	Combined	M	I	Combined
San Clemente Island	1109	5	1114	181.2	3.9	185.1
San Clemente Island outer coast				145.6	23.7	169.3
Santa Catalina Island	2507	42	2549	187.8	5.9	193.7
Santa Catalina Island outer coast				157.6	8.4	166.0
Whites Point (2)	960	32	992	85.0	68.6	153.6
Dana Point				139.1	21.6	160.7
San Diego	578	61	639	135.1	42.0	177.1
Santa Barbara Island	988	114	1102	107.5	16.9	124.4
Santa Cruz Island	366	178	544	121.3	34.8	156.1
Santa Cruz Island inner coast				132.1	12.3	144.4
Coal Oil Point	546	112	658	89.6	15.0	104.6
Malibu				137.0	15.6	152.6
San Nicolas Island	1399	56	1455	102.6	42.6	145.2
San Nicolas Island northwest coast				97.0	16.3	113.3
San Miguel Island	775	284	1059	111.2	31.7	142.9
San Miguel Island outer coast				108.5	63.1	171.6
Corona del Mar	369	46	414	102.0	3.3	105.3
Santa Rosa Island	2077	572	2649	113.0	67.1	180.1
Pt. Conception	1179	60	1239	117.6	28.9	146.5
Anacapa Island outer coast				115.8	35.8	151.6
Anacapa Island inner coast				123.4	40.0	163.4
Anacapa Island outer disturbed				117.6	46.7	164.3
Island mean	1317.3	178.7	1496.0	128.1	29.9	158.1
Mainland mean	726.4	62.2	788.4	115.1	27.9	142.9
Combined mean	1091.1	130.2	1201.2	124.0	29.3	153.3

$g\,m^{-2}$); comparable trends were apparent from the mean cover data (158% for island *vs.* 143% for mainland sites). In addition to the greater cover on islands, H' diversity (Shannon and Weaver, 1949) was 3% higher, D' richness (Margalef, 1968) was 19% greater and organic dry weight was 33% higher (organic dry weight in the lower intertidal was nearly double that of mainland means).

Much of the difference between the island and mainland communities is due to considerable differences in the standing stocks of lower intertidal macrophytes, that is, the large brown algae *Egregia menziesii, Eisenia arborea* and *Halidrys dioica*, and the surf grasses *Phyllospadix scouleri* and *P. torreyi*. These macrophytes, particularly the large brown algae, are often depauperate and patchy at mainland sites near cities. This apparent reduction in algal biomass is likely attributable to water quality (see discussion under Human Influences below).

In addition to the notable differences in the abundances of kelps and rockweeds between island and mainland habitats, there were also differences in the algal assemblages that dominated in the mid-intertidal zone (see Table 11.5). This region was occupied by tightly compacted articulated coralline algae and their algal epiphytes (see Stewart, 1983). While extensive algal turf communities were prevalent in the mid and low intertidal at nearly all sites, island habitats contained considerably larger and more robust forms. Conversely, mainland turf communities, particularly those near metropolitan areas, were characterized by smaller and simpler algal forms and more compact structure. These mainland turfs were often composed of fine filamentous epiphytes in conjunction with smaller thalli of articulated coralline algae. Such filamentous turf populations may in fact be highly useful in identifying intermediate successional communities maintained in subclimax by lack of environmental constancy or some form of physiological stress, as shown by Littler and Murray (1975) for a sewage-stressed system on San Clemente Island. Stewart (1983) has characterized this unique and taxonomically complex coralline–algal turf community of southern California, particularly in terms of its variability and role in trapping sediments.

PRIMARY PRODUCTIVITY AND FUNCTIONAL FORM RELATIONSHIPS

Productivity

The importance of intertidal seaweeds in fixing energy in coastal waters has been well documented (see, for instance, Blinks, 1955; Kanwisher, 1966; Mann, 1973; Littler et al., 1979). Standing stocks and net photosynthetic performances were determined concurrently (Littler et al., 1979) for the 13 most abundant intertidal macrophytes from a pristine habitat (San Clemente Island, southern California) over a four-season period. Highest net production rates were observed during summer for nine of the 13 species and minimum rates were recorded in spring and to a lesser extent during winter. Correlations between seasonal fluctuations in standing stock and net productivity were evident for only eight species. Total daily community production reached a peak in the fall (1.22 g C fixed per square meter of substratum per day) and declined sharply through winter to a spring low (0.47 $g\,C\,m^{-2}\,day^{-1}$), closely paralleling changes in ambient water temperature. Blue-green algae, *Corallina officinalis* var. *chilensis, Pterocladia capillacea* and *Egregia menziesii* contributed 76% of the total community primary productivity for the year. Seasonal patterns of photosynthetic performances were highly variable, with a tendency for most species to attain peak daily photosynthetic rates when temperatures were high and days long.

The impact of a low-volume discharge of domestic sewage on net community productivity and ecological energetics was studied (Littler and Murray, 1974) near Wilson Cove, San Clemente Island. The mean primary productivity of the macrophyte community at the outfall (127.1 mg $C\,m^{-2}\,h^{-1}$) was not appreciably different from that measured for nearby unpolluted controls (123.4 mg $C\,m^{-2}\,h^{-1}$), even though there was 11.7% less cover and fewer than half as many species in the sewage-affected area. The above rates are comparable to those reported for other intertidal systems. For example, a wave-exposed sea stack and a protected boulder beach at Cayucos Point, central California, were compared (Seapy and Littler, 1979); macrophytes contributed approximately one-third more to total community primary production on the boulder beach than did those of the sea stack (169.7 versus 116.5 net mg C m^{-2} h^{-1}), due mainly to the greater cover and concomitant production of Cyanophyta and Fucaceae.

Functional forms

Littler and Littler (1980) developed a functional-form model for understanding and predicting patterns of primary productivity (see also Littler, 1980b; Littler and Littler, 1981, 1983; Littler and Arnold, 1982; Littler and Kauker, 1984). Their paradigm was based on studies within diverse floras from a broad spectrum of rocky intertidal environments in southwestern North America, including several habitats within the Southern California Bight. Specifically, macroalgae having thin sheet-like thalli showed the highest productivity (mean apparent net photosynthetic performance = 5.16 mg C g^{-1} h^{-1}) with a reduction by a factor of about two between each of the following four groups: filamentous forms (2.47), coarsely branched forms (1.30), thick leathery forms (0.76) and jointed calcareous algae (0.45). The crustose group had by far the lowest mean net productivity of only 0.07 mg C g^{-1} h^{-1} (Littler and Arnold, 1982). The functional-form group approach is a promising tool for interpreting physiological and morphological co-evolved interrelationships. Specifically, it has demonstrated (Littler and Littler, 1984) considerable credibility in predicting the outcome of productivity-related ecological processes without being bound to a particular geographic region or phylogenetic line.

ENERGETIC AND FOOD-WEB RELATIONSHIPS

Calorific values

Littler and Murray (1977) examined the factors controlling food-web structure, and how natural intertidal communities in southern California deal with high-energy inputs from sewage-derived particulate matter. Much of the variation in the calorific values of intertidal macrophytes was related (Littler and Murray, 1977) to differences in life-form strategies. In agreement with data for colder north-temperate species (Paine and Vadas, 1969), those forms with relatively more structural tissues (presumably selected for by competition for space and light, predation or physical shearing stress) tended to contain fewer total calories per unit weight. On the other hand, fugitive or opportunistic species selected for rapid growth and high productivity and containing few predator or competitor defenses had relatively greater calorific values. Encrusting intertidal forms that are easily accessible to herbivores may have evolved reduced palatability through selection for thallus constituents with reduced calorific values. In nearly every case, macroinvertebrate populations near the outfall had higher energy contents than did unpolluted control populations. The potential ability of certain seaweeds to utilize and recycle organic materials back through the food web has been stressed (see Prince, 1974). Energy-rich sewage compounds appeared to enter the intertidal food web through populations of omnivores and suspension feeders (Fig. 11.7), which may explain the enhanced standing stocks of these organisms in the peripheral region of the effluent plume.

Food webs

There were two distinct food sub-webs characteristic of the rocky intertidal systems of the leeward coast of San Clemente Island (Fig. 11.7; Littler and Murray, 1978). The first is a macrophyte-based grazing sub-web near the outfall area

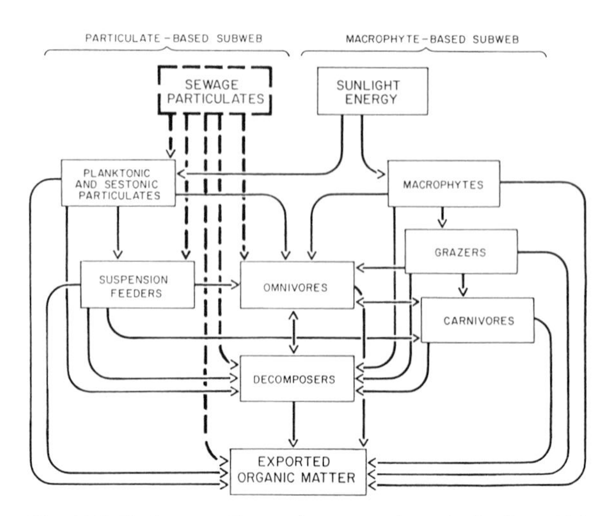

Fig. 11.7. Pathways of organic matter through the intertidal food web on San Clemente Island, southern California. Pathways unique to the San Clemente Island sewage outfall are shown by dashed lines. (From Littler and Murray, 1978.)

in which *Acmaea (Collisella)* spp., *Littorina planaxis, Lottia gigantea* and the omnivorous shore crab *Pachygrapsus crassipes* were the important herbivores. The same sub-web was also present on the unpolluted shoreline, where a similar guild of grazers was apparent, differing in the much lower numbers of *L. gigantea.* Hypothetically (Littler and Murray, 1978), the populations of grazers were enhanced energetically by the increased standing stocks of eurytolerant primary producers such as blue-green algae, *Gelidium pusillum* and *Ulva californica,* including their epiphytic filamentous brown algae, diatoms and bacteria.

The second food sub-web in the outfall and control areas (Fig. 11.7) utilized suspended and settled particulate materials as the primary energy source. In both the outfall and control areas, plankton and seston were available to suspension feeders. Populations of barnacles — *Chthamalus* spp. and *Tetraclita rubescens* — occurred commonly in the control areas and were the dominant suspension feeders.

However, in the outfall area, discharged organic particles served as the primary food source to the suspension-feeder sub-web, and organisms capable of utilizing sewage particulates predominated. These populations included *Pachygrapsus crassipes*, which was observed near the outfall to feed consistently on solid organic particles as well as algae during periods of low tide. The anemone *Anthopleura elegantissima* was also abundant in the area immediately fringing the point of discharge, but was rarely encountered in the unpolluted habitat. Also occurring in the outfall fringing area was an extremely dense population of very large individuals of the vermetid gastropod *Serpulorbis squamigerus*. This mollusc, which characteristically feeds upon detrital particles trapped from suspension, had formed an extensive bed at the lower intertidal margin of the outfall area and had been abundant in the sewage-impacted zone for many years, as evidenced by the reef-like build-up of dead calcareous tubes.

There was a much higher concentration of fishes in the vicinity of the San Clemente Island outfall than in a nearby unpolluted area (Horn, 1977). On average, gillnets in the outfall area trapped a larger number of fish species (1.4 times as great), higher densities (3 times as many individuals) and greater biomass (4.5 times) than were obtained in the control area. Horn (1977) concluded that the outfall served as both an attractant and food source for fishes that are generalized feeders.

Gut analyses

Pachygrapsus crassipes taken from the outfall area was found (see Littler and Murray, 1978 for details) to consume 18% more sewage particulates, as well as greater amounts of opportunistic primary producers, than *P. crassipes* from control areas, whose diet was more diverse and included calcareous algae as important constituents. Other grazers such as *Acmaea (Collisella) limatula, A. (Collisella) scabra* and *Lottia gigantea*, were found (Littler and Murray, 1978) to utilize preferentially high-energy ephemeral forms of algae in the outfall system. There was little qualitative difference in the gut contents of populations of *Littorina planaxis, L. scutulata* and *Serpulorbis squamigerus* in the outfall and control areas.

The omnivorous suspension feeder *Anthopleura elegantissima* occurred only rarely outside of the outfall fringe zone, where it regularly contained sewage in its gut. The outfall grazers consumed high-energy disclimax algal forms. For example, *Acmaea (Collisella) digitalis* utilized considerable amounts of *Enteromorpha* sp., while the diets of *A. (Collisella) limatula, A. (Collisella) scabra* and *Lottia gigantea* consisted of relatively large quantities of blue-green algae near the outfall. Specimens of *Acmaea (Collisella) digitalis*, a limpet occurring higher in the upper intertidal than other limpets and, consequently, nearer the outfall terminus, were found to be feeding considerably more on bacteria than were other gastropods.

ENVIRONMENTAL ASPECTS

Sand movement

Island sites tended to be dominated by the larger perennial species characteristic of mature communities. The habitat investigated on the western coast of San Nicolas Island near Dutch Harbor was unusual because of periodic episodes of sand movement and inundation, which occurred throughout the three years of the research program. In places where sand scouring was frequent (see Taylor and Littler, 1982; Littler et al., 1983 for detail), opportunistic organisms associated with early stages of community development (for example, seaweeds such as *Chaetomorpha linum, Cladophora columbiana, Enteromorpha intestinalis* and *Ulva lobata* and macroinvertebrates *Chthamalus dalli, Chthamalus fissus, Phragmatopoma californica* and *Tetraclita rubescens*) dominated the standing stocks. In contrast, elevated rocky substrata provided spatial escapes from sand inundation; these raised areas were occupied by relatively mature community assemblages, including long-lived molluscs such as *Haliotis cracherodii, Lottia gigantea* and *Mytilus californianus*. Santa Rosa Island also contained a highly variable zone below + 0.9 m, which appeared to be affected similarly by sand deposition. All three mainland areas that were subjected to a high degree of substratum instability (Corona del Mar, Whites Point and Coal Oil Point) also were distinguished by opportunistic species assemblages.

Wave exposure

There tended to be fewer macroinvertebrate taxa in samples from sites with heavy wave exposure (Fig. 11.8), while the greatest number of macroinvertebrates was encountered on Santa Catalina Island, a relatively sheltered habitat. Macrophyte species showed increased numbers at sites with prominent surge or swell, possibly due to less desiccation stress (Fig. 11.8), allowing a greater number of normally subtidal species to inhabit higher regions. The Ocean Beach community was relatively constant but lacked the large brown seaweeds usually found in other mature communities (see also Stephenson and Stephenson, 1972). The absence of these species is probably related to the high degree of wave shock that this site receives and to the friable nature of the soft sandstone substratum. It was observed frequently that the large seaweeds and barnacles were easily torn loose during periods of high wave energy.

Temporal variation

Seasonal patterns of distribution and abundance have often been described for populations inhabiting rocky intertidal habitats of temperate seas. For macrophytes, seasonality commonly has been

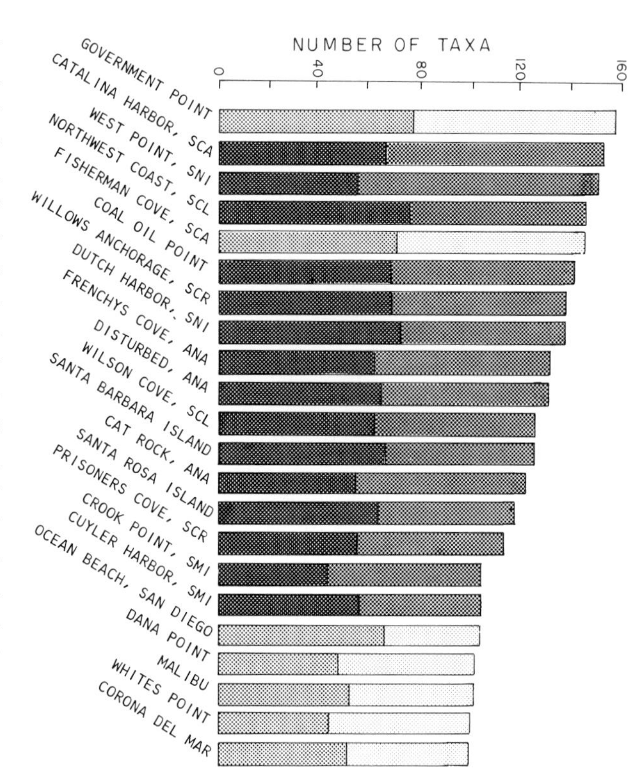

Fig. 11.8. Numbers of macrophyte taxa (left portions of histograms) and macroinvertebrate taxa (right portions of histograms) at each of the 22 study areas. Mainland stations are shaded lighter overall than island sites. The five stations below, located near large centers of human population, tended to have lower numbers of species. SCA: Santa Catalina Island; SNI: San Nicolas Island; SCL: San Clemente Island; SCR: Santa Cruz Island; ANA: Anacapa Island; SMI: San Miguel Island.

associated with fluctuations in environmental variables such as light, temperature and frequency of storms (see, for instance, Conover, 1958; Edwards, 1969; Doty, 1971; Horn et al., 1983). However, as demonstrated by several recent studies (Raffaelli, 1979; Lubchenco and Cubit, 1980; Dethier, 1981), biotic factors such as competition and predation may also contribute to seasonal patterns. For southern California, predictable seasonal cycles of certain abiotic features are apparent. For example, summer is characterized by longer daylight hours, warmer seawater and air temperatures, lower precipitation levels and fewer periods of stressful daytime aerial exposure than is the case during the winter.

Throughout 1975–1978 in southern California there were few widespread or consistent seasonal fluctuations of rocky intertidal populations except for (1) a slight lowering of most biological parameters at all stations following the daytime low-tide emersions preceding the winter months; (2) sporadic recruitment of barnacles and several other macroinvertebrates (such as *Aplysia*) at various sites during the winter through spring period; and (3) growth and accumulation of standing stocks of large seaweeds throughout the summer. Distinct seasonal patterns were most pronounced at the relatively protected, physically benign, southern island stations (for example, Santa Catalina Island, Fisherman Cove; San Clemente Island, Wilson Cove). At these sites, macrophyte standing stocks were greatest during the late summer and least during the winter when harsher environmental conditions probably resulted in lower growth rates. Reductions in intertidal standing stocks during periods of increased physical harshness have also been shown for New South Wales, Australia (Underwood, 1981) and central California (Horn et al., 1983). The overall lack of consistent temporal tendencies strongly suggests that local-scale or even site-specific conditions tend most often to predominate and obscure broad climatic effects. This agrees with other descriptions (Stephenson and Stephenson, 1972; Jernakoff, 1985) of rocky intertidal systems that also have a high degree of autonomy.

However, as is the case for many biological systems, major stochastic abiotic disturbances appear to be paramount in restructuring southern California intertidal communities. The most important of these during our study were (1) unusually heavy precipitation and flooding events that exposed intertidal organisms to reduced salinities and sediment loading; (2) conditions of extreme heat, low humidity and high winds that occurred during periods of daytime tidal emersion; (3) exceptionally large storm-generated waves; and (4) patterns of longshore sand transport that scoured and buried various organisms. The first two disturbances typically occur during the late fall and winter; however, large waves can strike the southern California coastline at any time, and the patterns of sand movement within stations appear to be quite variable, as well as differing from site to site (Littler, 1980b). Therefore, subtle seasonal patterns in community composition and standing stocks may be difficult to detect, particularly for systems that are subjected to high levels of such unpredictable disturbance.

Community responses to natural disturbances

During the three-year investigation, several of the sites experienced major natural physical disturbances. For example, the Corona del Mar study area became flooded by an unusually heavy rainfall (2.56 cm within 3 hours) in May 1977 (Littler and Littler, 1987). As a result, the sea urchin *Strongylocentrotus purpuratus* underwent a twenty-fold decrease in cover (from about 2.0% to less than 0.1%) in the lower intertidal zone (MLLW to +0.3 m) and disappeared entirely within the +0.3 to +0.6 m interval. However, the most dramatic effects were revealed by the density counts of *S. purpuratus* over relatively broad areas. Belt transects documented an average of 90.5% mortality, and a census of the total area between two permanent transect lines showed 93.6% of the *S. purpuratus* to be dead. A biotically similar area beyond the periphery of the region that was flooded (20 m north of the north transect line) experienced only 1.1% mortality of *S. purpuratus* (Littler and Littler, 1987). The serpulid worm *Eupomatus gracilis* decreased in mean density by 74 m^{-2}. Species consistently present previously, but absent following the flood, included four gastropods, the crab *Pugettia producta*, the bivalve *Glans carpenteri* and an unknown bryozoan. Ephemeral macrophytes characteristic of disturbed environments, the green algae *Enteromorpha* sp. and *Ulva californica* and Ectocarpaceae, increased markedly in mean overall cover (14.5% and 8.9%, respectively) following the rain-storm, as did newly recruited barnacles. However, the majority of long-lived macrophytes, such as *Hydrolithon decipiens*, blue-green algal crusts and *Gelidium* spp. (*G. coulteri* and *G. pusillum*), showed declines in overall mean cover (1.3%, 7.8%, and 5.0%, respectively).

During afternoon low tides in the winter of 1976, rocky intertidal organisms inhabiting the Santa Cruz Island site at Willows Anchorage were subject to prolonged exposure to hot, dry "Santa Ana" winds (Seapy and Littler, 1982). Numerous species exhibited pronounced die-back, including the barnacles *Chthamalus dalli* and *C. fissus* and the seaweeds *Codium fragile*, *Corallina officinalis* var. *chilensis, C. vancouveriensis, Cylindrocarpus rugosus, Endocladia muricata* and *Pelvetia fastigiata* f. *gracilis.* The die-backs were succeeded by blooms of morphologically simple seaweeds such as *Porphyra perforata* and *Ulva californica.*

During May 1976, an unusually severe storm devastated the Santa Catalina Island, Fisherman Cove site, which is normally sheltered in the lee of the prevailing swell patterns. This protected site was dominated by high standing stocks of large morphologically complex seaweeds and other organisms characteristic of mature stages of community development. The storm waves ripped up many kelp plants immediately seaward of the study area, in addition to breaking off a large portion of the three-meter high rocky bluff above the supralittoral fringe. Short-term effects of the storm were largely confined to decreases in the standing stocks of larger long-lived seaweeds, such as *Corallina officinalis* var. *chilensis, Egregia menziesii, Eisenia arborea* and *Gelidium purpurascens*, and increases in the abundance of morphologically simple opportunists, such as *Colpomenia sinuosa, Enteromorpha* sp. and *Scytosiphon dotyi.* These opportunistic species are known (Wilson, 1925; Emerson and Zedler, 1978; Murray and Littler, 1978) to readily occupy newly vacated patches within mature southern California communities. Where disturbance is periodic and not too severe, southern California rocky intertidal communities appear to exhibit an overall increase in diversity, owing to the coexistence of mixed patches of early and late successional species (Sousa, 1979a). For example, overall community *H'* diversity at Fisherman Cove increased following the 1976 storm from 2.80 to 3.04, due to the greater abundance of opportunists and decreased abundance of community dominants.

Drift seaweed biomass, consisting predominantly of the larger kelps, increases dramatically (ZoBell, 1971) on southern California beaches during the periods of most frequent storms (November through February). A strong positive correlation was noted (ZoBell, 1971) between the quantity of seaweed on shores and wave height, as well as wind velocity, thereby providing additional evidence of the significance of storm conditions to seaweed communities. Such natural events remove biomass and are recurrent physical factors in biological communities (Connell, 1978); their effects have been partially documented in the cases of sand scouring and inundation (Seapy and Littler, 1982; Taylor and Littler, 1982; Littler et al., 1983), substratum instability (Sousa, 1979a, 1980; Murray and Littler, 1984; Littler and Littler, 1984), extreme aerial exposure (Seapy and Littler, 1982) and unusually severe, storm-generated waves (Murray and Littler, 1978).

The unpredictable occurrence of storms or high waves has been correlated with subtidal standing-stock fluctuations in the tropics (Doty, 1971), where normal seasonality is obscured. Similarly, catastrophic events were related (Paine, 1979) to biomass fluctuations in the temperate sea palm *Postelsia.* Such stochastic events can have highly localized species-specific catastrophic effects on littoral populations; they may set in motion subsequent changes to overall community structure that would be difficult to understand from a program of infrequent sampling. Therefore, studies of the responses of intertidal communities to natural perturbations (Sousa, 1980; Littler and Littler, 1984) may give information on the ability of different components to recover, and thus provide one method of measuring community stability (resilience).

Recovery from disturbances

Successional studies have been conducted (Murray and Littler, 1979) throughout southern California to determine biological recovery rates following severe mechanical disturbances, that is, the removal of all fleshy and upright organisms. A primary objective was to identify sensitive populations that would be slow to re-establish. The resultant mortality suffered by mechanically harvesting populations provided free space such as that produced by natural physical disturbances.

Generally, most intertidal macrophytes were effective in rapidly re-establishing their overall cover on disturbed surfaces, while most macroinvertebrate populations tended to recover much more slowly. All of the sites (except Santa Cruz Island and Santa Catalina Island) regained their pre-harvest levels of mean macrophyte cover after 12 months, whereas none of the macroinvertebrate species attained preharvest levels of cover, and overall pre-harvest densities of macroinvertebrates were matched only at Whites Point. After 12 months, an average of only 55.4% of the pre-harvest densities and 64.4% Bray–Curtis cover similarity was recorded at the experimental sites.

This evidence is in agreement with that of Murray and Littler (1978) and Sousa (1980) in demonstrating that the successional stages of intertidal algal communities differ significantly in their responses to disturbance. Early successional communities tend to suffer more damage from a given level of perturbation but recover more quickly than either middle or late successional communities. The long-lived articulated coralline algae (such as, for instance, *Corallina officinalis* var. *chilensis*) show remarkably rapid recovery rates from remnant basal-crust portions (Littler and Murray, 1978; Littler and Kauker, 1984), particularly where they have previously formed a dense turf. In agreement, Stewart (1982) found that experimentally scraped quadrats could be distinguished from unscraped turfs after 12 to 14 months only because the genus *Corallina* remained less dense and in some instances, large clumps of *C. vancouveriensis* were still absent. After 24 months, turf regrowth areas were indistinguishable from the surrounding community, and *Corallina* was re-established completely on surfaces where it had formerly predominated. Certain of the frondose forms such as *Gigartina canaliculata* also showed rapid recovery following plot clearance (Sousa, 1979b). In our studies, *Egregia menziesii* was found to initiate recolonization of disturbed surfaces soon after harvesting; however, habitats subjected to continual disturbances (such as Whites Point, Coal Oil Point and Corona del Mar) were not readily recolonized by *Egregia*. Filamentous algae (such as rhodophycean turf at Coal Oil Point) showed rapid re-establishment to pre-harvest levels. Among the macroinvertebrates that appeared to be most rapid in recolonizing disturbed surfaces were the tube-worms *Dodecaceria fewkesi* and *Phragmatopoma californica*.

Several populations of macrophytes and macroinvertebrates were found to be exceptionally slow to recover, and are, thus, likely to be severely impacted by disturbance. The upper intertidal rockweeds *Hesperophycus harveyanus* and *Pelvetia fastigiata* consistently failed to re-establish. These species, even when abundant adjacent to disturbed plots, showed extremely slow vegetative encroachment or propagule recruitment. Other large frondose seaweeds, such as *Halidrys dioica* and *Prionitis lanceolata*, were also extremely slow to recruit or encroach. Populations of sessile bivalves were likewise quite slow to recover. For example, the mussels *Brachidontes adamsianus*, *Mytilus californianus*, *M. edulis* and *Septifer bifurcatus* all showed virtually negligible re-establishment on disturbed surfaces, as did the rock oyster *Pseudochama exogyra*.

HUMAN INFLUENCES

Intertidal habitats in southern California are unique in the U.S.A. owing to their large usage by an exceptionally numerous, recreation-oriented, human population. This intensive usage makes the intertidal zone particularly sensitive to additional forms of environmental stress. The effects of environmental deterioration and general lack of adequate baseline information have posed severe problems, particularly during the past two decades when attempts were made to assess the immediate effects of specific pollutants on coastal organisms in southern California. Such problems were obvious in the attempts to evaluate the impacts of the Santa Barbara oil spill of 1969 (see Foster et al., 1971; Nicholson and Cimberg, 1971; Straughan, 1971).

The most widely cited baseline information was that of Dawson (1959, 1965) for marine algae. Dawson noted reductions in algal species numbers ranging from 50 to 70% at sites near sewage outfalls. Because Dawson (1959, 1965) assessed only the "conspicuous" members of the seaweed flora, it has been difficult for subsequent researchers to make direct comparisons of richness (see Harris, 1980). Problems aside, other workers (Nicholson and Cimberg, 1971; Widdowson, 1971; Thom, 1976; Thom and Widdowson, 1978) have clearly substantiated further declines in macrophyte species numbers at many of the same areas studied by Dawson. Such declines were attributed to human influence but only circumstantial evidence was available. The declines do not seem to have been instantaneous (Nicholson, 1972), but probably are the result of human pressure that has been increasing markedly since the turn of the century.

Littler and Murray (1975) demonstrated considerable reductions in the standing stocks of large brown algae and seagrasses near a low-volume domestic sewage outfall on San Clemente Island. There were fewer species and less cover near the point of discharge (7 macroinvertebrates, 17.6% cover; 13 macrophytes, 91.7%) than in nearby "unpolluted" control areas (9 macroinvertebrates, 9.2%; 30 macrophytes, 103.4%). The outfall biota was less diverse than that of the controls, as shown by five different diversity indices. A great reduction in community stratification (spatial heterogeneity) and, hence, community complexity was noted near the outfall; this reduction in stratification was primarily due to the absence of large macrophytes such as *Egregia laevigata, Halidrys dioica, Phyllospadix torreyi* and *Sargassum agardhianum*. These were replaced in the mid intertidal near the outfall by a low turf of blue-green algae, *Gelidium pusillum, Ulva californica*, and small *Pterocladia capillacea*, and in the lower intertidal by *Serpulorbis squamigerus* covered with *Corallina officinalis* var. *chilensis*. As mentioned earlier, the enhancement of the suspension feeder *Serpulorbis squamigerus* and the omnivores *Anthopleura elegantissima, Ligia occidentalis* and *Pachygrapsus crassipes* in the outer fringe of the outfall plume appeared to be related to their ability to utilize sewage as a food source. A critical effect of the outfall was to decrease environmental stability, thereby favoring rapid colonizers and more sewage-tolerant organisms. The outfall macrophytes had relatively higher net primary productivity, smaller growth forms, and simpler and shorter life histories; most are components of early successional stages. The above findings (Littler and Murray, 1975) were in agreement with earlier observations (Widdowson, 1971; Thom and Widdowson, 1978) that there has been a shift from the more massive algal species towards turf and crustose taxa in southern California littoral communities since the surveys by Dawson (1959, 1965) in the 1950's. Additionally, Thom and Widdowson (1978) reported that these changes were most pronounced in habitats exposed to greatest disturbance, that is, localities near heavily populated metropolitan areas or in public parks subject to heavy human recreational usage.

Beauchamp and Gowing (1982) have shown that human trampling can cause serious degradation of littoral communities where usage is high. Murray and Littler (1984) have also documented a reduction of large brown and fleshy algae at Whites Point, a site exposed to extensive human traffic and inshore from the large (>300 million gal day^{-1}) sewage outfall operated by the Los Angeles County Sanitation District. Sousa et al. (1981) described reductions in standing stocks of long-lived, large brown algae for southern California intertidal habitats, and attributed their reduction to enhanced grazing pressure from sea urchins. However, our experience on the outer coastal habitats of the southern California islands is that abundant populations of the purple sea urchin *Strongylocentrotus purpuratus* commonly coexist with high standing stocks of large brown algae such as *Halidrys dioica* in the lower intertidal zone.

As a result of expansion of the human population in southern California, more of the marine communities on the relatively inaccessible offshore Channel Islands (for instance, Anacapa Island) are being altered (Littler, 1978b). By comparing an accessible station ("Anacapa-Disturbed"; Fig. 11.1; visited by approximately 50 000 boaters per year) with a relatively inaccessible habitat (Cat Rock) near Frenchys Cove, Anacapa Island, Littler (1978b) was able to show a community shift, from long-lived invertebrate and algal populations toward turfy opportunistic algae and the tube-worm *Phragmatopoma*, correlated with activities of visitors. Already most of southern California's rocky intertidal systems have been exposed to different degrees of anthropogenic influence; collecting, rock-turning, trampling, oil spills and sewage pollution have been so intensive that the conspicuous invertebrates and plant life have been severely reduced or eliminated near many of the heavily populated regions. Despite extensive research, human influence on intertidal coastal environments continues to outpace understanding of the ecological changes that are taking place.

REFERENCES

Anonymous, 1973. Physical and dynamic oceanography. In: *The Ecology of the Southern California Bight: Implications for Water Quality Management. Southern California Coastal Water Research Project (SCCWRP) Tech. Rep.*. 104, pp. 37–40.

Beauchamp, D.A. and Gowing, M.M., 1982. A quantitative assessment of human trampling effects on rocky intertidal community. *Mar. Environm. Res.*, 7: 279–293.

Blinks, L.R., 1955. Photosynthesis and productivity of littoral marine algae. *J. Mar. Res.*, 14: 363–373.
Bray, J.R. and Curtis, J.T., 1957. An ordination of the upland forest communities of southern Wisconsin. *Ecol. Monogr.*, 27: 325–349.
Briggs, J.C., 1974. *Marine Zoogeography*. McGraw-Hill, New York, 475 pp.
Bright, D.B., 1974. Benthic invertebrates. In: M.D. Dailey, B. Hill and N. Lansing (Editors), *A Summary of Knowledge of the Southern California Coastal Zone and Offshore Areas*. U.S. Department of the Interior, Bureau of Land Management, Washington, DC., pp. 10–1 to 10–291.
Brusca, R.C. and Wallerstein, B.R., 1979. Zoogeographic patterns of idoteid isopods in the Northeast Pacific, with a review of shallow water zoogeography of the area. *Bull. Biol. Soc. Wash.*, 3: 67–105.
Caplan, R.I. and Boolootian, R.A., 1967. Intertidal ecology of San Nicolas Island. In: R.N. Philbrick (Editor), *Proc. Symp. on the Biology of the California Islands*. Santa Barbara Botanic Garden, Santa Barbara, CA, pp. 203–207.
Cockerell, T.D.A., 1939. The marine invertebrate fauna of the Californian Islands. *Sixth Pac. Sci. Congr. Proc.*, 3: 501–504.
Connell, J.H., 1972. Community interactions in marine rocky intertidal shores. *Annu. Rev. Ecol. Syst.*, 3: 169–192.
Connell, J.H., 1978. Diversity in tropical rain forests and coral reefs. *Science*, 199: 1302–1310.
Conover, J.T., 1958. Seasonal growth of benthic marine plants as related to environmental factors in an estuary. *Publ. Inst. Mar. Sci. Univ. Tex.*, 5: 97–147.
Dawson, E.Y., 1949. Contributions toward a marine flora of the Southern California Channel Islands I–III. *Allan Hancock Foundation Publ., Univ. S. Calif.*, 8: 1–57.
Dawson, E.Y., 1959. A primary report on the benthic marine flora of southern California. In: *An Oceanographic and Biological Survey of the Continental Shelf Area of Southern California. Publ. Calif. St. Wat. Pollut. Contr. Bd.*, 20: 169–264.
Dawson, E.Y., 1965. Intertidal algae. In: *An Oceanographic and Biological Survey of the Southern California Mainland Shelf. Publ. Calif. St. Wat. Qual. Contr. Bd.*, 27: 220–231, 351–438.
Dawson, E.Y. and Neushul, M., 1966. New records of marine algae from Anacapa Island, California. *Nova Hedwigia*, 12: 173–187.
Dethier, M.N., 1981. Heteromorphic algal life histories: the seasonal pattern and response to herbivory of the brown crust, *Ralfsia californica*. *Oecologia (Berl.)*, 49: 333–339.
Doty, M.S., 1971. Physical factors in the production of tropical benthic marine algae. In: J. Costlow Jr. (Editor), *Fertility of the Sea, Vol. I.* Gordon and Breach, New York, pp. 99–121.
Edwards, P., 1969. Field and cultural studies on the seasonal periodicity of growth and reproduction of selected Texas benthic marine algae. *Contrib. Mar. Sci.* 14: 59–114.
Emerson, S.E. and Zedler, J.B., 1978. Recolonization of intertidal algae: an experimental study. *Mar. Biol. (Berl.)*, 44: 315–324.
Foster, M., Neushul, M. and Zingmark, R., 1971. The Santa Barbara oil spill, Part II. Initial effects on intertidal and kelp bed organisms. *Environm. Pollut.*, 2: 115–134.
Gunnill, F.C., 1979. *The effect of host distribution of the faunas inhabiting an intertidal alga*. Ph.D. Dissertation, Univ. of California, San Diego.
Gunnill, F.C., 1980a. Recruitment and standing stocks in populations of one green alga and five brown algae in the intertidal zone near La Jolla, California during 1973–1977. *Mar. Ecol. Prog. Ser.*, 3: 231–243.
Gunnill, F.C., 1980b. Demography of the intertidal brown alga *Pelvetia fastigiata* in Southern California, USA. *Mar. Biol.*, 59: 169–179.
Gunnill, F.C., 1983. Seasonal variations in the invertebrate faunas of *Pelvetia fastigiata* (Fucaceae): effects of plant size and distribution. *Mar. Biol.*, 73: 115–130.
Gunnill, F.C., 1985. Population fluctuations of seven macroalgae in southern California during 1981–1983 including effects of severe storms and an El Niño. *J. Exp. Mar. Biol. Ecol.*, 85: 149–164.
Harris, L.H., 1980. Changes in intertidal algae at Palos Verdes. In: *Coastal Water Research Project, Biennial Report for the years 1979–1980*. Southern California Coastal Water Research Project (SCCWRP), Long Beach, Ca., pp. 35–75.
Hayden, B.P. and Dolan, R., 1976. Coastal marine fauna and marine climates of the Americas. *J. Biogeogr.*, 3: 71–81.
Hendricks, T.J., 1977. Satellite imagery studies. In: *Coastal Water Research Project, Annual Report for the Year Ended 30 June 1977*. Southern California Coastal Water Research Project (SCCWRP), El Segundo, Ca., pp. 75–78.
Hewatt, W.G., 1946. Marine ecological studies on Santa Cruz Island, California. *Ecol. Monogr.*, 16: 187–208.
Hickey, B.M., 1979. The California Current system — hypotheses and facts. *Prog. Oceanogr.*, 8: 191–279.
Horn, M.H., 1977. Abundance and species composition of fishes. In: M.M. Littler and S.N. Murray (Editors), *Influence of Domestic Wastes on the Structure and Energetics of Intertidal Communities near Wilson Cove, San Clemente Island*. Tech. Pap. 396, Naval Undersea Center, San Diego, pp. 41–46.
Horn, M.H., Murray, S.N. and Seapy R.R., 1983. Seasonal structure of a central California rocky intertidal community in relation to environmental variations. *Bull. S. Calif. Acad. Sci.*, 82: 79–94.
Jernakoff, P., 1985. An experimental evaluation of the influence of barnacles, crevices and seasonal patterns of grazing on algal diversity and cover in an intertidal barnacle zone. *J. Exp. Mar. Biol. Ecol.*, 88: 287–302.
Jones, J.H., 1971. *General circulation and water characteristics in the southern California Bight*. Tech. Rep., 101, Southern California Coastal Water Research Project (SCCWRP), pp. 1–37.
Kanwisher, J.W., 1966. Photosynthesis and respiration in some seaweeds. In: H. Barnes (Editor), *Some Contemporary Studies in Marine Science*. Allen and Unwin, London, pp. 407–420.
Kimura, J.C., 1974. Climate. In: M.D. Dailey, B. Hill and N. Lansing (Editors), *A Summary of Knowledge of the Southern California Coastal Zone and Offshore Areas, Vol. I.* U.S. Department of the Interior, Bureau of Land Management, Washington, D.C., pp. 2-1 to 2-70.

Littler, M.M., 1978a. *The annual and seasonal ecology of Southern California rocky intertidal, subtidal, and tidepool biotas*. Southern California Baseline Study, Year Two, Final Rep., Vol. III, Rep. 1.1. Bureau of Land Management, U.S. Department of the Interior, Washington, D.C., pp. III-1.1.1-1 to III-1.1.1-6.

Littler, M.M., 1978b. *Assessments of visitor impact on spatial variations in the distribution and abundance of rocky intertidal organisms on Anacapa Island, California*. Final Rep. U.S. Nat. Park Service, Washington, D.C., 161 pp.

Littler, M.M., 1979. *Intertidal study of the Southern California Bight, 1977/1978 (third year)*. Southern California Baseline Study, Year Three, Vol. II, Rep. 1. Bureau of Land Management, U.S. Department of the Interior, Washington, D.C., pp. II-1.0-1 to II-1.0-9.

Littler, M.M., 1980a. Southern California rocky intertidal ecosystems: methods, community structure and variability. In: J.H. Price, D.E.G. Irvine and W.F. Farnham (Editors), *The Shore Environment, Vol. 2: Ecosystems. Systematics Association Special Vol. No. 17(b)*. Academic Press, London, pp. 565–608.

Littler, M.M., 1980b. Overview of the rocky intertidal systems of Southern California. In: D.M. Power (Editor), *The California Islands: Proceedings of a Multidisciplinary Symposium*. Santa Barbara Museum of Natural History, Santa Barbara, Ca., pp. 265–301.

Littler, M.M. and Arnold, K.E., 1982. Primary productivity of marine macroalgal functional-form groups from southwestern North America. *J. Phycol.*, 18: 307–311.

Littler, M.M. and Kauker, B.J., 1984. Heterotrichy and survival strategies in the red alga *Corallina officinalis* L. *Bot. Mar.*, 27: 37–44.

Littler, M.M. and Littler, D.S., 1978. Rocky intertidal island survey. In: M.M. Littler (Editor), *Intertidal Study of the Southern California Bight, 1977–1978 (Year Two). Vol. II, Rep. 5*. Bureau of Land Management, U.S. Department of the Interior, Washington, D.C., pp. II-5.0–1 to II–5.0–5.

Littler, M.M. and Littler, D.S., 1979. Mainland rocky intertidal aerial survey from Point Arguello to Point Loma, California. In: M.M. Littler (Editor), *Southern California Baseline Study, Intertidal, Year Three, Final Report*. Bureau of Land Management, U.S. Department of the Interior, Washington, D.C., pp. II-5.0-1 to II-5.0-7.

Littler, M.M. and Littler, D.S., 1980. The evolution of thallus form and survival strategies in benthic marine macroalgae: field and laboratory tests of a functional form model. *Am. Nat.*, 116: 25–44.

Littler, M.M. and Littler, D.S., 1981. Intertidal macrophyte communities from Pacific Baja California and the upper Gulf of California: relatively constant vs. environmentally fluctuating systems. *Mar. Ecol. Prog. Ser.*, 4: 145–158.

Littler, M.M. and Littler, D.S., 1983. Heteromorphic life-history strategies in the brown alga *Scytosiphon lomentaria* (Lyngb.) Link. *J. Phycol.*, 19: 425–431.

Littler, M.M. and Littler, D.S., 1984. Relationships between macroalgal functional form groups and substrata stability in a subtropical rocky-intertidal system. *J. Exp. Mar. Biol. Ecol.*, 74: 13–34.

Littler, M.M. and Littler, D.S., 1987. Effects of stochastic processes on rocky–intertidal biotas: an unusual flash flood near Corona del Mar, California. *Bull. S. Calif. Acad. Sci.*, 86: 95–106.

Littler, M.M. and Murray, S.N., 1974. Primary productivity of macrophytes. In: S.N. Murray and M.M. Littler (Editors), *Biological Features of Intertidal Communities Near the U.S. Navy Sewage Outfall, Wilson Cove, San Clemente Island, California*. Tech. Pap. 396, Naval Undersea Center, San Diego, pp. 67–79.

Littler, M.M. and Murray, S.N., 1975. Impact of sewage on the distribution, abundance and community structure of rocky intertidal macro-organisms. *Mar. Biol.*, 30: 277–291.

Littler, M.M., and Murray, S.N. (Editors), 1977. *Influence of domestic wastes on the structure and energetics of intertidal communities near Wilson Cove, San Clemente Island*. California Water Resources Center, Contribution No. 164, Univ. of California, Davis, Ca., 88 pp.

Littler, M.M. and Murray, S.N., 1978. Influence of domestic wastes on energetic pathways in rocky intertidal communities. *J. Appl. Ecol.*, 15: 583–595.

Littler, M.M., Murray, S.N. and Arnold, K.E., 1979. Seasonal variations in net photosynthetic performance and cover of intertidal macrophytes. *Aquat. Bot.*, 7: 35–46.

Littler, M.M., Martz, D.R. and Littler, D.S., 1983. Effects of recurrent sand deposition on rocky intertidal organisms: importance of substrate heterogeneity in a fluctuating environment. *Mar. Ecol. Prog. Ser.*, 11: 129–139.

Lubchenco, J. and Cubit, J., 1980. Heteromorphic life histories of certain marine algae as adaptations to variations in herbivory. *Ecology*, 61: 676–687.

Mann, K.H., 1973. Seaweeds: their productivity and strategy for growth. *Science*, 182: 975–981.

Margalef, R., 1968. *Perspectives in Ecological Theory*. Univ. of Chicago Press, Chicago, Ill., 111 pp.

Murray, S.N., 1974. Benthic algae and grasses. In: M. Dailey, B. Hill and N. Lansing (Editors), *A Summary of Knowledge of the Southern California Coastal Zone and Offshore Areas, Vol. II*. U.S. Department of the Interior, Bureau of Land Management, Washington, D.C., pp. 9–1 to 9–61.

Murray, S.N. and Littler, M.M. (Editors), 1974. *Biological features of intertidal communities near the U.S. Navy sewage outfall, Wilson Cove, San Clemente Island, California*. Technical Paper No. 396, Naval Undersea Center, San Diego, Ca., 85 pp.

Murray, S.N. and Littler, M.M., 1978. Patterns of algal succession in a perturbated marine intertidal community. *J. Phycol.*, 14: 506–512.

Murray, S.N. and Littler, M.M., 1979. Experimental studies of the recovery of populations of rocky intertidal macro-organisms following mechanical disturbance. In: M.M. Littler (Editor), *Southern California Baseline Study, Intertidal, Year Three, Final Report*. Bureau of Land Management, U.S. Department of the Interior, Washington, D.C., pp. II-2.0-1 to II-2.0-171.

Murray, S.N. and Littler, M.M., 1981. Biogeographical analysis of intertidal macrophyte floras of southern California. *J. Biogeogr.*, 8: 339–351.

Murray, S.N. and Littler, M.M., 1984. Analysis of seaweed communities in a disturbed rocky intertidal environment near Whites Point, Los Angeles, Calif., U.S.A. *Hydrobiologia*, 116/117: 374–382.

Murray, S.N., Littler, M.M. and Abbott, I.A., 1980. Biogeography of the California marine algae with emphasis on the Southern California Islands. In: D.M. Power (Editor), *The California Islands: Proceedings of a Multidisciplinary Symposium.* Santa Barbara Museum of Natural History, Santa Barbara, Ca., pp. 325–339.

Neushul, M., Clarke, W.D. and Brown, D.W., 1967. Subtidal plant and animal communities in the southern California islands. In: R.N. Philbrick (Editor), *Proc. Symp. on the Biology of the California Islands.* Santa Barbara Botanic Garden, Santa Barbara, Ca., pp. 37–55.

Newell, I.M., 1948. Marine molluscan provinces of western North America: a critique and a new analysis. *Proc. Am. Philos. Soc.*, 92: 155–166.

Newman, W.A., 1979. Californian transition zone: significance of short-range endemics. In: J. Gray and A. Boucot (Editors), *Historical Biogeography, Plate Tectonics, and the Changing Environment. The 37th Annual Biology Colloquium, April 23–24, 1976.* Oregon State University Press, Corvallis, Oreg., pp. 399–416.

Nicholson, N.L., 1972. The Santa Barbara oil spills in perspective. *Calif. Mar. Res. Comm. CalCOFI Rep.*, 16: 130–149.

Nicholson, N.L. and Cimberg, R.L., 1971. The Santa Barbara oil spills of 1969; a post-spill survey of the rocky intertidal. In: D. Straughan (Editor), *Biological and Oceanographic Survey of the Santa Barbara Channel Oil Spill 1969–1970, Vol. I.* Allan Hancock Foundation, Univ. of S. California, Los Angeles, Ca., pp. 325–399.

Paine, R.T., 1979. Disaster, catastrophe, and local persistence of the sea palm *Postelsia palmaeformis. Science.*, 205: 685–687.

Paine, R.T. and Vadas, R.L., 1969. Calorific values of benthic marine algae and their postulated relation to invertebrate food preference. *Mar. Biol.*, 4: 79–86.

Power, D.M. (Editor), 1980. *The California Islands: Proceedings of a multidisciplinary symposium.* Santa Barbara Museum of Natural History, Santa Barbara, Ca., 363 pp.

Prince, J.S., 1974. Nutrient assimilation and growth of some seaweeds in mixture of sea water and secondary sewage treatment effluents. *Aquaculture*, 4: 69–79.

Raffaelli, D., 1979. The grazer–algae interaction in the intertidal zone on New Zealand rocky shores. *J. Exp. Mar. Biol. Ecol.*, 38: 81–100.

Reid Jr., J.L. and Schwartzlose, R.A., 1962. Direct measurements of the Davidson Current off central California. *J. Geophys. Res.*, 67: 2491–2497.

Reid Jr., J.L., Roden, G.I. and Wyllie, J.G., 1958. Studies of the California Current system. *Calif. Mar. Res. Comm. CalCOFI Rep.*, 6: 27–56.

Ricketts, E.F., Calvin, J. and Hedgpeth, J.W., 1968. *Between Pacific Tides (4th ed.).* Stanford University Press, Stanford, Ca., 614 pp.

Schwartzlose, R.A., 1963. Nearshore currents of the western United States and Baja California as measured by drift bottles. *Calif. Mar. Res. Comm. CalCOFI Rep.*, 9: 15–22.

Seapy, R.R., 1974. Macroinvertebrates. In: S.N. Murray and M.M. Littler (Editors), *Biological features of intertidal communities near the U.S. Navy sewage outfall, Wilson Cove, San Clemente Island, California.* Tech. Pap. 396, Naval Undersea Center, San Diego, Ca., pp. 19–22.

Seapy, R.R. and Hoppe, W.J., 1973. Morphological and behavioral adaptations to desiccation in the intertidal limpet *Acmaea (Collisella) strigatella. Veliger*, 16: 181–188.

Seapy, R.R. and Littler, M.M., 1979. The distribution, abundance, community structure, and primary productivity of macroorganisms from two central California rocky intertidal habitats. *Pac. Sci.*, 32: 293–314.

Seapy, R.R. and Littler, M.M., 1980. Biogeography of rocky intertidal macroinvertebrates of the Southern California Islands. In: D.M. Power (Editor), *The California Islands: Proceedings of a Multidisciplinary Symposium.* Santa Barbara Museum of Natural History, Santa Barbara, CA, pp. 307–323.

Seapy, R.R. and Littler, M.M., 1982. Population and species diversity fluctuations in a rocky intertidal community relative to severe aerial exposure and sediment burial. *Mar. Biol.*, 71: 87–96.

Shannon, C.E. and Weaver, W., 1949. *The Mathematical Theory of Communication.* University of Illinois Press, Urbana, Ill., 117 pp.

Sheppard, F.P. and Emery, K.O., 1941. *Submarine topography off the California coast: canyons and tectonic interpretations.* Special Paper 31, Geological Society of America, 171 pp.

Sims, R.H., 1974. Macrophytes. In: S.N. Murray and M.M. Littler (Editors), *Biological Features of Intertidal Communities near the U.S. Navy Sewage Outfall, Wilson Cove, San Clemente Island, California.* Techn. paper 396, U.S. Naval Undersea Center, San Diego, pp. 13–17.

Smith, R., 1976. *Numerical Analysis of Ecological Survey Data.* Ph.D. Dissertation, Univ. of Southern California, Los Angeles, Ca., 401 pp.

Sousa, W.P., 1979a. Disturbance in marine intertidal boulder fields: the nonequilibrium maintenance of species diversity. *Ecology*, 60: 1225–1239.

Sousa, W.P., 1979b. Experimental investigations of disturbance and ecological succession in a rocky intertidal algal community. *Ecol. Monogr.*, 49: 227–254.

Sousa, W.P., 1980. The responses of a community to disturbance: the importance of successional age and species' life histories. *Oecologia (Berl.)*, 45: 72–81.

Sousa, W.P., Schroeter, S.C. and Gaines, S.D., 1981. Latitudinal variation in intertidal algal community structure: the influence of grazing and vegetative propagation. *Oecologia (Berlin)*, 48: 297–307.

Stephenson, T.A. and Stephenson, A., 1972. *Life Between Tidemarks on Rocky Shores.* Freeman, San Francisco, Ca., 425 pp.

Stewart, J.G., 1982. Anchor species and epiphytes in intertidal algal turf. *Pac. Sci.*, 36: 45–59.

Stewart, J.G., 1983. Fluctuations in the quantity of sediments trapped among algal thalli on intertidal rock platforms in southern California. *J. Exp. Mar. Biol. Ecol.*, 73: 205–211.

Stewart, J.G. and Myers, B., 1980. Assemblages of algae and invertebrates in southern California in *Phyllospadix*-dominated intertidal habitats. *Aquat Bot.*, 9: 73–94.

Straughan, D., 1971. *Summary of biological effects of oil*

pollution in the Santa Barbara Channel, Part I. Allan Hancock Foundation, Univ. S. Calif. Publ., pp. 1–11.

Taylor, P.R. and Littler, M.M., 1982. The roles of compensatory mortality, physical disturbance, and substrate retention in the development and organization of a sand-influenced, rocky intertidal community. *Ecology*, 63: 135–146.

Thom, R.M., 1976. *Changes in the intertidal flora of the Southern California mainland.* Thesis, California State University, Long Beach, Ca., 106 pp.

Thom, R.M. and Widdowson, T.B., 1978. A resurvey of E. Yale Dawson's 42 intertidal algal transects on the Southern California mainland after 15 years. *Bull. S. Calif. Acad. Sci.*, 77: 1–13.

Underwood, A.J., 1981. Structure of a rocky intertidal community in New South Wales: patterns of vertical distribution and seasonal changes. *J. Exp. Mar. Biol. Ecol.*, 51: 57–85.

Valentine, J.W. and Lipps, J.H., 1967. Late Cenozoic history of the Southern California Islands. In: R.N. Philbrick (Editor), *Proc. Symp. on the Biology of the California Islands.* Santa Barbara Botanic Garden, Santa Barbara, Ca., pp. 21–35.

Widdowson, T.B., 1971. Changes in the intertidal algal flora of the Los Angeles area since the survey by E. Yale Dawson in 1956–1959. *Bull. S. Calif. Acad. Sci.*, 70: 2–16.

Wilson, O.T., 1925. Some experimental observations of marine algal successions. *Ecology*, 6: 303–311.

ZoBell, C.E., 1971. Drift seaweeds on San Diego County beaches. *Beih. Nova Hedwigia*, 32: 269–314.

Chapter 12

LITTORAL ECOSYSTEMS OF TROPICAL WESTERN AFRICA

D.M. JOHN and G.W. LAWSON

INTRODUCTION

Someone once described the continent of Africa as a pear with a bite taken from it — a bite that extends from above Cape Palmas in Liberia to Moçâmedes in Angola! To continue the metaphor one might say that whoever took that bite must have removed with it a good deal of the marine flora and fauna of the region, for the tropical Atlantic coast of Africa has a very impoverished biota, making it virtually unique among such areas and posing the problem as to how this may have arisen. There is really no comparison, for instance, with the richness of plant and animal life on the tropical coast of East Africa, nor even with that of the shores of the western Atlantic. The western coast of Africa shares with the western shores of other continents, such as the Americas, the absence of coral reefs and consequently the rich and varied life associated with these biogenic structures. Similarly lacking are the extensive meadows of seagrasses that often occur in the lee of fringing reefs. However, it seems that even the absence of such major features is not enough to account for the present distribution of the biota, and other reasons need to be sought.

For the purpose of this chapter the coast of tropical West Africa (Fig. 12.1) is taken as extending from the Western Sahara–Morocco border (that is, just north of the Tropic of Cancer) to the Namibia–South African border (that is, some distance south of the Tropic of Capricorn). It thus covers more than twice the length of tropical coastline bordering what Lawson and John (1982, 1987) have loosely referred to as the Gulf of Guinea (Gambia to the Equator). We have attempted to bring together and synthesize the often incomplete information on the seaweed, seagrass, and animal communities associated with different substratum types along the open coast and, to a lesser extent, in lagoons, estuaries and tidal inlets. Wherever possible we have emphasized functional as well as structural aspects of the littoral communities. Though we have attempted to cover the topic as fully as possible, the emphasis in this chapter is on the findings of research carried out subsequent to an earlier review by Lawson (1966) covering much of the region considered here. Many of the references to the pre-1966 literature are not included as they can be found in that review.

THE PHYSICAL ENVIRONMENT

The coastline

The western coast of Africa is remarkably free of indentations, and this is partly accounted for by the deposition of sediments as sandbars and deltas during and since the last glaciation. Sandy beaches dominate most of the coastline; behind them are sometimes to be found extensive lagoon systems. Such sandy beaches are constantly changing their profiles due to the erosion and build-up of sand. Rocky outcrops occur at intervals (for example, headlands, low wave-cut platforms, boulder areas); some of the most extensive and continuous rocky shores occur in the Cap Vert area where they are volcanic in origin. The northern coast of Namibia and southern Angola (Bia dos Tigres) are particularly poorly endowed with rocky outcrops, and even those present are often exposed only at mean neap tide level (see Kensley and Penrith, 1980). In

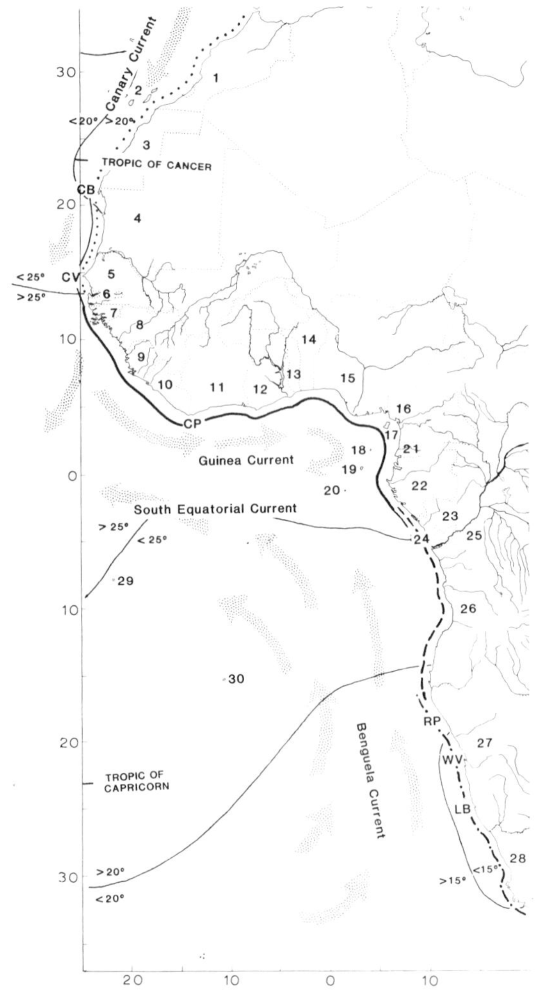

Fig. 12.1. Map covering the area of western Africa treated in this chapter showing main current systems, mean annual surface water temperatures (15, 20 and 25°C isotherms are shown) (after Ekman, 1953), and biogeographical regions (northern warm-temperate zone, dotted line; tropical zone, solid line; transition zone, broken line; southern temperate zone, dashed dotted line). The political entities are as follows: 1. Morocco, 2. Canary Islands, 3. Western Sahara (Spanish Sahara), 4. Mauritania, 5. Senegal, 6. Gambia, 7. Guinea-Bissau, 8. Guinea, 9. Sierra Leone, 10. Liberia, 11. Ivory Coast, 12. Ghana, 13. Togo, 14. Benin, 15. Nigeria, 16. Cameroun, 17. Bioko (formerly Fernando Póo), 18. Príncipé, 19. São Tomé, 20. Pagalu (=Annobón), 21. Equatorial Guinea (formerly Rio Muni), 22. Gabon, 23. Republic of the Congo, 24. Cabinda, 25. Zaïre, 26. Angola, 27. Namibia (South West Africa), 28. South Africa, 29. Ascension Island, 30. St. Helena. Also indicated are Cap Blanc (CB), Cap Vert (CV), Cape Palmas (CP), Rocky Point (RP), Walvis Bay (WB), and Lüderitzbucht (LB).

the northernmost part of the region under consideration, rocky outcrops are few except for the striking cliffs, the platforms, and the boulder areas of ill-consolidated Quaternary marine deposits forming the Cap Blanc peninsula in the north of Mauritania. Often, where cliffs occur, they are severely undercut, and the littoral may consist of a boulder beach formed of rocks fallen from above.

The coast is interrupted in places by river mouths, which often flow through wide estuaries or extensive deltas consisting of mangrove-fringed creeks (such as, for instance, the Niger delta). The larger estuaries and deltas are in the less arid parts of West Africa between Senegal and northern Angola. Mud flats are commonly very extensive in lagoons and in other wave-protected situations, though some estuaries also have rocky shores or banks. Harbour breakwaters or moles, bridge or pier supports, outfall pipes, and the girders of oil rigs, provide the only suitable surfaces for the attachment of benthic organisms along some parts of the coast. Mangrove vegetation is well-represented along the coast north of the Equator (especially from Gambia to the Sierra Leone–Liberian border, Nigeria and Cameroun) and is poorly developed to the south except at the mouth of the Zaïre river. This vegetation type is described elsewhere in this series (Chapman, 1977).

Some of the more extensive lagoon systems are to be found along the coastal region running almost parallel to the Equator. Lagoons associated with large river systems retain a permanent connection with the sea and may be termed "open" lagoons, whereas smaller, seasonally flowing rivers or streams are associated with lagoons cut off from the sea by sand barriers for shorter or longer periods and hence termed "closed" lagoons. A third type with an artificially restricted connection to the sea has been called "a semi-closed" lagoon (Pauly, 1975). The maximum salinity in open lagoons normally corresponds to that of seawater (*c.* 35‰). In contrast, closed or semi-closed lagoons may have salinities ranging from almost nil in the rainy season to more than twice that of seawater in the dry season. Further information on the formation and conditions in these lagoons has been given by Lawson (1966). To the south in Angola and Namibia there are a number of lagoon-like inlets or embayments. Each consists of an arm of the sea running roughly in a N–S or

NE–SW direction, the most westerly extending part of the coastline being a sandbar formed due to longshore drift. One such lagoon-like inlet, lying just to the south of Luanda in Angola, with a relatively narrow channel connecting it to the sea, has been studied by Lawson et al. (1975). Some similar inlets along the Namibian coast are becoming silted up as a result of a combination of the shifting shoreline and the addition of wind-borne sand (see Kensley, 1978). Any obstruction to the longshore drift of sand, whether a rocky headland or a man-made structure such as a harbour breakwater or mole, leads to the accumulation of sand on one side and the erosion of sand from the other (see John and Lawson, 1972).

Knowledge of the shallow seabed along the West African coast is still very incomplete. Most rocky headlands have underwater extensions, and a series of narrow rocky reefs or banks are reported lying immediately offshore between Ivory Coast and Nigeria (see Martin, 1971). Some of these offshore banks are formed of loosely consolidated beach rock, and represent wave-cut platforms formed when the sea was well below its present level. Two such banks have been intensively studied off the Ghanaian coast. In many places within the Gulf of Guinea (including the islands) the seabed is strewn with cobbles or nodules known as rhodoliths. Such areas have been reported off Sierra Leone, Ivory Coast, Ghana, and Togo (see Lieberman et al., 1979, for references). These areas have on occasion been misleadingly called "coral banks". In fact, true reef-building corals are only found in any abundance in a few protected, shallow-water coves (Laborel, 1974). Much of the shallow seabed consists of sand, gravel, or mud, which in places is strewn with shells and shell fragments. There is only very limited information on the soft-bottom communities, much of it derived from samples dredged off the Ghanaian coast and from stations sampled during the Danish *Atlantide* and *Galatea* expeditions to the coasts of tropical West Africa.

Oceanographic features

The northwest African coast is influenced by the cold southerly flowing Canary Current. This forms the eastern limb of the northern gyral circulation (see Fig. 12.1) and brings colder water conditions far into the tropics. In addition, cooling of the surface water is added to by the upwelling off the Mauritanian coast (centred from 20 to 22°N) of colder subsurface water. Though the average sea temperature is less than 18°C, it rapidly increases to the south where a branch of the Canary Current joins with an equatorial current (the Equatorial Counter Current) flowing in the calm doldrum zone lying between the North and South Equatorial currents. Throughout the year relatively warm water (usually >24°C) flows in an eastward direction along much of the Gulf of Guinea coast (the Guinea Current), though from time to time this direction may be temporarily reversed. Seasonal upwelling occurs between Cape Palmas (Liberia) and Cape Formosa (Nigeria), the principal one between June and October; a more limited upwelling is observed between December and March. During the upwelling period the surface water temperature may fall below 20°C. The relatively narrow boundary between the generally warmer so-called "Guinea water" and the colder water to the north is not constant, but moves northwards during the Northern Hemisphere summer and southwards during the winter months. This results in an alternating regime of warmer and colder water conditions along the coast between Cap Blanc and a point southwards of Cap Vert.

The southwest coast of Africa is influenced by the cold, north-flowing Benguela Current, which has been the subject of an extensive review by Shannon (1985). From this review it is apparent that this so-called Benguela Current is better regarded as a complex system under the control of local conditions, rather than merely as a branch of the southern Atlantic gyre. In this Benguela system offshore winds south of 15° are perennially favourable to upwelling. Such upwelling of cold subsurface water is especially evident in the Lüderitz region where it is associated with the narrow continental shelf, but it also occurs at places along the more northerly parts of the Namibian coast.

For the purposes of this chapter it is sufficient to know that to the south of the Equator the boundary between the warmer and colder inshore waters moves southwards during the Northern Hemisphere winter, and retreats northwards during the Southern Hemisphere winter. Thus the 24° sea-surface isotherm moves south from just below the Equator and reaches its southern limit in Angola (between Benguela and Moçâmedes) from January to April. During this period, according to Shannon (1985), warm saline water may reach as far south as Walvis Bay. Furthermore, the strong stratification that develops at this time restricts the amount of upwelling, and the average temperature in the region reaches about 16°C, as against an average of 13°C in the coldest months (August–November).

The coast of tropical Africa is for much of its length subject to more or less continuous wave action. There are few offshore shoals or banks and an absence of coral reefs to afford it protection. Only in a few isolated bays or embayments and, of course, in lagoons, harbours, and estuaries are there conditions that can be described as sheltered. Often the roughest seas with the greatest wave amplitudes coincide with the time of the local rainy season. Since wave height usually exceeds the tidal range (see below), it is an important factor determining the vertical extent of zone-forming organisms. Heavy seas during the rainy season, together with discharge from rivers and lagoons, result in much silting and high turbidity of the inshore water. Large rivers such as the Niger and Zaïre may well affect the turbidity of the water for considerable distances offshore and along the coast. Shore organisms are affected by salinity changes in those parts of the region receiving continuously or seasonally much inflow of freshwater (for example, Cameroun) with sublittoral algae especially influenced by the high turbidity. Salinities below 20‰ have been reported off the coasts at Sierra Leone and Guinea (see Lawson and John, 1982, 1987, for references).

The tides on the western coast of Africa are semi-diurnal, and their range is comparatively small, only exceeding 1.8 m at mean high water springs in a few river estuaries where they may attain 3 m. The two high tides and two low tides in each day often differ in amplitude. Along the Gulf of Guinea coast, for example, the lower low waters generally occur in the day during the Northern Hemisphere winter, but in the night-time during the summer months. This has ecological implications that are discussed under the section on seasonality (p. 317).

The coastal climate

Shore organisms are not influenced only by what might be termed the marine seasons but also by the climate, so both need to be considered together. Three main pressure systems affect western Africa: a subtropical high-pressure mass of air centred over the Sahara, a corresponding high-pressure air mass centred off the west coast of southern Africa, and a tropical low-pressure air mass (the doldrums) lying between the two high-pressure air masses. In this region the coastal climate is dominated by the seasonal displacement of these air masses, especially the northern high-pressure and tropical low-pressure air masses. Between these air masses is a zone of climatic instability (intertropical front, intertropical convergence zone, or monsoon front) separating the more southerly mass of moist equatorial air from the hot, dry, stable subtropical air. This front or zone lies roughly east–west and it moves north and south with the apparent migration of the sun so that its average position is furthest north in August (about 15°N on the coast) and farthest south in January (about 5 to 6°N). For a fuller general introduction to the climate of western Africa, see Griffiths (1972).

Rainfall varies widely in amount and shows seasonal differences along this coastal region, which covers 46 degrees of latitude. Precipitation is often insignificant over the desert regions bounding tropical western Africa, namely the Sahara (Western Sahara, Mauritania) in the north and the Namib (Namibia, southern Angola) in the south. For its latitude, the almost rainless coast of Namibia is very cool, though the environment along it is especially harsh for littoral organisms exposed during low water. To some extent it is moderated here and, to a lesser extent, in southern Angola by the occurrence of prolonged morning mists that are frequent at all times of the year (Walter, 1986). To the north of the Equator the visibility may be seriously impaired by the very hot, dry, and dust-laden Harmattan wind, which may have an effect on shore organisms. This wind, blowing from the north and north-east, comes from the African interior. It affects the coastal regions usually between December and March, reaching as far south as the Bight of Benin.

Southwards and northwards towards the Equator there is generally an increase in both the annual amount of precipitation and the length of the rainy season. The seasonal distribution also varies, with a single peak occurring along the coastal region to the north of Sierra Leone (above 8 to 9°N) and along the coast to the south of the Equator. Along most of the Gulf of Guinea coast there are two annual rainfall peaks. To the north of Libreville in Gabon the rainy season is during the months of the Northern Hemisphere summer, whereas southwards it is less prolonged with February to April often the wettest period. Libreville itself is almost on the Equator (0°27′N) and has a very short dry season (June–August) that occurs when the intertropical zone or front is farthest north. It is between here and Douala (4°04′N) in Cameroun that a dramatic change in the seasonal rainfall distribution takes place. July, which is the driest month at Libreville, is the wettest month of the year at Douala. The actual amount of rainfall, as well as the exact period of the rains, is also influenced by differences in the winds crossing the coast, the presence of coastal currents, and by coastal orientation and interior relief.

The winds produced in the centre of the northern subtropical air mass are the prevailing easterlies or trade winds (or Harmattan wind). In the more southerly tropical air mass the dominant winds are the southwesterlies, and these penetrate well to the north of the Equator following the northward migration of the intertropical zone or front. To the south of the Equator the coastal winds are generally southerly or southwesterly. These winds predominate, though when their strength is weak a diurnal land–sea breeze develops. From southern Angola southwards, warm, dry, easterly winds known as Berg winds sometimes blow from the interior. They cause exceptionally high temperatures and low humidities, and are often dust-laden. The significance of wind direction and strength on water loss from seaweeds on Ghanaian shores is discussed by Jeník and Lawson (1967).

GENERAL BIOGEOGRAPHICAL CONSIDERATIONS

The coastline of western Africa considered here corresponds to the geographical tropical zone, extending a short distance to the north and south of it. However, the distribution of the marine floras and faunas of this coast bears little relationship to mere latitude. This is because the cold Canary and Benguela currents running along the western coast of Africa from the north and south respectively restrict the occurrence of truly warm-water plants and animals to a relatively narrow band — very much narrower than that on the western side of the Atlantic. Even within this narrow band, local upwelling of cold water takes place, and surface water temperatures may fall below 20°C between July and September (see Houghton and Mensah, 1978). Ekman (1953) believes that such low temperatures, together with other factors such as the lack of suitable substrata, have contributed to the absence of coral reefs in the region. By thus bringing about a generally unstable temperature environment the upwelling may also account to some extent for the impoverished flora and fauna mentioned in the Introduction.

The exact northern and southern limits of tropical species in this zone have been set at a number of different places by various authors. There appears to be general agreement, however, that the northern limit as far as animals are concerned should be around or somewhat to the south of the Cap Vert peninsula. There is also general agreement that the southern limit should be at about the border between Angola and Namibia — or, more likely, in northern Namibia, according to Kensley and Penrith (1980). These limits are confirmed as far as marine algae are concerned by a quantitative study of seaweed floras by Lawson (1978). In this study, however, it appears that the Gulf of Guinea flora (Gambia to Gabon) is somewhat different from that of the region stretching from the Congo Republic to southern Angola, which is referred to as a "transition flora" since it is somewhat intermediate between tropical and temperate. Except for Bioko (=Fernando Póo), the other islands of the Gulf of Guinea [Príncipé, São Tomé, Pagalu (=Annóbon)] and the mid-Atlantic island of Ascension also fall within this transition zone and have a similar flora.

North of the Cap Vert peninsula there seems general agreement that the marine flora and fauna of the coastal region are of the warm-temperate type. South of the Angola–Namibia border there is some disagreement between various authors as to whether they are of the warm- or cold-temperate type; this is well discussed by Brown and Jarman (1978). Those who favour the former view (see, for example, Ekman, 1953; Briggs, 1974; Lüning, 1985) point to such features as the absence of certain species or groups supposed to be characteristic of Southern Hemisphere cold-water regions (such as, for instance, certain fish groups and the large brown alga *Durvillaea*) and, more indirectly, to the fact that the common *Laminaria* of the west coast of southern Africa (*L. pallida*) is apparently closely related to *L. ochroleuca*, which is a characteristic species of the northwestern Atlantic warm-temperate flora. The latter view (Hedgpeth, 1957; Stephenson and Stephenson, 1972) is mainly based on the fact that statistically there is a much greater preponderance of cold-water forms over warm-water forms in the area in question, and this applies more to algae than to animals. Whichever view is taken, it is clear that a major change does occur somewhere between southern Angola and northern Namibia. Furthermore, this change (certainly as far as the algae are concerned) is a more profound one than that which takes place at the northern limit of the tropical zone, where a gradual transition is more in evidence (Lawson, 1978).

With regard to the island groups lying off the coast of western Africa the Gulf of Guinea group and Ascension Island have already been mentioned, though it should be added that some authors [for example, Briggs (1974) on the basis of the fish fauna] believe that Ascension Island has more affinities with the American side of the Atlantic than the African. The Salvages, the Islas Canarias and St. Helena clearly are warm-temperate but the Cape Verde Islands are more controversial. Briggs (1974), on the basis of the fauna, regarded them as tropical, as also did Lüning (1985), whereas Lawson (1978), on the basis of algae, included them in the warm-temperate zone.

COMMUNITY STRUCTURE

Open rocky shores

The large majority of countries along the Atlantic seaboard of tropical western Africa have received some attention from shore ecologists. Some treatments have been more detailed than others, but only a few have extended downwards to include benthic associations on rocky areas in the shallow sublittoral. Most descriptions of littoral zonation are concerned with more or less continuous areas of rock rather than boulder shores. The general pattern of zonation described below refers to that found on the former type of shore. It is described using the scheme and terminology proposed by Lewis (1964).

The most important factor modifying littoral zonation along the western coast of Africa is wave action. This influences the width and vertical extent of the zones or belts of shore organisms, telescoping them in wave-sheltered situations and extending them vertically on wave-exposed shores, where the limits of zone-forming organisms are raised sometimes well above the normal tidal range (see Fig. 12.2). The composition of the zones is also altered by wave action, which is a very complicated factor varying with the slope and aspect of the rock surface (for example, see Lawson, 1957). With some variations, the general pattern of zonation shown in Fig. 12.2 is repeated along most parts of the western coast of Africa lying between Gambia in the north and Angola in the south. In the coastal regions to the north of Mauritania and to the south of Angola, colder-water organisms replace some of the tropical forms in the different zones.

Littoral zonation

Littoral fringe. This zone is influenced by sea spray and wave splash, and extends downwards from the first appearance of small gastropods (littorinids) to the upper limit of barnacles in quantity. Despite often being subject to wide variations in wave action, it is the zone on the shore showing the most constant species composition. Of the gastropods found the most characteristic is *Littorina punctata*, though it might on occasion have been mistaken for the possibly recently introduced *L. meleagris* (see Rosewater and Vermeij, 1972). *Littorina punctata* has been recorded as far north as southern Spain; its southern limit is somewhat problematical. According to Penrith and Kensley (1970a, b) and Kensley and Penrith (1980) it occurs all along the west coast of southern Africa, but Stephenson and Stephenson (1972) reported that the "supralittoral" fringe of the west coast of South Africa is inhabited by *Littorina knysnaensis*

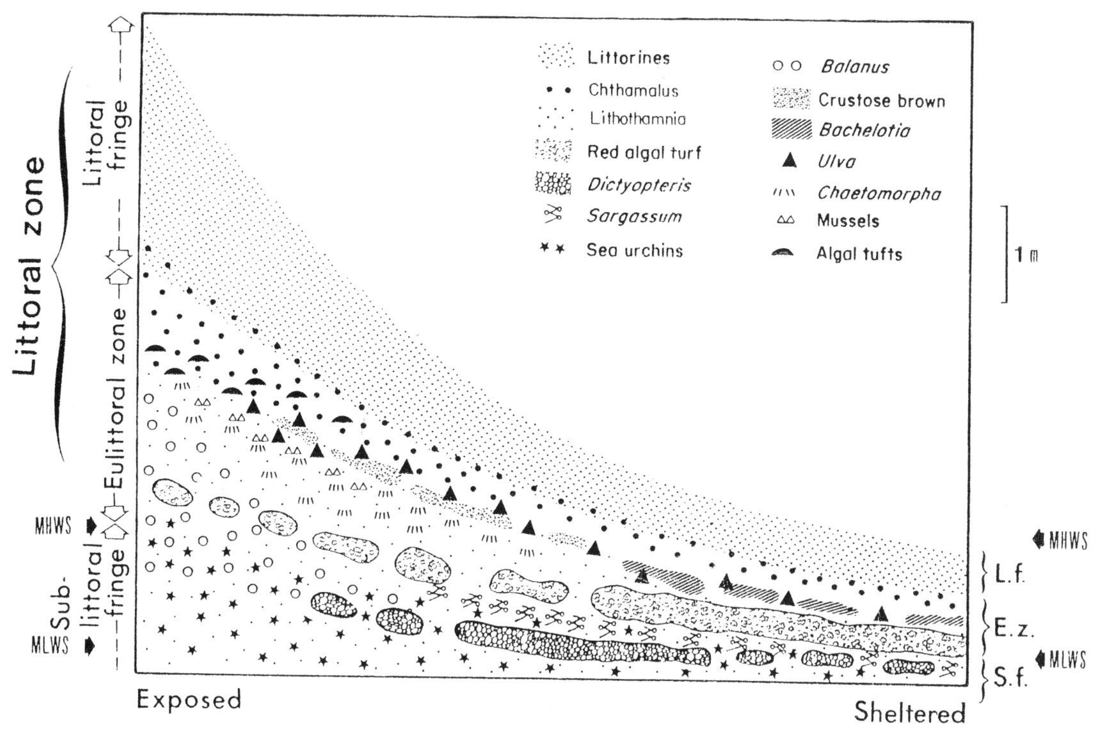

Fig. 12.2. Diagram showing the zones occupied by some of the more important littoral organisms on rocky shores in western Africa and their modification in relation to wave exposure. (After Lawson and John, 1982.)

and no other species of that genus. Rosewater (1981, p. 26) reported that, according to Kilburn, the southern limit of *L. punctata* is Rocky Point in northern Namibia, but that possible hybrids between *L. punctata* and *L. knysnaensis* may occur further south. A number of other littorinids are widely distributed on the littoral fringe of West African shores. *Littorina scabra angulifera*, for example, seems to be common on the shores of northern Gabon (John and Lawson, 1974b). Again *Littorina granosa* [= *Tectarius granosus*; not *Nodilittorina miliaris* as reported by Lawson and John (1982), see remarks by Rosewater (1981, p. 32)] extends from Cap Vert along the Gulf of Guinea coast to Gabon. On wave-exposed shores it occupies a somewhat higher level than *L. punctata* when the two occur together. Another secondary dominant at this level between Guinea in the north and northern Angola is *Littorina cingulifera*, especially in coastal areas influenced by freshwater. A colder-water *Littorina* (*L. neritoides*) is found on the Cap Blanc peninsula and extends northwards to the English and Norwegian coasts (see Rosewater, 1981).

The littoral fringe is comparatively barren of algae though crusts, cushion-like forms, or low mats are sometimes present and, like the littorinids in the uppermost part of this zone, are commonly confined to shaded cracks and crevices. These algal growths may be found in the open when rocks are shaded by trees or overhangs. The most common cushion or mat-forming algae are those forming the so-called *Bostrychia–Murrayella–Lophosiphonia* association, though this is not reported from Namibia or from the Cap Blanc peninsula. Its absence from the Cap Blanc peninsula has been attributed to the fact that the prevailing climatic conditions cause especially severe desiccation of algae growing in the higher zones during low water (see Lawson and John, 1977). Southwards from Cap Blanc blackish crusts of blue-green algae (mainly *Entophysalis deusta*) are found provided there is sufficient spray or shade to allow them to survive. In Angola such blue-green algal growths are often accompanied by patches of a black marine lichen and, in more southerly parts of that country, by a greenish layer tentatively referred to the alga *Epicladia* (= *Entocladia*) *testarum* (see

Lawson et al., 1975). To the immediate north of the tropics, the littoral fringe of Morocco is dominated by the black lichens *Verrucaria maura* and *Lichina pygmaea* (see Gayral, 1958).

Eulittoral zone. This zone (corresponding to the intertidal area of the shore) may, for convenience, be divided in most parts of western Africa into two subzones: an upper subzone characterized by barnacles, and a lower one characterized by the crustose calcareous red algae often collectively termed "lithothamnia". The upper subzone becomes more or less completely dried out during low water, whilst the lower one hardly ever dries out completely due to the almost continuous splash and spray encountered in all but the most wave-sheltered localities and, perhaps, to the water-holding capacity of the algal turf where this is well developed.

(a) *Barnacle subzone*. This upper subzone is so named because normally it is dominated by a balanoid belonging to the genus *Chthamalus*. On the Cap Vert peninsula and to the north of it the species is *C. stellatus* (see Sourie, 1954). To the south it becomes replaced by *C. dentatus*, and this is reported even at the southernmost site investigated by Penrith and Kensley (1970b) in Namibia (Lüderitzbucht), though it was absent from all but the most sheltered shores studied. They believed that its restriction to such shores was due to the higher water temperatures found within the sheltered bay as compared to the open coast. Most shores in the Lüderitzbucht area are dominated at this level by the limpet *Patella granularis*, a species endemic to southern Africa. Zonation on the shores of southern Namibia is essentially similar to that of the remainder of the west coast of southern Africa southwards to the Cape (see Penrith and Kensley, 1970b; Stephenson and Stephenson, 1972).

The oyster *Saccostrea* (= *Ostrea*, = *Crassostrea*) *cucullata* is reported from time to time as forming a belt in the lower part of this subzone on fairly flat rocks subject to moderate wave-exposure (see Gauld and Buchanan, 1959) and on more sheltered shores including the landward side of harbour breakwaters (John and Pople, 1973). On open shores oysters are usually confined to shallow pools, and so should not perhaps be regarded as a characteristic feature of rocky shores in the region (see Lawson, 1966). Though widespread in the region, this oyster is not reported further south than Angola (see Kensley and Penrith, 1973). Other common shore animals present at this level include species of the limpet *Siphonaria* and the gastropod *Nerita atrata* (= *N. senegalensis*); the latter is reported on shores from Senegal to Angola. *Nerita* is usually absent from wave-sheltered rocks, and in such conditions *Siphonaria* may largely replace *Chthamalus* (see Gauld and Buchanan, 1959). The genus *Siphonaria* is distributed throughout West Africa: *Siphonaria pectinata* is known from along the Gulf of Guinea coast, the Cap Vert peninsula (sometimes as *S. grisea*), and Angola (Lawson et al., 1975), whilst *S. capensis* is reported from farther south and has its northernmost limit at Rocky Point in Namibia (see Kensley and Penrith, 1980). Sometimes *Siphonaria* and *Nerita* are found higher on the shore in the littoral fringe (see Lawson and John, 1977); they extend downwards into the lower eulittoral or, in Sierra Leone, still further into the sublittoral fringe (Longhurst, 1958).

In general this subzone is animal-dominated and algae are usually not well represented, though occasionally extensive crusts or felt-like mats of blue-green algae may occur. Sometimes such algae as *Gelidium pusillum*, clumps of *Enteromorpha* and *Ulva* species, ectocarpoids, *Sphacelaria*, and tufts of *Chaetomorpha antennina* are found confined to cracks or crevices or on open rock surfaces in more wave-exposed situations. These algae are best developed in the lower part of the subzone, where the barnacles are still present but in lower numbers. One feature common to many shores from Cap Vert southwards along the Gulf of Guinea coast is a distinct belt of crustose brown algae at the junction of this subzone and the one below. In many places this belt is dominated by the regional endemic *Basispora africana* (see John and Lawson, 1974a). Often it is accompanied by another member of the Ralfsiaceae, namely *Ralfsia expansa*. In this position on the Angolan coast, on all but the most wave-sheltered shores, occurs a band of the red alga *Caulacanthus ustulatus* (see Lawson et al., 1975). Such a belt is also reported on the coast of Gambia (see John and Lawson, 1977a), though it is more pronounced along the Cap Vert (Sourie, 1954) and Cap Blanc peninsulas

(Sourie, 1954; Lawson and John, 1977). This belt is very well developed in moderately wave-exposed areas, and with increased shelter *Caulacanthus* often becomes just one of the components of the algal turf of the lower eulittoral subzone. To the north along the Moroccan coast (see Gayral, 1958) and on the Islas Canarias (see Lawson and Norton, 1971), it forms a distinct band at this level on the shore. A belt of *Porphyra* (probably *P. capensis*) is found on rocky shores in southern Angola (Lawson et al., 1975) and Namibia (Penrith and Kensley, 1970a, b) at this level, as well as in the littoral fringe above. With the exception of the report of *P. ledermannii* from Cameroun [see comments by Lawson and John (1982, p. 164)] this genus is unknown from the Gulf of Guinea. Sourie (1954) reported it in this subzone on the Cap Vert peninsula, and Lawson and John (1977) from more wave-exposed situations on the Cap Blanc peninsula:

(b) *The "lithothamnia" subzone*. This subzone is generally algal-dominated, with a wide range of growth forms present especially in more wave-sheltered localities. On wave-exposed rocks encrusting red algae ("lithothamnia") are the principal ground-cover species, and on occasion are accompanied by brown algae having a similar growth form (such as *Basispora africana* and *Ralfsia expansa*). In such situations other algae are usually diminutive, forming only low cushion-like clumps, ring-like patches, or low mats. The low-growing forms include species belonging to such widely distributed red-algal genera as *Gelidium*, *Herposiphonia*, *Laurencia*, *Polysiphonia*, *Taenioma* and *Wrangelia*. Conspicuous on most wave-exposed shores are the green tufts of *Chaetomorpha* and, less commonly, the membranaceous fronds of *Ulva*; these genera are found in this subzone and often in the one above. With increasing shelter, encrusting forms are less evident as most of the rock surface becomes overgrown by an almost closed algal turf made up of several species (for example, *Centroceras clavulatum*, *Ceramium* spp., *Gigartina acicularis*, *Herposiphonia* spp., *Hypnea* spp.) as well as clumps of coarser algae including both fleshy types and articulated corallines. Apart from the molluscs mentioned below the algal vegetation at this level offers shelter to many small cryptic animals, which are listed in some shore surveys (Penrith and Kensley, 1970a, p. 242). In southern Angola the coarse clumps of *Gelidium versicolor* form almost a belt at this level on the shore in somewhat wave-exposed situations. This large red alga is absent from the Gulf of Guinea, and to the north in Morocco the subzone is characterized either by another *Gelidium* (*G. spinulosum*) along with other red algae (such as *Gigartina acicularis*, *Laurencia pinnatifida*), or by the northerly distributed brown alga *Bifurcaria bifurcata*. In northern Namibia the subzone typically includes the limpet *Patella argenvillei*, and several large red algae such as species of *Aeodes*, *Chaetangium* and *Gigartina*.

Animals tend to dominate this subzone on wave-exposed shores where algae are poorly developed, except for "lithothamnia". These include the large mussel *Perna perna* (=*Mytilus perna*), limpets [mainly *Patella safiana* and *Fissurella* spp. (*F. coarctata* and *F. nabecula*: Gulf of Guinea; *F. mutabilis*: southern Namibia)], and carnivorous snails belonging to the genus *Thais* (mainly *T. nodosa*, *T. haemastoma*, *T. forbesi*). Most of these molluscs occur along the Gulf of Guinea, though colder-water forms often replace them to the north and south. For instance, *Patella safiana*, *Perna perna* and *Thais haemastoma* disappear along the Namibian coast and are replaced by other molluscs such as the large limpet *Patella argenvillei* and, on exposed shores, the mussel *Aulacomya magellanica* (see Penrith and Kensley, 1970b). Indeed, the zonation on the shore at Lüderitzbucht in southern Namibia closely resembles that described by Stephenson and Stephenson (1972) for the west coast of South Africa. Where wave action is considerable, the large barnacle *Balanus tintinnabulum* is often present in great abundance, except on Namibian shores, that are outside its southerly limit of distribution.

Other inhabitants of this subzone are often confined to wave-sheltered shores. These include the small mussel *Brachidontes puniceus*, which is reported from Senegal southwards to Gabon, especially on shores subject to brackish-water influence. Another local variant found in some wave-sheltered situations is the presence in some abundance of the fused calcareous tubes of *Vermetus*, which are sometimes partly covered by "lithothamnia". These tubes are reported to harbour a characteristic endofauna (see Sourie, 1954;

Gauld and Buchanan, 1959). In southern Namibia and on the west coast of South Africa, areas of rock at this level are occasionally covered by the coarse sandy tubes of *Gunnarea capensis*. These tubes form a narrow band or ridge between the *Patella* belts in more wave-exposed situations and in more protected sites, fuse to form a continuous mass (see Penrith and Kensley, 1970b; Stephenson and Stephenson, 1972). An interesting feature of wave-exposed shores along the Angola coast is the division of this "lithothamnia"-dominated subzone into an upper band of ridged and tortuous-surfaced lithothamnia and a lower one with the surface more or less smooth. Sometimes the upper band is yellowish due to blue-green algae growing over it, whilst the lower one remains pink. Ranging over the entire littoral in exposed situations is the crab *Grapsus grapsus*, which is common throughout the region except to the south of Rocky Point in Namibia (see Penrith and Kensley, 1970a). In more sheltered and landward positions a slimy greyish or often brownish-green zooanthid community (e.g. species of *Palythoa*) sometimes occupies this subzone or extends just above or below it.

Sublittoral fringe. Lawson (1966) has discussed at some length the difficulty of defining the junction of the sublittoral proper and the eulittoral zone along the Gulf of Guinea coast and northwards to the Cap Blanc peninsula. Though the upper limit of the brown algae *Dictyopteris delicatula* and *Sargassum vulgare* could in some places be used to define this fringe, a more constant feature of it is probably the upper limit of the sea urchin *Echinometra lucunter*. This forms a relatively narrow band over a wide range of wave-exposure, and extends southwards to southern Angola where it is the only organism (*Sargassum* being usually absent) defining this fringe (see Lawson et al., 1975). Sometimes along the Gulf of Guinea coast it is accompanied by other sea urchins, including the less conspicuous *Arbacia lixula*, and along harbour breakwaters by *Diadema antillarum*. Often *Echinometra* occupies deep depressions on very hard rocks such as granite. Lawson (1966) has questioned whether it is capable of making such holes on rock so hard, though its rock-boring activities are well-known in softer beachrocks in the West Indies (see Grunbaum et al., 1978).

To the north of Gambia and to the south of Angola brown algae other than *Dictyopteris* and *Sargassum* may be used in defining the sublittoral fringe, though generally they tend to occupy the sublittoral proper. Here the fringe is often occupied by upward extensions of populations of such "kelps" as *Ecklonia*, *Laminaria* and, possibly, *Saccorhiza* (northern coast of West Africa only). On the Cap Vert peninsula the *Sargassum* growing in this fringe is often accompanied by *Cystoseira* and *Ecklonia* (mainly *E. muratii*) in moderately wave-exposed situations (Sourie, 1954); these two latter brown algae are here at their southernmost limit along the north coast of western Africa (see Price et al., 1978). *Ecklonia* has only relatively recently been found growing on the western side of the Cap Blanc peninsula (Lawson and John, 1977) whereas only further north in Morocco do *Laminaria ochroleuca* and *Saccorhiza polyschides*, especially the former, form extensive populations at the low-water level and below (see Gayral, 1958). On the southern coast of western Africa (Namibia, South Africa) the fringe has been defined as the upper limit of another species of *Ecklonia* (*E. maxima* = *E. buccinalis*), *Laminaria pallida* and *Macrocystis pyrifera* (see, for example, Stephenson and Stephenson, 1972). According to Stephenson and Stephenson (1972) the kelp forest characterizing the "infralittoral fringe" (= sublittoral fringe) along this part of the African coast is inhabited by many species found in the zone above (*Patella argenvillei–P. cochlear* zone). *Ecklonia maxima* and *Laminaria pallida* extend northwards along this coast at least as far as Lüderitz, but at Rocky Point in northern Namibia the only kelp reported is a *Laminaria* (Penrith and Kensley, 1970a), though isolated fronds are washed ashore on beaches between False Cape Frio and the Kunene river (Kensley and Penrith, 1980). Kelp holdfasts, associated tufts of other algae, the undersides of rocks, and the spaces in *Gunnarea* tube masses provide shelter for a considerable cryptofauna [see list by Penrith and Kensley (1970b) for Lüderitzbucht].

Tide-pools

The conditions in tide-pools vary during the course of the tidal cycle, especially in small pools at higher levels on the shore. Those in the littoral fringe are not replenished with fresh seawater by the rise and fall of the tides, but are dependent for their existence on salt spray, wave splash, and rainfall. Such pools have been studied along the Gulf of Guinea coast and on the Cap Blanc peninsula (see Lawson and John, 1977, 1982). They are commonly lined by a mat of blue-green algae, especially if unshaded and contaminated (for example, by bird droppings). Often the algal mat is spongy, and detaches to float as surface flocks. The water is sometimes a turbid green due to the presence of large populations of microscopic algae. These pools and those in the upper part of the barnacle subzone often have a sparse fauna consisting of juvenile littorinids, and the shells of *Nerita* occupied by hermit crabs.

Rocky pools at somewhat lower levels on shores in the Gulf of Guinea are often lined by "lithothamnia", and these are commonly accompanied by the encrusting brown alga *Ralfsia expansa* and, less frequently, the prostrate form of *Lobophora variegata*. In such pools investigated in Ghana the "lithothamnia" crust may be elevated along the poolside as branched and anastomosing structures that form small shelf-like platforms level with the water surface. These crusts contain a burrowing and cryptozoic fauna including the date mussel *Lithophaga aristata*, *Pedalion anomioides*, and *Pachygrapsus* species (see Gauld and Buchanan, 1959). Such relatively wave-sheltered rocky pools also sometimes contain numerous colonies of the coral *Siderastrea radians*. Large pools in the "lithothamnia" subzone have a much richer fauna and flora than those at higher levels. Some rocky pools here are carpeted by green and brown zooanthids; those on the Cap Vert peninsula have been described in some detail by Sourie (1954). He studied large pools containing a variety of surfaces including boulders. Some of the pools were carpeted by species of the siphonaceous alga *Caulerpa* and by articulated corallines.

Deep pools, often with a restricted connection to the sea, may contain many algae normally found only in the shallow sublittoral. For example, such pools along the Gulf of Guinea coast and the Cap Vert peninsula usually contain dense populations of *Dictyopteris delicatula*, *Sargassum vulgare*, and species of *Dictyota*, *Galaxaura* and *Padina*. Those on the Cap Vert peninsula also contain in addition *Cystoseira senegalensis* and *Ecklonia muratii*. The lower littoral pools examined by Penrith and Kensley (1970b) much further to the south in Namibia (at Lüderitzbucht) also contained *Laminaria* and the southerly-distributed mussel *Choromytilus meridionalis*, which formed dense beds in these pools and in nearby channels. Sometimes they contained masses of the sandy tubes of *Gunnarea capensis*, and those pools containing loose rocks often had a dense growth of the colder-water green alga *Ulva atroviridis*. Many of the faunistic and floristic components mentioned by Penrith and Kensley are not found in northern Namibia, and some are southern African endemics.

The largely algal- and detrital-feeding fishes *Blennius cristatus* and *Ophioblennius atlanticus* are common in tide-pools along the Ghanaian coast. In pools studied in Namibia they are often accompanied by other blennies (such as, for example, *Blennius cornutus*, *B. fascigula*) seemingly endemic to southern Africa (see Penrith and Penrith, 1972). Observations on Ghanaian tide-pools show that adult *Ophioblennius* exhibit territoriality, defending an area in front of a hole, crevice, or depression from other fishes (see Sanusi, 1980). Such pools also contain the territorial damselfish *Microspathodon frontatus*, another damselfish *Abudefduf hamyi* (only juveniles), the sergeant-major fish *Abudefduf saxatilis*, and the parrotfish *Pseudoscarus hoefleri*. All these fishes are to a greater or lesser extent algivorous (Table 12.1), though they appear to take in incidentally or intentionally a wide variety of other food items. They probably account for the absence or diminutive growth of most algae other than crustose or calcareous forms in pools where these fish occur in large numbers. Generally their numbers are greatest in large pools having much surface relief (for example, boulder-lined). Often in such pools the only non-calcarous algae present are blue-greens, clumps of the red alga *Hypnea cervicornis*, or small and often creeping red algae confined to the territories of the very pugnaceous damselfish *Microspathodon*. Much of the algal material found in the stomachs of the fishes in these pools is of plants found growing on the open rocks or in other pools where there is little physical relief. This suggests that, when the pools are inundated by the tide, fish leave them to feed or else make use of drift plants. Other pools, mostly devoid of algae other than encrusting forms, possess large populations of *Echinometra lucunter*. These relatively

TABLE 12.1

Stomach contents of demersal fishes sampled from five habitats in Ghana. (Data from Sanusi, 1980.)

Frequency of groups (%)	Species[a]												
	1	*2*	*3*	*4*	*5*	*6*	*7*			*8*			
	TPE[b]	TPL	TPE	TPL	TPE	TPE	TPL	MB	TH	OR	TPE	MB	TH
Algae													
blue-greens	64 (4)[c]	91 (4)	100 (4)	65 (2)	69 (2)	65 (3)	78 (2)	–	50 (2)	50 (3)	74 (4)	56 (3)	75 (2)
diatoms	93 (2)	27 (1)	31 (1)	86 (2)	77 (2)	52.5 (3)	–	47 (1)	66 (1)	37.5 (1)	47 (3)	56 (2)	–
reds/browns/greens	100 (4)	100 (4)	100 (4)	100 (4)	100 (4)	100 (4)	80 (2)	100 (4)	100 (4)	100 (4)	100 (3)	100 (3)	100 (2)
Animals													
crustaceans	100 (4)	36 (1)	81 (1)	65 (1)	77 (3)	–	–	41 (1)	83 (4)	44 (4)	34 (3)	74 (2)	65 (3)
molluscs	9 (1)	9 (1)	56 (1)	17 (3)	47 (1)	1.5 (1)	–	17 (1)	50 (4)	43 (2)	–	26 (2)	–
bryozoans	–	–		12 (1)	45 (2)	–	–	–	41.5 (3)	44 (3)	–	60 (2)	–
No. of species of reds/browns/greens													
frequency > 50%	4	5	12	3	7	7	2	4	3	5	10	4	2
total no. of species	20	16	32	32	8	13	4	19	24	38	24	17	2

[a] *1* = *Microspathodon frontatus*; *2* = *Abudefduf saxatilis*; *3* = *Abudefduf hamyi*; *4* = *Ophioblennius atlanticus*; *5* = *Blennius cristatus*; *6* = *Pseudoscarus hoefleri*; *7* = *Stegastes imbricatus*; *8* = *Echinometra lucunter*.

[b] Key to habitats: TPL = tide-pool, littoral fringe; TPE = tide-pool, upper eulittoral sub-zone; MB = Miemia Bay; TH = Tema Harbour; OR = Offshore reef.

[c] Numbers in parentheses refer to maximum abundance: subjective scale from (1) — just present to (4) — very common.

sedentary animals also possess a variety of different algae in their stomachs (see Table 12.1) and it has been suggested that they might capture detached plants carried into the pools.

Harbour breakwaters and boulder shores

The breakwaters or moles of many harbours in western Africa show some resemblance to natural shores, as they consist of either large rough stone blocks or concrete tetrapods. Usually the pattern of zonation of shore organisms on these surfaces is somewhat akin to that found on nearby rocks subject to similar amounts of wave action. Comparisons have been made between the zonation patterns on harbour breakwaters along a stretch of the Gulf of Guinea (Benin to Ivory Coast) and any nearby rocky shores (see John and Pople, 1973).

Harbour breakwaters normally have an outer exposed wall and an inner very sheltered one. It might be expected, therefore, that the pattern of zonation on the outer walls would be similar to that found on natural rocky shores exposed to strong wave action and the pattern on the inner walls would approximate to the zonation on natural rocks in sheltered situations. This is indeed generally true, but local conditions with regard to such factors as the quantity of freshwater flowing into the harbour and the amount of pollution taking place mean that there is often considerable variation in zonation pattern, especially on the inner sheltered sides of such breakwaters. One way in which the outer parts of breakwaters may differ from natural beaches is that their slope below the low water mark may be much steeper. Again, the fact that harbour breakwaters in western Africa are usually constructed from loose, irregularly shaped rocks or concrete tetrapods means that there are many crevices to provide protection from predators for fish and other animals. Thus, there are often large numbers of demersal fish in the immediate sublittoral, including the black territorial damselfish *Stegastes imbricatus* and *Microspathodon frontatus*. Other fish commonly found are *Abudefduf saxatilis*, the surgeonfish *Acanthurus monroviae*, and *Ophioblennius atlanticus*. The presence in quantity of such fish, often mainly algivorous (see Table 12.1), means that the larger and more fleshy algae may be much reduced in the intertidal zone, leaving the largely inedible encrusting "lithothamnia" to dominate especially in wave-exposed areas. Only in the shallow sublittoral zone may be found clumps of algae such as *Hypnea cervicornis* and articulated corallines along with the "lithothamnia".

In the very sheltered inner harbour basins most littoral surfaces become covered by oil and fine sediment over which mats of blue-green algae frequently grow. Littorinids and the barnacle *Chthamalus dentatus*, constant features of rocky shores, may be absent here. *Balanus amphitrite* sometimes occupies the level normally taken up by the smaller *Chthamalus*. In moderately wave-sheltered parts of breakwater systems the oyster *Saccostrea cucullata* commonly forms a band extending from the littoral fringe into the shallow sublittoral. The sublittoral zone is here well defined by the presence of *Echinometra lucunter* and the longer-spined *Diadema antillarum*, the latter being confined to crevices and overhangs. It is interesting to note that *Diadema* exists only at much deeper levels (*c.* 10 m) on naturally occurring rocks.

The zonation of littoral organisms is less evident and less easy to discern on boulder shores than along harbour breakwaters. Such shores consisting of boulders with sand and gravel between them have been studied in Ghana by Bassindale (1961), Buchanan (1954), and Gauld and Buchanan (1959), and along the Cap Vert peninsula by Sourie (1954). If the boulders are unstable, littoral organisms are poorly represented, and even when they are stable the diversity of algae is low, and the animals tend to be especially abundant on shaded undersurfaces. The fauna and flora on shaded undersurfaces of boulders is, not unexpectedly, similar to that on harbour breakwaters or more continuous rocky surfaces but, in addition, there is a rich and varied "understone" fauna. Gauld and Buchanan (1959) recognized a distinct zonation of this understone fauna in some relatively sheltered boulder shores in Ghana, though this was often complicated by the boulders having sand or gravel between them and because they were sometimes lying in shallow pools. On the higher parts of the shore decapods were common (for instance, *Petrolisthes armata*, *Xanthodius* (= *Xantho*) *inaequalis* and the hermit crab *Pseudopagurus granulimanus*) as well as the gastropod *Nerita*. Other belts, including those of the gastropod *Planaxis lineatus* and of small hermit crabs and xanthids, were also described. Finally, at the lowest level of the shore various echinoderms and encrusting sponges seemed to be a ubiquitous feature of the understone fauna. Sourie (1954) recognized in Senegal various "mesobioses". One, where shading was only partial, was characterized by the red algae *Cryptonemia seminervis* and *Rhodymenia pseudopalmata*. Others beneath overhangs and under stones, where the algae had completely disappeared, were dominated by sponges and ascidians. Sourie also investigated the endolithic species associated with different surfaces (such as laterite, sandstone) including those in *Vermetus* tubes. The vertical distribution of these endoliths was compared to the distribution of plants and animals growing on the surfaces.

Little is known of the changes in the biota that take place in the immediate sublittoral zone. In a boulder-lined bay in western Ghana (Miemia Bay) the sublittoral fringe is defined by *Echinometra lucunter* (D.M. John, pers. observ.). This shallow bay has large populations of mainly algivorous fishes similar to those found along the moderately wave-sheltered sides of harbour breakwaters (see above). The main contrast with such artificial habitats is the absence of *Diadema antillarum* from the bay.

Smaller rocks in the lee of boulders, or protected by more continuous areas of littoral rock, are relatively stable and have a well-developed and rich cryptofauna. Usually on the undersides of these rocks are colonies of bryozoans. A detailed survey of bryozoans associated with this and other habitats in Ghana has been undertaken by Cook (1985). The major competitors of bryozoans growing on small rocks are sponges and foraminiferans, and "lithothamnia" also occur there. Such stones or small boulders may overlie mud, sand, or gravel. Gauld and Buchanan (1959) reported in Ghana large numbers of the polychaete *Eurythoe complanata* in a small bay where the stones lay on a mud–sand substratum. Animals associated with unstable sublittoral areas will be considered later. One of the most detailed investigations of the cryptofauna associated with loose stony bottoms was undertaken by Penrith and Kensley (1970b) in sheltered localities in southern Namibia. This fauna included a diversified community of chitons, small periwinkles, starfish and brittle-stars, pycnogonids, and a variety of polychaetes. Many of these animals have a southern African distribution, and are not found elsewhere in western Africa.

Open sandy beaches

Though sandy beaches dominate most of the coastline of tropical West Africa they have, partly because of their barren appearance, been more neglected by shore ecologists than the less extensive but more productive rocky areas. Littoral sandy areas, open or subject to tidal influences, are considered here whilst sand flats (including muddy areas) in estuaries or lagoons are discussed in a later section (p. 312). The principal accounts of the fauna of such unstable shores are those by Gauld and Buchanan (1956) for Ghana, Longhurst (1958) for Sierra Leone, McLachlan (1985) for Namibia, and Sourie (1957) for Senegal. In this last investigation an attempt was made to measure some of the possible factors affecting beach faunas. Special attention was given to the question of salinity, and the properties of sediments were also investigated with an attempt to correlate these with the distribution of the beach fauna. Except for the paper on Namibia (McLachlan, 1985) the studies mentioned above were discussed at some length by Lawson (1966).

West African beaches vary in form from very steep (10° slope) and narrow, to gently sloping (2–3° slope) and broader. In some places they are backed by high rocky cliffs, whilst elsewhere they may give way to low cliffs, earth, or dunes. Gauld and Buchanan (1959) recognized two types of sandy beach in Ghana: the narrow type dropping down steeply (1:6 slope) to low-water mark where a trench about 1 m deep and 5 m wide separates it from an almost flat bank about 40 m wide lying approximately at the low-water mark (very coarse sand, roughly stratified), and wider and less steep (1:20 slope) beaches with no trench or bank at the level of the low water (finer sand, unstratified). This latter type is similar to a long uninterrupted beach studied by Sourie (1957) to the north of the Cap Vert peninsula. The sand of this beach was light brown in colour, of medium particle size, and contained shell debris mostly of *Donax*. Small beaches on the rocky Cape itself are very variable, with the sand often greyish brown, heterogeneous in texture, and again often containing considerable amounts of shell debris. To the south the beaches all contain some grey sand though in other respects they are very different due to the varying degrees of wave action to which they are subject. For instance, one sheltered bay (Hann Bay) just to the north of Dakar was found by Sourie to contain coarser white and finer grey sand. The grey sand was coloured by organic matter, smelt of hydrogen sulphide, and was light yellow in colour when oxidized. On this and other beaches to the south the proportion of white sand varies from place to place, the white sand being predominant in the upper, often more steeply sloping, beach. Some beaches (e.g. Bargny) have a zone between the white and grey sand of very coarse particles accompanied by shell debris. It is of interest to note that these observations contrast with those of McLachlan (1985) for Namibia, where the coarsest particles are found near low water and the finest higher up the beach. Again, McLachlan apparently found that exposed beaches have a flatter profile than more wave-sheltered beaches, whereas the opposite holds for the Gulf of Guinea.

The narrow and steep beaches studied in Ghana possess a very impoverished fauna with the ghost crab *Ocypoda cursor* at the uppermost part of the shore and a sparse population of *Donax rugosus* near the low-water mark. Another more terrestrial ghost crab, *O. africana*, occurs in some places. Such beaches have a low water-retaining capacity due to the larger particle size of the sand. They therefore represent an inhospitable environment for burrowing animals during low water, when the interstitial solution may become concentrated by evaporation. For instance, Sourie (1957) has found that finer grey sands have a considerably greater (3 to 5 times) water-holding capacity than coarser sands which remain humid long after coarser sands have dried out. Steep sandy beaches are probably the norm along the wave-exposed coastline of western Africa rather than those described immediately to the south of Cap Vert, where they are relatively sheltered due to very shallow offshore areas.

Gently sloping beaches have a predominance of finer sand and a diversified macrofauna. Sourie (1957) found, with the notable exception of *Nerine cirratulus*, that many beach animals in Senegal may be restricted either to beaches with a predominance of white sand or to those with a predominance of finer sand. Some of the most commonly recorded beach animals have a zoned distribution, and Sourie believed this to be due mainly to the desiccation they face during low water. He did not think that salinity changes in themselves were the principal factors responsible for the zonation observed.

It is difficult to generalize concerning the pattern of zonation, though the northern Cap Vert beaches seem fairly similar in structure and faunistic composition to the broader and more gently sloping ones studied in Ghana (see Lawson, 1966, p. 433). Several differences are evident, including the fact that the most common polychaete (*Nerine cirratulus*) forms a zone in the lower half of the beach in Senegal rather than occurring in the ghost crab zone as in Ghana. In Senegal another *Nerine*, *N. agilis*, is often found in the upper half of the beach (above a *N. cirratulus* zone), or sometimes occupies the beach from the high water mark of neap tides to just below the level that remains constantly wetted by waves at low water. Also on some Ghanaian beaches *Donax puchellus* forms a littoral zone but apparently disappears before the sublittoral, whereas in Senegal this species is well represented below low water, especially just to the north of the Cap Vert peninsula. Longhurst (1958) recorded *Donax* (now *D. rugosus*) in the surf zone in Sierra Leone, but did not mention any *Nerine* species from its beaches. Buchanan (1954) considered *Donax* to characterize the surf zone along Ghanaian beaches, though Lawson (1966) pointed out that this mollusc is washed about by the waves, which makes its distribution somewhat erratic. Of the two *Donax* species recorded along the Gulf of Guinea coast and to the north in Senegal, one (*D. rugosus*) appears to be absent from beaches consisting of either very fine or very coarse sand.

In Namibia, McLachlan (1985) studied two beaches just to the north (moderately exposed) and south (very exposed) of Walvis Bay. Ocypods were not present in this region, but the upper part of the shore was occupied by the isopod *Tylos granulatus*. The middle zone of the beach was dominated by another isopod, *Excirolana natalensis* and a polychaete, *Scolelepis squamata*, in that order, from above downwards but with an appreciable overlap. Other animals at this level included *Talorchestia quadrispinosa* and *Pontogeloides latipes*. *Eurydice longicornis* was also present on the less exposed northern beach. At the lower level *Donax serra* was dominant on the very exposed shore and *Gastrosaccus* sp. on the moderately exposed shore. It should be noted that the genus *Gastrosaccus* is also found in some Ghanaian beaches, where it occurs below the main concentration of *Donax* (see Gauld and Buchanan, 1956). McLachlan (1985) also collected quantitative data and studied the meiofauna. The meiofauna and macrofauna have been studied by Tarr et al. (1985) on three beaches lying well to the north of Walvis Bay. These all lie along the so-called "Skeleton Coast" and the two most southerly ones receive inputs of drift kelp (*Laminaria schinzii*), unlike the most northerly beach (just south of the Kunene River). This latter beach has not only a relatively impoverished fauna due to the negligible amounts of drift weed compared to the other two, but it has species with tropical affinities as it lies beyond the influence of the Benguela upwelling system.

The demarcation of zones on sandy beaches is not as easy as on rocky shores, but a tentative generalization for West Africa is as follows:

(1) An upper zone, corresponding approximately to the littoral fringe, dominated by ocypod crabs in the tropical region and *Tylos granulatus* on the temperate south-west coast.
(2) A midlittoral zone, corresponding approximately to the eulittoral zone, characterized by isopods, especially species of *Excirolana*, and polychaetes (*Nerine*, *Scolelepis*).
(3) A lower zone, corresponding approximately to the sublittoral fringe and characterized by *Donax* or *Gastrosaccus*.

Lagoons, estuaries, and tidal inlets

The conditions in such habitats are very different from those found along the open coast, as organisms have to be adapted to little or no wave action, fluctuating salinities, high turbidity, smothering by sediment, and a preponderance of muddy or sandy substrata. Conditions for many "marine" organisms are least favourable in those lagoons periodically cut off from the sea (closed lagoons). Only considered here are open lagoons that are permanently connected to the sea by one or several channels. In open lagoons there is a genuine development of a littoral zone due to the rise and fall of the albeit often attenuated tide. Surfaces for the attachment of benthic organisms are few in estuaries and open lagoons, but include the roots of mangroves besides the rocky or lateritic banks that are to be found in some estuaries. The most extensive substratum type is mud, though sometimes sandy beaches occur towards the upper level of the shore. The typical mangrove vegetation of such habitats has been described elsewhere in this series (Chapman, 1977).

Soft substrata

Various studies have been concerned with particle size and other physical as well as chemical properties of the substrata. Some of the more complete of the earlier analyses were mentioned by Lawson (1966). Recent studies are few and widely separated. They include some ecological information and a faunal list by Kensley (1978) for a marine lagoon (Sandvis) in Namibia and a brief description of the macrofauna associated with the bloody cockle (*Anadara senilis*) in various Ghanaian and Nigerian lagoons. Little consideration has been given to estuaries south of the Equator except for Penrith (1970) who mentioned the fish collected from isolated pools in the Kunene River estuary (Namibia–Angola border).

The characteristic organism of the upper sandy beaches in such brackish-water habitats studied by Sourie (1957) in Senegal and by Longhurst (1958) in Sierra Leone is the fiddler crab *Uca tangeri*. This is often present in large numbers to the southernmost limit of its range in southern Angola (see Manning and Holthuis, 1981). A grapsid-dominated community occurs below the *Uca* in Sierra Leone, especially where plants other than mangroves help to consolidate the mud. On the muddy sand of the lower beach near the mouth of the Saloum River (at Djiferi) in Senegal, the most common animals are polychaetes such as *Cirratulus filiformis*, *Diopatra neapolitana* and *Macroclymene monilis*. *Arca senilis* is the dominant bivalve and is found just below the surface at the mean low-water level. Towards the open coast polychaetes become more abundant (see above, p. 310). In part of the Lagos lagoon system in Nigeria the lancelet *Branchiostoma nigeriense* is occasionally reported from littoral sands, though normally it is sublittoral (see Webb, 1958). Yoloye (1976) also found *Anadara senilis* in this lagoon system. It is restricted to areas of shallow water, and was also reported by him from a few of the numerous creeks of the Niger delta plus some of the lagoons and estuaries along the Ghanaian coast. Other bivalves commonly associated with it are *Aloidis* sp. and *Tellina nyphalis*. Sometimes the gastropod *Pugilina morio* was present in abundance, whilst *Thais haemastoma* was present in all the localities investigated, with large numbers of the hermit crab *Clibanarius africanus* almost invariably found in the shallows. Few of the species mentioned above were found by Kensley (1978) in the Sandvis Lagoon in Namibia. He recognized seven marine ecological areas in this lagoon based on the dominant organism(s) and the nature of the substratum. One of these is characterized in part by the same polychaete (*Diopatra neapolitana*) as reported by Sourie (1957) from a river mouth in Senegal (see above, p. 312).

Mangroves are generally the only plants to be found in any abundance fringing open lagoons and estuaries. Lawson (1966) mentioned a thin brown layer occurring on the mud surface beneath the mangrove *Rhizophora* in Ghana. This surface layer is covered by another with a green colour due to blue-green algae, which are widespread on mud in lagoons in western Africa. There is no doubt that the muddy floors of lagoons and estuaries present considerable barriers to the establishment of most other groups of benthic algae, though occasionally siphonaceous and other green algae may form a carpet over mud.

Rocky banks

Surfaces for the attachment of benthic organisms are few in lagoons and estuaries, if one excludes the roots of the mangroves. Often the only other surfaces available in estuaries are rocky or lateritic banks. Earlier accounts of the zonation of littoral organisms on such surfaces have been summarized by Lawson (1966). More recently observations have been made in Ivory Coast, Gabon, Gambia, and Liberia (see John, 1986).

The littoral organisms that develop on the often shaded rocky or lateritic banks of estuaries in western Africa are usually distinctly zoned. Often the vertical extent of the zones and the pattern of zonation change with increasing distance from the estuary mouth due to a reduction in marine influences including tides, wave action, and salinity. Thus littorinids characteristic of the littoral fringe of the open coast gradually become replaced up an estuary by *Littorina cingulifera*. The littorinid zone, as well as changing in composition, becomes lower and narrower and eventually disappears. When such surfaces are shaded by rocky overhangs or vegetation there may develop considerable growths of the green alga *Rhizoclonium*, accompanied by blue-green algae such as *Entophysalis deusta*. The belt of barnacles immediately beneath the zone of small gastropods is usually considered to demarcate the upper part of the eulittoral zone. It is generally more persistent than the gastropods of the littoral fringe, being found further up estuaries. Occasionally the red algae *Bostrychia* and *Caloglossa* crowd out the upper barnacles. The barnacle zone is occasionally absent in the Sierra Leone River estuary and replaced by a dirty grey covering of *Bostrychia radicans*. The oyster *Ostrea tulipa* characterizes what might be regarded as a middle eulittoral subzone, and is accompanied by the small limpet *Siphonaria pectinata*. Further up the estuaries so far studied, limpets eventually disappear and the oyster becomes replaced by the small mussel *Brachidontes puniceus*. The lowermost subzone of the eulittoral may be covered by a dirty grey felt or turf of small red and green algae, accompanied by blue-green algae. This subzone is often replaced in the Sierra Leone River estuary by animals including the serpulid worm *Pomatoleios*, the barnacle *Balanus amphitrite*, and sponges. In this estuary, Longhurst (1958) considered the upper limit of *Chama* as indicating the lower limit of the eulittoral zone. Below the *Chama* the sublittoral surfaces are blanketed with silt, and only a few individuals of *Balanus amphitrite* survive.

Seagrass communities

These have a very limited distribution in tropical western Africa, since they are absent from the Gulf of Guinea. Den Hartog (1970) mentioned just two species in the region, namely, *Halodule wrightii* (Angola, Mauritania, Senegal) and *Cymodocea nodosa* (Mauritania, Senegal). Sourie (1954) reported that the "Cymodocées formation" is not represented in the Dakar region but (p. 203) "il s'en trouvé, à l'état rudimentaire, quelques îlots qui peuplents les rentrants ensablés des plateformes rocheuses de la côte sud [there are some rudimentary islands in the sandy re-entrants of rocky platforms on the southern coast]". The only extensive seagrass meadows seem to occur in lagoon-like inlets and very wave-sheltered areas on the Bai de Lévrier side of the Cap Blanc peninsula in Mauritania (Lawson and John, 1977) and along the northern coast of Angola (Lawson et al., 1975). The seagrass-lined inlets along the Angolan coast are fringed by mangroves, whilst on the Cap Blanc peninsula similarly sheltered inlets contain salt marshes from which mangroves are entirely absent.

The seagrass meadows occur in the shallow sublittoral, though in a very sheltered inlet on the west side of the Cap Blanc peninsula they extend upwards to about the level of low-water neaps. In these inlets the dominant seagrass is *Halodule wrightii*, and immediately above it occurs a distinct zone of *Spartina maritima* with other halophytes (such as, for example, *Salicornia fruticulosa*, *Sesuvium portulacastrum*) occurring in the upper littoral. Similarly wave-sheltered inlets along the Angolan coast are lined by *Halodule wrightii*, but the littoral zone is dominated by the mangrove *Rhizophora mangle*, or, more rarely, by *Avicennia africana*. Along the western side of the Cap Blanc peninsula are large beds of the other seagrass, *Cymodocea nodosa*. Due to the heterogeneous nature of the sea bed immediately offshore, these are patchily distributed between rocky areas that are covered by larger algae. The seagrass beds contain a variety of algae, many of which grow epiphytically on the blades of the seagrass, on empty mollusc shells, and on other hard surfaces. Ascidians and sponges are reported as common in the seagrass beds studied in Angola, but otherwise there is little information on the associated fauna.

Sublittoral habitats

Much of the still very limited knowledge of sublittoral habitats along the coast of western Africa is based on dredge samples often collected below the depth of algal growth (*c.* −30 to −40 m). Dredge samples are especially useful when investigating soft-bottom communities; those obtained from hard bottoms are likely to give at best a very incomplete picture of the plants and animals present. Some of the most comprehensive investigations of soft-bottom communities were those carried out almost 30 years ago off the coast of Ghana by Bassindale (1961), Buchanan (1954, 1958), and Buchanan and Anderson (1955). The offshore fauna of unconsolidated areas of seabed is still virtually unknown elsewhere in the region, and the little information available is often based on the identification of animals associated with sediment samples collected by the Danish oceanographic expeditions of 1945–1946 and 1950–1952. Mainly taxonomic accounts of the animal groups collected on these expeditions are published in the volumes of *Atlantide Report* and the *Galathea Report*. Brief accounts of the shallow inshore vegetation (principally algal), and of the more obvious associated animals, exist for Gabon, Ghana and Sierra Leone (see Lawson and John, 1982, for references), Angola (Lawson et al., 1975), and the Mauritanian side of the Cap Blanc peninsula (Lawson and John, 1977). All these studies, and one by Laborel (1974) on West African corals (Senegal–Gabon), involved direct observations made by means of aqua-lung (SCUBA) diving. The description of the sublittoral vegetation along the Senegal coast to the north of Gambia is the result of a dredge survey (see Bodard and Mollion, 1974; Mollion, 1975). Much of the description that follows, however, is from a series of very detailed investigations carried out on shallow rocky reefs or banks and areas of cobbles lying off the Ghana coast. The studies by John et al. (1977), Lieberman et al. (1979, 1984), and Edmunds and Edmunds (1973) were carried out by diving in water shallower than about 30 m and by dredging at greater depths.

Rocky reefs

Rock surfaces exposed to considerable wave action are often covered with "lithothamnia" in the immediate sublittoral, and the most conspicuous animal present is *Echinometra lucunter*. On wave-sheltered shores along the Gulf of Guinea coast *Sargassum vulgare* is often present in the shallow sublittoral, though in deeper water it may become replaced by *Sargassum filipendula*. In very sheltered bays (in Sierra Leone, for example) where the bottom shows little surface relief, *Sargassum vulgare* often forms extensive beds down to a depth of 5 m (see John and Lawson, 1977b). If the rocky bottom is highly structured (e.g. boulder-lined), *Sargassum* may be absent and the community consists of cushion- or mat-forming algae, articulated coralline algae, and large clumps of a few species that are possibly resistant to grazing.

The dominant algae or animals on rocky areas studied below a depth of about 8 m off the Ghana coast vary depending on depth and the nature of the seabed. Shallower areas (*c.* −8 to −14 m) are often covered by the creeping net-like red alga *Dictyurus fenestratus* where the rocky reef abuts on the sand. In this area of low and little-structured rock, beds of *S. filipendula* covering several square metres of seabed may be present. Also growing in this area of intermittently sand-buried rocks are a number of brown and red algae. These little-studied sandy areas of the reef sampled off Ghana are almost devoid of molluscs except for *Alaba culliereti* (on *Sargassum*), *Terebra grayi*, *Turritella ungulina*, and the sea slugs *Aplysia dactylomela* and *A. fasciata*.

The algal vegetation is very inconspicuous in areas where the rocky reefs are raised to form cliffs, gullies, and ledges. In these areas the rocks are covered with crusts of "lithothamnia" and a low turf consisting of filamentous red and blue-green algae. Such a turf is best developed in the almost contiguous or even overlapping territories of *Stegastes imbricatus*. Solitary individuals of other algae are present including *Dictyopteris delicatula*, *Dictyota* spp., *Hypnea cervicornis*, *Laurencia majuscula*, *Predaea feldmannii* and purplish patches of blue-green algae. Often feeding on the bushy clumps of *Laurencia* is the sea slug *Aplysia winneba* (Edmunds and Edmunds, 1973). Sessile animals (particularly oysters and barnacles) are present in some abundance on the top of submarine cliff areas harbouring large populations of *Diadema antillarum*. Over these higher areas of the reef are many sponges, gorgonians, bryozoans and small tunicates. Large quantities of *Alcyonidium sanguineum* are found here growing on the alga *Dictyopteris*, on gorgonians, and on small stones. Another substratum for this bryozoan is a coelenterate, *Telesto*, which grows around gorgonians and eventually destroys their tissues. *Alcyonidium sanguineum* is able to invest the colonies of *Telesto*, and Cook (1985) found that the tubes and crevices of these joint colonies were inhabited by a brilliantly red-coloured polychaete and by a red ophiuroid. She also recorded a large number of other bryozoans growing on *A. sanguineum* or on the basal parts of gorgonians together with sea anemones, hydroids, ascidians, vorticellids, and several different types of sponges. The mollusc *Pteria* sp. was commonly attached to the gorgonian *Lophogorgia*.

The shallow rocky reef has many grazers, browsers, and carnivores which, along with ciliary feeders, have been recorded by Edmunds and Edmunds (1973, table 1). All the ciliary feeders are bivalves except the gastropod *Crepidula porcellana*. Though carnivorous molluscs are listed, they are not common when compared with deeper areas a few kilometres to the east. Small reef fishes, using rock crevices and the bryozoan and coelenterate colonies for protection, are abundant and many browse and graze on the algae and sedentary fauna. Some of the more common are the terrestrial damselfish, *Abudefduf saxatilis*, *Acanthurus monroviae*, *Ophioblennius atlanticus*, *Stegastes imbricatus*, and the large parrotfish, *Pseudoscarus hoefleri*. With increase in depth the diversity of algae is reduced, and Edmunds and Edmunds (1973) recorded a few molluscs not present in the shallower waters of the 10 m reef, such as *Cardium kobelti*, *Drillia pyramidata* and *Gari fervensis*.

Cobble areas

Cobbles or rhodoliths formed by the accretion of calcareous red algae are found strewn over the sandy bed in various places along the coast of western Africa. The biota associated with such areas has only been studied off the Ghanaian coast, where attention has been focused mainly on the plants (see Lieberman et al., 1979, 1984).

The cobbles provide a surface for the development of a diversified "algal plain community" during the calmer months of the dry season when cobble tumbling and sand burial are rare events. This algal plain community at 10 to 12 m depth is dominated by fleshy algae including *Gracilaria multipartita*, *Halymenia actinophysa*, *Laurencia brogniartii* and *Solieria filiformis* (see Lieberman et al., 1979). Many of the algae found here also occur on cobbles lying at about −20 m, where some of the plants are larger and better developed despite the increased depth. This is true for many red algae as well as for the stoloniferous green alga *Caulerpa taxifolia*. Over a depth range of about −3 to −10 m off the Senegalese coast, Bodard and

Mollion (1975) also found a community dominated by red algae. Many of the plants they recorded also occur off the Ghanaian coast, though others are apparently endemic to Senegal. The two most distinctive species off the northern Senegal coast to the south of Cap Blanc were *Anatheca montagnei* and *Halymenia senegalensis*. *Acrosorium uncinatum*, *Botryocladia senegalensis*, *Gracilaria camerunensis* and *Polyneura denticulata* were frequently abundant between Dakar and M'Bour, whereas further to the south the seabed was covered by a so-called "prairie" of *Hypnea* (mainly *H. cervicornis*) along with *Pseudobranchioglossum senegalense*.

Soft substrata

Some of the most detailed accounts of the fauna associated with soft bottoms have been of the seabed off the Ghanaian coast (see Buchanan, 1954, 1958; Buchanan and Anderson, 1955; Bassindale, 1961; Edmunds and Edmunds, 1973; Cook, 1985). The flora on such soft bottom areas off Ghana was mentioned by Lawson and John (1982; 1987) and John (1986). Similar soft-bottom associations are no doubt to be found elsewhere off the coast of western Africa, though this has still to be investigated.

The seabed off the Ghanaian coast may be divided into five zones as follows (see Buchanan, 1954; Bassindale, 1961):

(1) Surf zone: low water to a point beyond the breaking surf (*c.* −3.5 m), characterized by *Donax*.
(2) *Cultellus* zone: from the breaking surf to a depth of 14.5 m, *Cultellus tenuis* occurs throughout this zone, though a subzone characterized by *Dentalium* (elephant's tusk shell) occurs in the lower part.
(3) *Turritella* zone: from −11.5 to −36 m, with large numbers of this turret-shell occurring throughout, except for a narrow strip found at the upper and lower edge where only dead shells are present.
(4) Silt zone: outside the boundary of the *Turritella* zone and to a depth of 45 m, dead mollusc shells are encountered frequently throughout.
(5) Bryozoan zone: from about −45 to −90 m, a prolific growth of bryozoans and foraminiferans is encountered on the surface of the bottom deposit; this zone is particularly rich in molluscan species.

The *Turritella* zone is also characterized by large populations of small pagurid crabs. These live in shells of *Turritella*, which show borings indicating that carnivorous gastropods have removed the original occupants. There is, according to Edmunds and Edmunds (1973), an incredible number of predatory gastropods here (such as, for example, species of *Conus*, *Drillia*, *Murex*, *Terebra*) as well as various scavengers (e.g. *Nassa* spp.). They pointed out the striking richness of the molluscan fauna in this apparently uniform habitat, compared with the relatively low diversity of the shelled molluscan fauna in the more varied habitat of rocky reefs.

Few bryozoan species seem able to colonize shells inhabited by *Turritella* due to its burrowing habit which inhibits successful larval settlement. Empty shells also become buried by deposition, and so similarly are unavailable for settlement. Hermit crabs are surface feeders; when they occupy such shells they keep them at the surface, and bryozoans are therefore able to become established on them. Occasionally growing on these shells dredged from depths as great as 40 m are red algae including *Gracilaria multipartita*, *Hymenema* sp. and *Spyridia hypnoides*. Such shells are normally occupied by hermit crabs, suggesting that the same problems of settlement must apply equally well to the algae as to the bryozoans.

Zonation is often patchy and not as clear as the above scheme might indicate. For instance, some of the areas of grey mud found off the Ghanaian coast are overlain with pieces of silicious material formed by colonial Foraminifera, principally *Jullienella foetida* (referred to by Cook, 1985, as the *Jullienella* zone) and *Schizammina* spp. These are often at the same depth and on the same muddy bottom as the *Turritella*-dominated association, both sometimes occurring together to form the only solid surface available for many other species. Cook (1985) mentioned that the arenaceous test of the large foliaceous foraminiferan *Jullienella* is colonized by a very large bryozoan fauna. The free-living Cupuladriidae bryozoans, which are especially adapted to conditions of sandy seabed and high rates of deposition, occur in very large numbers in places, the colonies of *Discoporella umbellata* being of especially large size.

DYNAMIC ASPECTS

Seasonality

Most of the studies on littoral organisms along the coastline of tropical western Africa are descriptive and quantitative, and only few are concerned with functional aspects. One of the earliest studies to give some consideration to these aspects was undertaken by Sourie (1954). He observed the invasion of the barnacle belt from below by algae on a wave-exposed shore in Senegal during the rainy season. These algae persisted at this level for several months but became bleached and eventually disappeared during the dry season. Similar changes have been observed on Ghanaian shores by Lawson (1966) who found a correlation between the movement of the belts of certain shore organisms (particularly algae) and desiccation operating through the agency of the tides. During the period from September to March (dry-season months) the lower of the two low waters in any 24 hours occurs during the day time, whereas the position is reversed for the other six months of the year, which coincides roughly with the rainy season on the coast. Shore organisms in Ghana thus have to endure severe desiccation in the daytime during the dry season, and this leads to mass mortality of those that have become established at higher levels during the previous rainy season. Other factors are also involved in the seasonal development of shore organisms, especially those in tide-pools. John (1986) reported an increase in the ground cover of algae during the rainy season (May/June to October/November) in a tide-pool studied in western Ghana. He suggested that the decrease observed during the following dry season was due to high solar insolation and to various physico-chemical changes taking place because of the longer period the pools are isolated from the sea at low water. No doubt some of the differences in the vertical distribution of shore organisms reported for western Africa relate to the time of the year particular surveys were undertaken.

Very little is known of the functional aspects of benthic sublittoral associations along this coast. The only detailed investigations of the seasonality of such associations of algae have been confined to rocky reefs and cobble banks off the coast of Ghana (see John et al., 1977; Lieberman et al., 1979, 1984). On submarine banks of calcareous cobbles there is a rich and diversified development of algae. This is the algal plain community already mentioned, and it is found at depths between about 8 to 15 m during the two to four calmer months of the dry season. Most of these algae are removed during the rainy season due to surge causing cobble tumbling. The number of algal species per cobble has been found to be correlated with cobble movement or inertia (resistance to tumbling) during the disturbed conditions of the rainy season. With the return of calm weather during the dry season a rapid succession of algae takes place on the cobbles. Some of these algae probably arise from spores produced by those plants that survive the rainy season growing on nearby rocky banks. Nonetheless, many of the species on the cobbles are absent or very rare on these banks, and probably have arisen *in situ* from spores, germlings, juveniles, persistent holdfasts or stoloniferous bases. Ephemerals, annuals, pseudoannuals, and perennials appear sequentially during this short growing season. Algae with a greater size potential are in peak abundance late in the growing season, often never reaching their maximum size or reproducing. It is suggested that seasonal disturbance prevents the attainment of a true climax community on the cobbles and helps to maintain the high species diversity.

Offshore cobble areas are widespread, and yet little is known of the seasonality of the algae associated with them except for those in shallow waters off Ghana mentioned earlier. Some information has been given by Mollion (1975) for the algae growing mostly on small stones at depths ranging from 2 to 3.5 m on the seabed just to the south of the Cap Vert peninsula. Unfortunately the information obtained from dredge samples is insufficiently detailed to compare with that from the Ghanaian investigations. Sourie (1954) reported that algae are not usually thrown up on the northern Senegal beaches, but that they are frequent on the southern ones. Much of this cast-up material in southern Senegal is probably derived from the large offshore populations described by Mollion (1975) and Bodard and

Mollion (1975). Some kind of regular seasonal pattern is described by Lawson (1966), with green algae commoner in the drift during the dry season and reds more frequent during the rainy season. The reds are most probably derived from sublittoral associations growing on unstable substrata where they are often the dominant group.

With regard to the seasonality of sublittoral benthic associations growing on stable substrata, one of the few detailed investigations by SCUBA-diving has been off the coast of Ghana at depths of about 8 to 30 m (see John et al., 1977; Lieberman et al., 1984). Although these rocky reefs occur at depths comparable to those of the calcareous cobbles, their animal and plant communities not only differ in species composition and diversity but also with regard to patterns of growth-form, life-form, and longevity. Broadly similar patterns of seasonal change are observed on both substrata (see Lieberman et al., 1979, 1984) but are very much less variable on the rocky reef where many species persist throughout the year. Floristic differences between the two areas are due not only to differences in stability of the substratum but also to the greater effect of grazing on the rocky reefs by algivorous fishes, sea urchins, and probably other animals.

Recolonization and succession

The composition and extent of zones of organisms on western African shores may sometimes show aberrant patterns that are not readily explicable. These can sometimes be related to the past history of the site rather than to the normal zone-forming factors. For example, the absence of certain zone-forming algae on newly exposed rocks in Benin and Togo (see John and Lawson, 1972) is perhaps explicable in terms of differential recruitment rates by inocula carried in the prevalent west-to-east current. Sometimes very local differences may be due to some past destructive event such as sand movement, so what is observed is an association representing a phase in a successional sequence. Recolonization has been studied experimentally on shores in both Ghana (Lawson, 1966) and Senegal (Sourie, 1954). Three phases of recolonization have been recognized on Ghanaian shores if microscopic forms are ignored: denuded rocks are first quickly colonized by green algae (*Enteromorpha* quickly followed by *Ulva*), then by small spots of "lithothamnia," which later coalesce to give a crust that almost completely covers the originally cleared surface. Later follows a lengthy period of colonization by a number of other algae, some of which do not normally occur at the particular level of the shore cleared. This final phase has been estimated as taking perhaps six or seven years in Ghana, and Sourie (1954) found a 90% recolonization of the barnacle *Chthamalus* after two years on a Senegalese shore following denudation.

The seasonal succession of algae on cobble surfaces that takes place each year off the Ghanaian coast, following the wholesale removal of plants due to physical disturbance by rainy season storms, is not the same as takes place on newly exposed artificial surfaces. For instance, John (1986) found that the first colonizers (excluding a bacterial film) on roughened perspex (plexiglass) surfaces placed on the seabed off Ghana were filamentous brown algae rather than the filamentous red algae that appear on cobbles at the beginning of the dry season when tumbling ceases, or the green algae that first colonize denuded surfaces in the intertidal zone (see above). "Lithothamnia" usually appear as spots on artificial surfaces in just seven days, sometimes cover almost 40% of the total surface area after just a month, and thus resemble the events described by Lawson (1966) for the littoral zone. A mosaic of associations of different ages was often found on the plates after several months due to sloughing off of the original "lithothamnia"-covered surface which often took place where this crust had become overgrown by sessile animals (particularly barnacles).

The climax community off the Ghanaian coast on rocky reefs is probably one dominated by *Sargassum filipendula*. Grazing by fish and other animals on parts of such reefs where there is much physical relief seems to prevent this from being attained. In rocky areas lying adjacent to sand, a *Sargassum*-dominated algal association sometimes develops but this probably survives no more than a year or two due to burial by the constantly shifting sand. One of the few plants able to persist on subtidal rocks regularly buried by shifting sand is the stoloniferous red alga *Dictyurus fenestratus*. It thus appears that off the Ghanaian coast a stable climax community is rarely attained because of several reasons: grazing (principally by fish) of stable bottoms with much physical relief, periodic sand-burial of stable rocks abutting sandy areas, or seasonal instability of the substratum.

Plant–animal interactions

Another much neglected area of research in western Africa is that concerning the inter- and intraspecific interactions of benthic organisms. Large numbers of algal-feeding animals together with physical forces have been suggested as accounting for the diminutive size of fleshy and filamentous algae on wave-exposed shores (see John, 1986). Unfortunately there is no experimental evidence to substantiate this suggestion, except for a study by Gauld and Buchanan (1959) on rocks in Ghana. They observed a rapid and prolific growth of blue-green algae following the removal of *Siphonaria*. More recently, John (1986) has experimentally cleared a tide-pool in western Ghana of *Echinometra lucunter*, where most algae other than crustose forms were absent. This produced an increase in the cover of filamentous and fleshy algae, but the increase was associated with the creation of new, and the expansion of old, damselfish territories in addition to a reduction in sea urchin grazing. Indeed, in this pool there were also many small non-territorial fishes, which, following sea urchin removal, continued to graze, though outside the "algal farms" protected by the damselfish. Only in pools where evident grazers were absent was there a high cover of algae other than encrusting forms. Seasonal changes in such pools probably relate more to the physico-chemical environment than to grazing.

Grazing pressure by fish and invertebrates may well be one of the prime factors determining algal abundance and species composition in certain benthic habitats along the western coast of Africa. This seems to be the case in wave-sheltered situations in the shallow sublittoral zone where there is much physical relief (for example, rough stone harbour breakwaters, boulder-lined bays, broken-up offshore rocky banks). In such habitats the algal vegetation is species-poor, low in biomass, and dominated by crustose algae (principally "lithothamnia"), algae of very diminutive size, or larger algae possessing some defence against grazing whether structural (calcareous or gelatinous thalli) or, possibly, biochemical (see John, 1986). The important role of fish-grazing on the distribution of benthic algae has been demonstrated experimentally by John and Pople (1973) for parts of a breakwater system in Ghana. Similarly, large populations, often of the same demersal fishes as found along this breakwater system, also congregate around offshore rocky banks where holes, gullies, low cliffs, and crevices provide them with protection from marauding predators. Many of these small reef fishes are algivorous or omnivorous, as judged from an investigation of the stomach contents of individuals sampled from different habitats (such as rocky offshore banks, bays, or tide-pools) in Ghana (see Table 12.1). Large communities of benthic algae occur in habitats where cover for such small fishes is lacking. For instance, open areas of cobble bottom off the Ghanaian coast have few algivorous fish except for juveniles that have strayed away from the protection of nearby rocky banks. On these cobbles, as already indicated, there is found during the calmer months of the dry season a well-developed growth of algae up to 20 to 30 cm in height.

Information on other grazers along the western coast of Africa is very limited, though they are likely to influence profoundly the algal vegetation where they are present in large numbers. For instance, the shore crab *Grapsus grapsus*, common along most shores in the region, is an omnivore frequently to be seen eating algae. No quantitative data exist on its feeding activities, or on those of the sublittoral *Diadema antillarum* which is common especially along harbour breakwaters and on rocky banks off the Ghanaian coast. John (1986) has suggested that the occurrence of oysters and barnacles in *Diadema*-grazed areas is promoted by the removal by this sea urchin of competitively superior algae.

LITTORAL AND SUBLITTORAL ECOLOGY IN TROPICAL WESTERN AFRICA: PAST, PRESENT AND FUTURE

The question of the relative poverty of the marine flora and fauna of western Africa as compared with other tropical regions was mentioned in the Introduction. This was partly accounted for above (p. 301) by the suggestion that local upwelling of cold water on the Gulf of Guinea coast together with lack of firm substrata may have militated against the development of coral reefs and the many species associated with them. Van den Hoek (1975) has suggested another possible cause, namely a historical one. According to him, reduced sea temperatures during Pleistocene glaciations could have moved the 20°C winter isotherm latitudinally about 15 to 20° towards the Equator. This would have completely eliminated the tropical marine flora and fauna of the Gulf of Guinea, but left those of the western Atlantic relatively unaffected. Subsequently, the tropical eastern Atlantic would have been gradually recolonized by species from the west after the glaciations, but relatively little time would have been left for speciation to take place. This would explain the poverty, low endemism, and similarities in composition of the tropical American Atlantic with the tropical African Atlantic. Such suggestions are, of course, still only very speculative.

In his review of the littoral ecology of West Africa, Lawson (1966) ended with a consideration of some outstanding problems and possible future work. It is interesting to observe what has actually been accomplished in this respect in the 24 years that have elapsed since it was written. That review dealt with a rather more limited coastline, namely that part which is referred to loosely here as the Gulf of Guinea, and the Cap Vert peninsula in Senegal. In the present review the area of investigation has been considerably extended not only northwards but also to south of the Equator along shores that then had scarcely been looked at ecologically. Furthermore, some gaps in descriptive ecology specifically referred to in the earlier review, such as Liberia and Ivory Coast, are now filled, at least partially, by written accounts. Thus, although there is still much room for more detailed work, there is now fairly good coverage of the strictly littoral zone along most parts of the tropical coast of western Africa. This descriptive phase has been greatly enhanced by the many good taxonomic works on various groups of marine organisms that have been written during recent years. One may add, however, that good general accounts for the coasts of some countries such as Guinea-Bissau, Guinea, Republic of Congo, and Zaïre are not, so far as we have been able to ascertain, yet available. For the sublittoral, a zone not covered in the earlier review, a great deal of information is now available, but by far the greater part of this is from Ghana, and general remarks for the region often have to be extrapolated from this. Again, with regard to the more dynamic aspects of shore ecology, some progress has been made, especially with regard to plant–animal interactions and survival strategies, but many problems remain – for example, the occurrence in chronically fish-grazed areas of certain fleshy algae such as *Laurencia* which appear to have no structural adaptations for survival. It is perhaps in this field, bearing in mind the relatively sound basis of taxonomic work and descriptive ecology that is now in being, that the greatest potential for future research exists.

REFERENCES

Bassindale, R., 1961. On the marine fauna of Ghana. *Proc. Zool. Soc. London*, 137: 481–510.

Bodard, M. and Mollion, J., 1975. La végétation infralittorale de la petite côte sénégalaise. *Bull. Soc. Phycol. France*, 19: 193–221.

Briggs, J.C., 1974. *Marine Zoogeography*. McGraw-Hill, New York, 475 pp.

Brown, A.C. and Jarman, N., 1978. Coastal marine habitats. In: M.J.A. Wagner (Editor), *Biogeography and Ecology of Southern Africa. Monogr. Biol.*, 31: 1239–1277.

Buchanan, J.B., 1954. Marine molluscs of the Gold Coast, West Africa. *J.W. Afr. Sci. Ass.*, 1: 30–45.

Buchanan, J.B., 1958. The bottom fauna communities across the continental shelf off Accra, Ghana. *Proc. Zool. Soc. London*, 130: 1–56.

Buchanan, J.B. and Anderson, M.M., 1955. Additional records to the marine molluscan fauna of the Gold Coast. *J.W. Afr. Sci. Ass.*, 1: 57–61.

Chapman, V.J., 1977. Africa B. The remainder of Africa. In: V.J. Chapman (Editor), *Wet Coastal Ecosystems. Ecosystems of the World, Vol. 1.* Elsevier, Amsterdam, pp. 233–240.

Cook, P.L., 1985. Bryozoa: A preliminary survey. *Ann. Mus. Afr. Centr.*, 238: 1–276

Den Hartog, C., 1970. The sea-grasses of the world. *Verh. K. Ned. Akad. Wet. Reeks* 2, 59: 1–275.

Edmunds, J. and Edmunds, M., 1973. Preliminary report on the mollusca of the benthic communities off Tema, Ghana. *Malacologia*, 14: 371–372.

Ekman, S., 1953. *Zoogeography of the Sea.* Sidgwick and Jackson, London, 417 pp.

Gauld, D.T. and Buchanan, J.B., 1956. The fauna of sandy beaches in the Gold Coast. *Oikos*, 7: 293–301.

Gauld, D.T. and Buchanan, J.B., 1959. The principal features of rock shore fauna in Ghana. *Oikos*, 10: 121–132.

Gayral, P., 1958. *La Nature au Maroc 11. Algues de la Côte Atlantique Marocaine*. Société de Sciences Naturelle et Sciences Physiques de Maroc, Rabat, 524 pp.

Griffiths, J.F., 1972. General introduction. In: J.F. Griffiths (Editor), *World Survey of Climatology. Climates of Africa*, 10. Elsevier, Amsterdam, pp. 1–35.

Grunbaum, H., Bergman, G., Abbott, D.P. and Ogden, J.C., 1978. Intraspecific agonistic behavior in the rock-boring sea urchin *Echinometra lucunter* (L.) (Echinodermata: Echinoidea). *Bull. Mar. Sci.*, 28: 181–188.

Hedgpeth, J.W., 1957. Marine biogeography. In: J.W. Hedgpeth (Editor), *Treatise on Marine Ecology and Paleoecology. I. Ecology. Mem. Geol. Soc. Am.*, 67: viii + 1–1296.

Houghton, R.W. and Mensah, M.A., 1978. Physical aspects and biological consequences of Ghanaian coastal upwelling. In: R. Boje and M. Tomczak (Editors), *Upwelling Ecosystems*. Springer, Berlin, pp. 167–180.

Jeník, J. and Lawson, G.W., 1967. Observations on water loss of seaweeds in relation to microclimate on a tropical shore (Ghana). *J. Phycol.*, 3: 113–116.

John, D.M., 1986. Littoral and sub-littoral marine vegetation. In: G.W. Lawson (Editor), *Plant Ecology in West Africa: Systems and Processes*. Wiley, New York, pp. 215–246.

John, D.M. and Lawson, G.W., 1972. The establishment of a marine algal flora in Togo and Dahomey (Gulf of Guinea). *Bot. Mar.*, 15: 64–73.

John, D.M. and Lawson, G.W., 1974a. *Basispora*, a new genus of the Ralfsiaceae. *Br. Phycol. J.*, 9: 285–290.

John, D.M. and Lawson, G.W., 1974b. Observations on the marine algal ecology of Gabon. *Bot. Mar.*, 17: 249–254.

John, D.M. and Lawson, G.W., 1977a. The marine algal flora of the Sierra Leone peninsula. *Bot. Mar.*, 20: 127–135.

John, D.M. and Lawson, G.W., 1977b. The distribution and phytogeographical status of the marine algal flora of Gambia. *Feddes Reprium*, 88: 287–300.

John, D.M. and Pople, W., 1973. The fish grazing of rocky shore algae in the Gulf of Guinea. *J. Exp. Mar. Biol. Ecol.*, 1: 81–90.

John, D.M., Lieberman, D. and Lieberman, M., 1977. A quantitative study of the structure and dynamics of benthic subtidal algal vegetation off Ghana (Tropical West Africa). *J. Ecol.*, 65: 497–521.

Kensley, B., 1978. Interaction between coastal processes and lagoonal fauna between Walvis Bay and Lüderitzbucht, South West Africa. *Madoqua*, 11: 55–60.

Kensley, B. and Penrith, M.-L., 1973. The constitution of the intertidal fauna of rocky shores of Moçâmedes, southern Angola. *Cimbebasia Ser. A.*, 2: 113–123.

Kensley, B. and Penrith, M.-L., 1980. The constitution of the fauna of rocky intertidal shores of South West Africa. Part III. The north coast from False Cape Frio to the Kunene River. *Cimbebasia Ser. A.*, 5: 201–214.

Laborel, J., 1974. West African reef corals. An hypothesis on their origin. *Proc. Second Int. Coral Reef Symp.*, 1: 425–443.

Lawson, G.W., 1957. Seasonal variation of intertidal zonation on the coast of Ghana in relation to tidal factors. *J. Ecol.*, 45: 831–860.

Lawson, G.W., 1966. The littoral ecology of West Africa. *Oceanogr. Mar. Biol. Ann. Rev.*, 4: 405–448.

Lawson, G.W., 1978. The distribution of marine algal floras in the tropical and subtropical Atlantic Ocean: a quantitative approach. *Bot. J. Linn. Soc.*, 76: 177–193.

Lawson, G.W. and John, D.M., 1977. the marine flora of the Cap Blanc peninsula: its distribution and affinities. *Bot. J. Linn. Soc.*, 75: 99–118.

Lawson, G.W. and John, D.M., 1982. The marine algae and coastal environment of tropical West Africa. *Beih. Nova Hedwigia*, 70: 1–455.

Lawson, G.W. and John, D.M., 1987. The marine algae and coastal environment of tropical West Africa (2nd ed.). *Beih. Nova Hedwigia*, 93: vi + 1–415.

Lawson, G.W. and Norton, T.A., 1971. Some observations on littoral and sublittoral zonation at Teneriffe (Canary Isles). *Bot. Mar.*, 14: 116–120.

Lawson, G.W., John, D.M. and Price, J.H., 1975. The marine algal flora of Angola: its distribution and affinities. *Bot. J. Linn. Soc.*, 70: 307–324.

Lewis, J.R., 1964. *The Ecology of Rocky Shores*. The English Universities Press, London, 323 pp.

Lieberman, M., John, D.M. and Lieberman, D., 1979. Ecology of subtidal algae on seasonally devastated cobble substrates off Ghana. *Ecology*, 60: 1151–1161.

Lieberman, M., John, D.M. and Lieberman, D., 1984. Factors influencing algal species assemblages on reef and cobble substrata off Ghana. *J. Exp. Mar. Biol. Ecol.*, 75: 129–143.

Longhurst, A.R., 1958. An ecological survey of the West African marine benthos. *Fishery Publ. Colon. Off. London*, 11: 1–102.

Lüning, K., 1985. *Meeresbotanik. Verbreitung, Okophysiologie und Nutzung der marinen Makroalgen*. Georg Thieme, Stuttgart, New York, 375 pp.

McLachlan, A., 1985. The ecology of two sandy beaches near Walvis Bay. *Madoqua*, 14: 155–163.

Manning, R.B. and Holthuis, L.B., 1981. West African brachyuran crabs (Crustacea: Decapoda). *Smithson. Contrib. Zool.*, 306: xii + 1–379.

Martin, L., 1971. The continental margin from Cape Palmas to Lagos: bottom sediments and submarine morphology. In: F.M. Delany (Editor), *ICSU/SCOR Symp. Cambridge 1970: The Geology of the East Atlantic Continental Margin. Part 4. Africa. Rep. Inst. Geol. Sci.*, 70(16): 79–96.

Mollion, J., 1975. Étude quantitative d'une formation végétale marine de l'infralittoral supérieur au Sénégal. *Bull. Inst. Fond. Afr. Noire Sér. A*, 37: 537–554.

Pauly, D., 1975. On the ecology of a small West African lagoon. *Ber. Dt. Wiss. Komm. Meeresforsch.*, 24: 46–62.

Penrith, M.J. and Penrith, M.-L., 1972. The Blenniidae of western southern Africa. *Cimbebasia Ser. A*, 2: 65–90.

Penrith, M.-L., 1970. Report on a small collection of fishes from the Kunene river mouth. *Cimbebasia Ser. A*, 1: 165–176.

Penrith, M.-L. and Kensley, B., 1970a. The constitution of the fauna of rocky intertidal shores of South West Africa. Part II. Rocky Point. *Cimbebasia Ser. A*, 1: 243–268.

Penrith, M.-L. and Kensley, B.F., 1970b. The constitution of the intertidal fauna of rocky shores of South West Africa. Part I. Lüderitzbucht. *Cimbebasia Ser. A*, 1: 191–239.

Price, J.H., John, D.M. and Lawson, G.W., 1978. Seaweeds of the western coast of tropical Africa and adjacent islands: a critical assessment. II. Phaeophyta. *Bull. Br. Mus. Nat. Hist. (Bot.)*, 6: 87–182.

Rosewater, J., 1981. The Family Littorinidae in Tropical West Africa. *Atl. Rep.*, 13: 7–48.

Rosewater, J. and Vermeij, G.J., 1972. The amphi-Atlantic distribution of *Littorina meleagris*. *The Nautilus*, 86: 67–69.

Sanusi, S.S., 1980. *A study on grazing as a factor influencing the distribution of benthic littoral algae*. M.Sc. Thesis, University of Ghana, Ghana, 217 pp.

Shannon, L.V., 1985. The Benguela ecosystem. Part I. Evolution of the Benguela, physical features and processes. *Oceanogr. Mar. Biol. Ann. Rev.*, 23: 105–182.

Sourie, R., 1954. Contribution à l'étude écologique des côtes rocheuses du Sénégal. *Mém. Inst. Afr. Fr. Afr. Noire*, 38: 1–342.

Sourie, R., 1957. Étude écologique des plages de la côte Sénégalaise aux environs de Dakar (Macrofaune). *Ann. Ec. Sup. Sci. Dakar*, 2: 1–110.

Stephenson, T.A. and Stephenson, A., 1972. *Life between Tidemarks on Rocky Shores*. Freeman, San Francisco, 425 pp.

Tarr, J.G., Griffiths, C.L. and Bally, R., 1985. The ecology of three sandy beaches on the Skeleton Coast of South West Africa. *Madoqua*, 14: 295–304.

Van den Hoek, C., 1975. Phytogeographic provinces along the coasts of the northern Atlantic Ocean. *Phycologia*, 14: 317–330.

Walter, H., 1986. The Namib Desert. In: M. Evenari, I. Noy-Meir and D.W. Goodall (Editors), *Hot Deserts and Arid Shrublands. Ecosystems of the World, Vol. 12B*. Elsevier, Amsterdam, pp. 245–282.

Webb, J.E., 1958. The ecology of Lagos lagoon III. The life history of *Branchiostoma nigeriense* Webb in relation to its environment. *Philos. Trans. R. Soc. Ser. B*, 241: 335–353.

Yoloye, V., 1976. The ecology of the West African Bloody Cockle, *Anadara (Senilia) senilis* (L.). *Bull. Inst. Fond. Afr. Noire Ser. A*, 38: 25–56.

Chapter 13

LITTORAL AND SUBLITTORAL ECOSYSTEMS OF SOUTHERN AFRICA

J.G. FIELD and C.L. GRIFFITHS

INTRODUCTION

The southern African region is defined here as extending from the northern border of Namibia (South West Africa) (17°S) to the southern border of Mozambique (26°S). The overall length of the coastline between these points is approximately 4000 km, little more than the direct distance between them via Cape Agulhas, the southernmost tip of the continent (35°S). This exemplifies the very exposed, almost linear nature of the coastline. On the west coast the only significant shelter is afforded by relatively small lagoonal systems at Walvis Bay, Sandwich Harbour and the Saldanha Bay/Langebaan Lagoon system. The only other bay deep enough to provide significant shelter is False Bay, in the extreme southwest. The east coast incorporates a series of large "half-heart" bays, notably Algoa Bay at Port Elizabeth, although these afford little protection from wave action. A group of saline lakes and lagoons is located on the northern Natal coast, but while these are of considerable academic interest and conservation importance, they form a very small proportion of the coastal system.

Because of generally low and seasonally variable rainfall there are few major rivers, one result being that large, permanently open estuaries are rare, the best known being at Knysna on the south coast. Small, blind estuaries, on the other hand, are abundant in some areas, notably Transkei in the region of Port St. Johns (Fig. 13.1) and adjacent eastern Cape Province. The dry west coast supports very few estuarine systems, the only river of note being the Orange River.

The open shoreline that typifies this area can be apportioned into three habitat types: sandy beach, rocky shore and mixed shore (Bally et al., 1984). The last of these groupings includes cliffs with sandy bases and sand overlying wave-cut rocky platforms, as well as mixed boulder shores. Using these categories, and excluding southern Namibia, for which they were unable to obtain suitable data, Bally et al. (1984) calculated that 42% of the southern African coastline comprises open sandy beach and 27% rocky shore, with 31% in the mixed category (mostly sand overlying wave-cut rock). Of other types common elsewhere muddy shores are virtually absent outside the very few lagoonal areas, and cobble or pebble beaches constitute well under 1% of the coastline.

Southern African mixed and cobble shores have not been biologically investigated to date and are not considered further. The rocky and sandy shores are dealt with in separate sections below.

HYDROLOGY

Southern Africa is dominated by two current systems: the warm Agulhas Current, derived from the Mozambique Current in the east, and in the west the cool Benguela Current, which originates from water upwelled from subsurface layers along the west coast.

The Agulhas Current is a western boundary current of the south Indian Ocean and is formed by confluence of water derived from the South Equatorial Current, deflected southwards by the east coast of Madagascar, and the Mozambique Current, which flows southwards down the Mozambique Channel (Fig. 13.1). The current is strongest and warmest at the shelf break and reaches down to 2500 m (Heydorn et al., 1978).

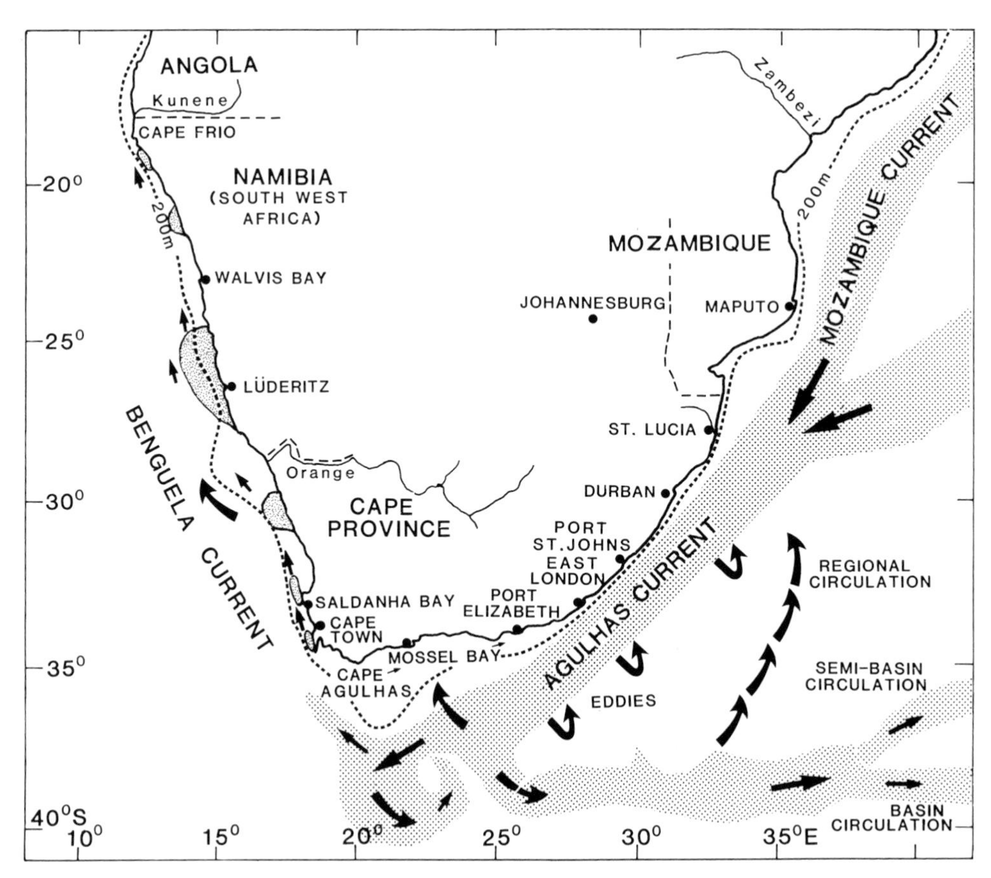

Fig. 13.1. Map of southern Africa showing major oceanographic features and sites mentioned in the text. The 200 m depth contour and principal ocean currents are shown, with arrows indicating currents, eddies and circulation patterns. Coarse stippling shows the warm-water currents. The main upwelling centres in the Benguela Current system are shown in fine shading.

Surface water flows at 2 m s^{-1} in the core, which ranges from 21 to 26°C and 35.4‰ salinity. It flows close to the coast north of Durban, meanders offshore where the shelf widens off Durban, hugs the shelf break close to East London and moves further offshore over the wide Agulhas Bank. Inshore of the core current the temperature is about 2°C cooler over the shelf, due to "dynamic upwelling" generated by the thrust and twist of the current. There are periodic "current reversals" with inshore pockets of cooler water flowing parallel to the coast in the opposite direction to the west-flowing Agulhas Current. The inshore counter currents are marked and frequent on the south coast between Cape Agulhas and Port Elizabeth, but less predictable as the shelf narrows further eastwards. The broad Agulhas Bank, which stretches from roughly Port Elizabeth to Cape Agulhas (Fig. 13.1) has a surface layer 20 to 50 m thick of warm (*c*. 18–22°C) water overlying cool bottom water (12–15°C). The stratification is marked in summer but tends to break down in winter storms. Periodically, parts of the south coast are subjected to local wind-generated upwelling of cool bottom water, which may even stun fish. On the seaward side of the Agulhas Current, the counter-clockwise recirculation patterns may be simplified into a series of four concentric gyres: (1) local eddies that recirculate within a few weeks on a spatial scale of a few hundred kilometres, (2) a regional circulation reaching as far east as the Madagascar Ridge (see Fig. 13.1), (3) the semi-basin circulation (mid-ocean ridge, 6 months), (4) the basin circulation (Australia, 1 year). Periodically, warm, mixed water from the Agulhas Bank intrudes into the South Atlantic Ocean carrying Indian Ocean plankton species and larvae up the west coast as far as Saldanha Bay.

The extent and frequency of this intrusion probably has a large influence on inter-annual variability of fish stock recruitment (L.V. Shannon, pers. commun.).

The Benguela Current influences the west coast from Cape Agulhas to southern Angola (Shannon, 1985). In the extreme southwest, the coast from Cape Agulhas to Cape of Good Hope may be regarded as a region of overlap between the warm-temperate south coast under the influence of the Agulhas Current and the cool-temperate Benguela region characterized by upwelling. In the southern Benguela region the upwelling is intermittent, being closely linked to the pulsing of southerly to southeasterly winds, which predominate in spring–summer (September–March). North of Saldanha Bay winds are more moderate and constant and upwelling is less seasonal. The major upwelling region is centred around Lüderitz, where near-gale-force southerly winds blow perennially alongside the Namib Desert, creating a semi-permanent plume of cool surface water. North of Walvis Bay the main upwelling season is in winter. Upwelling is concentrated into several main regions by local topography of mountains and bays, with three main plumes of cooler water, one south of the Orange River and two much larger ones off the Namibian coast. The upwelling regions are characterized by major fronts between cool (8–14°C) water inshore and warm, blue oceanic water (16–20°C). The front is marked by a north-flowing jet current which coincides approximately with the shelf edge but which meanders offshore and onshore according to the meteorological condition and strength of upwelling. The periodic clockwise passage of coastal low-pressure atmospheric conditions tends to be accompanied by an inshore counter-current transporting shelf waters southwards. This is often sun-warmed water in which phytoplankton blooms have had time to develop, so that plankton-rich water then bathes the shores. Large plankton blooms off Namibia result in massive decay in the sediments and oligoxic (<2 ml O_2 l^{-1}) bottom water is transported southwards in a deep counter-current as far south as Saldanha Bay, causing occasional inshore migration or even mortality of rock lobsters (*Jasus lalandii*).

Based on the currents, southern African shores may be divided into four main provinces (Brown and Jarman, 1978): a tropical east-coast province north of 26°S, which is characterized by coral reefs and is not discussed further here; the subtropical east-coast province from 26°S to about 31°S; the warm-temperate south coast reaching Cape Agulhas; and the cold-temperate west coast from Cape Agulhas to Angola. The distinction between west-coast and south-coast provinces is confined to the littoral zone, for below about 30 m depth the fauna has similar components (Brown and Jarman, 1978) and temperatures are remarkably constant at 10 to 14°C from Namibia to Port Elizabeth in the east.

Seasonality

On the east coast mean monthly sea surface temperatures range from 22° in winter to 27°C in summer, whilst on the south coast they range from 15 to 22°C, respectively. In False Bay on the eastern side of the Cape Peninsula, in the overlap between warm- and cold-temperature provinces, mean monthly values range from 14 to 19°C, whereas in Table Bay near Cape Town there is little difference between winter and summer means at 14 to 15°C. The constancy of monthly mean temperatures in Table Bay is misleading, because surface temperatures frequently fluctuate from 8 to 18°C during the summer upwelling season.

Tides and waves

The entire region is subjected to a simple semi-diurnal tidal regime, with a spring-tide amplitude of some 2 to 2.5 m and a neap-tidal range of about 1 m. The exposed nature of the coast, jutting south between the Atlantic and Indian Oceans towards the Southern Ocean, results in little tidal delay around the coast. Low water of spring tides occurs in the mornings between 08.00 and 10.00 and at night between 20.00 and 22.00 hours, resulting in relatively little heat stress and desiccation at low levels in the eulittoral, which are only exposed at spring tides. The exposed coastline is also subjected to strong wave action, particularly in the southwest, with maximum waves exceeding 6 m in height 10% of the time (Shillington, 1978). These are generated by storms up to 5000 km away in the South Atlantic Ocean.

ROCKY SHORES

Biogeographic patterns

Stephenson (1948) and Brown and Jarman (1978) have analyzed the biogeographic affinities of rocky-shore species from Durban to Port Nolloth, south of the Orange River. Figure 13.2 shows the four main components: a subtropical one declining westward, a warm-temperate one centred around the south coast, a cold-temperate component declining eastwards and a ubiquitous component. Thus the proportions of each component vary around the coast. The greatest species diversity and largest proportion of subtropical forms occur in the east-coast province, with decreasing numbers of species and increasing warm- and cold-temperate components as one moves to the south and west coasts (i.e., cooler waters).

Figure 13.3 summarizes the distribution of some of the commonest species, arranged approximately according to height up the shore. Several species at all levels have their western limit near Port St. Johns on the Transkei coast, marking the western limit of the subtropical east-coast region. Another group of species, including the barnacles, extend over both south and east coasts, whereas several sublittoral species are more or less confined to the cold-temperate west coast. Some species such as the supra-littoral *Littorina africana* and the sublittoral tunicate *Pyura stolonifera* span all three provinces. It should be mentioned that very little of the Namibian coast is rocky, so that most records come from only a few reasonably accessible rocky-shore sites such as at Lüderitz, and near Walvis Bay. The northern boundary of the cold-temperate west-coast province coincides approximately with the Namibian/Angolan boundary, the region between Cape Frio (Namibia) and Moçamedes at 15°10′S in Angola being an overlap region with subtropical biota. There are few rocky shores south of Moçamedes (Kensley and Penrith, 1970).

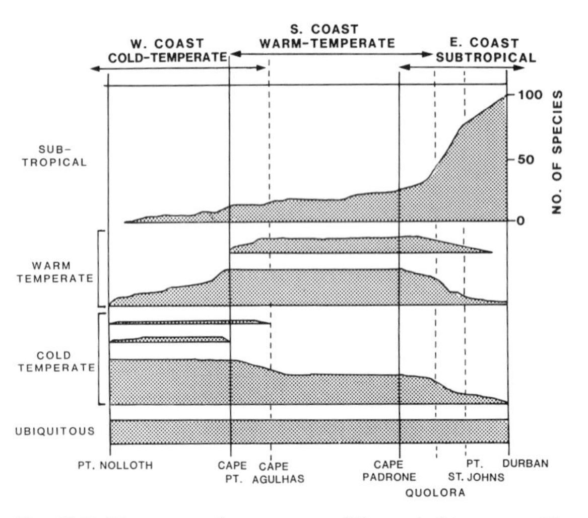

Fig. 13.2. Diagrammatic summary of the main biogeographic affinities of rocky-shore species from Port Nolloth on the west coast to Durban on the east coast. Biogeographic regions and overlap zones are indicated across the top of the diagram. The vertical axis shows the numbers of species. The four main biotic components are shown on the left of the diagram. (After Stephenson, 1948.)

Vertical zonation

Vertical zonation has been studied in southern African rocky shores since the 1930's when T.A. Stephenson and his team conducted qualitative surveys around the coast — summarized in work by Stephenson (1948) and Stephenson and Stephenson (1972). More recent discussions are given by Brown and Jarman (1978) and Branch and Branch (1981). Table 13.1 schematically summarizes the commonest species that characterize different vertical zones in each of the three biogeographic provinces. In all three biogeographic provinces the supra-littoral zone and supra-littoral fringe above the high water neap (HWN) tide level is called the *Littorina* zone because it is dominated by desiccation-resistant littorinids (principally *Littorina africana*, with a subspecies *L. africana* var. *knysnaensis* on the south and west coasts).

On the east coast (Table 13.1A) the upper and mid-littoral zones have an oyster belt of *Saccostrea* above a belt dominated by barnacles (*Chthamalus*, *Octomeris* and *Tetraclita*), the limpets *Patella* and *Cellana* and the polychaete worm *Pomatoleios*. The lower shore has a belt of anemones and mussels (*Zoanthus* and *Perna* respectively) while the infra-littoral fringe above the HWS mark has the tunicate *Pyura* plus the algae *Gelidium* and *Hypnea*.

On the warm-temperate south coast (Table 13.1B) typical shores have the supra-littoral and its fringe populated by *Littorina africana* and the algae *Bostrichia* and *Porphyra*. The upper mid-littoral is characterized by barnacles, limpets, the

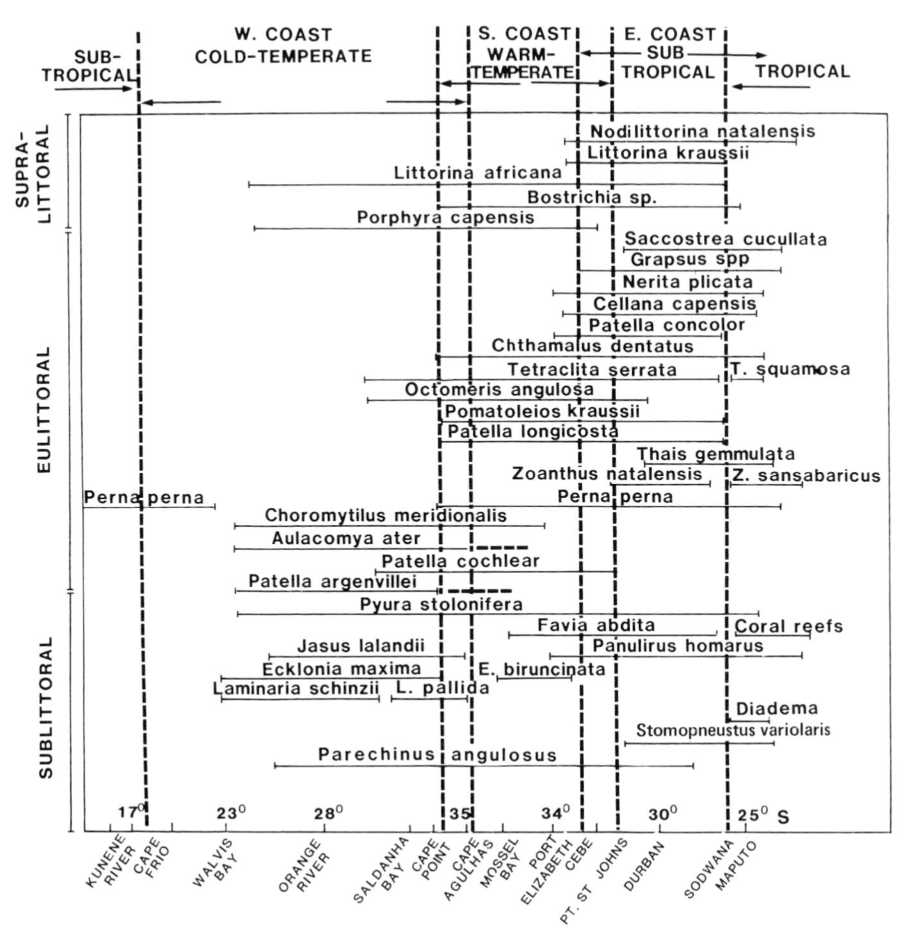

Fig. 13.3. Biogeographic distribution of some common littoral species from Mozambique in the east to Angola in the west, arranged approximately according to height up the shore. The main biotic regions are indicated at the top of the diagram from subtropical, cold-temperate, warm-temperate, to subtropical, and tropical. Overlaps between cold-temperate, warm-temperate and subtropical regions are also indicated. (After Brown and Jarman, 1978.)

gastropod *Oxystele* and the algae *Splachnidium* and *Ulva*. In the lower mid-littoral, the molluscs *Oxystele tigrina*, *Patella longicosta* and *Perna perna* dominate with the polychaete *Pomatoleios* and the alga *Gelidium*. The sublittoral fringe is very characteristic with a belt of pear-shaped limpets, *Patella cochlear*, with their algal gardens and encrusting grey coralline algae. The lower fringe just above LWS includes the alga *Hypnea* and the tunicate *Pyura*, which extends into the sublittoral. Among the common sublittoral forms are sea urchins, *Parechinus*, the rock crab *Plagusia*, and, in shallow gullies, the kelp *Ecklonia biruncinata*.

The cold-temperate west coast (Table 13.1C) has a *Littorina* zone similar to that of the south coast. The upper mid-littoral is typically reduced to two barnacles *Octomeris* and *Tetraclita* with *Patella granularis* and *Ulva* as other co-dominants. The lower mid-littoral has the limpet *P. granatina*, the alga *Splachnidium* and the colonial tube-building polychaete *Gunnarea*. The lowest levels of the eulittoral include a distinct belt of *Patella argenvillei* and *P. cochlear* with the anemone *Bunodactis* and the mussel *Choromytilus* dominating in patches. Recently *Mytilus galloprovincialis* has been found in the eulittoral zone, but it is not known when this species was introduced since it has been confused with *Choromytilus meridionalis* in the past (Grant and Cherry, 1985). Both the ribbed mussel *Aulacomya* and tunicate *Pyura* extend from the sublittoral just above LWS. Beds

TABLE 13.1

Schematic tabulation of vertical zonation patterns of the commonest species. Full vertical range of species is not shown. Tidal range is about 2 m. The eullitoral has been subdivided into a supra-littoral fringe, upper and lower mid-littoral zones and a sublittoral fringe. Standard abbreviations are used for tidal levels

Supra-littoral	Eulittoral				Sublittoral
	Supra-littoral fringe MHW–HWN	Upper mid-littoral HWN–MSL	Lower mid-littoral MSL–LWN	Sublittoral fringe LWN–LWS	below LWS
A. Subtropical east coast of southern Africa					
	Nodilittorina natalensis	*Saccostrea cucullata*	*Zoanthus natalensis*	*Hypnea* spp.	*Favia favus* (coral)
	Littorina natalensis	*Grapsus* spp.	*Perna perna*	*Gelidium amansii*	*Panulirus homarus*
	L. kraussii	*Cellana capensis*		*Pyura stolonifera*	*Stomopneustes variolaris*
		Patella concolor			
		Chthamalus dentatus			
		Octomeris angulosa			
		Tetraclita serrata			
		Pomatoleios kraussii			
B. Warm-temperate south coast of southern Africa					
	Littorina africana	*Oxystele variegata*	*Oxystele tigrina*	*Patella cochlear*	*Parechinus angulosus*
	Porphyra capensis	*Chthamalus dentatus*	*Gelidium pristoides*	Algal garden	*Ecklonia biruncinata*
	Bostrichia mixta	*Tetraclita serrata*	*Patella longicosta*	*Hypnea spicifera*	*Plagusia chabrus*
		Octomeris angulosa	*Perna perna*	*Pyura stolonifera*	
		Patella granularis	*Pomatoleios kraussi*		
		Splachnidium rugosum			
		Ulva sp.			
C. Cold-temperate west coast of southern Africa					
	Littorina africana	*Tetraclita serrata*	*Patella granatina*	*Patella cochlear*	*Aulacomya ater*
	Porphyra capensis	*Octomeris angulosa*	*Splachnidium rugosum*	*P. argenvillei*	*Pyura stolonifera*
	Bostrichia mixta	*Ulva* sp.	*Gunnarea capensis*	*Bunodactis reynaudi*	*Jasus lalandii*
		Patella granularis		*Choromytilus meridionalis*	*Laminaria* spp.
					Ecklonia maxima
					Parechinus angulosus

of *Laminaria* form a sub-canopy about 2 m above the bottom down to 10 or 20 m depth. The giant kelp *Ecklonia maxima* is more obvious from the shore because it floats at the surface by means of a distally swollen gas-filled stipe, forming a dense canopy of fronds at the surface. The kelp-bed community structure has been described by Field et al. (1980b) and includes a mosaic patch-work of understorey algae, mussels (*Aulacomya*), sea urchins (*Parechinus*) and holothurians, with the spiny lobster *Jasus lalandii*, a carnivore of commercial importance.

Relation to exposure and rock-type

The zonation described in the previous section was originally based on subjective assessment of the dominance of a few common species. It is thus not quantitative, and much information on the distribution of small or rarer species is ignored. In False Bay, the overlap region between south and west coast biogeographic provinces, Field and Robb (1970) compared 10 quantitative samples taken at three sites within 6 km of one another which were classified as exposed, semi-exposed and sheltered, with another 9 samples at different heights up the shore at the sheltered site. The results of multivariate analysis of the 79 species larger than 1 mm are reproduced in Fig. 13.4; they show that the biota as a whole varies fairly continuously with height up the shore, whereas classification into vertical zones based on dominant species gives the misleading impression that zones involve abrupt changes. The analysis also demonstrates that the biota changes more up the shore than with differing exposure to wave action over the restricted range studied, vector 1 extracting more of the variation than vector 2.

McQuaid and Branch (1984) compared 12 sites over some 100 km of coastline extending from the cold-temperate west coast to False Bay in the overlap region with the warm-temperate south-coast biota. Pooling results from 310 species at different levels at each site, they found that the biota was more affected by exposure to wave action than by the temperature regime, and that it varied little according to the geological type of substratum (coarse-grained granite, sandstone and shale). However, those sites with boulders that move around in storms had greatly reduced numbers of species and biomass. Sessile, attached filter-feeders were particularly affected.

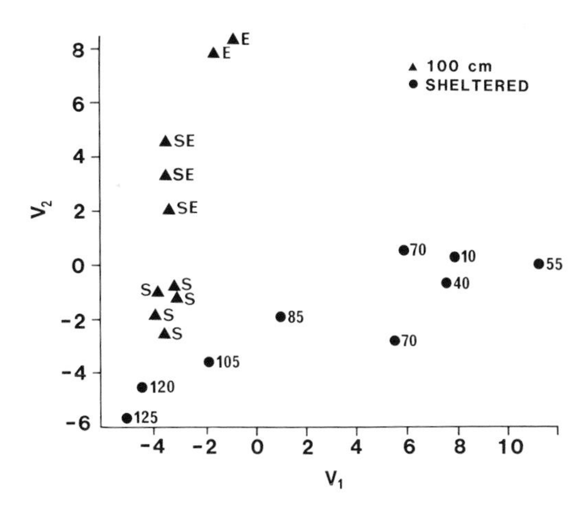

Fig. 13.4. Multivariate ordination of 19 rocky-shore samples from different heights up the eulittoral zone and different degrees of exposure to wave action, in False Bay, near Cape Town. Vector V_1 (horizontal axis) shows the major variation in log-transformed abundance data, and is strongly correlated with decreasing height down the shore (shown in cm). The vertical axis (V_2) is correlated with exposure to wave action, indicated by exposed (E), semi-exposed (SE) and sheltered (S) localities. Heights above datum (cm) are given for sheltered localities, others are all at 100 cm above datum. (Modified from Field and Robb, 1970.)

Rocky shores on the south and east coast have not been quantitatively analyzed with the same detail but Jackson (1976) attributes semi-quantitative changes in community structure in the east-coast region to differing exposure to wave action, amongst other factors. Similarly, on the south coast, McLachlan et al. (1981a) attributed the difference in community structure between two sites to contrasting exposure to wave action.

Rock pools

Brown and Jarman (1978) listed some species found in rock pools at different tidal levels in each of the biogeographic provinces. Huggett and Griffiths (1986) have analyzed the biota of 28 rock pools in the cold-temperate west coast in relation to fluctuating temperature and oxygen conditions in pools up the shore. Cluster analysis shows that the 33 species can be grouped into three zones in much the same way as the biota exposed to air. However, the factors limiting distribution are different in pools, since evaporative cooling is not possible and oxygen concentrations decrease in the dark. Species richness and biomass decrease with height up the shore. There is a complex but predictable interplay of physical and chemical factors with the biota. All pools experience two extremes daily: an oxygen minimum before dawn (or before flooding after a nocturnal low tide) and a temperature maximum. In summer at spring tides high shore pools heat up by over 12°C in 6 hours, with a drop from 28 to 16°C within a few minutes upon flooding. The largest oxygen fluctuations occur in lower mid-littoral pools, within regular diurnal ranges from <20% to >200%. Oxygen fluctuations are due mainly to metabolism of the pool biota. High on the shore, biomass in pools is reduced, but, as biomass increases down the shore, fluctuations in oxygen concentration also increase, until below LWN exposure time is so short that oxygen fluctuations are damped by regular inundation by the sea. Rock pools have not been observed to become anoxic; presumably the biota is regulated by oxygen conditions to prevent this.

Trophic relationships and energy-flow

It has been seen that exposure to wave action affects the species composition of the community. Although no quantitative data are available for east-coast rocky shores, biomass data collected by McQuaid and Branch (1985) at 12 sites around the Cape Peninsula and by McLachlan et al. (1981a) at 2 sites on the south coast show a relationship between biomass and trophic level at both exposed and sheltered sites (Fig. 13.5). At sheltered sites there is a semi-logarithmic relationship ($r^2=0.65$, $n=28$) between biomass and trophic level (with filter-feeders nominally at level 3). At exposed sites the biomass of algae, herbivores and carnivores is much the same, but there is a marked increase in filter-feeder biomass. The south-coast sheltered site has biomass values in the range of Cape Peninsula sites, but the exposed site has the lowest biomass values for all trophic categories except grazers. This bears out the conclusion of McQuaid and Branch (1985) that degree of exposure to wave action determines the trophic structure of a rocky-shore community, but biogeographic factors determine the species composition of the community.

Figure 13.6 depicts flows through a hypothetical generalized rocky-shore littoral system exposed to

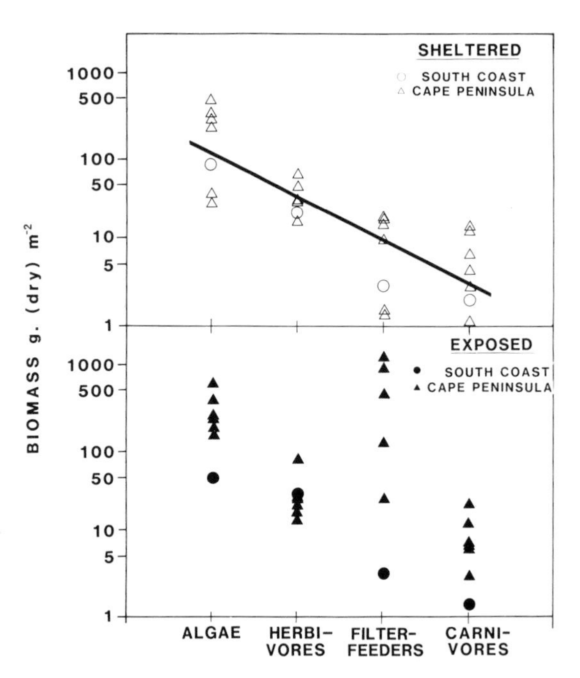

Fig. 13.5. Biomass of trophic components in the rocky eulittoral, at sites exposed to and sheltered from wave action along the south coast and the Cape Peninsula. At sheltered sites there is a semi-logarithmic relationship between dry biomass and trophic level, with filter-feeders at nominal trophic level 3 ($r^2=0.65$, $n=28$). At exposed sites, filter-feeder biomass is much increased, whereas other trophic categories have similar biomasses to sheltered sites. (Data from McLachlan et al. 1981a; McQuaid and Branch, 1985.)

wave action. Primary production is by phytoplankton, macroscopic seaweeds and also sporelings and micro-algae, which Branch (1981) has shown to be an important source of primary production for grazing molluscs in the eulittoral. The biomass of sporelings and micro-algae is negligible, but their turnover rate is fast and is maintained by the continued grazing and "gardening" activities of limpets. Thus their productivity is estimated to be about equivalent to that of the macrophytic seaweeds (Field, 1983). The productivity of phytoplankton is probably not as important to rocky-shore communities as the rate at which it is transported to and from filter-feeders by tide, waves and currents. Thus the "work-gates" in the diagram depict the importance of water movements in transporting phytoplankton and detritus, which is augmented by resuspension of animal faeces (not depicted). The rate of transport of material to rocky-shore communities has not been measured anywhere, yet it is important if one is to

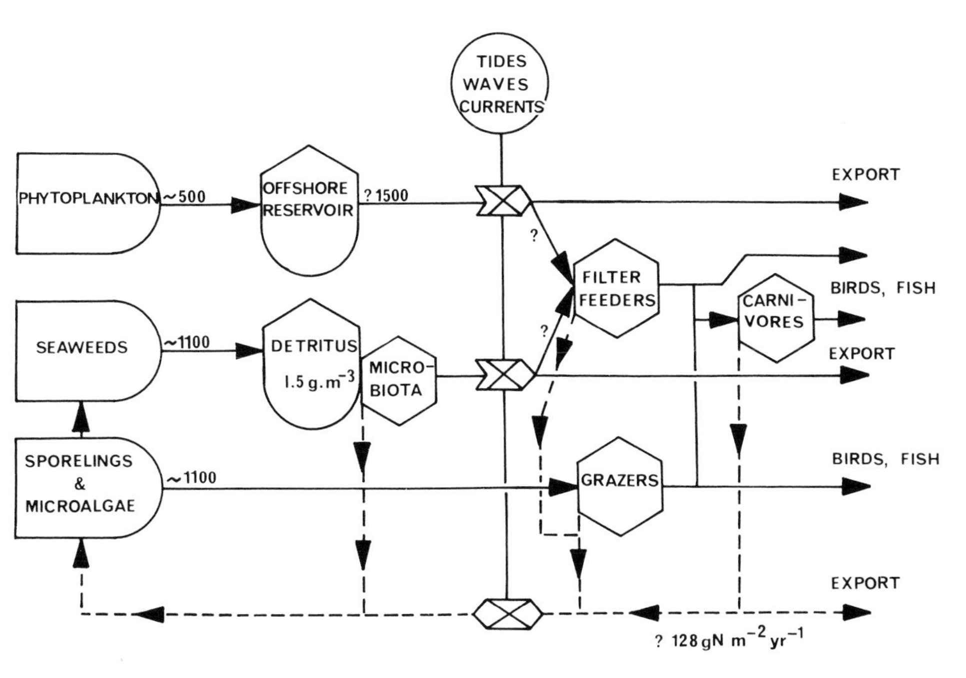

Fig. 13.6. Carbon and nitrogen flow through a hypothetical rocky-shore ecosystem exposed to wave action. Sources of primary production are shown by bullet symbols, and estimated carbon flows (g C m^{-2} yr^{-1}) by solid lines. The influence of external energy sources (tides, waves and currents) is depicted by uni- or bi-directional work-gate symbols, indicating their control in advecting and suspending phytoplankton, detritus and nutrients to filter-feeders or out of the system. Storages are indicated by "birdcage" symbols. Heterotrophic components are depicted by hexagonal symbols, whose excretions are indicated by broken arrows leading down and back to the seaweeds and micro-algae, or export from the system (g N m^{-2} yr^{-1}). (From Field, 1983, reproduced by permission of Wiley-Interscience.)

quantify the flow of materials through the system. From metabolic considerations one can estimate the food requirements of both filter-feeders and grazers, but it is difficult to apportion these amongst the various sources of food.

The rate of nutrient cycling can be roughly estimated if one assumes that all filter-feeders excrete ammonium at the same rate as mussels (5–40 μg NH_4-N $g^{-1}h^{-1}$; Bayne and Widdows, 1978; Kautsky and Wallentinus, 1980). Based on this assumption, the mean value for nitrogen released by all animals should support primary production of about 2200 g C m^{-2} yr^{-1}, the estimated production rate of small littoral algae in the Cape Peninsula (Field, 1983).

Biological interactions

Some of the biological interactions of intra- and inter-specific competition and predation that shape the structure of littoral communities have been investigated in southern Africa, but space permits only a few examples here. Thus intra-specific competition has been demonstrated in populations of the low-shore limpet *Patella cochlear*, which shows decreasing gonad output per unit biomass and decreasing maximum length with increasing density (Branch, 1975). Branch (1985) has also explained the mode of inter-specific competition among the 12 *Patella* species inhabiting southern African shores. Four mid- to upper-shore species are generalized browsers feeding opportunistically on an unpredictable and uncertain supply of macro-algae, diatoms and even spume left by the waves. These species compete by non-selective exploitation. On the other hand, infra-littoral and low-shore species tend to have specialized diets and compete for space on the shore, defending territories from intrusion by other limpet species. Thus *P. longicosta* and *P. tabularis* both defend their "gardens" of the alga *Ralfsia*, and *P. cochlear* individuals rotate on their scars and feed on a narrow band of algae around their scars that are defended territorially. Branch (1984) has shown that the clearly demarcated band of *P. cochlear* at the low-tide mark may exclude *P. longicosta* completely if densities of *P. cochlear* exceed 400 m^{-2}, eliminating *Ralfsia* and hence also the specialized *P. longicosta* that depends upon it. Branch (1984) has shown a close correlation between the number of limpet species co-existing on a shore and quantitative measures of both dietary and zonational specialization. On a geographic basis, Fig. 13.7 shows that the greatest degree of specialization occurs on the west and south coast, diminishing towards the subtropical regions further west and farther east (Branch, 1985). The incidence of interference competition (territorial defence of algal "gardens" or food plants, and physically pushing away neighbours) is strongly positively correlated with the number of co-existing species.

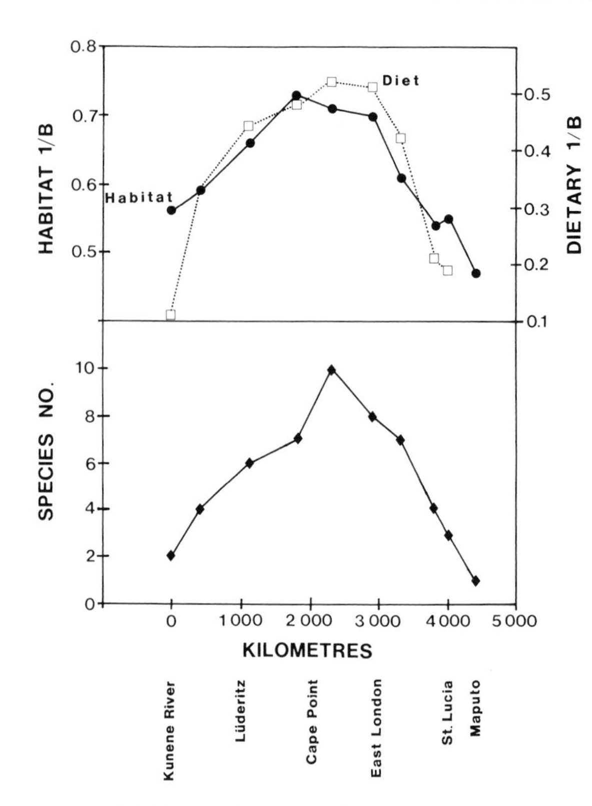

Fig. 13.7. Relationship between habitat specialization (1/*B*) (solid line), dietary specialization (1/*B*) (broken line), and species diversity (lower panel), in 12 limpet species at 10 localities around southern Africa. 1/*B* is the inverse of a quantitative measure of niche overlap. (After Branch 1984, 1985.)

Sublittoral rock

On the east coast, energy flow has been studied through a shallow reef ecosystem at "ORI Reef", located near the Oceanographic Research Institute at Durban. The reef is in the infralittoral fringe, being partly exposed at spring low tide. It is adjacent to shifting banks of sand and is exposed to wave action so that areas are alternately covered and uncovered by sand.

Figure 13.8 shows the estimated biomasses and some of the flows of carbon through the system (Berry et al., 1979; Schleyer, 1980; Berry, 1982). Accumulations of detached seaweeds and terrestrial plant material carried by currents and flooding rivers are characteristic of the area. Dense bacterial counts and high indices of bacterial activity, based on uptake of radio-labelled glucose and labelled algal exudate, lead to the conclusion that the animals depend heavily on detrital input, since phytoplankton primary productivity is relatively low. Seaweed cover and growth are intermittent and primary productivity of seaweeds has not been measured, so even relative contributions to primary production are difficult to estimate. The main grazers of seaweeds are fish, including the omnivorous blenny *Blennius cristatus*. The fish fauna is rich and diverse, 62 species having been recorded. Herbivores are poorly represented, and omnivores and carnivores dominate.

Suspension-feeders dominate the reef biomass. These include the brown mussel *Perna perna*, which has a large and variable biomass, fast growth rate and annual *P*/*B* (turnover) ratio of 4.8 (Berry, 1982). Other major suspension-feeders are the ascidian *Pyura stolonifera* and the oyster *Saccostrea margaretacea*, both of which also have turnover rates in excess of twice per annum, giving an annual average *P*/*B* of 3.54 and production of some 2.4 kg dry mass m^{-2} yr^{-1}. Carnivores include the spiny lobster *Panulirus homarus* and fish, such as the small *Blennius cornutus* and rock cod *Epinephelus andersoni*. None of the top carnivores are permanent residents, but they move offshore as they grow larger, or migrate periodically (for example, spiny lobster) or seasonally (for example, Cape cormorant, *Phalacrocórax capensis*). Thus there is export of energy at the highest trophic level, and import of detrital matter and plankton to support the filter-feeders. The system is thus a very open one, with physical forces such

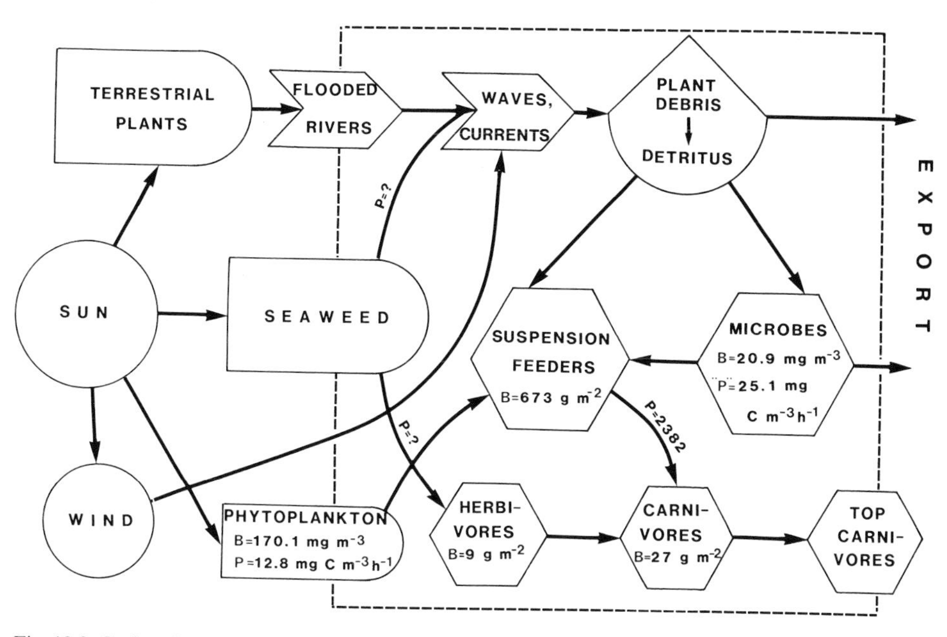

Fig. 13.8. Carbon flows through a sublittoral rocky reef community at Durban on the east coast. Symbols as in Fig. 13.6. Biomasses are given inside the symbols as mg or g m^{-2} or m^{-3} for suspended matter. Production values (*P*) are annual m^{-2}, unless indicated otherwise. Microbial production estimates are based on radio-labelled algal exudate (Schleyer, 1980). (Reproduced with permission from Schleyer, 1980; with additional data from Berry et al. 1979; Berry, 1982.)

as waves, tides, currents and flooding rivers playing an important role.

Figure 13.9 summarizes the factors that are believed to control the structure of the community at ORI Reef, and presumably other infra-littoral and shallow sublittoral reefs near sand banks in the east-coast subtropical region (Berry, 1982). Although the ascidian *Pyura stolonifera* and mussel *Perna perna* may co-exist for short periods, heavy settlements of *P. perna* occur occasionally, smothering *P. stolonifera* (and even adult mussels), with the result that mussels dominate the community. Normally however, mussel spat settle on established mussel beds and do not displace established *P. stolonifera*. After disturbance by waves or sand, the primary colonizers are algae, principally *Hypnea spicifera*, *Ulva* sp., *Enteromorpha* sp. and corallines such as *Cheilosporum cultratum* (Jackson, 1976).

Kelp beds dominate rocky subtidal areas on the west coast of southern Africa, which are bathed by the cool Benguela Current. The structure and biomass of kelp bed communities have been documented by Field et al. (1980b). Cluster analyses show that there are two principal types of biotic association within the kelp beds, areas

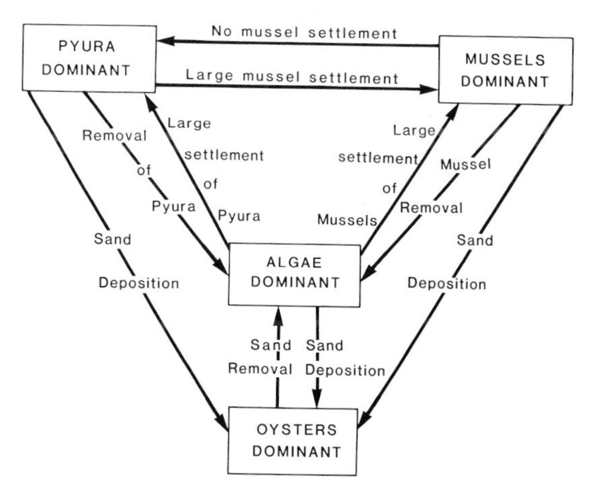

Fig. 13.9. Schematic representation of factors influencing community structure of a subtropical rocky sublittoral reef at Durban on the east coast. Four species that may dominate the community are represented in the boxes, and factors influencing transition to another dominant are represented on the arrows. (Reproduced with permission from Berry, 1982.)

dominated by algae (including kelps) and animals. These tend to form a mosaic, with inshore areas dominated by kelps and understorey algae down to 5 to 10 m depth, and dense patches of animals populating rock surfaces between the kelp in deeper water.

The kelps are of three species: *Ecklonia maxima*, reaching 10 m length and forming a canopy at the sea surface buoyed up by an expanded, gas-filled primary frond; *Laminaria pallida*, a smaller paler kelp that forms a sub-canopy 1 to 2 m off the bottom and extending deeper, depending upon water clarity; and in calmer protected waters, the much rarer *Macrocystis angustifolia*, with its elongate stipes and fronds reaching the surface. Under the kelps, one finds an understorey of red and brown algae such as *Botryocarpa prolifera*, *Epymenia obtusa*, *Gigartina radula*, and *Hymenena venosa* (Field et al., 1980b).

The animals are dominated by suspension-feeders, especially the ribbed mussel *Aulacomya ater*, sponges such as *Polymastia mamillaris* and *Tethya* spp., holothurians *Pentacta doliolum* and *Thyone aurea*, and the ascidian *Pyura stolonifera*. The barnacle *Notomegabalanus algicola* may also be common. These suspension-feeders vary in relative abundance from place to place and tend to form a dynamically changing mosaic of patches. The dynamics of patches is believed to be caused by a combination of storms uprooting kelp plants, patchy stochastic recruitment of animals and plants, and the sweeping effect of kelp fronds keeping areas of rock swept clear of many animals and plants (Velimirov and Griffiths, 1979; Velimirov, 1983). Grazers are much less common, but the specialized limpet *Patella compressa* lives on *Ecklonia maxima* stipes, its shell base being concave to match the convex stipe of the kelp, whereas the gastropod *Turbo cidaris* feeds largely on *Laminaria pallida*. Other grazers include the isopod *Paridotea reticulata*, the abalone *Haliotis midae*, and the sea urchin *Parechinus angulosus*. The latter two species probably subsist on sporelings and micro-algae on the rock surface, but get their main nutrition from drift weed broken off in the strong waves that occur frequently. The carnivores are dominated in biomass by the spiny lobster *Jasus lalandii*, which is also the target of a valuable commercial fishery. Their staple diet is believed to be the ribbed mussel *Aulacomya ater* (Velimirov et al., 1977), although the small, rapidly growing barnacle *Notomegabalanus algicola* has been found to be an important component in their diet in an area where mussels are rare (A. Barkai, pers. commun.). Other carnivores include various polychaete worms and the isopod *Cirolana imposita*, which although small, is very productive (Shafir and Field, 1980). Only one species of fish, *Pachymetopon blochii*, is common, feeding on mysids, and the epiphytes and cryptofauna on kelp.

Energy flow in west-coast kelp beds has been studied by Newell et al. (1982). Their figures for energy balance have been incorporated in Fig. 13.10, in which the kelp bed system is depicted within a broken line border. Kelps, understorey algae and epiphytes with a standing stock of 9208 kJ m^{-2} are estimated to produce an average of some 38 000 kJ m^{-2} yr^{-1}. The kelp fronds behave as moving belts of tissue, growing at the bases and eroding at the tips under the influence of strong wave action. The eroded particles form detritus, depicted as particulate organic matter (POM), dissolved organic matter (DOM), and the community of bacteria and other microbes which colonize the organic matter that nourishes them. The wave action also breaks free whole plants and large pieces of drift weed, which are fed on by herbivores such as the sea urchins and abalone; but their biomass is relatively small (300 kJ m^{-2}) and their food requirements modest at some 4650 kJ m^{-2} yr^{-1}. Jarman and Carter (1981) have estimated that some 12% of the biomass of kelp is uprooted in storms annually. Tidal and upwelling currents, influenced by the prevailing southerly and offshore winds, transport (export) quantities of drift weed out of the system. The remaining macrophyte production (33 554 kJ m^{-2} yr^{-1}) forms detritus, which may be kept in suspension by wave action and consumed by suspension-feeders such as mussels, barnacles and holothurians, totaling 5150 kJ m^{-2} yr^{-1} or 72% of the animal biomass. Another component of the diet of suspension-feeders is the phytoplankton, which has an annual estimated productivity of 23 986 kJ m^{-2} yr^{-1} inside the kelp bed. This is available mainly under calm winds or downwelling conditions, when phytoplankton from offshore is carried into the kelp beds. Under upwelling conditions both detritus and phytoplankton are transported out of kelp beds. Grazers and filter-feeders are preyed upon by carnivores of different sizes, pooled for simplicity into one compartment with a biomass of 1728 kJ m^{-2}. The production of lobsters and other carnivores is consumed by various top predators, including humans, seals,

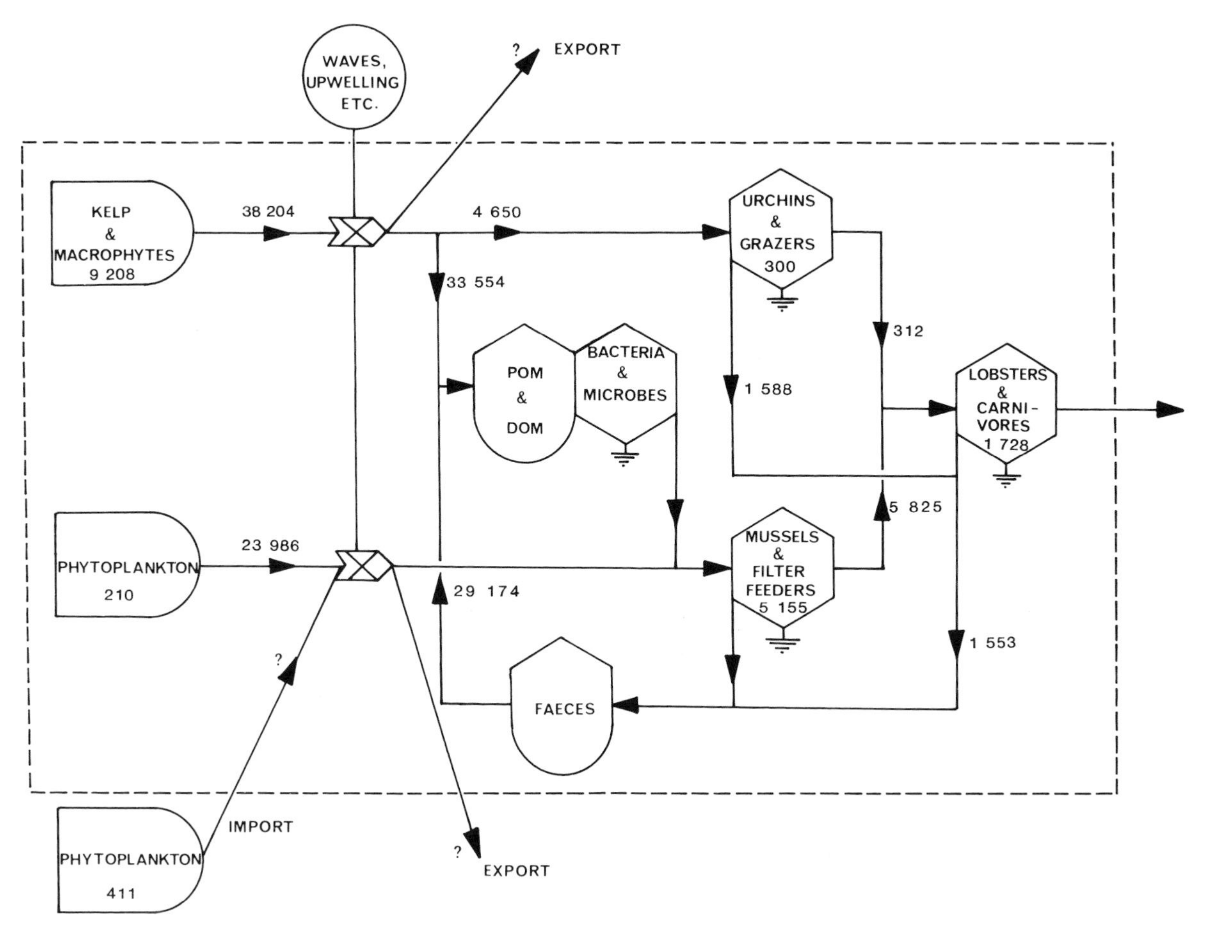

Fig. 13.10. Energy flow in a Benguela kelp bed ecosystem, using the symbols for primary producers, external energy sources, workgates, storage compartments, and heterotrophic organisms as in Fig. 13.6. Note the influence of waves, upwelling water movements, and other physical factors in controlling fragmentation of macrophytes, export of suspended matter and import of phytoplankton. Standing stocks are given inside the compartments as kJ m^{-2}, and average energy flows on the arrows between compartments as kJ m^{-2} yr^{-1}. The kelp bed is taken as having a mean depth of 10 m, and the boundaries of the system are represented as broken lines. The "faeces loop" of animal faeces is depicted as contributing substantially to the particulate and dissolved organic matter (POM and DOM, respectively). (Values after Newell et al., 1982.)

and sea birds, and is depicted by an arrow out of the system. Faeces from the animal populations form a source of detritus, depicted by arrows leading back to the POM/DOM compartment (in systems terms, a feed-back loop), which allows recolonization of refractory material by microbes and its re-enrichment in terms of nitrogen and protein (Newell and Field, 1983).

The energy-balance diagram is based on estimated annual average energy flows and, remarkably, Newell et al. (1982) found that the production of primary producers approximated to the estimated energy requirements of consumers; we deduced from this that imports of phytoplankton probably balance the export of detritus from the system, since it is recognized that the system is not a closed one. Measurements of water transport under upwelling and downwelling weather conditions showed that the water inside the study kelp bed at Oudekraal, near Cape Town, turned over up to seven times per day (Field et al., 1980a, 1981), throwing doubt upon the validity of the energy-balance treatment. However, it is in practice impossible to measure the rates of import of phytoplankton and export of detritus from the system. Accordingly, simulation models were developed in which a realistic but simplified kelp bed food web was modelled with ranges of biological electivity and assimilation efficiencies to test the effect of water transport under closed-system, upwelling and downwelling conditions (Wulff and Field, 1983; Wickens and Field, 1986). It is striking that the transport of phytoplankton, detritus and faecal matter by water movements had a greater effect on simulated filter-feeder feeding and growth rates than the full range of biological electivity and assimilation efficiencies. Surprisingly, it was also found that there was a shortage of food under conditions of continuous upwelling, when detritus is exported from kelp beds, whereas under downwelling conditions, the vast reservoir of phytoplankton from offshore provides a rich source of carbon and nitrogen for filter-feeders. When realistic weekly and seasonal rates of upwelling and downwelling were included in the models, filter-feeders fluctuated in scope for growth and reproduction, increasing during the winter downwelling season and decreasing during the upwelling season. These results are confirmed by analyses of lipid content of mussels and the fact that their main spawning activity is at the end of winter, after the downwelling season. The study thus emphasizes the importance of physical factors like the upwelling/downwelling patterns in influencing ecosystems in the sub-littoral zone of the Benguela region.

SANDY BEACHES

Since the southern African coastline is exposed to strong wave action, the few sheltered beaches are confined to estuary mouths or to the rare lagoonal systems. The proportion of fine sand in beach sediments is low. Thus, median grain sizes occur on the open shore (ranging from about 200 to 1000 μm with increasing wave action), and they decline to 120 μm in lagoonal systems. Since particle size, slope and oxygenation are closely related, such beaches tend to have moderate to steep slopes and to be fully oxygenated to at least 1 m below the sand surface. Because of the degree of sediment mobility few beaches can support macrofauna that construct stable burrows within the intertidal zone. Since exposure is the critical parameter controlling the biological processes occurring within beaches, an exposure scale from 1 to 20, as developed by McLachlan (1980a), is employed in the following account. On this scale almost all southern African oceanic beaches rate 12 (exposed) to 17 (very exposed).

Although pure sand beaches make up 42% of the southern African coastline (Bally et al., 1984), research into this habitat has lagged well behind that of rocky shores, beginning in earnest only with Brown's (1964) account of food relationships on the intertidal sandy beaches of the Cape Peninsula. A vigorous sandy-beach research programme has, however, subsequently developed, and southern Africa's sandy beaches are today undoubtedly the best known in the Southern Hemisphere.

Biogeographic patterns

Descriptive accounts of the ecology of sandy beaches are available for most regions around the southern African coast. Natal beaches have been surveyed by Dye et al. (1981), those of Transkei by Wooldridge et al. (1981), the eastern Cape by McLachlan (1977b, 1980b), the southern Cape by McLachlan et al. (1981b), the Cape Peninsula by Brown (1971) and Koop and Griffiths (1982), the western Cape by Bally (1983) and Namibia (South West Africa) by McLachlan (1985) and Tarr et al. (1986).

Northern Natal is characterized by long, moderately sloping beaches with wide surf zones, medium sand and exposure ratings of about 14. There is generally no drift line, and the sparse but moderately diverse macrofauna is characterized by tropical Crustacea, notably ghost crabs *Ocypode* spp. and mole crabs *Emerita austroafricana* and *Hippa adactyla*. Just north of Durban there is an abrupt change to very steep, coarse-grained beaches characterized by plunging waves. These beaches have very high exposure ratings of about 17.5, which preclude the establishment of macrofaunal communities within the eulittoral zone, the only species remaining being supralittoral *Ocypode* spp.

From the Transkei border to East London short pocket beaches confined between rocky headlands, and with exposure ratings of about 13, are common, many of them associated with small closed estuaries. The fauna here is more diverse, with gastropods and isopods increasingly important. The remainder of the south coast to the Cape Peninsula includes several long stretches of exposed to very exposed beach (rating 12–16), several of which fringe large half-heart bays. The fauna is moderately diverse and dominated in terms of biomass and numbers by dense populations of bivalves (*Donax*) and gastropods (*Bullia*).

The southwestern Cape coast incorporates numerous small pocket beaches that typically receive large inputs of kelp wrack from adjacent kelp beds. This material radically influences the macrofauna, which tends to consist predominantly of semi-terrestrial Crustacea and insects that feed on wrack along the drift-line.

Moving up the west coast there are long stretches of exposed to very exposed (rating 13–18) sand, usually high in calcium carbonate. The macrofauna is once again dominated by *Bullia* and *Donax*, although areas receiving significant wrack also support a rich drift-line fauna, including the giant isopod *Tylos granulatus*. An extensive area between St. Helena Bay (32°S) and Walvis Bay (22°S) remains unexplored, but at Walvis Bay the beaches rate 13–15.5 and consist of well-sorted medium quartz sands virtually devoid of calcium carbonate. There is a moderate drift-line fauna of scavenging isopods, amphipods and insects, but intertidal biomass is low and made up largely of isopods and polychaetes, with *Donax* and the mysid *Gastrosaccus* sp. becoming dominant at low water. In northern Namibia (South West Africa) beaches rating 15–17 support a macrofauna poor in molluscs and consisting mainly of eulittoral isopods and drift-line arthropods. Only in the most northerly site, at 17°S, are more tropical forms such as *Ocypode* spp. evident.

Longshore distribution ranges of major macrofaunal species found on southern African beaches are depicted in Fig. 13.11. The fauna clearly falls into biogeographic provinces similar to those described for rocky shores (Brown and Jarman, 1978). The subtropical component, exemplified by *Emerita*, *Hippa* and *Ocypode*, extends down the east coast to about Cebe, in Transkei. From here a relatively uniform temperate biota extends right around the coast to northern Namibia. Within this extensive zone a warmer south-coast component can, however, be distinguished from a cooler west-

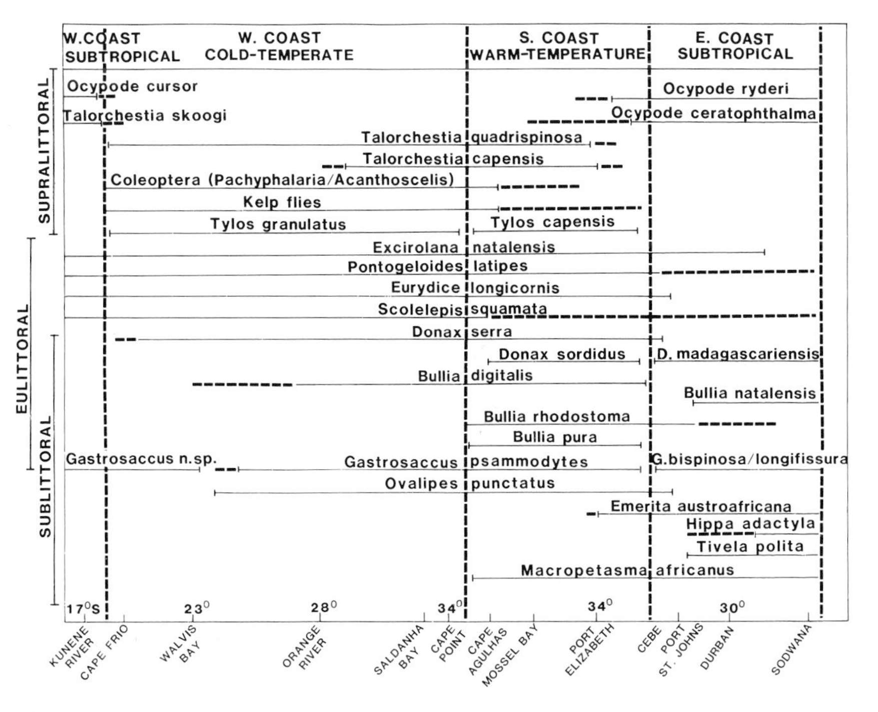

Fig. 13.11. Longshore distribution patterns of major sandy-beach macrofaunal species around the southern African coastline. The vertical axis depicts the height up the beach, whereas the horizontal axis represents the distance from west to east. Broken lines indicate unconfirmed distributions. (Modified after McLachlan et al., 1981b.)

coast one, the transition point occurring around Cape Point. This division is best characterized by the neat segregation of *Tylos granulatus* from *T. capensis*, but is also evident in the distribution of *Bullia pura*, *B. rhodostoma* and *Donax sordidus*.

The surprising northwards penetration of the cool-temperate fauna into Namibia is made possible by the cold northerly flowing Benguela Current, which reduces temperatures far below those experienced at equivalent latitudes on the east coast. It is thus only at 17°S that tropical elements, notably *Ocypode cursor* and *Talorchestia skoogi*, appear, whereas similar forms extend to 31°S on the east coast, which is warmed by the southerly flowing Agulhas Current.

Eulittoral zonation patterns

Macrofauna

Two universal zonation schemes for sandy-beach macrofauna have been proposed. The first of these, by Dahl (1952), recognizes three zones characterized primarily by their crustacean fauna. These are a supralittoral fringe, characterized by ocypodid crabs or talitrid amphipods, a midlittoral, typified by cirolanid isopods, and a sublittoral fringe, with a more diverse fauna in which crustaceans may be less important. Salvat (1964) subsequently proposed a scheme based on physical rather than biological parameters, and which recognizes four zones of drying, retention, resurgence and saturation.

Zonation on southern African beaches is specifically addressed by McLachlan (1980b), Bally (1983) and McLachlan et al. (1984), although many other papers refer to the topic in passing. McLachlan (1980b) found the intertidal macrofauna of a high-energy beach in the eastern Cape to be grouped into a lower belt, characterized by *Bullia rhodostoma*, *Donax sordidus* and *Gastrosaccus psammodytes*, and a midlittoral, dominated by *D. serra* and by isopods. The upper beach was devoid of macrofauna, probably because there was no distinct drift-line and the study area lay between the distribution limits of talitrid amphipods and ocypodid crabs. The zonation pattern thus corresponds to that proposed by Dahl (1952).

By contrast Bally (1983), using a more intensive sampling grid and sophisticated quantitative techniques, identified four zones on west-coast beaches, as proposed by Salvat (1964). An upper zone of drying was characterized by air-breathing Coleoptera, Diptera, Isopoda and Amphipoda. These were absent from the zone of retention, which was dominated by the spionid polychaete *Scolelepis squamata*, the isopods *Eurydice longicornis* and *Pontogeloides latipes*, and by juvenile *Donax serra*. In the zone of resurgence these were joined by other amphipods and cumaceans. The lowest zone of saturation was devoid of isopods but typified by dense populations of the mysid *Gastrosaccus psammodytes*, as well as *Bullia digitalis* and the nemertean *Cerebratulus fuscus*.

While both the above studies were limited to the intertidal, that of McLachlan et al. (1984) extended from the low water mark to 10 to 20 m depth. Numerical classification of these samples revealed three associations. The first, an inner turbulent zone, was restricted to very shallow water within the breakers and supported a low macrofaunal biomass comprised mostly of *Donax sordidus*, *Gastrosaccus psammodytes*, *Bullia* spp. and the prawn *Macropetasma africanus*. This is clearly an extension of the sublittoral fringe or the "saturation zone" of earlier authors. A second zone, just beyond the breakers, supported a very low biomass of *D. sordidus* and of species more abundant offshore and was termed the transition zone. The third grouping emerged as sediments became more stable at 5 to 13 m depth, resulting in a marked increase in both diversity and biomass. This outer turbulent zone was characterized by *Callianassa* spp., *Echinocardium cordatum*, the sipunculid *Golfingia capensis* and the bivalve *Phaxas decipiens*.

Coupling this zonation pattern with a similar series described off the west coast by Christie (1976) and with earlier intertidal systems, McLachlan et al. (1984) divided the littoral and sublittoral beach system into:

— A supralittoral zone characterized by air-breathing Crustacea.

— A midlittoral typified by cirolanid isopods.

— An inner turbulent zone (= sublittoral fringe or zone of saturation).

— A transition zone where extreme turbulence keeps the bottom almost devoid of macrofauna.

— An outer turbulent zone in which stability rapidly increases, resulting in increased biomass and diversity.

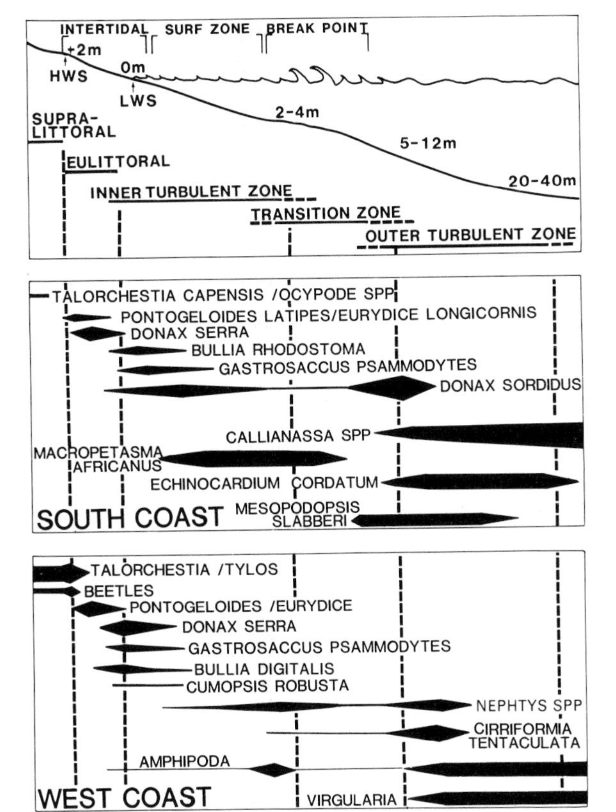

Fig. 13.12. Diagrammatic representation of zonation patterns of eulittoral and sublittoral macrofauna on exposed southern African sandy beaches. Zones and approximate heights in relation to LWS are given in the top panel. Thickness of the kites is proportional to relative abundance of species. (South coast data from McLachlan 1980b; McLachlan et al., 1984, west coast data from Christie, 1976; Bally, 1983.)

This scheme is depicted in Fig. 13.12 together with typical faunistic distribution patterns for the south and west coasts. The boundaries of these zones are not as distinct as they are on rocky shores but this is not surprising in view of the dynamic nature of the substratum, and because many species undertake both semi-lunar and tidal migrations up and down the shore (McLachlan et al., 1979b).

Meiofauna

Exposed, well-oxygenated beaches support a rich, often diverse meiofauna, which extends deep into the sediments, making it essential to consider zonation on a three-dimensional basis. Using data derived mainly from his earlier papers McLachlan (1977a, b, 1980b) divided the meiofauna of eastern Cape beaches into four fairly distinct strata, based on the degree of desiccation and oxygenation of the sediments. This system corresponds closely to the macrofaunal system proposed by Salvat (1964), consisting of the following in descending order:

(1) A dry sand stratum confined to the upper shore and extending to only about 15 cm depth. This zone may be subject to elevated temperatures, and water saturation may fall below 50% at low tide. The meiofauna is fairly rich (35–600 per 100 ml), consisting mainly of small nematodes and oligochaetes.

(2) A moist sand stratum extending from this point to the permanent water table and tapering to the surface near the low-water mark. The sand here remains at a fairly constant temperature and is never less than 50% water-saturated. Meiofaunal densities are very high (70–700 per 100 ml) with harpacticoid copepods and mystacocarids most abundant, followed by nematodes, oligochaetes and turbellarians.

(3) A water-table stratum, that may extend for some depth below the moist sand layer. Always water-saturated, this zone has oxygen saturations of 40 to 70%. Meiofaunal densities are moderate (20–350 per 100 ml) and divided fairly equally between nematodes and crustaceans.

(4) A low-oxygen layer that supports a sparse meiofauna (3–70 per 100 ml) dominated by nematodes, but which may extend for a metre or more in depth.

While this model appears to be of general application, it must be noted that meiofaunal composition and distribution are strongly influenced by local conditions. For example, increasing wave action, grain size and slope will tend to improve drainage, lowering the water table and hence reducing meiofaunal density in the dry zone, but depressing and often thickening the remaining zones. Grain size also influences meiofaunal composition, nematodes being most abundant in finer sediments and crustaceans in coarser ones. Organic loading may reduce drainage and lower oxygen tensions, pulling the zones upwards, but at the same time concentrations of food around the driftline attract a dense meiofauna to the upper layers (Koop and Griffiths, 1982). Zonation patterns will also vary with the migratory activity of the meiofauna. These may be associated with diurnal or tidal rhythms, or occur on a semilunar and seasonal basis (McLachlan et al., 1977).

Meiofaunal distribution patterns below the low-water mark essentially represent an extension of the deeper intertidal strata. The surface layer is water-saturated and well oxygenated to a depth dependent upon wave activity and particle size, and is underlain by a low oxygen layer. Malan and McLachlan (1985) show that in Algoa Bay the sediment remains well oxygenated and supports a significant meiofauna, consisting mostly of nematodes, to a depth of at least 50 to 60 cm in the surf zone. Variations in an offshore direction have not been conclusively established, although decreasing turbulence beyond the breakers should result in vertical compression of the zones.

Protozoa and bacteria

Few authors have incorporated estimates of bacterial or protozoan density and distribution into their surveys. McLachlan et al. (1979a) found that the density of Protozoa on Kings Beach, Port Elizabeth, declined from some 3000 g^{-1} at the surface to zero at 80 cm sediment depth; on the more exposed Maitland beach significant numbers were still present to the limit of core penetration, at 130 cm. Bacterial densities were variable, at 14 to 107×10^6 g^{-1}, but did not appear to decrease with depth into the sediment. Similar penetration of bacterial populations to at least 120 cm were recorded by Koop and Griffiths (1982) in the western Cape. These authors were also able to demonstrate a marked increase in bacterial abundance towards the low-water mark, where densities attained 252 x 10^6 cm^{-3}. Only one study, that of Malan and McLachlan (1985) records bacterial densities in sublittoral sediments. These ranged from 12 to 130 x 10^6 cm^{-3}, again with little distinct variation with depth into the sediment, although there was an apparent increase in numbers with increasing water depth over the range 0.5 to 8 m.

Community structure

Macrofaunal species richness, as well as density and biomass estimates for all intertidal faunal components on a variety of southern African beaches are given in Table 13.2, all figures being expressed per running metre of beach, a standard measure adopted to incorporate changes in beach profile and migratory movements of the biota.

It is immediately apparent that the structure and composition of the intertidal macrofauna is extremely variable, species richness ranging from 1, on the very exposed coarse beaches of Natal, to 18 in the Western Cape (far more if species of terrestrial origin are counted). Density and biomass figures are even more variable, covering 3 to 4 orders of magnitude. McLachlan et al. (1981a) attempted to correlate macrofaunal diversity and abundance with exposure and hence particle size. These relationships are reproduced in Fig. 13.13, with the addition of more recent data. The coarsest, most exposed beaches clearly can only support a low biomass and only few highly mobile species, while the most sheltered ones can harbour a much richer and more diverse fauna that includes delicate, relatively immobile and tube-forming species. While this results in significant correlations it is equally clear that a great deal of scatter occurs over the particle-size range 200 to 400 μm, where most of the data points in fact lie. These variations can be attributed to a number of other factors, including the nature and amount of food available (see next section) and geographical location relative to the ranges of potential resident species.

The density and biomass of the intertidal meiofaunal community appear to be much more stable than that of the macrofauna. Density averages about 10^8 m^{-1} and biomass ranges from about 20 to 200 g dry weight m^{-1}, which is similar to the mean for macrofauna. Bacterial biomass estimates have seldom been attempted, but it is clear that bacterial biomass may often exceed those of the macro- and meiofauna combined.

Given that the productivity to biomass ratios (P/B) for macrofauna, meiofauna and bacteria are probably of the order of 2.5, 10 and 30 respectively (Koop and Griffiths, 1982), it is clear that in terms of annual production bacteria will be by far the most important element on most beaches and macrofauna probably the least so.

Relatively few comparable data are available for sublittoral systems, although McLachlan et al. (1984) have shown that macrofaunal species diversity in the surf zone of Algoa Bay is far greater (28–51 species) than in the adjacent intertidal. Meiofaunal biomass is, however, relatively low, probably because of poorer depth penetration into the sediment, resulting from inadequate oxygenation (Malan and McLachlan,

TABLE 13.2

Physical and biological characteristics of 25 beaches around the southern African coastline. Where species of terrestrial origin (mostly insects) were recorded these have been omitted from subsequent analysis and the number of marine species is listed, with total number (including insects) following in brackets. Other figures in brackets were not directly cited by the authority listed, but could be estimated from their descriptive accounts, or from other sources

Site	Exposure (1–20)	Grain size (μm)	Beach width (m)	Macrofauna			Meiofauna		Bacterial biomass (g dry wt m^{-1})	Ref.[a]
				No. of species	No. m^{-1}	Biomass (g dry wt m^{-1})	No. m^{-1} ($\times 10^7$)	Biomass (g dry wt m^{-1})		
Natal										
Sodwana	14	400	100	5	201	87	3.5	21	–	[1]
St. Lucia	14	289	50	9	250	9	0.8	10	–	[1]
Blythedale	17.5	864	60	1	3	18	2.4	26	–	[1]
Kelso	17.5	992	35	1	5	33	6.1	106	–	[1]
Transkei										
Thompsons	13.5	346	50	3	105	9	10.0	111	–	[2]
Mpande	12.5	221	80	7	2790	99	17.0	90	–	[2]
Cebe	12	218	100	8	2206	56	10.0	65	–	[2]
Gulu	13.5	247	135	7	3041	19	20.0	156	–	[2]
South Cape										
Maitland	16	302	80	11	(9000)	6600	(13.0)	76	(1000–2000)	[3]
Struisbaai	15	367	100	12	2555	108	38.0	243	–	[4]
Stillbay	13.5	223	100	11	624	23	26.0	233	–	[4]
Wilderness	15	281	100	8	1309	37	24.0	148	–	[4]
Keurboumstrand	14	305	100	7	1510	36	20.0	117	–	[4]
West coast										
Kommetjie	14	238	63	6 (15)	28 838	241	9.0	200	663	[5]
Melkbosstrand	15	215	90	19 (23)	95 066	324	10.0	625	–	[6]
Ysterfontein	15	205	60	18 (22)	153 945	682	20.2	1277	–	[6]
Rocherpan	16	320	49	14 (18)	7283	92	18.9	571	–	[6]
Lynch Point	(15)	*c.* 800	32	3 (5)	*c.* (20)	(< 10)	–	–	–	[7]
Langebaan	(7)	120	80	13 (16)	(> 1000)	(< 100)	–	–	–	(Lagoonal)
Oesterwal	(7)	173	390	32	(> 10 000)	(5000)	–	–	–	(Lagoonal)
Namibia										
Platjies	15.5	294	45	7	4501	123	29.0	193	–	[8]
Langstrand	13	272	22	8 (10)	5847	47	9.0	77	–	[8]
Toscannini	16	580	39	7 (22)	8638	82	–	–	–	[9]
Hoarusib	15	275	35	11 (30)	4660	66	–	–	–	[9]
Bosluisbaai	17	350	49	8 (11)	1830	20	–	–	–	[9]
Mean	14	356	86	9.4	13 809	556.8	14.9	116		
Range	7–17.5	120–992	32–390	1–32	3–154 000	9–6600	10–1277	663–(2000)		

[a]References:
[1] Dye et al. (1981)
[2] Wooldridge et al. (1981)
[3] McLachlan (1977b)
[4] McLachlan et al. (1981a, b)
[5] Koop and Griffiths (1982)
[6] Bally (1981, 1983)
[7] Day (1959)
[8] McLachlan (1985)
[9] Tarr et al. (1985)

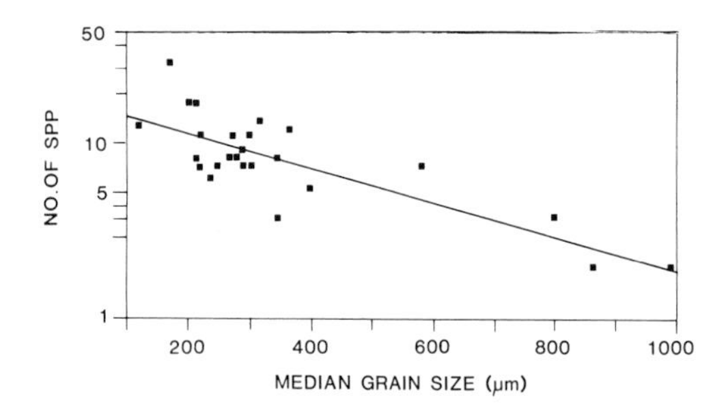

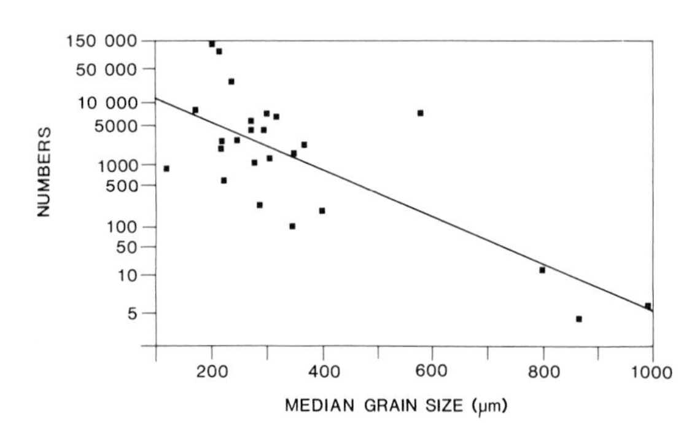

Fig. 13.13. Macrofaunal diversity and abundance in relation to particle size of sandy-beach sediments. The number of macrofaunal species is plotted against median grain size in the top panel:

$\log_e$ (no. of species) = 2.996 − 0.002 (grain size)
(F = 41.931; df = 1, 23; P < 0.001).

The number of macrofaunal individuals m^{-1} is plotted in relation to median grain size in the lower panel:

$\log_e$ (no. of individuals) = 10.609 − 0.009 (grain size)
(F = 30.93; df = 1, 23; P < 0.001).

Note the logarithmic *y*-axes. (After McLachlan et al. 1981b, with additional data from R. Bally, pers. commun.)

1985). Bacterial density, being less sensitive to oxygen tension, remains high, bacterial biomass approximating that of the macrofauna.

Material and energy flows

Outside the few lagoonal and estuarine systems, southern African beaches are essentially devoid of primary producers. The fauna is thus entirely dependent upon imported food resources. These can orginate from land (Brown, 1964), but the vast bulk undoubtedly derives from the sea, taking the form either of larger items of algal or animal debris, or of finely divided detrital or phytoplankton material. The quantities and proportions of these food materials will vary with the geographic location and topography of the beach, as well as the proximity and productivity of adjacent rocky areas. The nature of the available food in turn determines the biomass, composition and distribution pattern of the fauna. Two extreme situations are represented by the small pocket beaches of the southwestern Cape, which are often flanked by highly productive kelp beds, and the sweeping expanses of clean sand in areas such as Algoa Bay.

The community structure and patterns of materials and energy flow in a short beach receiving large inputs of kelp wrack have been described in a series of papers summarized by Griffiths et al. (1983) and Koop and Lucas (1983). The study area, Kommetjie Beach on the Cape Peninsula, was estimated to receive an input in excess of 2000 kg wet mass m^{-1} yr^{-1} as kelp wrack. This enormous supply of food, calculated per running metre of beach, was concentrated along the drift line, attracting a rich and diverse macrofauna (35 species, totalling 2256 g m^{-1} dry weight) dominated by semi-terrestrial amphipods, Diptera and Coleoptera. By contrast the lower reaches of the beach were virtually devoid of macrofauna, resulting in the unusual situation in which both biomass and species diversity declined markedly downshore.

The herbivores amongst the macrofauna, mainly amphipods of the genus *Talorchestia* and kelp fly (*Fucellia capensis*) larvae, were together estimated to consume 71% of the kelp deposited on the beach. They in turn were fed upon by several species of birds, isopods (*Eurydice longicornis* and *Exosphaeroma truncatitelson*) and carnivorous Coleoptera.

Much of the kelp eaten by the herbivores was returned to the beach in the form of faeces and entered the sand column, together with the decomposed remains of any uneaten kelp. This material supported a dense meiofauna (624 g m^{-1}) comprised mostly of nematodes and oligochaetes, and again concentrated in the area of maximum food availability below the driftline. Bacterial biomass was 961 g m^{-1} dry weight, with maximum densities occurring on the lower shore, possibly because meiofaunal predation pressure was minimal in this area.

Using a microcosm experiment set up in the

field, Koop et al. (1982a, b) showed that over 90% of the carbon leached into a sand column beneath decomposing wrack was absorbed by bacteria and/or meiofauna within 1 m, and that 28% of this was incorporated into bacterial biomass. Most of the bacterial production was thought to be consumed by the meiofauna, which had an annual food requirement of about 50 times its biomass, or 31 kg dry weight m^{-1} yr^{-1}. This suggests that the interstitial environment is a fairly closed system in which most of the organic input is absorbed by bacteria, which in turn are eaten by meiofauna. While the majority of the carbon deposited on the beach is probably respired as it passes up the food chain, the nitrogen is efficiently retained (Koop and Lucas, 1983) and must ultimately find its way back into the sea. This nutrient return is, however, insignificant compared to the considerable requirements of adjacent kelp beds and hence plays little role in influencing subsequent wrack input to the beach.

A quite different pattern is evident on the long unbroken expanses of beach in the southern and eastern Cape. There is little input of macrophyte debris or carrion to such systems, but their broad, shallow surf zones tend to support cellular circulation patterns that retain nutrients leaching from the beach, encouraging the development of surf zone phytoplankton blooms. These in turn provide a rich source of food for dense populations of macrofaunal filter-feeders and their associated fish and bird predators. These beaches and their surf zones thus form semi-enclosed ecosystems, with surf zone phytoplankton the primary producers, the beach macrofauna the consumers and the interstitial fauna the main decomposers (McLachlan, 1980c).

Surf zone phytoplankton blooms have been observed over at least 800 km of the southern Cape coastline, from False Bay to Algoa Bay, consisting mainly of the diatom *Anaulus birostratus*, although other species also occur, particularly *Aulacodiscus kittoni*. Blooms generally accumulate within 100 m of the beach, forming in the early morning over rip currents and persisting throughout the day. In late afternoon the cells apparently lose buoyancy and sink, to be dispersed outside the breaker zone overnight, where they are possibly less vulnerable to zooplankton grazing (McLachlan and Lewin, 1981). It has been estimated that the average annual phytoplankton biomass within the breaker zone is 264 g C m^{-1} with a further 382 g C m^{-1} within another 250 m offshore. Annual primary production within 500 m of the shore would thus be of the order of 99 kg C m^{-1} yr^{-1}, of which 20 to 25% is in the form of DOM (McLachlan and Bate, 1984).

This material, plus lesser amounts of faeces and macrophyte debris, is consumed by benthic filter-feeders. The most important of these are bivalves of the genus *Donax* (*D. serra* dominating in the intertidal and *D. sordidus* in the surf zone), and zooplankton crustaceans, particularly the prawn *Macropetasma africanus* and the mysid *Mesopodopsis slabberi*. Higher trophic levels are represented by scavenging whelks of the genus *Bullia*, by

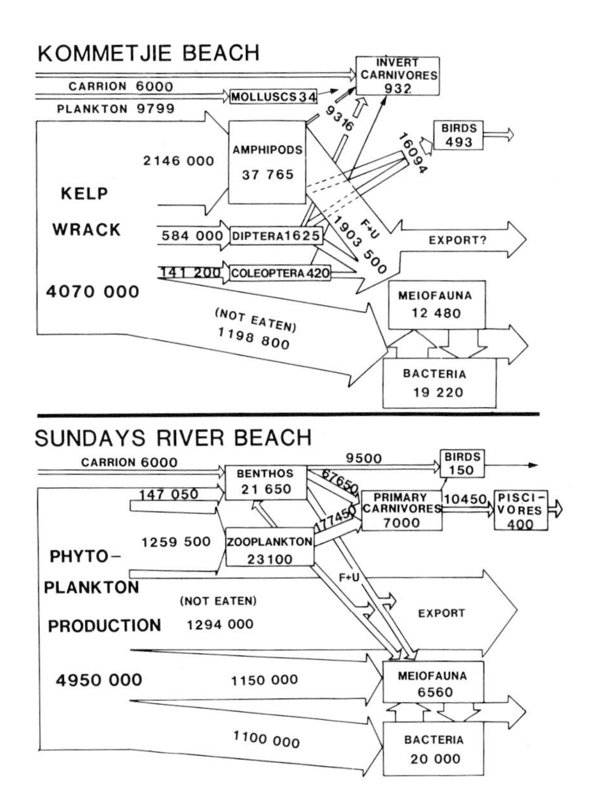

Fig. 13.14. Diagrammatic energy flow diagrams for Kommetjie Beach (above) and Sundays River Beach (below). Arrows show annual flows in kJ m^{-1} yr^{-1} and boxes standing stocks in kJ m^{-1}. Conversions have been made assuming 1 g dry mass = 20 kJ; 1 g C = 50 kJ. The Kommetjie model incorporates the eulittoral zone only (63 m width), whereas the Sundays River model incorporates both beach and surf zone, together 500 m wide. (Modified after Griffiths et al., 1983; McLachlan and Bate, 1984; with additional data from Malan and McLachlan, 1985.)

the swimming crab *Ovalipes punctatus*, which preys mostly on *Donax* and *Bullia*, and by a number of fish, of which the sandshark *Rhinobatos annulatus* is the most important. Birds may be significant predators in the intertidal, but are of minor importance overall.

Using these and other estimates of macrofaunal biomass and consumption, together with similar figures for the interstitial and microscopic food chains, McLachlan and Bate (1984) have produced a carbon budget for a high energy surf zone, which is reproduced in a simplified form in Fig. 13.14. Unlike the kelp-subsidized system of the western Cape, for which a similar diagram is provided, these areas appear to function as self-sustaining units. Surf zone diatoms provide ample primary production to supply not only the macroscopic food chain but also the interstitial fauna and microbial loop. The system may indeed act as a significant exporter of energy and carbon to nearshore waters, this taking the form of either phytoplankton, detritus or DOM.

While the two systems process surprisingly similar amounts of energy these are derived from totally different sources, illustrating the enormous variations in community structure to be found in sandy beaches around southern Africa. Both systems, however, emphasize the importance of the interstitial and microbial fauna, which have only recently begun to receive the attention they deserve.

ACKNOWLEDGEMENTS

This chapter was written while one of us (CLG) was on study leave in the School of Animal Biology, University College of North Wales; we are indebted to Prof. E. Naylor and his staff for placing the facilities of that department at our disposal. We also thank R. Bally, S. Tolosana, D. Gianakouras and P. Wickens for help in preparing the manuscript.

REFERENCES

Bally, R., 1981. *The ecology of three sandy beaches on the west coast of South Africa.* Ph.D. Thesis, University of Cape Town, Cape Town, South Africa, unpubl.

Bally, R., 1983. Intertidal zonation on sandy beaches of the west coast of South Africa. *Cah. Biol. Mar.*, 24: 85–103.

Bally, R., McQuaid, C.D. and Brown, A.C., 1984. Shores of mixed sand and rock: an unexplored marine ecosystem. *S. Afr. J. Sci.*, 80: 500–503.

Bayne, B.L. and Widdows, J., 1978. The physiological ecology of two populations of *Mytilus edulis* (L.). *Oecologia*, 37: 137–162.

Berry, P.F., 1982. Biomass and density of detritivores on a littoral rocky reef on the Natal coast, with an estimate of population production for the ascidian *Pyura stolonifera. Investig. Rep. Oceanogr. Res. Inst.* (Durban, South Africa), 53: 1–15.

Berry, P.F., Hanekom, P., Joubert, C., Joubert, M., Schleyer, M., Smale, M. and Van der Elst, R., 1979. Preliminary account of the biomass and major energy pathways through a Natal nearshore reef community. *S. Afr. J. Sci.*, 75: 565.

Branch, G.M., 1975. Intraspecific competition in *Patella cochlear* (Born). *J. Animal Ecol.*, 44: 263–282.

Branch, G.M., 1981. The biology of limpets: physical factors, energy flow and ecological interactions. *Oceanogr. Mar. Biol. Annu. Rev.*, 19: 235–380.

Branch, G.M., 1984. Competition between marine organisms: ecological and evolutionary implications. *Oceanogr. Mar. Biol. Annu. Rev.*, 22: 429–593.

Branch, G.M., 1985. Competition: its role in ecology and evolution in intertidal communities. In: E.S. Vrba, (Editor), *Species and Speciation, Transvaal Museum Monograph 4.* Transvaal Museum, Pretoria, pp. 97–104.

Branch, G.M. and Branch, M.L., 1981. *The Living Shores of Southern Africa.* C. Struik, Cape Town, 272 pp.

Brown, A.C., 1964. Food relationships on the intertidal sandy beaches of the Cape Peninsula. *S. Afr. J. Zool.*, 16: 210–218.

Brown, A.C., 1971. The ecology of the sandy beaches of the Cape Peninsula, South Africa. Part 1. Introduction. *Trans. R. Soc. S. Afr.*, 39: 247–277.

Brown, A.C. and Jarman, N., 1978. Coastal marine habitats. In: M.J.A. Werger (Editor), *Biogeography and Ecology of Southern Africa.* Junk, The Hague, pp. 1239–1277.

Christie, N.D., 1976. A numerical analysis of the distribution of a shallow sublittoral sand macrofauna along a transect at Lamberts Bay, South Africa. *Trans. R. Soc. S. Afr.*, 42: 149–172.

Dahl, E., 1952. Some aspects of the ecology and zonation of the fauna on sandy beaches. *Oikos*, 4: 1–27.

Day, J.H., 1959. The biology of Langebaan Lagoon: a study of the effects of shelter from wave action. *Trans. R. Soc. S. Afr.*, 35: 475–548.

Dye, A.H., McLachlan, A. and Wooldridge, T., 1981. The ecology of sandy beaches in Natal. *S. Afr. J. Zool.*, 16: 200–209.

Field, J.G., 1983. Coastal ecosystems: flow patterns of energy and matter. In: O. Kinne (Editor), *Marine Ecology, Vol. 5.* Wiley-Interscience, Chichester, pp. 758–794.

Field, J.G. and Robb, F.T., 1970. Numerical methods in marine ecology. 2. Gradient analysis of rocky shore samples from False Bay. *Zool. Afr.*, 5: 191–210.

Field, J.G., Griffiths, C.L., Linley, E.A., Carter, R.A. and Zoutendyk, P., 1980a. Upwelling in a nearshore marine

ecosystem and its biological implications. *Estuarine Coastal Mar. Sci.*, 11: 133–150.
Field, J.G., Griffiths, C.L., Griffiths, R.J., Jarman, N., Zoutendyk, P., Velimirov, B. and Bowes, A., 1980b. Variation in structure and biomass of kelp communities along the south-west Cape coast. *Trans. R. Soc. S. Afr.*, 44: 145–203.
Field, J.G., Griffiths, C.L., Velimirov, B., Zoutendyk, P. and Carter, R.A., 1981. Wind-induced water movements in a Benguela kelp bed. In: F.A. Richards (Editor), *Coastal Upwelling*. American Geophysical Union, New York, pp. 507–513.
Grant, W.S. and Cherry, M.I., 1985. *Mytilus galloprovincialis* Lmk. in southern Africa. *J. Exp. Mar. Biol. Ecol.*, 90: 179–191.
Griffiths, C.L., Stenton-Dozey, J.M.E. and Koop, K., 1983. Kelp wrack and the flow of energy through a sandy beach ecosystem. In: A. McLachlan and T. Erasmus (Editors), *Sandy Beaches as Ecosystems*. Junk, The Hague, pp. 547–556.
Heydorn, A.E.F., Bang, N.D., Pearse, A.F., Flemming, B.W., Carter, R.A., Schleyer, M.H., Berry, P.F., Hughes, G.R., Bass, A.J., Wallace, J.H., Van der Elst, P.P., Crawford, R.J.M. and Shelton, P.A., 1978. Ecology of the Agulhas Current region: an assessment of biological responses to environmental parameters in the Southwest Indian Ocean. *Trans. R. Soc. S. Afr.*, 43: 151–189.
Huggett, J. and Griffiths, C.L., 1986. Some relationships between elevation, physico-chemical variables and biota of intertidal rock pools. *Mar. Ecol. Prog. Ser.*, 29: 189–197.
Jackson, L.F., 1976. Aspects of the intertidal ecology of the east coast of South Africa. *Investig. Rep. Oceanogr. Res. Inst.* (Durban, South Africa), 46: 1–72.
Jarman, N.G. and Carter, R.A., 1981. The primary producers of the inshore regions of the Benguela. *Trans. R. Soc. S. Afr.*, 44: 321–326.
Kautsky, N. and Wallentinus, I., 1980. Nutrient release from a Baltic *Mytilus*-red algal community and its role in benthic and pelagic productivity. *Ophelia*, 1 (Suppl.): 17–30.
Kensley, B.F. and Penrith, M.-L., 1970. The constitution of the intertidal fauna of rocky shores of South West Africa. 1. Lüderitzbucht. *Cimbebasia A*, 1: 191–239.
Koop, K., and Griffiths, C.L., 1982. The relative significance of bacteria, meio- and macrofauna on an exposed sandy beach. *Mar. Biol.*, 66: 295–300.
Koop, K. and Lucas, M.I., 1983. Carbon flow and nutrient regeneration from the decomposition of macrophyte debris in a sandy beach microcosm. In: A. McLachlan and T. Erasmus (Editors), *Sandy Beaches as Ecosystems*. Junk, The Hague, pp. 249–262.
Koop, K., Newell, R.C. and Lucas, M.I., 1982a. Biodegradation and carbon flow based on kelp (*Ecklonia maxima*) debris in a sandy beach microcosm. *Mar. Ecol. Prog. Ser.*, 7: 315–326.
Koop, K., Newell, R.C. and Lucas, M.I., 1982b. Microbial regeneration of nutrients from the decomposition of macrophyte debris on the shore. *Mar. Ecol. Prog. Ser.*, 9: 91–96.
Malan, D.E. and McLachlan, A., 1985. Vertical gradients of meiofauna and bacteria in subtidal sandy sediments from two high-energy surf zones in Algoa Bay, South Africa. *S. Afr. J. Mar. Sci.*, 3: 43–53.
McLachlan, A., 1977a. Studies on the psammolittoral meiofauna of Algoa Bay, South Africa. II. The distribution, composition and biomass of the meiofauna and macrofauna. *Zool. Afr.*, 12: 33–60.
McLachlan, A., 1977b. Composition, distribution, abundance and biomass of the macrofauna and meiofauna of four sandy beaches. *Zool. Afr.*, 12: 279–306.
McLachlan, A., 1980a. The definition of sandy beaches in relation to exposure: a simple rating system. *S. Afr. J. Sci.*, 76: 137–138.
McLachlan, A., 1980b. Intertidal zonation of macrofauna and stratification of meiofauna on high energy sandy beaches in the Eastern Cape, South Africa. *Trans. R. Soc. S. Afr.*, 44: 213–223.
McLachlan, A., 1980c. Exposed sandy beaches as semi-closed ecosystems. *Mar. Environm. Res.*, 4: 59–63.
McLachlan, A., 1985. The ecology of two sandy beaches near Walvis Bay. *Madoqua*, 14: 155–163.
McLachlan, A. and Bate, G., 1984. Carbon budget for a high energy surf zone. *Vie et Milieu*, 34: 67–77.
McLachlan, A. and Lewin, J., 1981. Observations on surf phytoplankton blooms along the coasts of South Africa. *Bot. Mar.*, 24: 553–557.
McLachlan, A., Erasmus, T. and Furstenberg, J.P., 1977. Migrations of sandy beach meiofauna. *Zool. Afr.*, 12: 257–277.
McLachlan, A., Dye, A.H. and Van der Ryst, P., 1979a. Vertical gradients in the fauna and oxidation of two exposed sandy beaches. *S. Afr. J. Zool.*, 14: 43–47.
McLachlan, A., Wooldridge, T. and Van der Horst, G., 1979b. Tidal movements of the macrofauna on an exposed sandy beach in South Africa. *J. Zool.* (London), 188: 433–442.
McLachlan, A., Lombard, H.W. and Louwrens, S., 1981a. Trophic structure and biomass distribution on two East Cape rocky shores. *S. Afr. J. Zool.*, 16: 85–89.
McLachlan, A., Wooldridge, T. and Dye, A.H., 1981b. The ecology of sandy beaches in southern Africa. *S. Afr. J. Zool.*, 16: 219–231.
McLachlan, A., Cockcroft, A.C. and Malan, D.E., 1984. Benthic faunal response to a high energy gradient. *Mar. Ecol. Prog. Ser.*, 16: 51–63.
McQuaid, C.D. and Branch, G.M., 1984. The influence of sea temperature, substratum and wave exposure on rocky intertidal communities: an analysis of faunal and floral biomass. *Mar. Ecol. Prog. Ser.*, 19: 145–151.
McQuaid, C.D. and Branch, G.M., 1985. Trophic structure of rocky intertidal communities: response to wave action and implications for energy flow. *Mar. Ecol. Prog. Ser.*, 22: 153–161.
Newell, R.C. and Field, J.G., 1983. Relative flux of carbon and nitrogen in a kelp-dominated system. *Mar. Biol. Lett.*, 4: 249–257.
Newell, R.C., Field, J.G. and Griffiths, C.L., 1982. Energy balance and significance of micro-organisms in a kelp bed community. *Mar. Ecol. Prog. Ser.*, 8: 103–113.
Salvat, B., 1964. Les conditions hydrodynamiques interstitielles des sédiments meublés intertidaux et la répartition verticale de la faune endogée. *C.R. Acad. Sci.*, 259: 1576–1579.

Schleyer, M.H., 1980. *The role of microorganisms and detritus in the water column of a subtidal reef of Natal, South Africa.* Ph.D. Thesis, University of Natal, Durban, South Africa, unpublished.

Shafir, A. and Field, J.G., 1980. Importance of a small carnivorous isopod in energy transfer. *Mar. Ecol. Prog. Ser.*, 3: 203–215.

Shannon, L.V., 1985. The Benguela Ecosystem, Part. 1. Evolution of the Benguela, physical features and processes. *Oceanogr. Mar. Biol. Annu. Rev.*, 23: 105–182.

Shillington, F.H., 1978. Surface waves near Cape Town: measurements and statistics. *Civ. Eng. S. Afr.*, 20: 203–206.

Stephenson, T.A., 1948. The constitution of the intertidal fauna and flora of South Africa, III. *Ann. Natal Mus.*, 11: 207–324.

Stephenson, T.A. and Stephenson, A., 1972. *Life between Tidemarks on Rocky Shores.* Freeman, San Francisco, 425 pp.

Tarr, J.G., Griffiths, C.L. and Bally, R., 1985. The ecology of the sandy beaches on the skeleton coast of South West Africa. *Madoqua*, 14: 295–304.

Velimirov, B., 1983. Succession in a kelp bed ecosystem: clearing of primary substrate by wave-induced kelp sweeping. In: *Proc. 17th Europ. Marine Biological Symp., Brest. Oceanologia Acta*, 201–206.

Velimirov, B. and Griffiths, C.L., 1979. Wave-induced kelp movement and its importance for community structure. *Bot. Mar.*, 22: 169–172.

Velimirov, B., Field, J.G., Griffiths, C.L. and Zoutendyk, P., 1977. The ecology of kelp bed communities in the Benguela upwelling system. Analysis of biomass and spatial distribution. *Helgol. Wiss. Meeresunters.*, 30: 495–518.

Wickens, P.A. and Field, J.G., 1986. The effect of water transport on nitrogen flow through a kelp-bed community. *S. Afr. J. Mar. Sci.*, 4: 79–92.

Wooldridge, T., Dye, A.H. and McLachlan, A., 1981. The ecology of sandy beaches in Transkei. *S. Afr. J. Zool.*, 16: 210–218.

Wulff, F.V. and Field, J.G., 1983. Importance of different trophic pathways in a nearshore benthic community under upwelling and downwelling conditions. *Mar. Ecol. Prog. Ser.*, 12: 217–228.

Chapter 14

LITTORAL AND SUBLITTORAL COMMUNITIES OF CONTINENTAL CHILE

BERNABÉ SANTELICES

INTRODUCTION

The coast of mainland Chile is about 4200 km long, extending from Cape Horn in subantarctic waters (55°58′S) to Arica in the Tropics (18°20′S). The first descriptions of eulittoral marine communities along this extended coastline were performed about 35 years ago (Dahl, 1953; Guiler, 1959a, b) and less than 50 reports have been produced thereafter. Emphasis was first placed on qualitative and then on quantitative descriptions of eulittoral communities. In recent years emphasis has switched to quantitative explorations of the sublittoral zone and to experimental evaluations of interactions among the ecological factors determining the organization of these communities. Most of these studies have focused on rocky eulittoral communities of central Chile (30°–40°S). Therefore, there are extended areas along this coast that have been scarcely explored (for example, 18–25°S; 42–55°S), and whole ecosystems that have not yet been scientifically studied. This review is intended to give an integrated view of the patterns of community structure and organization studied so far.

The accumulated information indicates that coastal topography and climate, oceanographic factors and the biogeographic attributes of the biotic components are of paramount importance in understanding the structural patterns of the commonest communities. Therefore, the first part of this chapter will provide a general characterization of the coastal zone, while the second and third parts will review the general patterns of community structure and organization found. When reviewing patterns of community structure, the main accent will be on physiognomically dominant biota such as epilithic algae and sessile invertebrates on rocky shores, and benthic soft-bottom animals in estuaries and sandy shores. When reviewing patterns of community organization, emphasis will be on organisms which apparently play the most important roles as structuring species of these communities.

CHARACTERIZATION OF THE COASTAL ZONE

Coastal topography and climate

Coastal topography clearly defines two very different regions along mainland Chile (Guiler, 1959b; Stephenson and Stephenson, 1972; Castilla, 1976b, c; Viviani, 1979; Santelices, 1980a, b). North of Chiloé (41°29′S) the coastline is very regular, almost straight-lined, fully exposed to the prevailing winds and waves, with open sandy beaches and few sheltered bays (Fig. 14.1). South of Chiloé the coast is dismembered and precipitous, with mountains along the shore rising above 3000 m and with a string of off-shore islands protecting this coast from the frequent storms outside.

Topography and climate further differentiate various types of marine habitats within these two regions (Figs. 14.1A, 14.1B). The coastal range between Arica (18°20′S) and Punta Rincón, to the south of Antofagasta (25°S), consists of Jurassic and Cretaceous volcanic rocks with sedimentary intrusions elevated during the Quaternary (Araya-Vergara, 1976). Here, the rainy season is restricted to a few weeks every summer (Di Castri and Hajek, 1976), with especially intense rainfall in the Cordillera de los Andes, less intense in the central

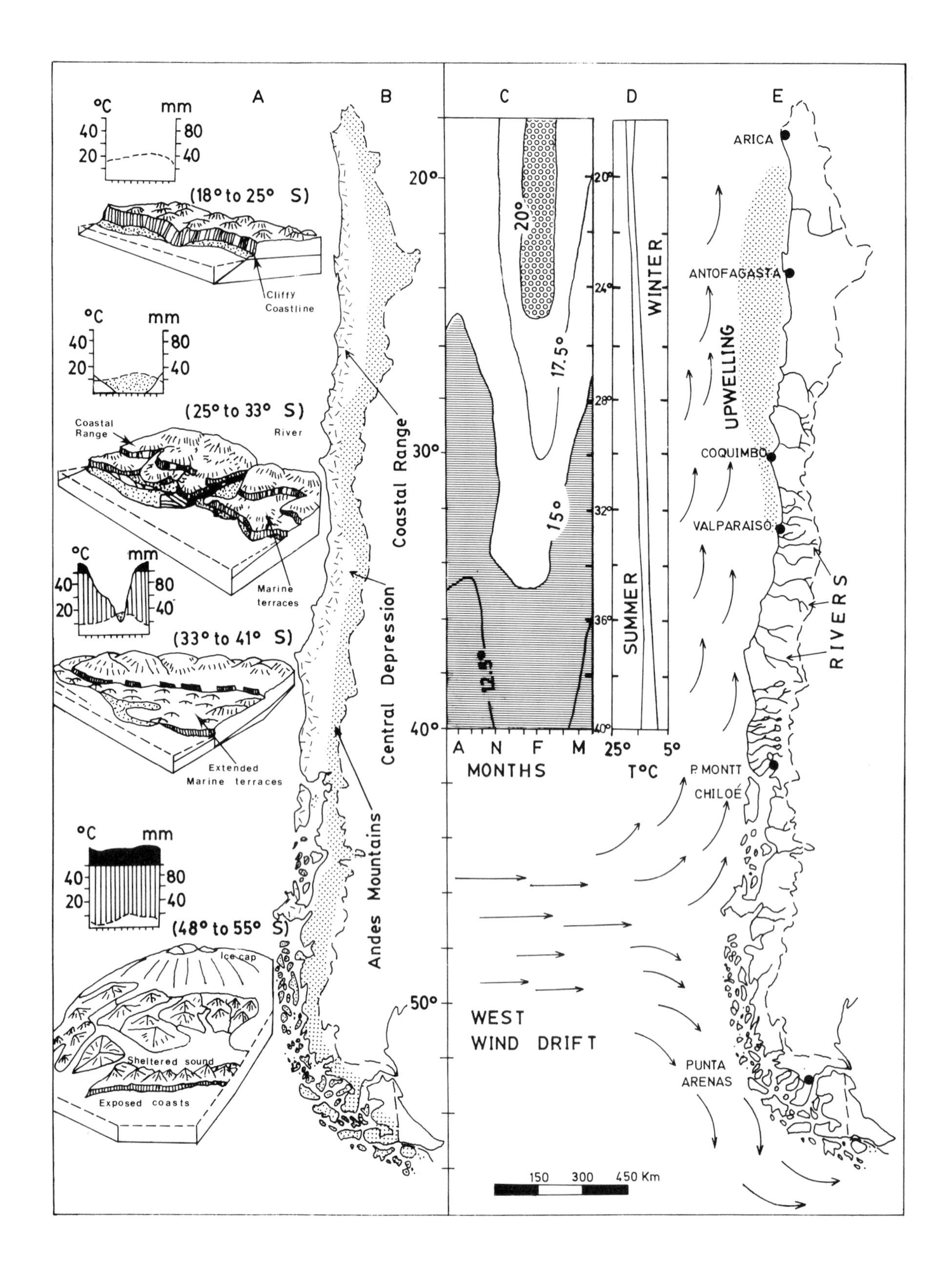
A
B
C
D
E
°C
mm
40
80
20
40
(18° to 25° S)
Cliffy Coastline
(25° to 33° S)
Coastal Range
River
Marine terraces
(33° to 41° S)
Extended Marine terraces
(48° to 55° S)
Ice cap
Sheltered sound
Exposed coasts
Coastal Range
Central Depression
Andes Mountains
20°
30°
40°
50°
17.5°
15°
12.5°
A N F M
MONTHS
WINTER
SUMMER
25° 5°
T°C
ARICA
ANTOFAGASTA
UPWELLING
COQUIMBO
VALPARAISO
RIVERS
P. MONTT
CHILOÉ
WEST WIND DRIFT
PUNTA ARENAS
150 300 450 Km

depression and of limited importance in the coastal range. In addition, with the exception of the narrow and shallow Rio Loa (normal flow about 1.5 to 2 $m^3 s^{-1}$), all rivers originating in the Andes disappear in the central depression without reaching the coastal range. Thus, this stretch of coast is very cliff-like, it has limited access, a desert-like climate and is homogeneously exposed to prevailing southwesterly winds and waves. By setting offshore currents in motion, the winds cause coastal upwelling, bringing cool, nutrient-rich water to the surface (Fig. 14.1). In addition, these southwesterly sea breezes can whip up short, steep waves on top of the usual oceanic swell, producing turbulent seas and modifying zonation patterns. At these latitudes only the north-facing sides of rocky outcrops projecting into the sea or the northern sides of nearshore islands constitute constantly sheltered habitats. On warmer shores fog is common, affording some protection to eulittoral and upper littoral organisms from intense solar radiation.

South of Antofagasta (25°S) and extending up to the vicinity of Navidad (33°26′S), the coastal range is composed primarily of granitic rock, has lower elevation than further north and leaves somewhat extended marine terraces. The monotonously homogeneous rocky coastline is interrupted now and then by a few open sandy beaches and, in increasing numbers south of 27°S, by estuaries of rivers draining into the sea the melting snow from the Andes. The direction of the prevailing onshore winds changes seasonally, being southwest in summer and northwest in winter. In this area, therefore, the north-facing as well as the south-facing sides of beaches, rocky outcrops and islands can be alternately exposed to winds and waves.

From Navidad (33°26′S) to Chiloé (41°29′S) the coastal range is made up of Precambrian metamorphic shales of low elevation, leaving extended coastal terraces. A number of rivers drain into the sea either melt-water from the snows of the Andes, or the rain-water that is increasingly abundant southward, adding fair quantities of sediments and terrigenous materials to the coastal zone. In addition to rocky formations, a number of sandy beaches, several important estuaries and extensive mud flats can also be found in this part of the country.

South of Chiloé (41°29′S), the coastal geomorphology typically corresponds to an erosional tectonic pattern of fjords with glaciated and unglaciated hinterlands (Araya-Vergara, 1976). The central depression between the Cordillera de los Andes and the coastal range is here covered by the sea. Between 40° and 48°S numerous rivers drain the abundant rainfall into the sea, while between 48° and 52°S some large fjords with glaciated hinterlands reach the sea; the volumes of water pouring down their ravines substantially modify salinity regimes. Between this coastline and the open sea a number of offshore islands provide protected waters, forming sheltered channels normally used for inshore navigation (Patagonian Channels). Due to the obvious differences in wave impact, type of bottoms, salinity and temperature between the open coast of these islands and the sheltered sounds, a pronounced habitat diversification occurs south of 42°S as compared to the rest of the coastline.

Oceanographic conditions

Surface currents and temperature regimes

Traditionally, the coast of Chile has been thought to be mainly under the influence of the Humboldt Current (Gunther, 1939; Knox, 1960; Brandhorst, 1963; Wyrtki, 1963, 1964, 1966, 1967; Robles, 1966; Stephenson and Stephenson, 1972). The West Wind Drift reaches Pacific South America near 45°S latitude, branching into two unequal masses of water flowing in opposite directions (Johnson et al., 1980). Southwardly deflected, the Cape Horn Current follows the

Fig. 14.1. Coastal topography, land and oceanic climate determining different types of coastal systems along mainland Chile. (A) Coastal topography (after Araya-Vergara, 1976), rainfall (solid line) and air temperature (dashed line) at four localities along Chile (after Di Castri and Hajek, 1976). (B) Relative position of the Cordillera de los Andes (Andes Mountains) the central depression and the coastal range along Chile. (C) and (D) Monthly mean temperatures and temperature ranges of surface coastal waters off Chile (81–82°W) between 18° and 40°S (after Wooster and Sievers, 1971). (E) General direction of the surface coastal currents (Coastal Branch of the Humboldt Current), presence of upwelling areas (shaded area), and of the main fresh-water rivers along Chile.

southernmost coast of Chile, rounds the tip of the continent and projects itself through Drake Passage. The northwardly deflected mass, the Humboldt Current, flows along the east coast of South America up to approximately 4°27′S in northern Peru, and then turns west-northwest across the Pacific Ocean (Fig. 14.1E). More recent information, however, indicates that the Humboldt Current flows 300 to 400 km off the coast, and that it is only one of several currents and counter currents found north of 40°S (Wooster and Sievers, 1971; Robles et al., 1976, 1980; Silva and Sievers, 1981; Bernal et al., 1982). Indeed, two water masses, the Subantarctic Water and the Subtropical Surface Water, dominate the oceanographic regime in this sector. The Subantarctic Water originates in the eastern branch of the anticyclonic gyre of the southeastern Pacific, flowing northward. At about 1000 km off the South American coast, and south of 35°S, this is the dominant element throughout the year, determining temperatures of 10 to 18°C and salinities of 32 to 34.8‰ in the uppermost 100 m of water. The Subtropical Surface Water is a water mass approximately 30 m deep flowing southward over the Subantarctic Water, reaching 25 to 27°S in summer and 23 to 25°S in winter, with temperatures above 18°C and salinities above 34.9% (Fig. 14.1C).

The circulation in the region is complex, and as one approaches the coast from the open ocean several currents (flowing to the north) and counter currents can be distinguished (Bernal et al., 1982). They include the Chile–Peru Ocean Current, centered about 1000 km off the coast; the Peruvian Oceanic Counter Current, between 76° and 77°W; the Humboldt Current, 300 to 400 km off the coast; the Chile Coastal Counter Current centered between 100 to 200 km off the coast; and the Fjord Current (that is, the Coastal Branch of the Humboldt Current, which is a surface flow with its axis lying 100 km off the coast). This last current, together with the Subtropical Surface Water, determine the basic oceanographic conditions found off the coast at several latitudes along Chile. Thus, south of 39°S the uppermost 50 m correspond to Fjord Current Waters, with temperatures between 11.5 and 13.5°C and salinities of 32.7 to 33.4‰. Farther north (37–28°S), the uppermost 10 to 17 m are Fjord Current waters (13.5–17.5°C) flowing over the Subantarctic Water, which has temperatures between 11.5 and 13.5°C, down to 50 m. Between 18° and 28°S the uppermost 12 m of water correspond to the Subtropical Surface Water, with temperatures above 18.5°C and salinities around 34.9‰. Underneath and down to a depth of 45 m, temperatures (13–14.5°C) and salinities (about 34.8‰) are characteristic of the Subantarctic Water. These data indicate that a summer temperature difference of 7 to 8°C and a winter difference of 5 to 6°C should be expected between 40° and 18°S (Fig. 14.1D). Yet, the intense coastal upwelling occurring along central and northern Chile reduces temperatures of these coastal waters, resulting in a remarkably homogeneous temperature regime along most of Chile's coast. For example, surface temperatures during the winter and summer were as follows: 5 to 6°C and 8 to 9°C, respectively in a shallow (8 m deep) *Macrocystis* bed at 53°S (Santelices and Ojeda, 1984b), 9 to 10°C and 15 to 16°C in a *Lessonia* bed at 33°S (Santelices et al., 1981), and 14°C and 16.5°C at 20°S (Viviani, 1979).

The general oceanographic characteristics of northern Chile are substantially modified at irregular intervals due to the "El Niño" phenomenon. This is a massive southerly penetration of warm oceanic subtropical and equatorial subsurface water into the Chile–Peru Current system (Robles et al., 1976, 1980; Bernal et al., 1982; Barber and Chavez, 1983). Increased surface temperatures are accompanied by heavy rains and, among other oceanographic changes, by reduced upwelling and increased downwelling close to shore. All these changes have catastrophic effects on fish and coastal biota. The frequency and severity, as well as the southward extension of the phenomenon, varies in an as yet unpredictable way. However, during strong "El Niño" events, abnormally high temperatures have been recorded as far south as 35°S.

Salinity variations

The normal variations of salinities discussed previously are much less than the tolerance ranges of coastal invertebrates and algae. However, significant salinity reductions may occur in surface waters of enclosed bays during the rainy season (Ahumada and Chuecas, 1979), in estuaries, which are increasingly common south of 28°S, and in the Patagonian sounds that are under the influence of rainfall and melting ice from the glaciers (Pickard, 1971).

Nutrient concentrations

Data on nutrient concentrations are available for a few localities in Chile. On average, coastal sea water around Valparaíso (33°S) contains 0.05 ± 0.01 μmol P l^{-1} and 0.41 ± 0.1 μmol N l^{-1} (Alveal, 1981). In upwelling areas, coastal concentrations of 1.7 μmol l^{-1} of phosphate and 5.6 to 10 μmol l^{-1} of nitrate have been recorded (Díaz, 1984). As stated previously, the coast of Chile north of 35°S experiences persistently favourable upwelling winds during spring, summer and autumn.

Tides

For the stretch of coast north of Chiloé (18° to 40°S), tides are semidiurnal with an amplitude of about 2 m (Guiler, 1959b). The dismembered coastal topography south of Chiloé significantly reduces the predictability of tidal amplitude and regularity. In the vicinity of Puerto Montt and Chiloé (40°–41°S) and around the Canal Beagle area (53°S), tidal amplitudes can reach up to 8 to 12 m, permitting the development of extended sand flats and pebble beaches respectively. As will be discussed further below, the Chiloé area is one of the few places along Chile, other than a few estuarine situations, with extensive mud-flat systems.

The littoral and sublittoral biota

The biogeographic affinities of the marine macrobenthos of mainland Chile have been explored for well over a century (Forbes, 1856 in Briggs, 1974). A few generalizations so far reached seem of paramount importance with respect to community structure and interactions. Therefore they will be briefly analyzed below.

A high degree of eco-geographic isolation seems to be a general characteristic of these intertidal–shallow subtidal biota. There are low floristic and faunistic similarities (especially of endemic species) with marine biotas in Ecuador, the Galápagos and the Juan Fernández Archipelagos, suggesting that the low temperatures and the generally northward flow of the subantarctic waters have allowed only a limited biotic exchange through surface waters with the western Pacific, the central Pacific islands and even with the eastern tropical Pacific (Santelices, 1980a). The sub-superficial layers of water (deeper than 50 m) in northern Chile have anaerobic conditions (Gallardo, 1963), further limiting invasions of deep-water forms from the north (Gallardo, 1963; Castilla, 1979; Viviani, 1979), while the extremely low salinities prevailing in southern Patagonia also limit faunistic exchange from the south (Viviani, 1979). It is not surprising, then, that this marine fauna and flora show a high degree of endemism. For example, 55.6% of the littoral molluscs (Dall, 1909), 53.3% of the anomuran decapods (Haig, 1965), 23.3% of the brachyuran crabs (Garth, 1957) and 32.3% of the benthic algal flora (Santelices, 1980a) are endemic species. Several of these endemic species have become ecologically important parts of the local communities, often occupying unique ecological niches with no known parallels in comparable habitats elsewhere. This is the case for the gastropod *Concholepas concholepas* (Castilla, 1976a; Castilla et al., 1979), the clingfish *Sicyases sanguineus* (Paine and Palmer, 1978), the eulittoral kelp *Lessonia nigrescens* (Santelices et al., 1980), and the dominant mid-eulittoral encrusting green alga *Codium dimorphum* (Santelices et al., 1981). Furthermore, experimental manipulations of some of these populations indicate these organisms may not show convergence in ecological properties with morphological or physiognomic analogs elsewhere. Thus, the ecological role of the South American sea otter *Lutra felina* seems related to the structuring of intertidal rather than subtidal marine communities, and it should not be considered an ecological equivalent of the North American sea otter *Enhydra lutris* (Castilla and Bahamondes, 1979). Common sea urchins do not over-exploit live plants of *Macrocystis pyrifera* in kelp forests of southernmost Chile, as has frequently been reported in California (Castilla and Moreno, 1982). Contrary to results in the Northern Hemisphere the second algal stratum in these southern kelp forests decreases rather than increases in biomass after removal of the floating canopy of *M. pyrifera* (Santelices and Ojeda, 1984a, b). The middle eulittoral mussel *Perumytilus purpuratus* increases its cover upon removal of the starfish *Heliaster helianthus*; but it does not escape in size as other mussels do, as shown in similar experiments conducted in Washington State and New Zealand (Paine et al., 1985).

The isolated character of the marine benthic biota seems also to have some effects on species richness, at least for some groups (Santelices, 1980a). It is true that the limited biological exploration of the region so far could account, at least in part, for fewer species having been recorded. However, the northern flow and the low temperature of the cool Subantarctic Water could limit migration or colonization by the biotas from the eastern tropical Pacific or the central Pacific islands. If this is correct, one should expect that the marine communities from southern Chile were richer in species than farther north and that species richness in temperate Pacific South America was reduced, as compared with climatically equivalent regions having contact with more effective routes of migration (such as the Pacific coast of North America). The comparative data so far published tend to support this hypothesis. Thus, Alveal et al. (1973) found more species and documented more complex *Macrocystis* communities in the Magellanic region (*c.*55°S) than in central Chile (30°S). Dawson et al. (1964) commented that a collection of benthic algae in central and southern Peru yielded scarcely half the diversity of species that a comparable effort revealed in the similar environment of southern California. Dayton et al. (1973) and Santelices and Ojeda (1984a) have noted the much simpler community structure of the South American *Macrocystis* beds, emphasizing the absence of ecological equivalents (other than *Lessonia flavicans*) of the several genera of Laminariales, which in the North Pacific form the first and second vegetational strata. The ichthyofauna (Moreno and Jara, 1984) and the invertebrate fauna (Ojeda and Santelices, 1984a), associated with *Macrocystis pyrifera* in the Beagle Channel, were much less diversified than those occurring with *M. pyrifera* in California (Ghelardi, 1971). According to Marincovich (1973), Olsson (1961) estimated that about three times as many species of shell-boring molluscs occurred in North America, as in a region of similar size and environmental diversity in South America. The only published exception to this trend seems to be the family Fissurellidae which, according to Guiler (1959b), reaches a considerable degree of diversification in Chile.

Two major faunistic components have been traditionally recognized along the temperate coast of Pacific South America: a warm temperate component in Peru and northern Chile and a cold-temperate component in southern Chile. However, the way in which these faunistic components intergrade along Chile, the number of biogeographic provinces, and the provincial boundaries in the area, are still in dispute. Woodward (1856), Dall (1909), Olsson (1961) and Marincovich (1973) consider the Isla de Chiloé (*c.* 42°S) as a clear boundary between the so-called Magellanic and Peruvian Provinces; others (Balech, 1954; Knox, 1960; Dahl, 1960; Dell, 1971; Viviani, 1970, 1979; Castilla, 1979) recognize a transition zone of cold-temperate mixed waters between 30° and 42°S. Brattström and Johanssen (1983) have recognized a northern warm-temperate and a southern cold-temperate region with a border at about 42°S but with a transitional area (30–40°S) within the warm-temperate region. Knox (1960) characterized this transition area as a clearly defined province and Viviani (1979) extended its northernmost limit to Pisco (*c.*13°S) in central Peru. However, Ekman (1953), Dahl (1960), Dell (1971) and Castilla (1979) have insisted on the transitional character of this area. Some faunistic groups such as Asteroidea (Madsen, 1956), Bryozoa (Viviani, 1970, 1979), Polyplacophora and endemic *Fissurella* [Posada and McLean, in personal communications to Castilla (1979)] show increased species richness along this transitional region (20°–40°S), which results from the overlapping of the Peruvian and Magellanic faunistic components, added to the presence of numerous species endemic to the transitional region. Other groups, however, do not show this pattern. For example, species richness of littoral fishes along Chile increases to the north, due to the latitudinal addition of either herbivorous fishes, or species (Labridae) capable of feeding on sea urchins (Moreno et al., 1979). Species richness among anthozoans is maximum around 40°S, the region where the partially enclosed sounds and complex archipelagos begin, with concomitant habitat diversification (Sebens and Paine, 1979). The constraints on community organization imposed by these biogeographic patterns of distribution are only now coming to light. However, they might be of major importance. For example, the geographic distribution of two important consumers of primary space-users (the starfish *Heliaster helianthus* and the black sea urchin *Tetrapygus niger*) have their southern geographic limit at about 38°S, probably leading to conspicuous differences in the biotic interactions organizing rocky eulittoral communities north and south of this latitude.

The only extensive biogeographic analysis conducted on the macroalgae of temperate Pacific South America (Santelices, 1980a) distinguished three areas: one encompassing the region from Magellan Strait to Cape Horn (53°–55°S), characterized by numerous subantarctic and endemic species of restricted distribution; a second, very broad, intermediate area (5°–53°S) containing many endemic and bipolar species and very few tropical species, and characterized by a gradual northwards decrease of its numerous subantarctic species; and a third area (4°–5°S) characterized by the presence of a few tropical species, a scarcity of endemic elements and the complete absence of subantarctic species. The macroalgae along Chile, therefore, include two floristic components with a major discontinuity at 53°S, produced by the addition, south of 53°S, of a number of subantarctic species occurring in the sheltered sounds and archipelagos of that area. Probably many of these species also occur between 53° and 42°S, but sampling there has been far from complete. Contrary to the geographic patterns of species richness described for several groups of invertebrates and fishes, species richness of the marine macroalgae shows a definite latitudinal decrease towards the Equator in the stretch of coast between 55° and 4°S. This is the result of the northward reduction in subantarctic species without an equivalent replacement by tropical species. Indeed, of 130 species (34.4%) with subantarctic affinities found in southernmost Chile, about 50 are found at 40°S, 8 at 23°S and none at 5°S. By contrast, the contribution of eastern tropical Pacific elements is restricted to 13 species found at 4°S; six of these extend beyond 12°S, but none occur south of Coquimbo (30°S). Because macroalgae are among the dominant biota in eulittoral and sublittoral rocky communities, the decrease in species richness northwards has a profound effect on the structure of these communities (see below).

DISTRIBUTIONAL AND ZONATION PATTERNS

Rocky shores

Arica–Chiloé (18°–42°S)

Wave-exposed rocky surfaces are common habitats in the coastline between 18° and 42°S, and the zonation patterns have been repeatedly described at several latitudes (Guiler, 1959a, b; Alveal, 1970, 1971; Romo and Alveal, 1977; Santelices et al., 1977, 1980, 1981; Santelices and Castilla, 1977; Castilla, 1979, 1981; Montalva and Santelices, 1981; Ruiz and Giampoli, 1981; Ojeda and Santelices, 1984b).

In central Chile (*c.* 30°S) *Durvillaea antarctica* and *Lessonia nigrescens* form a boundary fringe between the eulittoral and the sublittoral zones (Fig. 14.2A). The extent of the fringe and the specific relative abundance are determined by rocky slope and wave impact (Santelices et al., 1980). Steep, almost vertical wave-exposed rocks are monopolized by *L. nigrescens*. Exposed shores with more gradual slopes have kelp beds of *L. nigrescens* with scattered individuals of *D. antarctica.* In more sheltered habitats *D. antarctica* displaces the wave-resistant *L. nigrescens*. When the *Lessonia–Durvillaea* belt is absent, two other algal formations can be found at this level, depending on the type of opening left by the absence of kelp (Ojeda and Santelices, 1984b). In small vegetational openings surrounded by individuals of *Lessonia nigrescens*, and in the absence of large grazers, large patches of *Gelidium chilense* can be found. In vegetational openings with abundance of grazers, patches of calcareous crusts and bare rock normally occur. Indeed, a pink cover of *Mesophyllum* sp. normally extends into the sublittoral from the lower limit of the kelp holdfasts. Hordes of the black sea urchin *Tetrapygus niger*, large individuals of the chiton *Acanthopleura echinata* and the black snail *Tegula atra* can be found on the pink crusts.

The zone of 0.5 to 1.5 m immediately above *Lessonia nigrescens* is dominated by various algal associations whose relative importance is correlated with changes in rocky slope, wave exposure and incident light (Fig. 14.2A). On vertical walls, and especially in shaded places, *Codium dimorphum* forms expanded green crusts often covering other mid-eulittoral organisms and supporting a

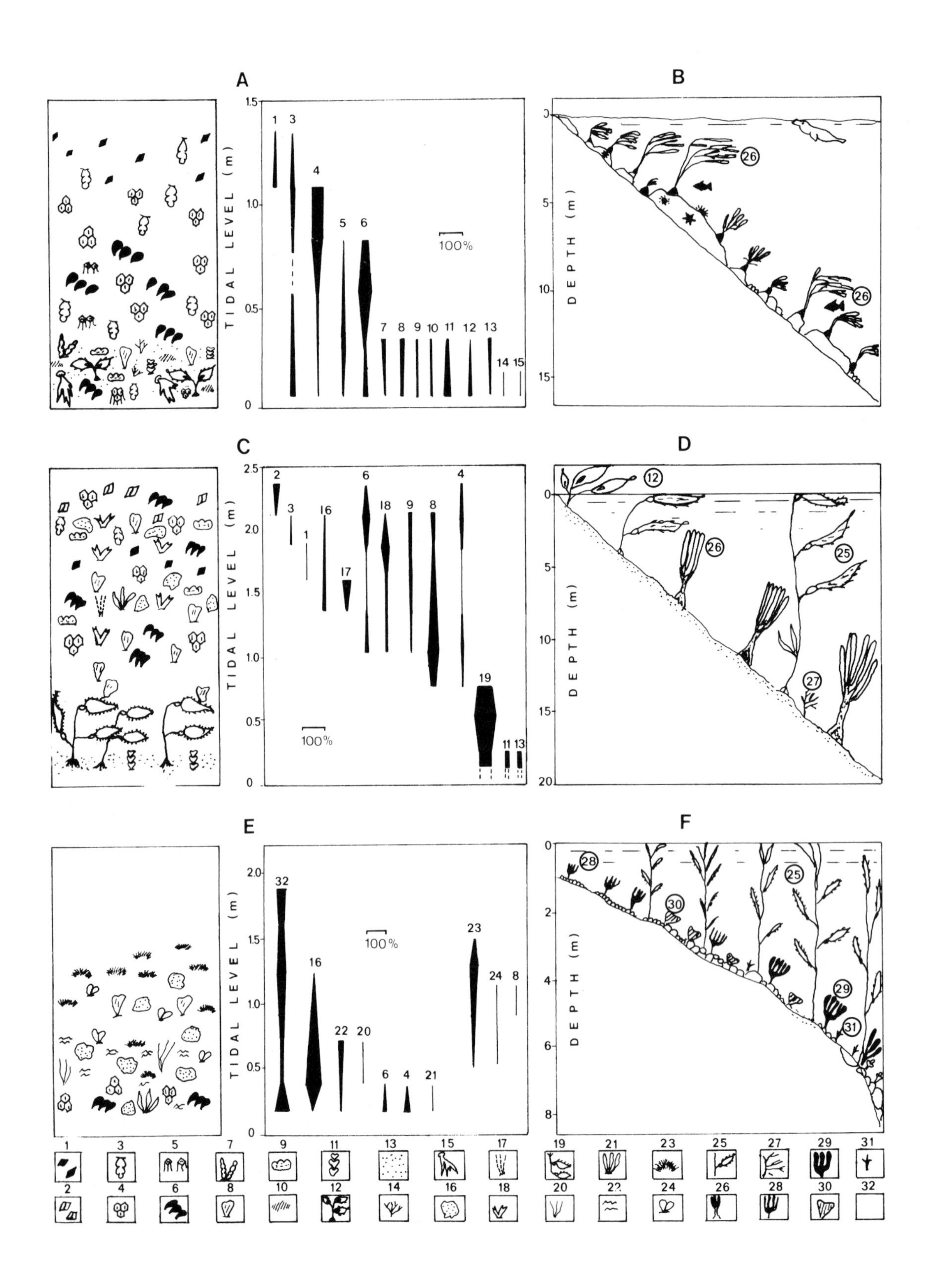

A
B
C
D
E
F
TIDAL LEVEL (m)
DEPTH (m)
100%

diversity of small limpets (such as *Collisella ceciliana*, *C. zebrina*, *Siphonaria lessoni*). On more horizontal platforms with good water exchange and higher levels of light intensity, *Gelidium chilense*, *G. lingulatum*, *Laurencia chilense* and tufts of *Corallina officinalis* form the dominant vegetation. *Fissurella crassa* and *F. limbata* are common grazers at the lower levels of the eulittoral, while *Chiton granosus* extends farther up into the middle and upper eulittoral.

The middle eulittoral normally contains a band of the mussel *Perumytilus purpuratus*, its width varying as a function of slope and wave exposure. The uppermost eulittoral, and up to 2 to 2.5 m above mean low-water level (MLWL) are dominated by pure or mixed stands of the chthamaloid barnacles *Chthamalus scabrosus* and *Jehlius cirratus*. Several filamentous or frondose algae, such as *Centroceras clavulatum*, *Enteromorpha compressa*, *Iridaea laminarioides*, *Ulva rigida*, and several species of *Polysiphonia* can be found among the mussels, extending slightly up towards the barnacle zone.

Three meters above tidal datum the cover values of *Jehlius cirratus* are much reduced and they co-occur with patchy fronds of *Porphyra columbina* and aggregations of *Littorina araucana* and *L. peruviana*. Rocky places exposed to heavy sunlight show, at these tidal elevations, patches of a complex of dark-red crusts, the commonest of which is *Hildebrandia lecannellieri*.

A comparison of wave-exposed, rocky eulittoral habitats in the latitudes between Arica and northern Chiloé indicates an essentially similar pattern of distribution of eulittoral organisms, modified by a northward reduction in the number of belt-forming species of algae. In Mar Brava (*c.* 41°S), Maicolpué and Pucatrihue (*c.* 38°S), and Cocholgue Bay (*c.* 36°S), the eulittoral–sublittoral boundary contains *Macrocystis pyrifera* in addition to *Durvillaea antarctica* and *Lessonia nigrescens*. A belt of *Ahnfeltia plicata* and *Gymnogongrus furcellatus* can locally co-occur with or replace *Codium dimorphum* and *Gelidium* spp. A conspicuous belt of *Iridaea ciliata* is found immediately above, while a mixed belt of *Bostrychia scorpioides*, *Gelidium pusillum* and *Prasiola stipitata* is evident at the uppermost eulittoral levels. *Macrocystis pyrifera* disappears north of Concepción (36°S), and *A. plicata*, *G. furcellatus* and *I. ciliata* cease to be important as belt-forming species around Valparaíso (33°S). *Iridaea laminarioides* also disappears further north (32°S), while north of Coquimbo (30°S) the relative importance of *Gelidium lingulatum* is restricted to extremely wave-exposed habitats. Between this latitude and Antofagasta (25°S), *Durvillaea antarctica* drops off. Thus, the wave-exposed eulittoral habitats around Antofagasta have a eulittoral–sublittoral fringe dominated by *Lessonia nigrescens* only, while the zone 1.0 to 1.5 m above contains very patchy populations of *Codium dimorphum*, *Enteromorpha* spp. and *Ulva rigida*. Rocky habitats around Iquique (20°S) show an essentially similar pattern of distribution, with locally abundant populations of *Colpomenia sinuosa* in the mid-eulittoral zone.

Fig. 14.2. Structural patterns of some eulittoral and sublittoral rocky communities in Chile. (A) Qualitative and quantitative representation of a wave-exposed eulittoral community from central (30°S) Chile (after Santelices et al., 1977). (B) Qualitative representation of exposed sublittoral kelp beds from central Chile, dominated by *Lessonia trabeculata* (after Villouta and Santelices, 1984). (C) Qualitative and quantitative representation of wave-protected eulittoral communities from central Chile (after Santelices et al., 1977). (D) A wave-protected sublittoral community from central Chile dominated by *Macrocystis* and with a second canopy of *Lessonia trabeculata* (after Alveal et al., 1973). (E) Qualitative and quantitative structure of wave-protected eulittoral rocky communities from the Canal Beagle area (53°S). (F) Wave-protected sublittoral kelp communities from the Canal Beagle area (53°S) dominated by *Macrocystis pyrifera* and with a second canopy of *Lessonia flavicans* and *L. vadosa*.

All quantitative data are cover values (%). The key for the species is as follows:

(1) *Littorina araucana*
(2) *Littorina peruviana*
(3) *Porphyra columbina*
(4) *Jehlius cirratus*
(5) *Chthamalus scabrosus*
(6) *Perumytilus purpuratus*
(7) *Scytosiphon lomentaria*
(8) *Ulva rigida*
(9) *Colpomenia sinuosa*
(10) *Codium dimorphum*
(11) *Corallina officinalis*
(12) *Lessonia nigrescens*
(13) *Mesophyllum* sp.
(14) *Polysiphonia-Ceramium*
(15) *Durvillaea antarctica*
(16) *Hildenbrandia* sp.
(17) *Chaetomorpha linum*
(18) *Nothogenia fastigiata*
(19) *Macrocystis integrifolia*
(20) *Pilayella littoralis*
(21) *Enteromorpha* sp.
(22) *Bostrychia scorpioides*
(23) *Spongomorpha* sp.
(24) *Adenocystis utricularis*
(25) *Macrocystis pyrifera*
(26) *Lessonia trabeculata*
(27) *Glossophora kunthii*
(28) *Lessonia vadosa*
(29) *Lessonia flavicans*
(30) *Gigartina skottsbergii*
(31) *Callophyllis variegata*
(32) Bare rock

Most generalizations about sublittoral rocky communities between Arica and Chiloé have assumed that the coastline has kelp beds dominated by *Macrocystis* (Fritsch, 1945; Mann, 1973; Michanek, 1979). However, these coasts are generally exposed to heavy waves, strong surge and storms, and are unlikely habitats for species of *Macrocystis*. The sublittoral communities most commonly found in recent explorations between Osorno (40°28′S) and Antofagasta (23°26′S) are kelp beds formed by *Lessonia trabeculata*, extending down to a depth of 20 m and seemingly occurring in all except very sheltered sublittoral places (Villouta and Santelices, 1984). Normally the beds start at a depth of about 1 to 2 m (Fig. 14.2B) and their lowest limit seems to be set by substratum and light availability. In places with a low rocky slope, these beds can be up to 300 m wide.

Although detailed, quantitative studies of these beds have been performed only at two localities in central Chile (both around 32°S), results suggest that plant densities, as well as species composition of the understorey algal community, change with depth (Villouta and Santelices, 1984). In shallow depths (−2 to −7 m), high densities (2 plants m^{-2}) of rather small (1 m long) plants are common. The substratum between the holdfasts of *Lessonia trabeculata* is dominated by calcareous crusts, and by the polychaete *Idanthyrsus armata*. Non-calcareous crusts such as *Ralfsia*, *Hildenbrandia* and *Peyssonnelia*, as well as barnacles and bryozoa, each may have cover values of 5 to 10%. In deeper waters (−15 to −17 m), kelp plants are larger (up to 2.5 m long) and less dense (1 plant m^{-2}). A second vegetational stratum develops here, formed mainly by *Bossiella orbigniana*, *Gelidium chilense*, *Halopteris funicularis*, *H. scoparia*, *Glossophora kunthii* and *Plocamium secundatum*. Sponges and bryozoans are infrequent. Taking both depths, a total of 47 taxa of invertebrates occur within and around the holdfasts of *L. trabeculata*. At least seven species of fishes and two of marine mammals (*Otaria byronia* and *Lutra felina*) also visit the beds. Barren grounds containing high densities of several species of *Fissurella* and *Tetrapygus niger* also occur within these beds.

Descriptions of eulittoral rocky sheltered communities exist for a few places along the coastline between Arica and Chiloé (Guiler, 1959a, b; Alveal et al., 1973; Romo and Alveal, 1977; Ruiz and Giampoli, 1981; Moreno and Sutherland, 1982). Individuals of *Macrocystis integrifolia* in the localities north of Valparaíso, and *M. pyrifera* in places further south, mark the lower limit of the eulittoral zone (Fig. 14.2C). In a few places, scattered individuals of *Lessonia trabeculata* can reach this level which is exposed during low tides, but this species normally does not form a conspicuous belt at this level. Mixed or monospecific belts of *Corallina chilensis*, *Glossophora kunthii*, *Nothogenia fastigiata*, *Plocamium pacificum* and tufts of *Griffithsia chilensis* occur in the lower eulittoral zone (Fig. 14.2B). *Adenocystis utricularis*, *Colpomenia sinuosa* and *Scytosiphon lomentaria* can be seasonally abundant within and on top of dense beds of *Perumytilus purpuratus* shells that cover the middle eulittoral levels. The species represented and the patterns of distribution of the organisms at the uppermost eulittoral levels are quite similar to those described for exposed surfaces. The sublittoral communities normally show a first stratum formed by the floating canopy of *Macrocystis* (Fig. 14.2D) and, sometimes, a second stratum formed by the lower canopy of *Lessonia trabeculata*. Calcareous crusts and tufts of ectocarpoids cover the bottom (Alveal et al., 1973). When *Macrocystis* canopy is sparse and the second canopy is lacking, individuals of *Glossophora kunthii* and *Plocamium violaceum*, 5 to 10 cm long, can also be found (Alveal et al., 1973). In Bahía Corral, in the vicinity of Valdivia (39°51′S), a kelp bed of *Macrocystis pyrifera* studied by Moreno and Sutherland (1982) extended from a depth of 1 to 8 m; *L. trabeculata* was absent and encrusting algae covered 80 to 90% of the bottom. Here, winter storms removed the canopy every year, but the surviving holdfasts produced new fronds the following winter. Upright forms of perennial and ephemeral algae forming the understorey vegetation were abundant only in late winter and spring, when the *Macrocystis* canopy was reduced.

Quantitative descriptions of rocky sheltered communities for habitats in northern Chile are lacking, except for Guiler's (1959a) notes on the patterns of eulittoral biota in Bahia Antofagasta (23°42′S). Guiler thought that shore topography in the area had considerable bearing on the type of zonation. Here the rocks are of a shale that gently dips away from the sea forming a wide erosion platform. In this place, immediately below the barnacle belt, there is a *Pyura–Corallina* association formed by a closely packed, dense cover of *Pyura praeputialis*, with *Corallina chilensis* using the ascidian as substratum. At lower intertidal elevations *Ulva* replaces *Corallina* as the seaweed with highest cover values. Among the ascidian tests, as well as underneath and around the fronds of *Ulva* and *Corallina*, several other algae and invertebrates occur. The large barnacle *Austramegabalanus psittacus* occurs in the lowest part of the *Pyura* belt, and both organisms are frequently covered with *Ectocarpus confervoides* and *Halopteris hordacea*. *Pyura praeputialis* only forms this belt at Antofagasta; the closely related species *Pyura chilensis* is widespread along the entire coast of Chile but does not form conspicuous belts in northern Chile (Guiler, 1959a; Paine and Suchanek, 1983).

Sheltered shores south of Valparaíso might withstand the influence of periodic sand intrusion. Zonation patterns of eulittoral organisms in these habitats are slightly different. In general, most species of the perennial lower and mid-eulittoral habitats without sand influence are missing here (for instance, *Austramegabalanus psittacus*, *Codium dimorphum*, *Lessonia trabeculata*, *Macrocystis pyrifera*). The number of grazers and predators in general is also reduced, and hundreds of individuals are covered by sand every time sand invasion reaches the lower eulittoral. The most conspicuous vegetation at these levels consists of successive belts of *Gymnogongrus furcellatus*, *Iridaea laminarioides* and *Nothogenia chilensis*, all species with a crustose base or crustose alternate phase that allows them to survive sand invasion.

Chiloé to Cape Horn (42°–55°S)

As stated previously, south of Chiloé the coast becomes more and more precipitous and wild. The weather is almost always rigorous, particularly during summer, and the winds blow almost continually from the west. Offshore islands shelter some portions of these coasts from storms outside, thus generating habitat diversity. However, due to the prevailing weather conditions and limited access, only limited information of patterns of community structure is available for these latitudes. Currently no one has described the eulittoral and sublittoral communities in these wave-exposed open coasts. Likewise, there is no quantitative information on communities occurring in the euryhaline, low-temperature conditions in the channels under the influence of the fjords. Most of the available information refers to sheltered eulittoral and sublittoral islands around the Estrecho de Magallanes (Alveal et al., 1973), the Canal Beagle and the Bahía de Nassau (55°S) in the southernmost tip of South America (Guzmán and Rios, 1981; Rios and Guzmán, 1982; Santelices and Ojeda, 1984a, b; Castilla, 1985). These beaches are often characterized by the presence of boulder fields or low rocky outcrops. Ice and snow are common in the supralittoral and uppermost eulittoral zones during winter. In summer, the organisms might receive direct sunlight, and experience mid-day temperatures up to 9 to 10°C.

In the Canal Beagle the uppermost individuals of *Lessonia vadosa* mark the lowest limit of the eulittoral (Fig. 14.2E). The zone of 20 to 40 cm immediately above is covered by pink calcareous crusts and high densities of *Nacella magellanica* and *N. mytilina*. The next zone of 50 cm has mussels and barnacles, and, especially underneath stones and boulders, there are high densities of *Acmaea*, *Collisella*, *Nacella* and *Siphonaria*. In winter at this level macroalgae are completely absent. The next zone of 30 to 40 cm has a mixed cover of *Bostrychia mixta*, *Hildenbrandia lecannellieri* and *Pilayella littoralis*. During spring and summer, *Adenocystis utricularis* and species of *Enteromorpha*, *Porphyra* and *Spongomorpha* also occur here. The uppermost eulittoral as well as the supralittoral contains several bands of lichens.

The sublittoral communities in these habitats consist of narrow coastal belts, 50 to 100 m wide, of *Macrocystis pyrifera* that surround the sheltered and semi-sheltered sides of these islands. The upper limit of *Macrocystis pyrifera* appears sharply bounded below a belt of *Lessonia vadosa* 1 to 2 m wide and extends down to 10 to 15 m (Fig. 14.2F). In a transectional view the first 10 m of kelp bed are characterized by small-sized, densely packed individuals of *M. pyrifera* living on consolidated rocky substrata or boulders. The population extends down to 15 m where the substratum is seemingly limiting (muddy). In some areas, consolidated rocky substrata can be found carrying stands of *M. pyrifera* mixed with *Lessonia flavicans*.

These kelp forests of *Macrocystis pyrifera* in southern Chile have several vegetational strata. The first stratum is represented by the floating canopy of *M. pyrifera*. The second is formed by plants of *Lessonia flavicans*, 2 to 3 m long, that extend across most of the forest. Underneath this canopy a third stratum is formed by fleshy and frondose algae about 50 cm in height, among which *Epymenia falklandica* and *Gigartina skottsbergii* are the commonest and most important as producers of biomass, covering extensive portions of the forest. Occasionally, elongated individuals of *Enteromorpha prolifera*, *Scytosiphon lomentaria*, tufts of *Ballia callitrichia*, *Halopteris hordacea*, *Plocamium secundatum* and *Spongomorpha pacifica*, as well as elongated blades of *Callophyllis variegata*, *Hymenena laciniata*, *Myriogramme* sp., *Phycodrys* sp. and *Ulva rigida* can be found forming associations and covering localized patches in the bed. In areas where the canopy of *L. flavicans* or *M. pyrifera* is less dense, a cover of short, filamentous members of the genera *Ceramium*, *Ectocarpus*, *Griffithsia* and *Lophurella* can be found as well. Most of the primary substratum across the bed is covered by crustose algae that occur underneath the strata mentioned above.

Sandy shores

Sandy shores have considerable development in Chile, and become increasingly more frequent as one moves from central to southern Chile. The first observations on community structure of the macrofauna in these habitats were reported by Dahl (1953) when proposing his general zonation scheme for sandy shores. Dahl studied the distributional patterns of larger Crustacea in Montemar (32°57′S), in the vicinity of Puerto Montt (41°29′S) and in Punta Arenas (53°22′S). Based on these and numerous other studies, Dahl (1953) distinguished for temperate latitudes an uppermost, subterrestrial fringe characterized by talitrid amphipods, a midlittoral zone occupied by cirolanid isopods, and a sublittoral fringe characterized by several burrowing amphipods. Thereafter, several authors have made qualitative (Osorio et al., 1967; Nuñez et al., 1974; Epelde-Aguirre and López, 1975) and quantitative descriptions (Castilla et al., 1977; Jaramillo, 1978, 1982; Sánchez et al., 1982; Varela, 1983) of these communities. All of these studies have been conducted in shores located within 30° and 42°S, and most authors have compared their results with the zonation schemes proposed by Dahl (1953), often pointing out local differences. Only Sánchez et al. (1982) have attempted latitudinal comparisons of zonation patterns.

The information accumulated indicates that three faunistic fringes (*sensu* Trevallion et al., 1970) can be found in the eulittoral levels of these sandy beaches (Fig. 14.3A). The uppermost fringe is characterized by the amphipod *Orchestoidea tuberculata*. In Playa Morrillos (Fig. 14.3A), the isopod *Tylos* sp., the beetle *Phalerisida maculata,* and larvae and adult individuals of several insect species can also be found at this level. *Tylos* sp. is absent from Quilliruca and Mehuín while *P. maculata* is also absent from Mehuín. In this last locality the cirolanid isopod *Excirolana braziliensis* extends up to this uppermost fringe.

Cirolanid isopods of the genus *Excirolana* seem to be characteristic members of the middle fringe. *E. braziliensis* was the predominant organism at this level in Morrillos, *E. hirsuticauda* was the numerically important one at Quilliruca, while these two species together with *E. monodi* defined this fringe in Mehuín. Polychaetes are also important at this level. *Euzonus heterocirrus* and *Scolelepis* sp. were quite conspicuous in Quilliruca, and *Hemipodus* sp. characterized this fringe in Morrillos. Although no species of Polychaeta were typical of this level in Mehuín, Jaramillo (1978) suggested they may be found in sandy shores with less wave exposure, and on muddier bottoms.

In all three shores the lower fringe is characterized by *Emerita analoga* (Anomura), *Mesodesma donacium* (Bivalvia), and different species of Polychaeta, which can be locally important. *Nephtys impressa* is quite abundant in Morrillos and Mehuín, while *Scolelepis* sp. is numerically important in Quilliruca. Some decapods (*Lepidopa chilensis*), cirolanid isopods and amphipods (*Bathyporeiapus magellanicus*) can also be locally common at this level.

A comparison of these results with the scheme proposed by Dahl (1953) suggests at least two sources of variation. Owing to local habitat variations, several species can modify their expected representation. This could be the case of the cirolanid isopods found at the uppermost fringe in Mehuín, whose presence is unexpected under

Dahl's scheme. Likewise, in Bahía Corral (37°03′S), Epelde-Aguirre and López (1975) found cirolanid isopods together with *Emerita analoga* in the lower fringe probably due to the strong wave impact in the mid-eulittoral levels. In Playa La Herradura, less than 10 km away from Playa Quilliruca, Castilla et al. (1977) found a significant modification in vertical zonation patterns, reduced species richness and selective absence of seemingly more sensitive species in places affected by oil pollution.

A second source of variation arises because Dahl's scheme is based only on Crustacea. Various authors have remarked on the presence of various other macrofaunistic groups useful to characterize vertical zones. In the sandy shores so far studied in Chile, Polychaeta in the middle and lower fringes and Bivalvia in the lower fringe consistently appear as numerically important organisms.

Quantitative variations are also common in these communities and should be borne in mind when attempting general characterization. Jaramillo (1978) simultaneously studied four different sites in Playa Mehuín, finding significant differences both in species composition and density values. Sánchez et al. (1982) quantified seasonal differences in diversity and vertical distribution of several numerically important members of the macrofaunal community of Playa Morrillos, while Bertrán (1984) and Jaramillo et al. (1984) reported macrofaunal changes due to variations in salinity, wave exposure and sediment type (see Estuaries section, p. 361).

Many species found in the lower eulittoral seemingly extend their distributional range into the sublittoral. However, at present qualitative or quantitative descriptions of such communities are lacking. Several protected sandy shores have commercially important beds of *Gracilaria* spp. extending down to depths of 7 to 10 m. A few studies have described quantitative variations in standing stock and have measured production rates (Santelices and Fonck, 1979; Romo et al., 1979; Westermeier, 1980; Black and Fonck, 1981; Santelices et al., 1984; Westermeier et al., 1984). Although a few seaweed species have been occasionally included in these studies, no consideration of the whole macrobenthic community occurring in these habitats has been made.

Sand and mud flats

Sand and mud flats are specially well developed between 38°S and 45°S, although they may also be abundant further south. They form extensive platforms protected from wave impact that remain uncovered during low tides. Several species of *Enteromorpha*, *Gracilaria* and *Ulva* as well as different types of mussels and polychaetes, can be found here. Because they contain abundant *Gracilaria* beds, these flats have supported heavy commercial exploitation during the last 15 years. Yet there have been no qualitative or quantitative studies of the macrobenthic communities found here.

Estuaries

Central and southern Chile are characterized by numerous estuaries. Physico-chemical characterizations of these habitats have been performed (Campos et al., 1974) and the patterns of ecological distribution of selected taxonomic groups have been established [for instance, polychaetes: Bertrán (1980); bivalves: Stotz (1981); barnacles: Arenas (1971); crabs: Retamal (1979); fishes: Fischer (1963), Campos (1973), Pequeño (1979)]. Less frequently, however, has the community structure of the littoral and sublittoral macro-infauna been analyzed. Only Bertrán (1984) and Jaramillo et al. (1984) have attempted such analyses in estuaries in the vicinity of Valdivia (*c.* 40°S), comparing faunistic assemblages found at different distances from the mouth of the estuaries. In Río Lingue, Bertrán (1984) found that the two inner stations (Fig. 14.3B) had muddy sediments, a high content of organic matter and a notorious black reduction layer. The main faunistic components at these stations were Polychaeta, making up 47.7 to 87.3% of the whole macro-infauna. *Perinereis gualpensis* and *Boccardia polybranchia* were the most abundant species, with densities ranging from 686 to 1132 and from 107 to 1480 ind. m^{-2} respectively. Amphipods (*Paracorophium* sp.), bivalves (*Kingiella chilenica*) and insect larvae were the other abundant members in this community. The two stations near the mouth of the estuary had clean sand and a low content of organic matter and lacked a black reduction layer. Crustacea here constituted almost the whole of the macro-infauna;

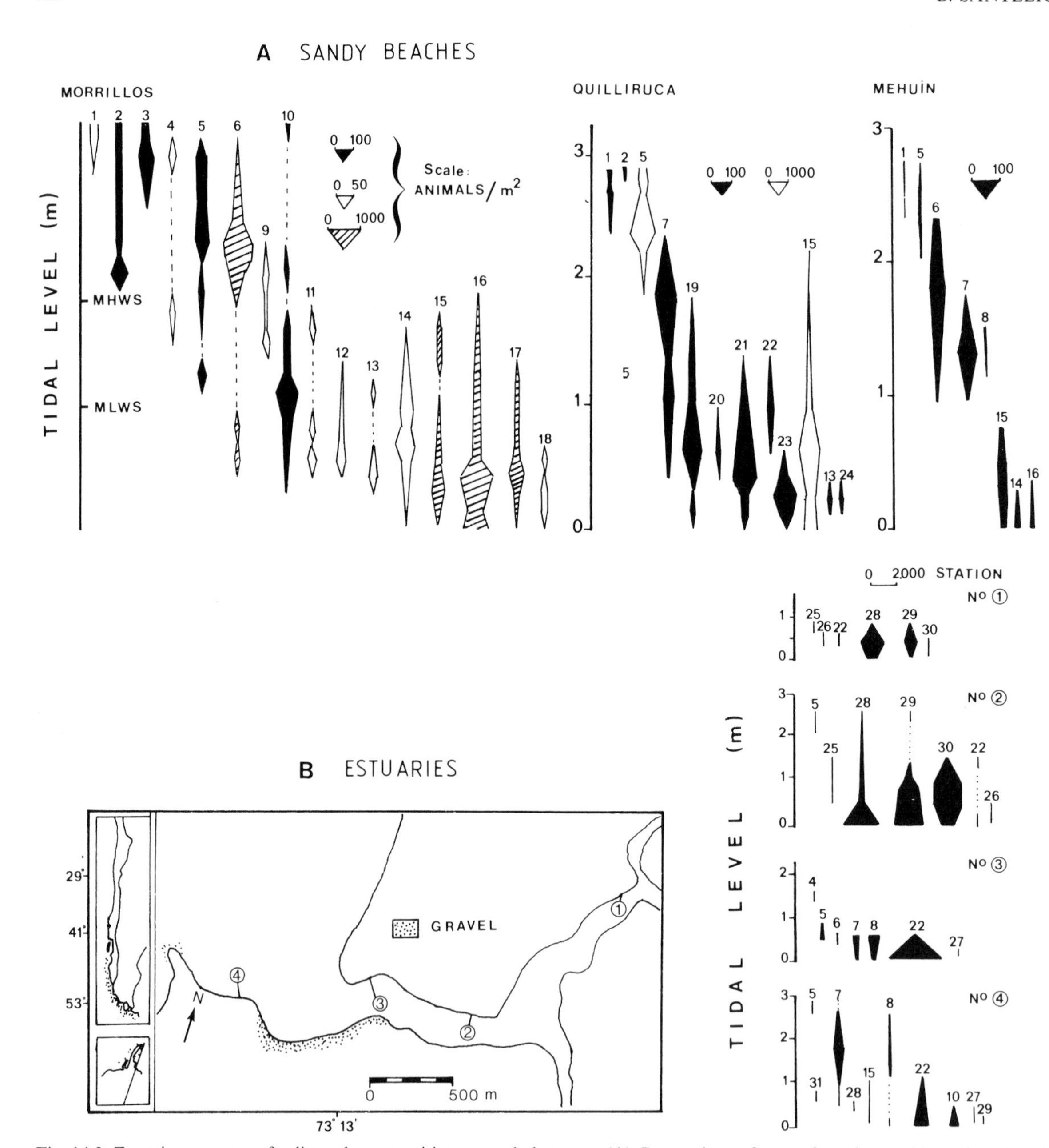

Fig. 14.3. Zonation patterns of eulittoral communities on sandy bottoms. (A) Comparison of macrofaunal assemblages found at three sandy beaches along Chile; Playa Morrillos (30°09'S; after Sánchez et al., 1982); Quilliruca (32°42'S; after Castilla et al., 1977); and Mehuín (39°23'S; after Jaramillo, 1978). The key for the species is as follows:

(1) *Phalerisida maculata*	(9) *Hemipodus* sp.	(17) *Polydora* sp.	(25) Insect larvae
(2) *P. maculata*, Larvae	(10) *Bathyporeiapus magellanicus*	(18) *Bellia picta*	(26) *Kingiella chilenica*
(3) *Tylos* sp.	(11) *Chaetilia paucidens*	(19) *Euzonus heterocirrus*	(27) *Phoxocephalopsis* sp.
(4) *Oligochaeta* indet.	(12) Nemertinea indet.	(20) *Lumbrineris* sp.	(28) *Paracorophium* sp.
(5) *Orchestoidea tuberculata*	(13) *Nephtys monilibranchiata*	(21) *Scolelepis* sp.	(29) *Perinereis gualpensis*
(6) *Excirolana braziliensis*	(14) *N. impressa*	(22) *Phoxocephalidae*	(30) *Boccardia polybranchia*
(7) *E. hirsuticauda*	(15) *Emerita analoga*	(23) *Scoloplos* sp.	(31) *Exosphaeroma lanceolata*
(8) *E. monodi*	(16) *Mesodesma donacium*	(24) *Sthenelais* sp.	

(B) Pattern of zonation of macrofaunal assemblages at four stations with different types of bottom, salinities and wave exposure in an estuary at 39°S (redrawn from Bertrán, 1984).

the amphipod *Paracorophium* sp. and the isopods *Excirolana monodi* and *Emerita analoga* were numerically important members of this assemblage.

Bertrán (1984) has emphasized the relationship between faunistic assemblages and the type of substrate. Furthermore, when comparing the zonation patterns found here with nearby sandy shores, Bertrán (1984) noticed that the only significant difference was the general absence of bivalves near the mouth of the estuary. A similar relation between substratum type and faunistic assemblages was found also by Jaramillo et al. (1984) in the estuary of the Río Queule. Thus, a group of sandy stations near the outlet was numerically dominated by suspension feeders (Bivalvia and Amphipoda) while the stations in the middle and upper parts of the estuary, with muddy and sandy bottoms and a higher percentage of organic matter, were dominated by deposit feeders. The bivalve *Mesodesma donacium* represented the dominant faunal assemblage near the sandy outlet (up to 13 000 ind. m^{-2}) while the polychaete *Minuspio chilensis* was the numerically dominant deposit feeder (up to 22 000 ind. m^{-2}) in the middle and upper parts of the estuary. Boundaries between the two faunal assemblages were not clear however, and several species showed a continuous distribution. When comparing these findings with other temperate estuaries, especially in the Northern Hemisphere, Jaramillo et al. (1984) noticed a reduced species richness and a general absence of cosmopolitan estuarine taxa in southern Chile.

PATTERNS OF COMMUNITY ORGANIZATION

Experimental studies on community organization began less than ten years ago and have been restricted mainly to exposed and semi-exposed rocky eulittoral habitats around 30° and 39°S. A few experimental studies have been performed also in sheltered eulittoral and sublittoral habitats of the Canal Beagle area (54°S), but integrated ideas on community organization have not yet been provided.

Macroalgal belts and competitive interactions

Experimental analysis of the interacting factors determining floristic patterns in the lower and middle rocky eulittoral habitats of central Chile have pointed to the widespread importance of interference competition. In these habitats Santelices (1981) has noticed zonation of algal morphologies rather than of species, recognizing three permanent algal belts and one or several temporary algal belts. From above down, the three permanent components include a wide belt of *Mesophyllum* sp. and other calcareous crusts, a belt of kelp-like seaweeds (*Lessonia-Durvillaea*) and a narrow belt of crustose or cushion-forming algae (*Gelidium* spp. and/or *Codium dimorphum*) (Fig. 14.4).

Much of the physiognomy of the lower eulittoral zone seems to be determined by factors regulating juvenile recruitment of kelps (Fig. 14.4). Recruitment can be completely inhibited either by interference from adult kelps or grazing by large sublittoral herbivores (Santelices and Ojeda, 1984c). Because both factors are mutually exclusive in the field, juvenile recruitment is maximal in patchy vegetational openings which are large enough for disturbance by adult plants to be reduced and small enough for grazing pressure to be reduced. Larger vegetational openings suffer intense grazing pressure, resulting in bare rock patches (Ojeda and Santelices, 1984b). Less intense grazing allows a pink cover of calcareous crusts in which specific composition seems to be regulated by light intensity and water movement.

The relative importance of *Durvillaea antarctica* and *Lessonia nigrescens* within the belt depends on the degree of wave impact (Santelices et al., 1980). *Lessonia nigrescens* seems to be adapted to strong wave impact. Yet complete space monopolization is prevented by a series of adaptations of *D. antarctica*. Certain morphological forms of this species are less affected by wave action, allowing a population stock to persist even at the most exposed places. Boring into algal holdfasts by invertebrates weakens the mechanical resistance of old, eroded plants, providing open space where juveniles of either species could settle. *Durvillaea antarctica* seems to take greater advantage of this primary space by means of a fugitive life history. Thus, in wave-exposed habitats *D. antarctica* is constantly removed by water movement but persists because of higher colonization rates. In more sheltered habitats, competitive displacement by *L. nigrescens* might occur, but *D. antarctica* is heavily

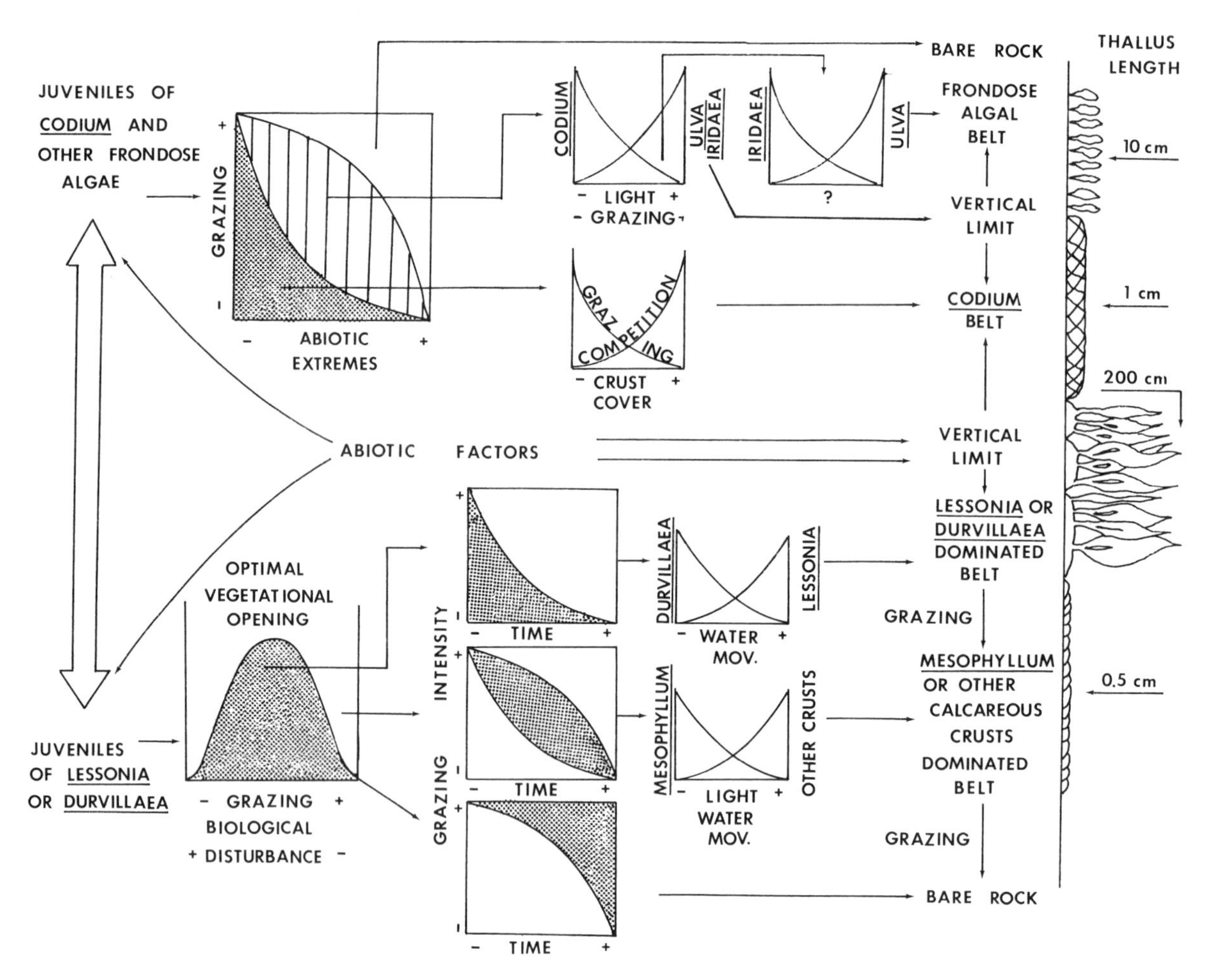

Fig. 14.4. Ecological interactions resulting in three permanent belts of algae (*Mesophyllum* sp., *Lessonia–Durvillaea* and *Codium*) in eulittoral–shallow sublittoral rocky habitats of central Chile.

collected by fishermen for human consumption as food.

Pure or mixed stands of *Codium dimorphum* and of species of *Gelidium* occur at the levels immediately above. Experimental removal of *L. nigrescens* at some seasons allows expansion of the lower vertical limit of these species (Ojeda and Santelices, 1984b), suggesting that interspecific competition is important here. By contrast, removal of *C. dimorphum* or *Gelidium* does not result in upward expansion of *L. nigrescens*, suggesting that the abiotic extremes limit its uppermost distribution (Santelices and Ojeda, 1984c).

The relative importance and the upper limit of *Codium dimorphum* are, however, determined by abiotic climate and grazing. Summer bleaching of the crustose, expanded colonies of *C. dimorphum* results in reduction of cover by this species in the lower eulittoral zone and its disappearance from the mid-eulittoral zone. This, in turn, results in the expansion of the lower limit of several mid- and upper-eulittoral species, among which *Iridaea laminarioides* and *Ulva rigida* are important. Herbivore exclusion experiments indicate that grazing increases cover reduction of *C. dimorphum* after initial summer bleaching of the colony. During winter, *C. dimorphum* is able to overgrow, exclude and, therefore, limit the downward extent of most other intertidal species expanding into the middle eulittoral. Thus, while growing, the relative importance of grazing on the structure of this belt decreases while the importance of interspecific competition increases. The opposite seems true for summer conditions.

Equivalent data for the species of *Gelidium* that also occur at this level are lacking. However, interspecific interference does occur among these species (Montalva and Santelices, 1981), and they have significant differences at least in tolerance for light intensity (Oliger and Santelices, 1981).

The factors regulating the abundance of the uppermost eulittoral vegetational belts have not yet been as thoroughly studied. The abundance of *Iridaea laminarioides* and *Ulva rigida* is strongly affected by the herbivore *Fissurella picta* (Moreno and Jaramillo, 1983) – a grazer that is also heavily collected for human consumption (Moreno et al., 1984). Herbivore removal in the rocky eulittoral of Mehuín, in the vicinity of Valdivia, allowed high cover (80%) of *I. laminarioides* which, in turn, precluded further invasion by ephemerals (such as *Ulva rigida*). Herbivore addition caused a drastic decline of the erect portion of *Iridaea*; the substrate become dominated by barnacles and crustose algae, allowing recruitment of ephemeral species (Jara and Moreno, 1984).

It is interesting to note the general importance of interspecific interactions between algae of similar morphology within each vegetational belt. This seems to explain the occurrence of zonation of external morphologies rather than strict species zonation patterns. Furthermore, this lower eulittoral and shallow sublittoral vegetation is characterized by strongly trimodal size distribution (Santelices, 1981). The shallow sublittoral calcareous crust (*Mesophyllum* spp.) is replaced vertically by large kelp-like seaweeds (*Lessonia* and *Durvillaea*) above which expanded and short non-calcareous crusts and cushions (*Codium* and *Gelidium*) are found. Even though several factors determine the presence of these species, many have morphologies adapted to escape or resist heavy grazing pressure. This could perhaps explain why interspecific interference is apparently so important in this system.

Predators and predation

Focusing his studies from a different point of view, Castilla (1981) has identified at least nine predators of high trophic levels, each with a trophic sub-web in these same habitats (Fig. 14.5). These can be grouped into two types. The first group, represented by the seagull *Lárus dominicanus*, the seabird *Haemátopus ater* and the South American sea otter *Lutra felina*, visit the eulittoral mainly for feeding activities, and spend most of their time elsewhere. Due to logistic problems of experimental manipulation, the structuring role of these species on eulittoral communities is poorly known. Yet, direct field observations, analysis of food remains, and faeces analysis indicate that *L. dominicanus* preys on two species of *Acanthocyclus*, which, in turn, prey on mussels and barnacles. Therefore, the gull may play an important role in structuring these communities (Bahamondes and Castilla, 1986). Likewise, *Lutra felina* consumes (Castilla and Bahamondes, 1979) seven species of herbivore gastropods, in addition to *Concholepas concholepas* and *Sicyases sanguineus*, both of which are important predators on primary space-users. Therefore, the sea otter may also play a role in structuring eulittoral communities. However, after decades of heavy hunting it is now in endangered status, and its densities in central Chile do not seem important enough to affect these communities significantly.

The second group of predators includes the starfish *Heliaster helianthus*, the decapods *Acanthocyclus gayi* and *A. hassleri*, the gastropods *Crassilabrum crassilabrum* and *Concholepas concholepas*, and the clingfish *Sicyases sanguineus*. All these species occupy eulittoral habitats on a permanent basis, at least during some part of their life history.

The community effects of the starfish *Heliaster helianthus* have been studied more extensively than those of other predators. This species is a generalized consumer of marine invertebrates, including barnacles, a wide range of molluscs and solitary tunicates (Paine et al., 1985). Yet the major component (59–63%) of its observed adult diet is made up of *Perumytilus purpuratus*, a common mussel of exposed mid-eulittoral levels along most of the coastline between Arica and Chiloé (18° to 42°S). Experimental removal of *H. helianthus* leads to an increase in the abundance of mussel and barnacles (Paine et al., 1985) and to an expansion of the lower vertical limit of the mussel. Castilla (1981) suggested that the ecological role played by all other predators combined (except *Crassilabrum crassilabrum*) could be similar to *H. helianthus*, as they also remove primary space users. But this may not be the case of *Sicyases sanguineus*. The clingfish is abundant, quite voracious, and preys

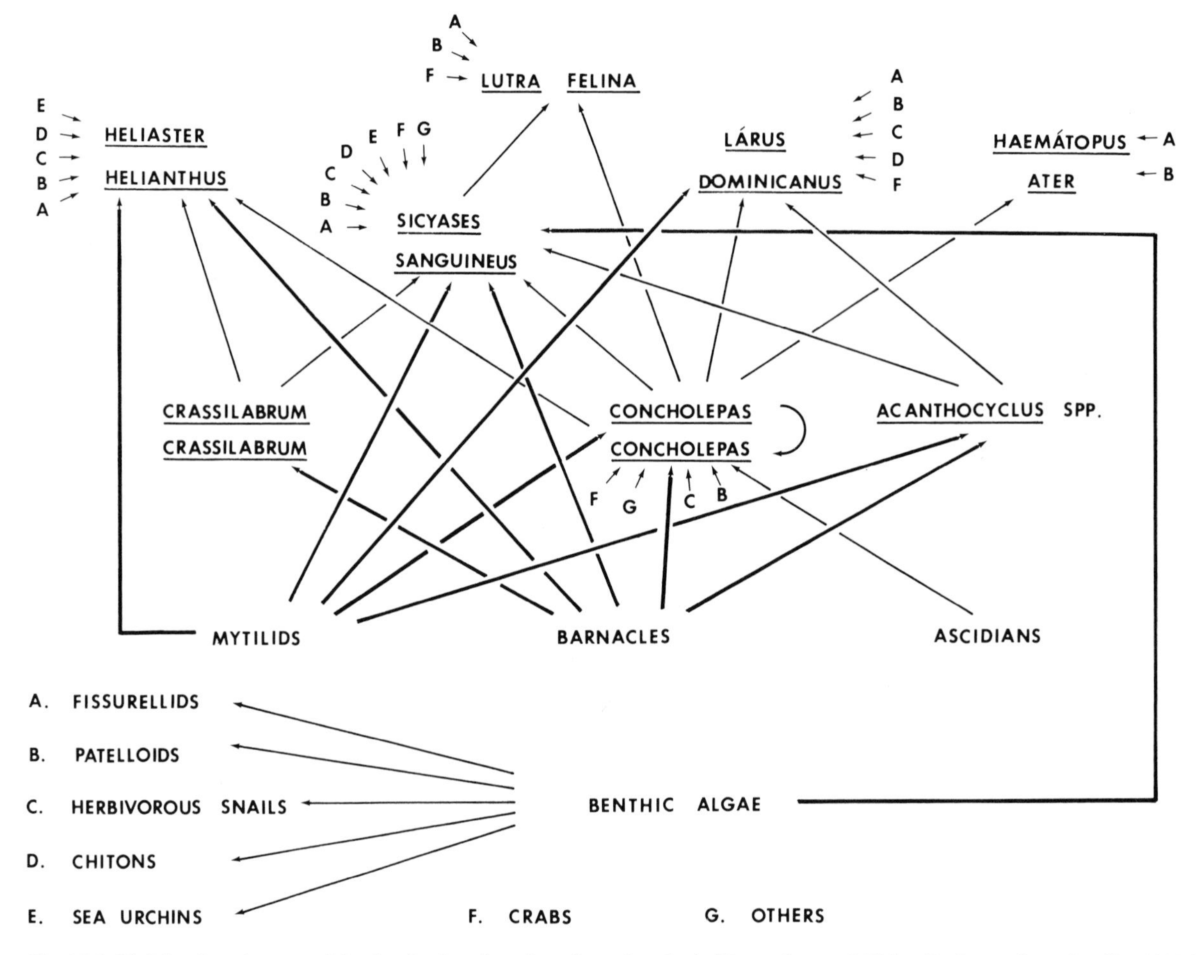

Fig. 14.5. High-level predators and food webs they form in eulittoral rocky habitats of central Chile. (Redrawn from Castilla, 1981.)

on most species on vertical walls throughout the eulittoral zone with little evidence for selectivity. However, *S. sanguineus* is limited by size of the mouth and it cannot control the distribution and abundance of prey species becoming too large to be consumed (Paine and Palmer, 1978). In addition, many less-generalized invertebrate consumers coexist with *S. sanguineus* in the middle and lower zone and their combined foraging pressure seemingly compensates for the absence of *Sicyases* (Paine and Palmer, 1978).

The community effects of *Concholepas concholepas* have been experimentally evaluated just recently. This species is normally subjected to strong fishing pressure along the coast. Establishment of a marine reserve in Mehuín (39°24′S) in 1978 resulted in a striking change of the population structure of *C. concholepas*, by the addition of larger size classes (Moreno et al., 1986). Simultaneously, the width of the *Perumytilus purpuratus* belt decreased, becoming restricted to a narrow fringe at upper intertidal levels. Moreno et al. (1986) noticed that *Heliaster helianthus*, which also preys on *P. purpuratus*, is lacking south of 33°S, and that the response of *P. purpuratus* to increased snail predation on the marine reserve seems more striking than the effect described in central Chile by Paine et al. (1985) after the removal of *H. helianthus*. Equivalent data (Castilla and Durán, 1985) from a second reserve established in 1982 in Las Cruces (Central Chile) indicate that similar phenomena are also occurring there. It should be borne in mind, however, that both of these *C. concholepas* repopulations are taking place in the absence of the endangered *Lutra felina*, one of the top predators in the system, which under natural densities perhaps would be able to regulate the density of *C. concholepas*.

Several of these studies point to the ecological role of human predation in the coastline between 18° and 42°S. Indeed, it is difficult to define what is "natural" in the ecological history of these communities. It is probably true that the Chilean shoreline at present is being heavily overexploited, and a rich and productive shoreline community is being altered at an increasing pace (Castilla, 1976b). However, early inhabitants of these coastal areas also were intertidal harvesters (Viviani, 1979; Moreno et al., 1986). Those inhabiting southern Chile also had the possibility of hunting and plant gathering on land, but those living north of 30°S had to face a desert land, and to depend almost entirely on the intertidal zone for protein supplies. Shell middens along the coast attest to important human impact as early as 8000 B.P., and recent data suggest that inland settlers could have exerted important predatory pressures on these communities as early as 12 000 B.P. (Moreno et al., 1986). Invertebrate species common in some of these shell middens (such as, for instance, *Choromytilus chorus*) are now locally extinct; it is at present unknown if this resulted from human predation or from some other causes (Viviani, 1979). Perhaps, as Moreno et al. (1986) suggested, anthropological factors have been more important than previously thought in shaping and maintaining the structure of these coastal communities.

CONCLUSIONS

Although clearly incomplete and unevenly distributed, the present information on coastal Chilean ecosystems allows some preliminary conclusions on distributional patterns for a few habitats in some geographic areas and a partial understanding of the interacting effects of several ecological factors in organizing these communities. So far the emerging picture suggests that coastal topography, land and oceanographic climate have produced a highly isolated biota, organized into structurally simple communities, where several ecologically important species are endemic and have unique ecological niches with no parallel development in comparable habitats elsewhere. Thus, the historical components of the system are of paramount importance to understand its structure and organization. Future descriptive research will be especially rewarding if oriented to the exploration of coastal communities in little-studied habitats such as mud and sand flats, and sand and rocky systems on the exposed and protected Patagonian shores. Further experimental studies of all these systems will be useful, not only for a general understanding of the ecological interactions shaping these communities, but also for testing general ecological theory in an isolated and seemingly quite unique coastline.

ACKNOWLEDGEMENTS

This review has benefited from several years of research on the rocky communities especially of central Chile. It is a pleasure to acknowledge the continuous support on these activities from Dirección de Investigación, Pontificia Universidad Católica de Chile through several grants. I extend my gratitude to J.C. Castilla, J. Cancino, F. Jaksić, A. Hoffmann, D. Lopez, C. Moreno and P. Sánchez for reviewing, criticizing and improving the manuscript and to G. Espinosa, R. Ugarte and E. Martínez for much help with the illustrations.

REFERENCES

Ahumada, R. and Chuecas, L., 1979. Algunas condiciones hidrográficas estacionales de la Bahía de Concepción y áreas adyacentes. *Gayana Bot. Misc.*, 8: 1–56.

Alveal, A., 1981. *Fluctuaciones del fitoplancton en la Bahía de Valparaíso en el periodo de primavera-vernao.* Thesis, Universidad de Valparaíso. Valparaíso, unpublished.

Alveal, K., 1970. Estudios ficoecológicos en la región costera de Valparaíso. *Rev. Biol. Mar.* (*Valparaíso*), 14: 7–84.

Alveal, K., 1971. El ambiente costero de Montemar y su expresión biológica. *Rev. Biol. Mar.* (*Valparaíso*), 14: 85–119.

Alveal, K., Romo, H. and Valenzuela, J., 1973. Consideraciones ecológicas de las regiones de Valparaíso y de Magallanes. *Rev. Biol. Mar.* (*Valparaíso*), 15: 1–29.

Araya-Vergara, J.F., 1976. Reconocimiento de tipos e individuos geomorfológicos regionales en la costa de Chile. *Inf. Geogr.* (*Chile*), 23: 9–30.

Arenas, J., 1971. Distribución de *Elminius kingii* (Gray) (Cirr.) en el estuario del Río Valdivia. *Beitr. Neotrop. Fauna*, 6: 199–206.

Bahamondes, I. and Castilla, J.C., 1986. Predation of marine invertebrates by the kelp gull *Larus dominicanus* in an undisturbed intertidal rocky shore of Central Chile. *Rev. Chil. Hist. Nat.*, 59: 65–72.

Balech, E., 1954. División zoogeográfica del litoral Sudamericano. *Rev. Biol. Mar.* (*Valparaíso*), 4: 184–195.

Barber, R.T. and Chávez, F.P., 1983. Biological consequences of El Niño. *Science*, 222: 1203–1210.

Bernal, P.A., Robles, F.L. and Rojas, O., 1982. Variabilidad física y biológica en la región meridional del sistema de corrientes Chile–Perú. *Monogr. Biol.*, 2: 75–102.

Bertrán, C., 1980. Análisis taxonómico de *Perinereis gualpensis* Jeldes y *Perinereis vallata* Grube (*Annelida, Polychaeta*) en el estuario del Río Lingue, Chile. *Stud. Neotrop. Fauna Environm.*, 15: 81–89.

Bertrán, C., 1984. Macroinfauna intermareal en un estuario del sur de Chile (estuario del Río Lingue, Valdivia). *Stud. Neotrop. Fauna Environm.*, 19: 33–46.

Black, H.Y. and Fonck, E., 1981. On the vegetation dynamics of *Gracilaria* sp. in Playa Changa, Coquimbo, Chile. *Proc. Int. Seaweed Symp.*, 10: 223–228.

Brandhorst, W., 1963. *Descripción de las condiciones oceanográficas en las aguas costeras entre Valparaíso y Golfo de Arauco, con especial referencia al contenido de oxígeno y su relación con la pesca.* Ministerio de Agricultura, Dirección de Agricultura y Pesca, Chile, 55 pp.

Brattström, H. and Johanssen, A., 1983. Ecological and regional zoogeography of the marine benthic fauna of Chile. *Sarsia*, 68: 289–339.

Briggs, J.C., 1974. *Marine Zoogeography*. McGraw-Hill Series in Population Biology. McGraw-Hill, New York, 475 pp.

Campos, H., 1973. Migration of *Galaxias maculatus* (Jenyns) (Galaxiidae, Pisces) in Valdivia estuary, Chile. *Hydrobiologia*, 43: 301–312.

Campos, H., Bucarey, E. and Arenas, J., 1974. Estudios limnológicos del lago Riñihue y Río Valdivia (Chile). *Bol. Soc. Biol. Concepción*, 48: 47–67.

Castilla, J.C., 1976a. A unique mollusc. *Sea Frontiers*, 22: 302–304.

Castilla, J.C., 1976b. Parques y reservas marítimas chilenas, necesidad de creación, probables localizaciones y criterios básicos. *Medio Ambiente*, 2: 70–80.

Castilla, J.C., 1976c. Ecosistemas marinos de Chile: principios generales y proposición de clasificación. In: F. Orrego (Editor), *Preservación del Medio Ambiente Marino*. Instituto de Estudios Internacionales, Universidad de Chile, Santiago, pp. 22–37.

Castilla, J.C., 1979. Características bióticas del Pacífico Sur-Oriental, con especial referencia al sector chileno. *Rev. Com. Perm. Pacífico Sur*, 10: 167–182.

Castilla, J.C., 1981. Perspectivas de investigación en estructura y dinámica de comunidades intermareales rocosas de Chile Central. II. Depredadores de alto nivel trófico. *Medio Ambiente*, 5: 190–215.

Castilla, J.C., 1985. Food webs and functional aspects of the kelp, *Macrocystis pyrifera*, community in the Beagle Channel, Chile. In: W.R. Siegfried, P.R. Condy and R.M. Laws (Editors), *Antarctic Nutrient Cycles and Food Webs*. Springer, Berlin, pp. 407–414.

Castilla, J.C. and Bahamondes, I., 1979. Observaciones conductuales y ecológicas sobre *Lutra felina* (Molina 1782) (Carnivora: Mustelidae) en las zonas Central y Centro-Norte de Chile. *Arch. Biol. Med. Exp.*, 12: 119–132.

Castilla, J.C. and Durán, R., 1985. Human exclusion from the rocky intertidal zone of Central Chile: the effects on *Concholepas concholepas* (Gastropoda). *Oikos*, 45: 391–399.

Castilla, J.C. and Moreno, C., 1982. Sea urchin and *Macrocystis pyrifera*. Experimental test of their ecological relations in southern Chile. In: J.M. Lawrence (Editor), *International Echinoderm Conference, Tampa Bay*. Balkema, Rotterdam, pp. 257–263.

Castilla, J.C., Sánchez, M. and Mena, O., 1977. Estudios ecológicos en la zona costera afectada por contaminación del "Northern Breeze". I. Introducción general y communidades de playas de arena. *Medio Ambiente*, 2: 53–64.

Castilla, J.C., Guisado, Ch. and Cancino, J., 1979. Aspectos ecológicos y conductuales relacionados con la alimentación de *Concholepas concholepas* (Mollusca: Gastropoda: Muricidae). *Biol. Pesq.*, 12: 91–97.

Dahl, E., 1953. Some aspects of the ecology and zonation of the fauna on sandy beaches. *Oikos*, 4: 1–27.

Dahl, E., 1960. The cold temperate zone in Chilean seas. *Proc. R. Soc. London Ser. B.*, 152: 631–633.

Dall, W.H., 1909. Report on a collection of shells from Peru with a summary of the littoral marine Mollusca of the Peruvian zoological province. *Proc. U.S. Nat. Mus.*, 37: 147–294.

Dawson, E.Y., Acleto, C. and Foldvik, N., 1964. The seaweed of Peru. *Nova Hedwigia Beih.*, 13: 1–111, 80 pls.

Dayton, P., Rosenthal, R.J. and Mahan, L.C., 1973. Kelp communities in the Chilean Archipelago: R/V *Hero Cruise* 72–5. *U.S. Antarct. J.*, 8: 34–35.

Dell, R.K., 1971. The marine mollusca of the Royal Society Expedition to southern Chile, 1958–1959. *Rec. Dom. Mus.*, 7: 155–233.

Díaz, M., 1984. Distribución de fosfatos, nitratos y nitritos en una sección frente a Iquique (20°16′S), diciembre 1982. *Invest. Pesq.*, 31: 103–108.

Di Castri, F. and Hajek, E.R., 1976. *Bioclimatología de Chile*. Vicerrectoría Académica, Universidad Católica de Chile, Santiago de Chile, 129 pp.

Epelde-Aguirre, A. and López, M., 1975. Zonación en el sustrato arenoso de Playa Blanca, Bahía de Coronel y observaciones sobre crustáceos poco frecuentes. *Bol. Soc. Biol. Concepción*, 49: 161–170.

Ekman, S.P., 1953. *Zoogeography of the Sea*. Sidgwick and Jackson, London, 417 pp. [translated from Swedish by E. Palmer].

Fischer, W., 1963. Die Fische des Brackwassersgebietes Lenga bei Concepción (Chile). *Int. Rev. Gesamten Hydrobiol.*, 48: 419–511.

Forbes, E., 1856. Map of the distribution of marine life. In: A.K. Johnston (Editor), *The Physical Atlas of Natural Phenomena*. Johnston, Edinburgh, pl. No. 31. [cited from Briggs (1974)].

Fritsch, F.E., 1945. *The Structure and Reproduction of Algae*, Vol. 2. Cambridge University Press, Cambridge, 939 pp.

Gallardo, V.A., 1963. Notas sobre la densidad de la fauna bentónica en el sublitoral del norte de Chile. *Gayana (Zool.)*, 10: 3–15.

Garth, J.S., 1957. The Crustacea Decapoda Brachyura of Chile. *Reports of the Lund University Chile Expedition 1948–1949. Lunds Univ. Arsskr. Avd. 2*, 53(7): 1–130.

Ghelardi, R.J., 1971. Species structure of the holdfast community. In: W.J. North (Editor), *The biology of giant kelp beds (Macrocystis) in California. Nova Hedwigia*, 32: 381–420.

Guiler, E.R., 1959a. Intertidal belt forming species on the rocky coast of northern Chile. *Pap. Proc. R. Soc. Tasmania*, 93: 33–58.

Guiler, E.R., 1959b. The intertidal ecology of the Montemar area, Chile. *Pap. Proc. R. Soc. Tasmania*, 93: 165–183.

Gunther, E.R., 1939. A report on oceanographical investigations in the Perú coastal current. *Discovery Rep.*, 13: 107–276.

Guzmán, L. and Ríos, C., 1981. Estructura del conjunto de macroorganismos de una playa de bloques y cantos de Isla Wollaston, archipiélago del Cabo de Hornos. *An. Inst. Patagonia* Punta Arenas (Chile), 12: 257–271.

Haig, J., 1965. The Crustacea Anomura of Chile. *Reports of the Lund University Chile Expedition 1948–1949. Lunds Univ. Arsskr. Avd. 2*, 51: 1–68, 13 figs.

Jara, H.F. and Moreno, C.A., 1984. Herbivory and structure in a midlittoral rocky community: a case in southern Chile. *Ecology*, 65: 28–38.

Jaramillo, E., 1978. Zonación y estructura de la comunidad macrofaunística en playas de arena del sur de Chile (Mehuín, Valdivia). *Stud. Neotrop. Fauna Environm.*, 13: 71–92.

Jaramillo, E., 1982. Taxonomy, natural history and zoogeography of sand beach isopods from the coast of southern Chile. *Stud. Neotrop. Fauna Environm.*, 17: 175–194.

Jaramillo, E., Mulsow, S., Pino, M. and Figueroa, H., 1984. Subtidal benthic macroinfauna in an estuary of south Chile: distribution pattern in relation to sediment types. *Mar. Ecol.*, 5: 119–133.

Johnson, D.R., Fonseca, T. and Sievers, H., 1980. Upwelling in the Humboldt Coastal Current near Valparaíso, Chile. *J. Mar. Res.*, 38: 1–16.

Knox, G.A., 1960. Littoral ecology and biogeography of the southern oceans. *Proc. R. Soc. London Ser. B.*, 152: 577–624.

Madsen, F.J., 1956. Asteroidea. *Reports of the Lund University Chile Expedition 1948–1949. Lunds Univ. Arsskr. Avd. 2*, 52: 1–53.

Mann, K.H., 1973. Seaweeds: their productivity and strategy for growth. *Science*, 182: 975–981.

Marincovich Jr., L., 1973. Intertidal molluscs of Iquique, Chile. *Sci. Bull. Nat. Hist. Mus. Los Angeles*, 16: 1–49.

Michanek, G., 1979. Phytogeographic provinces and seaweeds distribution. *Bot. Mar.*, 22: 375–391.

Montalva, S. and Santelices, B., 1981. Interspecific interference among species of *Gelidium* from Central Chile. *J. Exp. Mar. Biol. Ecol.*, 53: 77–88.

Moreno, C.A. and Jara, H.F., 1984. Ecological studies on fish fauna associated with *Macrocystis pyrifera* belts in the south of Fueguian Islands, Chile. *Mar. Ecol. Prog. Ser.*, 15: 99–107.

Moreno, C.A. and Jaramillo, E., 1983. The role of grazers in the zonation of intertidal macroalgae of the Chilean coast near Valdivia. *Oikos*, 41: 73–76.

Moreno, C.A. and Sutherland, J.P., 1982. Physical and biological processes in a *Macrocystis pyrifera* community near Valdivia, Chile. *Oecologia* (Berlin), 55: 1–6.

Moreno, C.A., Duarte, W.E. and Zamorano, J.H., 1979. Variación latitudinal del número de especies de peces en el sublitoral rocoso: una explicación ecológica. *Arch. Biol. Med. Exp.*, 12: 169–178.

Moreno, C.A., Sutherland, J.P. and Jara, H.F., 1984. Man as a predator in the intertidal zone of southern Chile. *Oikos*, 42: 155–160.

Moreno, C.A., Lunecke, K.M. and López, M.I., 1986. The response of an intertidal *Concholepas concholepas* (Gastropoda) population to protection from Man in southern Chile and the effects on benthic sessile assemblages. *Oikos*, 46: 359–364.

Nuñez, J., Aracena, O. and López, M., 1974. *Emerita analoga* en Llico, Provincia de Curicó (Crustacea, Decapoda, Hippidae). *Bol. Soc. Biol. Concepción*, 48: 11–22.

Ojeda, F.P. and Santelices, B., 1984a. Invertebrate communities in holdfasts of the kelp *Macrocystis pyrifera* from southern Chile. *Mar. Ecol. Prog. Ser.*, 16: 65–73.

Ojeda, F.P. and Santelices, B., 1984b. Ecological dominance of *Lessonia nigrescens* (Phaeophyta) in Central Chile. *Mar. Ecol. Progr. Ser.*, 19: 83–91.

Oliger, P. and Santelices, B., 1981. Physiological ecological studies on Chilean Gelidiales. *J. Exp. Mar. Biol. Ecol.*, 53: 65–75.

Olsson, A.A., 1961. *Mollusks of the Tropical Eastern Pacific. Panamic-Pacific Pelecypoda.* Paleontol. Res. Inst., Ithaca, New York, 574 pp.

Osorio, C., Bahamondes, N. and López, M., 1967. El limanche *Emerita analoga* (Stimson) en Chile. *Bol. Mus. Nacional Hist. Nat. Chile*, 29: 61–116.

Paine, R.T. and Palmer, A.R., 1978. *Sicyases sanguineus*: a unique trophic generalist from the Chilean intertidal zone. *Copeia*, 1978: 75–81.

Paine, R.T. and Suchanek, T.H., 1983. Convergence of ecological processes between independently evolved competitive dominants: a tunicate-mussel comparison. *Evolution*, 37: 821–831.

Paine, R.T., Castilla, J.C. and Cancino, J., 1985. Perturbation and recovery patterns of starfish-dominated intertidal assemblages in Chile, New Zealand and Washington State. *Am. Nat.*, 125: 679–691.

Pequeño, G., 1979. Antecedentes alimentarios de *Eleginops maclovinus* (Valenciennes, 1830) (Teleostomi: Nototheniidae) en Mehuín, Chile. *Acta Zool. Lilloana*, 35: 207–230.

Pickard, G.L., 1971. Some physical oceanographic features of inlets of Chile. *J. Fish. Res. Board Can.*, 28: 1077–1106.

Retamal, M., 1979. *Hemigrapsus crenulatus* (H. Milne Edwards, 1937) en el estero Lenga (Crustacea, Decapoda, Grapsidae). *Bol. Soc. Biol. Concepción*, 41: 281–302.

Ríos, C. and Guzmán, L., 1982. Reevaluación de la estructura de la comunidad en una playa de bloques y cantos de alta latitud. Archipiélago de Cabo de Hornos. *An. Inst. Patagonia*, 13: 211–224.

Robles, F., 1966. *Descripción gráfica de las condiciones oceanográficas frente a la provincia de Tarapacá en base a los datos de la Operación Oceanográfica Mar Chile II.* Instituto Hidrográfico de la Armada, Valparaíso, Chile, 68 pp.

Robles, F., Alarcón, E. and Ulloa, A., 1976. Las masas de agua en la región Norte de Chile y sus variaciones en un período frío (1967) y en períodos cálidos (1969, 1971–1973). In: *Reunión de Trabajo sobre el Fenómeno de "El Niño".* Comisión Oceanográfica Intergubernamental, COI. Guayaquil, Ecuador, 9–12 Diciembre, 1974, 68 pp.

Robles, F.L., Alarcón, E. and Ulloa, A., 1980. Water masses in the northern Chilean zone and their variations in the cold period (1967) and warm periods (1969, 1971–1973). In: *Proc. Workshop on the phenomenon known as "El Niño".* UNESCO, pp 83–174.

Romo, H. and Alveal, K., 1977. Las comunidades del litoral rocoso de Punta Ventanilla, Bahía de Quintero, Chile. *Gayana* (Miscelánea), 6: 1–41.

Romo, H., Alveal, K. and Dellarossa, V., 1979. Biología de *Gracilaria verrucosa* (Hudson) Papenfuss en Chile Central. In: B. Santelices (Editor), *Actas del Primer Simposio sobre Algas Marinas Chilenas.* Ministerio de Economía, Fomento y Reconstrucción, Santiago, Chile, pp. 155–163.

Ruiz, E. and Giampoli, L., 1981. Estudios distribucionales de la flora y fauna costera de Caleta Cocholgue, Bahía de Concepción, Chile. *Bol. Soc. Biol. Concepción,* 52: 145–166.

Sánchez, M., Castilla, J.C. and Mena, O., 1982. Variaciones Verano-Invierno de la macrofauna de arena en Playa Morrillos (Norte Chico, Chile). *Stud. Neotrop. Fauna Environm.,* 17: 31–49.

Santelices, B., 1980a. Phytogeographic characterization of the temperate coast of Pacific South America. *Phycologia,* 19: 1–12.

Santelices, B., 1980b. Muestreo cuantitativo de comunidades intermareales de Chile Central. *Arch. Biol. Med. Exp.,* 13: 413–424.

Santelices, B., 1981. Perspectivas de investigación en estructura y dinámica de comunidades intermareales rocosas de Chile Central. I. Cinturones de macroalgas. *Medio Ambiente,* 5: 175–189.

Santelices, B. and Castilla, J.C., 1977. Estudios ecológicos en la zona costera afectada por la contaminación del "Northern Breeze". III. Informe de daños ecológicos y destrucción de recursos. *Medio Ambiente,* 2: 84–91.

Santelices, B. and Fonck, E., 1979. Ecología y cultivo de *Gracilaria lemanaeformis.* In: B. Santelices (Editor), *Actas del Primer Simposio sobre Algas Marinas Chilenas.* Ministerio de Economía, Fomento y Reconstrucción, Santiago, Chile, pp. 165–200.

Santelices B. and Ojeda, F.P., 1984a. Effects of canopy removal on the understory algal community structure of coastal forests of *Macrocystis pyrifera* from southern South America. *Mar. Ecol. Prog. Ser.,* 14: 165–173.

Santelices, B. and Ojeda, F.P., 1984b. Population dynamics of coastal forests of *Macrocystis pyrifera* in Puerto Toro, Isla Navarino, southern Chile. *Mar. Biol. Prog. Ser.,* 14: 175–183.

Santelices, B. and Ojeda, F.P., 1984c. Recruitment, growth and survival of *Lessonia nigrescens* (Phaeophyta) at various tidal levels in exposed habitats of Central Chile. *Mar. Biol. Prog. Ser.,* 19: 73–82.

Santelices, B., Cancino, J., Montalva, S., Pinto, R. and González, E., 1977. Estudios ecológicos en la zona costera afectada por contaminación del "Northern Breeze". II. Comunidades de playas de rocas. *Medio Ambiente,* 2: 65–83.

Santelices, B., Castilla, J.C., Cancino, J. and Schmiede, P., 1980. Comparative ecology of *Lessonia nigrescens* and *Durvillaea antarctica* (Phaeophyta) in Central Chile. *Mar. Biol.,* 59: 119–132.

Santelices, B., Montalva, S. and Oliger, P., 1981. Competitive algal community organization in exposed intertidal habitats from Central Chile. *Mar. Ecol. Prog. Ser.,* 6: 267–276.

Santelices, B., Vásquez, J., Ohme, U. and Fonck, E., 1984. Managing wild crops of *Gracilaria* in Central Chile. *Hydrobiologia,* 116/117: 77–89.

Sebens, K.P. and Paine, R.T., 1979. Biogeography of anthozoans along the west coast of South America: habitat, disturbance and prey availability. In: *Proc. Int. Symp. on Marine Biogeography and Evolution in the Southern Hemisphere, Auckland, New Zealand. Vol. 1* N.Z. Dept. of Scientific and Industrial Research Information Series 137, pp. 219–237.

Silva, N. and Sievers, H., 1981. Masas de agua y circulación en la región de la rama costera de la Corriente de Humboldt, latitudes 18°S–33°S (Operación Oceanográfica MAR-CHILE 10 — ERFEN 1). *Ciencia Tecnol. Mar.,* 5: 5–50.

Stephenson, T.A. and Stephenson, A., 1972. *Life Between Tidemarks on Rocky Shores.* Freeman, San Francisco, 425 pp.

Stotz, W., 1981. Aspectos ecológicos de *Mytilus edulis chilensis* (Huje, 1854) en el estuario del Río Lingue (Valdivia, Chile). *Rev. Biol. Mar. Inst. Oceanogr. Valparaíso,* 17: 335–377.

Trevallion, A., Ansell, A.D., Sivadas, P. and Narayanan, B., 1970. A preliminary account of two sandy beaches in South West India. *Mar. Biol.,* 6: 268–279.

Varela, C., 1983. Anfípodos de las playas de arena del sur de Chile (Bahía de Maiquillahue, Valdivia). *Stud. Neotrop. Fauna Environm.,* 18: 25–52.

Villouta, E. and Santelices, B., 1984. Estructura de la comunidad submareal de *Lessonia* (Phaeophyta, Laminariales) en Chile Norte y Central. *Rev. Chil. Hist. Nat.,* 57: 111–122.

Viviani, C.A., 1970. Die Bryozoen (Ento- und Ectoprocta) des chilenischen Litorals. *Chemoprint-Giessen,* 1,2: 1–307.

Viviani, C.A., 1979. Ecogeografía del litoral chileno. *Stud. Neotrop. Fauna Environm.,* 14: 65–123.

Westermeier, R., 1980. Explotación de *Gracilaria verrucosa* (Hudson) Papenfuss en los estuarios de los ríos Maullín y Quenuir, X Región y antecedentes para su manejo racional. *Medio Ambiente,* 4: 90–94.

Westermeier, R., Steubing, L., Rivera, P. and Wenzel, H., 1984. *Gracilaria verrucosa* (Hudson) Papenfuss en la X Región (Maullín y Quenuir - Provincia Llanquihue) Chile. *Mem. Asoc. Latinoamericana Acuicultura,* 5: 419–430.

Woodward, S.P., 1856. *A Manual of the Mollusca, or Rudimentary Treatise of Recent and Fossil Shells. Part 3.* John Weale, London, 486 pp.

Wooster, W. and Sievers, H.A., 1971. Seasonal variations of temperature drift and heat exchange in surface waters off the west coast of South America. *Limnol. Oceanogr.,* 16: 595–605.

Wyrtki, K., 1963. The horizontal and vertical field of motion in the Perú Current. *Bull. Univ. Calif. Scrips Inst. Oceanogr.*, 8: 313–346.

Wyrtki, K., 1964. The thermal structure of the Eastern Pacific Ocean. *Dtsch. Hydrogr. Erganzungshelft Reihe A (8°)*, 6: 1–88.

Wyrtki, K., 1966. Oceanography of the Eastern Equatorial Pacific Ocean. *Oceanogr. Mar. Biol. Annu. Rev.*, 4: 33–68.

Wyrtki, K., 1967. Circulation and water masses in the Eastern Equatorial Pacific Ocean. *Int. J. Oceanol. Limnol.*, 1: 117–147.

Chapter 15

SEAGRASS ECOSYSTEMS IN THE TROPICAL WEST PACIFIC

JOOP J.W.M. BROUNS and FRANCISCA M.L. HEIJS

INTRODUCTION

This chapter deals with a part of the largest archipelagic region in the world. The approximately 21 000 islands of the archipelago are the territory of three states: Indonesia, Papua New Guinea and the Philippines. In this area many groups of animals and plants reach a peak in diversity (McCoy and Heck, 1976; Sale, 1980). The variability of the ecosystems in general is reflected in the diversity of the marine coastal habitats. Extensive estuarine areas exist in Indonesia and Papua New Guinea, particularly with the high rainfall zones and gently sloping coastal plains (Polunin, 1983).

Rocky shores are rare in the area described. The largest part of the coastline of Indonesia and New Guinea is bordered by mangrove swamps, particularly the relatively unpopulated shores of the largest islands, namely Sumatra, Borneo and New Guinea. Sandy beaches are widely distributed throughout the area. The generalized zonation pattern, first proposed by Stephenson and Stephenson (1949), applies well to the exposed tropical shores:

(1) the maritime zone, characterized by the "pes-caprae" community, is named after the dominant species *Ipomoea pes-caprae*;

(2) the littoral fringe shows a predominance of littorinids, neritids and patellid limpets;

(3) the eulittoral zone is divided into an upper eulittoral zone, with a sparse algal film and mainly neritids and limpets, a mid-eulittoral zone with attached bivalves and barnacles, and a lower-eulittoral zone, which can be recognized by pink and brown-red algal turfs, scoured by sea urchins;

(4) the sublittoral fringe, only briefly emersed at extreme low tides, shows a profusion of algae near the low water mark and numerous corals (such as *Acropora, Montipora, Pocillopora, Porites*).

Another important aspect to consider in the ecology of living reefs is the degree of wave action, which is interrelated with the topography. Wide fringing reefs with either beaches of sand or debris or sandy to muddy areas, appear at sites with slight to moderate wave action. Usually the eulittoral zone is poorly represented in these localities. Under heavy wave action the reef is much narrower, with a marked eulittoral zone. Under varying degrees of wave action the following classes can be distinguished:

(a) Wave-swept slopes or beaches under constant surf attack. Corals are seldom abundant, and may be virtually absent. The main debris-cementing organisms are the calcareous Rhodophyceae (such as *Neogoniolithon* and *Porolithon* spp.).

(b) Exposed reefs with a seaward rampart breaking the full attack of the waves. The seaward side shows a high abundance of calcareous red algae, often with *Sargassum* and *Turbinaria* spp. on top, and a rich algal turf. The landward side shows a wide moat with living and/or dead corals, micro-atolls, and in the landward reaches seagrasses. The faunal community of the moat consists of molluscs, echinoderms (such as *Echinometra, Linckia, Holothuria, Stichopus, Tripneustes* and several ophiuroids), a wide range of gastropods (for example, *Cypraea, Strombus, Terebra*) and numerous Alcyonaria. Common moat algae are *Actinotrichia, Amphiroa, Boodlea, Caulerpa, Galaxaura, Gracilaria, Halimeda, Hypnea, Laurencia, Peyssonnelia, Struvea, Tolypiocladia* and *Valonia.*

(c) Sheltered shores in calm localities consist of a wide moat area (up to 1 km) with, on the seaward side, a slightly raised rim dominated by corals (usually *Montipora* or massive *Porites* micro-atolls). The landward reaches consist of a sandy beach or sand–mud flats with seagrass meadows and/or mangroves (for example, *Avicennia, Bruguiera* and *Rhizophora*). Algal species of the mangrove zone belong to the *Bostrichia–Caloglossa* community, and the algal component characteristic for seagrass meadows consists of *Avrainvillea, Boodlea, Caulerpa, Halimeda, Hypnea, Padina* and *Valonia*. Among the macrofauna are holothurians (such as *Metriatyla* and *Synapta*), echinoids (*Tripneustes*), asteroids (*Archaster*), sponges, gastropods (*Conus, Lambis, Mitra, Strombus, Terebellum* and *Vexillum*), polychaetes (for example, Ariciidae, Clyceridae, Nereidae), and crustaceans (such as *Calappa, Callianassa* and stomatopods). This chapter focusses on those communities that are dominated by seagrasses.

The data and particularly the descriptions of the eulittoral and sublittoral communities dominated by seagrasses are from personal observations. The observations and surveys are all from an area between 2° and 11°S latitude and between 118° and 152°E longitude, being approximately the Austro-Malayan subregion (Wallace, 1876). The localities are presented in Fig. 15.1. The climate of the area is monsoonal. Notably the Lesser Sunda Islands and the south coast of Papua New Guinea have a distinct dry and wet season. Figure 15.2 shows the climatogram of the Port Moresby region, from 1960–1980 (Brouns and Heijs, 1985). The total yearly insolation of the sites lies between 5.4 and 5.9 GJ m^{-2} (Landsberg et al., 1966). The temperature of the seawater varies from 24.4°C in August to 30.8°C in March for the Port Moresby region, Papua New Guinea.

The seagrass flora of the western tropical Pacific is rather rich in species (Den Hartog, 1970), as illustrated by several reports on the distribution of seagrasses in the Indo-Pacific and West Pacific region: in Micronesia 10 species (Tsuda et al., 1977); in Papua New Guinea 10–13 species (Johnstone, 1978a, b, 1982; Brouns and Heijs, 1985; Heijs and Brouns, 1986); in the Philippines 11 species (Meñez et al., 1983); in Palau 9 species (Kock and Tsuda, 1978; Ogden and Ogden, 1982); and in eastern Indonesia 12 species (pers. observ. during the Snellius-II expedition, 1984).

The seagrass beds in the Indo-Pacific region are generally of a mixed nature (Kock and Tsuda; 1978; Ogden and Ogden, 1982; Meñez et al., 1983). In Papua New Guinea as well as in eastern Indonesia seagrass beds consisting of six to seven species are found. In these multi-species meadows *Thalassia hemprichii* is the dominant seagrass, followed by *Enhalus acoroides* and *Cymodocea rotundata. Cymodocea serrulata, Halophila ovalis, Halodule uninervis* and *Syringodium isoetifolium* are generally less important, but may locally be as abundant as *Enhalus acoroides* or *Cymodocea rotundata*. This hierarchy in dominance has also

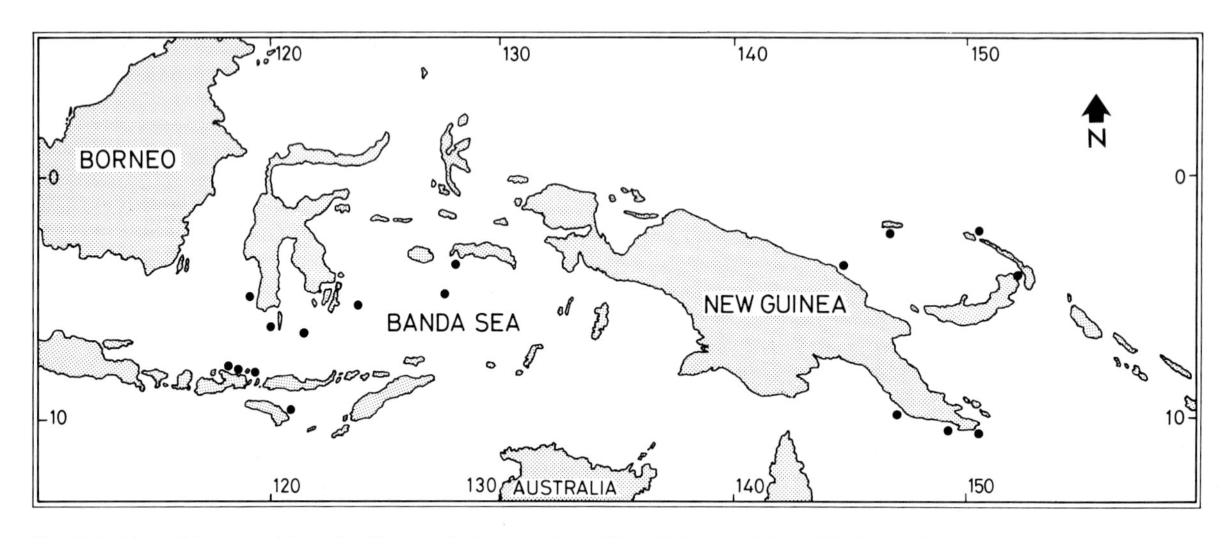

Fig. 15.1. Map of the central Indo-Pacific area: Indonesia, Papua New Guinea and the Philippines. The sites that have been surveyed by the authors are indicated by dots.

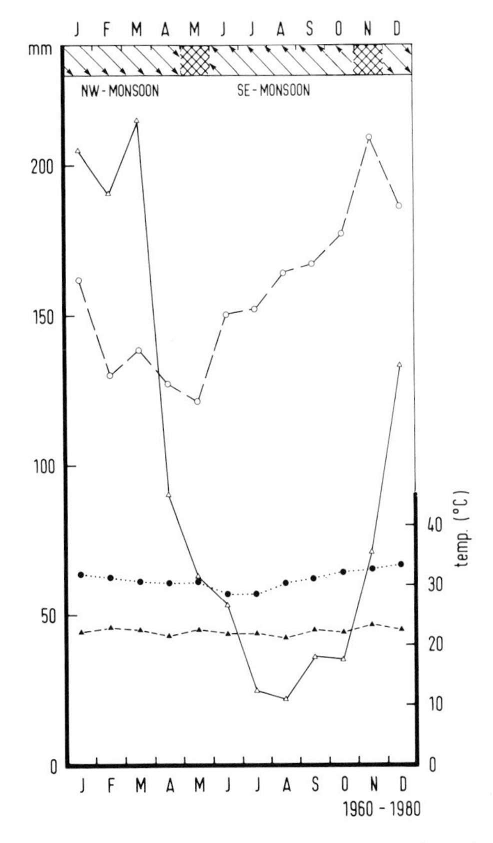

Fig. 15.2. Climatogram from the Port Moresby region in Papua New Guinea. Open triangles: precipitation (mm); open circles: evaporation (mm); solid circles: maximum air temperature (°C); solid triangles: minimum air temperature (°C).

been observed in Palau (Ogden and Ogden, 1982) and in the Philippines (Meñez et al., 1983). Apart from the multi-species seagrass beds, any combination of two, three or four seagrass species can be found, as well as monospecific patches.

SEAGRASS COMMUNITIES

Monospecific seagrass beds

Monospecific seagrass beds of all above-mentioned species can be found. However, it is important to realize that generally the occurrence of a monospecific seagrass bed is temporary and should be considered as an intermediate phase towards a more stable situation (that is, the mixed seagrass meadow). Furthermore, monospecific seagrass beds are observed only occasionally.

Thalassodendron ciliatum

Thalassodendron ciliatum is the only species in the area that forms mainly pure communities with many accompanying algae. In Papua New Guinea, *Thalassodendron ciliatum* shows a rather disjunct distribution pattern (Heijs and Brouns, 1986), whereas in eastern Indonesia it is quite common (Brouns, 1985c). It has been found from the upper sublittoral zone down to 10 m depth, where it may form large meadows on substrate generally consisting of coral rubble mixed with coral sand.

Because of their growth form, these communities are vertically more complex than other tropical monospecific seagrass communities. Several layers can be recognized: a thick rhizome-mat, partly above the substrate; the relatively long vertical woody stems; and the canopy formed by the seagrass leaves. The space between the leaf canopy and the bottom usually harbours an extremely rich and abundant algal and faunal community.

The sciaphilic algal growth consists of Rhodophyceae (the genera *Amphiroa, Gracilaria, Haloplegma, Jania, Laurencia, Liagora* and *Peyssonnelia*), Chlorophyceae (species of the genera *Bryopsis, Caulerpa, Chaetomorpha* and *Halimeda*) and Phaeophyceae (mainly species belonging to the genera *Dictyopteris, Dictyota* and *Padina*). The photophilic community on the leaves is almost exclusively composed of Rhodophyta (such as Ceramiales), of which the encrusting coralline algae (for instance, *Fosliella farinosa, F. lejolisii* and *Melobesia membranacea*) dominate the community quantitatively.

Thalassia hemprichii

This seagrass species is the dominant representative of the marine angiosperms of the Indo-Pacific. It occurs on all kinds of substrata, varying from fine silty mud to coarse coral rubble, and is found from the lower eulittoral down to *c.* 5 m depth. In and around pools left behind by the receding tide, growth can be very luxuriant, whereas on the tidal flats densities may vary considerably. *Thalassia hemprichii* can form a dense monospecific community on coarse sand. It has also been observed growing in mats formed by the green alga *Halimeda opuntia* (*Halimeda–Thalassia* community, abundance ratio 2:1) and in association with *H. opuntia* and the coral *Montipora*. The rhizome layer of *Thalassia hemprichii* can be very compact and up to 20 cm thick.

The algal component of the *Thalassia hemprichii* community consists generally of *Avrainvillea erecta*, and the genera *Ceratodictyon, Champia* and *Hypnea*. In the *Halimeda–Thalassia* community and the *Halimeda–Thalassia–Montipora* association the algal community can be very rich, due to the availability of additional hard substrata (Heijs, 1985b). The photophilic algal community consists mainly of Ceramiales and encrusting epiphytes (Heijs, 1985a).

Enhalus acoroides

Enhalus acoroides is the second most abundant seagrass in Papua New Guinea and eastern Indonesia. It is the largest seagrass species, and is easily recognized by the long strap-like leaves. It occurs on muddy and sandy substrates, and is found from the intertidal zone down to 6 m depth. In the lower eulittoral *Enhalus acoroides* occupies the small pools that retain water during low tide. Here the leaves remain short (20–50 cm), whereas in the sublittoral the leaves may reach a length of 180 cm.

Enhalus acoroides often grows close to mangroves, where it can form extensive meadows in shallow areas. In deeper water (3–5 m depth) small circular patches may be formed, each consisting of one plant with many shoots, or large beds, bordering the seaward edge of a mixed seagrass meadow.

In the intertidal zone the algal community mainly consists of *Caulerpa* spp. and *Halimeda* spp. In deeper parts no macroalgae have been observed associated with this seagrass bed. The epiphytic algae are poorly represented in the monospecific *Enhalus* beds (Brouns and Heijs, 1986). In shallow muddy areas, leaves of *Enhalus acoroides* accumulate detritus upon which a variety of invertebrates may be grazing.

Cymodocea

Two species belonging to this genus have a wide distribution in the Indo-Pacific:

Cymodocea rotundata. Monospecific fringes of this species usually are found in the lower eulittoral zone on sandy substrate. Here *Cymodocea rotundata* is often exposed during low water of spring-ebb tides. In pools growth may be luxuriant, whereas on the tidal flats the seagrass leaves may have a "burned" appearance after several days of exposure.

Cymodocea serrulata. This species occurs mainly in the subtidal down to 5 m depth, where it may form extensive monospecific beds. Occasionally this species is found intertidally. Substrata varying from silty mud to coarse coral rubble are suitable for this seagrass species. The highest densities are usually found on mud-covered coral rubble.

The macroalgal growth associated with the two *Cymodocea* species is almost completely composed of green algae (the genera *Caulerpa, Halimeda* and *Neomeris*). Occasionally a few Rhodophyta (for example, *Acanthophora spicifera* and *Spyridia filamentosa*) and Phaeophyta (such as *Dictyota* spp.) may be found (Heijs, 1985b).

The epiphytic algal community on the leaves of *Cymodocea rotundata* and *Cymodocea serrulata* is very similar in species composition; however, it is more abundant on the leaves of *Cymodocea serrulata* (Heijs, 1985c).

Syringodium isoetifolium

Syringodium isoetifolium grows from ELWS down to 5 m depth. Seagrass communities consisting of *Syringodium isoetifolium* only are found just below the lowest low water mark. Apparently this species can tolerate only very short periods of exposure. The cylindrical leaves are usually covered with epiphytic algae (Heijs, 1985c). Hydroids are frequently observed on the leaf sheaths.

Halodule

Halodule uninervis and *Halodule pinifolia* are both found in the Indo-Pacific. *Halodule uninervis* is more common than *Halodule pinifolia*. *Halodule uninervis* shows distinct morphological differences in different habitats: a narrow-leaved form in the higher parts of the intertidal; and a wide-leaved form, from the lower eulittoral down to 7 m depth. The narrow-leaved form develops mainly pure fringes on sandy substrata, whereas the wide-leaved form occurs on a variety of substrates. *Halodule uninervis* is a typical pioneer species. *Halodule pinifolia* grows in the lower eulittoral and upper sublittoral where it forms a pure vegetation on sandy and muddy substrata, but these monospecific beds have been observed infrequently.

Halophila

Four species belonging to this genus are found in the area (see also Den Hartog, 1970): *Halophila ovalis, H. ovata, H. decipiens* and *H. spinulosa*. The latter two species are the least common, and occur only as pure stands predominantly in the sublittoral zone.

Halophila decipiens may form the lower edge of a seagrass bed at *c.* 10 m depth, but it has been observed at a depth of 50 m on the Great Barrier Reef. *Halophila spinulosa* forms pure communities on white coral sand in very clear water down to 45 m. This community occurs only sporadically. *Halophila ovata* is less abundant than *Halophila ovalis*. Due to its small size, *Halophila ovata* is often completely buried in sand and silt. This species grows mainly in the middle eulittoral zone. *Halophila ovalis* occurs on a wide range of substrata from the lower eulittoral down to 22 m depth. The species has a wide ecological range, and is often seen as a pioneer colonizing newly available sandy substrates. The leaf morphology may vary between locations, probably as a consequence of environmental conditions.

Epiphytic algae are less common on the *Halophila* species than on the previously-mentioned seagrasses, particularly in the monospecific seagrass beds, due to the short lifetime of the leaves.

Zostera capricorni

This seagrass species just reaches Papua New Guinea as a northern extension of its Australian range. So far *Zostera capricorni* has only been collected from Daru in Papua New Guinea, and it has not yet been recorded from eastern Indonesia. At Daru it occurs on muddy bottoms around the low water mark.

Associations of two/three seagrass species

Any combination of two or three seagrass species can be found. Mixed stands are indeed more frequently encountered than strictly monospecific seagrass beds. Several combinations can be considered abundant:

(a) An association of *Enhalus acoroides* and *Thalassia hemprichii* (and occasionally *Halophila ovalis*). This community occurs usually in the intertidal zone and just below ELWS level. The abundance of each species varies between the locations. Close to mangroves, *Enhalus acoroides* is dominant (up to 80%), whereas in the subtidal areas *Thalassia hemprichii* usually reaches a higher density (70–80%). In shallow lagoons this community may form extensive meadows, with large plants. Intertidally, the plants are generally smaller and adapted to the tidal regime. The *Enhalus acoroides–Thalassia hemprichii* (*–Halophila ovalis*) associations occur on muddy substrates as well as on calcareous mud and sand.

The macroalgae associated with this community include particularly *Caulerpa* spp. and *Halimeda* spp., with occasionally loose-lying or washed-in algae (genera *Dictyota, Hydroclathrus, Hypnea, Tolypiocladia*) (Heijs and Brouns, 1986; Heijs, 1987a). The epiphytic community on the leaves of the large *Enhalus acoroides* plants and *Thalassia hemprichii* are similar to those found in the monospecific seagrass beds. However, at times an abundance of epiphytes may completely smother the leaves in part of the seagrass bed.

(b) An association of *Cymodocea rotundata–Halodule uninervis* or *Cymodocea rotundata–Halophila ovalis* or *Halodule uninervis–Halophila ovalis*. These combinations of seagrass species are mainly observed in the mid- to lower eulittoral zone. Which combination is found, as well as the respective abundance of the species, depends on the local habitat conditions, hence may vary readily. All three seagrass species are well adapted to exposure, and are considered pioneer species. They all are quick colonizers with a rapid growth of the rhizome and a relatively short life-span of the shoots.

The association of *Halodule uninervis* and *Halophila ovalis* is not only observed in the intertidal area but also occurs subtidally. In the upper sublittoral zone this community grows frequently on sand mounds scattered throughout the seagrass bed. These sand mounds are deposited by burrowing thalassinid shrimps. The top layer is therefore regularly disturbed and will be recolonized by the pioneer species.

(c) An association of *Thalassia hemprichii* and *Cymodocea rotundata*. This community often forms a narrow zone between the previously described zone (type b) and the mixed seagrass meadow (see below). The *Thalassia hemprichii–Cymodocea rotundata* association occurs between +0.5 and +0.2 m ELWS on a variety of substrata. The density of the respective seagrass species varies between locations and is determined by environmental factors.

Mixed seagrass communities

The mixed seagrass meadows generally consist of any association of at least 4 of the following 7 seagrass species: *Cymodocea rotundata, C. serrulata, Enhalus acoroides, Halodule uninervis, Halophila ovalis, Syringodium isoetifolium* and *Thalassia hemprichii.* The framework is nearly always composed of a *Thalassia hemprichii–Enhalus acoroides* association (dominant seagrass species), with a lesser abundance of the other seagrass species, rather homogeneously mixed. The abundance of the accompanying seagrass species is variable within the mixed meadow and is caused by biotic and abiotic factors (such as water depth, substrate characteristics). Figure 15.3 represents a transect of a mixed seagrass bed on a sandy slope near Port Moresby, Papua New Guinea. The species composition and abundance change with increasing depth. Several distinct associations can be distinguished:

Thalassia hemprichii–Enhalus acoroides–Cymodocea rotundata –Syringodium isoetifolium community (Fig. 15.4).

This community is found from +0.3 m ELWS down to −2.0 m ELWS on gently sloping sandy beaches. *Thalassia hemprichii* is the dominant representative (cover, 40%) followed by *Enhalus acoroides* (cover, 25%), *Cymodocea rotundata* and *Syringodium isoetifolium* (both covering 10%). Occasionally *Halodule uninervis* and/or *Halophila ovalis* are also present, although their abundance is negligible. *Syringodium isoetifolium* is most abundant around chart datum (mean sea-level), and *Cymodocea rotundata,* rather common in the lower eulittoral, may be replaced by *C. serrulata* in the sublittoral zone. A co-occurrence of the *Cymodocea* species, however, has also been observed. *Enhalus acoroides* is interspersed among the other seagrass species in this type of seagrass meadow.

Thalassia hemprichii–Cymodocea rotundata–C. serrulata–Syringodium isoetifolium community.

This community generally inhabits the sand accumulations of the "inner" reef flat. The mixed seagrass meadow, which is well protected from incoming ocean swell, is composed of *Thalassia hemprichii* (dominant seagrass, cover 50%), *Cymodocea rotundata, Syringodium isoetifolium* (both covering 20%) and *Cymodocea serrulata* (cover up to 10%). The shallower parts of this community show a *Thalassia hemprichii–Cymodocea rotun-*

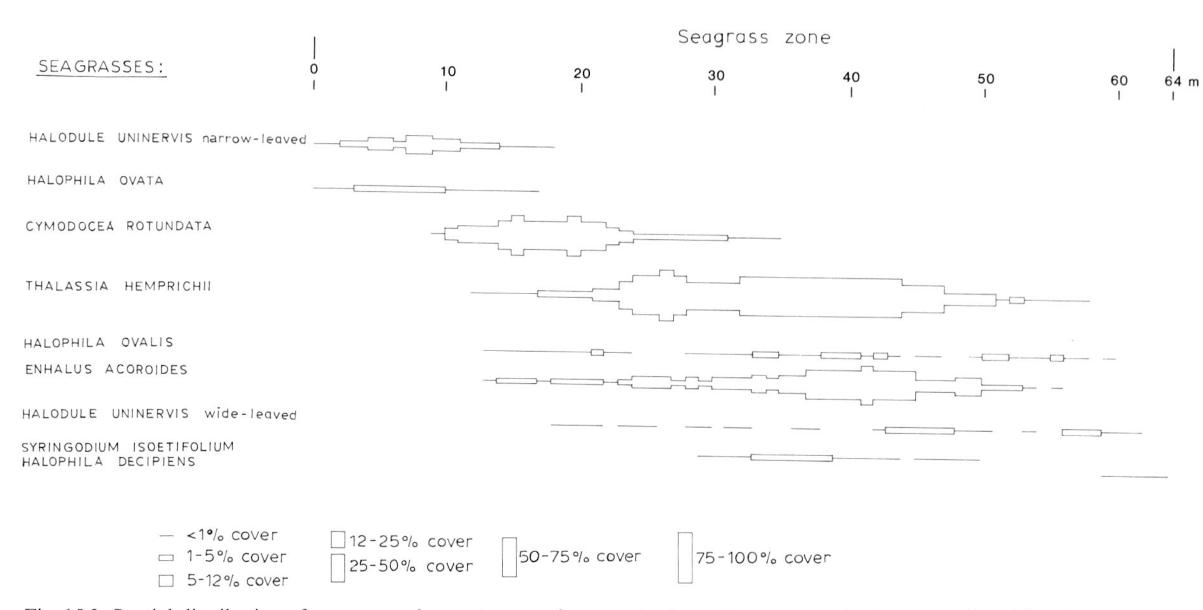

Fig. 15.3. Spatial distribution of seagrasses along a transect from a mixed meadow on a sandy slope near Port Moresby.

Fig. 15.4. Mixed seagrass bed on a sandy substrate. The tall seagrass is *Enhalus acoroides*. The mixed undergrowth is composed of *Thalassia hemprichii* and *Cymodocea serrulata*.

data–Halodule uninervis–Syringodium isoetifolium association (all species equally abundant). Towards the reef crest the mixed meadow is replaced by a zone of coral rubble and living coral colonies. In the absence of a reef crest, the character of the mixed seagrass meadow changes with increasing depth. At the lower edge *Thalassia hemprichii* and *Syringodium isoetifolium* disappear and *Halodule uninervis* and *Halophila ovalis* are increasing in abundance. The distribution of *Enhalus acoroides* in this mixed meadow is typical. It generally occurs in small isolated patches, which generally consist of a single plant with several to many shoots. In the undergrowth *Halodule uninervis, Halophila ovalis* and *Cymodocea rotundata* are often seen, particularly on the edges of these patches.

The distribution of macroalgae in these mixed seagrass meadows is rather variable between locations. They can be quite abundant, and species diversity may be high (n >20 species; cover >100%) (Fig. 15.5). The algal layer, in the shallow areas, can locally be more than 10 cm thick. These algal communities commonly include the rhizophytic *Caulerpa* spp. (mainly *C. cupressoides, C. racemosa, C. sertulariodes, C. serrulata*), *Halimeda cylindracea, H. macroloba, H. opuntia, Chaetomorpha crassa* and *Dictyota* spp. that grow entangled among the seagrasses; in addition, a mixture of *Actinotrichia fragilis, Amphiroa* spp., *Jania* spp., *Valonia aegagropila* and *V. utricularis*, and several *Gracilaria* spp. (*G. eucheumoides, G. salicornia*) occur among the seagrasses. Depending on the availability of hard stabilized substrata, other macroalgal species (for example, *Codium, Neomeris* and *Padina*) may occur (Heijs and Brouns, 1986; Heijs, 1987a).

The epiphytic algal community found on the leaves in a mixed seagrass meadow is similar in species composition and abundance to that observed on leaves in monospecific seagrass beds. The epiphytic component is largely composed of

Fig. 15.5. A dense cover of the macroalga *Chaetomorpha crassa* in a meadow of *Enhalus acoroides* and *Thalassia hemprichii*.

Ceramiales (mainly *Ceramium*, *Herposiphonia* and *Polysiphonia*), Cryptonemiales (*Fosliella*) and a few Nostocales (for instance, *Calothrix*, *Lyngbya*) (Heijs, 1987b).

SPATIAL PATTERN

Seagrass beds are generally found on the inner reef flat and often bordered by coral growth and/or coral rubble areas that determine the zonation pattern. The variability in seagrass communities is considerable, being determined by local biotic and abiotic circumstances (exposure, substrate characteristics, etc.). It is therefore difficult to give an overall zonation pattern characteristic for seagrasses. The general pattern can be described as follows:

(1) Mid eulittoral
– sheltered muddy habitats. *Enhalus acoroides* with occasionally *Halophila ovalis* or *Thalassia hemprichii*. Monospecific patches of *Cymodocea rotundata*, *Halophila ovata*, *Halophila ovalis* or *Halodule uninervis*.
– sheltered and exposed habitats with coarse substrate. Any monospecific or bispecific growth of *Cymodocea rotundata*, *Halodule uninervis*, *Halophila ovalis* and *H. ovata*.

(2) Lower eulittoral to upper sublittoral.
The monospecific and bispecific seagrass patches are invaded by other seagrass species, eventually becoming a mixed seagrass meadow in which the seven most common species are found.

(3) Lower sublittoral.
The mixed meadow changes through tri- and bispecific seagrass communities into a monospecific seagrass bed. The seagrass species found will depend on local circumstances (water depth, substrate, etc.). This type of seagrass community is only found when the seagrass bed is not bordered by a zone of living and/or dead coral. When this is the case, the zonation pattern will end with the mixed seagrass meadow.

This overall zonation scheme for seagrasses, also

described by Den Hartog (1970), appears to be applicable for monospecific and mixed seagrass communities throughout Papua New Guinea and eastern Indonesia, and possibly for the whole Indo-West Pacific.

Although quite a number of macroalgal species are found associated with seagrass beds, only very few are characteristic species. The species composition of the macroalgal component in the seagrass beds is mainly determined by the substrate characteristics. Truly rhizophytic algae (such as *Avrainvillea, Caulerpa, Halimeda*) are common to abundant in seagrass beds, since they all favour sandy and/or muddy substrata. The haptophytic macroalgae are dependent on suitable hard substrata for colonization and subsequent growth, and are, therefore, not characteristic for the seagrass beds.

The different algal groups associated with the seagrass zonation can be described as rhizophytic algal species using the same substrate as seagrasses (*Avrainvillea, Caulerpa, Halimeda*), haptophytic algae on stabilized substrata (for example, *Dictyota, Neomeris, Padina*), and loose-lying/washed-in algae, which may continue their growth in the seagrass bed after being torn loose (for instance, *Gracilaria, Hypnea, Laurencia*).

Most species on the reef flat are widely distributed. Some show either gradients in abundance across the reef flat or are locally abundant or dominant, whereas others exhibit no such trends. Apart from substrate characteristics, the presence and abundance of these macroalgae are also determined by other biotic and abiotic factors (growth characteristics of each alga, competition, sedimentation, insolation, temperature, etc.). In the eulittoral, mainly *Avrainvillea erecta, Halimeda macroloba* and *H. opuntia* are likely to occur in association with seagrasses, whereas in the upper sublittoral many more species may be present.

Figure 15.6A illustrates a pattern typical for seagrass communities in sheltered, shallow localities, often on muddy substrata. The seagrass beds are generally dominated by *Enhalus acoroides* with *Thalassia hemprichii* in the undergrowth. On exposed shores, often characterized by a reef platform that can be very wide, a different zonation pattern is found (Fig. 15.6B). Here the "inner" reef flat is partly protected by the reef crest. The substrate may consist of sand, coral rubble or a mixture of these. Locally depressions may be filled by foraminiferan skeletons. Consolidated limestone is generally present below the top layer of loose sediment. Fig. 15.6C shows a pattern typical for the seagrass vegetation on sandy slopes in semi-exposed areas.

The epiphytic component on the leaves of the various seagrass species shows the same characteristics in monospecific seagrass beds and in mixed seagrass meadows (Heijs, 1985a, c, 1987b; Brouns and Heijs, 1986):

(1) the number of algal species and their abundance increase with increasing leaf age;

(2) the number of algal epiphytes and their abundance is of the same order on corresponding leaves of shoots of the various seagrasses;

(3) a significant difference exists in number of epiphytes, their abundance and reproduction density on the upper (oldest) and lower (youngest) leaf parts of corresponding leaves;

(4) no difference is found in number and abundance of epiphytic algae on the inner and outer face of each leaf;

(5) the Ceramiales dominate the epiphytic community in number of species whereas the encrusting coralline algae are quantitatively the most dominant group;

(6) algal species present as initial colonizers are still present on the older leaves. During the lifetime of the host, colonizing epiphytes are joined by other algal species, hence the composition of the epiphytic community alters on subsequent older parts of the host. True succession, where one group of epiphytic algae is replaced by a different group of epiphytes, does not take place. New species are able to settle and grow; however, they cannot oust the pioneer community.

TEMPORAL PATTERN

When sampling the various communities, no clear seasonal trend becomes apparent with respect to the species composition of the marine angiosperm component. Changes are relatively slow and take probably more than a year.

The temporal pattern of macroalgae may vary considerably between locations. Species present throughout the year in one location may show a strong seasonal distribution in another location. This is a consequence of different environmental and topographical characteristics at the various

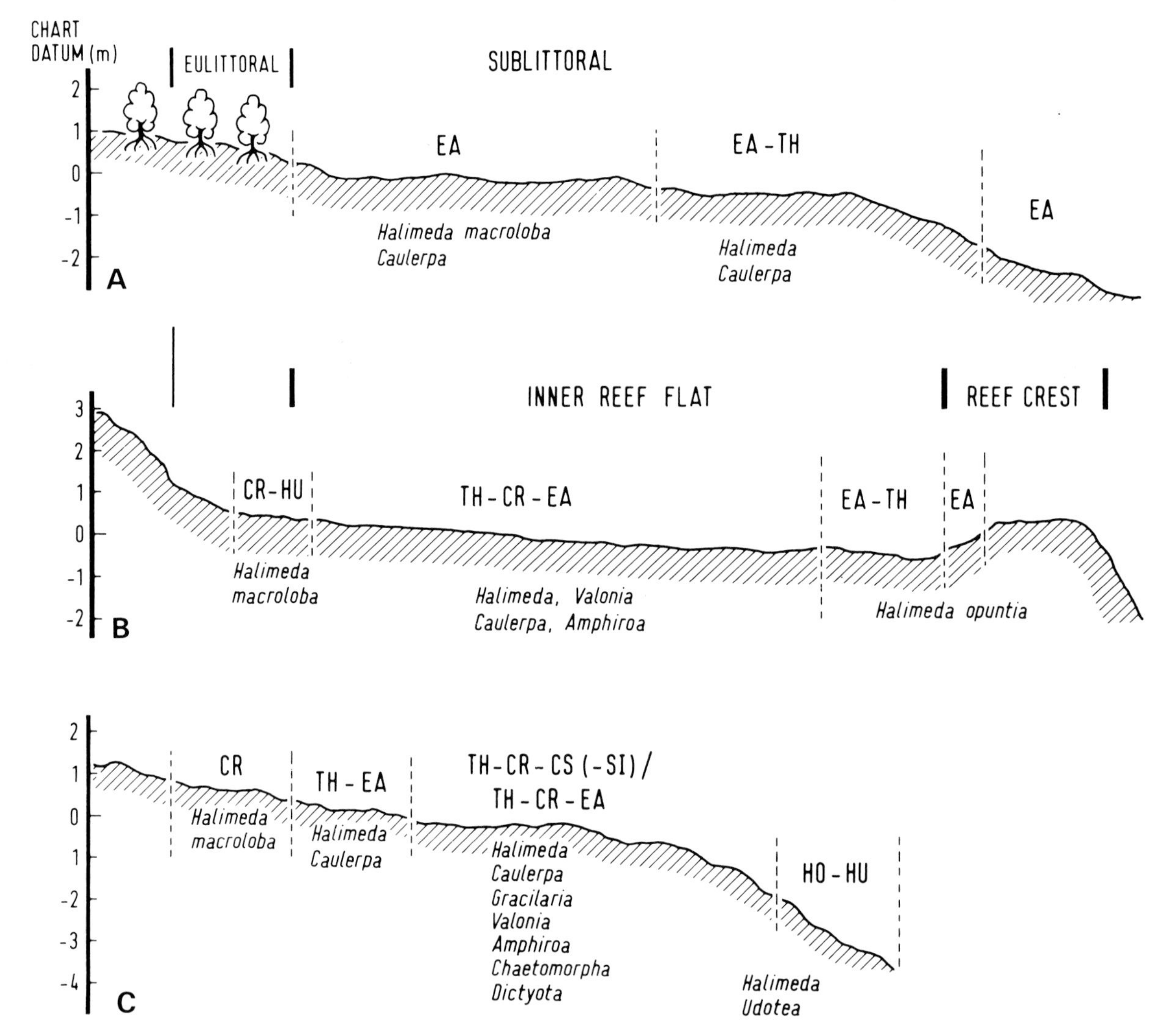

Fig. 15.6. Spatial pattern of seagrasses and macroalgae characteristic of different types of shores. (A) Shallow muddy lagoons in sheltered locations. (B) "Inner" reef flat on exposed shores. (C) Sandy slopes on semi-exposed shores. CR = *Cymodocea rotundata*, CS = *Cymodocea serrulata*, EA = *Enhalus acoroides*, HO = *Halophila ovalis*, HU = *Halodule uninervis*, SI = *Syringodium isoetifolium*, TH = *Thalassia hemprichii*

locations. Furthermore, biotic and abiotic factors (water temperature, illumination, variation in tidal patterns, substrate characteristics, etc.) may be important for the distribution and occurrence of macroalgae (Lawson, 1957, 1966; Doty, 1971; Santelices, 1977; Heijs, 1985b, 1987a).

The epiphytic component, however, shows a clear pattern. In Fig. 15.7 a generalized temporal pattern of epiphytic algae is presented for monospecific seagrass beds as well as mixed meadows. It appears that the same epiphytes are common to abundant on the various seagrass species in monospecific stands and in mixed vegetation (Heijs, 1985a, c, 1987b; Brouns and Heijs, 1986). From this table a group of algae can be recognized that may be considered characteristic for the seagrasses throughout Papua New Guinea. This is also clear from the biological importance of the various algal epiphytes (Table 15.1). This characteristic group of algal epiphytes on seagrass leaves consists of encrusting algae, Cyanophyta, *Audouinella* spp., *Ceramium gracillimum, Enteromorpha flexuosa* and/or *Polysiphonia savatierii.* Some epiphytic algae in Fig. 15.7 show a different gradient

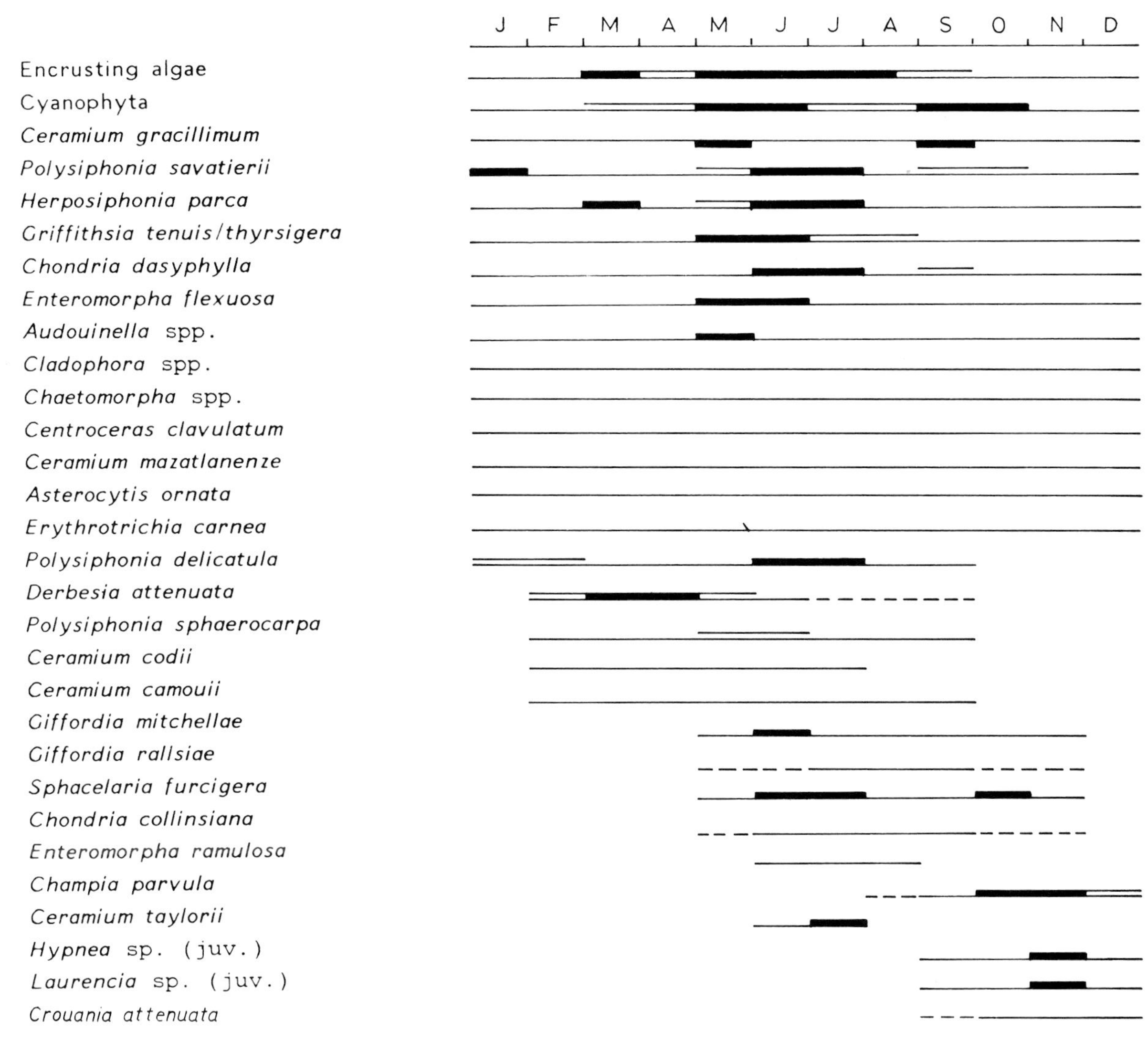

Fig. 15.7. Temporal pattern of epiphytes characteristic for the various seagrass species in monospecific and mixed situations. Dashed line: cover <5%; solid line: cover 5–10%; open bar: cover 10–20%; solid bar: cover >20%.

of abundance on the various seagrass species during the year, whereas others may be dominant only on one or two seagrass species, due to locally different microhabitat conditions. The majority of epiphytic algae, however, reach their maximum cover from May to August (Heijs, 1987b).

FUNCTIONAL ASPECTS

Seagrasses

From the most important species of the area (viz. *Cymodocea rotundata, C. serrulata, Enhalus acoroides, Halodule uninervis, Syringodium isoetifolium* and *Thalassia hemprichii*) some parameters of production and biomass are presented in Table 15.2. The mean annual value of the plastochrone interval of the leaves (PIL), or the time interval between the appearance of two successive leaves for most species is of the same magnitude (10.0–12.2 days), except for *Enhalus acoroides* (Brouns, 1985a, b, 1987a, b; Brouns and Heijs, 1986).

The lowest PIL as well as the lowest annual mean PIL, has been calculated for *Thalassia hemprichii*, both under eulittoral and sublittoral conditions (Brouns, 1985a, b). The highest annual

TABLE 15.1

Biological rank of importance of the 10 most abundant algal epiphytes in the different monospecific seagrass beds

Seagrass species	*Thalassia hemprichii*		*Cymodocea rotundata*	*Cymodocea serrulata*	*Syringodium isoetifolium*	*Halodule uninervis*	*Enhalus acoroides*
	eulittoral	sublittoral					
Encrusting algae	1	1	1	1	1	1	1
Cyanophyta	2	2	2	2	2	2	2
Ceramium gracillimum	3	3	3	3	3	3	3
Cladophora spp.	4	4	–	–	–	–	6
Polysiphonia savatierii	5	5	5	5	5	4	4
Audouinella spp.	7	7	4	4	6	5	5
Enteromorpha flexuosa	6	6	7	7	–	9	7
Chaetomorpha spp.	8	8	6	9	8	6	–
Griffithsia tenuis/thyrsigera	9	9	–	8	9	10	8
Chondria dasyphylla	10	–	–	–	7	–	10
Centroceras clavulatum	–	10	–	6	4	–	9
Ceramium taylorii	–	–	8	10	–	–	–
Derbesia attenuata	–	–	9	–	–	–	–
Herposiphonia secunda	–	–	10	–	10	8	–
Ceramium mazatlanenze	–	–	–	–	–	6	–

TABLE 15.2

Production and biomass (annual mean) of seagrass leaves and epiphytes in monospecific and mixed seagrass beds

Species	Biomass (g ADW m^{-2})		Production (g ADW m^{-2} day^{-1})	
	monospecific seagrass bed	mixed seagrass meadow	monospecific seagrass bed	mixed seagrass meadow
Seagrasses				
Cymodocea rotundata	62	9	2.0	0.3
Cymodocea serrulata	98	12	3.6	0.4
Enhalus acoroides	77	–	1.9	–
Halodule uninervis	68	1	2.8	0.04
Syringodium isoetifolium	194	22	5.5	0.7
Thalassia hemprichii				
eulittoral	70	–	4.2	–
sublittoral	40	56	2.6	2.5
Thalassodendron ciliatum	113	–	3.8	–
	Biomass (g ADW m^{-2} leaf surface)		Production (g ADW m^{-2} leaf surface day^{-1})	
Epiphytes				
Cymodocea rotundata	2.39 (22%)[a]	1.84 (19%)	0.16 (38%)	0.14 (25%)
Cymodocea serrulata	2.97 (24%)	1.94 (22%)	0.18 (39%)	0.12 (33%)
Enhalus acoroides	0.52 (3%)	2.68 (17%)	0.01 (6%)	0.05 (9%)
Halodule uninervis	1.72 (23%)	1.93 (24%)	0.14 (44%)	0.15 (31%)
Syringodium isoetifolium	4.77 (23%)	6.16 (21%)	0.25 (35%)	0.27 (19%)
Thalassia hemprichii				
eulittoral	1.31 (15%)	–	0.12 (37%)	–
sublittoral	1.59 (14%)	1.55 (14%)	0.10 (19%)	0.11 (16%)

[a]In parentheses the contribution of the epiphytic component to the total above-ground biomass/production.

mean PIL has been found for *Enhalus acoroides* (PIL= 31 days) (Brouns and Heijs, 1986).

Annual mean biomass values, in Table 15.2, are presented for seagrass leaves and the associated epiphytes. Density and biomass of the various seagrass species are highly variable and depend on local environmental factors (substrate characteristics and level, turbidity, etc.). The variability of the density in monospecific seagrass beds is roughly by a factor of 2 for most seagrasses. The exceptions are *Thalassia hemprichii*, under sublittoral conditions, and *Halodule uninervis*. *Thalassia hemprichii* occurs frequently as the only seagrass species in coarse coral rubble or interspersed between living coral. The variability in leaf biomass is often a consequence of the number of shoots per m^2 (density) and the total leaf area per leaf cluster. The biomass of a shoot and the density of the plant cover are negatively correlated in most seagrass species considered. For instance, the phenotypical variation in *Halodule uninervis* is of such an order that a dense meadow (cover, 75–100%) may be composed of 1500 large shoots, or of 18 300 small shoots. The mean maximum biomass of the leaves of a single shoot in a sample can amount to 18 mg ash-free dry weight (ADW) in meadows with a low density but dense cover; or become as little as 5 mg ADW in meadows with a maximum observed shoot density (Brouns, 1987a). Part of this variation is seasonal.

The range in biomass of the rhizomatous plant parts is wider, depending on the respective seagrass species. The variability reflects the relative age of the seagrasses in the sampling quadrat.

The rhizome mat of *Thalassodendron ciliatum* has not been sampled. The exposed rhizome layer on a study site in Indonesia reached a thickness of 75 cm (Brouns, 1985c). This observation, and the measured biomass of the rhizome of this seagrass in the Red Sea (Lipkin, 1979), indicates that the rhizome biomass can reach values higher than 50 g ADW per shoot. This seagrass is the only species with persistent rhizomes. The calculated mean turnover time for the rhizomes of *Enhalus acoroides* was 1580 days (Brouns and Heijs, 1986). The longevity of the rhizomes in the other species is considerably less. In Table 15.2 the total annual mean standing crop of the seagrasses studied in Papua New Guinea is presented. The daily mean production ranged from 1.9 g ADW m^{-2} day^{-1} (*Enhalus acoroides*) to 5.5 g ADW m^{-2} day^{-1} (*Syringodium isoetifolium*) (Brouns and Heijs, 1986; Brouns, 1987a).

Macroalgae

The seasonal variability of algal biomass and species composition is considerable (Fig. 15.8). Several papers clearly indicate how variable and aggregated macroalgal densities may be in localized areas. For instance, Virnstein and Carbonara (1985) reported that the highly aggregated benthic macroalgae in Florida could reach 15 000 g dry weight (DW) m^{-2}. Cowper (1978) found that macroalgae contributed up to 30% of the total above-ground phytomass of benthic vegetation in a Texas lagoon. Thorne-Miller et al. (1983), studying the distribution and biomass of submerged macrophytes in five neighbouring coastal lagoons in Rhode Island, found that the algal portion ranged from 13 to 46%. Heijs (1987a) reported that, when present, the macroalgal component of mixed seagrass beds in Papua New Guinea contributed 18 to 32% of the total above-ground biomass. Direct correlations, however, between the presence and abundance of macroalgae and environmental variables (illumination, tidal pattern, water temperature, etc.) are difficult to assess. It seems likely, therefore, that combinations of, or complex interactions between, these environmental variables are responsible for local differences and possibly also year-to-year variations (Santelices, 1977; Heijs, 1985b; Virnstein and Carbonara, 1985).

Epiphytes

The structure and function of the epiphytic community are largely determined by the biotic (for example, characteristics of the host, interactions between host and epiphytes, inter- and intra-specific interactions between epiphytes, life-cycles of epiphytes) and abiotic factors (such as temperature, water depth, insolation, exposure, mechanical wear and tear).

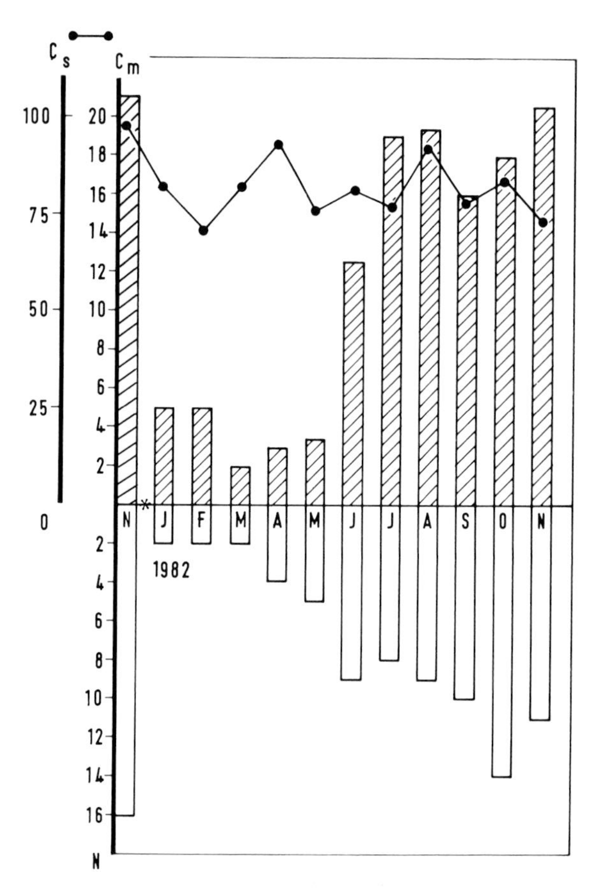

Fig. 15.8. Temporal abundance (% cover) of seagrasses (C_s) and macroalgae (C_m) and the number of macroalgal species (N) from a mixed seagrass meadow in Papua New Guinea.

The range and the mean annual values for biomass and production of epiphytes on the leaves of the various seagrass species are presented in Table 15.2. In monospecific seagrass beds, the highest annual mean biomass of the epiphytic component has been observed on the leaves of *Syringodium isoetifolium* and *Cymodocea serrulata*, namely, 4.8 and 3.0 g ADW m^{-2} leaf surface, respectively. The lowest annual mean epiphyte biomass has been found on the leaves of *Enhalus acoroides* (0.5 g ADW m^{-2} leaf surface). The values for epiphyte biomass on leaves of the other seagrass species are intermediate.

In a mixed seagrass meadow, annual mean

epiphyte biomass was highest for *Syringodium isoetifolium* and *Enhalus acoroides*. Epiphyte biomass on the leaves of the other seagrass species was approximately 1.8 g ADW m^{-2} leaf surface (Brouns and Heijs, 1986; Heijs, 1987b).

The contribution of the epiphytic community to the annual mean above-ground plant biomass (seagrass leaves and epiphytes, expressed as ash-free dry weight) ranged from 3% on leaves of *Enhalus acoroides* to 24% on leaves of *Cymodocea serrulata* and *Halodule uninervis*. When comparing the different seagrass species in the various monospecific and mixed seagrass communities, the biomass of epiphytes was virtually the same (Heijs, 1984, 1985a, c, 1987b; Brouns and Heijs, 1986).

When expressing epiphyte biomass in dry weight, 10 to 55% of the total above-ground plant biomass can then be attributed to the epiphytic component.

These values are of the same magnitude for other seagrass species, being 24 to 35% for *Zostera marina* (Penhale, 1977; Borum and Wium-Andersen, 1980); and 47% for *Halodule wrightii* (Morgan and Kitting, 1984).

Net production of epiphytes in Table 15.2 is defined as gross production minus respiration, losses due to grazing, mechanical wear and tear and senescence. Therefore, biomass-increment per unit time can be used as a minimum net production value, since the influence of these losses is minimal over short-term periods.

In the monospecific seagrass beds, the annual mean increase in epiphyte biomass (absolute growth rate) ranges from 9 mg ADW m^{-2} leaf surface per day on leaves of *Enhalus acoroides* to 250 mg ADW m^{-2} leaf surface per day on leaves of *Syringodium isoetifolium*. The production values for epiphytes on leaves of *Enhalus acoroides* are considerably lower than those found for other tropical Indo-Pacific seagrass species (Table 15.2), which is a consequence of differing topographic and hydrodynamic conditions.

Bulthuis and Woelkerling (1983), studying epiphyte production on leaves of *Heterozostera tasmanica* during the growing season only, have found values ranging from 0.06 to 1.04 g ADW m^{-2} leaf surface per day.

Penhale (1977) and Borum and Wium-Andersen (1980) studied the production of epiphytes on the leaves of *Zostera marina*. They found that the epiphytic component contributed 20 to 50% to the total above-ground plant production. Of the total above-ground plant production of *Halodule wrightii*, 48 to 56% can be attributed to the epiphytes (Morgan and Kitting, 1984). Jones (1968) estimated that production of epiphytes amounted to 20% of the total above-ground plant production of *Thalassia testudinum*.

Epiphyte production values observed in the mixed seagrass meadow are comparable to those calculated for the respective monospecific seagrass communities. This is also illustrated by the contribution of the epiphytic component to the annual mean above-ground production (seagrass leaves and epiphytes). An estimated 6% (*Enhalus acoroides*) to 44% (*Halodule uninervis*) of the total above-ground production is contributed by the epiphytic community of the tropical Indo-Pacific seagrasses.

The variability in structural and functional aspects of the algal component (for example, macroalgae and epiphytes) between monospecific and mixed seagrass communities is likely the result of (a) structure and growth characteristics of the dominant plant species (seagrasses and/or macroalgae); (b) differing topographic and physiographic conditions (water depth, substrate characteristics, etc.); (c) different complexities of the two types of seagrass beds.

The animal community associated with the different seagrass communities, the arrangement in time and space, and the interactions within the community and with the surrounding environment, will change depending on the phenology of the dominant plant species (Heck and Orth, 1980). Furthermore, besides providing additional substrate, food, shelter and refuge for small invertebrates (Den Hartog, 1979; Orth and Van Montfrans, 1984), the macroalgae and epiphytic algae contribute significantly to the net productivity of seagrass communities (Conover, 1964; Dawes et al., 1974; Cowper, 1978; Den Hartog, 1979; Heijs, 1984, 1985a, c, 1987a, b; Virnstein and Carbonara, 1985), the decomposition (Zieman, 1975), the dissolved organic carbon (DOC) pool (Sieburth, 1969) and the total standing crop (Cowper, 1978; Virnstein and Carbonara, 1985; Heijs, 1984, 1985a, c, 1987a, b; Brouns and Heijs, 1986).

A possible negative effect of large amounts of macroalgae and/or a dense cover of epiphytic algae in seagrass beds may be competition for light (shading), nutrients and space between the seagrasses and algae (Conover, 1964; Mathieson and Dawes, 1974; Josselyn, 1977; Sand-Jensen, 1977; Cowper, 1978; Borum and Wium-Andersen, 1980; Bulthuis and Woelkerling, 1983; Orth and Van Montfrans, 1984).

MACROFAUNA OF THE SEAGRASS COMMUNITIES

The composition and the occurrence of macrofauna in the seagrass communities along the shores of Papua New Guinea and Eastern Indonesia reflect the density of the human population and the pressure that is exerted on the community by fishing and collecting of invertebrates for consumption. Generally animals without nutritional value, or which are considered unpalatable, remain. Nevertheless, during the surveys in Indonesia, large areas of the seagrass beds appeared to be completely devoid of macrofauna.

Few animals feed directly on the seagrasses. Of the vertebrates the potentially most important group of feeders on seagrasses are the dugongs (*Dugong dugon*) and some species of marine turtles. These animals are severely endangered and are extinct in many localities.

Few invertebrates feed directly on seagrasses. The grazers are mainly sea urchins. Their main food is not the leaf but the associated algal flora (Kikuchi, 1980). Most marine invertebrates are not capable of digesting structural carbohydrates (Kristensen, 1972). The sessile fauna includes hydrozoans, actinians, bryozoans, foraminiferans and tube-dwelling polychaetes. The mobile fauna consists of gastropods, turbellarians, nemertines, crustaceans and echinoderms. The feeding types of the epifaunal community associated with seagrasses consist of herbivores, and detritivores including nematodes, polychaetes, amphipods, harpacticoids, ostracods, tardigrades, prosobranchs and opisthobranchs (Lewis and Hollingworth, 1982). The herbivorous fauna is capable of directly scraping the epiphytic microflora and the entangled detritus off the seagrass blades (Randall, 1965; Carr and Adams, 1973). Predators are less abundant, and include turbellarians, nematodes, polychaetes, opisthobranchs, hydroids and sea anemones.

The epifauna itself serves as a food source for vagile larger invertebrates and vertebrates. Fish, polychaetes, crabs and shrimps are the main predators (Hagerman, 1966; McIntyre, 1969; Kikuchi, 1974).

Most leaf material is exported out of the seagrass bed or deposited on the bottom. In many seagrass species the leaves break up in fragments. The leaf parts, buoyant because of the gases in the leaves or trapped between the epiphytic algae, rise to the surface. Larger parts can become entangled in dense seagrass beds. The plant tissue is fragmented by detritus feeders, and the organic matter is decomposed by micro-organisms and utilized through the detritus food chain. Quite a few organisms may, depending on the locality, occur in large numbers in the seagrass beds.

The most abundant echinoderm in the seagrass beds from Papua New Guinea is *Tripneustes gratilla* (family Toxopneustidae). This species, living on the surface, covers itself with seagrass leaves, mainly of *Thalassia hemprichii*, coral debris and bivalve shells, probably as protection against insolation or as camouflage. The influence of this behaviour and the grazing of this animal has a marked effect on the standing crop of this seagrass in particular. Stomach contents of *T. gratilla* have shown large quantities of fragmented seagrass leaves. Other members of the class Echinoidea are *Echinometra mathaei*, common in the mid- and lower eulittoral, *Heterocentrotus mammillatus* in the lower eulittoral and sublittoral fringe, and *Diadema setosum*, locally found throughout the seagrass beds, where it may form aggregations during daytime. *Echinometra mathaei* is often found in deep cavities beneath coral blocks.

The most conspicuous group of invertebrates in seagrass meadows are the holothurians. Several species of four families (Cucumariidae, Holothuroiidae, Stichopodidae and Synaptidae) are found. The largest invertebrate is the synaptid *Synapta maculata*, rather common on seagrass flats. The most abundant sea-cucumber in Papua New Guinea, however, is the black *Halodeima atra*, which locally reaches a high density on open patches of sand in the lower eulittoral. *Actinopyga mauritiana, Bohadschia argus, Halodeima atra* and *Metriatyla scabra* are frequently observed in seagrass meadows.

Some of the Holothuroiidae are locally used as a food source for human consumption (bêche-de-mer or trepang), and therefore have commercial value. *Microthele nobilis* is the most valuable species, but species of the genera *Halodeima, Stichopus* and *Thelenota* are locally consumed and hence collected. Tropical holothurians feed on an array of food items and are relatively efficient in processing sediment. By ingesting sediment and through burrowing activities, holothurians work over the substrate and destroy the initial stratification. The resulting bare areas are rapidly colonized by seagrasses.

The Asteroidea are not collected by the local inhabitants because they are considered inedible. Many species favour a sandy substrate, often with seagrasses, though some are found on rocky substrates. *Protoreaster nodosus* is very common in mixed seagrass meadows and monospecific seagrass beds in Papua New Guinea as well as in eastern Indonesia. This species can be considered the most abundant and conspicuous invertebrate. Less abundant and very variable in appearance is *Culcita novaeguineae*. Other locally common species, particularly in the eulittoral zone, just below the sand surface, are *Archaster typicus* and *Astropecten polyacanthus*. The bright blue *Linckia laevigata* is found mainly on coral debris bordering the seagrass beds.

The more common species in the class Gastropoda (phylum Mollusca) belong to the families Conidae, Cypraeidae and Strombidae. Most species occur in the eulittoral as well as in the sublittoral belt. *Strombus gibberulus* and *S. luhuanus* usually form aggregations with sometimes up to 20 to 30 individuals per m^2. Juveniles in particular are epiphytic on seagrass leaves. The spider shell *Lambis lambis* and several species of *Strombus* are collected for consumption. These species are lacking in some heavily utilized seagrass beds. *Cypraea annulus* and *C. moneta* are very characteristic for the seagrass beds and occur in great numbers. Frequently encountered Conidae are *Conus leopardus, C. litteratus* and *C. marmoreus*, well camouflaged by the overgrowth of green and calcareous algae. *Malleus malleus* and *Pinna nobilis* belong to the molluscan infauna of the intertidal and sublittoral zone and are frequently encountered in seagrass meadows, where they live in a vertical position, almost wholly buried in the sediment, with the gaping valve edges protruding above the surface.

Only one qualitative and quantitative inventory of the fish fauna associated with seagrass meadows in the area (Java, Indonesia) is known (Hutomo and Martosewojo, 1977). The authors collected 78 species in a seagrass meadow dominated by *Halodule uninervis*, the smallest of the tropical seagrasses with strap-like leaves. Apogonidae occupied the first rank in relative as well as total abundance. Pollard (1984) reviewed ecological studies on seagrass–fish communities. In a numerical classification analysis the three Indo-Pacific sites (Java, Guam and Madagascar) grouped together, and the Java and Guam sites were closely associated. The Madagascar site is characterized by the diversity of multi-species seagrass meadows adjacent to coral reefs. The structural complexity of the area is reflected in the relatively large number (189) of fish species (Vivien, 1974).

Fishes use the seagrass beds for various reasons: (1) among the herbivores feeding directly on the seagrasses and epiphytes are the parrot-fishes (Scaridae), the surgeon-fishes (Acanthuridae) and the halfbeaks (Hemirhamphidae); (2) juveniles of the surgeon-fishes, snappers (Lutjanidae) and puffers (Tetraodontidae) use the seagrass leaves as food and shelter (Ogden, 1980); (3) residents, using the seagrass beds mainly for shelter, include pipefishes (Syngnathidae), eels and wrasses (Labridae). Of the fishes only the halfbeak (*Hemirhamphus far*) is frequently observed feeding. The animal selectively takes the leaves of the seagrass *Syringodium isoetifolium* from among the broken-off leaves floating on the surface.

Few herbivores are exclusive to seagrass areas. They often migrate between the reef areas and the seagrass beds. They are capable of channelling plant material out of the seagrass beds to the surrounding communities, where it becomes available to higher trophic levels. The movements of these macroconsumers therefore form an important link between the seagrass bed and the adjacent communities such as coral reefs and mangroves (Randall, 1965; Ogden and Zieman, 1977; Hay, 1981).

CONCLUSIONS

Finally, a few recommendations for future research may be offered:

– Structural and functional aspects.

Although some qualitative and quantitative data have been collected on the macrofauna of the seagrass beds, the epifauna, the infauna and the microfauna on the bottom have yet to be studied quantitatively. Furthermore, information on their spatial and temporal distribution, as well as on their interrelations, has to be sought. Attention should also be addressed to productivity, nutritional relationships, decomposition and biomass.

The floral component has been investigated rather intensively, in particular the structural aspects. The functional aspects of the macroalgae associated with the seagrass beds, however, should receive more attention. To learn more about the specificity of the epiphytes on seagrass leaves, further research is needed on the epiphytic component on other living (for example, macroalgae) and non-living substrata (for example, rocks).

– Dynamic aspects and autecology.

Various aspects of the monospecific and mixed seagrass community have been studied; namely, spatial and temporal distribution, productivity and biomass of seagrasses, epiphytes and macroalgae, and growth patterns of the various seagrass species. However, to get more insight into the dynamic relations within the communities and the autecology of the various seagrass species, data should be gathered on phenological aspects, competition, recolonization, tolerances to environmental conditions, etc. For instance, the intraspecific variation in the growth rates of the rhizomes was remarkable. It is however unknown whether this growth rate is related to the size of the plant, or simply related to the natural variation within the population. The first possibility would indicate a transport of nutrients from older shoots to rhizome meristems. Changes in time in the chemical composition of the rhizomes could then give insight into these processes. The incorporation of radio-tracers into a single shoot, in order to study translocation processes, appears feasible and may elucidate the functioning of the plant as a single unit (Allessio and Tieszen, 1978; Tietema, 1981).

– Interactions with the surrounding environment.

Continuous measurements have been made of several abiotic factors (for instance, insolation, precipitation, water temperature and tidal patterns). However, more research is needed on the nutrient content of the interstitial and overlying water layer of the seagrass beds, and the physical and chemical properties of the sediment.

Furthermore, the position of the seagrass community relative to other shallow-water ecosystems in the coastal zone should be assessed, that is, the interactions of seagrass communities with surrounding coral and/or mangrove communities, or "micro-atolls," which are scattered throughout the seagrass beds and reef platforms, should be evaluated. Such investigations, which could be carried out intensively in one location, would make it easier to compare similar communities in other areas.

REFERENCES

Allessio, M.L. and Tieszen, L.L., 1978. Translocation and allocation of C-photoassimilate by *Dupontia fisheri*. In: L.L. Tieszen (Editor), *Vegetation and Production Ecology of an Alaskan Arctic Tundra. Ecological Studies, Vol. 29.* Springer, Berlin, pp. 393–413.

Borum, J. and Wium-Andersen, S., 1980. Biomass and production of epiphytes on eelgrass (*Zostera marina* L.) in the Øresund, Denmark. *Ophelia* (Suppl.), 1: 57–64.

Brouns, J.J.W.M., 1985a. The plastochrone interval method for the study of the productivity of seagrasses; possibilities and limitations. *Aquat. Bot.*, 21: 71–88.

Brouns, J.J.W.M., 1985b. A comparison of the annual production and biomass in three monospecific stands of the seagrass *Thalassia hemprichii* (Ehrenb.) Aschers. *Aquat. Bot.*, 23: 149–175.

Brouns, J.J.W.M., 1985c. A preliminary study of the seagrass *Thalassodendron ciliatum* (Forssk. den Hartog) from Eastern Indonesia. *Aquat. Bot.*, 23: 249–260.

Brouns, J.J.W.M., 1987a. Aspects of production and biomass of four seagrass species (Cymodoceoideae) from Papua New Guinea. *Aquat. Bot.*, 27: 333–362.

Brouns, J.J.W.M., 1987b. Quantitative and dynamic aspects of a mixed seagrass meadow in Papua New Guinea. *Aquat. Bot.*, 29: 33–47.

Brouns, J.J.W.M. and Heijs, F.M.L., 1985. Tropical seagrass ecosystems in Papua New Guinea. A general account of the environment, marine flora and fauna. *Proc. K. Ned. Akad. Wetensch.*, C88: 145–182.

Brouns, J.J.W.M. and Heijs, F.M.L., 1986. Production and biomass of the seagrass *Enhalus acoroides* (L.f.) Royle and its epiphytes. *Aquat. Bot.*, 25: 21–45.

Bulthuis, D.A. and Woelkerling, Wm. J., 1983. Biomass accumulation and shading effects of epiphytes on leaves of *Heterozostera tasmanica*, in Victoria, Australia. *Aquat. Bot.*, 16: 137–148.

Carr, W.E.S. and Adams, C.A., 1973. Food habits of juvenile marine fishes occupying seagrass beds in the estuarine zone near Crystal River, Florida. *Trans. Am. Fish. Soc.*, 102: 511–540.

Conover, J.T., 1964. The ecology, seasonal periodicity and

distribution of benthic plants in some Texas lagoons. *Bot. Mar.*, 7: 4–41.

Cowper, S.W., 1978. The drift algal community of seagrass beds in Redfish Bay, Texas. *Contrib. Mar. Sci.*, 21: 125–132.

Dawes, C.J., Mathieson, A.C. and Cheney, D.P., 1974. Ecological studies of Floridan *Eucheuma* (Rhodophyta, Gigartinales). *Bull. Mar. Sci.*, 24: 235–273.

Den Hartog, C., 1970. Seagrasses of the world. *Verh. K. Ned. Akad. Wetensch., Natuurk, Reeks 2*, 59(1): 1–275.

Den Hartog, C., 1979. Seagrasses and seagrass ecosystems, an appraisal of the research approach. *Aquat. Bot.*, 7: 105–117.

Doty, M.S., 1971. Antecedent event influence on benthic marine algal standing crops in Hawaii. *J. Exp. Mar. Biol. Ecol.*, 6: 161–166.

Hagerman, L., 1966. The macro- and microfauna associated with *Fucus serratus* L. with some ecological remarks. *Ophelia*, 3: 1–43.

Hay, M.E., 1981. Spatial patterns of grazing intensity on a Caribbean barrier reef: Herbivory and algal distribution. *Aquat. Bot.*, 11: 97–109.

Heck Jr., K.L. and Orth, R.J., 1980. Seagrass habitats: the roles of habitat complexity, competition and predation in structuring associated fish and motile macroinvertebrate assemblages. In: V.S. Kennedy (Editor), *Estuarine Perspectives*. Academic Press, New York, pp. 449–464.

Heijs, F.M.L., 1984. Annual biomass and production of epiphytes in three monospecific seagrass communities of *Thalassia hemprichii* (Ehrenb. Aschers.). *Aquat. Bot.*, 20: 195–218.

Heijs, F.M.L., 1985a. The seasonal distribution and community structure of the epiphytic algae on *Thalassia hemprichii* (Ehrenb. Aschers.) from Papua New Guinea. *Aquat. Bot.*, 21: 295–324.

Heijs, F.M.L., 1985b. The macroalgal component in monospecific seagrass beds from Papua New Guinea. *Aquat. Bot.*, 22: 291–324.

Heijs, F.M.L., 1985c. Some structural and functional aspects of the epiphytic component on four seagrass species (Cymodoceoideae) from Papua New Guinea. *Aquat. Bot.*, 23: 225–247.

Heijs, F.M.L., 1987a. Community structure and seasonality of macroalgae in mixed seagrass meadows from Papua New Guinea. *Aquat. Bot.*, 27: 139–158.

Heijs, F.M.L., 1987b. Qualitative and quantitative aspects of the epiphytic component in a mixed meadow from Papua New Guinea. *Aquat. Bot.*, 27: 363–384.

Heijs, F.M.L. and Brouns, J.J.W.M., 1986. A survey of seagrass communities around the Bismarck Sea, Papua New Guinea. *Proc. K. Ned. Akad. Wetensch.*, C89: 11–44.

Hutomo, M. and Martose Wojo, S., 1977. The fishes of seagrass community on the west side of Burung Island (Pari Islands, Seribu Islands) and their variations in abundance. *Mar. Res. Indonesia*, 17: 147–172.

Johnstone, I.M., 1978a. The ecology and distribution of Papua New Guinea seagrasses. I. Additions to the seagrass flora of Papua New Guinea. *Aquat. Bot.*, 5: 229–233.

Johnstone, I.M., 1978b. The ecology and distribution of Papua New Guinea seagrasses. II. The Fly Islands and Raboin Island. *Aquat. Bot*, 5: 235–243.

Johnstone, I.M., 1982. Ecology and distribution of the seagrasses. *Monogr. Biol.*, 42: 497–152.

Jones, J.A., 1968. *Primary productivity of the tropical marine turtle grass Thalassia testudinum König, and its epiphytes*. Thesis, Miami University, Miami, Fla., 196 pp.

Josselyn, M.N., 1977. Seasonal changes in the distribution and growth of *Laurencia poitei* (Rhodophyceae, Ceramiales) in a subtropical lagoon. *Aquat. Bot.*, 3: 217–229.

Kikuchi, T., 1974. Japanese contributions on consumer ecology in eelgrass (*Zostera marina* L.) beds, with special reference to trophic relationships and resources in inshore fisheries. *Aquaculture*, 4: 145–160.

Kikuchi, T., 1980. Faunal relationships in temperate seagrass beds. In: R.C. Phillips and C.P. McRoy (Editors), *Handbook of Seagrass Biology: An Ecosystem Perspective*. Garland STPM Press, New York, pp. 153–172.

Kock, R.L. and Tsuda, R.T., 1978. Seagrass assemblages of Yap, Micronesia. In: R.T. Tsuda (Editor), *Marine Biological Survey of Yap Lagoon. Univ. Guam Mar. Lab. Report*, 45: 16–20.

Kristensen, J.H.M., 1972. Carbohydrates of some marine invertebrates with notes on their food and on the natural occurrence of the carbohydrates, studied. *Mar. Biol.*, 14: 130–142.

Landsberg, H.E., Lippmann, H., Paffen, K. and Troll, C., 1966. *World Maps of Climatology*. Springer, Berlin, 27 pp.

Lawson, G.W., 1957. Seasonal variation of intertidal zonation on the coast of Ghana in relation to tidal factors. *J. Ecol.*, 45: 831–860.

Lawson, G.W., 1966. The littoral ecology of West Africa. *Oceanogr. Mar. Biol. Ann. Rev.*, 4: 405–448.

Lewis, J. and Hollingworth, C.E., 1982. Leaf epifauna of the seagrass *Thalassia testudinum*. *Mar. Biol.*, 71: 41–49.

Lipkin, Y., 1979. Quantitative aspects of seagrass communities, particularly of those dominated by *Halophila stipulacea*, in Sinai (Northern Red Sea). *Aquat. Bot.*, 7: 119–128.

Mathieson, A.C. and Dawes, C.J., 1974. Ecological studies on Floridan *Eucheuma* (Rhodophyta, Gigartinales). II. Photosynthesis and respiration. *Bull. Mar. Sci.*, 24: 274–285.

McCoy, E.D. and Heck, K.L., 1976. Biogeography of corals, seagrasses, and mangroves: an alternative to the center of origin concept. *Syst. Zool.*, 25: 201–210.

McIntyre, A.D., 1969. Ecology of marine meiobenthos. *Biol. Rev.*, 44: 245–290.

Meñez, E.G., Phillips, R.C. and Calumpong, H.P., 1983. Seagrasses from the Philippines. *Smiths. Contr. Mar. Sci.*, 21: 1–40.

Morgan, D. and Kitting, C.L., 1984. Productivity and utilization of the seagrass *Halodule wrightii* and its attached epiphytes. *Limnol. Oceanogr.*, 29: 1066–1076.

Ogden, J.C., 1980. Faunal relationships in Caribbean seagrass beds. In: R.C. Phillips and C.P. McRoy (Editors), *Handbook of Seagrass Biology: An Ecosystem Perspective*. Garland STPM Press, New York, London, pp. 173–198.

Ogden, J.C. and Ogden, N.B., 1982. A preliminary study of two representative seagrass communities in Palau, Western Caroline Islands. *Aquat. Bot.*, 12: 229–244.

Ogden, J.C. and Zieman, J.C., 1977. Ecological aspects of coral

reef-seagrass contacts in the Caribbean. *Proc. Third Int. Coral Reef Symp.*, 1: 377–382.
Orth, R.J. and Van Montfrans, J., 1984. Epiphyte-seagrass relationships with emphasis on the role of micrograzing: a review. *Aquat. Bot.*, 18: 43–69.
Penhale, P.A., 1977. Macrophyte-epiphyte biomass and productivity in an eelgrass (*Zostera marina* L.) community. *J. Exp. Mar. Biol. Ecol.*, 26: 211–224.
Pollard, D.A., 1984. A review of ecological studies on seagrass-fish communities, with particular reference to recent studies in Australia. *Aquat. Bot.*, 18: 3–42.
Polunin, N.V.C., 1983. The marine resources of Indonesia. *Oceanogr. Mar. Biol. Ann. Rev.*, 21: 455–531.
Randall, J.E., 1965. Grazing effects on seagrasses by herbivorous reef fishes in the West Indies. *Ecology*, 46: 255–260.
Sale, P.F., 1980. The ecology of fishes on coral reefs. *Oceanogr. Mar. Biol. Ann. Rev.*, 18: 367–421.
Sand-Jensen, K., 1977. Effects of epiphytes on eelgrass photosynthesis. *Aquat. Bot.*, 3: 55–63.
Santelices, B., 1977. Water movement and seasonal algal growth in Hawaii. *Mar. Biol.*, 43: 225–235.
Sieburth, J. McN., 1969. Studies on algal substances in the sea. III. The production of extracellular organic matter by littoral marine algae. *J. Exp. Mar. Biol. Ecol.*, 3: 290–309.
Stephenson, T.A. and Stephenson, A., 1949. The universal features of zonation between tide-marks on rocky coast. *J. Ecol.*, 37: 289–305.
Thorne-Miller, B., Harlin, M.M., Thursby, G.B., Brady-Campbell, M.M. and Dworetzky, B.A., 1983. Variations in the distribution and biomass of submerged macrophytes in five coastal lagoons in Rhode Island, U.S.A. *Bot. Mar.*, 26: 231–242.
Tietema, T., 1981. *Ecophysiology of the sand sedge Carex arenaria L.* Thesis, State University of Utrecht, 106 pp.
Tsuda, R.T., Fosberg, F.R. and Sachet, M.H., 1977. The distribution of seagrasses in Micronesia. *Micronesica*, 13: 191–198.
Virnstein, R.W. and Carbonara, P.A., 1985. Seasonal abundance and distribution of drift algae and seagrasses in the mid-Indian River lagoon, Florida. *Aquat. Bot.*, 23: 67–82.
Vivien, M.L., 1974. Ichthyofaune des herbiers de phanerogames marines du Grand Récif de Tuléar (Madagascar) 1. Les peuplements et leur distribution écologique. *Téthys*, 5: 425–436.
Wallace, A.R., 1876. *The Geographical Distribution of Animals*, Reprinted 1962, Hafner. New York, London.
Zieman, J.C., 1975. Quantitative and dynamic aspects of the ecology of turtle grass *Thalassia testudinum*. *Estuarine Res.*, 1: 541–562.

Chapter 16

LIFE IN THE LITTORAL OF THE RED SEA (with remarks on the Gulf of Aden)

YAACOV LIPKIN

INTRODUCTION

The Red Sea as an environment

The Red Sea, geologically an ocean in infancy, is a relatively young branch of the Indian Ocean to which it is connected via the Gulf of Aden. It extends in a SE-NW direction from the tropics at its connection with the Gulf of Aden at Bab al Mandab (*c*. 12°40′N) to subtropical latitudes at its northernmost end, the tip of the Gulf of Suez (*c*. 30°N) separating Arabia on its Asian side from Africa. The narrow (24 km wide) passage of Bab al Mandab is partially blocked by a shallow sill (to 137 m deep), strongly affecting properties of Red Sea water. The Red Sea, especially the northern half, is located in a hot and extremely arid region surrounded by very dry and hot deserts. Evaporation being very high, exchange of water with the ocean rather limited, and incoming runoff water practically non-existent, the Red Sea is the warmest and most salty body of water with a direct open connection to the ocean. The Red Sea environment and oceanography have been described elsewhere in this series (Ross, 1983), so only a brief outline is necesary.

Girdler (1983) proposed a model for the evolution of the Gulf of Aden and the Red Sea in three stages:

(1) In the Oligocene, about 42 million years ago, an early rift system propagated from the Indian Ocean through the Gulf of Aden, then the southern Red Sea, the northern Red Sea, and finally the Gulf of Suez, reaching the latter region about 38 million years ago. This rift was a response to a NE-SW tensional stress related to the rotation of the Arabian plate anticlockwise with respect to the African plate about a pivot in the Mediterranean off North Africa at *c*. 33°N, 24°E.

(2) In the beginning of the Miocene the rift system was reactivated, resulting in major movements throughout the region that started about 24 million years ago and lasted through the lower Miocene until about 16 million years ago. This was a period of active separation and widening, with the propagation of a new crack, the early Gulf of Elat (‘Aqaba)–Dead Sea rift system, in the beginning of the period. The Gulf of Aden widened first, the Red Sea a little later. At the end of the period the Red Sea had closed to become an isolated basin in which evaporites *c*. 5000 m thick were deposited.

(3) In the earliest Pliocene (about 5 million years ago) a further major reactivation occurred. The sea broke through the straits of Bab al Mandab terminating the deposition of evaporites, and sea-floor spreading was rejuvenated, lasting to the present.

Rainfall around the Red Sea ranges between 11 mm yr^{-1} in the north to less than 200 mm yr^{-1} in the south. Years without rains are not a rarity, especially in the northern half. No rivers flow into this sea; only occasional floods, resulting from local rains in the desert, reach the sea through wadis that are usually dry. Average evaporation is 2150 mm yr^{-1}, increasing from south to north (Table 16.1). Air and sea surface temperatures are high, decreasing from south to north (Table 16.1). In the Gulf of Suez winter air temperatures are lowest; the mean minimum of January, the coolest month, is only 6°C, and the shallow water of the Gulf quickly cools down as well. Salinities are also high, being lowest at the entrance to the sea (Table 16.1), increasing to nearly 43‰ in semi-closed northerly habitats in the Gulf of Elat (Gulf

TABLE 16.1

Evaporation, air and sea-surface temperatures, and surface salinities in open waters in the Red Sea

Part of Red Sea	Evaporation (mm yr^{-1})	Air temperatures (°C)				Sea-surface temperatures (°C)		Salinity (‰)
		January[a]		August[b]				
		mean minimum	mean maximum	mean minimum	mean maximum	winter	summer	
Southern	1820	24	32	27–29	43	28	*c.* 31	37
Northern	*c.* 4000	10	25	25	39	22	28	40–41.5

[a]Coolest month.
[b]Hottest month.

of ‘Aqaba) or to nearly 44‰ in the Gulf of Suez. Throughout the year, prevailing winds in the northern part are from the northwest along the axis of the Red Sea. In the southern part, however, the winds change seasonally. During the SW monsoon period (April to September) they blow towards the southeast, changing direction by 180° during the NE monsoon period (October to March). Surface currents agree with prevailing wind directions. During the SW monsoon period they flow in a southeast direction along the entire Red Sea and out through Bab al Mandab to the Gulf of Aden. During the NE monsoon period a surface current flowing south meets another flowing north at about 20°N, where they collide and the surface water sinks to the depths. Nutrients are low, resulting in a paucity of plankton and good penetration of light to the depths. Nutrient concentrations decrease progressively from south to north, especially in the upper 100 m layer. North of latitude 22°N they are only 0.1 to 0.2 of the concentrations near Bab al Mandab. In the open Red Sea, the concentration of nitrate-nitrogen in surface waters was less than 1 $\mu mol\, l^{-1}$ during the SW monsoon period. Phosphate-phosphorus was then 0.3 $\mu mol\, l^{-1}$ in the Gulf of Aden; it dropped to near 0.1 $\mu mol\, l^{-1}$ at Bab al Mandab and was even less in most of the Red Sea. Nearshore waters in the central Red Sea had 0.01 to 0.02 $\mu mol\, l^{-1}$ NO_3–N and 0.03 to 0.4 $\mu mol\, l^{-1}$ PO_4–P. During the SW monsoon period, nutrient poor surface water of the Red Sea flows out into the Gulf of Aden, whereas during the NE monsoon period rich Gulf of Aden surface water flows into the Red Sea. Right over the sill, denser Red Sea water flows out into the Gulf of Aden year-round. During the NE monsoon period only this Red Sea water and the inflowing surface current occur above the sill. During the SW monsoon period, when the surface flow reverses, an intermediate inflowing current develops, bringing Gulf of Aden water into the Red Sea to compensate for the losses. The stronger incoming flow during the NE monsoon period raises the mean level of the Red Sea in winter by about 0.5 m. This affects the intertidal zone, causing consequent seasonal shifts in spatial distribution of its organisms and communities. Generally, tidal ranges are widest at both ends; in the south it is *c.* 1.2 m and in the north 1.5 to 2 m. In the central parts it is smaller, being only 0.3 to 0.4 m on the Arabian coast near 23°N, and practically nil at Bur Sudan. However, there are local deviations from this general pattern (Clark and Gohar, 1953; Neumann and McGill, 1962; Oren, 1962; Gorgy, 1966; Nir, 1971; Por, 1972; McGill, 1973; Baeshin and Aleem, 1978; Nasr, 1982; Ketchum, 1983; Ross, 1983; Reiss and Hottinger, 1984).

Rocky shores along the Red Sea coast vary greatly in type and nature of substrate, topography, and degree of exposure to wave action. They harbour a very wide range of communities. Igneous or sedimentary rocks are sometimes exposed within the intertidal zone. Recently formed beach-rock slabs also sometimes occur within this zone, usually at the upper levels. Commonest of all are fossil coral-reef limestones that occur over the entire range of the intertidal zone. Highly sheltered rocky shores lined at the intertidal levels by convex walls, and undercuts in such limestone [the so-called Entedebir Reef Limestone (Nir, 1971)], were studied on both sides of the southern Red Sea (Klausewitz, 1967; Lipkin, 1987c). In Sinai in the north, a variety of substrates under different wave-exposure regimes was also examined. Rocky shores there amounted to only 15% of the coastline within the intertidal, the rest being a variety of unconsolidated substrates (cf. Safriel and Lipkin, 1964, 1975; Ayal and Safriel, 1980). The types of rocky shores studied in the northern Red Sea were as follows:

(1) Vertical and often convex walls characterize some shores where no platforms occur (see below); in some cases these have a deep undercut within fossil coral limestone (cf. Wainwright, 1965, Fig. 4; Klausewitz, 1967, Fig. 15).

(2) Blocks and boulders of different rock types also occur along shores devoid of platforms.

(3) Beach-rock slabs occur on some shores overlying unconsolidated substrates. They are fine- to coarse-grained sandstones when formed on more sheltered beaches versus conglomerates on less sheltered ones. In both cases they dip gently towards the sea, usually inclined at 4.5 to 12%.

(4) Commonest are sub-horizontal platforms on fossil reef limestone that are formed by abrasion and usually situated near mean sea level. They fringe a large portion of the Red Sea coast, being interrupted by wadi mouths (forming deltas built of terrigenous sediments) or by sandy coves. These "fringing reefs" either start on the beach or are separated from it by shallow lagoons; they are usually 2 to 5 m deep and rarely to 10 m. The platforms, as well as the lagoons, range from a few metres to several hundred metres wide; the lagoons may even be 1 to 2 km wide. When contiguous with the beach the platforms start on the landward side below a low sandy or muddy beach, or they continue any of the rocky beach types mentioned above. Usually they are inclined at 0.25 to 1% towards the sea. The upper surfaces of the platforms are exposed during regular ebbs, except for the seaward edges that only become denuded rarely; they usually remain covered with a layer of water up to 1 m thick during flows.

As the coast is dissected by many faults into separate blocks moving tectonically independent of one another, some blocks may have horizontal platforms and others may be slightly inclined towards the coast — the seaward margin being the highest part of the platform. The upper, abraded surface of the platform has many little projections and depressions of various sizes and depths. On wider platforms the depressions closer to the coastline are filled with sediment, usually coralligenous sand, and are usually rather shallow. They hold water during low tide when the surrounding rocky surface of the platform dries up. The seaward parts of platforms are exposed to considerable wave action, which gradually decreases to the leeward. On very wide platforms the parts nearest to the beach are rather sheltered. In some places the platforms drop abruptly to great depths on the seaward side near vertical rocky walls. However, these rocky structures usually end in vertical walls or steep slopes only a few metres high. Coral reefs develop narrow belts on the surface of these platforms towards the seaward margins. At the foot of the reefs lies a sandy to muddy area several metres deep, which slopes gently to 10 to 200 m at the edge of the shelf, depending on its width. Where the shelf is very narrow and ends at shallow depths, some of the coastal communities occur on the rather steep continental slope. On wider shelves, barrier reefs reaching the surface occur in some places. They form platforms at about mean sea level, enclosing large lagoons about 40 m deep between them and the nearshore fringing reefs. In other places coral knolls develop on the shelf and do not reach the surface. Reefs are more numerous on the Arabian side. In the southern part of the Red Sea the shelf is very wide on both sides. Numerous islands, islets, and rocks form the Dahlak Archipelago on the Ethiopian side and the Farasan Islands off the Arabian coast. In the Gulf of Elat in the north, the shelf is often very narrow and sometimes even altogether nonexistent, with the steep slopes of mountains descending directly to the depths of the Gulf.

Shores covered by unconsolidated substrates also vary considerably, but, as their topography is much simpler than that of rocky shores, they embrace fewer habitats and consequently harbour fewer communities. Nevertheless, some of the most productive communities occur in these situations.

Sands are the most common substrates within the intertidal zone of the Red Sea. They may be of different grain sizes, with coarser sands on the more exposed beaches. In some places the sand is covered with pebbles. The beach may be covered by an arkose sand with a mixture of unrounded grains of many minerals coming from nearby granites. In other places it is covered by a sand with better rounded grains of fewer, partially weathered minerals, like that with the golden glitter covering the beach at Dahab ("gold" in Arabic) in Sinai, which contains much vermiculite. Still other beaches may be covered by sand with well-rounded quartz grains or ooliths. In sheltered coves clay may occur on the beach. In the shallow subtidal the most common sediment is white calcareous sand composed of tiny fragments of corals, coralline algae, shells of invertebrates, foraminifers, and others. Off wadi deltas, terrigenous sediments occur in the entire depth range discussed here, with finer grain sizes farther offshore.

Biological study of the Red Sea

Red Sea biota have been studied for the last two centuries. The history of marine algal, seagrass, and zoological research in this sea has been reviewed by Papenfuss (1968), Lipkin (1975a), and Mergner (1984). In earlier years the studies were carried out by visiting naturalists and scientists, mostly from Europe, who usually visited the area for short periods. This early period was characterized, by and large, by the collection of specimens and subsequent structural studies of the organisms collected. With a few exceptions, the accounts published during this period were of a floristic and faunistic nature, sometimes including a few remarks of ecological significance (but see Kluzinger, 1872).

In the 1930s, with the establishment of the Marine Research Station at Al Ǵhurdaqah (Egypt), a slow change started to take place. More and more resident scientists from countries bordering the Red Sea took part in the biological research, and the research itself involved increased ecological aspects, including a continuously growing proportion of experimental field and laboratory work. Most of the research of this kind has been carried out in the northern part of the Red Sea because most marine research laboratories are located here (for details, see Mergner, 1984). Since the establishment of the H. Steinitz Marine Biological Laboratory at Elat in 1968, a period of intensive research started, centred mainly on the Gulf of Elat (Gulf of 'Aqaba); this resulted in many papers, including two books (Reiss and Hottinger, 1984; Friedman and Krumbein, 1985) summarizing various aspects of this research. Only relatively recently have quantitative data on different communities started to accumulate, much of them on the coral reefs.

The origin of the Red Sea biota is essentially from the Indo-West Pacific. Only after the Suez Canal had been opened could considerable exchange of species between the Red Sea and the Mediterranean take place through migration, resulting in the occurrence of Atlanto–Mediterranean species in the Red Sea. The migration of organisms from one basin to the other has been by no means symmetrical, that of Red Sea organisms into the Mediterranean being by far the more extensive (Steinitz, 1967; Por, 1978). The diversity of taxa in the Red Sea decreases from south to north. Many tropical genera and species present in the southern part do not reach the northern end; for instance, of 56 Indian Ocean Cypraeidae, 38 were reported from the Gulf of Aden, 32 from the southern Red Sea, 26 from its central part and only 21 from the north. A large group of tropical organisms is restricted to the south, including the benthic algae *Caloglossa* and *Chlorodesmis*, plus the mangrove *Ceriops*, and the gobies *Glanogobius*, *Glossogobius*, and *Gnatholepis*, along with the green algae *Caulerpa ambigua*, *C. fastigiata*, and *C. selago* and the fiddler crabs, *Uca lactea albimana*, and *U. vocans hesperiae* (Lipkin, 1968, 1974; Foin, 1972; Lewinsohn, 1977b; Goren, 1979).

A variety of habitats is occupied by animal and plant life along the Red Sea coasts and in its shallows. In this chapter benthic communities of the intertidal and upper subtidal to *c.* 30 m water depth will be described, with the exception of those of coral reefs and mangal stands. Quantitative data, when available, will be emphasized. Coral reefs and mangals are subjects of separate volumes in this series (Dubinsky, 1990; Chapman, 1977) and will not be described here. Coral reefs are narrowly interpreted as including those parts of the marine benthic ecosystem where actively growing hermatypic corals dominate the scene, forming actual reefs. In the Red Sea, corals often occur on fossil or subfossil formations, which are much more widely distributed than the actual reefs. They grow either on the seaward side of fossil reef structures of fringing reefs along the shores, or around offshore barrier reefs. It should, however, be borne in mind that the reef and the mangal are inseparable parts of the intertidal and shallow submerged system interacting with other parts in many ways; therefore, reference to them and to their role in the ecosystem will sometimes be unavoidable.

LIFE ON ROCKY SHORES

Important factors determining the nature of a community at a certain niche on the rocky shore are the level at which it is located in relation to sea level, degree of exposure to water movement, and, in the subtidal, amount and nature of the light reaching it. Location and topography of the rocky surface on which organisms live determine these and associated factors.

Intertidal zonation: supralittoral and eulittoral

The system for the subdivision of the intertidal, based on physical and biological boundaries between the zones, is less adequate for describing the ecological distribution of communities than the so-called "universal" one proposed by Stephenson and Stephenson (1949, 1972). This system is based on eco-biodistributional criteria, which has been used in most descriptions of the intertidal zonation of the Red Sea. The former system is used here for purposes of uniformity in this Volume.

The upper belts of the intertidal, the so-called black zone of the supralittoral, and the barnacle belt of the uppermost eulittoral zone appear on vertical or nearly vertical walls, on vertical or nearly vertical sides of blocks, or on the landward parts of beach-rock slabs located high on the beach. Platforms, as a rule, are located at mid- to lower eulittoral levels, harbouring algal-dominated communities. Only occasionally, when rocky projections (pebbles or boulders) occur on the platform surfaces, do barnacles cover the portions protruding into the upper eulittoral. Taller platforms that reach supralittoral levels even harbour blue-green algae and periwinkles.

The supralittoral zone

The supralittoral zone roughly corresponds to the so-called black or dark zone. In the Red Sea it is characterized, as on other coasts world-wide, by a belt dominated by littorinid snails and blue-green algae (Cyanobacteria of many recent works). The latter usually occur endolithic in the upper parts and epilithic in the lower, especially in humid crevices. The zone is most conspicuous on light-coloured rocks (limestones and, to a lesser extent, sandstones) and especially on vertical walls (cf. Wainwright, 1965, figs. 2, 3). Usually it does not include any other belt. In the northern Gulf of Elat, on seaward-dipping beach-rock slabs, the littorinids–blue-green algae belt is characterized by the periwinkles *Nodolittorina subnodosa* and *N. millegrana* (only the former in the Gulf of Suez), which feed on endolithic blue-greens, and by *Planaxis sulcatus*. *Nodolittorina millegrana* and *P. sulcatus* increase in abundance with exposure to wave action. The herbivorous *P. sulcatus* migrates up and down, following the fluctuations of the tide. On shaded vertical surfaces at this level the much less common *Littorina scabra* and *Nerita undata* occur. The isopod *Ligia* is sometimes also very common at this level, *Ligia exotica* being reported from the southern Red Sea. On vertical walls and undercuts in the fossil reef a distinct dark belt appears along the entire Red Sea, with the endolithic blue-greens and usually a few littorinids — extremely few in sheltered areas of the Dahlak Archipelago (Safriel and Lipkin, 1964; Lewinsohn and Fishelson, 1968; Ayal and Safriel, 1980; Lipkin, 1987c).

The eulittoral zone, which comes below, usually contains a considerable number of communities arranged in distinct belts on vertical surfaces and inclined beach-rock slabs. Indistinct belts occur on the platforms, where they are often arranged in a mosaic pattern as a result of patchiness of ecological niches.

Upper eulittoral zone

A barnacle belt occurs below the dark belt (in the upper eulittoral zone) on rocks exposed to enough wave action to support them. Four barnacles occur here and are listed in their order of decreasing dependence on wave action: *Tetraclita squamosa rufotincta*, *Tetrachthamalus oblitteratus*, *Chthamalus barnesi*, and *Balanus amphitrite*. *Tetraclita* and *Tetrachthamalus* compete in the upper eulittoral. The latter has a wider vertical range, whereas the former is favoured by wave exposure. Zonation between the two, as described by Safriel and Lipkin (1964), occurs only under exposure conditions favourable for *Tetraclita*. Achituv (1981) provided evidence that under such conditions *T. squamosa* competitively excludes *T. oblitteratus* from the central part of the latter's vertical range of distribution. This part eventually becomes the *Tetraclita* zone. Density of *T. squamosa* on beach-rock ranges between nil to *c.* 8500 m^{-2}, according to the degree of wave exposure. On vertical rocks and south-facing surfaces it may reach *c.* 25 000 m^{-2}. This belt is richer in life than the former. A variety of snails accompanies the barnacles, such as *Monodonta canilifera*, which prefers to stay above water level and migrates vertically with tidal fluctuations. *Thais savignyi* hides during low water in cavelets and crevices. It becomes active when submerged by incoming tides, when it bores into *T. squamosa* shells and feeds on its soft parts, as does also *Morula granulata*. The herbivorous limpets *Cellana eucosmia* (their density according to one report is 55–58 m^{-2}) and *Siphonaria kurracheensis*, the fuzzy chiton *Acanthopleura haddoni*, and the prosobranch *Nerita sanguinolenta* are also active during high water, when they feed on algae. Other snails in this belt are *Clypeomorus moniliferus*,, *Drupa morum morum*, *D. ricinus*, and *Engina mendicaria*. Bivalves occur here as well, such as *Chama rueppelli*. The goose barnacle *Ibla cumingi* occurs abundantly among the much larger *Tetraclita*, and the sea anemone *Anthopleura elatensis*, characteristic of lower levels, is often found in empty shells of *Tetraclita*. In crevices there appears the mussel *Brachidontes variabilis*, which in sheltered conditions in the Gulf of Elat may cover portions of slabs, platforms, or solution basins at a high density (to 3000 m^{-2}). *Ibla cumingi* is concentrated mainly in the central part of the *Tetraclita* belt, where its density is around 2000 m^{-2}, with local concentrations up to 7500 m^{-2}. It lives in close association with *T. squamosa*, protected from predation among its shells. In the lower part of the *Tetraclita* belt, small brown patches of the encrusting brown alga *Petroderma maculiforme* join the blackish to dark olive-green specks of encrusting blue-greens on beach-rock and shells of barnacles. In this lower part larger algae appear in winter, like *Enteromorpha clathrata*, *E. compressa*, *Porphyra umbilicalis*, and *Ulva lactuca*. The bivalve *Saccostrea cucculata* also occurs at this level. On undercut surfaces it forms dense porous masses. In the cavities of these surfaces, *I. cumingi* individuals concentrate in densities twice those in the *Tetraclita* belt. The blennies *Alticus kirki magnusi*, *Istiblennius rivulatus*, and *Salarias fasciatus* often occur on wave-washed slabs in shallow basins, or even on shaded vertical walls at this level. *Octopus*, most often *O. cyaneus*, but also *O. aegina* and *O. macropus*, are very active at these levels during the night, hiding in crevices below during the day (Safriel and Lipkin, 1964; Safriel, 1969; Achituv, 1972; Ayal and Safriel, 1980; Safriel et al., 1980a, b; Achituv and Klepal, 1981; Fishelson, 1983; Hullings, 1985).

In the inner parts of the great archipelagos of the southern Red Sea, the Dahlak Archipelago, and the Farasan Islands, exposure conditions differ markedly from those in the Gulf of Elat. Located on a very wide, shallow shelf, the shores of the inner islands in both groups are highly sheltered from wave action. This results in an almost complete absence of barnacles in the intertidal and a scarcity of living coral reefs in the subtidal, although solitary colonies and small groups of hermatypic corals do exist. On vertical surfaces of highly sheltered coasts, instead of barnacles, a narrow belt of a thin coralline red alga covers the rock surface right below the dark belt. Below, a *Cladophoropsis membranacea* belt appears (with the green alga sparsely scattered in pits 2 to 4 cm in diameter) on the vertical surface, which is otherwise covered by endolithic and epilithic bluegreens. A few limpets and *Nerita* also occur. Somewhat lower, the wall flattens into a sub-horizontal rock-flat and another belt occurs. It is occupied by an *Enteromorpha compressa* community, with *E. compressa*, *Cladophora lehmanniana* plus *Chondria dasyphylla* occurring in basins on the flat surface at the foot of the cliff (Wainwright, 1965; Klaueswitz, 1967; Lipkin, 1987c).

Mid-eulittoral zone

Below the *Enteromorpha compressa* belt two belts dominated by algae (*Chondria dasyphylla* and Laurencia papillosa) cover much of the flat at Dahlak. From similar flats on the opposite (Arabian) coast, in the Farasan Islands, the following gastropods were reported: *Cerithium erythraeonense*, *Conus textile*, *Cypraea turdus*, *Latirus polygonus*, *Strombus fasciatus*, and *Tectus noduliferus*. Also present were the bivalve *Chama rueppelli*, the polychaete *Galeodes paradisiaca*, the blennies *Antennablennius hypenetes*, *Istiblennius edentulus* and I. flaviumbrinus, and the gobies *Acentrogobius meteori* and *A. ornatus*. Some of these species were most abundant in the lower eulittoral. On slabs in Sinai a belt of turf-forming algae occurs below the barnacle. In this belt *Sphacelaria tribuloides* often dominates, but *Centroceras clavulatum*, *Gelidium pusillum*, *Herposiphonia tenella* and other similarly tiny algae are important and may even dominate in some sites. A few large algae like *Laurencia papillosa*, *Padina gymnospora*, or *Spyridia filamentosa* may also occur. *Chama rueppelli* is also an important component in this belt, as well as the gastropods *Cerithium* cf. *caeruleum*, *C. columna*, and *Thericium scabridum*. The mussel *Modiolus auriculatus* is very common in sheltered locations where it may be covered with sediment almost to the tips. In such habitats it may reach densities of *c.* 800 m^{-2}. A number of organisms common in higher and lower belts occurs as well. Some species that reach their uppermost vertical limits here are the polychaete *Branchiomma luculana*, the small abalone *Sanhaliotis* cf. *pustulata*, small individuals of the giant clam *Tridacna maxima*, the sea urchin *Echinometra mathaei*, and the common dromiid crabs *Cryptodromia canaliculata* and *C. granulata*. To a much lesser extent, *C. hilgendorfi* and *Dromidia unidentata* occur plus the common grapsids *Grapsus albolineatus* and *G. granulosus*, which migrate with tidal fluctuations so as to remain mostly at the water level and just above. Less common is *Plagusia tuberculata* (Safriel and Lipkin, 1964; Klausewitz, 1967; Holthuis, 1977; Lewinsohn, 1977a; Ayal and Safriel 1982b; Lipkin, 1987c).

The lower eulittoral zone

The lower eulittoral levels less commonly become exposed to air. On slabs in the Gulf of Elat large algae dominate, such as the common *Caulerpa serrulata*, *Codium dwarkense*, *Cystoseira myrica*, *Dictyosphaeria cavernosa*, *Digenea simplex*, *Gelidiella acerosa*, *Halimeda discoidea*, *Laurencia papillosa*, *Sargassum dentifolium*, *Spyridia filamentosa*, *Turbinaria elatensis* and *Ventricaria ventricosa*, and the less common *Caulerpa peltata*, *Lobophora variegata*, *Neomeris annulata*, *Polyphysa parvula*, and *Udotea argentea*. During the winter and early spring, *Colpomenia sinuosa*, *Hydroclathrus clathratus* or *Stypopodium zonale* occur as well. Amongst algae and in crevices in the belt the following animals are common: *Modiolus auriculatus*, the vermetid gastropod *Serpulorbis inopertus*, the cerithiid gastropod *Clypeomorus petrosa gennesi*, the brittle star *Ophiocoma scolopendrina*, the octocorallinio *Xenia umbellata*, and the sea urchins *Diadema setosum* and *Tripneustes gratilla*. Small, isolated colonies of hermatypic corals also appear, commonest among them is the pinkish *Stylophora pistillata*. In the thickets of the algae, especially in shaded locations where sediments accumulate around thalli, a rich fauna occurs, especially in the sediment-loaded algal masses. It includes amphipods, isopods, polychaetes, ostracods, copepods, kinorhynchs, and, in the larger cavities, also hydroids, calcareous sponges, or larvae of opisthobranchs. In such masses 35 polychaete and 23 ostracod species were found, compared with only 5 and 2 to 4, respectively, in algal masses without sediments. Some of the many inhabitants of such masses are the sponge *Sycandra*, the copepod *Pseudocyclops gohari*, the amphipod *Elasmopus steinitzi*, and the mites *Litarachna denhami* and *Pontarachna punctulum*. At lower levels the brown alga *Turbinaria elatensis* may dominate in the north, with scattered vermetids, *Dendropoma maxima*, faithfully accompanying it. *Cerithium echinatum* and *Sargassum* spp., as well as scattered hermatypic corals of several species, also occur (Safriel and Lipkin, 1964; Remane and Schulz, 1964; Safriel et al., 1980b; Ayal and Safriel, 1982b).

On wide platforms in the Gulf of Elat differences in heights of rocky surfaces are slight. They are, however, very important as exposure regimes in this harsh environment vary markedly on surfaces, even with only a few cm difference in height. Such small differences in level therefore result in different communities. Another important factor determining the distribution of parts of the communities on the platforms is the distance of a site from the seaward edge of a platform, which determines its exposure to waves. As a result, communities are arranged in belts parallel to the platform margins and subject to modifications due to topographical details of the platform. The part of the platform closest to the shoreline holds fine sediment in its depressions and is often covered by diatom films and blue-green algae, with a few snails feeding on them. On rocky exposures hardly any growth of algae can be seen other than microscopic forms — mostly blue-greens. They harbour a few snails, such as *Cerithium* cf. *caeruleum*, *C. columna*, and *Thericium scabridium*, the chiton *Acanthopleura haddoni*, the crab *Metopograpsus messor*, and little else. To the seaward side of this belt another habitat covered by a *Digenea simplex* community occurs, with *Dictyosphaeria cavernosa* and *Modiolus auriculatus* as important components. Conspicuous companions are the octocorals *Clavularia hamra*, *Sinularia compressa*, or *S. polydactyla* and, in crevices, *Tubipora musica*. Greatest densities of cerithiids in this community were $131 \pm 102\ m^{-2}$ *Clypeomorus petrosa gennesi*, $73 \pm 30\ m^{-2}$ *T. scabridum*, and $65 \pm 35\ m^{-2}$ *C.* cf. *caeruleum*. Usual densities of all were 30 to 60 m^{-2}. Small shallow depressions in this belt hold water during ebb tide. In these depressions *T. scabridium* reached $250 \pm 38\ m^{-2}$ (in the Gulf of Suez) and *C.* cf. *caeruleum* $300 \pm 180\ m^{-2}$. Ayal and Safriel (1982a, b) suggested that the capabilities of different species to hide from predatory fishes coming from nearby reefs determine their spatial distribution in the different intertidal niches of the platform. A thin film of sediment usually covers plants in this area. The bottoms of the depressions are covered with a thin layer of coralligenous sand, sometimes with the seagrass *Halodule uninervis* or, in larger ones, *Halophila stipulacea*. The rock around the perimeter of the depressions supports richer algal vegetation, outstanding in which are clumps of light green, clavate vesicles of *Boergesenia forbesii*. In crevices in this area *Ophiocoma scolopendrina* hide, coming out by the thousands when the tide goes out; they turn partially upside down combing the surface film of the water with their arms for floating microorganisms. Especially abundant among the latter are detached epiphytic diatoms. They float in small masses of slime buoyed by tiny oxygen bubbles (resulting from their photosynthetic activity) that are carried away by the outgoing tide. Such depressions, when about 15 cm deep, harbour the sea urchin *Echinometra mathaei* in considerable numbers. Deeper, larger depressions, *c.* 40 cm deep, are inhabited by the black, long-spined urchin *Diadema setosum*, that grazes much more effectively than *E. mathaei*, resulting in destruction of the vegetation in the neighborhood.

In the northern Red Sea *E. mathaei* often lives symbiotically with the alpheid shrimp *Athanas indicus*. One or two shrimps, and often more, are found on large urchins (> 15 mm test diameter) among spines near the mouth where they are protected from predators. The shrimp is specific to the urchin and is not found on other urchin species. In the northern Red Sea region the urchin is sometimes also associated with tiny gastropods. Pairs of *Monogamus entopodia* live in the ambulacra around the mouth, while *Robillardia cernica* lives in small depressions on the body. The small cardinalfish *Siphamia permutata*, young of *Apogon* (*Nectamia*) *cyanosoma* and *Cheilodipterus bipunctatus*, and the tiny squid (2 to 5 cm long) *Sepia gibba*, swim protected among the long spines of *Diadema setosum*. The squid only occurs during the night, while during the day it hides under the clusters of the urchins. When the latter move to feed at night, the fish and squid move with them. The urchin also harbours a few taeniacanthid copepods. Of these, *Echinosocius pectinatus* is common around the mouth, probably feeding on food brought in by the host. *Echinosocius elatensis* is common around the anus, probably feeding on wastes, while *Echinirus diadematis* is so rare that its biology is unknown. All three are highly specific to their host (Klausewitz, 1966; Karplus and Masry, 1972; Jacob-Judah, 1976; Lützen, 1976; Pomeranz and Tsurnamal, 1976; Fishelson, 1983).

Towards the sea a wide area is occoupied by a *Cystoseira myrica* community. Here water energy is strong enough to prevent settlement of sediment on plants but is still too low to support a luxuriant algal growth. The grapsid crab *Pachygrapsus minutus* is abundant in this belt, hiding in crevices during high water and climbing on boulders on the platform during low tides. Diversity is still rather low, and fronds of the dominant species are dwarfed. Octocorals, such as *Clavularia hamra*, *Lobophytum pauciflorum*, and *Parerythropodium fulvum*, are common. This community in turn is replaced by a *Laurencia papillosa* community accompanied by *Padina gymnospora*. The latter stretches to a few metres from the seaward edge of the platform.

Next come several relatively narrow belts, most

commonly a *Dictyota* belt and a *Gelidiella acerosa* belt. The few metres of platform closest to the seaward edge harbour either a coral reef or a highly diverse and luxuriant growth of algae, both on the surface and in holes and cavelets in the rock. Dominant on the surface are species of *Laurencia* containing iodo- and bromophenols, which protect them from the many grazing fishes that are very active on this part of the platform during high water. The fish roam in mixed schools containing especially surgeonfishes (*Acanthurus nigrofuscus*, *A. sohal*, *Ctenochaetus striatus*, *Zebrasoma veliferum* and *Z. xanthurum*) but also triggerfishes (*Balistapus undulatus* and *Pseudobalistes fuscus*). The unicorn-fish *Naso lituratus* and others that feed on fleshy algae also occur, plus parrotfish (*Scarus gibbus*, *S. niger*, or *S. sordidus*) that bite into rock surfaces, taking chunks of living tissue from encrusting calcareous red algae and other organisms. Density of grazing fishes was greatest at the seaward edge of the platform and on knolls in front of it (their densities in different habitats are indicated below, p. 418). At the edge of the platform in wave-beaten sites the spiny lobster *Panulirus penicillatus* is active at night, feeding on various invertebrates, mainly slow creatures like molluscs, sea urchins, and invalid fish. During the day it hides in crevices in the subtidal. This part of the platform also has an abundance of the red alga *Champia irregularis*, painting the platform with patches of fluorescent light-blue colour when the sun is low.

On the Sinai coast (northern Gulf of Elat) platforms are so narrow that even their landward parts are subjected to some influence of wave action and are thus devoid of fine sediments. Higher levels are occupied by a *Sphacelaria tribuloides* community, whereas the lower parts have a *Turbinaria elatensis* community. The former is characterized by an almost continuous cover of a short-haired algal carpet, the latter by the sparse vegetation of robust brown algae, 20 to 30 cm tall, all moving back and forth as one with the waves. Towards the seaward edge, reef organisms become more common, including small colonies of hexacorals. The margin itself and the vertical wall beyond are occupied, as in the wide platforms, by an actual coral reef. Similar platforms on the Sinai coast of the Gulf of Suez support a much richer vegetation at all levels. This may be attributed in part to the smaller population of grazing fish (due to the scarcity of reefs which in other parts of the Red Sea provide shelter) and to the greater exposure of these platforms to wave action, that limits the duration of grazing.

Seaward edges of narrow platforms or beach-rock slabs in the southern Gulf of Elat are exposed to extensive wave action and are fringed by an actively growing hard and porous biogenic crust of the vermetid *Dendropoma* sp. (a new undescribed species close to *Dendropoma meroclista* see Safriel and Hadfield, 1988). Dominant among the infauna in this crust were polychaetes. Of these, *c.* 54% were carnivorces, *c.* 30% detritivores, *c.* 16% omnivores and less than 0.5% herbivores (Ben Eliahu, 1975; Safriel and Lipkin, 1975; Ayal and Safriel, 1980; Ben Eliahu and Safriel, 1982).

The sublittoral zone

Rocks in the shallow subtidal are usually occupied by reef-building corals and associates of the coral reef. They are easily accessible by grazers and therefore usually devoid of vegetation other than a few species (*Asparagopsis taxiformis*, and *Dasya baillouviana*) containing repellent substances, or the stone-hard crustose coralline reds. On the wide shallow banks of the southern Red Sea coral-reef development is rather poor, although reef-building corals occur, usually as meagre, sparse growths. A rich growth of sponges may take their place, accompanied by a rich infauna and epifauna composed mainly of crustaceans (cirripeds and decapods), polychaetes, and molluscs. The commonest sponges are *Cinachyra alba tridens*, *Fasciospongia cavernosa*, *Heteronema erecta*, *Spirastrella inconstans*, and *Tethya seychellensis*. Fishelson (1966) reported 35 species associated with three individuals of the greenish brown *S. inconstans* in the Dahlak Archipelago. Most prominent among them was *Balanus longirostrum*, 250 individuals of which inhabited a single sponge with a surface area of *c.* 500 cm^2.

Petrolisthes leptocheles and the tiny ophiurid *Ophiactis savignyi* were also rather common, with 60 and 380 individuals, respectively, in a sponge head. An apogonid fish, *Apogon* (*Nectamia*) *spongicolus*, was described from this habitat. In a similar habitat on the central Arabian coast, Weitlauf

(1978b) found the cowries *Cypraea kieneri* and *C. walkeri*, and some crustaceans.

On rock flats in sheltered areas around the islands of the Dahlak Archipelago, a *Laurencia papillosa* community occurred in the uppermost levels to 50 cm deep at low water. A *Caulerpa racemosa* community occupied the levels 50 to 100 cm deep and was replaced by a *C. racemosa-Halimeda opuntia* variant at that depth. In deeper water a *Sargassum forsskalii* community and a *Turbinaria triquetra* community appeared. Individuals of the giant clam *Tridacna* were scattered in these communities. Interesting cyclopoid copepods live in their pallial cavities — *Anthessius amicalis* in *Tridacna maxima* and *A. alatus* in *T. noae*. Humes and Stock (1965) also found the latter copepod on the starfish *Acanthaster planci* and the orange sponge *Acanthella aurantiaca*, occurrences they regarded as accidental. Colonies of octocorals, among them *Xenia blumi*, *X. hicksoni*, and *X. umbellata* also occurred; in the latter the pipefish *Siokunichthys bentuviai* hides during the day. At the other end of the Red Sea, in the Gulf of Suez, coral-reefs are also limited and other communities take their place. On rock flats or bottoms covered with pebbles and boulders, communities dominated by the large brown algae of the Red Sea occupy wide patches. *Cystoseira trinodis*, *Hormophysa cunaeiformis*, *Turbinaria triquetra*, and a few species of *Sagassum* are the commonest community builders at these depths (Humes and Stock, 1965; Levi, 1965; Wainwright, 1965; Clark, 1966; Lewinsohn and Fishelson, 1968; Lipkin, 1987b, c).

In crevices, cavelets, or hollows under submerged rocks throughout the Red Sea a rich sciaphilic algal community develops where grazing allows. Among the commonest species are *Boodlea composita*, *Botryocladia skottsbergii*, *Bryopsis plumosa*, *Caulerpa lentillifera*, *C. peltata*, *C. webbiana*, *Codium arabicum*, *Dictyopteris delicatula*, *Gracilaria foliifera*, *Haloplegma duperrei*, and *Valonia macrophysa*. Among the less common, one finds *Martensia elegans*, *Rhipiliopsis aegyptiaca*, *Struvea anastomosans*, and the epiphytic *Gymnothamnion elegans*, *Taenioma perpusillum*, or *Tolypiocladia glomerulata*. Some species, such as *Coelothrix irregularis*, *Gelidiopsis acrocarpa*, *G. capitata* and *Sarconema scinaioides* were found in this community only in the south (Lipkin, 1987b, c).

LIFE ON BEACHES WITH UNCONSOLIDATED SUBSTRATES

The displacement of sediment particles by waves continuously beating the beach causes severe difficulties to potential inhabitants of the intertidal and shallow submerged zones on unconsolidated bottoms. In such conditions sessile organisms may be buried by deposited sediment or lose their footing by its removal. As a result, a basic difference between the intertidal of rocky shores and that of the unconsolidated sediments is that most organisms of the latter are mobile, whereas many of those of the former are sessile. Many inhabitants of the intertidal of soft bottoms keep moving with the fluctuations of the tide trying to hold the same position in relation to the changing water level. Only when the beach is rather sheltered from wave action, with resulting relative stability of the substrate, do plants and sessile invertebrates develop on these bottoms. Under such circumstances, although zonation occurs on the soft bottoms, the vertical division of the intertidal zone between different communities is much less defined than on rocks. The different communities on such coasts tend to divide the beach between them horizontally rather than vertically, mostly according to the nature of substrate and exposure to wave action. Another feature of unconsolidated bottoms is that many of their occupants live within the sediments (either completely underneath the surface or in burrows) in order to escape the harsh conditions of the intertidal zone and extensive predation in the subtidal.

Intertidal zone: supralittoral and eulittoral

Supralittoral zone

The uppermost part of the intertidal zone on sandy beaches is characterized by the very common ghost crab *Ocypode saratan* and the land hermit-crab *Coenobita scaevola*. In undisturbed areas the males of the former may often be encountered on the mounds they build (Fig. 16.1), sometimes rather close to one another and in front of their burrows that reach to the subsand water level. In the burrows they spend the hot hours of the day. In the same habitat the smaller and much less common *Ocypode cordimanus* also occurs. *Coenobita scaevola* may be active during the day or

Fig. 16.1. Mounds of male ghost crab *Ocypode saratan* on a low sandbar separating a lagoon from the open sea (El Bilaiyim, Sinai coast of the Gulf of Suez).

night, probably depending on the season; they may be observed in large, dense crowds scavenging just above the waterline. When not active they either dig in the moist sand to hide a few centimetres below the surface or invade rocky shores from nearby sandy stretches and aggregate in cavelets and other shaded niches, sometimes in very great numbers. They may wander hundreds of metres from the coastline, especially on flat beaches. The gills of *C. scaevola* are often infected by the parasitic mite *Andregamasus steinitzi*, probably throughout the Red Sea. In the Gulf of Elat another mite, *Askenasia sinusarabicus*, was found on the same host. All three crabs are scavengers and are active mostly at night; they are present all around the Red Sea and in the Gulf of Aden (Costa, 1965, 1972; Lewinsohn, 1969, 1977b, 1983).

Above the subsand water level interstitial communities inhabit wide sectors of the sandy beaches. The most characteristic group in this habitat is the harpacticoid copepod, both with respect to the number of species and individuals, although most other marine invertebrate groups are also represented. The interstitial habitat in the Red Sea, in contrast with the situation on other beaches studied, is rather stable. It is continuously hypersaline, warm throughout the year, and with very little (if any) continental water flowing through it. Well-sorted sand of medium-size grains (0.25–0.5 mm) is preferred by most organisms and contains the richest communities. In sand harpacticoid copepods dominate, whereas in mud nematodes are most prominent. The upper layer (0 to 25 cm) is influenced by air conditions and is therefore warmest, saltiest, and driest. Here the terrestrial component is most abundant; the mite *Rhodacaroides aegyptiacus* is a representative. The marine component also includes oligochaetes (*Enchytraeus* sp., *Michaelsena* sp., *Rhyacodrilus* sp., and *Tubifex* cf. *costatus*), nematodes, turbellarians, tardigrades, copepods (such as *Kliopsyllus* sp.), ostracods, and gastrotrichs. The lowest layer (35 to 50 cm) below the surface is closest to the subsand water and is influenced by conditions in the water; it fluctuates with a considerable lag behind the tidal fluctuations. The intermediate layer, 25 to 35 cm below the surface, is the most stable, with an annual range of temperatures of less than 5°C and of salinity less than 20‰. This layer is therefore the richest in fauna. Among others, it harbours a yet undescribed species of the copepod *Arenopontia*. Experiments with *Kliopsyllus* and *Arenopontia* from the Gulf of Elat showed that the former was more resistant than the latter to higher temperatures and salinities. Optimal ranges were 25 to 30°C and 38 to 49‰ for the former and 20 to 25°C and 34 to 40‰ for the latter (Remane and Schulz, 1964; Darom, 1974).

Upper and central eulittoral zone

In the upper part of the swash region all around the Red Sea a *Hippa* community occurs on sandy beaches that are relatively exposed to wave action and have steep slopes of around 12 to 20%. These beaches are often located on deltas at wadi mouths. The community from the northern Gulf of Elat has been described as containing *Hippa celaeno* and *H. picta* intermingled in about equal numbers. These small crustaceans penetrate at least 5 cm into the sand and follow the rising and receding tides, with the younger individuals staying closer to the waterline. They dig deepest in sand of medium-sized grains but are also found in sand of different grain sizes. *Hippa picta* seems to prefer a sediment with larger particles, which is not very well sorted, whereas *H. celaeno* seems indifferent to sorting level. The crustaceans exhibit a wide variability in colour, with the colour of the population matching that of the substrate. Animals living in white coralligenous sand are white, whereas those living in variegated granitic sand show a range of different patterns and combinations of white, red and black. The two species are taken by predatory fishes and birds and are affected by periodic floods rushing through their habitat, causing mass mortalities. Both species are carnivorous and detritivorous, but *H. celaeno* feeds mostly on small bivalves. The same grounds harbour the clam *Atactodea glabrata* and the worm *Polygordius* sp. Other communities of the swash region occur on somewhat sheltered beaches. These include the *Mactra olorina* — *Cladophora rupestris* community and the *Impages hectica* — *Enteromorpha* community. In both, the molluscs are buried in the sand with only their shell tips protruding; green algae grow attached to them. The individual pairs are sparsely scattered *c.* 50 to 100 cm apart. These molluscs, as well as *A. glabrata*, follow the waterline and fluctuate with the tides (Por and Lerner-Segev, 1966; Klausewitz, 1967; Lewinsohn, 1969; Lipkin, 1973; Turko, 1976).

On the sheltered coast of the El Kura lagoon in Sinai a peculiar oyster reef develops at the upper eulittoral on the sand flat. *Saccostrea cucculata* larvae seem to have attached here and there to empty shells or little stones, developing into adults on which later generations of larvae settle, eventually constructing a reef structure (Fig. 16.2) (Lipkin, unpubl.). Another attached mollusc, *Chama rueppelli*, also occurs nearby in the same habitat. These bivalves grow singly or in small clusters of 2 to 3 individuals (10 to 20 cm apart) and do not form reef structures. They are buried in the sandy mud and also attach to small shell fragments, stones, or dead *Chama* valves (Hottinger, 1972).

In relatively sheltered coves of the northern Red Sea brown patches of diatoms (e.g. *Pleurosigma*) develop in the eulittoral zone on gently sloping sandy beaches. Microfauna also live on and mostly within the sand. The microfauna is composed of ciliates, such as *Frontonia* sp., *Trachelocerca* sp., a black *Stentor* sp. or the euplotid *Discocephalus rotatorius*, the gastrotrichs *Aspidiophorus* sp. or *Macrodasys* sp., the kinorhynchan *Pycnophyes* sp., and polychaetes like *Armandia leptocirris*, *Fabricia acuseta*, *Fabriciola ghardaqa*, *Neanthes* sp., *Nereis* sp., *Oriopsis armandi*, *Pisionidens indica*, *Polyophthalmus pictus* or *Staurocephalus*. In sand grains of medium-size the nemertean *Ototyphlonemertes* occurs in the lower eulittoral. Conversely, in coarse sand the polychaete *Saccocirrus papillocercus* is replaced by *Goniadides aciculata* in the uppermost sublittoral (Remane and Schulz, 1964).

On beaches sheltered from severe wave action fine particles settle, and consequently the bottoms are clayey. These bottoms are more stable than sandy ones merely because of the weaker wave motion affecting the sediments; in addition, the sediment particles stick together more strongly. Here mudflats are formed, sometimes with underlying rock at various depths beneath the surface. During low tide wide areas of the eulittoral are exposed. Such areas are populated in the Red Sea and the Gulf of Aden by the small burrowing sand-crab *Dotilla sulcata*. It occupies a belt from the lowest waterline to 30 to 40 cm above. Its width depends on the slope of the beach. Density of the crabs in this community is usually greater in the south, 240 m^{-2} on average in Dahlak (170–420 m^{-2}), and smaller in the north, 165 m^{-2} being the average in Sinai (75–200 m^{-2}).

Fig. 16.2. A patch of oyster reef (*Saccostrea cucculata*) in the upper eulittoral zone on a sheltered sandy beach (El Kura' lagoon, Sinai coast of the Gulf of Elat).

The more sandy the bottom, the sparser the population. In the *D. sulcata* belt other crabs, such as *Ebalia abdominalis* and the hermit crab *Diogenes avarus*, occur in smaller numbers. On more sheltered beaches with a resulting higher proportion of clay in the sediment, species of the colourful fiddler crab *Uca* share in the community, becoming dominant in extremely sheltered habitats such as the mangal stands. The largest and most beautiful species in the region, *Uca tetragonon*, occurs throughout the Red Sea and the Gulf of Aden together with *Uca inversa*. The other two species in the basin, *U. lactea albimana* and *U. vocans hesperiae*, are restricted to the southern part, the former occurring in the Gulf of Aden as well. In these communities much organic matter is available for the scavengers. The crabs, which are the most prominent scavengers, hide during high water when the area is flooded and accessible to predators. When the tide is out and the bottom is exposed they clear the openings of their burrows and emerge by the hundreds, feeding on diatoms and other microscopic organisms covering the sediment particles. In this process they form the typical small sediment pellets that soon cover the surface around their burrows, giving the area a a peculiar appearance. In many colonies the whole surface is densely covered by pellets before the next tide comes in. *Dotilla* pellets are small, being 2 to 4 mm in diameter, whereas those of *Uca* are 5 to 8 mm in diameter, making it easy to assess the proportion of each in the community. In the sand of this habitat Fishelson (1984) found nematode densities of over 10 000 m^{-2}, somewhat fewer crustaceans (mostly harpacticoid copepods and ostracods), *c.* 1100 mm^{-2} small polychaetes, and more than 3000 m^{-2} foraminifers. All of these groups were represented in the stomach contents of *D. sulcata*. In general, organic matter in the sand decreases with distance from the low tide mark. In the Gulf of Elat there were 160 to 180 mg g^{-1} of organic matter in the sand at the lowest part of the *Dotilla* belt and only *c.* 60 mg g^{-1} in the uppermost part. *Dotilla sulcata* is taken by birds, such as the mangrove heron (*Egretta schistacea*) or large plovers (*Dromas*), which pick out the larger crabs (Fishelson, 1971, 1984; Lewinsohn, 1977b).

Lower eulittoral zone

On pebble beaches lower eulittoral levels are characterized by an algal cover on the upper surfaces of pebbles (usually small and turf-forming species) and by many animal species hiding under the pebbles. The commonest algae in this habitat are *Laurencia papillosa*, *Padina boryana*, *P. gymnospora*, *Sphacelaria tribuloides*, and *Spyridia filamentosa*. Porcellanid crabs are very common under stones in the northern Red Sea, particularly *Petrolisthes leptocheles*. Grapsid crabs, such as *Pseudograpsus elongatus*, also abound under stones in the Red Sea, while in muddy areas *Metopograpsus messor* is abundant. The boxer crab *Lybia leptochelis* commonly carries a sea anemone on each claw for protection — usually a small Triactis producta. The small abalone *Haliotis unilateralis* sometimes occurs, while the sea anemone *Anthopleura elatensis* and the gorgonarian *Acabaria pulchra* are very common. The common orange sponge *Tethya seychellensis*, plus the less common white sponge *Leuconia bathybia* and the polychaetes *Eurythoe complanata* and *Hermodice carunculata*, also occur under stones. The gastropod *Nerita polita* is also characteristic of such habitats; it prefers pebbles 5 to 10 cm in diameter, under which it hides during the day. The mean density of this species in the northern Gulf of Elat was 3.2 (0.8–4.4) m^{-2}. *Nerita sanguineolenta* also occurs on such beaches, but it is less common than on slabs (Safriel, 1969; Fishelson, 1970; Talmadge, 1971; Holthuis, 1977; Lewinsohn, 1977a; Tsurnamal, 1983).

Although typically subtidal, seagrasses start in the lower eulittoral. Until recently, nine seagrasses have been known from the Red Sea: *Cymodocea rotundata*, *C. serrulata*, *Enhalus acoroides*, *Halodule uninervis*, *Halophila ovalis*, *H. stipulacea*, *Syringodium isoetifolium*, *Thalassia hemprichii*, and *Thalassodendron ciliatum*. A tenth species, *Halophila ovata*, has recently been reported from the Arabian coast of the central Red Sea; it only occurs in the central and southern parts. In previous records from the Red Sea it has been called *H. ovalis*. The commonest species in the northern Red Sea are *H. uninervis* and *H. stipulacea*. These two also reach the highest level on the shore — the mid-eulittoral. To the south *H. stipulacea* is less common, and it is rather rare within the tropical region. Most seagrasses extend from the subtidal to the lower eulittoral levels, and some only extend to the lower limit of this zone. Most seagrasses usually form monospecific communities in the intertidal zone, as in the subtidal, but mixed populations are also present (Den Hartog, 1970; Lipkin, 1975a; Baeshin and Aleem, 1978; Aleem, 1979).

Intertidal seagrass communities usually occur on mudflats that are highly sheltered from wave action as well as in shallow, sediment-filled depressions on the leeward side of wide rocky platforms. However, *H. stipulacea* stands are also encountered on sandy, even gritty, beaches. The highest of all seagrass communities occurs within the mid-eulittoral zone and is represented by a very sparse community of filiform-leaved *H. uninervis*. The total standing crop was only about 10 to 50 g dry weight m^{-2}, with about 10% consisting of leaves. This community grows on a very shallow substrate (10 to 12 cm thick) deposited on fossil reef limestone. At the same levels on thicker sediment the community is much denser and has standing crops 5 to 10 times greater. The leaves here were also filiform. The former might represent ephemeral communities germinating and establishing each year anew, whereas the latter are more stable. In shallow depressions on the flats (*c.* 20 cm deep) dense populations of *Halodule uninervis* and *Halophila stipulacea* occur. In these depressions the bottom is never exposed to air, as tidal water is trapped during ebb tide. Nevertheless, the plants are subjected to considerable fluctuations in environmental conditions, such as temperature, concentrations of oxygen, carbon dioxide and other gases, plus pH. Leaves of *H. uninervis* in this habitat were 2 to 3 mm wide, similar to subtidal leaves. *Halophila stipulacea* leaves were shorter (*c.* 4 cm long) and narrower when compared with those from deeper levels. Their tips often get "burnt" as a result of prolonged exposures, yet the plant flowers abundantly and sets seeds normally. In the northern Red Sea the entire populations in this extreme habitat may in certain years be wiped out when the winter temperatures are too low. In spring the species is re-established from seeds. In small depressions the plants are often heavily covered with epiphytic algae: blue-greens, diatoms and macroalgae like *Centroceras*, *Chondria*, *Hypnea cornuta*, and *Spyridia filamentosa*. Many foraminifers also occur, mostly *Sorites orbiculus* on *Halodule uninervis* and Miliolidae on *Halophila stipulacea*. In thickets, amphipods like *Siphonoecetes erythraeus*, the gastrotrichan *Urodasys*, and polychaetes such as *Hesionides* sp. or *Mystides* sp. occurred, as well as many other animals. Grazing fish can only reach this habitat for a rather short period during high water and cannot crop off the entire production of epiphytes. In wider depressions farther from the shoreline local grazers of microalgae occur, such as the large strombs *Strombus tricornis* and the smaller *S. gibberulus albus*; *Tectus dentatus* also occurs near the edges close to rocky exposures. In these depressions the black sea-cucumber *Halodeima atra* is very common. In somewhat deeper depressions the spider shell *Lambis truncata sebae* dwells (Remane and Schulz, 1964; Lipkin, 1979; Reiss and Hottinger, 1984). Berner et al. (1986) reported symbiotic zooxanthellae inside heavy shells of living *Strombus tricornis*.

At lower eulittoral levels on the platforms (in the parts closer to deeper water of sheltered coves) a variety of intertidal seagrass communities occur, which are, in fact, the uppermost extensions of the subtidal populations. They include populations of all seagrasses and are very similar to the shallow subtidal communities.

The sublittoral zone

A variety of communities occupy the uppermost levels of the subtidal zone on unconsolidated substrates. Seagrass communities are the commonest feature of the soft bottoms in sheltered areas. In the upper subtidal they are usually dense. Exceptions are *Halophila ovalis*, which is almost always sparse. Sparse populations of other seagrasses occur towards the depth limit of these communities. Large areas of seagrass communities descend far below the levels discussed here — to at least 80 m. In the shallower beds a variety of animals find refuge among the leaves of the seagrass, and a diverse animal life occurs above the sediment surface and in its upper layers. Other areas harbour communities of invertebrates and fishes. In sites devoid of seagrass the bottom looks barren, with predatory fish roving above. Most of the indigenous fauna dig into the sediment seeking protection. Only occasionally a head or an upper part of the body of a creature protrudes from its burrow or, here and there, a mound of a larger burrower is encountered. Otherwise, only organisms extremely well camouflaged or those equipped with special means of deterring predators may permanently stay above the ground with a chance of surviving.

Sublittoral seagrass beds

Most seagrass communities appear in shallow water (0.5 to 5 m deep), except for *Halophila stipulacea* and *Thalassodendron ciliatum* that continue into deeper water. Some seagrasses, such as *Halodule uninervis*, occur with others in mixed populations. *Syringodium isoetifolium* almost always occurs with other seagrasses and only rarely forms pure stands. All seagrass beds harbour a relatively rich associated fauna and flora: a bottom infauna dominated by cirratulid and terebellid polychaetes, plus clams; epiphytic algae and sessile invertebrates on its leaves; and mobile animals on and amongst the seagrass. Some of these organisms may become very conspicuous and even share dominance of the community. Most of the species mentioned in similar habitats in the lower eulittoral zone also occur. Some additional subtidal species are the razor clam (*Pinna muricata*) that occurs almost completely buried in the sediment and the crinoid *Heterometra savignyii*, which dwells in coral reefs — like all other Red Sea crinoids. Many other animals normally inhabiting the reef occur occasionally, especially on scattered rubble. Even alcyonarians like *Clavularia hamra*, so typical of the coral reef, are among them. Prominent among the animals inhabiting seagrass beds are young stages of various reef dwellers or pelagic organisms, plus species typical of other habitats when adults. Schools of grazing fishes often nibble seagrasses, cleaning them of epiphytes.

Therefore, in seagrass habitats accessible to grazing fish the standing stock of epiphytic algal vegetation is usually very small.

In the northern Red Sea seagrass communities in which *Halophila stipulacea* or *Halodule uninervis* are either the only components or the dominant ones, cover large areas. Those based on the former species cover a wide ecological range from the low intertidal zone to the lowest limits of seagrass vegetation, whereas those based on *H. uninervis* range only between the low intertidal zone to about 5 m depth. To the south *H. uninervis* remains very common, but *H. stipulacea* becomes progressively more scarce. In the central Red Sea *H. stipulacea* is an uncommon species, remaining so in the southern part as well. These two seagrasses, although rather different from one another, have in common the same growth form. Thus, they form vegetation of a similar structure, comparable to most other seagrasses. In such vegetation the leaves emerge near the bottom, taking on an upright position that allows easy access by animals (including grazers on epiphytes) to most parts of the leaf surface and its epibiota (Aleem, 1979; Lipkin, 1979; Reiss and Hottinger, 1984).

In the shallow subtidal zone, as in the intertidal, *Halodule uninervis* and *Halophila stipulacea* communities are either composed of these two seagrasses alone or include others, such as *Cymodocea rotundata, Halophila ovalis*, or *Syringodium isoetifolium*. Below 3 m, *H. stipulacea* almost always forms pure populations, whereas *H. uninervis* rarely descends below 2 m. Lipkin (1979) found at Marsa abu Zabad (on the Sinai coast) that the vegetation in *H. stipulacea* beds was densest (total plant cover often 100% or nearly so) and standing crops were largest in shallow waters no deeper than 10 m. Both cover and standing crop decreased progressively with depth, although plant cover was still 60 to 70% at 20 m. Reiss and Hottinger (1984), however, reported maximum densities at depths of 15 to 25 m near Jezirat el Far'un, *c.* 150 km north-northeast. The water in the latter site is clearer than the embayment at Marsa abu Zabad, which may be the reason for this difference. Standing crops in beds of *H. stipulacea* and *H. uninervis* in the northern Red Sea quickly increase with depth to a maximum at *c.* 2 m; they then gradually decrease. The largest standing crop of the former plant was 550 g dry weight m^{-2}, while the total for all seagrasses in this community was 800 g dry weight m^{-2} (Fig. 16.3). In *H. uninervis* beds the standing crop was *c.* 1250 g dry weight m^{-2}. In beds of *Thalassia hemprichii* in Sinai, the standing crop was *c.* 2700 (950–4050) g dry weight m^{-2}, and in the central Red Sea (near Jeddah) it was about 950 g dry weight m^{-2}. In *Cymodocea rotundata* and *Syringodium isoetifoliurn* beds in Sinai the standing crop was 1150 and 1000 g dry weight m^{-2}, respectively. In the Gulf of Aden, pure beds of *Cymodocea serrulata* and mixed beds of *C. serrulata* and *Syringodium isoetifolium* (both having 100% plant cover) had standing crop of only 290 and 420 g dry weight m^{-2}, respectively (Hirth et al., 1973; Lipkin, 1977a, 1979; Aleem, 1979).

While standing crop in *Halophila stipulacea* beds in Sinai decreases with depth below 3 m (after an increase from sea level to that depth), other quantitative parameters of the community continue to increase with depth before decreasing. For example, the density of leaves reached a maximum of 15 200 m^{-2} around 10 m (Fig. 16.4). Plant biometric parameters also vary with depth. The most striking change was in leaf size. Leaf length increased continuously with depth without a decline; it almost doubled from 4.05±0.5 cm in the uppermost subtidal to 7.9±0.6 cm at 20 to 30 m

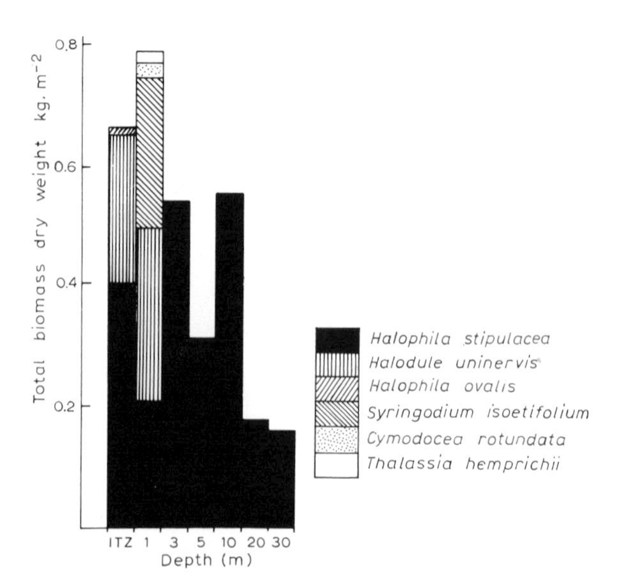

Fig. 16.3. Standing crop of seagrass communities dominated by *Halophila stipulacea* at E'Mteine (Marsa abu Zabad, Sinai coast of the Gulf of Elat) ITZ = Intertidal zone. (From Lipkin, 1979.)

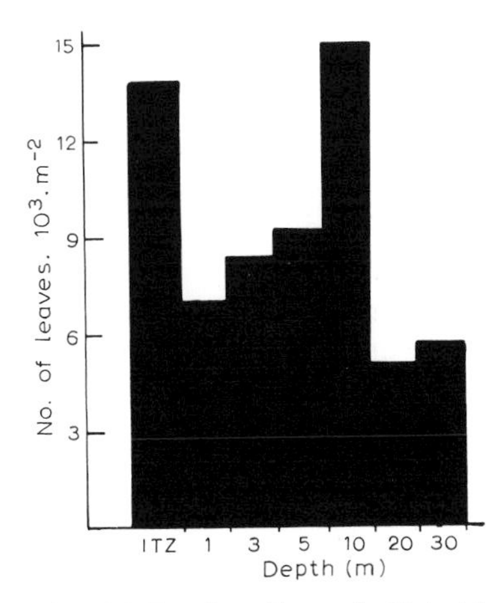

Fig. 16.4. Density of leaves in *Halophila stipulacea* community at different depths in E'Mteine (Marsa abu Zabad, Sinai coast of the Gulf of Elat). (From Lipkin, 1979.)

where individual leaves were often 12 cm long (Fig. 16.5). Tips of most *Halodule uninervis* leaves, except the youngest, were chopped off, probably by grazers; their undisturbed lengths could therefore not be determined. Within the same depth range leaf width almost doubled. The largest leaf surface area per unit area of sea bottom was at

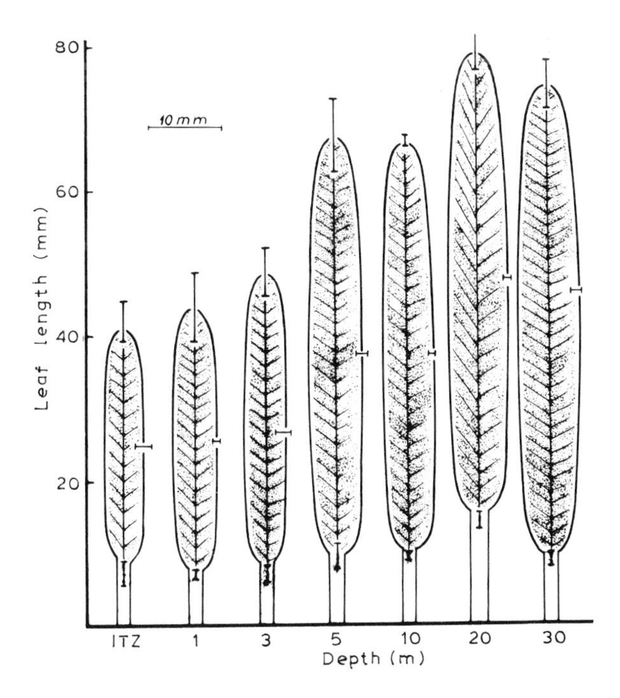

Fig. 16.5. Size of *Halophila stipulacea* leaves (length and width) at different depths in E'Mteine (Marsa abu Zabad, Sinai coast of the Gulf of Elat). (From Lipkin, 1979.)

about 10 m, where it was *c.* 8 $m^2 m^{-2}$. The ratio between the plant parts buried in the sediment to those above the surface in *Halophila stipulacea* changed with depth as well. In the intertidal and shallow subtidal the former contained a larger portion of the biomass, more than one half to two thirds the latter. The proportion of plant parts occurring above the surface increased with depth, reaching two thirds of the biomass at 10 to 20 m. By contrast, the ratio for *H. uninervis* was about the same at all depths (Lipkin, 1979).

The biomass of epibiota in the subtidal zone, especially epiphytic algae, is controlled by grazing fish. Only a few grazer-resistant epiphytes are apparent on seagrass leaves, among them large foraminifers that are symbiotic with various micro-algae. The most common species to 5 m in the northern Red Sea are *Amphisorus hemprichii* and *Sorites orbiculus*, which are symbiotic with *Amphora tenerrima*, *Chlorella* sp., and *Symbiodinium microadriaticum*. *Amphisorus hemprichii* was more common in *H. stipulacea* beds versus *Sorites orbiculus* in *H. uninervis* beds. At 4 m the density of *A. hemprichii* was *c.* 1500 to 3500 m^{-2}. Less common in the beds were other foraminifera, such as the thick-shelled *Amphistegina lessonii* and *A. lobifera*, that inhabit beds in shallow waters. Both foraminifera live symbiotically with diatoms (*Amphora tenerrima*, *Fragilaria shiloii*, *Nitzschia frustulum* var. *subsalina*, and *Nitzschia panduriformis* var. *continua*), while *A. lessonii* is also associated with the green alga *Chlorella* sp. The algae are located individually within the "cells" of the honeycomb-structured inner surface of the foraminiferan shells. *Amphistegina lessonii* composed some 75% of the *Amphistegina* population at depths down to 50 m, below which it was gradually replaced by thinner-shelled species (Hansen and Buchardt, 1977; Hottinger, 1977; Lee et al., 1979; Saks, 1981).

The small ahermatypic coral *Heterocyathus* (1 cm in diameter) forms an interesting association in these shallow beds, growing attached to an empty shell of a gastropod (usually a *Turritella*) and having a sipunculid worm in its cavity. The sipunculid is, according to Reiss and Hottinger (1984), a species of *Aspidosiphon*. Gastropods like *Cerithium erythraeonense*, *Cerithium spathuliferum*, and *Rhinoclavis sinensis* are common in seagrass beds. The razor-fish *Aeoliscus punctulatus* may often be encountered hovering in a vertical position, usually head down among the seagrass leaves. The small brownish fish mimics decomposing leaves of *Halophila stipulacea* (plentiful in the bed) not only in its colour but also in the size and shape of its strongly compressed body, which is narrow at the down-pointing snout and wide above. Other fishes like the pipefish *Halicampus macrorhynchus* or species of sea horse (mostly *Hippocampus fuscus*, *Hippocampus hystrix*, and *Hippocampus kuda*) also occur in the beds. At similar depths, where there is no clay in the sediment, *Asymmetron* cf. *lucayanum* is often common within the seagrass beds and in surrounding areas. The density of *Asymmetron* in the one location where quantitative data was obtained was 46 m^{-2}, but this should be regarded as a casual value rather than a representative one (Gohar, 1948; Steinitz, 1962; Klausewitz, 1964; Fishelson, 1971; Scheidegger, 1972; Ayal and Safriel, 1983).

Among the most spectacular companions in the *Halophila stipulacea* beds are the garden eels, *Gorgasia sillneri*, colonies of which sometimes occur in the beds or at their edges to depths of 3 to 45 m — usually to 20 m. Large colonies may be about 100 m long and contain over a thousand individuals. The fish live in sinusoidal tubes in the sea bottom. The usual density in a colony is 1 to 2.5 eels m^{-2}, with the highest density being 5 m^{-2}. The same feeding grounds are shared by the sand-diving fish *Trichonotus nikii*, which is about 5 cm long. Hundreds of these fish occupy the water layer at the level of their heads when the latter are hidden in their tubes — both during midday hours and at night. The sand-diving fish hang suspended in a peculiar oblique position of *c*. 50°, exhibiting undulating movements of their elongated bodies and diving into the sand head-first when approached. Other sand-diving fishes, such as *Limnichthys*, often occur by the thousands; the labrid *Xyrichtys pavo* often inhabit the garden-eel colonies as well. The latter, as well as *Callechelys marmorata* and *Synodus variegatus*, may devour young garden eels and larvae (Klausewitz, 1962; Clark and Von Schmidt, 1966; Fricke, 1970; Clark, 1971, 1972).

Fishelson (1971) remarked that the eels preferred muddy flats adjacent to steep slopes. By contrast, Clark (1972) states that the colonies occur 2 to 30 m away from reefs, off the mouths of wadis, where the texture of the sand is right for them. The big stingray *Taeniura lymma* usually lives in shallow waters (to 10 m deep) of lagoons and sandy areas among coral colonies. Periodically it visits sheltered sandy coves, which are otherwise suitable for the establishment of garden-eel colonies. During the breeding season it moves into these habitats for longer periods. It disturbs the sediment either by waving its wing-like flanks to expose the buried polychaetes (which are the main staple of its diet), or lies just under the surface of the sand with only its eyes, spiracles, and spines protruding. The leopard ray *Aëtobatus narinari* that ploughs the bottom with its longer lower jaw seeking bivalves, crustaceans, and other bottom invertebrates, also causes a similar disturbance. Other fishes, such as the sparid *Rhabdosargus haffara*, also dig into the sand for clams and crustaceans. (Klausewitz, 1959; Fishelson, 1983). Clark (1972) believes that the disturbance of the sediment by these and other fishes with similar behaviour is a major factor restricting the distribution of garden-eel colonies.

The *Thalassodendron ciliatum* community is unique among seagrass communities of the Red Sea as it almost always occurs in pure stands. Unlike the leaves of most other seagrasses that are borne on very short vertical stems situated on horizontal rhizomes near the bottom (being orientated vertically or obliquely to the substrate), those of *T. ciliatum* occur at the end of long, woody vertical stems. Each such stem has a bunch of falcate leaves at the tip, being orientated more or less horizontally. The canopy of the usually dense beds thus confines a sheltered, shaded space underneath — between the sea bottom and the close layer of leaves. The *T. ciliatum* community is very common in the Red Sea, extending from the uppermost sublittoral zone to at least 30 m. It occurs on horizontal and nearly horizontal bottoms, which are composed of coarse coralligenous sand containing coral and shell rubble. In the shallows it often occurs near coral reefs without the wide halo zone (several metres) that typically separates other seagrass communities from the reef. The variability of plant biometric and quantitative community parameters were similar to those of the *Halophila stipulacea* community. As in the *H. stipulacea* community, the standing crop of *T. ciliatum* is lowest in the uppermost subtidal zone where the tips of leaf bunches extend into the lowermost intertidal zone and become exposed during exceptionally low water. Its standing crop was highest at *c.* 2 m (Fig. 16.6), averaging about 2800 g dry weight m^{-2} but often reaching 5000 g dry weight m^{-2} or more. In the closed space underneath the canopy a rich sciaphilic community develops — both epiphytic upon the vertical stems and upon other hard objects. It contains elements of the rocky sciaphilic community, as well as calcareous red algae like *Haliptilon* and *Peyssonnelia*. Interestingly, grazing fish do not penetrate the canopy to consume algae in this community, although they check the photophilic epiphytic community on its leaves. Some animals from other habitats lay their eggs in this sheltered space, with their young spending their early stages here protected from predators. Most conspicuous is the squid *Sepioteuthis lessoniana*, which is often seen hovering close over the canopy with its body bent over to lay its eggs on the seagrass stems or other hard objects beneath. With increasing depth the length of the vertical stems elongates, determining the height of the vegetation. From less than 20 cm in the uppermost sublittoral zone it increases to more than a metre at 15 to 30 m. Similarly, leaf length almost doubled within the same range from 6.2 ± 0.3 to 12.1 ± 1.2 cm. The number of leaves on a vertical stem increased from about 5 to 8 between 1 and 15 m depth, decreasing with further depth. The density of leaves was highest at 5 m, being *c.* 11 000 m^{-2} (Fig. 16.7). Vertical stems were densest at 8 m, numbering *c.* 2000 m^{-2} (Lipkin, 1977a, 1979, 1988).

Fringing lagoons

In lagoons separating the beach from the fringing or patch reefs the bottom is usually sandy, with some larger pieces of coral rubble. Here and there, patches of seagrass, clayey sediment, or pebbles occur within the intertidal and uppermost sublittoral zones. The vegetation is extremely poor because of many grazers, and the fauna is typical of soft bottoms. The black sponge *Biemma fortis* and the razor clam *Pinna muricata* are buried in the coralligenous sand with only their tips protruding. Clams (*Laevicardium orbita* and *Tellina pharaonis*), gastropods (*Conus arenatus*, *C. striatus*, or *Terebra* spp.), and sea-cucumbers (*Holothuria arenicola*) also live under the surface. The sea anemone *Radianthus koseiriensis* is also typical here. It is buried in the sediment, attached to a stone or rock below, with its tentacles spread to capture the tiny creatures it feeds upon. It may hold young fish (*Amphiprion bicinctus* and *Dascyllus trimaculatus*) that, when adults, move to the reef community where they live symbiotically with larger sea

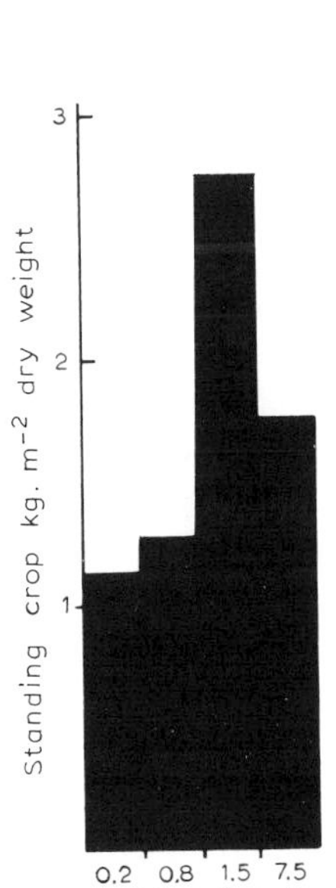

Fig. 16.6. Standing crop of stands of the *Thalassodendron ciliatum* community at different depths in E'Mteine (Marsa abu Zabad, Sinai coast of the Gulf of Elat). (From Lipkin, 1988.)

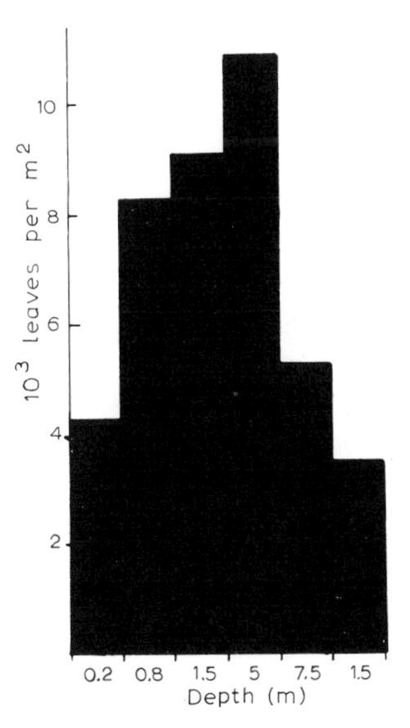

Fig. 16.7. Density of leaves in stands of the *Thalassodendron ciliatum* community at different depths in E'Mteine (Marsa abu Zabad, Sinai coast of the Gulf of Elat). (From Lipkin, 1988.)

anemones, such as *Gyrostoma*. On the slightest threat (at night), *R. koseiriensis* contracts into the sediment together with its symbionts. The strange-looking *Stoichactis tapetum*, which occurs a little deeper, also harbours young *A. bicinctus* and *D. trimaculatus*. When *R. koseiriensis* retracts in the evening, it is replaced by the much less common *Macrodactyla aspera* that emerges then to catch small fish and crustaceans. At a depth of 1 m, and partially overlapping with *R. koseiriensis*, other sea anemones are encountered: nocturnal *Megalactis hemprichii* and *Cerianthus maua*. The former forms rather sparse populations from 2 to 8 m on coarse sandy bottoms, having few detrital particles and many stones; the latter occurs from 1 to 20 m. In parts of the lagoon where the sediment is rather fine and thick it forms dense populations. On stones, octocorals like *Xenia macrospiculata* and *Xenia membranacea* grow, while *Anthelia fishelsoni* occurs under stones. As in the intertidal zone, life under stones within the Red Sea is very rich, colourful and diverse. It includes, among others, the red sponge *Tethya seychellensis*, the opisthobranch *Dolabrifera dolabrifera*, the gastropod *Conus musicus*, and a wealth of cowries (*Cypraea carneola*, *Cypraea grayana*, *Cypraea isabella*, *Cypraea lynx*, and *Cypraea nebrites*). *Cypraea pantherina* occurs within the same habitats under larger boulders. On muddy sand the octocoral *Sarcophyton glaucum* sometimes occurs, although it is more common on platforms and coral reefs (Cohen, 1970; Fishelson, 1970, 1983; Weitlauf, 1978a; Heinemann and Heinemann, 1982; Dafni, 1983; Tsurnamal, 1983).

Sublittoral sandy areas devoid of seagrass

In areas 0.2 to 1.5 m deep where a dense monolayer of small stones and pebbles covers a sandy substrate, most intertidal elements of similar habitats occur. In addition, the nocturnal sea slugs *Berthellina citrina*, *Discodoris erythraeensis*, *Discodoris vanicorum*, and *Nembrotha affinis* are very common; during the day they hide under stones. Many of these sea slugs are very conspicuous when present on stones because of their bright warning colours and magnificent patterns, which place them among the most beautiful creatures of the region, even against the background of this highly colourful tropical marine life. *Berthellina citrina* feeds on sponges non-selectively; spines of at least seven different sponges may occur in their stomachs during a single study. Upon contact it defends itself by the secretion of a highly acidic fluid. It appears that other opisthobranchs do the same, as predators in the area avoid them. In algal thickets (on stones and blocks of clayey sand) in highly sheltered habitats the nudibranch *Notarchus indicus* is plentiful between 1 and 3 m, with densities to 50 m^{-2}. It feeds on epiphytic microalgae growing on larger algae or seagrasses as well as on microalgae covering smooth mud. Its occurrence and population size correspond with the larger algae, which flourish in spring and decline later during summer (Gohar and Aboul Ela, 1959; Schuhmacher, 1973; Marbach and Tsurnamal, 1973; Fishelson, 1983).

In the uppermost sublittoral zone of sandy, sheltered coves a *Lovenia elongata* community occupies patches. The sea urchins dig into the upper layers of the bottom. They prefer clean medium-fine sand (particles with a diameter of 0.15 to 0.3 mm and sorting coefficient, $S_o = 1.3$) on stoneless and grassless bottoms, including clearings in the seagrass bed. *Lovenia elongata*, which is the dominant component in the community, occurs from the lowermost eulittoral zone to *c.* 95 m. The urchins stay most of the time under the surface of the sand where they crawl in search of food. At Taba, in the northern Gulf of Elat, they were found to occupy a sandy belt at 3 to 5.5 m where the average grain diameter was 0.28 mm. The density of the population there was 62.5 ± 24 m^{-2}, quickly decreasing in adjacent shallow areas (2.7 to 3 m) with coarse sand (grain diameter 0.48 mm) or in deeper depths (5.5 m) with fine sand (grain diameter 0.21 mm). Based on experience elsewhere the authors nevertheless concluded that the nature of the substrate and lack of turbulence, rather than depth, determined the distribution of this community. The overall density of *Lovenia elongata* in the subtidal communities ranged between 2.6 ± 0.5 m^{-2} (adults only) and 109 ± 18 m^{-2} (adults and young). In its typical habitat within the upper sublittoral zone *L. elongata* is accompanied by several other echinoids, such as *Clypeaster humilis*, *Echinodiscus auritus*, *Laganum depressum*, *Metalia spatagus*, *M. sternalis*, *Moira stygia*, *Nudechinus scotiopremnus*, and *Schizaster lacunosus*. All these echinoids, like *L. elongata*, ingest sand to extract nourishment from the organic detritus within it. Other accompanying species are the gastropods *Casmaria*, *Nassarius albescens*, *Polinices tumidus*, *Strombus fasciatus*, *S. gibberulus albus*, and *Terebra* spp., the bivalves *Cardium* and *Mactra olorina* that in places may be so common as to share dominance of the community, and the very common crab *Calappa hepatica*. The latter, plus the gastropod *Matuta* (family Cassididae), the starfish *Luidia maculata*, and some fish like *Gerres oyena* are possible predators of *Lovenia*. The mixed population of *Lovenia elongata* and *Mactra olorina* described by Por and Lerner-Segev (1966) from the northermost Gulf of Elat had almost completely disappeared by 1970 probably because of excess silt resulting from construction activities (Por and Lerner-Segev, 1966; Lawrence and Ferber, 1971; Ferber, 1973, 1983; Ferber and Lawrence, 1976).

All species accompanying these sea urchins also live in similar sandy habitats without urchins, with many of them being more abundant in the latter habitat. The same type of habitat also harbours burrowing gastropods and clams. The most common ones are *Terebra dimidiata*, *Terebra maculata*, *Conus arenatus*, *Conus tessulatus*, *Codakia tigrina*, *Polinices tumidus*, *Tellina virgata*, and *Turritella maculata*; below 5 m *Glycymeris pectunculus* also occurs. The worm-like chordate *Ptychodera flava*, which burrows deep into the sand, and the big synaptid *Synapta maculata* also share this habitat. At Al Ǵhurdaqah the meiofauna of these bottoms is rich, mainly composed of syllid polychaetes, hydracarines and harpacticoid copepods. Commonest among the latter are a species of *Canuellina* and *Stenhelia polluta*. In the northern Gulf of Elat nematodes dominate, composing 37 to 68% of the total meiofauna; harpacticoid copepods come next (19.5–23.5%), then ostracods (8.75–15%), and polychaetes (10.8–11%). At 1 to 5 m the density is high, with the total for individuals of all groups being about $2 \times 105\ m^{-2}$, the density decreases below 10 m but increases again at greater depths (50–250 m). A typical interstitial fauna at Al Ǵhurdaqah was found to contain, among other taxa, the chordate *Branchiostoma*, the gastrotrichan *Protodrilus*, the gastropod *Sapha amicorum*, and species of the rotatorian *Encentrum*. In algal thickets amphipods like *Guernea petalocera* and *Perioculodes* cf. *longimanus* abounded, along with tanaidaceans, ostracods and copepods, such as *Neodactylopus cyclopoides* and *Syngastes pietschmanni* (Remane and Schulz, 1964; Por and Lerner-Segev, 1966; Fishelson, 1971; Schmalbach and Por, 1977; Dafni, 1983).

In shallow (to −2.5 m), sandy, areas throughout the Red Sea the blenniid *Xiphasia setifer* occurs. It lives solitarily in tubes (with densely woven walls) and resembles in general appearance a single garden eel. It is somewhat shorter than the eels, being 50 to 60 cm long (Fridman and Masry, 1971; Dor, 1984).

Several plant communities dominated by green algae occupy patches of sandy bottoms in shallow waters (1 to 15 m deep). The calcareous *Halimeda cylindracea* and *Halimeda macroloba* and the non-calcareous *Avrainvillea erecta* and *Chlorodesmis papenfussii* form sparse communities. The first three occur in all parts of the Red Sea, while *C. papenfussii* is restricted to the south. Distance between individual plants in these communities is usually 20 to 80 cm. Similarly, a *Tydemania expeditionis* community occurs at depths below 10 m, usually next to coral-reef knolls. Robust species of *Caulerpa* sometimes form communities on soft bottoms. As a rule the total plant cover in these communities is rather small, usually being less than 15% of the bottom sea. A *Caulerpa serrulata* community appears on coarse sand, usually in small patches where the vegetation is not very dense. A *Caulerpa racemosa* community sometimes occurs on rather coarse sand, but it seems to prefer fine soft mud on which the vegetation is often very dense. In the southern Red Sea *Caulerpa selago* and *Caulerpa sertularioides* communities occur on medium-size sand protected from waves and currents. The dominant species in these communities also occur in mixed populations with the seagrasses *Cymodocea rotundata* and *Thalassia hemprichii* and with *Caulerpa racemosa* (Lipkin, 1974, 1987b).

On sandy bottoms in the upper sublittoral zone a number of interesting burrow-inhabiting snapping shrimp–goby associations occur, some of them throughout the Red Sea while others only occur locally. In the northern Gulf of Elat at least a dozen fish and a similar number of shrimp are involved, with even more occurring within the Red Sea as a whole (Luther, 1958a, b; Abel, 1960; Klausewitz, 1960, 1964, 1967, 1969, 1970, 1974; Magnus, 1967; Karplus et al., 1972a, b, 1974, 1981;

TABLE 16.2

Distribution and relative abundance of some shrimp–goby associations in habitats of the northern Gulf of Elat. (After Karplus et al., 1981; and Goren and Karplus, 1983, generalized.)

Shrimp	Goby usually associated	Relative abundance in				Sediment preferred (grain size)
		lagoon (0–2 m)	shallow fore-reef (4–8 m)	patch reef (8–14 m)	deep fore-reef (14–20 m)	
?	*Tomiyamichthys randalli*	+				fine
?	*Vanderhorstia delagoae*	++				fine
Alpheus djiboutensis[a]	*Cryptocentrus cryptocentrus* and *C. lutheri*	++				medium
A. rubromaculatus	*Lotilia graciliosa*	++	+			fine
A. purpurilenticularis	*Amblyeleotris steinitzi*	+++	++	+		fine to medium
A. rapax	*Ctenogobiops maculosus*	+	++	+		coarse
A. sp.	?		+[b]	++	+	
A. bellulus	?		+[b]	++	+++	
A. ochrostriatus	?		+	+	+	

[a]Juveniles near shore where there are many stones; the only association in the near-shore belt.
[b]From 6 m depth.

Goren and Karplus, 1983; Karplus, 1987). See Table 16.2 for a listing of these assessments.

In the same type of habitat other burrow-inhabiting fish occur that are not associated with shrimps, including the gobies *Acentrogobius ornatus*, *Amblycentrus arabicus*, *Amblycentrus magnusi*, *Amblygobius albimaculatus*, and *Valenciennea sexguttata*, plus the branchiostegid *Malacanthus brevirostris* (Klausewitz, 1967, 1968; Clark and Ben Tuvia, 1973; Goren, 1979, 1983). The latter carries in its mouth small stones, coral, or shell rubble to build mounds over its burrows for further protection. The gobies *Ptereleotris evidus* and *Ptereleotris microlepis* also dig burrows; during daylight hours they swim high above the bottom, though not too far from their burrows. Others either swim solitarily (*Yongeichthys nebulosus* and *Yongeichthys pavidus*) or in schools like the mullets (*Mulloides flavolineatus*, *Pseudupeneus macronema*, and *Upeneus asymmetricus*) above the sandy bottom (Tortonese, 1968; Goren, 1983; Dor, 1984).

On sandy bottoms a number of predatory fish wait for their prey, being camouflaged by their colours or colour-patterns or being partially buried in the sediment. These include the synodontids *Synodus variegatus* and *Trachinocephalus myops*, the platycephalid *Sorsogona prionota*, the parapercid *Parapercis hexophthalma*, the gerrid *Gerres oyena* and sometimes the grouper *Epinephelus areolatus* or the flatfish *Bothus pantherinus* (Klausewitz, 1967; Tortonese, 1968). The flatfish *Pardachirus marmoratus*, which lies colour-camouflaged on the bottom, has developed an additional means of defense. It secretes a highly potent acidic proteinaceous toxin (pardaxin) into the water from glands at the bases of its dorsal and anal fin-rays; the toxin strongly affects gill adenosine triphosphatase and has been proven to deter potential aggressors. The amount of the toxin varies between individual fish, being 20 to 80% of the total proteins in the crude secretion (Clark and Chao, 1972; Primor et al., 1977, 1980; Primor and Tu, 1980). Similarly equipped with both protective toxin and camouflaged morphology are the stonefish. One of these, *Synanceia verrucosa*, hardly moves, being mostly buried in the sand among stones with only its eyes and enormous mouth above. It is often overgrown with algae, bryozoans, and other animals that occur on stones. The two other smaller stonefish (*Synanceia nana* and *Inimicus filamentosus*) also occur (Eschmeyer and Rama Rao, 1973; Fishelson, 1973, 1983).

THALASSOHALINE HABITATS SEPARATED FROM OR BARELY CONNECTED TO THE SEA

Lagoons barely connected to the sea

Coastal lagoons varying in nature, depth, and degree of connection with the open sea occur all around the Red Sea. Some have a wide shallow entrance over a bar. Others have a narrow channel or opening to the sea: Rhizophora Lagoon on the southern part of Museri Island; the central part of the lagoon described by Kassas (1957) north of Marsa Halout and 25 km south of Bur Sudan; and El Bilaiyim Lagoon on the Sinai coast of the northern Gulf of Suez. All of them have higher temperatures and salinities (especially in summer) than in the open sea, so that they are often well within the metahaline range (cf. Por, 1972). Many of these lagoons are surrounded by mangroves, and for most of them little else has been reported. Species diversity in the lagoons is lower than in open habitats — the more sheltered they are, the lower the diversity.

One of the most characteristic organisms in such lagoons is the scyphomedusa *Cassiopea andromeda*. It occurs solitarily or in groups in shallow lagoons (0.5 to 9 m) throughout the Red Sea and in the Gulf of Aden. The more sheltered the habitat, the larger the groups. The medusas feed on plankton but cannot survive on plankton alone; they also depend upon the photosynthates produced by symbiotic algae they contain. The symbionts are also important for the asexual reproduction of *C. andromeda*. Strobilation rate and the development of the ephyrae are considerably higher in animals with symbionts. The only enemies of *C. andromeda* within the Red Sea seem to be the leatherback sea turtle *Dermochelys coriacea* and the crab *Grapsus albolineatus* (Gohar and Eisawy, 1961; Adar, 1980).

Land-locked nearshore water-bodies

Pools located in cracks

Two typical Red Sea habitats occurring within cracks of elevated fossil reefs have been studied in some detail. One of them occurs in the south, the Devil's Crack on Entedebir Island, Dahlak Archipelago, while the other occurs in the north, the Ras Muhammad cracks in southern Sinai. Both habitats are located in deep, narrow cracks where hardly any direct sunlight penetrates. The cracks are situated some 150 m from the waterline and are fed subterraneously by seawater seeping through highly porous rock. They are inhabited by light-avoiding and shade-preferring organisms of marine origin.

The bottom in Devil's Crack is muddy; in the Ras Muhammad Elongated Crack an accumulation of sand occurs at one end, but the rest of the crack consists of almost vertical rocky walls descending below observable depths. Salinity in the former crack was 36‰, a little below that of the water in the nearby sea (38‰); in the latter crack it was almost the same as the nearby seawater, being about 41‰.

The fauna and flora in these habitats is rather limited. Devil's Crack has the blind red atyid shrimp *Antecaridina lauensis*, a water strider, *Halobates* sp., two ostracods (one of them a species of *Bairdia*), two nematodes, a very rich fauna of copepods, including a dominant calanoid, *Pseudocyclops steinitzi*, four harpacticoids, *Nitokra affinis stygia*, *Typhlamphiascus latifurca*, and the less common *Cletocamptus xenuus*, and *Schizoperoides expeditionis*; the latter are all new species described from the crack by Por (1968). Of the algae, only blue-greens and diatoms occur. A similar habitat in the same region is Avicennia Pool on Museri Island, which is located about 250 m from the nearest coast and is surrounded by a narrow belt of the mangrove *Avicennia marina* (Holthuis, 1963; Por, 1968; Lipkin, unpubl.).

The Ras Muhammad Elongated Crack has a more diverse fauna and flora than Devil's Crack. Blue-greens dominate the walls in the intertidal levels, along with a chiton having densities of 20 to 30 m^{-2} at the lower levels. Several shade algae, like *Botryocladia skottsbergii*, *Codium arabicum*, and *Valonia aegagropila* abound on the ledge and upper parts of the vertical walls below. Two shrimps dominate the submerged fauna, the bright red, blind *Calliasmata pholidota* and *Periclimenes pholeter*, both described from this crack. In addition, sponges, ascidians, polychaetes (a terebellid and a serpulid), the key-hole limpet *Diodora rueppelli*, and a Cardita-like bivalve are common on the walls, along with the algae, at about the water level and a little below. A benthic calanoid belonging to a yet undescribed species of *Pseudocyclops* is present. The isopod *Cymodoce* occurs in the algal mass, and the goby *Monishia ochetica* is also present. The globose, vesicular green alga *Ventricaria ventricosa* dominated for a few years after the crack was first opened by an earthquake in 1968. It grew in great numbers mostly on tin cans dumped into the crack shortly after its discovery, then decreased in numbers as the cans rusted and slowly disintegrated. Two other cracks in the vicinity had a similar though poorer biota. Both had *Antecaridina lauensis*, which was missing from the elongated crack. A similar crack reported from Jezirat Sinafir also had a red, blind shrimp of the *Antecaridina* type (Holthuis, 1973; Por and Tsurnamal, 1973; Por and Dor, 1975; Tsurnamal, 1975; Klausewitz, 1975; Goren, 1985; Por, 1985).

Pools located on flat coasts

Land-locked seawater pools occur near the coastline on low flat land. These are usually large, unshaded, and hypersaline, with their salinities ranging from only slightly higher than the open sea to highly hypersaline. Several pools were studied in detail on the Sinai coast of the Gulf of Elat. Two representative examples are described here: a slightly hypersaline one with mostly marine organisms and relatively few hypersaline elements and a highly hypersaline pool populated by hypersaline elements and with only a few marine species.

The Di Zahav Pool, situated *c*. 4 km south of the oasis of Dahab, was described by Por and Dor (1975). This is an oval bowl-shaped small pool ($110 \times 70 \times 4.5$ m^3) separated from the sea by a 30 m wide sand bar. Salinities in the pool range between 45 and 60‰; thus it may be classified as metahaline (Por, 1972). Tidal range is about a tenth of that in the open sea. Temperature fluctuations are considerably wider than in the nearby sea, winter minimum being 14°C (and occasionally even 9°C) compared to 21°C in the sea. Beachrock covers most of the perimeter of the pool above and below the water line. The fauna is basically composed of hypersaline marine species with no marked seasonal changes in composition. On beachrock in the shallows the serpulid *Vermiliopsis pygidialis* grows in groups, forming rounded masses of interwoven tubes as large as a human's head. Boring sponges (*Tethya* cf. *globostellata* and a clavellinid sponge) penetrate the bases of these masses, gradually eroding them until the "heads" detach and roll down to the bottom. Where the substrate is soft blue-green algae occur; the gastropods *Pirenella cauilliaudi* feed on them not allowing any mats to form. In deeper parts of the pool the gravelly bottom is cemented into a 5 to 6 cm thick elastic crust by several sponges — *Hymeniacidon sanquinea*, *Spirastrella* sp., and an unidentified clionid species. On this crust the stunted red alga *Laurencia papillosa* grows sparsely, accompanied by some growth of the blue-green *Microcoleus tenerrimus*. Along steep, sandy slopes descending to the bottom *Cassiopea andromeda* occur in considerable numbers.

Boulders, composed of reef-like structures from serpulids, lie on the bottom of the pool and are the main habitat for macroscopic life in the pool. Their upper surfaces and sides were covered by several centimetres of two-layered thick crusts composed of three green algae that dominated among these life forms. Directly on the boulders grew *Dictyosphaeria cavernosa* and *Valonia aegagropila*, which were covered by a light green layer of *Siphonocladus rigidus*. The bright red ascidian *Ecteinascidia* cf. *thurstoni* subdominated, forming dense growth around the bases of boulders wherever the algal crust was broken. It was also interwoven with *V. aegagropila* under overlaying *S. rigidus* where the ascidian looked pale or colorless. Around the bases of boulders the key-hole limpet *Diodora rueppelli* abounded as well as the isopod *Cymodoce truncata*; sponges and serpulids inhabited the undersides of boulders. The meiobenthos of the muddy bottom was very rich and included representatives of most meiobenthic groups. The dominant species among the ostracods was *Cyprideis torosa*, accompanied by a species of *Bairdia*. On the sedimentary bottom *Heterolaophonte quinquespinosa* dominated among the copepods and *Typhlamphiascus confusus* abounded. *Orthopsyllus linearis* was common in masses of algae, accompanied by *Tisbe* sp.; *Ridgewayia typica* and *Euryte* sp. were also very common. Three new copepods were described from this pool: *Dizahavia halophila*, *Cletodes dorae* and *Cyclopuella sinaitica*. Two fishes, the euryhaline *Aphanius dispar*, common in brackish to hypersaline waterbodies on land, and a marine goby also inhabited the pool (Por and Dor, 1975; Por, 1975, 1979).

The Solar Lake (9.5 km southwest of Elat) has received much attention. This is an oval pool 5 m deep, 140 m long and 50 m wide, separated from the sea by a sand-bar 45 to 60 m wide and 3 m high; it is sheltered from the prevailing winds by nearby hills. The water level in summer is about 1 m below that of winter. The pool has a gently sloping "shelf" to 1.5 m deep in winter and a steep slope reaching to the bottom. It is stratified during most of the year with brines of different concentrations — the lower layer with salinities of 160 to 180‰ and the upper with 45 to 100‰. In summer, holomixis occurs, resulting in salinities of 140 to 180‰ throughout the entire water column. The water-body is therefore classified as hypersaline (Por, 1972). When stratified it functions as a heat trap, its lower layers (the hypolimnion) warming up in summer to about 50°C (maximum measured was 60.5°C in November, 1974). During the stratification period these layers also become anaerobic, following the activity of *Desulfovibrio* and other anaerobic fermentation processes, with accumulation of up to 40‰ of hydrogen sulphide. In the anaerobic metalimnion and hypolimnion, plates of photosynthetic bacteria develop. In the upper levels, which lack oxygen but have a little hydrogen sulphide, a *Chromatium* plate also forms; deeper, where hydrogen sulphide concentrations are around 15 ppm, a *Chlorobium* plate is also found, while on the bottom where a high concentration of hydrogen sulphide occurs, a dense bloom of the blue-greens *Oscillatoria lacus-solaris* and *Phormidium hypolimneticum* forms a mat. The blue-greens disappear and partially disintegrate during the holomixis. In cultures these blue-greens did not release oxygen during photosynthesis, but instead granular sulfur settled on algal filaments, implying that they may have used hydrogen sulphide rather than water as an electron donor. The upper layer, the epilimnion, remains between 14 and 28°C and contains most of the life in the pool (Eckstein, 1970; Por, 1972; Cohen et al., 1975; Por and Dor, 1975).

The poor fauna and flora in this pool (only about twenty animal species were listed below the water level) is typical of that within hypersaline water-bodies rather than marine. The composition of the biota changes markedly seasonally. Blue-green algal mats (also referred to as microbial mats), black in the shallows and becoming brown in deeper waters, covered the bottom of the "shelf" area. In shallow water the mat was composed of polygons 0.3 to 1 m across with raised rims, becoming flat and on the average larger towards the edge of the shelf. There the mats were torn, with greenish pieces hanging down loosely over the edge. The mats, which were dominated by *Oscillatoria lacus-solaris* and *Phormidium hypolimneticum*, had a surface layer of coccoid blue-greens (*Aphanocapsa littoralis*, *Aphanothece halophytica*, and *Entophysalis deusta*) and diatoms (*Amphora coffeaeformis*, *Navicula* sp., and *Nitzschia thermalis*) as well as flexibacteria. During the end of winter and early spring purple photosynthetic bacteria also occurred. Just belows the thermocline, in a layer with temperatures of 35 to 40°C, the planktonic blue-green *Dactylococcopsis* sp. abounded. *Artemia* feeding on them had to dive for short periods into this inhospitable layer. Nine species of halophilic Coleoptera were charactertistic inhabitants, with the most important being *Anacaena* and *Paraberosus melanocephalus*, along with *Artemia* sp. and *Nitokra lacustris*. The only marine elements were the copepod *Robertsonia salsa* and the ostracod *Cyprideis torosa*. The turbellarian *Macrostoma* sp., larvae of the coleopterans *Ochthebius* cf. *auratus* and *Bledius* spp., and the isopod *Halophiloscia* were also important inhabitants.

At the bottom of the pool *Halophiloscia* feed on decaying plant material. *Robertsonia salsa* nauplii eat diatoms, while *Nitokra lacustris* nauplii and the nematode *Monhystera* sp. feed on filamentous blue-greens. Mayfly larvae also eat these nauplii, as well as diatoms and filamentous blue-greens. Both the larvae and adults of *Paraberosus* feed on blue-greens and diatoms, while the swimming larvae also consume, in addition, *Artemia* and larvae of *Robertsonia salsa* and *Ochthebius*. Anacaena larvae feed on the resting eggs (cysts) of *Artemia*, on *Halophiloscia*, and on *Bledius* larvae. An algal biomass of 100 g dry weight (organic) was sufficient to support over a thousand larvae and adults of *Paraberosus* but only about 450 larvae of mayfly.

Primary production in the pool was highest in the metalimnion and hypolimnion during the stratification period. The highest value obtained was 4960 mg C m^{-3} day^{-1} at 4 m depth on 20 November 1970 versus only 50 mg C m^{-3} day^{-1} in the epilimnion. The mats are 35 to 200 cm in thickness. The subfossil lower layers, which are over 2000 years old, overlay carbonate mud (about 4500 years old) with shells of *Pirenella cauilliaudi* and a mytilid. The mollusc shells, suggesting lagunar conditions, testify to lower salinities that eabled their development. The occurrence of *P. cauilliaudi*, which feeds on blue-greens, would not allow development of mats. The former lagoon was separated from the sea about 2400 years ago (Por, 1972, 1977, 1985; Friedman et al., 1973; Krumbein and Cohen, 1974; Cohen et al., 1975; Krumbein, 1975; Por and Dor, 1975; Zalcman and Por, 1975; Dimentman and Spira, 1982; Friedman and Foner, 1982; Spira and Van Rijn, 1982; Campbell, 1985; Campbell and Golubic, 1985; Gerdes et al., 1985). Similar water-bodies in the region with higher salinities do not contain any marine species (Por, 1975; Friedman and Krumbein, 1985).

PRODUCTIVITY AND CONTRIBUTION TO THE ECOSYSTEM OF INTERTIDAL AND SHALLOW SUBTIDAL COMMUNITIES

The southern Red Sea is richer than the northern, while the Gulf of Aden is even richer. Halim (1969) reported that the total pigment in the water column (to a depth of 200 m) was less than 30, 40 to 50, and 100 mg m^{-2}, respectively, in the waters of the northern Red Sea, its southern part and the Gulf of Aden during the SW monsoon period. During the NE monsoon period the northern part was richer, with values of *c*. 60 mg m^{-2}. Similarly, Kimor (1973) reported that both phyto- and zooplankton in the northern Red Sea were poorer and less diverse than in the southern part, and that the Gulf of Aden was richer. According to Levanon-Spanier et al. (1979), primary productivity in the Gulf of Elat is no more than 3.38 μg C m^{-3} h^{-1} in winter and only 0.05 to 0.27 μg C m^{-3} h^{-1} in summer, with diatoms composing 97% of the winter phytoplankton and only 33.4% of the summer phytoplankton (49% dinoflagellates, 17.6% blue-greens–*Trichodesmium*). Beckmann (1984) reported 3.8–32.8 g FW m^{-2} (to a depth of 100 m) as typical biomass figures for mesozooplankton. He also remarked that the Gulf of Aden was richer than the Red Sea but that the area above the sill of Hanish, at the entrance to the Red Sea, was the richest spot in the entire region. Steinitz (1973) noted that the transition between the richer and poorer parts is at about 20°N, which applies to all groups that are eventually dependent on the marine primary producers, including porpoises, sharks, sea turtles, seabirds, and fish. Nevertheless, estimates of fisheries catches in the two parts seem similar (Ben Tuvia, 1968), although data from some countries in the region are inaccurate.

Benthic productivity and production

The primary productivity of a few seagrasses (Wahbeh, 1983) and several common benthic algae (Dor and Levy, 1984) was estimated under ambient conditions within the northern Gulf of Elat (Table 16.3). In addition, the productivity of *Halophila stipulacea* was determined along a depth gradient between 3 and 30 m (Table 16.4). Most algae were studied on rocky bottoms surrounding mangal stands in southern Sinai where a measurement of *in situ* gross primary productivity of the entire community was attempted by the oxygen exchange method, using glass bells placed on the bottom over intact portions of the community. Hourly values, obtained from daily figures averaged over 24 hours, were 375, 146 and 67 mg O^2 per square metre of bottom, respectively, for the *Spyridia filamentosa*, *Digenea simplex*, and blue-green mat communities (Dor and Levy, 1984). Lack of quantitative information on plant communities in the region (area covered by each, extent of standing crops etc.) does not allow use of these data for estimation of potential contribution of algal communities to the entire ecosystem. *In vitro* productivity of two algal symbionts of epiphytic foraminifers from *H. stipulacea* leaves (in the absence of the host) was 6.14 and 8.78 μmol C (mg dry weight)$^{-1}$ h^{-1}. They consumed about 60% through respiration, retained 35 to 40% within their cells, and released to the media only 0.3 to 0.5% (Saks, 1981). Higher light intensities than those used in the experiments could cause higher

TABLE 16.3

Gross primary productivity, dark respiration [both in mg O_2 (g dry weight)$^{-1}$h^{-1}] and P/R ratio of some seagrasses and benthic algae in the northern Gulf of Elat

Plant species	Productivity	Respiration	P/R
Seagrasses[a]			
Halophila ovalis	2.1 (1.5–3.2)	0.9 (0.7–1.1)	2.3
Halophila stipulacea	1.5 (0.9–2.9)	0.3 (0.2–0.4)	6.1
Halodule uninervis	3.2 (1.0–4.5)	0.4 (0.3–0.5)	9.4
Algae[b]			
Valonia aegagropila	5.6±3.4	2.9±0.4	1.9
Mats of blue-greens	1.8±0.5	0.9±0.1	2.1
Cystoseira trinodis	5.0±2.5	1.7±0.1	2.9
Hormophysa cunaeiformis	8.7±4.5	2.7±0.2	3.2
Sargassum subrepandum	10.0±2.1	3.0±0.2	3.3
Caulerpa racemosa	28.0±10.0	7.6±0.7	3.7
Digenea simplex	6.1±2.5	1.6±0.1	3.7
Cystoseira myrica	9.3±3.2	2.4±0.1	3.8
Padina pavonica	12.0±7.0	3.0±0.2	4.0
Laurencia papillosa	13.8±4.2	3.5±0.2	4.0
Spyridia filamentosa	17.0±5.1	4.3±0.4	4.0

[a]From Wahbeh (1983); mean values for productivity, with range of values, rounded to nearest decimal; P/R based on unrounded values.

[b]Based on data presented by Dor and Levy (1984); mean maximal values for productivity rounded to nearest decimal; respiration calculated, not measured directly; P/R based on unrounded values.

TABLE 16.4

Primary productivity, respiration [both in mg O_2 (g dry weight)$^{-1}$h^{-1}] and P/R ratio of *Halophila stipulacea* along a depth gradient in the northern Gulf of Elat. (After Wahbeh, 1983.)

Depth (m)	Productivity	Respiration	P/R
3	1.3	0.2	5.8
10	1.1	0.3	4.1
20	1.0	0.3	4.0
30	0.8	0.5	1.8

Values rounded to nearest decimal, P/R based on unrounded values.

production, and in the presence of the host a larger portion of the production is possibly released and available to the host forams. With *in situ* experiments, photosynthesis of *H. stipulacea* was maximal at 100 W m^{-2}, whereas the maximum for *Halophila ovalis* occurred at 150 W m^{-2}. Halodule uninervis did not reach its maximum values even at 160 W m^{-2}, which was the most intensive irradiance used. The average production of organic matter by *H. stipulacea* at Al 'Aqabah was 2.2 g dry weight m^{-2} day^{-1} for a period betwewen mid-August and early January, with a maximum in November (3.2 g dry weight m^{-2} day^{-1}) and a minimum in late December (Wahbeh, 1983, 1984).

Production of living material in wide seagrass beds in the Red Sea is rather high. In the entire Red Sea it may be estimated to be at least 7500 to 8000 t dry weight yr^{-1}, which is probably a very conservative estimate and does not include the production of epiphytes. Seagrasses contribute to the ecosystem mostly through the organic detritus they produce and the epiphytic algae they carry.

Consumption, grazing and predation

Reef-dwelling herbivorous fish and sea urchins are responsible for the halo zones several metres wide between the reefs and the beds of all seagrasses, except for *Thalassodendron ciliatum*. The width of this zone depends on predation pressure on the populations of these herbivores — the lighter the pressure the farther from the protective reef they dare to reach. In seagrass beds the production of fast-growing epiphytic algae is cropped by schools of grazing fish, such as rabbitfishes (*Siganus* spp.), that frequently visit the beds and are often observed nibbling at the leaves. Under very heavy grazing pressure some algae of this epiphytic community have even developed special means to cope with it, such as the missile-shaped propagules of epiphytic *Centroceras* that markedly enhance its propagation capabilities (Lipkin, 1977b). By contrast, direct consumers of seagrasses are very few but include some gastropods. Sea cows or dugongs (*Dugong dugon*) and green turtles (Chelonia mydas), which are present throughout the Red Sea and the Gulf of Aden, feed almost exclusively on seagrasses, especially delicate species (Gohar, 1957; Lipkin, 1975b). However, this is certainly not the case for *Posidonia oceanica*, which is reported by Hirth and Carr (1970) and cited by Mortimer (1982) as food of the green turtle in the Gulf of Aden since this seagrass does not occur in the region. Every dugong consumes large amounts of seagrass, preferring the more delicate forms. Even with their grazing pressure concentrated on delicate species, the extremely low density of dugong populations in the Red Sea accounts for the negligible impact their grazing has on seagrass beds. Although the green turtle has extensive feeding grounds on the coast of South Yemen on the Gulf of Aden and in the Red Sea, its populations in the latter are also small. Only 600 are estimated for the entire coast of South Yemen, and they do not affect seagrass beds markedly. However, on the coast of North Yemen, which is only 2.4 times longer than that of Yemen, the population was estimated at 30 000, but their impact on seagrass beds has not been reported. Caloric values of Gulf of Aden seagrasses range between 3.01 and 3.12 kcal (g dry weight)$^{-1}$. The nesting grounds of turtles in the Red Sea, as elsewhere, are continuously shrinking and being destroyed by human activities (Hirth et al., 1973; Ross and Barwani, 1982; Sella, 1982). According to Wahbeh and Mahasneh (1985), the major consumers of *Halophila stipulacea* at Al 'Aqabah are the sea urchin *Tripneustes gratilla* and the rabbitfish *Siganus rivulatus*. Wahbeh (1984) reported fish bites on *H. stipulacea* leaves in Al 'Aqabah. On the Sinai side of the Gulf of Elat, however, *H. stipulacea* forms only a small portion of the stomach contents of this and other *Siganus* species (Lundberg and Lipkin, 1979), which consume mostly seaweed. Only one of 141 specimens studied had a stomach exclusively filled with seagrass. In others, the relatively few seagrass fragments in their stomachs were probably casually ingested together with the epiphytic algae that grew on them. In addition, small, green gastropods of the species *Smaragdia rangiana* also feed solely on the seagrasses. They cut through the epidermis on one side of the leaf to get to the soft tissues inside, creating little "glassed" windows with the epidermis on the other side of the leaf. Their populations also are too small to have a considerable impact on seagrass beds. A small portion of seagrass production enters the food web in the form of dissolved organic matter, which may be utilized directly by bacteria attached to leaf surfaces. These may in turn be consumed by micrograzers. The largest density of such bacteria on seagrass leaves in the northern Gulf of Elat occurs during summer, the period of most intensive photosynthetic activity (Wahbeh and Mahasneh, 1984).

The bulk of the seagrass biomass thus enters the food web *via* the decomposers. Little information is available as to the rate of decomposition of seagrass remains in the Red Sea. Wahbeh and Mahasneh (1985) reported that when leaves of *Halophila stipulacea* were exposed to decomposers, nonsporogenous forms of bacteria quickly broke down labile organic compounds. Subsequently, the numbers of bacteria on the decomposing leaves decreased as they were consumed by detritivorous meiofauna. Most important among these detritivores at Al 'Aqabah were nematodes (32–91%). Later, a second peak of bacteria built up when populations of micro-organisms that break down structural carbohydrates occurred. Among these, species of *Arthrobacter*, *Bacillus*, and *Pseudomonas* were the most important. Starting with experiments with bleached leaves that had already lost most of their labile organic compounds, the further loss of weight was as follows: 40% after 100 days, 55% after 200 days, and 60% after 300 days. The largest density of micro-organisms on the leaves of the second group occurred some 250 days after initiation of the experiments.

Other seagrasses decompose even slower than Halophila stipulacea, especially *Thalassodendron ciliatum*. Leaves of *T. ciliatum* contain large numbers of "tannin cells" and are extremely resistant to decomposition. As a result they accumulate in great numbers near the beds, are carried onto the beach, and, after drying, are driven by the winds inland to great distances. This is the only seagrass with leaves found in large amounts well away from the coastline in Sinai (Lipkin, 1987a). The woody stems of the plant, the only seagrass in the Red Sea with heavily lignified stems, are even more difficult to decompose than its leaves. The stems usually remain *in situ* and die after becoming buried by accumulation of sediments in the bed; they very slowly decompose, eventually becoming carbonized (Lipkin, 1977a).

Algal production is mostly harvested by fish and, to a lesser extent, by sea urchins (mostly *Diadema setosum*) and large gastropods (*Tectus dentatus*). Surgeonfish (mostly *Acanthurus nigrofuscus*, plus *Ctenochaetus striatus*) and parrotfish (*Scarus soridus* and two other unidentified *Scarus* spp.) formed 98.4% of the grazing fish on platforms and adjacent habitats in the northern Gulf of Elat and were common in all habitats.

Zebrasoma xanthurum occurred largely in fringing lagoons, *Zebrasoma veliferum* mostly on the platform, and *Siganus argenteus* and *Siganus luridus* mostly at the seaward edge of the platform but no deeper than 5 m. The highest density and diversity of herbivores occur at the active, living reef — i.e. the seaward margin of the platform (fore reef) and nearby knolls at 2 m depth (Table 16.5). Surgeonfish dominated among the grazing fishes to a depth of 5 m, and at greater depths parrotfish took over. Surgeonfish comprised about 63% and parrotfish 35% of the grazers at the seaward edge; on the platform the former comprised 54–65%, increasing to 85% in the fringing lagoon. In shallow waters (about 3 m deep) in the southern Red Sea the density of all fish was estimated at 0.05 m^{-2}. The ratio of surgeonfish to parrotfish was estimated at 1.07, compared with 1.8 in the north. The standing crop was estimated conservatively at 10 g dry weight m^{-2} (Clark et al., 1968; Bouchon-Navaro and Harmelin-Vivien, 1981).

Some surgeonfish (*Acanthurus nigrofuscus* and *Acanthurus sohal*) have symbiotic micro-organisms in their gut, including bacteria, trichomonadids, flagellates, and a peculiar unicellular organism. The latter represents an undescribed group of protists, which is very common in the gut contents ($2–10 \times 10^4$ cells mm^{-3}). These various protists presumably help the fish break down compounds in their algal diet. Other surgeonfish (*Ctenochaetus striatus*, *Zebrasoma veliferum*, and *Zebrasoma xanthurum*) and herbivorous fish (*Siganus luridus*), which feed on these algae in the same area (often in the same schools) do not contain such symbionts (Fishelson et al., 1985b). Rabbitfish, which represent only 2% of the herbivores on the platform, are important grazers of algal epiphytes of seagrass beds. The three species present in the Red Sea (*Siganus argenteus*, *Siganus luridus*, and *Siganus rivulatus*) differ from one another in their preferences for food components and thus in the composition of their diet. In the Gulf of Elat, *S. argenteus* and *S. rivulatus* took reds (27–32%) and browns (33–42%) in about the same proportion and took fewer greens (18–20%); the rest of their diet consisted of blue-greens and seagrass fragments. By contrast, *S. luridus* consumed mostly browns (83%), mainly three robust species: *Lobophora variegata* (19%), *Cystoseira myrica* (18%), and *Sargassum* spp. (14.5%) — see Lundberg and Lipkin (1979). Fishelson et al. (1985a) showed that the accumulation of fat in special bodies associated with the gonads of surgeonfish (especially *Acanthurus nigrofuscus*) was seasonal and correlated with variations in green algal growth (in the northern Gulf of Elat) preceding spawning.

Some predators at different trophic levels, mostly specialists, occur in the Red Sea; their staple foods are listed in Table 16.6. Their activity determines the composition and spatial distribution of communities in which their prey takes part. Thus, there are marked differences in ecological ranges and niche inhabitance of different species of cerithiid gastropods and mussels between the Gulf of Elat and the Gulf of Suez; these differences were attributed by Ayal and Safriel (1982a, b) to the more numerous fish predators in the Gulf of Elat. At higher trophic levels in the Red Sea one finds a wide variety of predators of different kinds. Uppermost in the food web are the large marine creatures (fishes, sharks, and porpoises) living within the ecosystem, as well as terrestrial animals (seabirds or man) that take their prey out of the ecosystem onto land. Quantitative studies on the food of predators of the highest tropic levels in the Red Sea are rare. One study analysed the diet of the osprey, *Pandion haliaëtus*, from the island of

TABLE 16.5

Fish populations in the Gulf of Elat[a]

Habitat	Number of species	Density (m^{-2})
Lagoon	7	0.13
Landward and central parts of platform	10	0.15
Seaward part of platform	14	0.18
Knolls (2 m depth)	17	0.23
Gentle sandy slope		
5 m	15	0.08
10 m	12	0.11
20 m	12	0.06
30 m	9	0.05
40 m	7	0.04

[a]Based on data presented by Bouchen-Navaro and Harmelin-Vivien (1981).

TABLE 16.6

Some predators at various trophic levels, permanent or visiting, in the shallow waters of the Red Sea and their staple foods. (Based mainly on Baranes, 1983; Dafni, 1983; Fishelson, 1983; Lewinsohn, 1983; Sella, 1983.)

Predator	Staple food	Remarks
Sea anemones		
Gyrostoma helianthus	Fish and invertebrates	The largest sea anemone in the Red Sea
Macrodactyla aspera	Small fish and crustaceans	
Radianthus ritteri	Small creatures (including fish)	
Crustaceans		
Calappa hepatica	Sand invertebrates	
Hymenocera picta	Starfish (including *A. planci*)	Living in pairs cooperating in feeding
Panulirus penicillatus	Mainly molluscs and sea urchins	
Scyllarides tridacnophaga	Clams (including giant clams)	
Gastropods		
Charonia tritonis	Clams, starfish (including *A. planci*)	Largest gastropod in the Red Sea
Chicoreus ramosus	Clams, including giant clams	Reef dweller, largest Red Sea muricid
Conus arenatus	Sand-dwelling worms and crustaceans	In sand
Conus textile	Fish and invertebrates	Most poisonous cone, among stones
Nassa francolina	Worms, soft invertebrates	Under stones in the sublittoral
Nassarius albescens	Sand worms and crustaceans	In sand
Scabricola fissurata	Sand worms and clams	In sand
Thais savignyi	Barnacles (*Tetraclita*), other gastropods, clams	Burrows in shells
Tonna perdix	Clams, also sea-cucumbers	Burrows in shells, on sandy bottom
Sea slugs		
Berthellina citrina	Mainly sponges	Among the commonest Red Sea sea slugs
Chromodoris quadricolor	Sponges, including the poisonous *Sigmosceptrella magnifica*	Nocturnal
Hexabranchus sanguineus	Sponges, invertebrates	Nocturnal, in reefs and rocks
Cephalopods		
Octopus cyaneus	Invertebrates and fish	Nocturnal
Sepioteuthis lessoniana	Pelagic fish	Approaching coast to spawn
Starfishes		
Acanthaster planci	Corals	Nocturnal
Astropecten polyacanthus	Gastropods, clams, small sea urchins	Nocturnal
Choriaster granulatus	Mostly corals	Nocturnal
Luidia maculata	Other starfishes, sea urchins, brittle stars	Sand-dweller, rather rare

Tirân in the northern Red Sea. It was shown that its diet was greatly influenced by the nest's location. With nests located near the open sea and separated from it only by a narrow platform, 92% of the fish taken were carnivores and only 8% herbivores. Where nests were located farther from the sea, behind a wide platform, the ratio was nearly 3:2. Diversity of prey fish in the diet of birds nesting near the open water was rather limited, with the open-water fish *Fistularia petimba* composing most of the food. Birds nesting farther from open waters had more diverse diets without preference for a certain fish; only 28% of the specimens taken were *F. petimba* (Ben Hur, 1982).

Fishing in the Red Sea is not very intensive. In the poorer northern part it is mostly limited to coastal fishing by the local sparse population of nomadic bedouin. In the richer south a limited

Table 16.6 (*continued*)

Sharks		
Carcharhinus melanopterus	Reef fish (parrot fish, labrids), also squids and octopuses	Most common shark in the Red Sea
Galeocerdo cuvier	Various fish including rays, sharks, large rock and reef fish	Large, pelagic, occasionally near shore
Mustelus mosis	Clams, also bites bottom fish	Approaches shore at night
Negaprion acutidens	Fish occurring near reefs and on the bottom	In shallows, near the bottom
Sphyrna mokarran	Small sharks, groupers, parrot fish, squids, lobsters etc.	Large, sometimes approaches the shore
Stegostoma varium	Bottom fish and invertebrates	Bottom dweller in shallows
Triaenodon obesus	Reef fish, also crustaceans	In shallows near coral reefs
Rays		
Aëtobatus narinari	Clams, crabs, a few worms and other bottom invertebrates	In groups
Rhina ancylostomus	Bottom invertebrates: crustaceans, squids, sometimes bottom fish: flatfish, goatfish	Nocturnal, buried during day with only eyes and spiracles above sand
Rhinobatos halavi	Crabs, small fish	
Rhynchobatus djiddensis	Crabs, bottom fish	Mostly nocturnal
Taeniura lymma	Bottom worms, shrimps	
Torpedo sinuspersici	Crustaceans, worms, small fish	Buried in sand with eyes and spiracles exposed
Other fish		
Acanthopagrus bifasciatus	Hard-shell invertebrates, mainly clams and gastropods	Also eats strongly attached species like *Chama* or *Modiolus*
Adioryx diadema	Small fish and invertebrates	Nocturnal
Cephalopholis miniata	Small fish and crustaceans	This and other species of the group compose most of the predator biomass in the shallows
Cheilinus trilobatus	Gastropods, a few crabs and hermit crabs	
Diplodus noct	Molluscs, worms, crustaceans	Near the bottom
Fistularia petimba	Small fish	
Hemigymnus fasciatus	Gastropods, a few crabs and hermit crabs	
Coris aygula	Crabs, a few sea urchins	
Lethrinus mahsena	Crustaceans, clams, squids, sea urchins, algae	Mostly nocturnal rock fish
Lutjanus bohar	Small fish and invertebrates	
Mulloides flavolineatus	Burrowing bottom invertebrates	In small to large schools
Pterois volitans	Small fish and invertebrates	In rocks and reefs, poisonous
P. radiata	Prefers invertebrates to fish	
Sphyraena chrysotaenia	Small fish and crustaceans	
Synanceia verrucosa	Fish near the bottom	On soft bottom, poisonous
Sea turtles		
Caretta caretta	Small benthic invertebrates, invalid fish	Uncommon
Dermochelys coriacea	Mostly medusae	Usually pelagic

offshore fishery has developed as well, in which a few trawlers are involved. Estimates of the annual catch were *c.* 40 000 metric tonnes for the entire Red Sea in the mid-sixties (Ben Tuvia, 1968). Today the proportion of production taken by man from the marine system cannot be dramatically higher.

REFERENCES

Abel, E.F., 1960. Zur Kenntnis des Verhaltens und der Ökologie von Fische an Korallenriffen bei Ghardaqa (Rotes Meer). *Z. Morphol. Ökol. Tiere*, 149: 430–503.

Achituv, Y., 1972. The zonation of *Tetrachthamalus oblitteratus* Newman, and *Tetraclita squamosa rufotincta* Pilsbry in the Gulf of Elat, Red Sea. *J. Exp. Mar. Biol. Ecol.*, 8: 73–81.

Achituv, Y., 1981. Interspecific competition between *Tetrachthamalus oblitteratus* and *Tetraclita squamosa. Mar. Ecol.*, 2: 241–244.

Achituv, Y. and Klepal, W., 1981. Distribution and diversity of *Ibla cumingi* Darwin (Crustacea, Cirripedia) from the Gulf of Elat (Red Sea). *Mar. Ecol.*, 2: 295–305.

Adar, O., 1980. The effect of the endosymbiotic zooxanthellae on asexual reproduction and growth of *Cassiopea andromeda* (Coelenterata). *Isr. J. Zool.*, 29: 203–204 (abstract).

Aleem, A.A., 1979. A contribution to the study of seagrasses along the Red Sea coast of Saudi Arabia. *Aquat. Bot.*, 7: 71–78.

Ayal, Y. and Safriel, U.N., 1980. Intertidal zonation and key-species associations of the flat rocky shores of Sinai, used for scaling environmental variables affecting cerithiid gastropods. *Isr. J. Zool.*, 29: 110–124.

Ayal, Y. and Safriel, U.N., 1982a. Species diversity of the coral reef — a note on the role of predation and of adjacent habitats. *Bull. Mar. Sci.*, 32: 787–790.

Ayal, Y. and Safriel, U.N., 1982b. Role of competition and predation in determining habitat occupancy of Cerithiidae (Gastropoda: Prosobranchia) on the rocky, intertidal, Red Sea coasts of Sinai. *Mar. Biol.*, 70: 305–316.

Ayal, Y. and Safriel, U.N., 1983. Does a suitable habitat guarantee successful colonization? *J.Biogeogr.*, 10: 37–46.

Baeshin, N.A. and Aleem, A.A., 1978. Littoral vegetation at Rabegh (Red Sea Coast) Saudi Arabia. *Bull. Fac. Sci. K.A.U., Jeddah*, 2: 123–130.

Baranes, A., 1983. Selachii. In: L. Fishelson (Editor), *Aquatic Life. Plants and Animals of the Land of Israel, Vol. 4*, Ministry of Defense, The Publishing House and Society for Protection of Nature, Tel Aviv, pp. 130–134 [in Hebrew].

Beckmann, W., 1984. Mesozooplankton distribution on a transect from the Gulf of Aden to the central Red Sea during the winter monsoon. *Oceanol. Acta*, 7: 87–102.

Ben Eliahu, N.M., 1975. Polychaete cryptofauna from rims of similar intertidal vermetid reefs on the Mediterranean coast of Israel and in the Gulf of Elat: Sabellidae (Polychaeta Sedentaria). *Isr. J. Zool.*, 24: 54–70.

Ben Eliahu, M.N. and Safriel, U.N., 1982. A comparison between species diversities of polychaetes from tropical and temperate structurally similar rocky intertidal habitats. *J. Biogeogr.*, 9: 371–390.

Ben Hur, Y., 1982. *The diet of an ultimate predator in the coral reef — diet of the osprey (Pandion haliaetus) population on Tiran Island.* M.Sc. thesis, Department of Zoology, The Hebrew University, Jerusalem, 99 pp. [in Hebrew].

Ben Tuvia, A., 1968. Report on the fisheries investigation of the Israel South Red Sea Expedition, 1962. *Sea Fish. Res. Stn. Haifa Bull.*, 52: 21–55.

Berner, T., Wishkovsky, A. and Dubinsky, Z., 1986. Endozoic algae in shelled gastropods — a new symbiotic association in coral reef? I. Photosynthetically active zooxanthellae in *Strombus tricornis. Coral Reefs*, 5: 103–106.

Bouchon-Navaro, Y. and Harmelin-Vivien, M.L., 1981. Quantitative distribution of herbivorous reef fishes in the Gulf of Aqaba (Red Sea). *Mar. Biol.*, 63: 79–86.

Campbell, S.E., 1985. *Oscillatoria limnetica* from Solar Lake, Sinai is a *Phormidium* (Cyanophyta or Cyanobacteria). *Arch. Hydrobiol. Suppl.*, 71: 175–190.

Campbell, S.E. and Golubic, S., 1985. Benthic cyanophytes (cyanobacteria) of Solar Lake (Sinai). *Arch. Hydrobiol. Suppl.*, 71: 311–329.

Chapman, V.J. (Editor), 1977. *Wet Coastal Ecosystems. Ecosystems of the World, Vol. 1.* Elsevier, Amsterdam. 428 pp.

Clark, E., 1966. Pipefishes of the genus *Siokunichthys* Herald in the Red Sea with description of a new species. *Sea Fish. Res. Stn. Haifa Bull.*, 41: 3–6.

Clark, E., 1971. Observations on a garden eel colony at Elat. *The Hebrew Univ. Jerusalem Mar. Biol. Lab. Sci. Newsletter*, 1: 5.

Clark, E., 1972. The Red Sea's gardens of eels. *Nat. Geogr. Mag.*, 142: 724–735.

Clark, E. and Ben Tuvia, A., 1973. Red Sea fishes of the family Branchiostegidae with a description of a new genus and species *Asymmetrurus oreni. Sea Fish. Res. Stn. Haifa Bull.*, 60: 63–74.

Clark, E. and Chao, S., 1972. A toxic secretion from the Red Sea flatfish, *Pardachirus marmoratus* (Lacepede). *The Hebrew Univ. Jerusalem Mar. Biol. Lab. Sci. Newsletter*, 2: 14 (reprinted with slight changes and illustrations in *Sea Fish. Res. Stn. Haifa Bull.*, 60: 53–56, 1973).

Clark, E. and Gohar, H.A.F., 1953. The fishes of the Red Sea: order Plectognati. *Publ. Mar. Biol. Stn. Ghardaqa*, 8: 3–80.

Clark, E. and Von Schmidt, K., 1966. A new species of *Trichonotus* (Pisces, Trichonotidae) from the Red Sea. *Sea Fish. Res. Stn. Haifa Bull.*, 42: 29–36.

Clark, E., Ben Tuvia, A. and Steinitz, H., 1968. Observations on a coastal fish community, Dahlak Archipelago, Red Sea. *Sea Fish. Res. Stn. Haifa Bull.*, 49: 15–31.

Cohen, J., 1970. *Octocorallia from the Gulf of Elat and ecological observations on them.* M.Sc. thesis, Department of Zoology, The Hebrew University, Jerusalem, 93 pp. [in Hebrew].

Cohen, Y., Krumbein, W.E. and Shilo, M., 1975. The Solar Lake: limnology and microbiology of a hypersaline, meromictic heliothermal heated sea-marginal pond (Gulf of Aqaba, Sinai). *Rapp. Comm. Int. Mer Mediterr.*, 23 (3): 105–107.

Costa, M., 1965. *Andregamasus* n. gen., a new genus of mesostigmatic mites associated with terrestrial hermit crabs. *Sea Fish. Res. Stn. Haifa Bull.*, 38: 6–14.

Costa, M., 1972. Notes on mites (Acari) associated with the hermit crab *Coenobita scaevola* Forskal. *Isr. J. Zool.*, 21: 41–48.

Dafni, J., 1983. Gastropoda, Bivalvia. In: L. Fishelson (Editor), *Aquatic Life. Plants and Animals of the Land of Israel, Vol 4.* Ministry of Defense, The Publishing House and Society for Protection of Nature, Tel Aviv, pp. 184–191, 195–198 [in Hebrew].

Darom (Masry), D., 1974. *A systematic and ecological study of some interstitial Crustacea from sandy beaches along the*

Gulf of Elat (Aqaba). Ph.D. Thesis, The Hebrew University, Jerusalem, 100 pp. [in Hebrew].

Den Hartog, C., 1970. The sea-grasses of the world. *Verh. K. Ned. Akad. Wet. Afd. Natuurkd. Reeks 2*, 59 (1): 1–275.

Dimentman, C. and Spira, J., 1982. Predation of *Artemia* cysts by water tiger larvae of the genus *Anacaena* (Coleoptera, Hydrophilidae). *Hydrobiologia*, 97: 163–165.

Dor, I. and Levy, I., 1984. Primary productivity of the benthic algae in the hard-bottom mangal of Sinai. In: F.D. Por and I. Dor (Editors), *Hydrobiology of the Mangal*. W. Junk, The Hague, pp. 179–191.

Dor, M., 1984. *CLOFRES — Checklist of the Fishes of the Red Sea*. Israel Academy of Sciences and Humanities, Jerusalem, 437 pp.

Dubinsky, Z. (Editor), 1990. *Coral Reefs. Ecosystems of the World, Vol. 25*. Elsevier, Amsterdam.

Eckstein, Y., 1970. Physicochemical limnology and geology of a meromictic pond on the Red Sea shore. *Limnol. Oceanogr.* 15: 363– 372.

Eschmeyer, W.N. and Rama Rao, K.V., 1973. Two new stonefishes (Pisces, Scorpenidae) from the Indo-West Pacific, with a synopsis of the subfamily Synanceiinae. *Proc. Pacif. Acad. Sci.* 4th Ser., 39 (18): 337–382.

Ferber, I., 1973. *Contributions to the biology of Lovenia elongata (Gray) (Echinoidea: Spatangoidea) in the Gulf of Elat, Red Sea*. M.Sc. Thesis, Oceanography, The Hebrew University, Jerusalem, 78 pp.

Ferber, I., 1983. Echinoidea. In: L. Fishelson (Editor), *Aquatic Life. Plants and Animals of the Land of Israel, Vol. 4*. Ministry of Defense, The Publishing House and Society for Protection of Nature, Tel Aviv, pp. 169–174 [in Hebrew].

Ferber, I. and Lawrence, J.M., 1976. Distribution, substratum preference and burrowing behaviour of *Lovenia elongata* (Gray) (Echinoidea: Spatangoida) in the Gulf of Elat ('Aqaba), Red Sea. *J. Exp. Mar. Biol. Ecol.*, 22: 207–225.

Fishelson, L., 1966. *Spirastrella inconstans* Dendy (Porifera) as an ecological niche in the littoral zone of the Dahlak Archipelago (Eritrea). *Sea Fish. Res. Stn. Haifa Bull.*, 41: 17–25.

Fishelson, L., 1970. Littoral fauna of the Red Sea: the population of non-scleractinian anthozoans of shallow waters of the Red Sea (Elat). *Mar. Biol.*, 6: 106–116.

Fishelson, L., 1971. Ecology and distribution of the benthic fauna in the shallow waters of the Red Sea. *Mar. Biol.*, 10: 113–133.

Fishelson, L., 1973. Observations on skin structure and sloughing in the stonefish *Synanceja verrucosa* and related fish species as a functional adaptation to their mode of life. *Z. Zellforsch.*, 140: 497–508.

Fishelson, L. (Editor), 1983. *Aquatic Life. Plants and Animals of the Land of Israel, Vol. 4*. Ministry of Defense, The Publishing House and Society for Protection of Nature, Tel Aviv, 342 pp.

Fishelson, L., 1984. Population ecology and biology of *Dotilla sulcata* (Crustacea, Ocypodidae) typical for sandy beaches of the Red Sea. In: A. McLachlan and T. Erasmus (Editors), *Sandy Beaches as Ecosystems*. Junk, The Hague, pp. 643–654.

Fishelson, L., Montgomery, W.L. and Myrberg, Jr., A.A., 1985a. A new fat body associated with the gonads of surgeonfishes (Acanthuridae: Teleostei). *Mar. Biol.*, 86: 109–112.

Fishelson, L., Montgomery, W.L. and Myrberg, Jr., A.A., 1985b. A unique symbiosis in the gut of tropical herbivorous surgeonfish (Acanthuridae: Teleostei) from the Red Sea. *Science*, 229: 49–51.

Foin, T.C., 1972. The zoogeography of the Cypraeidae in the Red Sea basin. *Argamon*, 3: 5–16.

Fricke, H.W., 1970. Ökologische und verhaltensbiologische Beobachtungen an den Röhrenaalen *Gorgasia sillneri* und *Taenioconger hassi* (Pisces, Heterocongridae). *Z. Tierpsychol.*, 27: 1076–1099.

Fridman, D. and Masry, D., 1971. *Xiphasia setifer* Swainson, a blenniid fish new for the Gulf of Elat. *Hebrew Univ. Jerusalem Mar. Biol. Lab. Sci. Newsletter*, 1: 6.

Friedman, G.M. and Foner, H.A., 1982. pH and Eh changes in sea-marginal algal pools of the Red Sea: and their effect on carbonate precipitation. *J. Sediment. Petrol.*, 52: 41–46.

Friedman, G.M. and Krumbein, W.E. (Editors), 1985. *Hypersaline Ecosystems. The Gavish Sabkha. Ecological Studies, Analysis and Synthesis, Vol. 53*. Springer, Berlin, 484 pp.

Friedman, G.M., Amiel, A.J., Braun, M. and Miller, D.S., 1973. Generation of carbonate particles and laminites in algal mats — example from sea-marginal hypersaline pool, Gulf of 'Aqaba, Red Sea. *Am. Ass. Petrol. Geol. Bull.*, 57: 541–557.

Gerdes, G., Spira, J. and Dimentman, C., 1985. The fauna of the Gavish Sabkha and the Solar Lake — a comparative study. In: G.M. Friedman and W.E. Krumbein (Editors), *Hypersaline Ecosystems. The Gavish Sabkha. Ecological Studies, Analysis and Synthesis, Vol. 53*. Springer, Berlin, pp. 322–345.

Girdler, R.W., 1983. The evolution of the Gulf of Aden and Red Sea in space and time. *Deep Sea Res.*, 31: 747–762.

Gohar, H.A.F., 1948. A description and some biological studies of a new alcyonarian species "Clavularia hamra Gohar". *Publ. Mar. Biol. Stn. Ghardaqa*, 6: 3–33.

Gohar, H.A.F., 1957. The Red Sea dugong. *Publ. Mar. Biol. Stn. Ghardaqa*, 9: 3–49.

Gohar, H.A.F. and Aboul Ela, I.A., 1959. On the biology and development of three nudibranchs (from the Red Sea). *Publ. Mar. Biol. Stn. Ghardaqa*, 10: 41–69.

Gohar, H.A.F. and Eisawy, A.M., 1961. The biology of *Cassiopea andromeda* (from the Red Sea). *Publ. Mar. Biol. Stn. Ghardaqa*, 11: 3–42.

Goren, M., 1979. The Gobiinae of the Red Sea. *Senckenberg. Biol.*, 60: 13–64.

Goren, M., 1983. Gobiidae. In: L. Fishelson (Editor), *Aquatic Life. Plants and Animals of the Land of Israel, Vol. 4*. Ministry of Defense, The Publishing House and Society for Protection of Nature, Tel Aviv, pp. 162–165 [in Hebrew].

Goren, M., 1985. A review of the gobiid fish genus *Monishia* Smith, 1949, from the western Indian Ocean and Red Sea, with description of a new species. *Contr. Sci.*, 360: 1–9.

Goren, M. and Karplus, I., 1983. *Tomiyamichthys randalli* n. sp., a gobiid associated with a shrimp, from the Red Sea. *Senckenberg. Biol.*, 63: 27–31.

Gorgy, S., 1966. Contribution à l'étude du milieu marine et de la pêche en mer Rouge (secteur de la Republique Arab Unie). *Rev. Trav. Inst. Pêches Maritim.*, 30 (1): 93–112.

Halim, Y., 1969. Plankton of the Red Sea. *Oceanogr. Mar. Biol. Ann. Rev.*, 7: 231–275.

Hansen, H.J. and Buchardt, B., 1977. Depth distribution of *Amphistegina* in the Gulf of Elat, Israel. *Utrecht Micropaleont. Bull.*, 15: 205–224.

Heinemann, H. and Heinemann, H., 1982. Remarks on Cypraeidae from the Red Sea. *Levantina*, 41: 475–485.

Hirth, H. and Carr, A., 1970. The green turtle in the Gulf of Aden and the Seychelles Islands. *Verh. K. Ned. Akad. Wetenschap. Afd. Natuurk. Reeks 2*, 58 (5): 1–44.

Hirth, H.F., Klikoff, L.G. and Harper, K.T., 1973. Seagrasses of Khor Umaira, Peoples Democratic Republic of Yemen with reference to their role in the diet of the green turtle, *Chelonia mydas*. *Fish. Bull.*, 71: 1093–1097.

Holthuis, L.B., 1963. On red coloured shrimps (Decapoda, Caridea) from tropical land-locked saltwater pools. *Zool. Meded. Rijksmus. Nat. Hist. Leiden*, 38: 261–279.

Holthuis, L.B., 1973. Caridean shrimps found in land-locked saltwater pools at four Indo-west Pacific localities (Sinai Peninsula, Funafuti Atoll, Maui and Hawaii Islands), with the description of one new genus and four new species. *Zool. Verh. Rijksmus. Nat. Hist. Leiden*, 128: 1–48.

Holthuis, L.B., 1977. The Grapsidae, Gecarcinidae and Palicidae (Crustacea, Decapoda, Brachyura) of the Red Sea. *Isr. J. Zool.*, 26: 141–192.

Hottinger, L., 1972. *Pseudochama cornucopiae* (Linne) in Dahab, a model of rudist "reef"? *The Hebrew Univ. Jerusalem Mar. Biol. Lab. Sci. Newsletter*, 2: 6–7.

Hottinger, L., 1977. Distribution of larger Peneroplidae, *Borelis* and Nummulitidae in the Gulf of Elat, Red Sea. *Utrecht Micropaleont. Bull.*, 15: 35–110.

Hullings, N.C., 1985. Activity pattern and homing in two intertidal limpets, Jordan Gulf of 'Aqaba. *Nautilus*, 99: 75–80.

Humes, A.G. and Stock, J.H., 1965. Three new species of *Anthessius* (Copepoda, Cyclopoida, Myicolidae) associated with *Tridacna* from the Red Sea and Madagascar. *Sea Fish. Res. Stn. Haifa Bull.*, 40: 49–74.

Jacob-Judah, S., 1975. Commensal taeniacanthid copepods from the sea urchin *Diadema setosum* at Elat (Red Sea) with a description of two new species. *Isr. J. Zool.*, 24: 1–15.

Karplus, I., 1987. The association between gobiid fishes and burrowing alpheid shrimps. *Oceanogr. Mar. Biol. Ann. Rev.*, 25: 507–562.

Karplus, I. and Masry, D., 1972. A cephalopod associated with *Diadema setosum*. *The Hebrew Univ. Jerusalem Mar. Biol. Lab. Sci. Newsletter*, 2: 5.

Karplus, I., Szlep, R. and Tsurnamal, M., 1972a. Associative behaviour of the fish *Cryptocentrus cryptocentrus* (Gobiidae) and the pistol shrimp *Alpheus djiboutensis* (Alpheidae) in artificial burrows. *Mar. Biol.*, 15: 95–104.

Karplus, I., Tsurnamal, M. and Szlep, R., 1972b. Analysis of mutual attraction in the association of the fish *Cryptocentrus cryptocentrus* (Gobiidae) and the shrimp *Alpheus djiboutensis* (Alpheidae). *Mar. Biol.*, 17: 275–283.

Karplus, I., Szlep, R. and Tsurnamal, M., 1974. The burrows of alpheid shrimp associated with gobiid fish in the northern Red Sea. *Mar. Biol.*, 24: 259–268.

Karplus, I., Szlep, R. and Tsurnamal, M., 1981. Goby-fish partner specificity. I. Distribution in the northern Red Sea and partner specificity. *J. Exp. Mar. Biol. Ecol.*, 51: 1–19.

Kassas, M., 1957. On the ecology of the Red Sea coastal land. *J. Ecol.*, 55: 187–203.

Ketchum, B.H., 1983. Enclosed seas — introduction. In: B.H. Ketchum (Editor), *Estuaries and Enclosed Seas, Ecosystems of the World Vol. 26*, Elsevier, Amsterdam, pp. 209–218.

Kimor, B., 1973. Plankton relations of the Red Sea, Persian Gulf and Arabian Sea. In: B. Zeitzschel and S.A. Gerlach (Editors), *The Biology of the Indian Ocean. Ecological Studies, Analysis and Synthesis, Vol. 3*. Springer, Berlin, 549 pp.

Klausewitz, W., 1959. Fische aus dem Roten Meer. I. Knorpelfische (Elasmobranchii). *Senckenberg. Biol.*, 40: 43–50.

Klausewitz, W., 1960. Fische aus dem Roten Meer. IV. Einige systematisch und ökologisch bemerkenswerte Meergrundeln (Pisces, Gobiidae). *Senckenberg. Biol.*, 41: 149–162.

Klausewitz, W., 1962. *Gorgasia sillneri* ein neuer Rohrenaal aus dem Roten Meer (Pisces, Apodes, Heterocongridae). *Senckenberg. Biol.*, 43: 433–435.

Klausewitz, W., 1964. Fische aus dem Roten Meer. VI. Taxonomische und ökologische Untersuchungen an einigen Fischarten der Küstenzone. *Senckenberg. Biol.*, 45: 123–144.

Klausewitz, W., 1966. Fische aus dem Roten Meer. VII. *Siphamia permutata* n. sp. (Pisces, Perciformes, Apogonidae). *Senckenberg. Biol.*, 47: 217–222.

Klausewitz, W., 1967. Die physiographische Zonierung der Saumriffe von Sarso. 4. Beitrag der Arbeitsgruppe Litoralforschung. *"Meteor" Forschungserg. Reihe D*, 2: 44–68.

Klausewitz, W., 1968. Fische aus dem Roten Meer. VIII. *Biat magnusi* n. sp., eine neue Meergrundel (Pisces, Ostichytes, Gobiidae). *Senckenberg. Biol.*, 49: 13–17.

Klausewitz, W., 1969. Fische aus dem Roten Meer. XI. *Cryptocentrus sungami* n. sp. (Pisces: Gobiidae). *Senckenberg. Biol.*, 50: 41–46.

Klausewitz, W., 1970. Wiederfund von *Lotilia graciliosa* (Pisces: Gobiidae). *Senckenberg. Biol.*, 51: 177–179.

Klausewitz, W., 1974. Fische aus dem Roten Meer. XIII. *Cryptocentrus steinitzi* n. sp., ein neuer "Symbiose-Gobiide" (Pisces: Gobiidae). *Senckenberg. Biol.*, 55: 69–76.

Klausewitz, W., 1975. Fische aus dem Roten Meer. XV. *Cabillus anchialinae* eine neue Meergrundel von der Sinai-Halbinsel (Pisces: Gobiidae: Gobiinae). *Senckenberg. Biol.*, 56: 203–207.

Klunzinger, C.B., 1872. Zoologische Excursion auf ein Korallenriff des Rothen Meeres bei Kossēr. *Z. Ges. Erdkunde Berlin*, 7: 20–56.

Krumbein, W.E., 1975. The biogeochemistry of algal mats of the Solar Lake. *Rep. H. Steinitz Mar. Biol. Lab. Elat*, 4: 29 (abstract).

Krumbein, W.E. and Cohen, Y., 1974. Biogene, klastische und evaporitische Sedimentation in einem mesothermen monomiktischen ufernahen See (Golf von Aqaba). *Geol. Rundsch.*, 63: 1035–1065.

Lawrence, J.M. and Ferber, I., 1971. Substrate particle size and the occurrence of *Lovenia elongata* (Echinodermata: Echinoidea) at Taba, Gulf of Elat (Red Sea). *Isr. J. Zool.*, 20: 131–138.

Lee, J.J., McEnery, M.E., Shilo, M. and Reiss, Z., 1979. Isolation and cultivation of diatom symbionts from larger Foraminifera (Protozoa). *Nature*, 280: 57–58.

Levanon-Spanier, I., Padan, E. and Reiss, Z., 1979. Primary production in a desert-enclosed sea — the Gulf of Elat (Aqaba), Red Sea. *Deep Sea Res.*, 6A: 673–685.

Levi, C., 1965. Spongiaires recoltés par l'expédition israélienne dans le sud de la mer rouge en 1962. *Sea Fish. Res. Stn. Haifa Bull.*, 40: 3–27.

Lewinsohn, C., 1969. Die Anomuren des Roten Meeres (Crustacea Decapoda: Paguridea, Galatheidea, Hippidea). *Zool. Verhand. Rijksmus. Nat. His. Leiden.*, 104: 1–213.

Lewinsohn, C., 1977a. Die Dromiidae des Roten Meeres (Crustacea Decapoda, Brachyura). *Zool. Verhand. Rijksmus. Nat. His. Leiden*, 151: 1–41.

Lewinsohn, C., 1977b. Die Ocypodidae des Roten Meeres (Crustacea Decapoda, Brachyura). *Zool. Verhand. Rijksmus. Nat. His. Leiden*, 152: 43–84.

Lewinsohn, C., 1983. Crustacea Decapoda. In: L. Fishelson (Editor), *Aquatic Life. Plants and Animals of the Land of Israel, Vol. 4.* Ministry of Defense, The Publishing House and Society for Protection of Nature, Tel Aviv, pp. 200–209 [in Hebrew].

Lewinsohn, C. and Fishelson, L., 1968. The second Israel South Red Sea Expedition, 1965 (general report). *Isr. J. Zool.*, 16: 59–68.

Lipkin, Y., 1968. Red Sea vegetation. I. *Encyclopedia Hebreica, 19.* Encyclopedia Publ. Co., Jerusalem, p. 899 [in Hebrew].

Lipkin, Y., 1973. Vegetation of the Bitter Lakes in the Suez Canal water system. *Isr. J. Zool.*, 21: 447–457.

Lipkin, Y., 1974. Ecological distribution of *Caulerpa* in the Red Sea. *J. Mar. Biol. Ass. India*, 15: 160–167.

Lipkin, Y., 1975a. A history, catalogue and bibliography of Red Sea seagrasses. *Isr. J. Bot.*, 24: 89–105.

Lipkin, Y., 1975b. Food of the Red Sea *Dugong* (Mammalia: Sirenia) from Sinai. *Isr. J. Zool.*, 24: 81–98.

Lipkin, Y., 1977a. Seagrass vegetation of Sinai and Israel. In: C.P. McRoy and C. Helfferich (Editors), *Seagrass Ecosystems. A Scientific Perspective.* Dekker, New York, pp. 263–293.

Lipkin, Y., 1977b. *Centroceras*, the "missile"-launching marine red alga. *Nature*, 270: 48–49.

Lipkin, Y., 1979. Quantitative aspects of seagrass communities, particularly of those dominated by *Halophila stipulacea*, in Sinai (northern Red Sea). *Aquat. Bot.*, 7: 119–128.

Lipkin, Y., 1987a. Seagrasses on the coasts of Sinai. In: A. Shmueli and G. Gvirtzman (Editors), *Sinai Book.* Dept. of Geography, Tel Aviv University, pp. 495–504 [in Hebrew].

Lipkin, Y., 1987b. Marine algal vegetation of Sinai. In: A. Shmueli and G. Gvirtzman (Editors), *Sinai Book.* Dept. of Geography, Tel Aviv University, pp. 505–514 [in Hebrew].

Lipkin, Y., 1987c. Marine vegetation of the Museri and Entedebir Islands (Dahlak Archipelago, Red Sea). *Isr. J. Bot.*, 36: 87–99.

Lipkin, Y., 1988. *Thalassodendretum ciliati* in Sinai (northern Red Sea) with special reference to quantitative aspects. *Aquat. Bot.*, 31: 125–139.

Lundberg, B. and Lipkin, Y., 1979. Natural food of the herbivorous rabbitfish (*Siganus* spp.) in northern Red Sea. *Bot. Mar.*, 22: 173–181.

Luther, W., 1958a. Symbiose von Fischen (Gobiidae) mit einem Krebs (*Alpheus djiboutensis*) im Roten Meer. *Z. Tierpsychol.*, 15: 175–177.

Luther, W., 1958b. Symbiose von Fischen mit Korallentieren und Krebsen im Roten Meer. *Nat. Volk*, 88: 141–146.

Lützen, J., 1976. On a new genus and two new species of Prosobranchia (Mollusca), parasitic on the tropical sea urchin *Echinometra mathaei*. *Isr. J. Zool.*, 25: 38–51.

Magnus, D.B.E., 1967. Zur Ökologie sedimentbewohnender *Alpheus*-Garnelen (Decapoda, Natantia) des Roten Meeres. *Helgoländer Wiss. Meeresunters.*, 15: 506–522.

Marbach, A. and Tsurnamal, M., 1973. On the biology of *Berthellina citrina* (Gastropoda: Opistobranchia) and its defensive acid secretion. *Mar. Biol.*, 21: 331–339.

McGill, D.A., 1973. Light and nutrients in the Indian Ocean. In: B. Zeitzschel and S.A. Gerlach (Editors), *The Biology of the Indian Ocean. Ecological Studies, Analysis and Synthesis, Vol. 3.* Springer, Berlin, pp. 53–102.

Mergner, H., 1984. The ecological research on coral reefs of the Red Sea. In: M.V. Angel (Editor), *Marine Science of the North-west Indian Ocean and Adjacent Waters, Proc. Mabahiss/John Murray Int. Symp., Egypt, 3–6 Sept., 1983. Deep Sea Res.*, 31: 855–884.

Mortimer, J.A., 1982. Feeding ecology of sea turtles. In: K.A. Bjorndal (Editor), *Biology and Conservation of Sea Turtles, Proc. World Conf. Sea Turtle Conserv., Washington, D.C., 26–30 Nov., 1979*, Smithsonian Inst. Press and World Wildlife Fund, pp. 103–109.

Nasr, D.H., 1982. Observations on the mortality of the pearl oyster, *Pinctada margaritifera*, in Dongonab Bay, Red Sea. *Aquaculture*, 28: 271–281.

Neumann, A.C. and McGill, D.A., 1962. Circulation of the Red Sea in early summer. *Deep Sea Res.*, 8: 223–235.

Nir, Y., 1971. Geology of Entedebir Island and its recent sediments, Dahlak Archipelago — southern Red Sea. *Sea Fish. Res. Stn. Haifa Bull.*, 58: 9–42.

Oren, O.H., 1962. The Israel South Red Sea Expedition. *Nature*, 194: 1134–1137.

Papenfuss, G.F., 1968. A history, catalogue and bibliography of Red Sea benthic algae, *Isr. J. Bot.*, 17: 1–118.

Pomeranz, E., and Tsurnamal, M., 1976. The symbiotic association between the shrimp, *Athanas indicus*, and the sea urchin, *Echinometra mathaei*, in the Gulf of Elat. *Isr. J. Zool.*, 25: 204–205 (abstract).

Por, F.D., 1968. Copepods of some land-locked basins of the islands of Entedebir and Nocra (Dahlak Archipelago, Red Sea). *Sea Fish. Res. Stn. Haifa Bull.*, 49: 32–50.

Por, F.D., 1972. Hydrobiological notes on the high-salinity waters of the Sinai Peninsula. *Mar. Biol.*, 14: 111–119.

Por, F.D., 1975. A typology of the nearshore seepage pools of Sinai. *Rapp. Comm. Int. Mer. Mediterr.*, 23 (3): 103.

Por, F.D., 1977. Animal food chains of Solar Lake and the predominance of Coleoptera. *Rep. H. Steinitz Mar. Biol. Lab. Elat*, 6: 25 (abstract).

Por, F.D., (Editor) 1978. *Lessepsian Migration. The Influx of Red Sea Biota into the Mediterranean by way of the Suez Canal, Ecological Studies, Analysis and Synthesis, Vol. 23.* Springer, Berlin, 228 pp.

Por, F.D., 1979. The Copepoda of Di Zahav pool (Gulf of Elat, Red Sea). *Crustaceana*, 37: 13–30.

Por, F.D., 1985. Anchialine pools — comparative hydrobiology. In: G.M. Friedman and W.E. Krumbein (Editors). *Hypersaline Ecosystems. The Gavish Sabkha. Ecological Studies, Analysis and Synthesis, Vol. 53.* Springer, Berlin, pp. 136–144.

Por, F.D. and Dor, I., 1975. Ecology of the metahaline pool of Di Zahav, Gulf of Elat, with notes on the Siphonocladacea and the typology of near shore marine pools. *Mar. Biol.*, 29: 37–44.

Por, F.D. and Lerner-Segev, R., 1966. Preliminary data about the benthic fauna of the Gulf of Elat (Aqaba), Red Sea. *Isr. J. Zool.*, 15: 38–50.

Por, F.D. and Tsurnamal, M., 1973. Ecology of the Ras Muhammad crack in Sinai. *Nature*, 241: 43–44.

Primor, N. and Tu, A.T., 1980. Conformation of pardaxin, the toxin of the flatfish *Pardachirus marmoratus*. *Biochem. Biophys. Acta*, 626: 299–306.

Primor, N., Parness, J. and Zlotkin, E., 1977. Studies on the chemistry and action of the toxic skin secretion of the Red Sea flatfish *Pardachirus marmoratus*. *Rep. H. Steinitz Mar. Biol. Lab. Elat*, 6: 24 (abstract).

Primor, N., Sabnay, I., Lavie, V. and Zlotkin, E., 1980. Toxicity to fish, effect on gill ATPase and gill ultrastructural changes induced by *Pardachirus* secretion and its derived toxin pardaxin. *J. Exp. Zool.*, 211: 33–43.

Reiss, Z. and Hottinger, L., 1984. *The Gulf of 'Aqaba, Ecological Micropaleontology, Ecological Studies, Analysis and Synthesis, Vol. 50.* Springer, Berlin, 354 pp.

Remane, A. and Schulz, E., 1964. Die Strandzonen des Roten Meeres und ihre Tierwelt. *Kiel Meeresforsch.*, 20 (Sonderheft): 5–17.

Ross, D.A., 1983. The Red Sea. In: B.H. Ketchum (Editor), *Estuaries and Enclosed Seas, Ecosystems of the World, Vol. 26*, Elsevier, Amsterdam, pp. 293–307.

Ross, J.P. and Barwani, M.A., 1982. Review of sea turtles of the Arabian area. In: K.A. Bjorndal (Editor), *Biology and Conservation of Sea Turtles. Proc. World Conf. Sea Turtle Conserv., Washington, D.C., 26–30 Nov., 1979.* Smithsonian Inst. Press and World Wildlife Fund, Washington, D.C. pp. 373–383.

Safriel, U., 1969. Ecological segregation, polymorphism and natural selection in two intertidal gastropods of the genus *Nerita* at Elat (Red Sea, Israel). *Isr. J. Zool.*, 18: 205–231.

Safriel, U. and Lipkin, Y., 1964. On the intertidal zonation of the rocky shores at Eilat (Red Sea, Israel). *Isr. J. Zool.*, 13: 187–190.

Safriel, U. and Lipkin, Y., 1975. Intertidal zonation on the Red Sea rocky shores of Sinai — preliminary report. Dept. Zool., The Hebrew Univ. Jerusalem, (mimeographed Hebrew MS of limited circulation), 19 pp.

Safriel, U.N. and Hadfield, M.G., 1988. Sibling speciation by life-history divergence in *Dendropoma* (Gastropoda; Vermetidae). *Biol. J. Linn. Soc.*, 35: 1–13.

Safriel, U.N., Felsenburg, T. and Gilboa, A., 1980a. The distribution of *Brachidontes variabilis* (Krauss) along the Red Sea coasts of Sinai. *Argamon*, 7: 31–43.

Safriel, U.N., Gilboa, A. and Felsenburg, T., 1980b. Distribution of rocky intertidal mussels in the Red Sea coasts of Sinai, the Suez Canal and the Mediterranean coast of Israel, with special reference to recent colonizers. *J. Biogeogr.*, 1: 39–62.

Saks, N.M., 1981. Growth, productivity and excretion of *Chlorella* spp. endosymbionts from the Red Sea. *Bot. Mar.*, 24: 445–449.

Scheidegger, G., 1972. *Heterocyathus*, an ahermatypic madreporarian in the *Halophila* environment of the Gulf of Elat, Red Sea. *The Hebrew Univ. Jerusalem Mar. Biol. Lab. Sci. Newsletter*, 2: 8–9.

Schmalbach, A.E. and Por, F.D., 1977. Granulometry and distribution of the meiobenthos in the north of the Gulf of Elat (Red Sea). *Rep. H. Steinitz Mar. Biol. Lab. Elat*, 6: 27 (abstract).

Schuhmacher, H., 1973. Notes on occurrence, feeding and swimming behavior of *Notarchus indicus* and *Melibe bucephala* at Elat, Red Sea (Mollusca: Opistobranchia). *Isr. J. Zool.*, 22: 13–25.

Sella, I., 1982. Sea turtles in the eastern Mediterranean and northern Red Sea. In: K.A. Bjorndal (Editor), *Biology and Conservation of Sea Turtles, Proc. World Conf. Sea Turtle Conserv., Washington, D.C., 26–30 Nov., 1979*, Smithsonian Inst. Press and World Wildlife Fund, Washington, D.C., pp. 417–423.

Sella, I., 1983. Marine mammals; Sea turtles. In: L. Fishelson (Editor), *Aquatic Life. Plants and Animals of the Land of Israel, Vol. 4.* Ministry of Defense, The Publishing House and Society for Protection of Nature, Tel Aviv, p. 129 [in Hebrew].

Spira, J. and Van Rijn, J., 1982. Feeding of *Artemia* sp. in the Solar Lake. *Isr. J. Zool.*, 31: 70–71 (abstract).

Steinitz, H., 1962. On the occurrence of *Asymmetron* in the Gulf of Eylath (Aqaba). *Sea Fish. Res. Stn. Haifa Bull.*, 30: 35–38.

Steinitz, H., 1967. A tentative list of immigrants via the Suez Canal. *Isr. J. Zool.*, 16: 166–169.

Steinitz, H., 1973. Fish ecology of the Red Sea and Persian Gulf. In: B. Zeitzschel and S.A. Gerlach (Editors), *The Biology of the Indian Ocean*. Springer, Berlin, pp. 465–466 (abstract).

Stephenson, T.A. and Stephenson, A., 1949. The universal features of zonation between tide marks on rocky shores. *J. Ecol.*, 37: 289–305.

Stephenson, T.A. and Stephenson, A., 1972. *Life between tide marks on rocky shores.* Freeman, San Francisco, 425 pp.

Talmadge, R.R., 1971. Notes on Israeli haliotids. *Argamon*, 2: 81–85.

Tortonese, E., 1968. Fishes from Eilat (Red Sea). *Sea Fish. Res. Stn. Haifa Bull.*, 51: 6–30.

Tsurnamal, M., 1975. Ras Muhammad Pleistocene reef cracks, their fauna and flora. *Rep. H. Steinitz Mar. Biol. Lab. Elat*, 4: 39–40 (abstract).

Tsurnamal, M., 1983. Porifera. In: L. Fishelson (Editor), *Aquatic Life. Plants and Animals of the Land of Israel, Vol. 4.* Ministry of Defense, The Publishing House and Society for Protection of Nature, Tel Aviv, pp. 243–245 [in Hebrew].

Turko, A., 1976. *The biology of Hippa (Decapoda: Anomura: Hippidae)* in the northern Gulf of Elat, Red Sea. M.Sc.

Thesis, The Hebrew University, Jerusalem, 36 pp. [in Hebrew].
Wahbeh, M.I., 1983. Productivity and respiration of three seagrass species from the Gulf of Aqaba (Jordan) and some related factors. *Aquat. Bot.*, 15: 367–374.
Wahbeh, M.I., 1984. The growth and production of the leaves of the seagrass *Halophila stipulacea* (Forsk.) Aschers from Aqaba, Jordan. *Aquat. Bot.*, 20: 33–41.
Wahbeh, M.I. and Mahasneh, A.M., 1984. Heterotrophic bacteria attached to leaves, rhizomes and roots of three seagrass species from Aqaba (Jordan). *Aquat. Bot.*, 20: 87–90.
Wahbeh, M.I. and Mahasneh, A.M., 1985. Some aspects of decomposition of leaf litter of the seagrass *Halophila stipulacea* from the Gulf of Aqaba. *Aquat. Bot.*, 21: 237–244.
Wainwright, S.A., 1965. Reef communities visited by the Israel South Red Sea Expedition, 1962. *Sea Fish. Res. Stn. Haifa Bull.*, 38: 40–53.
Weitlauf, G., 1978a. Shallow water shelling in the Red Sea. *Conchol. Am. Bull.*, 13: 5–6 (1980, reprinted in *Levantina*, 24: 272–276).
Weitlauf, G., 1978b. Red Sea shelling. *Conchol. Am. Bull.*, 14: 2 (1980, reprinted in *Levantina*, 25: 285–287).
Zalcman, D. and Por, F.D., 1975. The food web of Solar Lake (Sinai coast, Gulf of Elat). *Rapp. Comm. Int. Mer Mediterr.*, 23 (3): 133–134.

Chapter 17

SOUTHEASTERN AUSTRALIA

R.J. KING, P.A. HUTCHINGS, A.W.D. LARKUM and R.J. WEST

INTRODUCTION

The marine biota of the temperate coast of southeastern Australia is especially rich and diverse. This review outlines the major environmental features of the littoral zone in the region and introduces the characteristic biota of the major habitats: rocky shores; sandy beaches; and coastal lagoons, estuaries and embayments. Early accounts of shoreline ecology in Australia as elsewhere were largely descriptive and restricted to rocky shores, but studies are now quantitative and/or experimental, and often concerned with sheltered shores and estuaries. Recent publications have also stressed the role of biological interactions, rather than simply the effects of physical and chemical factors. There have been no satisfactory studies addressing the agents that might bring about the species richness and diversity of the region.

ENVIRONMENT

Geology and geomorphology

Much of the open coast of southeastern Australia consists of sandy bays or long stretches of sandy beaches alternating with rocky headlands. A striking feature of these headlands is the development of shore platforms (Bird, 1984). On the New South Wales coast platforms occur on a variety of substrata (Griffiths, 1982) but commonly on sandstones; as the result of water layer weathering, large flat platforms are formed at about the high tide level. Intertidal and high-tide platforms also occur in Victoria, for example at Flinders, and Phillip Island (Western Port), where they are developed on basalt and associated pyroclastic rocks of Cenozoic age (Jutson, 1950) (Fig. 17.1a). In Victoria and parts of South Australia there are extensive lengths of coastline with low-tide platforms developed on Pleistocene aeolian calcarenite. The platforms are broad, often up to 100 m or more across, and are much dissected with numerous pools and overhangs, providing a wealth of microhabitats (Fig. 17.1b). Where massive outcrops of granite and quartzite occur, weathering is less effective and the headlands are not bordered by shore platforms (Bird, 1984) (Fig. 17.1c).

The physical aspects of the mobile sandy beaches between the rocky headlands have been studied extensively (see for example Bird, 1983, 1984; Chapman, 1983; Short, 1983). These habitats are practically devoid of macroscopic plants and animals, and the microscopic biota is virtually unstudied. In marked contrast are the seagrass-dominated sandy and muddy tidal flats, which extend into the sublittoral zone on wave-sheltered shores. These occur in protected bays and inlets and especially in the coastal saline lagoons and barrier estuaries which are so characteristic of the southeastern Australian coast (Barnes, 1980; Roy, 1984).

Nutrients

Except in specific localities where high nutrient levels are associated with pollution, the nutrient levels in southeastern Australian coastal waters are low. Nitrate concentrations are around 1.0 μg-at. NO_3 N l^{-1} or less, and phosphate concentrations are around 0.2 μg-at. PO_4 P l^{-1} or less (Jeffrey, 1981; Rochford, 1984). Minor upwellings have been reported for three localities in southeastern

Fig. 17.1. Rocky coast shores in southeastern Australia: (a) High-tide platform, Bismark Reef, Flinders, Victoria. (b) Low-tide platform on dune limestone, Glaneuse Reef, Point Lonsdale, Victoria. (c) Plunging granite cliffs, Deal Island, Kent Group, Bass Strait.

Australia (Rochford, 1972, 1975, 1977a) as well as near Robe in South Australia (Rochford, 1977b). The deep waters of the East Australian Current have nutrient levels some 2 to 3 times greater than the surface water (Rochford, 1980) so that upwelling can be a source of enrichment.

The low nutrient levels have been attributed to (1) generally low nutrient levels in the coastal soils; (2) the isolation of coastal waters from the rich sub-Antarctic waters to the south; and (3) the dominance, in the region, of subtropical waters with limited nutrient reserves down to 100 to 200 m (Jeffrey, 1981).

Temperature

On open coasts along the southern Australian mainland the mean summer surface water temperature is mostly 18 to 20°C (Fig. 17.2a). Along the New South Wales coast there are marked changes associated with latitude, and waters are warmed due to the southward movement of eddies that comprise the so-called East Australian Current (Fig. 17.2b).

Despite the importance of seawater temperature in limiting the distribution of marine eulittoral organisms there are comparatively few data for onshore localities. King (1970) summarized three years of continuous records for Point Lonsdale at Port Phillip Heads, Victoria and recorded an absolute summer maximum of 20.6°C and an absolute minimum of 10.1°C. Differences as great as 1.9°C were recorded for the same month in different years and up to 3.8°C between the absolute maximum and minimum for any particular month. The temperature in a rock pool at the

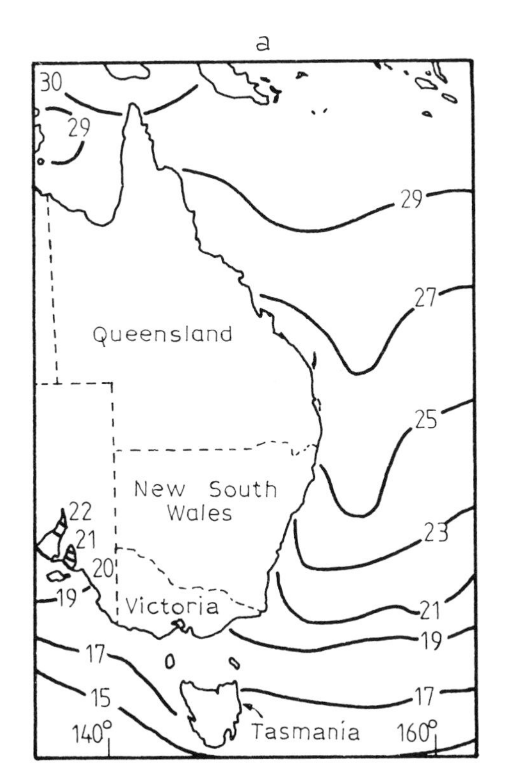

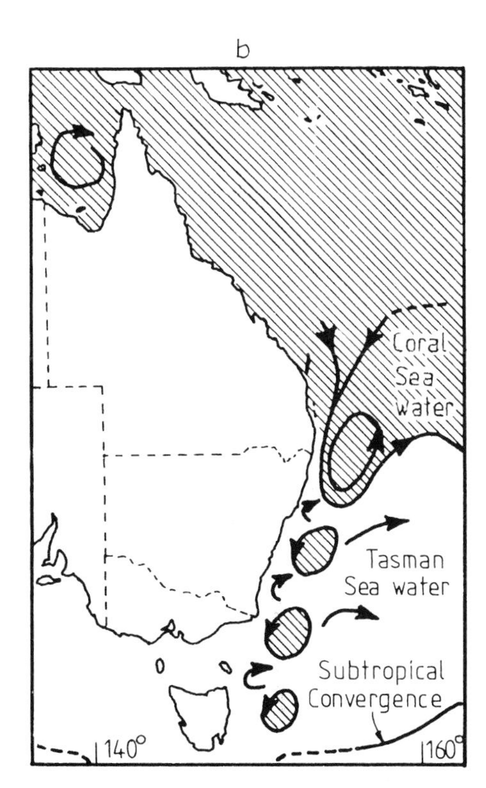

Fig. 17.2. (a) Mean summer surface seawater temperatures in eastern Australia (after Jeffrey, 1981). (b) Ocean circulation features in eastern Australia (after Jeffrey, 1981).

same locality showed a much wider range (Fig. 17.3), with the absolute temperature depending on the air temperature and the state of tide (Fig. 17.4). Organisms in the eulittoral zone are subjected to even greater fluctuations in temperature. The combination of high air temperatures and lower humidity has been regarded as the main cause of the lack of algae in summer at upper shore levels in southern Australia (Womersley, 1984).

Tides

Tides in southeastern Australia are semi-diurnal with two spring and two neap tides per lunar cycle. There is a marked diurnal inequality (Fig. 17.5).

Extreme tidal ranges at open coast localities are generally less than 2 m, though higher ranges occur (3.6 m at Western Port). Marked local variations are associated with channels and embayments.

Tidal emersion is regarded as perhaps the most important factor in the establishment of the zonation pattern on open coasts. The tidal emersion curve (Fig. 17.6) shows only gradual changes from high to low water with the greatest change occurring around mean low water and mean high water.

The tidal level is affected by winds and barometric pressure so that there are seasonal differences. Changes in sea level due to barometric conditions can be as much as 0.3 m (Australian Government, 1985), and on coasts with a small tidal range this can be important when combined with strong winds or storm surges. It is likely that freak conditions exercise greater control over zonation patterns than do average conditions. An example is provided by the exceptionally low tides experienced at Point Lonsdale in March 1971 (Fig. 17.6). Lower zone algae suffered desiccation and even death during this period.

Light

Kirk (1976, 1977, 1979) and Phillips and Kirk (1984) have provided detailed information on underwater light regimes in coastal water and inland lakes of southeastern Australia. Kirk (1976)

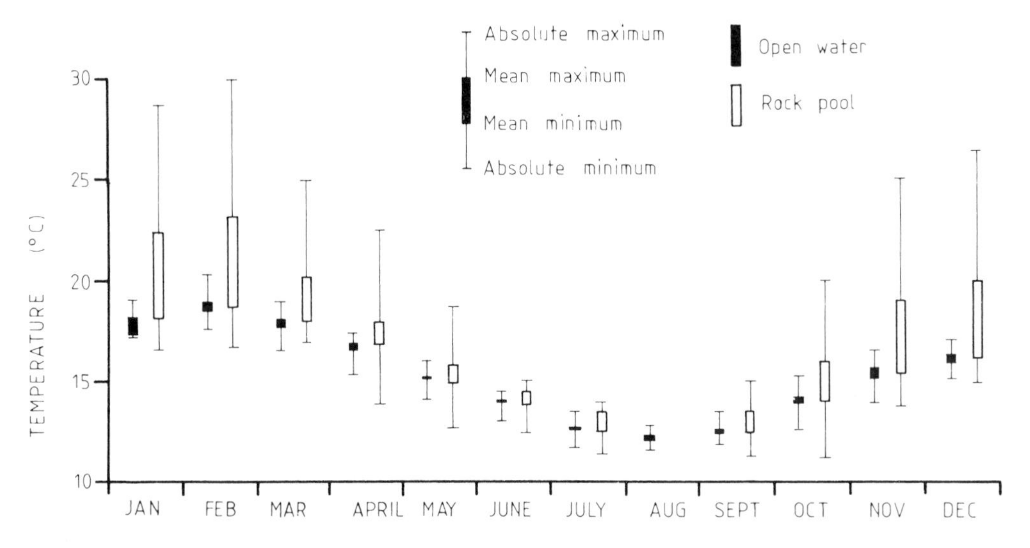

Fig. 17.3. Comparison of temperature data for a rock pool and adjacent inshore waters, Point Lonsdale, Victoria, 1967.

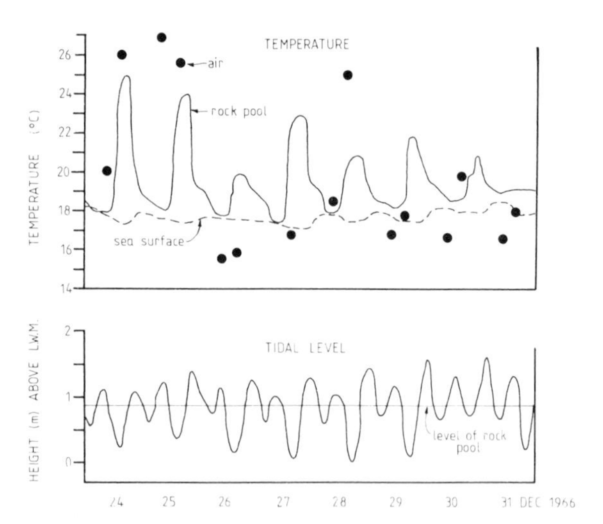

Fig. 17.4. Relationship between rock pool water temperatures, surface sea temperatures, air temperatures and tidal emersion. (Data from Point Lonsdale, Victoria.)

found that yellow substance ("Gelbstoff") contributed strongly to light reduction, especially in the blue region of the spectrum within inland waters. In coastal waters the yellow substance is often at low concentrations (Kirk, 1976, 1979; Phillips and Kirk, 1984). Two opposing factors come into play: the proximity of clear oceanic water close to the southeast Australian coast and the drain-off of fresh waters with high concentrations of yellow substance. In shallow estuaries and mangrove flats the concentrations of yellow substance may reach high levels, but the underwater light quality in such areas has not been measured to date. Phytoplankton and suspended inorganic particles are generally low in coastal waters (Revelante, 1976–77). Thus in most coastal waters the peak of quantum irradiance is 550–580 nm (Kirk, 1979; Phillips and Kirk, 1984). Coastal lagoons constitute a different type of water body. The only work on the light climate of such bodies is a study of the Gippsland Lakes, Victoria (Hickman et al., 1984). There the greatest attenuation of light was caused by suspended inorganic particles, and the contribution of yellow substance was much less; this led to a quantum irradiance spectrum with a broad maximum between 550 and 680 nm, and a slight peak at 570 nm.

ZONATION ON ROCKY SHORES IN SOUTHEASTERN AUSTRALIA

Stephenson and Stephenson published their account of the universal features of zonation between tide marks on rocky coasts in 1949, but earlier studies on the New South Wales coast (Pope, 1943; Dakin et al., 1948) had already described these shores in terms that were almost directly compatible. In the 1950s a number of authors discussed the applicability of the zonation terminology to Australasian shores (Womersley and Edmonds, 1952; Chapman and Trevarthen, 1953; Guiler, 1953b). In 1961, Lewis first pub-

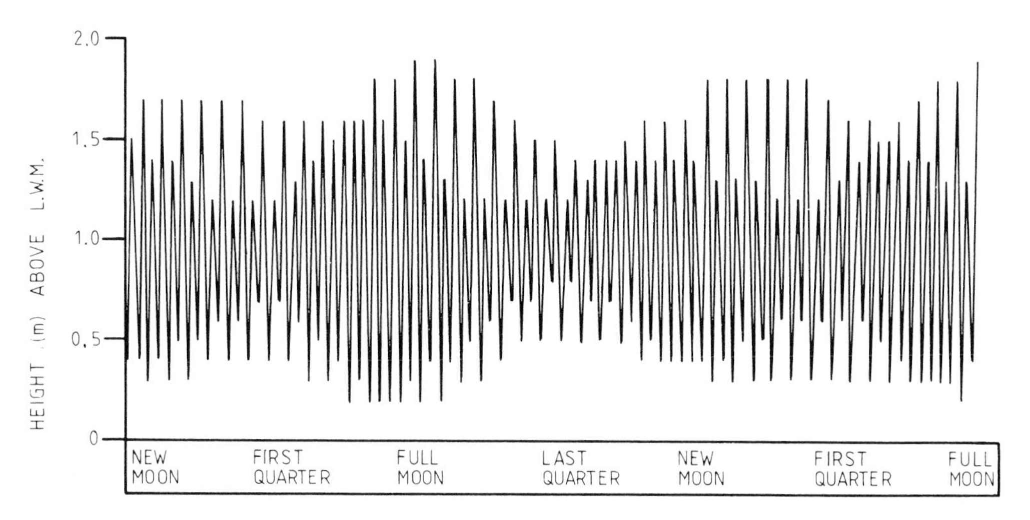

Fig. 17.5. Tidal pattern and range, based on predicted tide values for Sydney (Fort Denison).

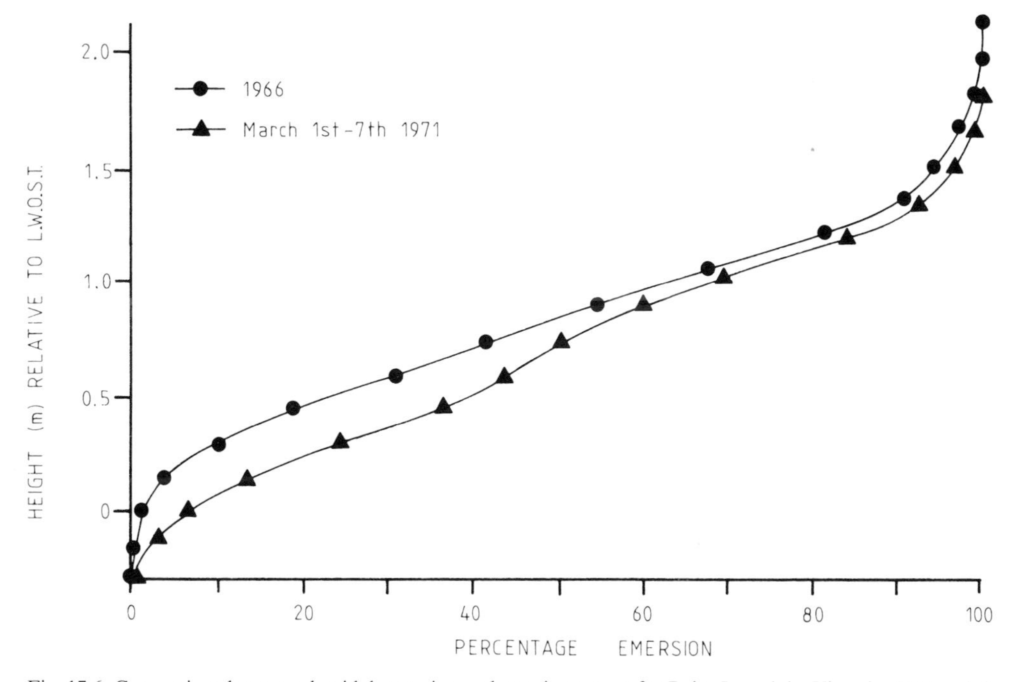

Fig. 17.6. Comparison between the tidal emersion–submersion curves for Point Lonsdale, Victoria, 1966 and the emersion–submersion curve during a period of unusually low tides associated with strong off-shore winds (March, 1971). (Drawn from original data supplied by Public Works Department, Victoria.)

lished the scheme that is now so widely accepted, and that is adopted here. This scheme emphasizes the biological nature of the zones, and the importance of water movement in determining the relative vertical extent of the major zones (see also Fig. 17.7).

The general pattern of major zone forming shore organisms in southeastern Australia is covered in a number of review articles (Stephenson and Stephenson, 1972; Chapman and Chapman, 1973; Womersley, 1981a; Barson and King, 1982). Table 17.1 lists the dominants for each zone on coasts of moderate to strong wave action. Some of the detailed accounts of specific areas,

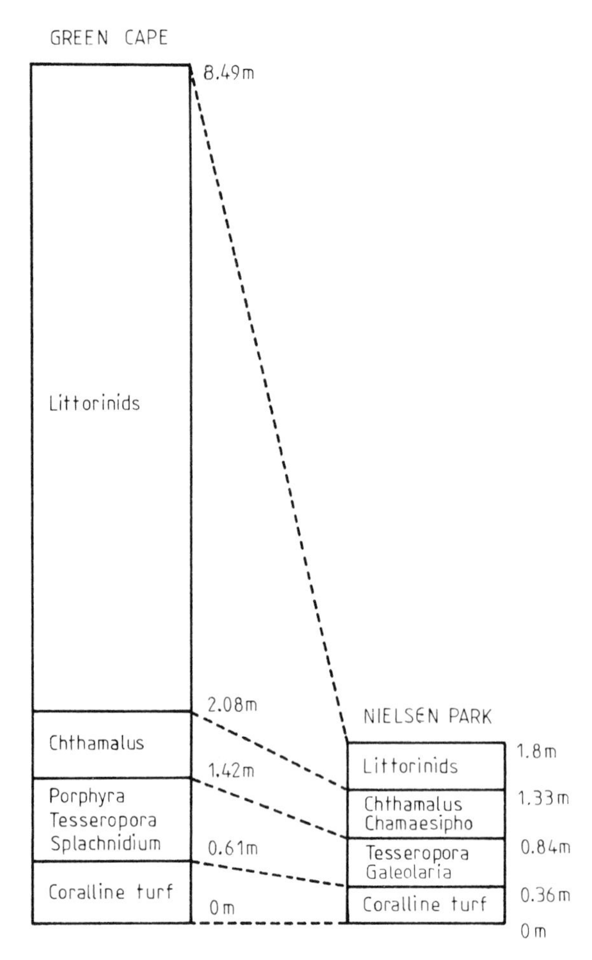

Fig. 17.7. Comparison of the vertical extent of the major zones at two localities with similar tidal range on the New South Wales coast. Green Cape is a rough-water locality south of Eden, southern New South Wales; Nielsen Park is a sheltered locality within Sydney Harbour.

which form the basis of these reviews, are listed in Table 17.2.

Although zonation terminology was developed for use on rocky shores with at least moderate wave action it has been successfully applied to calm water coasts in southern Australia (Womersley 1981a, 1984). It is, however, difficult to equate the zones with those on rocky shores, as the biota of sandy and muddy shores include none of the indicator species characteristic of zones on rocky shores. In Table 17.1 the dominants for New South Wales salt-marsh and mangrove coasts with slight wave action are listed, along with species from rocky shores.

Zonation of organisms continues into the sublittoral, and in South Australia upper, mid- and lower sublittoral zones have been recognized (Shepherd and Womersley, 1970, 1971, 1976; Womersley, 1972; Shepherd and Sprigg, 1976). The zones are not clear-cut but grade with depth due to the decrease in both light intensity and water movement (Womersley, 1981a). The three zones recognized by Shepherd and Womersley are (1) an upper sublittoral zone characterized by coralline and other turf forming algae; (2) a mid-sublittoral zone dominated by *Ecklonia radiata* and other large brown algae; and (3) the lower sublittoral zone dominated by delicate and diverse red algae. In the Sydney region the distribution of subtidal organisms has been described in terms of these three zones (Farrant and King, 1982; Van der Velde and King, 1984). In Botany Bay (State Pollution Control Commission, 1978) the three zones could not be recognized. On calm water coasts, Womersley (1981a) reported that the upper sublittoral zone of a coralline algal turf does not occur, and the zone of brown algae comes to low-tide level. This appears to be the case within Jervis Bay (May and Larkum, 1981).

Coralline algal flats (urchin barren grounds) occur frequently in the mid-sublittoral zone on the coast of New South Wales, but apparently not in other parts of Australia, despite the fact that similar algal assemblages, especially *Ecklonia radiata* forests, occur in these other regions. The flats seem to have a similar structure to those observed in many other parts of the world (Mann, 1982), with the urchin *Centrostephanus rodgersii* and several species of limpet interacting in their grazing patterns to restrict recolonization by macrobenthic algae (Fletcher, 1987). These flats have been described briefly by Shepherd and Womersley (1970) and May and Larkum (1981), and in greater detail by Fletcher (1987).

BIOGEOGRAPHY

Hedley (1904) was the first author to recognize and name the marine zoogeographic provinces in southeastern Australia. This he did on the basis of mollusc distributions. He considered that there was a similarity in the southern molluscan fauna from Geraldton, Western Australia to Western Port, Victoria and this he named the Adelaidian [this province was subsequently named the Flindersian by Cotton (1930)]. The fauna of New South Wales was distinct and this he named the Peronian. In 1926 Hedley recognized in addition the Maugean Province in Tasmania, following Iredale and May (1916). Other workers have made contributions based on distributions of particular taxa (for example, Ashby, 1926: chitons; Clarke, 1946: echinoderms; Kott, 1952, 1957, 1962, 1963: ascidians), but Bennett and Pope (1953) were the first to take into account more than one group in the delimitation of biogeographic provinces. This approach was subsequently followed by Womersley and Edmonds (1958), Womersley (1959), Bennett and Pope (1960) and Knox (1963). In all of these studies emphasis was placed on intertidal and uppermost subtidal organisms.

Womersley (1981b) summarized the situation for southeastern Australia. The warm-temperate coast is referred to as the *Peronian Province*. This extends from southern Queensland along the New South Wales coast, which becomes cool temperate near the Victorian border. It corresponds to the area under the influence of the East Australian Current (Fig. 17.2b). The *Flindersian Province* comprises the southern coast of Australia with the southeast coasts (Victoria and Tasmania) showing cold-temperate affinities and referred to sometimes as the *Maugean Province* or *Subprovince*.

Despite the considerable work done on the marine flora and fauna we do not have the data to critically evaluate the validity of such provinces in terms of the total biota. The marine biota of southeastern Australia is especially rich. For the algae, Womersley (1981b, 1984) provided data to show that southern Australia is the outstanding region (within a limited temperature range) of the world for richness of benthic algae, the number of species being 50 to 80% greater than for other comparable regions. Many taxonomic groupings require extensive revision before they can be included in a re-evaluation of biogeography. The present status of some of the major groups is summarized here.

FLORA

Macroalgae

The marine algal flora of southern Australia has been well documented (for references see Clayton and King, 1981; and Womersley, 1984). The southern flora, extending east to include the Victorian and Tasmanian coasts, has been noted not only for its richness but also for the high degree of endemism (Womersley, 1981b, 1984). Womersley (1984) has calculated the percentage endemism in the three divisions as:

Chlorophyta	genera	5%	species	*c.* 30%
Phaeophyta		19%		70%
Rhodophyta		30%		75%

Since the reviews by May (1938, 1939) for Chlorophyta and Phaeophyta there have been essentially no comprehensive taxonomic works on the algae of the New South Wales coast. Two exceptions are papers on some marine algae of Lord Howe Island: Dictyotales (Allender and Kraft, 1983), and *Codium* (Jones and Kraft, 1984) and the Dictyotales of the mainland coast (Farrant and King, 1989). The handbook to the green and brown algae of New South Wales (Borowitzka et al., 1982) only summarized pre-existing data. Many of the taxa that are noteworthy in southern Australia are absent from New South Wales coasts. Conspicuous larger brown algae such as *Macrocystis* (Laminariales), *Durvillaea* and many species of *Cystophora* (Fucales) are absent from the upper sublittoral zones. Tropical species such as *Caulerpa brachypus*, *Halimeda discoidea* and *Valoniopsis pachynema* extend from the north to varying extents, but the few warmer water species do not compensate for the loss of the southern flora, so that in comparison with the latter the diversity of the New South Wales flora is considerably lower.

Since the review by Saenger et al. (1977) there have been several studies specifically concerned with the algae of mangroves (Davey and Woelkerling, 1980, 1984; King 1981a, b, d; Beanland and Woelkerling, 1982, 1983; King and Wheeler, 1985; King and Puttock, 1989). These studies emphasize the uniformity of the mangrove associated algal flora at the generic level, but at the species level there are latitudinal differences along the New South Wales coast.

Seagrasses

The seagrass flora of Australia, especially southwestern Australia, is particularly rich (McComb et

TABLE 17.1

The intertidal zones and their dominants[a] on southeastern Australian shores subject to strong to moderate wave action

	Tasmania		Victoria	New South Wales		
	Rough water (west and south)	More sheltered (north and east)		Rocky coasts with extreme–strong wave action	Rocky coasts with moderate wave action	Mangrove/saltmarsh coasts with slight wave action
Littoral fringe	***Nodolittorina praetermissa***	***Nodolittorina unifasciata***	***Nodolittorina unifasciata*** ***N. praetermissa*** *Lichina confinis* [*Verrucaria*]	***Nodolittorina tuberculata*** ***Nodolittorina unifasciata***	***Nodolittorina unifasciata***	Cyperaceae
Eulittoral Upper	*Lichina confinis* ***Chthamalus antennatus*** ***Chamaesipho columna***	*Lichina confinis* ***Chthamalus antennatus*** ***Chamaesipho columna***	***Chthamalus antennatus*** ***Chamaesipho columna***	***Chthamalus antennatus*** ***Chamaesipho columna***	***Chthamalus antennatus*** ***Chamaesipho columna***	*Sarcocornia quinqueflora* *Sporobolus virginicus* ***Assiminea*** spp. ***Ophicardelus*** spp. ***Salinator*** spp. ***Tatea*** spp. ***Austrocochlea constricta*** ***Bembicium auratum*** [*Aegiceras corniculatum*] *Avicennia marina* *Bostrychia/Caloglossa* association ***Saccostrea commercialis*** ***Littoraria*** spp. ***Batillaria australis*** ***Teredo*** spp.
Mid	***Patelloida latistrigata*** various **limpets** [***Catomerus polymerus***] blue-green algae [*Porphyra columbina*] ***Austromytilus rostratus*** ***Galeolaria caespitosa***	***Patelloida latistrigata*** various **limpets** blue-green algae [*Porphyra columbina*] ***Austromytilus rostratus*** ***Galeolaria caespitosa*** ***Patelloida alticostata***	***Catomerus polymerus*** [*Splachnidium rugosum*] **limpets** ***Austromytilus rostratus*** blue-green algae [*Porphyra columbina*] ***Galeolaria caespitosa***	***Tesseropora rosea*** ***Catomerus polymerus*** [*Splachnidium rugosum*][b] [*Porphyra columbina*]	***Tesseropora rosea*** various **molluscs** including ***Cellana tramoserica*** ***Galeolaria caespitosa***	

Lower	*Lithophyllum hyperellum*	*Lithophyllum hyperellum*	coralline mat	coralline mat	coralline mat	numerous **crab** species *Avicennia marina* *Bostrychia/ Caloglossa* association *Hormosira banksii*
	Patella peroni coralline mat **Plaxiphora costata** *Xiphophora gladiata* [**Pyura stolonifera**]	**Patella peroni** coralline mat **Plaxiphora costata** *Hormosira banksii* [**Pyura stolonifera**]	*Hormosira banksii* [**Pyura stolonifera**] [**Megabalanus nigrescens**]	**Megabalanus nigrescens** **Pyura stolonifera** *Pterocladia capillacea*	*Hormosira banksii* **Pyura stolonifera**	
Upper sublittoral	*Durvillaea potatorum*	*Cystophora torulosa*	**Plaxiphora** spp. [*Cystophora intermedia*] *Durvillaea potatorum*	*Durvillaea potatorum*[b] *Phyllospora comosa*	*Ecklonia radiata* mixed algae including Dictyotales	*Zostera capricorni* *Posidonia australis* **Alpheus** spp. and other burrowing **prawns** **Anadara trapezia** **Conuber** spp. **Nassarius** spp. **Pyrazus ebeninus**
	"lithothamnia" *Cystophora* spp. *Macrocystis pyrifera*	*Cystophora* spp. *Phyllospora comosa* *Ecklonia radiata* *Macrocystis angustifolia*	*Phyllospora comosa* *Macrocystis angustifolia*	*Ecklonia radiata* *Sargassum* spp. *Cystophora* spp.[b]		

[a]Plant names are in light-face, animal names in bold-face type. Names in brackets are either inconspicuous or seasonal taxa, or locally prominent.
[b]Temperate species characteristic of southern New South Wales.

TABLE 17.2

General descriptive accounts of rocky shore ecology and zonation studies (including subtidal studies) in southeastern Australia

Geographic area	Reference
Tasmania	
General	Bennett and Pope (1960); Guiler (1950, 1952c); Edgar (1983a, b, 1984)
Specific localities	
Port Arthur/Eaglehawk Neck	Cribb (1954b); Guiler (1952b)
Pipe Clay Lagoon	Guiler (1951a)
Freycinet Peninsula	Guiler (1951b, 1953a)
D'Entrecasteaux Channel	Guiler (1952a); Sanderson and Thomas (1987)
South Cape Bay and First Lookout Point	Guiler (1954)
Trial Harbour	Guiler (1960b)
Bass Strait	King (1973)
Fisher Island	Gillham (1965); Guiler et al. (1958)
Victoria	
General	Bennett and Pope (1953)
Specific localities	
Port Phillip Bay	King et al. (1971)
Port Phillip Bay (subtidal)	Port Phillip Survey (1966, 1971); Womersley (1966); Melbourne Metropolitan Board of Works and Fisheries and Wildlife Dept. (1973); Lewis (1983)
Western Port	Smith (1971)
New South Wales	
General	Dakin et al. (1948); Guiler (1960a)
Specific localities	
Sydney region	Pope (1943); May et al. (1970); State Pollution Control Commission (1981a)
Sydney region (subtidal)	Farrant and King (1982); Van der Velde and King (1984); Jones (1977); Watson (1979)
Jervis Bay (subtidal)	May (1981); May and Larkum (1981)
Southern NSW	Owen (1978)

al., 1981; Larkum et al., 1989). The species present in southeastern Australia have been monographed by E. L. Robertson (1984).

FAUNA

Fish

About 800 species of marine fish are known from southeastern Australia. Much of the fish fauna of New South Wales has closer affinities with that of tropical regions than with the temperate southern fauna.

Approximately 20 to 25% of the rocky reef fish of southeastern Australia are endemic to New South Wales and eastern Victoria. Estuarine and other marine species tend to be more widespread. In the Victorian and Tasmanian region there is also a high percentage of endemism, in all environments except the deep sea. The southern temperate fish fauna has some affinities with the fish fauna of tropical regions, but the greatest affinities are with those of other temperate areas, such as New Zealand and South Africa (D.F. Hoese, pers. commun., 1988). Although the fish of the region are well described and documented (see review by Whitley, 1964), there are few quantitative studies apart from those in mangrove and seagrass areas (see pp. 448ff.).

Echinoderms

Approximately 400 species of echinoderms have been recorded from southeastern Australia (New South Wales, Tasmania, Victoria). The majority are known from shallow water (less than 30 m). In terms of biogeography, three groups can be recognized: some 50% of species are tropical, extending south to about Montague Island (New South Wales); 12% are southern Australian species extending north along the New South Wales coast and into southern Queensland; the remaining 38% are endemic to southeastern Australia (though 25 to 30% of these extend across the Tasman Sea to Lord Howe Island, Norfolk Island, Kermadec Island and northern New Zealand). As the number of tropical species diminishes rapidly with distance southwards along the New South Wales coast, the degree of endemism increases. The high degree of endemism in Tasmanian echinoderms has been documented by Rowe and Vail (1982).

Molluscs

There is a northern Australian molluscan fauna and a southern Australian molluscan fauna, with regions of overlap on both the east and west coasts of the continent (Wilson and Gillett, 1974). A large number of endemic species occurs in New South Wales, particularly on the continental shelf but also to some extent on the continental slope, although the fauna of the latter is less well known. Many endemic species occur intertidally, and these may extend to Mallacoota Inlet or Wilsons Promontory, Victoria. The molluscan fauna in the Bass Strait region is a very complex one, probably as a result of the land bridge that connected mainland Australia to Tasmania during the Pleistocene. Many species occur at either the eastern or western end of Bass Strait only (W.F. Ponder, pers. commun., 1988), whereas others extend across the Strait, and some are restricted to this region. The molluscan fauna of Tasmania has been discussed by Dartnall (1974).

Crustaceans

The Australian decapod fauna is relatively well known (Griffin and Yaldwyn, 1968). The decapod fauna can be divided into a larger northern (tropical) fauna and a southern (temperate) fauna. There is a broad transition zone between the two on both the east and west coasts of Australia, with few species extending along the whole eastern or western coasts.

Many of the species in the tropical decapod fauna are widespread throughout the Indo-Pacific. In contrast, the species in the temperate fauna are largely restricted to Australia, and they show a variety of distribution patterns (for details see Griffin and Yaldwyn, 1968). Some species have widespread southern distributions extending from New South Wales or Victoria to southwestern Australia.

There are also transition zones in the north around the Torres Strait area, and in southeastern South Australia. Several species which are otherwise common in tropical areas occur in Spencer Gulf, South Australia. Poore and Griffin (1979) found a similar pattern for the southeastern Australian Thalassinidae.

There is a warm-temperate isopod fauna on the southeastern coast. This fauna, which has a very high degree of endemism, probably extends to southwestern Australia. A few species from the Victorian Shelf have been found in the Great Barrier Reef region and on the North West Shelf (Poore, 1985).

The amphipod fauna is not yet well enough known to make any generalizations (J.K. Lowry, pers. commun., 1988).

Polychaetes

To date about 500 species of polychaetes have been recorded from Australia. A large percentage of these have been recorded from southeastern Australia (Day and Hutchings, 1979), the area where most collecting has been done. It is premature to comment on the degree of endemism of Australian polychaetes because of the present state of taxonomic knowledge. Many genera are worldwide, with species restricted to Australia or part of Australia. Southeastern Australia has numerous species restricted to the general region, but no clearly defined biogeographic barriers are apparent. It is probable that each species has a fairly well-defined distribution determined by a number of species-specific factors. In a study of the polychaete fauna of *Posidonia* seagrass beds, salinity and sediment characteristics were more important in determining distributions than was latitude (Collett et al., 1984).

PLANT BIOMASS AND PRIMARY PRODUCTIVITY

Algae

The few studies on algal biomass and primary production in southeastern Australia are summarized in Table 17.3. In the case of *Ecklonia radiata*, which has a high biomass, primary production is high but there is no reason to expect a correlation between biomass and production. Harada (1980) studied a number of sublittoral benthic algae with low to medium biomass but high primary production. The low values for biomass and primary production of *Macrocystis pyrifera* shown in Table 17.3 were obtained by surface harvesting (Cribb, 1954a); greater values of both would be expected if complete plants had been harvested.

Inshore phytoplankton production, where measured, is moderately low, being 0.04 g C m^{-3} day^{-1} for Port Hacking (Scott, 1978) and 0.05 g C m^{-3} day^{-1} for Sydney Harbour (Revelante and Gilmartin, 1978). Nevertheless in many embayments and estuaries phytoplankton may contribute a large share of the primary production. In South West Arm (Port Hacking) it was established that phytoplankton produced 400 mg C m^{-2} day^{-1}, seagrasses 102 mg C m^{-2} day^{-1} and benthic microorganisms 186 to 310 mg C m^{-2} day^{-1} over an area of 78 ha (Cuff et al., 1983). In Botany Bay, phytoplankton is thought to contribute 35% of the primary production (Larkum, 1981). It is probable that seagrasses covered a large proportion of Botany Bay before European settlement (Larkum, 1976), and under these conditions the proportional contributions of phytoplankton would have been much reduced. This illustrates the point that in shallow estuaries with large beds of seagrasses and mangroves the proportional contribution from sources other than phytoplankton is high. This in turn would lead to food chain systems dominated by detritivores, since herbivory accounts for very little secondary production in seagrass beds (Conacher et al., 1979; Kirkman and Reid, 1979; Nicholls

TABLE 17.3

Algal biomass and production data

Algal species/ community	Geographic area	Depth (m)	Biomass g DW m^{-2}	Production g DW m^{-2} yr^{-1}	Reference
Ecklonia radiata	Sydney, NSW	1.0	740–2230	2900–3100	Larkum (1986)
Macrocystis pyrifera	Port Arthur, Tasm.	4–10	20–50	60–150[a]	Cribb (1954a)
Hormosira banksii	Botany Bay, NSW	intertidal	280–638	360–440	King (1981b)
Sublittoral benthic community	Long Reef, Sydney, NSW	0.5	670	5400	Harada (1980)
Sublittoral benthic community	Sydney Harbour, NSW	0.1–5	80–1000	–	Farrant and King (1982)
Seagrass epiphytes	Western Port, Vict.		20–150	40–190[b]	Bulthuis and Woelkerling (1983)
Sand microalgae	Port Hacking, NSW	6–10	0.5	110–270[c]	Giles (1983)

[a]Based on wet weight: dry weight ratio of 10:1 (see Cribb, 1954a).
[b]Based on leaf area index of 2–5.
[c]Based on C content of 47% and ash weight of 10%.

TABLE 17.4

Productivity of seagrasses in southeastern Australia

Seagrass	Locality	Standing crop (g DW m^{-2})	Leaf production (g DW m^{-2} day^{-1})	Reference
Heterozostera tasmanica	Western Port	27–173	0.34–4.20	Bulthuis and Woelkerling (1983)
Posidonia australis	Port Hacking	140	0.57–1.44	Kirkman et al. (1979)
	Botany Bay	101–280	1.0–2.3	West and Larkum (1979)
	Jervis Bay	294–453	0.9–5.2	West and Larkum (1979)
Zostera capricorni	Lake Macquarie	19–187		King and Barclay (1986)
	Tuggerah Lake	10–68		
	Port Hacking	50–70	0.56–1.82	Kirkman et al. (1982)
	Botany Bay	64–331	1.37–8.92	Larkum et al. (1984)
	Smiths Lake	12–60		Robinson et al. (1983)
Zostera muelleri	Western Port	17–151		Bulthuis (1981)

et al., 1985; Larkum et al., 1989). In deep estuaries, such as drowned river valleys, phytoplankton dominates and may support herbivore-dominated secondary production (Cuff et al., 1983).

Seagrasses

Seagrass primary production was reviewed by West and Larkum (1983), with an emphasis on Australian examples (see also Larkum et al., 1989). Table 17.4 summarizes data for biomass, primary production and turnover for seagrasses in southeastern Australia. Standing crop and production rates are usually seasonal, with a slowing of growth during colder months, and occasionally a "winter dieback". Localized conditions, such as exposure, turbidity, and wave action, lead to large variations in seagrass cover, standing crop and consequently primary production on an areal basis.

The most extensive studies of environmental factors in relation to productivity are those of Bulthuis and Woelkerling (1981) and Bulthuis (1983a–c), which document the effects of light, temperature, epiphyte load, and nutrient addition on the growth of *Heterozostera tasmanica* in Western Port. Tyerman et al. (1984) have studied the effect of salinity on the growth and osmotic relations of *Posidonia australis*.

The majority of studies on growth in seagrasses have concerned the rate of growth of leaves. There are very few data concerning rhizome growth of Australian seagrasses. West and Larkum (1983) stated that rhizome production reflects the growth habit of individual species and the stability of a particular stand. For example, *Posidonia australis* rhizomes show a variable growth rate depending upon site, with some erect shoots being over 20 years in age and others showing active elongation. The inability of *Posidonia* to develop active apical meristems, when growing in stable situations, results in slow regrowth of denuded areas (Larkum and West, 1983). King and Holland (1986) have estimated seasonal variation in starch content of the rhizome in *Zostera capricorni*. Values of up to 22.6% of the dry weight of the rhizome occurred in summer, but this value fell rapidly during autumn and winter, presumably due to mobilization and use of reserves by the plant.

DECOMPOSITION AND NUTRIENTS

Decomposition

Little work on decomposition and detritus formation in algae has been carried out. *Ecklonia radiata*, the dominant alga in many open coast sublittoral sites in southeastern Australia has a high rate of production (Table 17.4) and, since it supports few herbivores, it gives rise to much detritus and presumably dissolved organic carbon (DOC) (Kirkman, 1984; Larkum, 1986). Considerably more work has been done on seagrasses and mangroves. Bavor and Millis (1976) and Bavor (1978) studied decomposition of the seagrass *Zostera muelleri*. During the first three weeks there was a rapid rise in the population of heterotrophic bacteria and a rapid loss of organic carbon and organic nitrogen. In *Posidonia australis* beds in Port Hacking less than 5% of leaf production is consumed by herbivores (Kirkman and Reid, 1979), compared with less than 1% in similar beds in Botany Bay (Conacher et al., 1979). Detached leaves, which decomposed *in situ*, accounted for nearly 60% of production, while the remaining 35% was lost to the seagrass bed either as particulate organic carbon (POC), DOC or as leaf fragments. *Posidonia australis* leaf-litter bags have been shown to lose 40% of their dry weight in 50 days (West, 1980). Van der Valk and Attiwill (1983, 1984) presented data on *in situ* decomposition of leaf and root litter and mineral loss in the mangrove *Avicennia marina*. Leaf litter was broken down almost completely after 230 days; fibrous roots decomposed less rapidly. Crabs ate 50% of the leaf litter retained in the mangrove zone, but it was estimated that about 40% of the leaf litter was exported from the mangrove zone either as whole leaves, POC or DOC. Similar rates of decomposition were obtained for the same species in Sydney (Goulter and Allaway, 1979) and in New Zealand (Albright, 1976).

Nutrient cycling

Nutrient cycling has been little studied in southeastern Australia. Seagrasses (Kirkman et al., 1979; McComb et al., 1981; Bulthuis and Woelkerling, 1981), mangroves (Van der Valk and Attiwill, 1984) and phytoplankton (Bulleid, 1983) have been shown to be large reservoirs of nitrogen and phosphorus and presumably this also applies to benthic marine algae. In several instances the water column associated with plant communities has been shown to be enriched for nitrogen and phosphorus (Bulthuis and Woelkerling, 1981; Kirkman et al., 1979; Bulleid, 1983). Remineralization of nitrogen and phosphorus *in situ* is, therefore, probably a very important process, especially considering the low levels of those nutrients in the upper layers of the Tasman Sea, and the low input of freshwater runoff. In a study of nitrogen recycling in South West Arm (Port Hacking) above an unvegetated sandy bottom, about 40% of nitrogen turnover occurred at the sediment surface (Bulleid, 1983), and the remainder in the water column. Cycling of nitrogen and phosphorus in communities of benthic algae, seagrasses and mangroves remains an outstanding question. Presumably decomposition *in situ* leads to much recycling but the finding that 40% of the leaf litter is lost from *Avicennia marina* communities in Victoria (Van der Valk and Attiwill, 1984) suggests that there must be other inputs of nitrogen and phosphorus into this system. This view is strengthened by the fact that in the *A. marina* vegetation in Botany Bay, approximately 55% of the total nitrogen pool is located within the mangroves (Clarke, 1985). Clarke suggested that, as well as nitrogen input from precipitation and fixation, there may be significant ground water enrichment from adjacent saltmarsh and terrestrial communities.

Nutrient levels and absorption by marine plants

The nutrient levels in ocean water in the region are low (see p. 429). With the close proximity of the 100 m depth line to the coast, the poor nutrient status is characteristic of the inshore waters and for many of the estuaries that are subject to little fresh water discharge (Rochford, 1951; Bulleid, 1983). River flood waters may cause localized increases in nitrogen from one to as much as 20 μg-at. NO_3 N l^{-1} and in phosphorus in the northern part of the region only (where richer soils occur) from 0.2 to as much as 5 μg-at. PO_4 P l^{-1} (Rochford, 1951). Also upwellings and pollution may cause localized increases in nutrient levels.

An important unresolved question is whether growth of marine plants is limited by nitrogen or phosphorus or indeed by any other mineral nutrient. Borowitzka (1972) presented evidence indicating that the growth of some algae might be nitrogen-limited and not phosphorus-limited. On the other hand, Bulthuis and Woelkerling (1981) found no effect of nitrogen or phosphorus on the leaf growth of the seagrass *Heterozostera tasmanica*, even when intertidal sediment concentrations were experimentally increased 100-fold. This result may merely reflect the sufficiency of nitrogen and phosphorus in established seagrass beds. Support for this conclusion comes from work by Boon et al. (1986), which indicates a rapid turnover and uptake of ammonium in beds of *Zostera capricorni* at Cleveland, Queensland. Also Kirkman et al. (1979) recorded a net loss of reactive phosphate from a bed of the seagrass *Posidonia australis* in the Port Hacking estuary. McComb et al. (1981) suggested from data on seagrass beds in Cockburn Sound (Western Australia) that the sediments were rich in phosphates. Net losses of phosphorus from seagrass beds for any sustained period are unlikely, since these marine sand sediments are generally low in exchangeable phosphates (Jitts, 1959). Phosphorus may be taken up by sediments during periods of seasonally high concentrations associated with stratified deoxygenated water (Rochford, 1951; Bulleid, 1983). Detrital input of organic phosphorus may be mineralized later (Scott, 1978; Bulleid, 1983).

Nitrogen fixation has been shown to be an important source of nitrogen in *Avicennia marina* stands in Victoria (Van der Valk and Attiwill, 1984). Larkum (unpubl.) measured high rates of nitrogen fixation on the leaves of *Posidonia australis*, in line with work elsewhere (Capone and Taylor, 1977). No work has been carried out on rocky shores but the abundance of heterocystous cyanobacteria in the upper littoral zones suggests this may be another important site for fixation of nitrogen.

ROCKY SHORES

Initial studies on the biota of rocky shores in southeastern Australia essentially describe the fauna or the general zonation of organisms (see pp. 438–440). The few papers with a quantitative approach have generally considered distributions in terms of physical environmental factors and often dealt with only one group of organisms. Examples are the papers by Meyer and O'Gower (1963) and O'Gower and Meyer (1965, 1970). Since the 1970s there has been an increasing number of experimental and/or quantitative studies, especially the works by Underwood and his colleagues, which have addressed the general issues of structure and dynamics in intertidal communities.

In addition there has been a number of quantitative studies concerned specifically with sewage disposal and with more general aspects of "pollution". Borowitzka (1972) determined species diversity of sublittoral macroalgae at three sewage outfalls in the Sydney region and recorded a reduction in species numbers (especially Phaeophyta and Rhodophyta) in the vicinity. Algal species diversity was reduced with increased height on the shore, distance from the platform edge, and proximity to the outfalls. May (1981) observed the long-term variation in intertidal floras at two outfalls, one with strong wave action (Ulladulla), the other within Jervis Bay and relatively sheltered from waves (Plantation Point). Naturally occurring variation in the floras was greater than any induced by the low volume of well-treated sewage, except in the immediate vicinity of the outfall. In a continuation of this study, May (1985) recorded a decline in species in the vicinity of the outfalls, whereas a few species, namely *Gelidium pusillum*, *Ralfsia* sp. and *Ulva lactuca*, were thriving adjacent to the sewage flow. Changes were greatest at Ulladulla where the volume of sewage was greater and its pretreatment less thorough.

A survey on the effects of primary treated sewage at an outfall off North Head, Sydney, recorded increased turbidity of the water, and a change from an algal-dominated benthic community to one dominated by filter-feeding sponges and ascidians, plus associated species (Jones, 1977).

May (1976) has recorded the changing dominance of the green alga *Caulerpa filiformis*. This species was first recorded in the Sydney region in 1923 and noted as rare. It is now abundant at many localities in the Sydney region though whether this is due to acclimatization of an introduced species, or response to increased local pollution has not been determined. Similarly the record of the brown alga *Striaria attenuata* near Eden in New South Wales (King and Wheeler, 1985) may represent a recent introduction, possibly due to shipping as Skinner and Womersley (1983) suggested was the case in South Australia.

Underwood (1981) provided a detailed and quantitative account of the structure of rocky intertidal communities in New South Wales, so as to provide a background for the experimental studies undertaken by his group.

The lower levels of the shore were dominated by foliose macroalgae with 100% cover, and few animals were present. Mid-shore levels were dominated by grazing molluscs, sessile animals (mainly barnacles and tubeworms) and/or encrusting algae. At the upper levels of the shore there was a zone of littorinid gastropods. In the mid-shore areas, foliose algae were sparse, except in pools, and there was a positive correlation with the abundance of sessile animals. On exposed rocky shores the upper limits of the vertical distribution of dense foliose algae, the height of peak abundance of mid-shore grazers, and the upper limits of these grazers were higher than on less wave-exposed shores. Within a zone, considerable patchiness in the dominant organisms was apparent, and no clear trends of diversity of species along a gradient of exposure to wave action were evident.

The factors that bring about this pattern are exceedingly complex, and hence there is the need for carefully planned experiments (Underwood, 1985) which take into account seasonal variation and differences between sites (Underwood et al., 1983). Physical factors associated with different heights on the shore and varying periods of emersion during low tide are inadequate to explain many patterns of littoral zonation of groups of species, or as factors determining the upper and/or lower limits of distribution of individual species (Underwood, 1978; Underwood and Denley, 1984). On rocky shores there is generally an abrupt upper limit to foliose macroalgae so that only low shore areas are dominated by plants (Underwood, 1980, 1981). With the exception of species that are seasonal or have a sporadic occurrence, above this limit foliose algae are generally scarce, limited to pools, or occur amongst barnacles. Temporal and small scale variability in abundance of intertidal algae have been discussed in detail by Jernakoff (1985c). At these upper levels grazing gastropods and encrusting algae are abundant (Underwood, 1981; Jernakoff, 1983). The vertical distribution of algae varies seasonally, tending downshore in warmer periods of the year. Also there is a general trend for algae to extend to higher levels on wave-exposed shores than on more sheltered shores. These observations suggest that physical factors associated with periods of emersion during low tides are important factors influencing the upper limit of algal distribution. Carefully controlled experiments have shown that littoral grazing (mainly by molluscs) plays a major role in determining the vertical distribution of foliose macroalgae in New South Wales, and the upper limit of algal distribution (Underwood, 1979, 1980; Jernakoff, 1983; Underwood and Jernakoff, 1984). At the same time, however, physical factors associated with emersion (probably desiccation, sunlight and temperature) greatly influence the growth and abundance of plants. This is clearly shown if grazers are removed. The algae can then extend to higher levels on the shore, especially if shade is provided. But what limits the distribution of those littoral grazers? Studies by Underwood and Jernakoff (1981) suggest that limpets (an important group of grazers) are prevented from invading dense stands of macroalgae, as they are swept away by waves if they move over the plants. If areas of shore are cleared of algae, and limpets placed in this cleared land, the clearings are invaded by rapidly growing algae and the limpets eventually starve to death as they cannot eat fully grown plants. The limpets are therefore kept out of low-shores areas by the rapid colonization and growth of algae, and this results in domination at these levels by algae. Among the grazers themselves competitive interactions have been demonstrated (Creese, 1982; Creese and Underwood, 1982).

Barnacles can also influence algal diversity and cover (Jernakoff, 1985a). Jernakoff (1983) showed that on a barnacle-dominated shore, grazing molluscs were still important and that neither crevices between barnacles nor the barnacles themselves influenced algal recruitment. The distribution of algae reflected the past "escapes" of spores. Jernakoff (1985b) has also studied the effect of algal overgrowth of the barnacle *Tesseropora rosea* and shown that, at least over one year, there were no significant differences in the mortality, weight, and fecundity compared with barnacles free of plants. He suggested that competition for space between algae and barnacles was important at recruitment stages when an algal layer could prevent the settlement of barnacle spat. Spatial and temporal variation in settlement and recruitment vary greatly and such factors need to be considered in community studies (Caffey, 1985).

Underwood (1984a), in studies at Cape Banks, Sydney, showed that the abundance of intertidal microalgal food (microalgae and algal spores) could be estimated by measurement of chlorophyll on and in the surface of the substratum. Except during summer there was a marked increase in microalgae towards the bottom of the shore. The availability of microalgal food, and the growth of intertidal gastropods at various shore heights, have also been investigated (Underwood, 1984b). The intensity of competition between intertidal gastropods varies temporally and spatially according to the densities and relative abundance of the grazers and according to the availability of microalgal food (Underwood, 1984c).

A further series of experiments has been concerned with predation by the whelk *Morula marginalbaoe* (Moran et al., 1984; Fairweather et al., 1984). Predation was shown to affect the structure of the community. Highly preferred prey (oysters and the limpet *Patelloida latistrigata*) were eliminated by whelks, whereas less preferred species such as the barnacle *Chaemosipho columna* were reduced in density when alternative prey were also less preferred. The differential predation on alternative prey is further discussed by Fairweather (1985); see also Fairweather and Underwood (1983). Effects of prey density, prey size and predator size were discussed by Moran (1985). Jernakoff and Fairweather (1985) analyzed some interactions between barnacles, carnivorous whelks, gastropods and algae and reached a conclusion which could apply with equal validity to many of the studies listed above: "that the observed pattern is due to a number of different interactions among these organisms". This is not an implied criticism of such studies, but rather a recognition of the state of knowledge in this area, and the fact that no simple generalizations can be drawn from the results so far available.

On the basis of experimental analysis of the structure and dynamics of mid-shore communities, Underwood et al. (1983) concluded that the studies had no predictive capacity. This was because of the variation in timing and intensity of recruitment of reproductive units of all components of the ecosystem, and in the density and activity of predatory whelks in different localities. In the discussion above, grazing and predation have been emphasized but even within the studies cited they are by no means the only important interactions between intertidal organisms that were considered. Further examples are found in the studies on commensal relationships between intertidal gastropods *Patelloida mufira* and *Austrocochlea constricta* (Mapstone et al., 1984) and studies on the relationship between the minute gastropod *Littorina acutispira* and barnacles, which suggest that the barnacles provide a refuge from wave-shock rather than protection from desiccation or high temperatures (Underwood and McFadyen, 1983).

Subtidal studies may address the same fundamental ecological questions as those discussed here but are further complicated by methodological problems. Some of these can be overcome by specially constructed samplers (Kennelly and Underwood, 1985) or use of an underwater microscope (Kennelly and Underwood, 1984). Nonetheless, many subtidal studies still make use of settlement plates and such a technique has been extended to study colonization of subtidal algae in an *Ecklonia* kelp community in Sydney Harbour (Kennelly, 1983; Kennelly and Larkum, 1983). These studies documented marked seasonal changes in total algal cover, diversity of colonizers and relative cover of different algal species. In an ancillary study (Kennelly, 1983), the amount of algae initially present was shown to be reduced by invertebrate grazers, the abundances of which were in turn affected by predation. Other studies on *Ecklonia radiata* communities include those of Larkum (1986) on productivity (see pp. 440ff.) and studies of the epifauna on *Ecklonia* plants at Portsea (Victoria) by Fletcher and Day (1983). In the latter it was shown that disturbance rather than competition was the major factor influencing the community.

Burchmore et al. (1985) were responsible for one of the few quantitative studies of rock-reef fish communities in this region.

SANDY BEACHES

Until 1983 there were no published accounts of the sandy beach fauna in New South Wales (Dexter, 1983a, b). Dexter (1983a) listed the fauna of sandy beaches in central New South Wales; of the 118 species collected at least half were undescribed. The range in species diversity on New South Wales sandy beaches is comparable to that reported for boreal, temperate and tropical sandy beaches. Subsequently, Dexter (1984) described the community structure of the fauna of four sandy beaches in southeastern New South Wales. Densities changed during the year but not in a predictable way and no seasonal changes in intertidal zonation patterns were found. Distribution and abundance of species were strongly influenced by tidal level. The number of species increased down the beach from high to low tidemarks, and also increased with decreasing exposure to wave action. Crustaceans dominated the fauna at the most wave-exposed site and decreased in abundance with increasing protection. One or two dominant species could be used to indicate the degree of wave action. Dexter (1985) described the life histories of abundant crustaceans on these beaches. The crustacean communities were dominated by peracarids belonging to the same or related families of isopods and amphipods as those in sandy sediments of temperate and tropical beaches elsewhere in the world. Clearly related species occurring either in different habitats or when present in the same habitat, showed increased niche diversification. The crustacean fauna also shows parallel convergence in life history patterns to those described for the fauna of other sandy beaches.

The surf zone of sandy beaches in New South Wales, Queensland and South Australia are characterized by a group of polychaetes (family Onuphidae) which until recently were regarded as one species *Australonuphis teres*. Paxton (1979) reviewed this taxon and found it to consist of several (5) species (already distinguished by beach fishermen). These beachworms are the largest polychaetes known and can attain lengths of 1 m or more. They live in fragile tubes or are naked burrowing into the sediment in the surf zone, and regularly moving along the beach as the beach profile changes during a storm.

Hutchings and Recher (1974) and Day and Hutchings (1984) have provided faunal lists for intertidal estuarine sandy beach habitats.

Data on rates of turnover and secondary production are not available.

COASTAL LAGOONS, ESTUARIES AND EMBAYMENTS

The flora of estuaries and coastal lagoons in southeastern Australia is generally dominated by seagrasses, and such communities have been especially well studied in the Sydney region. Seagrass communities are dealt with in the following section.

Coastal lagoons

The southeastern coast of Australia is characterized by a series of coastal lagoons, some of which are permanently open to the sea while others are closed by sand bars for varying periods (Barnes, 1980; Roy, 1984).

The fauna of several such lagoons has been described. Atkinson et al. (1981) carried out a detailed survey of Myall Lakes, a series of shallow brackish-water lagoons on the central coast of New South Wales. The diversity of the fauna was closely related to the salinity regime operating within the lagoon and the time of year when the lagoon was open to the sea, and hence recruitment could occur. In some coastal lagoons, probably no reproduction occurs and recruitment is entirely by pelagic larvae when the lagoon is tidally flushed. Many coastal lagoons on the New South Wales coast have been modified by construction work, dredging, and pollution especially from septic tanks (Weate and Hutchings, 1977).

Poore (1982) studied the benthos of the Gippsland Lakes (Victoria) and discussed the likely effects on the fauna resulting from changes in water usage in the catchment area. The fauna of the Gippsland Lakes is similar to that of other estuaries in southeastern Australia. Several species with widespread distributions were recognized, notably the bivalves *Spisula trigonella*, *Tellina deltoidalis*, and *Theora fragilis*, the polychaetes *Australonereis ehlersi*, *Barantolla lepte*, *Ceratonereis aequisetis*, *Nephtys australiensis*, and *Prionospio cirrifera*, the isopods *Colanthura peroni* and *Syncassidina aesturia*, and the shrimps *Callianassa arenosa* and *Macrobrachium intermedium*. No brachyuran crabs, which are characteristic of estuaries in New South Wales, occurred. The Gippsland Lakes have a rich fauna of amphipods in particular with at least 15 species, the genera *Gammaropsis*, *Melita* and *Paracorophium* being represented by two or more species each.

Recently, Day and Hutchings (1984) described two coastal lagoons at Merimbula and Pambula on the south coast of New South Wales. These are basically marine, being permanently open to the sea. The fauna in these lagoons resembles that recorded for Careel Bay, Pittwater, Sydney (Hutchings and Recher, 1974; Hutchings and Rainer, 1979) and other sheltered but fully marine environments.

Data on rates of turnover and secondary productivity of the fauna of these coastal lagoons are unavailable.

Estuaries and bays

Information on the benthos and epibenthic fauna of estuaries of southeastern Australia is scanty. Information on the fauna of seagrass beds and intertidal sandy beaches is discussed elsewhere (pp. 445, 448). Detailed long term studies of the benthos of the Hawkesbury River and central New South Wales are underway. Jones (1987) dealt with the effects of dredging and spoil disposal on the macrobenthos. A characteristic of all the rivers in southeastern Australia is the relatively low volumes of freshwater being discharged during most of the year; hence the marine/estuarine fauna extends a long way upstream, but one suspects that much of the fauna is washed out during flood conditions. Thus the fauna of estuaries is determined by rainfall patterns, rather than seasonal factors as is generally the case in Europe. Obviously the timing of a flood will be crucial to the survival of a population since replacement after the flood will only be by recruitment of larvae from breeding populations in nearby estuaries. Information on rates of turnover and secondary production are sparse for any southeastern Australian estuary (Rainer, 1984).

Poore and Kudenov (1978a) investigated the benthos of the Yarra River, which runs through the heart of Melbourne (Victoria). This river has been subject to pollution, developed as a river port, the shores reclaimed and the original sand bar across its entrance removed. The fauna of the lower marine dominated regions was compared with that of the central muddy basin of Port Phillip Bay, the bay into which the Yarra flows. The faunal diversity was lower than that of equivalent areas in the Northern Hemisphere, and no seasonality in species composition or diversity was seen. The dominant river species was the bivalve *Theora fragilis*, which also occurred, but in lower numbers, in Port Phillip Bay. Poore and Kudenov discussed the diversity, stability and predictability of these benthic communities and concluded that the river sites had a benthos of low diversity and highly predictable species composition.

Taxonomy is still a major problem in studies of estuarine fauna. Hutchings and Murray (1984) have described the polychaete fauna of estuarine areas and coastal embayments in central and south New South Wales. Over 180 species were recorded, many of which were undescribed. A similar pattern would emerge in a study of the small crustaceans, and to a lesser extent for molluscs, especially micro-molluscs, of estuarine areas.

The major embayments in southeastern Australia have been the subject of a series of investigations — for instance the Botany Bay Study (State Pollution Control Commission, 1978, 1979, 1981a–c); the Western Port Bay Environmental Study (Brand et al., 1974; Littlejohn et al., 1974), and the Port Phillip Bay Study (Port Phillip Survey, 1966, 1971; Melbourne Metropolitan Board of Works etc., 1973; Poore and Kudenov, 1978b).

A review of the benthos of the muddy environments in Port Phillip Bay (Poore and Rainer, 1979) showed that the benthos exhibited small scale spatial patchiness, with the fauna at all stations dominated by the deposit feeding bivalve *Theora fragilis*, and the decapod shrimp *Callianassa limosa*. Only minor species were restricted to single stations. Density of a few species varied seasonally, and several were more abundant in some years than others. Use of the Bray–Curtis coefficient grouped some samples by station and some by season, and it was concluded that the main differences between stations were in overall species composition, not in densities of dominant species. Temporal changes in the composition of muddy bottom benthic communities are common only in cool-temperate latitudes (Boesch et al., 1976). Changes in community structure of the benthos in Port Phillip Bay were non-seasonal, with irregular fluctuations in the density of the common species and in the presence of the minor species. Poore and Rainer (1974) also examined the distribution of soft-bottom molluscs in Port Phillip Bay, in terms of abundance, feeding types and patterns of diversity, in relation to both depth and sediment type. Particular feeding categories dominated particular environments, with infaunal suspension feeders dominating in marginal sandy substrata, and surface deposit feeders dominating in the silt and clay sediments. The main factors determining the distribution of molluscs were substratum and food supply, with purely hydrological effects being limited to nearshore areas.

Detailed studies in the vicinity of the Werribee sewage treatment farm, Port Phillip Bay, have been carried out by Poore and Kudenov (1978b) and Dorsey (1982), and off Black Rock sewage outfall, Connewarre, Victoria, by Dorsey and Synnot (1980). In these studies opportunistic species, such as certain groups of polychaetes and amphipods, were dominant in the communities closest to the outfall. The effect of the drains on the macrobenthos is exerted through particulate organic matter, dissolved nutrients and freshwater inputs.

SEAGRASS ECOSYSTEMS

Distribution of seagrasses

One result of the high-energy wave climate along the open coastline, is that the establishment of seagrass meadows in southeastern Australia is almost entirely restricted to the protected waters of estuaries and semi-enclosed embayments.

Bentham (1878) and later Ostenfeld (1916) provided lists of seagrass species and their distributions in Australian waters, but it was not until Wood's studies (1959) that the seagrass communities of southeastern Australia were described. More recently, Den Hartog (1970), Larkum (1977), McComb et al. (1981), King (1981c), Hutchings (1982), Robertson (1984) and Larkum et al. (1989) have all reviewed various aspects of Australian seagrass biology.

There are at least seven recorded species of seagrasses in southeastern Australia (E.L. Robertson, 1984), and, in addition, three species of *Ruppia* (Jacobs and Brock, 1982). *Ruppia* species are not considered to be true seagrasses, since although they grow in inland lakes (fresh and saline) and coastal saline lagoons, they are usually not found in fully marine conditions.

Apart from Wood's general account (Wood, 1959), the description of botanical aspects of seagrass communities in southeastern Australia has been largely site-specific. For example, distribution studies have been carried out within the following areas (from north to south): Smiths Lake (Robinson et al., 1983), Lake Macquarie and Tuggerah Lakes (Higginson, 1965, 1971; King, 1986; King and Barclay, 1986; King and Hodgson, 1986; King and Holland, 1986), Botany Bay, Pittwater and Jervis Bay (Larkum, 1976), Lake Illawarra (Harris et al., 1980), the Gippsland Lakes (Ducker et al., 1977a), and Western Port (Bulthuis, 1981). Very recently, West et al. (1985) have mapped the distribution of seagrasses in New South Wales. In this survey of 111 estuaries the total area of seagrass was approximately 155 km^2. King and Hodgson (1986) have presented quantitative data for six years, both for the area of all seagrass species and for the total biomass of *Zostera* in Lake Macquarie and Tuggerah Lakes; they showed that there were marked, and unpredictable, changes in seagrass meadow composition and distribution over this time.

The most widespread seagrass throughout southeastern Australia is *Zostera capricorni* (Wood, 1959; West et al., 1985). This species is replaced by *Zostera muelleri* and *Heterozostera tasmanica* to the south of Mallacoota Inlet (Jacobs and Williams, 1980). As E.L. Robertson (1984: 122) has stated in relation to the Australian Zosteraceae: "a broad spectrum of intergrades occurs" and "further work is required to elucidate the *Z. mucronata–Z. muelleri–Z. capricorni* complex". As well, *Heterozostera tasmanica* is only distinguishable from *Zostera* species on the basis of a larger number of cortical vascular bundles, and is recognized as a separate genus only because this is consistent with current usage (Jacobs and Williams, 1980). Records of *Heterozostera* from the Sydney region were considered to be drift specimens (Jacobs and Williams, 1980), but the species, though uncommon, has now been collected as far north as Port Stephens. Similar problems exist in the field identification of *Ruppia* species (Jacobs and Brock, 1982). To a lesser extent they also occur with *Halophila*, due to the variability in vegetative material, particularly leaves (E.L. Robertson, 1984). *Posidonia australis* occurs in southeastern Australian waters south of Wallis Lake, but in southern Australia there are several other species of the genus which form more extensive meadows in the sandy, marine environments of the South Australian gulfs (Womersley,

1984) and along the southwestern Australian coast (Cambridge, 1975, McComb et al., 1981).

In southeastern Australia *Zostera capricorni* grows in a wide range of habitats, including coastal rivers, lakes, lagoons, creeks, and open embayments (West et al., 1985) and on a wide range of sediment types (Harris et al., 1980). The distribution of *Zostera* is considered dependent upon light penetration and wave action (Higginson, 1965; Harris et al., 1980; Robinson et al., 1983; Larkum et al., 1984). *Zostera* occurs in the estuarine portion of coastal rivers on mudflats and shoals and, in such cases, is usually limited to depths of approximately two metres (see, for example, Wood, 1959; for the Macleay River). In coastal lakes *Zostera* and *Halophila* generally fringe the shoreline and may occur to depths of four metres (Harris et al., 1980; Robinson et al., 1983).

In comparison, *Posidonia australis* occurs in fewer estuaries and is restricted to open embayments and a small number of lagoons and inlets. *Posidonia* is absent from coastal rivers and from lakes and lagoons that occasionally close (West et al., 1985). This absence may be related partly to salinity tolerance although Tyerman et al. (1984) stated that "some factor other than the tolerance of the mature plant . . . operates to limit the habitat to stenohaline conditions". The distribution of *Posidonia* is also influenced by sediment. Wood (1959) commented that *Posidonia* occurs on sandy shores with a positive redox potential (Eh). This is verified if one compares the seagrass distribution maps of West et al. (1985) with the distribution of sediment types (Roy, 1984). *Posidonia australis* is found on coastal barrier and tidal-delta sands, and not on estuarine muds and fluvial sands.

Ruppia is found predominantly on muddy substrata in closed or semi-closed coastal lagoons and in the cut-off lagoons and creeks of coastal rivers (Jacobs and Brock, 1982; West et al., 1985). Many lakes and lagoons in New South Wales are subject to periodic and variable entrance-closures dependent on rainfall. In such situations, the seagrasses appear to undergo cyclic changes, being dominated by *Ruppia* in times of low salinity and by *Zostera* in periods of high salinity (see, for example, Robinson et al., 1983). This is consistent with the observation that *R. megacarpa* will only germinate at low salinities (Brock, 1982). In rare situations, such as Wallis Lake, Lake Macquarie and St. Georges Basin, *Zostera, Posidonia* and *Ruppia* may occur in close association (Wood, 1959; West et al., 1985).

In open embayments and marine-dominated lakes and inlets, *Zostera* can occur from the intertidal sand flats to a depth of about 10 m (below mean low water, MLW) but in stable sublittoral conditions is often replaced by *Posidonia* (Wood, 1959; Larkum, 1976). In such cases, *Zostera* (or *Heterozostera*) and *Halophila* generally occur at the upper and lower limits of the *Posidonia* meadow (Larkum, 1976; West et al., 1985). The situation is reversed when tidal range is negligible, as in St. Georges Basin and Wallis Lake. *Posidonia* here occupies the landward fringe, and *Zostera* inhabits the deeper waters (West et al., 1985).

The distribution of the genus *Halophila* is not as well documented, though the species is widely distributed in terms of estuary type and sediment.

The distribution of seagrasses in Western Port has been described by Bulthuis (1981). Here *Amphibolis antarctica* occurs near the entrance in sandy exposed areas, while *Heterozostera* occupies the intertidal mudflat and *Zostera muelleri* the higher littoral zones.

There have been many recorded losses in the distribution of seagrass beds, principally adjacent to large cities. The southeastern Australian coast is particularly vulnerable in this respect, as seagrasses are confined to semi-enclosed estuarine waters. Larkum and West (1983) have reviewed the causes of seagrass depletion. Prominent examples are the loss of *Posidonia* from Botany Bay due to erosion and turbidity (Larkum, 1976), the loss of *Heterozostera* from Western Port (Bulthuis, 1985), and the loss of *Zostera* from the northern New South Wales rivers.

Seagrass epiphytes

On sandy and muddy coasts seagrasses are often the only stable substrata for algal colonization. Ducker et al. (1977b) provided a list of epiphytes occurring on *Amphibolis*, and Ducker and Knox (1978, 1984) have investigated the biological relationships between algal epiphytes and their seagrass hosts.

Harlin (1975) suggested that epiphytic algae would provide a means of monitoring environmental change. May et al. (1978) and May (1982) used such an approach in a study of sites in Botany Bay and Jervis Bay, but while differences could be recognized, the absence of appropriate environmental data severely limits the interpretation of such results.

Infauna of seagrass beds

The fauna of Australian seagrass beds was reviewed by Hutchings (1982). Much of the information reviewed was purely qualitative, especially that dealing with epifauna (animals living on the surface of the mud at the base of the seagrass) and sessile communities on the blades of the seagrass. Quantitative studies have been restricted to infauna and fish populations (see below). Hutchings concluded that seagrass faunal studies were at a stage when experimental, manipulative studies were required. Since that time, Bell et al. (1985) have experimented, using artificial seagrass beds.

A quantitative survey of the infauna of *Posidonia* beds along the New South Wales coast (Collett et al., 1984) failed to detect a characteristic fauna associated with this vegetation. Extensive sampling revealed only two "constant" species, and these were widely distributed in the nine sites sampled. No species or genera were found to be obligately dependent on, or unique to, *Posidonia* beds. Instead, sediment and salinity regimes determined the species composition of the community. Long-term studies of Careel Bay (Hutchings and Recher, 1974, and unpublished data) have documented the variability in composition of animal communities in seagrass beds. This presumably results from the pelagic recruitment of most species, so that local current and weather patterns will determine species settlement. This is especially true of species that breed once and then die. Poore (1982), working on the benthic communities of the Gippsland Lakes, found no correlation between temporal changes in the benthic fauna and seasonal changes in temperature or any other environmental factor.

Rainer (1981) and Rainer and Fitzhardinge (1981) studied in detail the temporal patterns of benthic communities in Cabbage Tree Basin, Port Hacking, over 18 months, including the fauna associated with seagrass beds. Approximately 50% of the common fauna exhibited strong seasonal variations in abundance, whereas the remaining species exhibited no seasonal variation or episodic variations in numbers. No strong synchrony in peaks of seasonal abundance was evident. Rainer discussed in detail the stability of the individual communities, in relation to stress caused by fluctuating temperatures, salinities and periods of deoxygenation at the deeper sites. The species composition of the intertidal and shallow-water sites (including seagrass beds) was more stable than that of the deeper sites, due to the presence of short-lived opportunistic species present after periods of deoxygenation. The stable community structure and species composition at the intertidal and shallow-water sites indicate that greater environmental harshness does not necessarily imply less faunal stability. These results may have widespread applicability since many other estuarine areas or coastal lagoons in southeastern Australia have similarly "harsh" environments, or even harsher, especially those with entrances periodically closed during drought periods.

No detailed studies on rates of secondary production of seagrass faunas have been made except by Rainer (1982) who studied the macrobenthic communities of Gunnamatta Bay, Port Hacking. Estimated production varied from 10 kJ m^{-2} yr^{-1} (in an intertidal area of well sorted sand) to 210 to 450 kJ m^{-2} yr^{-1} (in shallow subtidal areas), with an average of 160 kJ m^{-2} yr^{-1} for the estuary as a whole. Such figures are relatively low for estuaries, which may reflect the relative paucity of molluscs, resulting from excessive harvesting. Polychaetes dominated the production (45% of the total), crustaceans (20%) and molluscs and echinoderms (12%). Detritus feeders dominated each community, contributing 61% of the total estimated net production; carnivores contributed 32%, and suspension feeders 7%. Production estimated for carnivores was higher than could be supported by the non-carnivore populations, suggesting that some of the species considered to be carnivores have additional feeding modes. Rainer (1984) examined the possibility of using polychaetes alone to characterize the secondary production of a benthic community but concluded that all of the major taxa needed to be considered.

The ways in which primary production of the seagrass beds is transferred to the faunal community has hardly been investigated in Australia. Brand (1977) experimentally investigated the production of detritus from the breakdown of *Zostera* by amphipods (*Parhyalella* spp.). This was to quantify part of the model of the carbon budget proposed by Brand et al. (1974) for estuarine ecosystems in Western Port Bay, Victoria. The amphipods ingested the epiphytes growing on the blades of *Zostera*. Brand did not provide evidence that the amphipods actually obtained nutrients from the plant matter, but implied that nutrition was obtained mainly from the attached biota. The amphipods produce nitrogen-enriched faecal pellets, which facilitate further breakdown of the *Zostera* fragments. In addition to amphipods, some polychaetes and a few fish in Australian seagrass beds are herbivores. Leatherjackets have been shown to obtain some nutrients from ingested *Posidonia australis* fragments (Conacher et al., 1979). Nonetheless most of the fish occurring in seagrass beds are carnivores (Littlejohn et al., 1974) that feed on the detritus-feeding crabs, molluscs and polychaetes (Robertson, 1980). Although Robertson and Howard (1978) have documented the importance of zooplankton for several fish species in Western Port Bay, fish diets may shift during the year as prey species fluctuate (Littlejohn et al., 1974).

Seagrasses and fish ecology

A major study of fish utilization of estuarine habitats carried out in Botany Bay (State Pollution Control Commission, 1981b, c) found that inshore estuarine wetland communities (seagrasses and mangroves) were nursery areas for many economically important species of fish.

Pollard (1984) reviewed ecological studies on seagrass–fish communities with particular reference to Australian studies. Major studies have been carried out in Botany Bay (see Middleton et al., 1984) and Western Port, Victoria (see E.L. Robertson, 1984). Pollard (1984) has summarized the role of seagrasses in maintaining fisheries resources:

> "The significance of seagrass habitats in many parts of the world as nursery areas for juvenile and sub-adult stages of fishes important to commercial and recreational fisheries is stressed. The utilization of such seagrass habitats as 'fish nurseries' appears to be based on their provision of both adequate shelter for small fishes from predators and an abundant food source, particularly in the form of small epibenthic crustaceans which in turn depend on the seagrass detritus cycle as the basis for their food resources."

On the basis of the few studies undertaken in adjacent mangrove habitats, Bell et al. (1984) concluded that mangrove communities in temperate Australia were important as nursery areas for fishes inhabiting adjacent estuarine and inshore marine habitats as adults.

CONCLUDING REMARKS

At the completion of this overview of ecological studies on the southeastern Australian coast it is perhaps reasonable to attempt a comparison with eulittoral communities in other parts of the world. Despite differences in the species involved, the high species diversity and degree of endemism are characteristic of the region, and there are marked similarities in the types and distributions of eulittoral plants and animals: the so-called "universal features" first elaborated by Stephenson and Stephenson (1949). Certainly descriptive studies of the local distributions of eulittoral organisms in southeastern Australia fit well within this broad descriptive framework, and the work of Womersley and co-workers on sublittoral communities in southern Australia has shown that such an approach may be extended usefully into the sublittoral zones. One might ask how similar in functional respects are Australian marine communities? Underwood and Fairweather (1985) attempted to answer that question when they compared experiments in different geographical regions with the aim of deducing rules governing the organization of rocky intertidal communities. In three specific cases, results from the New South Wales coast were compared with similar studies by ecologists in other countries. In each instance difficulties resulted due to the fact that the aims, hypotheses and approaches used were not comparable, even in superficially similar studies. This lack of comparability between studies makes it impossible to presently answer the question: "are Australian marine ecosystems different."

ACKNOWLEDGEMENTS

We wish to thank colleagues for their help and advice, but in particular Des Griffin and Frank

Rowe (Australian Museum) for comments on animal groups, and Penny Farrant (University of New South Wales) for drawing the diagrams.

REFERENCES

Albright, L.J., 1976. *In situ* degradation of mangrove tissues. *N.Z. J. Mar. Freshwater Res.*, 10: 611–647.

Allender, B.M. and Kraft, G.T., 1983. The marine algae of Lord Howe Island (New South Wales): the Dictyotales and Cutleriales (Phaeophyta). *Brunonia*, 6: 73–130.

Ashby, E., 1926. The regional distribution of Australian chitons (Polyplacophora). *Rep. Aust. Assoc. Advanc. Sci.*, 17: 366–393.

Atkinson, G., Hutchings, P.A., Johnson, M., Johnson, W.D. and Melville, M.D., 1981. An ecological investigation of the Myall Lakes region. *Aust. J. Ecol.*, 6: 299–327.

Australian Government, 1985. *Australian National Tide Tables 1986*. A.G.P.S., Canberra, 160 pp.

Barnes, R.S.K., 1980. *Coastal Lagoons*. CUP, Cambridge, 106 pp.

Barson, M. and King, R.J., 1982. Australia — Coastal ecology. In: M.L. Swartz (Editor), *Encyclopaedia of Beaches and Coastal Environments*. Dowden, Hutchinson and Ross, Penns., pp. 103–110.

Bavor, J., 1978. *Microbiological studies in Westernport Bay. 1. Seagrass biodegradation*, Report to Westernport Bay Environmental Study, Ministry for Conservation, Melbourne.

Bavor, J. and Millis, N.F., 1976. *Bacteriological studies in Westernport Bay*. Report to Westernport Bay Environmental Study, Ministry for Conservation, Melbourne.

Beanland, W.R. and Woelkerling, W.J., 1982. Studies on Australian mangrove algae: composition and geographic distribution of communities in Spencer Gulf, South Australia. *Proc. R. Soc. Victoria*, 94: 89–106.

Beanland, W.R. and Woelkerling, W.J., 1983. *Avicennia* canopy effects on mangrove algal communities in Spencer Gulf, South Australia. *Aquat. Bot.*, 17: 309–313.

Bell, J.D., Pollard, D.A., Burchmore, J.J., Pease, B.C. and Middleton, M.J., 1984. Structure of a fish community in a temperate tidal mangrove creek in Botany Bay, New South Wales. *Aust. J. Mar. Freshwater Res.*, 35: 33–46.

Bell, J.D., Steffe, A.S. and Westoby, M., 1985. Artificial seagrass: how useful is it for field experiments on fish and macroinvertebrates? *J. Exp. Mar. Biol. Ecol.*, 90: 171–177.

Bennett, I. and Pope, E.C., 1953. Intertidal zonation of the exposed rocky shores of Victoria together with a rearrangement of the biogeographical provinces of temperate Australian shores. *Aust. J. Mar. Freshwater Res.*, 4: 105–159.

Bennett, I. and Pope, E.C., 1960. Intertidal zonation of the exposed rocky shores of Tasmania and its relationship with the rest of Australia. *Aust. J. Mar. Freshwater Res.*, 11: 182–221.

Bentham, G., 1878. *Flora Australiensis: a description of the plants of the Australian Territory. Vol. 7*. Reeve. London.

Bird, E.C.F., 1983. Provenance of beach sediments in south-eastern Australia. In: A. McLachlan and T. Erasmus (Editors), *Sandy Beaches as Ecosystems*. Junk, The Hague, pp. 87–95.

Bird, E.C.F., 1984. *Coasts — An Introduction to Coastal Geomorphology*. ANU Press, Canberra, 320 pp.

Boesch, D.F., Wass, M.L. and Virnstein, R.W., 1976. The dynamics of estuarine benthic communities. In: M. Wiley (Editor), *Estuarine Processes Vol. 1. Uses, Stresses and Adaptation to the Estuary*. Academic Press, New York, pp. 177–196.

Boon, P.I., Moriarty, D.J.W. and Saffigna, P.G., 1986. Rates of ammonium turnover, and the role of amino-acid deamination, in seagrass (*Zostera capricorni* Aschers.) beds of Moreton Bay, Australia. *Mar. Biol.*, 91: 259–268.

Borowitzka, M.A., 1972. Intertidal algal species diversity and the effect of pollution. *Aust. J. Mar. Freshwater Res.*, 23: 73–84.

Borowitzka, M.A., King, R.J. and Larkum, A.W.D., 1982. *A field guide to the marine plants of N.S.W. I. Chlorophyta, Phaeophyta and Seagrasses*. University of New South Wales, Botany School, Kensington, 83 pp. and 37 figs.

Brand, G.W., 1977. *Epifaunal activity in seagrass breakdown*. Report to Westernport Bay Environmental Study. Ministry for Conservation, Victoria, 17 pp.

Brand, G.W., Bulthuis, D. and Cuff, W.R., 1974. *Energy flow and its regulation in the Westernport Bay ecosystem*. Report to Westernport Bay Environmental Study 1973–74. Project Report No. 4.3.11 Ministry for Conservation. Victoria, 12 pp.

Brock, M.A., 1982. Biology of the salinity tolerant genus *Ruppia* L. in saline lakes in South Australia. I. Morphological variation within and between species and ecophysiology. *Aquat. Bot.*, 13: 219–248.

Bulleid, N.C., 1983. The nutrient cycle of an intermittently stratified estuary. In: W.R. Cuff and M. Tomczak, Jr. (Editors), *Synthesis and Modelling of Intermittent Estuaries*. Springer, Berlin, pp. 55–57.

Bulthuis, D.A., 1981. Distribution and summer standing crop of seagrass and macro-algae in Westernport, Victoria. *Proc. R. Soc. Victoria*, 92: 107–112.

Bulthuis, D.A., 1983a. Biomass accumulation and shading effects of epiphytes on leaves of the seagrass *Heterozostera tasmanica* in Victoria, Australia. *Aquat. Bot.*, 16: 137–148.

Bulthuis, D.A., 1983b. Effects of *in situ* light reduction on density and growth of the seagrass *Heterozostera tasmanica* in Westernport, Victoria, Australia. *J. Exp. Mar. Biol. Ecol.*, 67: 91–103.

Bulthuis, D.A., 1983c. Effects of temperature on the photosynthesis- irradiance curve of the Australian seagrass *Heterozostera tasmanica*. *Mar. Biol. Lett.*, 4: 47–57.

Bulthuis, D.A., 1985. Demise of seagrasses in Westernport, Victoria 1973–84. *Aust. Mar. Sci. Bull.*, 89: 24–25.

Bulthuis, D.A. and Woelkerling, W.J., 1981. Effects of *in situ* nitrogen and phosphorus enrichment of the sediments on the seagrass *Heterozostera tasmanica* (Martens ex Ascher.) Den Hartog in Western Port, Victoria, Australia. *J. Exp. Mar. Biol. Ecol.*, 53: 193– 207.

Bulthuis, D.A. and Woelkerling, W.J., 1983. Seasonal variation in standing crop, density and leaf growth rate of the seagrass *Heterozostera tasmanica* in Westernport and

Port Phillip Bay, Victoria, Australia. *Aquat. Bot.*, 16: 111–136.

Burchmore, J.J., Pollard, D.A., Bell, J.D., Middleton, M.J., Pease, B.C. and Mathews, J., 1985. An ecological comparison of artificial and natural rocky reef fish communities in Botany Bay, New South Wales, Australia. *Bull. Mar. Sci.*, 37: 70–85.

Caffey, H.M., 1985. Spatial and temporal variation in settlement and recruitment of intertidal barnacles. *Ecol. Monogr.*, 55: 313–332.

Cambridge, M.L., 1975. Seagrasses of southwestern Australia with special reference to the ecology of *Posidonia australis* Hook f. in a polluted environment. *Aquat. Bot.*, 1: 149–161.

Capone, D.G. and Taylor, B.F., 1977. Nitrogen fixation (acetylene reduction) in the phyllosphere of *Thalassia testudinum*. *Mar. Biol.*, 40: 19–28.

Chapman, D.M., 1983. Sediment reworking on sandy beaches. In: A. McLachlan and T. Erasmus (Editors), *Sandy Beaches as Ecosystems*. Junk, The Hague, pp. 45–61.

Chapman, V.J. and Chapman, D.J., 1973. *The Algae*, 2nd ed. MacMillan, London, 497 pp.

Chapman, V.J. and Trevarthen, C.B., 1953. General schemes of classification in relation to marine coastal zonation. *J. Ecol.*, 41: 198–204.

Clarke, H.L., 1946. *The echinoderm fauna of Australia*. Publ. Carnegie Inst. No. 566, 567 pp.

Clarke, P.J., 1985. Nitrogen pools and soil characteristics of a temperate estuarine wetland. *Aquat. Bot.*, 23: 275–290.

Clayton, M.N. and King, R.J. (Editors), 1981. *Marine Botany: An Australasian Perspective*. Longman Cheshire, Melbourne, 468 pp.

Collett, L.C., Hutchings, P.A., Gibbs, P.J. and Collins, A.J., 1984. A comparative study of the macro-benthic fauna of *Posidonia australis* seagrass meadows in New South Wales. *Aquat. Bot.*, 18: 111–134.

Conacher, M.J., Lanzing, W.J.R. and Larkum, A.W.D., 1979. Ecology of Botany Bay. II. Aspects of the feeding ecology of the fanbellied leatherjacket, *Monacanthus chinensis* (Pisces: Monacanthidae), in *Posidonia australis* seagrass beds in Quibray Bay, Botany Bay, New South Wales. *Aust. J. Mar. Freshwater Res.*, 27: 117–127.

Cotton, B.C., 1930. Fissurellidae from the "Flindersian" region, southern Australia. *Rec. S. Aust. Mus.*, 4: 219–222.

Creese, R.G., 1982. Distribution and abundance of the acmaeid limpet *Patelloida latistrigata*, and its interaction with barnacles. *Oecologia*, 52: 85–96.

Creese, R.G. and Underwood, A.J., 1982. Analysis of inter- and intra-specific competition amongst limpets with different methods of feeding. *Oecologia*, 53: 337–346.

Cribb, A.B., 1954a. *Macrocystis pyrifera* (L.) Ag. in Tasmanian waters. *Aust. J. Mar. Freshwater Res.*, 5: 1–34.

Cribb, A.B., 1954b. The algal vegetation of Port Arthur, Tasmania. *Pap. Proc. R. Soc. Tasmania*, 88: 1–44.

Cuff, W.R., Sinclair, R.E., Parker, R.R., Tranter, D.J., Bulleid, N.C., Giles, M.S., Godfrey, J.S., Griffiths, F.B., Higgins, H.W., Kirkman, H., Rainer, S.F. and Scott, B.D., 1983. A carbon budget for south west arm, Port Hacking. In: W.R. Cuff and M. Tomczak, Jr. (Editors), *Synthesis and Modelling of Intermittent Estuaries*. Springer, Berlin, pp. 193–222.

Dakin, W.J., Bennett, I. and Pope, E.C., 1948. A study of certain aspects of the ecology of the intertidal zone of the New South Wales coast. *Aust. J. Sci. Res. B*, 1: 176–230.

Dartnall, A.J., 1974. In: W.D. Williams (Editor), *Biogeography and Ecology in Tasmania. Monographiae Biologicae, Vol. 25*. Junk, The Hague, pp. 171–194.

Davey, A. and Woelkerling, W.J., 1980. Studies on Australian mangrove algae. I. Victorian communities: composition and geographic distribution. *Proc. R. Soc. Victoria*, 91: 53–66.

Davey, A. and Woelkerling, W.J., 1984. Studies on Australian mangrove algae. III. Victorian communities: structure and recolonization in Western Port Bay. *J. Exp. Mar. Biol. Ecol.*, 85: 177–190.

Day, J.H. and Hutchings, P.A., 1979. An annotated checklist of Australian and New Zealand Polychaeta, Archiannelids and Myzostomida. *Rec. Aust. Mus.*, 32: 80–161.

Day, J.H. and Hutchings, P.A., 1984. Descriptive notes on the fauna and flora of Merimbula, Pambula and Back Lakes, New South Wales. *Aust. Zool.*, 21: 269–289.

Den Hartog, C., 1970. *The Seagrasses of the World*. North-Holland, Amsterdam, 275 pp.

Dexter, D.M., 1983a. Community structure of intertidal sandy beaches in New South Wales, Australia. In: A. McLachlan and T. Erasmus (Editors), *Sandy Beaches as Ecosystems*. Junk, The Hague, pp. 461–472.

Dexter, D.M., 1983b. A guide to sandy beach fauna of New South Wales. *Wetlands (Aust)*, 3: 94–104.

Dexter, D.M., 1984. Temporal and spatial variability in the community structure of the fauna of four sandy beaches in southeastern New South Wales. *Aust. J. Mar. Freshwater Res.*, 35: 663–672.

Dexter, D.M., 1985. Distribution and life histories of abundant crustaceans of four sandy beaches of southeastern New South Wales. *Aust. J. Mar. Freshwater Res.*, 36: 281–290.

Dorsey, J.H., 1982. Intertidal community offshore from the Werribee sewage treatment farm: an opportunistic infaunal assemblage. *Aust. J. Mar. Freshwater Res.*, 33: 45–54.

Dorsey, J.H. and Synnot, R.N., 1980. Marine soft bottom benthic community offshore from Black Rock sewage outfall, Connewarre, Victoria. *Aust. J. Mar. Freshwater Res.*, 31: 155–162.

Ducker, S.C. and Knox, R.B., 1978. Alleloparasitism between a seagrass and algae. *Naturwissenschaften*, 65: 391–392.

Ducker, S.C. and Knox, R.B., 1984. Epiphytism at the cellular level with special reference to algal epiphytes. In: H.-F. Linskens and J. Heslop-Harrison (Editors), *Encyclopedia of Plant Physiology. Vol. 17. Cellular Interactions*, Springer, Berlin, pp. 113–133.

Ducker, S.C., Brown, V.B. and Calder, D.M., 1977a. *An identification of aquatic vegetation in the Gippsland Lakes*. School of Botany. University of Melbourne, 111 pp.

Ducker, S.C., Foord, N.J. and Knox, R.B., 1977b. Biology of Australian seagrasses, the genus *Amphibolis* C. Agaradh (Cymodoceaceae). *Aust. J. Bot.*, 25: 67–95.

Edgar, G.J., 1983a. The ecology of southeast Tasmanian phytal animal communities. I. Spatial organization on a local scale. *J. Exp. Mar. Biol. Ecol.*, 70: 129–157.

Edgar, G.J., 1983b. The ecology of south-east Tasmanian phytal animal communities. II. Seasonal change in plant and animal populations. *J. Exp. Mar. Biol. Ecol.*, 70: 159–179.

Edgar, G.J., 1984. General features of the ecology and biogeography of Tasmanian subtidal rocky shore communities. *Pap. Proc. R. Soc. Tasmania*, 118: 173–186.

Fairweather, P.G., 1985. Differential predation on alternative prey, and the survival of rocky intertidal organisms in New South Wales. *J. Exp. Mar. Biol. Ecol.*, 89: 136–157.

Fairweather, P.G. and Underwood, A.J., 1983. The apparent diet of predators and biases due to different handling times of their prey. *Oecologia*, 56: 169–179.

Fairweather, P.G., Underwood, A.J. and Moran, M.J., 1984. Preliminary investigations of predation by the whelk *Morula marginalba*. *Mar. Ecol. Prog. Ser.*, 17: 143–156.

Farrant, P.A. and King, R.J., 1982. The subtidal seaweed communities of the Sydney region. *Wetlands (Aust)*, 2: 51–60.

Farrant, P.A. and King, R.J., 1989. The Dictyotales (Algae: Phaeophyta) of New South Wales. *Proc. Linn. Soc. NSW*, 110: 369–406.

Fletcher, W.J., 1987. Interactions among subtidal Australian sea urchins, gastropods, and algae: effects of experimental removals. *Ecol. Monogr.*, 57: 89–109.

Fletcher, W.J. and Day, R.W., 1983. The distribution of epifauna on *Ecklonia radiata* (C. Agardh) J. Agardh and the effect of disturbance. *J. Exp. Mar. Biol. Ecol.*, 71: 205–220.

Giles, M.S., 1983. Primary production of benthic microorganisms in south east arm, Port Hacking, New South Wales. In: W.R. Cuff and M. Tomczak Jr. (Editors), *Synthesis and Modelling of Intermittent Estuaries*. Springer, Berlin, pp. 147–166.

Gillham, M.E., 1965. IV. Vegetation: Additions and Changes — The Fisher Island Field Station. *Pap. Proc. R. Soc. Tasmania*, 99: 71–73.

Goulter, P.F.E. and Allaway, W.G., 1979. Litter fall and decomposition in a mangrove stand, *Avicennia marina* (Forssk.) Vierh. in Middle Harbour, Sydney. *Aust. J. Mar. Freshwater Res.*, 30: 541–546.

Griffin, D.J.G. and Yaldwyn, J.C., 1968. The constitution, distribution and relationships of the Australian decapod crustacea. A preliminary review. *Proc. Linn. Soc. NSW*, 93: 164–183.

Griffiths, O., 1982. *Coastal headlands survey. Pt. 4, A preliminary geomorphological and biological survey of the intertidal rock platforms of the headlands along the New South Wales coast*. Nat. Trust. Aust., Sydney, 108 pp.

Guiler, E.R., 1950. The intertidal ecology of Tasmania. *Pap. Proc. R. Soc. Tasmania*, 84: 135–201.

Guiler, E.R., 1951a. The intertidal ecology of Pipe Clay Lagoon. *Pap. Proc. R. Soc. Tasmania*, 85: 29–52.

Guiler, E.R., 1951b. Notes on the intertidal ecology of the Freycinet Peninsula. *Pap. Proc. R. Soc. Tasmania*, 85: 53–70.

Guiler, E.R., 1952a. The ecological features of certain sheltered intertidal areas in Tasmania. *Pap. Proc. R. Soc. Tasmania*, 86: 1–11.

Guiler, E.R., 1952b. The intertidal ecology of the Eaglehawk Neck area. *Pap. Proc. R. Soc. Tasmania*, 86: 13–29.

Guiler, E.R., 1952c. The nature of intertidal zonation in Tasmania. *Pap. Proc. R. Soc. Tasmania*, 86: 31–61.

Guiler, E.R., 1953a. Further observations on the intertidal ecology of the Freycinet Peninsula. *Pap. Proc. R. Soc. Tasmania*, 87: 93–95.

Guiler, E.R., 1953b. Intertidal classification in Tasmania. *J. Ecol.*, 41: 381–384.

Guiler, E.R., 1954. The intertidal zonation at two places in southern Tasmania. *Pap. Proc. R. Soc. Tasmania*, 88: 105–117.

Guiler, E.R., 1960a. Notes on the intertidal ecology of Trial Harbour, Tasmania. *Pap. Proc. R. Soc. Tasmania*, 94: 57–62.

Guiler, E.R., 1960b. The intertidal zone-forming species on rocky shores of the east Australian coast. *J. Ecol.*, 48: 1–28.

Guiler, E.R., Serventy, D.L. and Willis, J.H., 1958. The Fisher Island field station — with an account of its principal fauna and flora. I. General description of Fisher Island and its mutton-bird rookeries. *Pap. Proc. R. Soc. Tasmania*, 92: 165–167.

Harada, K., 1980. *A preliminary study of colonization by benthic macro-algae at an exposed coastal site off Sydney*. M.Sc. Thesis, University of Sydney, Sydney.

Harlin, M.M., 1975. Epiphyte-host relations in seagrass communities. *Aquat. Bot.*, 1: 125–131.

Harris, M. McD., King, R.J. and Ellis, J., 1980. The eelgrass *Zostera capricorni* in Illawarra Lake, New South Wales. *Proc. Linn. Soc. NSW*, 104: 23–33.

Hedley, C., 1904. The effect of the Bassian isthmus upon the existing marine fauna: a study in ancient geography. *Proc. Linn. Soc. NSW*, 28: 876–883.

Hedley, C., 1926. Zoogeography. In: *Australian Encyclopedia, Vol. 2*. Angus and Robertson, Sydney, pp. 743–744.

Hickman, N.J., McShane, P.E. and Axelrad, D.M., 1984. Light climate in the Gippsland Lakes, Victoria. *Aust. J. Mar. Freshwater Res.*, 35: 517–524.

Higginson, F.R., 1965. The distribution of submerged aquatic angiosperms within the Tuggerah Lake system. *Proc. Linn. Soc. NSW*, 90: 328–334.

Higginson, F.R., 1971. Ecological effects of pollution in Tuggerah Lakes. *Proc. Ecol. Soc. Aust.*, 5: 143–152.

Hutchings, P.A., 1982. The fauna of Australian seagrass beds. *Proc. Linn. Soc. NSW*, 106: 181–200.

Hutchings, P.A. and Murray, A., 1984. Taxonomy of polychaetes from the Hawkesbury River and the southern estuaries of New South Wales, Australia. *Rec. Aust. Mus. Suppl.*, 3: 1–118.

Hutchings, P.A. and Rainer, S.F., 1979. The polychaete fauna of Careel Bay, Pittwater, New South Wales, Australia. *J. Nat. Hist.*, 13: 745–796.

Hutchings, P.A. and Recher, H.F., 1974. The fauna of Careel Bay with comments on the ecology of mangrove and seagrass communities. *Aust. Zool.*, 18: 99–128.

Iredale, T. and May, W.L., 1916. Misnamed Tasmanian chitons. *Proc. Malac. Soc. Lond.*, 12: 94–117.

Jacobs, S.W.L. and Brock, M.A., 1982. A revision of the genus *Ruppia* (Potamogetonaceae) in Australia. *Aquat. Bot.*, 14: 325–337.

Jacobs, S.W.L. and Williams, A., 1980. Notes on the genus *Zostera s. lat* in New South Wales. *Telopea*, 1: 451–455.

Jeffrey, S.W., 1981. Phytoplankton ecology — with particular reference to the Australasian region. In: M.N. Clayton and R.J. King (Editors), *Marine Botany: an Australasian Perspective*. Longman Cheshire, Melbourne, pp. 241–291.

Jernakoff, P., 1983. Factors affecting the recruitment of algae in a midshore region dominated by barnacles. *J. Exp. Mar. Biol. Ecol.*, 67: 17–31.

Jernakoff, P., 1985a. An experimental evaluation of the influence of barnacles, crevices and seasonal patterns of grazing on algal diversity and cover in an intertidal barnacle zone. *J. Exp. Mar. Biol. Ecol.*, 88: 287–302.

Jernakoff, P., 1985b. The effect of overgrowth by algae on the survival of the intertidal barnacle *Tesseropora rosea* Krauss. *J. Exp. Mar. Biol. Ecol.*, 94: 89–97.

Jernakoff, P., 1985c. Temporal and small-scale spatial variability of algal abundance on an intertidal rocky shore. *Bot. Mar.*, 28: 145–154.

Jernakoff, P. and Fairweather, P.G., 1985. An experimental analysis of interactions among several intertidal organisms. *J. Exp. Mar. Biol. Ecol.*, 94: 71–88.

Jitts, H.R., 1959. The adsorption of phosphate by estuarine bottom deposits. *Aust. J. Mar. Freshwater Res.*, 10: 7–21.

Jones, A.R. (Editor), 1977. *An ecological survey of nearshore waters. East of Sydney, NSW 1973–1975*. The Australian Museum, Sydney, 320 pp.

Jones, A.R., 1987. Temporal patterns in the macrobenthic communities of the Hawkesbury Estuary, N.S.W. *Aus. J. Mar. Freshwater Res.*, 38: 607–624.

Jones, R. and Kraft, G.T., 1984. The genus *Codium* (Codiales, Chlorophyta) at Lord Howe Island (N.S.W.). *Brunonia*, 7: 253–276.

Jutson, J.T., 1950. The shore platforms at Flinders, Victoria. *Proc. R. Soc. Victoria*, 60: 57–73.

Kennelly, S.J., 1983. An experimental approach to the study of factors affecting algal colonization in a sublittoral kelp forest. *J. Exp. Mar. Biol. Ecol.*, 68: 257–276.

Kennelly, S.J. and Larkum, A.W.D., 1983. A preliminary study of temporal variation in the colonization of subtidal algae in an *Ecklonia radiata* community. *Aquat. Bot.*, 17: 275–282.

Kennelly, S.J. and Underwood, A.J., 1984. Underwater microscopic sampling of a sublittoral kelp community. *J. Exp. Mar. Biol. Ecol.*, 76: 67–78.

Kennelly, S.J. and Underwood, A.J., 1985. Sampling of small invertebrates on natural hard substrata in a sublittoral kelp forest. *J. Exp. Mar. Biol. Ecol.*, 89: 55–67.

King, R.J., 1970. Surface sea-water temperatures at Port Phillip Bay Heads, Victoria. *Aust. J. Mar. Freshwater Res.*, 21: 47–50.

King, R.J., 1973. The distribution and zonation of intertidal organisms in Bass Strait. *Proc. R. Soc. Victoria*, 85: 145–162.

King, R.J., 1981a. The free-living *Hormosira banksii* (Turner) Decaisne associated with mangroves in eastern Australia. *Bot. Mar.*, 24: 569–576.

King, R.J., 1981b. The macroalgae of mangrove communities in eastern Australia. *Phycologia*, 20: 107–108.

King, R.J., 1981c. Marine angiosperms: seagrasses. In: M.N. Clayton and R.J. King (Editors), *Marine Botany: an Australasian Perspective*. Longman Cheshire, Melbourne, pp. 201–210.

King, R.J., 1981d. Mangroves and saltmarsh plants. In: M.N. Clayton and R.J. King (Editors), *Marine Botany: An Australian Perspective*. Longman Cheshire, Melbourne, pp. 308–328.

King, R.J., 1986. Aquatic angiosperms in coastal saline lagoons of New South Wales. 1. The vegetation of Lake Macquarie. *Proc. Linn. Soc. NSW*, 109: 11–23.

King, R.J. and Barclay, J.B., 1986. Aquatic angiosperms in coastal saline lagoons of New South Wales. 3. Quantitative assessment of *Zostera capricorni* Ascherson. *Proc. Linn. Soc. NSW*, 109: 41–50.

King, R.J. and Hodgson, B., 1986. Aquatic angiosperms in coastal saline lagoons of New South Wales. 4. Long term changes. *Proc. Linn. Soc. NSW.*, 109: 51–60.

King, R.J. and Holland, V.M., 1986. Aquatic angiosperms in coastal saline lagoons of New South Wales. 2. The vegetation of Tuggerah Lakes with specific comments on the growth of *Zostera capricorni* Ascherson. *Proc. Linn. Soc. NSW.*, 109: 25–39.

King, R.J. and Puttock, C.F., 1989. Morphology and taxonomy of *Bostrychia montagne* and *Stictosiphona* J.D. Hooker and Harvey (Rhodomelaceae/Rhodophyta). *Aus. Syst. Bot.*, 2: 1–73.

King, R.J. and Wheeler, M.D., 1985. Composition and geographic distribution of mangrove macroalgal communities in New South Wales. *Proc. Linn. Soc. NSW*, 108: 97–117.

King, R.J., Black, J.H. and Ducker, S.C., 1971. Intertidal ecology of Port Phillip Bay with systematic lists of plants and animals. *Mem. Nat. Mus. Victoria*, 32: 93–128.

Kirk, J.T.O., 1976. Yellow substance (Gelbstoff) and its contribution to the attenuation of photosynthetically active radiation in some inland and coastal south-eastern Australian waters. *Aust. J. Mar. Freshwater Res.*, 27: 61–71.

Kirk, J.T.O., 1977. Attenuation of light in natural waters. *Aust. J. Mar. Freshwater Res.*, 28: 497–508.

Kirk, J.T.O., 1979. Spectral distribution of photosynthetically active radiation in some southeastern Australian waters. *Aust. J. Mar. Freshwater Res.*, 30: 81–91.

Kirkman, H., 1984. Standing stock and production in *Ecklonia radiata* (Turn.) J. Agardh. *J. Exp. Mar. Biol. Ecol.*, 76: 119–130.

Kirkman, H. and Reid, D.D., 1979. A study of the role of the seagrass *Posidonia australis* in the carbon budget of an estuary. *Aquat. Bot.*, 7: 173–183.

Kirkman, H., Griffiths, F.B. and Parker, R.R., 1979. The release of reactive phosphate by a *Posidonia australis* seagrass community. *Aquat. Bot.*, 6: 329–337.

Kirkman, H., Cook, I.H. and Reid, D.D., 1982. Biomass and growth of *Zostera capricorni* Aschers., in Port Hacking, NSW, Australia. *Aquat. Bot.*, 12: 57–67.

Knox, G.A., 1963. The biogeography and intertidal ecology of the Australasian coasts. *Oceanogr. Mar. Biol. Ann. Rev.*, 1: 341–404.

Kott, P., 1952. The Ascidians of Australia. I. Stolidobranchiata

Lahille and Phlebobranchiata Lahille. *Aust. J. Mar. Freshwater Res.*, 3: 205–333.

Kott, P., 1957. The Ascidians of Australia. II. Aplousobranchiata Lahille: Clavelinidae Forbes and Hanly and Polyclinidae Verrill. *Aust. J. Mar. Freshwater Res.*, 8: 64–109.

Kott, P., 1962. The Ascidians of Australia. III. Aplousobranchiata Lahille: Didemnidae Giard. *Aust. J. Mar. Freshwater Res.*, 13: 265–334.

Kott, P., 1963. The Ascidians of Australia. IV. Aplousobranchiata Lahille: Polyclinidae Verrill (continued). *Aust. J. Mar. Freshwater Res.*, 14: 70–118.

Larkum, A.W.D., 1976. Ecology of Botany Bay. I. Growth of *Posidonia australis* (Brown) Hook. f. in Botany Bay and other bays of the Sydney Basin. *Aust. J. Mar. Freshwater Res.*, 27: 117–127.

Larkum, A.W.D., 1977. Some recent research on seagrass communities in Australia. In: C.P. McRoy and C. Helfferich (Editors), *Seagrass Ecosystems: A Scientific Perspective*. Marcel Dekker, New York, pp. 247–262.

Larkum, A.W.D., 1981. An introduction to the physiology of marine plants. In: M.N. Clayton and R.J. King (Editors), *Marine Botany: an Australasian Perspective*. Longman Cheshire, Melbourne, pp. 346–348.

Larkum, A.W.D., 1986. A study of growth and primary production of *Ecklonia radiata* (C. Ag.) J. Agardh (Laminariales) at a sheltered site in Port Jackson, New South Wales. *J. Exp. Mar. Biol. Ecol.*, 96: 177–190.

Larkum, A.W.D. and West, R.J., 1983. Stability, depletion and restoration of seagrass beds. *Proc. Linn. Soc. NSW*, 106: 201–212.

Larkum, A.W.D., Collett, L.C. and Williams, R.J., 1984. The standing stock, growth and shoot production of *Zostera capricorni* Aschers., in Botany Bay, NSW, Australia. *Aquat. Bot.*, 19: 307–327.

Larkum, A.W.D., McComb, A.J. and Shepherd, S.A. (Editors), 1989. *The Biology of Seagrasses*. Elsevier, Amsterdam, 841 pp.

Lewis, J.A., 1983. Floristic composition and periodicity of subtidal algae on an artificial structure in Port Phillip Bay (Victoria, Australia). *Aquat. Bot.*, 15: 257–274.

Lewis, J.R., 1961. The littoral zone on rocky shores — a biological or physical entity? *Oikos*, 12: 280–301.

Littlejohn, M.J., Watson, G.F. and Robertson, A.I., 1974. The ecological role of macrofauna in eelgrass communities. In: M.A. Shapiro (Editor), *A Preliminary report Westernport Bay environmental study*. Ministry for Conservation, Victoria, pp. 1–16.

Mann, K.H. (Editor), 1982. *Ecology of Coastal Waters. A Systems Approach. Studies in Ecology, Vol. 8*. Blackwell, Oxford, 322 pp.

Mapstone, B.D., Underwood, A.J. and Creese, R.G., 1984. Experimental analyses of the commensal relation between intertidal gastropods *Patelloida mufria* and trochid *Austrocochlea constricta*. *Mar. Ecol. Prog. Ser.*, 17: 85–100.

May, V., 1938. A key to the marine algae of New South Wales. Pt. 1. Chlorophyceae. *Proc. Linn. Soc. NSW*, 63: 207–218.

May, V., 1939. A key to the marine algae of New South Wales. Pt. 2. Melanophyceae (Phaeophyceae). *Proc. Linn. Soc. NSW*, 64: 191–215.

May, V., 1976. Changing dominance of an algal species (*Caulerpa filiformis* (Suhr) Hering). *Telopea*, 1: 136–138.

May, V., 1981. Long-term variation in algal intertidal floras. *Aust. J. Ecol.*, 6: 324–435.

May, V., 1982. Short note — the use of epiphytic algae to indicate environmental changes. *Aust. J. Ecol.*, 7: 101–102.

May, V., 1985. Observations on algal floras close to two sewerage outlets. *Cunninghamia*, 1: 385–394.

May, V. and Larkum, A.W.D., 1981. A subtidal transect in Jervis Bay, New South Wales. *Aust. J. Ecol.*, 6: 439–457.

May, V., Bennett, I. and Thompson, T.E., 1970. Herbivore–algal relationships on a coastal rock platform (Cape Banks, N.S.W.). *Oecologia*, 6: 1–14.

May, V., Collins, A.J. and Collett, L.C., 1978. A comparative study of epiphytic algal communities on two common genera of seagrasses in eastern Australia. *Aust. J. Ecol.*, 3: 91–104.

McComb, A.J., Cambridge, M.L., Kirkman, H. and Kuo, J., 1981. The biology of Australian seagrasses. In: J.S. Pate and A.J. McComb (Editors), *The Biology of Australian Plants*. University of Western Australia Press, Perth, pp. 258–293.

Meyer, G.R. and O'Gower, A.K., 1963. The ecology of six species of littoral gastropods I. Associations between species and associations with wave action. *Aust. J. Mar. Freshwater Res.*, 14: 176–193.

Melbourne Metropolitan Board of Works and Fisheries and Wildlife Department of Victoria, 1973. *Environmental Study of Port Phillip Bay. Report on Phase One 1968–1972*, 372 pp.

Middleton, M.J., Bell, J.D., Burchmore, J.J., Pollard, D.A. and Pease, B.C., 1984. Structural differences in the fish communities of *Zostera capricorni* and *Posidonia australis* seagrass meadows in Botany Bay, New South Wales. *Aquat. Bot.*, 18: 89–109.

Moran, M.J., 1985. Effects of prey density, prey size and predator size on rates of feeding by an intertidal predatory gastropod *Morula marginalba* Blainville (Muricidae), on several species of prey. *J. Exp. Mar. Biol. Ecol.*, 90: 97–105.

Moran, M.J., Fairweather, P.G. and Underwood, A.J., 1984. Growth and mortality of the predatory intertidal whelk *Morula marginalba* Blainville (Muricidae): the effects of different species of prey. *J. Exp. Mar. Biol. Ecol.*, 75: 1–17.

Nicholls, P.D., Klumpp, D.W. and Johns, R.B., 1985. A study of food chains in seagrass communities. III. Stable carbon isotope ratios. *Aust. J. Mar. Freshwater Res.*, 36: 683–690.

O'Gower, A.K. and Meyer, G.R., 1965. The ecology of six species of littoral gastropod molluscs. II. Seasonal variations in the six populations. *Aust. J. Mar. Freshwater Res.*, 16: 205–218.

O'Gower, A.K. and Meyer, G.R., 1970. The ecology of six species of littoral gastropods. III. Diurnal and seasonal variations in densities and patterns of distribution in three environments. *Aust. J. Mar. Freshwater Res.*, 22: 35–40.

Ostenfeld, C.H., 1916. Contributions to West Australian Botany. I. The seagrasses of West Australia. *Dansk. Bot. Ark.*, 2: 1–44.

Owen, P.C., 1978. Intertidal and sublittoral zones. In: R.H. Gunn (Editor), *Land use on the South Coast of New South Wales. A Study in Methods of Acquiring and Using*

Information to Analyse Regional Land Use Options, Vol. 2, Bio-physical background studies, CSIRO, Melbourne, Ch. 7, pp. 90–99.

Paxton, H., 1979. Taxonomy and aspects of the life history of Australian beachworms (Polychaeta: Onuphidae). *Aust. J. Mar. Freshwater Res.*, 30: 265–294.

Phillips, D.M. and Kirk, J.T.O., 1984. Study of the spectral variation of absorption and scattering in some Australian coastal waters. *Aust. J. Mar. Freshwater Res.*, 35: 635–644.

Pollard, D.A., 1984. A review of ecological studies on seagrass–fish communities with particular reference to recent studies in Australia. *Aquat. Bot.*, 18: 3–42.

Poore, G.C.B., 1982. Benthic communities of the Gippsland Lakes, Victoria. *Aust. J. Mar. Freshwater Res.*, 33: 901–915.

Poore, G.C.B., 1985. Diversity distribution and biogeographic affinities of some crustacea in Bass Strait. Abstract AMSA Conf., Dec. 1984. *Aust. Mar. Sci. Bull.*, 89: 36–37.

Poore, G.C.B. and Griffin, D.J.G., 1979. The Thalassinidae (Crustacea: Decapoda) of Australia. *Rec. Aust. Mus.*, 32: 217–321.

Poore, G.C.B. and Kudenov, J.D., 1978a. Benthos of the Port of Melbourne: the Yarra River and Hobson's Bay, Victoria. *Aust. J. Mar. Freshwater Res.*, 29: 141–155.

Poore, G.C.B. and Kudenov, J.D., 1978b. Benthos around an outfall of the Werribee sewage-treatment farm: Port Phillip Bay, Victoria. *Aust. J. Mar. Freshwater Res.*, 29: 157–167.

Poore, G.C.B. and Rainer, S., 1974. Distribution and abundance of soft bottom molluscs in Port Phillip Bay, Victoria, Australia. *Aust. J. Mar. Freshwater Res.*, 25: 371–411.

Poore, G.C.B. and Rainer, S., 1979. A three year study of benthos of muddy environments in Port Phillip Bay, Victoria. *Estuarine Coast. Mar. Sci.*, 9: 477–497.

Pope, E.C., 1943. Animal and plant communities of the coastal rock platforms at Long Reef, N.S.W. *Proc. Linn. Soc. NSW*, 68: 221–254.

Port Phillip Survey, 1966. Port Phillip Survey 1957–1963. Part 1. *Mem. Nat. Mus. Victoria*, 27: 1–385.

Port Phillip Survey, 1971. Port Phillip Survey 1957–1963. Part 2. *Mem. Nat. Mus. Victoria*, 32: 1–172.

Rainer, S.F., 1981. Temporal patterns in the structure of macrobenthic communities of an Australian estuary. *Estuarine Coast. Shelf Sci.*, 13: 597–620.

Rainer, S.F., 1982. Trophic structure and production in the macrobenthos of a temperate Australian estuary. *Estuarine Coast. Shelf Sci.*, 15: 423–441.

Rainer, S.F., 1984. Polychaeta as a representative taxon, their role in function and secondary production of benthic communities in three estuaries. In: P.A. Hutchings (Editor), *Proc. First Int. Polychaete Conf., Sydney. Proc. Linn. Soc. NSW*, pp. 370–382.

Rainer, S.F. and Fitzhardinge, R.C., 1981. Benthic communities in an estuary with periodic deoxygenation. *Aust. J. Mar. Freshwater Res.*, 32: 227–243.

Revelante, N., 1976–1977. Tropical phytoplankton community diversity. *Aust. Inst. Mar. Sci. Res. Rep.*, 1976–1977: 24–26.

Revelante, N. and Gilmartin, M., 1978. Characteristics of the microplankton and nanoplankton communities of an Australian coastal plain estuary. *Aust. J. Mar. Freshwater Res.*, 29: 9–18.

Robertson, A.I., 1980. The structure and organisation of an eelgrass fish fauna. *Oecologia*, 47: 76–82.

Robertson, A.I., 1984. Trophic interactions between fish fauna and macrobenthos of an eelgrass community in Western Port, Victoria. *Aquat. Bot.*, 18: 135–153.

Robertson, A.I. and Howard, R.K., 1978. Diel trophic interactions between vertically migrating zooplankton and their fish predators in an eelgrass community. *Mar. Biol.*, 48: 207–213.

Robertson, E.L., 1984. Seagrasses. In: H.B.S. Womersley (Editor), *The Marine Benthic Flora of Southern Australia, Part I.* Government Printer, South Australia, pp. 57–122.

Robinson, K.I.M., Gibbs, P.J., Barclay, J.B. and May, J.L., 1983. Estuarine flora and fauna of Smiths Lake, New South Wales. *Proc. Linn. Soc. NSW*, 107: 19–34.

Rochford, D.J., 1951. Studies in Australian estuarine hydrology. I. Introductory and comparative features. *Aust. J. Mar. Freshwater Res.*, 2: 1–116.

Rochford, D.J., 1972. *Nutrient enrichment of East Australian coastal waters I. Evans Head upwelling.* Tech. Rep. 33, CSIRO Aust. Div. Fish. Oceanogr., Cronulla.

Rochford, D.J., 1975. Oceanography and its role in the management of aquatic ecosystems. *Proc. Ecol. Soc. Aust.*, 8: 67–83.

Rochford, D.J., 1977a. *The present evidence for summer upwelling off the northeast coast of Victoria.* Rep. 77, CSIRO Aust. Div. Fish. Oceanogr., Cronulla.

Rochford, D.J., 1977b. *A review of a possible upwelling situation off Port MacDonell, S.A.* Rep. 81, CSIRO Aust. Div. Fish. Oceanogr.

Rochford, D.J., 1980. *The nutrient status of the oceans around Australia.* Ann. Rep. 1977–79, CSIRO Aust. Div. Fish. Oceanogr., Cronulla, pp. 9–20.

Rochford, D.J., 1984. Nitrates in eastern Australian coastal waters. *Aust. J. Mar. Freshwater Res.*, 35: 385–397.

Rowe, F.W.E. and Vail, L.L., 1982. The distribution of Tasmanian echinoderms in relation to southern Australian biogeographic provinces. In: J.M. Lawrence (Editor), *Echinoderms: Proc. Int. Conf., Tampa Bay.* Balkema, Rotterdam. pp. 319–325.

Roy, P.S., 1984. New South Wales estuaries: their origin and evolution. In: P. Thom (Editor), *Coastal Geomorphology in Australia.* Academic Press, Sydney, pp. 99–121.

Saenger, P., Specht, M.M., Specht, R.L. and Chapman, V.J., 1977. Mangal and coastal salt-marsh communities in Australasia. In: V.J. Chapman (Editor), *Wet Coastal Ecosystems, Ecosystems of the World, Vol. 1.* Elsevier, Amsterdam, pp. 293–345.

Sanderson, J.C. and Thomas, D.P., 1987. Subtidal macroalgal community distribution in the D'Entrecasteaux Channel, Tasmania. *Aust. J. Ecol.*, 12: 41–51.

Scott, B.D., 1978. Nutrient cycling and primary production in Port Hacking, N.S.W. *Aust. J. Mar. Freshwater Res.*, 29: 307–315.

Shepherd, S.A. and Sprigg, R.C., 1976. Substrate sediments and subtidal ecology of Gulf St. Vincent and Investigator Strait. In: C.R. Twidale, M.J. Tyler and B.P. Webb

(Editors), *Natural history of the Adelaide Region*. R. Soc. S. Aust., Adelaide, pp. 161–174.

Shepherd, S.A. and Womersley, H.B.S., 1970. The sublittoral ecology of West Island, South Australia: 1. Environmental features and algal ecology. *Trans. R. Soc. S. Aust.*, 94: 105–138.

Shepherd, S.A. and Womersley, H.B.S., 1971. Pearson Island expedition 1969. 7. The sub-tidal ecology of benthic algae. *Trans. R. Soc. S. Aust.*, 95: 155–167.

Shepherd, S.A. and Womersley, H.B.S., 1976. The subtidal algal and seagrass ecology of St. Francis Island, South Australia. *Trans. R. Soc. S. Aust.*, 100: 177–191.

Short, A.D., 1983. Sediments and structures in beach-nearshore environments, southeast Australia. In: A. McLachlan and T. Erasmus (Editors), *Sandy Beaches as Ecosystems*. Junk, The Hague, pp. 145–155.

Skinner, S. and Womersley, H.B.S., 1983. New records (possibly introductions) of *Striaria, Stictyosiphon* and *Arthrocladia* (Phaeophyta) for southern Australia. *Trans. R. Soc. S. Aust.*, 107: 59–68.

Smith, B.J. (Editor), 1971. *Littoral survey of Western Port Bay*. Interim Rep. Marine Study Group of Victoria, Melbourne, 57 pp.

State Pollution Control Commission, 1978. *Seagrasses of Botany Bay. Environmental Control Study of Botany Bay*. Rep. BBS3, SPCC, Sydney, 55 pp.

State Pollution Control Commission, 1979. *Effects of dredging on macrobenthic infauna of Botany Bay. Environmental Control Study of Botany Bay*. Rep. BBS10, SPCC, Sydney, 55 pp.

State Pollution Control Commission, 1981a. *Rocky shores of Botany Bay and their benthic flora and fauna*. Rep. No. 22 SPCC, Sydney, 22 pp.

State Pollution Control Commission, 1981b. *The ecology of fish in Botany Bay. Environmental Control Study of Botany Bay*. Rep. BBS23, SPCC, Sydney, 78 pp.

State Pollution Control Commission, 1981c. *The ecology of fish in Botany Bay. Environmental Control Study of Botany Bay*. Rep. BBS23A, SPCC, Sydney, 127 pp.

Stephenson, T.A. and Stephenson, A., 1949. The universal features of zonation between tide-marks on rocky coasts. *J. Ecol.*, 37: 289–305.

Stephenson, T.A. and Stephenson, A., 1972. *Life Between Tidemarks on Rocky Shores*. Freeman, San Francisco, 425 pp.

Tyerman, S.D., Hatcher, A.I., West, R.J. and Larkum, A.W.D., 1984. *Posidonia australis* growing in altered salinities: leaf growth, regulation of turgor and the development of osmotic gradients. *Aust. J. Plant Physiol.*, 11: 35–47.

Underwood, A.J., 1978. A refutation of critical tidal levels as determinants of the structure of intertidal communities on British shores. *J. Exp. Mar. Biol. Ecol.*, 33: 261–276.

Underwood, A.J., 1979. The ecology of intertidal gastropods. *Adv. Mar. Biol.*, 16: 111–210.

Underwood, A.J., 1980. The effects of grazing by gastropods and physical factors on the upper limits of distribution of intertidal macroalgae. *Oecologia*, 46: 201–213.

Underwood, A.J., 1981. Structure of a rocky intertidal community in NSW: patterns of vertical distribution and seasonal changes. *J. Exp. Mar. Biol. Ecol.*, 51: 57–86.

Underwood, A.J., 1984a. The vertical distribution and seasonal abundance of intertidal microalgae on a rocky shore in New South Wales. *J. Exp. Mar. Biol. Ecol.*, 78: 199–220.

Underwood, A.J., 1984b. Microalgal food and the growth of the intertidal gastropods *Nerita atramentosa* Reeve and *Bembicium nanum* (Lamarck) at four heights on a shore. *J. Exp. Mar. Biol. Ecol.*, 79: 277–291.

Underwood, A.J., 1984c. Vertical and seasonal patterns in competition for microalgae between intertidal gastropods. *Oecologia*, 64: 211–222.

Underwood, A.J., 1985. Physical factors and biological interactions: the necessity and nature of ecological experiments. In: P.G. Moore and R. Seed (Editors), *The ecology of rocky coasts*. Hodder and Stoughton, London, pp. 372–390.

Underwood, A.J. and Denley, E.J., 1984. Paradigms, explanations and generalizations in models for the structure of intertidal communities on rocky shores. In: D.R. Strong, D.S. Simberloff, L.G. Abele and A.B. Thistle (Editors), *Ecological Communities: Conceptual Issues and the Evidence*, Princeton, University Press, N.J., pp. 151–180.

Underwood, A.J. and Fairweather, P.G., 1985. Intertidal communities: do they have different ecologies or different ecologists? *Proc. Ecol. Soc. Aust.*, 14: 7–16.

Underwood, A.J. and Jernakoff, P., 1981. Interactions between algae and grazing gastropods in the structure of a low-shore algal community. *Oecologia*, 48: 221–233.

Underwood, A.J. and Jernakoff, P., 1984. The effects of tidal height, wave-exposure, seasonality and rock-pools on grazing and the distribution of intertidal macroalgae in New South Wales. *J. Exp. Mar. Biol. Ecol.*, 75: 71–96.

Underwood, A.J. and McFadyen, K.E., 1983. Ecology of the intertidal snail *Littorina acutispira* Smith. *J. Exp. Mar. Biol. Ecol.*, 66: 169–197.

Underwood, A.J., Denley, E.J. and Moran, M.J., 1983. Experimental analyses of the structure and dynamics of mid-shore rocky intertidal communities in New South Wales. *Oecologia*, 56: 202–219.

Van der Valk, A.G. and Attiwill, P.M., 1983. Above- and below-ground litter decomposition in an Australian saltmarsh. *Aust. J. Ecol.*, 18: 441–447.

Van der Valk, A.G. and Attiwill, P.M., 1984. Decomposition of leaf and root litter of *Avicennia marina* at Western Port Bay, Victoria, Australia. *Aquat. Bot.*, 18: 205–221.

Van der Velde, J.T. and King, R.J., 1984. The subtidal seaweed communities of Bare Island, Botany Bay. *Wetlands (Aust.)*, 4: 7–22.

Watson, J.E., 1979. Biota of a temperate shallow water reef. *Proc. Linn. Soc. NSW*, 103: 227–235.

Weate, P.B. and Hutchings, P.A., 1977. Gosford lagoons — environmental study — the benthos. *Operculum*, 5: 137–143.

West, R.J., 1980. *A study of growth and primary production of the seagrass Posidonia australis Hook. f.* M.Sc. Thesis, University of Sydney, Sydney, 128 pp.

West, R.J. and Larkum, A.W.D., 1979. Leaf productivity of the seagrass, *Posidonia australis*, in eastern Australian waters. *Aquat. Bot.*, 7: 57–65.

West, R.J. and Larkum, A.W.D., 1983. Seagrass primary

production — a review. *Proc. Linn. Soc. NSW*, 106: 213–223.

West, R.J., Thorogood, C.A., Walford, T.R. and Williams, R.J., 1985. *An estuarine inventory for New South Wales, Australia.* Fisheries Bulletin 2. Dept. Agric. NSW.

Whitley, G.P., 1964. A survey of Australian ichthyology. *Proc. Linn. Soc. NSW*, 89: 11–127.

Wilson, B.R. and Gillett, K., 1974. *Australian shells, illustrating and describing 600 species of marine gastropods found in Australian waters.* A.H. and A.W. Reed, Sydney, rev. ed., 168 pp.

Womersley, H.B.S., 1959. The marine algae of Australia. *Bot. Rev.*, 25: 545–614.

Womersley, H.B.S., 1966. Port Phillip survey, 1957–1963: algae. *Mem. Nat. Mus. Victoria*, 27: 133–156.

Womersley, H.B.S., 1972. Aspects of the distribution and ecology of subtidal marine algae on Southern Australian coasts. *Proc. 7th Int. Seaweed Symp. (1971)*. Univ. Tokyo Press, Tokyo, pp. 52–54.

Womersley, H.B.S., 1981a. Marine ecology and zonation of temperate coasts. In: M.N. Clayton and R.J. King (Editors), *Marine Botany: An Australasian Perspective.* Longman-Cheshire, Melbourne, pp. 211–240.

Womersley, H.B.S., 1981b. Biogeography of Australasian marine macroalgae. In: M.N. Clayton and R.J. King (Editors), *Marine Botany: An Australasian Perspective.* Longman-Cheshire, Melbourne, pp. 292–307.

Womersley, H.B.S., 1984. *The marine benthic flora of Southern Australia. Part I.* Government Printer, South Australia, 329 pp.

Womersley, H.B.S. and Edmonds, S.J., 1952. Marine coastal zonation in southern Australia in relation to a general scheme of classification. *J. Ecol.*, 40: 84–90.

Womersley, H.B.S. and Edmonds, S.J., 1958. A general account of the intertidal ecology of South Australian coasts. *Aust. J. Mar. Freshwater Res.*, 9: 217–260.

Wood, E.J.F., 1959. Some East Australian seagrass communities. *Proc. Linn. Soc. NSW*, 84: 218–226.

Chapter 18

REMOTE SENSING OF THE NEAR-SHORE BENTHIC ENVIRONMENT

MAHLON G. KELLY

INTRODUCTION

Distributional patterns of shallow-water benthos, and the factors causing those patterns, are poorly understood. This is not surprising because, in contrast to intertidal and terrestrial systems, the patterns cannot be easily seen from either the surface or by diving. Sampling by dredging, trawling or diving gives only point observations that are hard to synthesize into patterns. However, remote sensing and imaging from aircraft and satellites allow the patterns to be discerned. Remote sensing is a valuable but little-used tool both for basic research into factors causing the distribution of biota and for inventoring coastal community types. In this chapter the available techniques for remote sensing and validation will be reviewed, some examples of what may be learned with such techniques will be given, and a discussion of how the further use of remote sensing may lead to a better understanding of littoral ecosystems will be given. The work that has been done on determining phytoplanktonic chlorophyll, will not be discussed nor will studies on the intertidal or marshes. Chlorophyll determination is a very complex topic presently under development, with satellite instrumentation being dedicated to the task. Because of these complexities it cannot be covered adequately here. The work by Trees and El-Sayed (1986), Kim and Van der Piepen (1986), and Venable et al. (1986) are examples of three main approaches being developed (remote colorimetry, sunlight-induced fluorescence, and laser-induced fluorescence). The approaches and problems associated with intertidal and marsh work are very similar to those for work on land (as discussed by Colwell, 1983), since there is no intervening water. Examples of the use of remote sensing in the intertidal are the work by Coulson et al. (1980) and by Meulstee et al. (1986). Freshwater work has largely focused on emergent vegetation (see, for instance, Riatala and Lampinen, 1985; and Riatala et al., 1985).

Aerial photography of the submerged bottom was suggested during World War I (Lee, 1922) and extensively used by the U.S. Navy for bathymetric reconnaissance in the Pacific during World War II. Early scientific application was mostly by geologists — for example, Cloud's (1962) studies of carbonate deposition on the Bahama Banks, Kumpf and Randall's (1961) work in the Virgin Islands, and the study of Ball et al. (1967) on the effects of a hurricane in southern Florida. I know of only two biological studies prior to 1968. Newell et al. (1959) used aerial photography to study the distribution of communities on the Bahama Banks, and Thomas et al. (1961) examined the effects of a hurricane on *Thalassia* beds in Biscayne Bay, Florida. During the Mercury and Gemini space missions, astronauts noted striking patterns in coastal areas, which led to interest by the National Aeronautics and Space Agency (NASA) in finding the causes of these patterns; this led to the work of Conrod and the present author in developing methods for identifying and studying the patterns by aerial photography (Conrod et al., 1968; Kelly and Conrod, 1969; Kelly, 1969a, b; Nichols et al., 1972). Unfortunately most of this work is in contract reports or publications with limited circulation. Surprisingly, since then the methods have been little applied to submerged benthos (but see below), although there have been several unpublished studies for resource inventory by local and national agencies in the United States. Landsat and SPOT (Satellite pour l'Observation de la Terre) images now make the use of satellite imagery feasible (Bour et al., 1986). Modern rapid computing techniques, such as those used to enhance and analyze astronomical photography, should make new methods of image enhancement and identification possible, but they have not been tried yet.

TECHNIQUES

The techniques for obtaining remote images may be classified as to type of platform (aircraft or satellite) and type of imaging device (camera or scanner). Techniques for image interpretation include photointerpretation, image enhancement and interpretation using a computer, or a combination of both. Image interpretation depends on field surveys to provide so-called ground truth information, so that the biota seen in the patterns may be identified.

Types of imagery

Photography

Photography is primary limited to use with aircraft, except for shots from the Gemini and Apollo satellite series and photographs from military satellites that are generally unavailable. It is of three basic types: hand-held, usually oblique "snapshots"; mapping-quality photogrammetric images; and collimated multispectral images.

Hand-held photographs are the least expensive to obtain, and are commonly most useful for providing initial orientation to a field site. Simple visual reconnaissance from the air, even without photography, is often useful. Oblique hand-held photographs are usually the most easily understood by untrained workers, and they are useful for instructing a field crew or seminar audience. Oblique photographs, preferably including some of the shoreline, are intuitively understood more easily than vertical images. In cases where the objective is simply to locate a seagrass bed or a reef, hand-held oblique photographs may be preferable to higher-quality photogrammetric images. Such photographs may point out features of the bottom communities that would otherwise be missed by surface investigation. The author was told that seagrasses were not to be found in a particular location, only to see them easily from a low-flying light aircraft.

Obtaining high-quality hand-held images requires some care. Either a 35 mm or 70 mm film-format, single-lens reflex camera is most commonly used. The former has the advantage of providing easily projected images, while the latter is better for high-quality enlargements. A lens of the shortest possible focal-length should be used (although avoiding a fish-eye effect); it allows photographs to be taken from lower altitudes, minimizing the effects of haze and blurring due to camera shake, which is always a problem in a light aircraft. A camera should never be steadied by pressing against the airframe; vibration will blur the image. It may help to attach a weight to the camera to increase its mass; a shoulder stock or chest pad is usually too cumbersome in the confined space of a light plane. The highest possible shutter speed and largest lens opening should be used, again to avoid blurring, and because depth of field is of no concern. Exposure is best determined by using a spot-meter aimed at the area of interest. If there is negligible haze, the exposure should be slightly greater than that indicated by the meter so as to maximize penetration. With haze or considerable light scattering due to suspended sediment the indicated exposure (or even less) should be used to avoid overestimating exposure because of scattered light. If true colour rendering is unimportant (as is usually the case) a yellow filter or haze filter will increase contrast and reduce the effects of light scattering caused by haze and suspended sediment. In that case the exposure should also be determined through the filter. Colour film should always be used, for the photointerpretative cues provided by colour are very important. All positive and negative colour films are suitable, but a high speed-rating is to be preferred. Film grain is not the limiting factor in resolution, colour rendering is generally unimportant, and the high shutter-speed that is allowed will minimize effects of camera shake. Camera orientation and angle are also important. Reflection of the sun must be avoided, while light penetration of the water should be maximized. A high sun-angle assures the latter, while aiming away from the sun avoids reflection. Reflection will also be minimized if the sea-surface is very smooth; of course, surface spray and foam must be avoided. Cloudy days are unsuitable because of reflection of diffuse sky light by the surface.

High-quality photogrammetric images have several advantages. Because of their large format (usually 22 cm square) and the use of lenses of short focal length (usually 15 cm) and high quality, large amounts of information are contained in a single photograph. The photograph or a portion thereof may be enlarged by a factor of 8 or more, still retaining image resolution better than that of the human eye — that is, there is no blurring. Images from 10 000 m altitude will typically cover an area 15 km on a side with resolution of about 5 m. At other altitudes the coverage and resolution are increased or decreased linearly. This means that very large areas may be seen in a relatively few frames. A second advantage of photogrammetric images is that they have little distortion from edge to edge, allowing fairly simple creation of maps of benthic biota. There is some distortion, however, in part due to the lens, and in part because the camera mount may not be aimed perfectly on the vertical at the time of exposure. If true mapping-quality is needed there must be precisely located landmarks or targets to provide reference points; a contractor then can produce corrected, distortion-free images. A further advantage of photogrammetric images is that if there is sufficient overlap between frames (70% overlap is usually used) bottom relief can be seen with a stereoscope. However, unlike terrestrial photographs, photographs of the bottom cannot be used to map relief because refraction by the water distorts the vertical scale. A final advantage of photogrammetric shots is that they can be used for navigation in the field while interpreting the images.

Because of the expense of photogrammetric equipment mounted in an adequate aircraft, photogrammetric images will usually be obtained by a contractor or perhaps a governmental agency. However, since the problems of photographing the shallow marine bottom are different from those in photographing the land, and since contractors will have little or no experience, the user must work closely with the contractor, and preferably the air crew. The air crew must be aware of several factors. Weather is more of a problem than in terrestrial work, and contingency dates are very important. First, a mission should not be flown soon after a storm as suspended sediment will produce too much light-scattering. Second, a clear sky is important to obtain maximum light penetration and minimum surface reflection. Third, the sea surface should be calm, in order to prevent reflection. The depth of penetration of photography is not limited by available light, but rather the contrast obtainable for deep objects. Light scattering and surface reflection reduce contrast, with scattering acting in two ways: producing a "fog" effect that interferes with the image and obscuring light reflected from images on the bottom. The same factors that determine the depth of visibility of a Secchi disk are acting here, and the problem is analogous to, but much more severe than, the problem of photographing through haze. Plankton also produces scattering. Better penetration is generally available in the tropics. The author obtained usable images from 60 m depth on the west edge of the Bahama Banks, while the deepest useful images in the Chesapeake Bay or the southern bays of Long Island were at 12 m. The sun angle, and thus the time of day, is also important. With a nearly vertical sun, reflection from the surface will completely obscure the bottom. There must be no direct reflection of the sun. The area of sun reflection will be much smaller if there are no capillary waves or short swells. However, the sun angle should be as high as possible while avoiding reflection, because there will be less light penetration and more scattering with low sun angles. Finally, of course, a season must be chosen in which the vegetation is well developed.

Altitude selection is important, usually being a compromise between area of coverage and resolution. The following discussion assumes a 22 cm format and 15 cm focal-length lens. Below 1000 m (giving 1.5 km coverage and about 30 cm resolution) there is rarely an increase in resolution because of sea-surface irregularities and scattering, so 1000 m is usually the lowest altitude that should be used. Since the objective is usually to identify types of bottom cover and patterns, rather than discrete objects, resolution of 60 cm, obtained at an altitude of 2000 m, is usually more than adequate. The author has most commonly used 4000 m altitudes; photography of the south shore of Long Island from a U-2 aircraft at 23 000 m allowed detailed mapping of beds of *Zostera*. The major seagrass beds of the Bahama Banks could be identified in hand-held photographs from the Gemini spacecraft at 200 km.

The choice of film, filter, and exposure is different from that of terrestrial photography. Colour film should be used because of the importance of colour for photointerpretation. The only reason to use panchromatic film is if numerous inexpensive copies or enlargements must be made. While colour-infrared film (which gives a false red image of radiation in the near infrared) is usually chosen for photography of terrestrial vegetation, due to the reflection from chlorophyll in the near infrared, it should normally be avoided for marine work because there is no penetration and therefore no image from the near infrared in the water. Color-infrared is very useful if accurate delineation of the shoreline or emergent vegetation (marshes or mangroves, for example) is important, because the land–water margin is conspicuous. Water penetration by light absorbed by the two pigments sensitive to visible radiation will still be obtained. There are also "minus-blue" films available that are specially designed for photography through the water. They attempt to avoid backscattered light by having only two pigment layers; the blue-sensitive layer is not present, since scattering usually is maximal in blue wavelengths. Minus-blue films may provide slightly better contrast in deep waters, but lose important information in shallow waters. It was found that photography with a yellow filter gave just as good contrast in deep water (Conrod et al., 1968). Exposures should be judged as described above for hand-held photography. Although it is not possible to obtain more than one exposure in a single flight path, it is usually cost-effective to re-fly the path using a second exposure. This is particularly important if the path covers water of very different depths and therefore requires different exposures.

If it is not feasible to obtain special photogrammetric imagery, satisfactory photographs may be obtained from a variety of sources. Mapping photography of coastal areas often covers near-shore waters as well, although the factors described above, such as sun-angle, water clarity, and film and exposure, will not be optimized. In most countries, national and/or local agencies photograph most land areas every few years. Such sources are particularly valuable if change in community distribution is being studied; in some locations photography is available from as long ago as the 1930's. Coastal photography from the early 1940's is available for many areas in the Pacific from the U.S. Army and Navy. Orth and Moore (1983) reviewed the use of historical photography for analyzing the distribution of seagrasses, giving several examples.

Multispectral photography is a much more specialized technique that will be described only briefly. If several photographs are obtained on panchromatic film of an identical area but with different narrow-band filters, the photographs may be combined to form false-colour images that may aid in photointerpretation. It is most common to use a cluster of 4 or 9 collimated cameras, dividing the spectrum into 3 or 8 "slices" with an additional colour photograph. If, then, a particular type of object has a specific wavelength distribution of reflectance (often called a "spectral signature") it may be enhanced or identified by using careful selection of filters and careful combination of the images in forming a colour enlargement or projection. These techniques have been used for mapping such things as groundwater distribution, crop types, and forest cover. Unfortunately, the reflectance of submerged objects depends on several factors unrelated to the type of object, such as water colour, depth, and turbidity. This makes multispectral photography of limited use in the submerged near-shore environment. An attempt was made, for example, to identify *Thalassia* beds in this way, and to judge their biomass using microdensitometry of the images (Kelly and Conrod, 1969), but variation in epiphyte cover and plant density had too much effect on the spectral signatures. These problems are shared by multispectral scanner imagery, and are further discussed below.

Scanner imagery

Multispectral scanner imagery may be obtained from either aircraft or satellites, and is the only practical imagery presently available from satellites (Landsat and SPOT imagery is obtained by multispectral scanners). A multispectral scanner scans a line perpendicular to the flight path with a rocking prism or mirror. The light from the scanned line is then broken into a spectrum by a prism or grating, and sensors sample portions of the spectrum, with the number of sensors or spectral slices depending on the particular device. The signals are digitized (usually to 8 bits, giving 256 intensity levels) at frequent intervals and recorded. In an aircraft, recording is on magnetic tape. In a satellite it is in random-access memory, which is periodically "dumped" to ground stations where tapes are made. Image resolution (the size of the picture element or pixel) is determined by the scanning rate and sampling rate, the former giving the resolution along the flight path, and the latter determining the within-line resolution. The advantages of multispectral scanning are that more information about spectral composition is gained than with colour photography, and that the information is in a form that may be examined with a computer for various purposes (although photographs may also be digitized). The disadvantages are lower resolution than mapping photographs and the cost of equipment and processing.

Scanner images are reconstructed by selecting three wavelength bands and displaying them on a colour video-display terminal or a similar device. The video image may then be photographed. Since any possible combination of three wavelength bands may be used, this produces "false-colour" images — unless, of course, the wavelength bands are the same as the retinal sensitivity of the human eye. Images may be enhanced by a variety of techniques as an aid to image interpretation. For example, the image may be produced not by the intensity of the signal but by its first derivative — that is, the variation in intensity from area to area. Otherwise a single wavelength may be selected and different intensities displayed as different colours or regions of rapid change in intensity (usually the edges of objects) may be emphasized. Of more specialized but very powerful use is the identification and mapping of areas with specific spectral signatures. One or more "calibration targets" are chosen and the ratios of intensities at the recorded wavelengths are determined. Other areas are then located that have the same ratios (within certain error bounds) and these are mapped. Such techniques have been used to map, for example, types of crop cover, forest types, crop and forest disease, and soil moisture.

Unfortunately, multispectral scanner techniques have not proven useful for automated identification or mapping of submerged vegetation. For example, unpublished maps of vegetation in Biscayne Bay, Florida, produced by the U.S. Geological Survey contained gross errors, with more than 50% of the vegetation misidentified. The difficulty is that the spectral signatures of submerged vegetation are much more variable than for terrestrial vegetation. This is due to four factors: (1) The cover of submerged vegetation is variable, making the colour of bottom sediment a factor; (2) nearly all submerged vegetation carries epiphytes, but to varying degrees so that the colours change; (3) the colour depends on overlying suspended sediment; and (4) water absorbs some wavelengths more rapidly than others, making colours change with depth. Only a trained interpreter can make the best use of scanner imagery.

Although multispectral techniques have not been successful for objective identification of benthic types, Bour et al. (1986) used an airborne radiometer to simulate the imagery that would be received from the SPOT satellite over New Caledonia to see if it could be used to identify the reef biotopes in which the snail *Trochus niloticus* would be found. They argued that, by using sophisticated statistical analysis of spectral signatures, reef biotopes and facies could be reliably identified and mapped. However, if the reef types they were examining had the appearance of typical coral environments, differences in colour would be more pronounced and consistent (in part because of the very clear water) than differences between vegetation types.

A particular problem is the estimation of biomass of submerged vegetation. Conrod et al. (1968) tried to estimate biomass of *Thalassia testudinum* from imagery of the Bahama Banks; they concluded that variation in colour, because of the overlying water, epiphyte cover, and variable colour of sediments would make reliable measurements impossible. They were working primarily with areas of complete coverage, in which biomass increases would be due to length of blades and density. Orth and Moore (1983) published maps of change in percent coverage of *Zostera marina* in the Rappahannock River estuary from 1937 to 1978, but of course they could not calibrate the historical photography, and they gave no figures as to the reliability of the estimates of coverage. There are fewer problems in the intertidal. Meulstee et al. (1986) showed a good correlation between density

($g\,m^{-2}$) of *Z. marina* and microdensitometric measurements of infrared photography in the intertidal of the Oosterschelde estuary (The Netherlands). Even there the biomass variation was associated with change in percentage of cover, and very careful field calibration was required. Coulson et al. (1980) gave some promising results in estimating biomass of *Fucus*, but there too the variation was in percentage of cover, and there was no field calibration. There have been several attempts at estimating biomass of emergent freshwater macrophytes (see, for instance, Adams et al., 1977) but the author knows of no successful ones. It appears that, at best, remote sensing must serve as an adjunct to careful field measurements when estimating biomass.

Photointerpretation and field interpretation

Accurate photointerpretation is an iterative process. That is, the interpreter should tentatively identify objects, perhaps with an overlay of the images, then check the identification in the field, change the identification criteria as needed, map and check again, change again, and so on until interpretation is accurate. The process is highly intuitive, and is a skill gained by experience. The criteria used by a skilled interpreter (although sometimes not consciously) are colour (hue and saturation), texture and shape, regularity, and sharpness, or diffuseness of edges. Repeated or regular geometric patterns often provide important clues. For example, the contact between the Miami oolite and Key Largo limestone formations in the Florida Keys may be identified by a change in algae, soft corals, and sponges growing on the rocks. This results in a change in colour and texture along a sharp, smooth edge, allowing the contact to be very accurately mapped where the rocks are not covered by sediment. In wave- and current-washed areas the edges of seagrass beds are usually sharp (where the grass is undercut), with strong differences in saturation between the sediment and seagrass, and a uniform tone in the seagrass. In still areas the edges of seagrass beds are usually diffuse with a mottled texture due to variable cover — this is true for both *Thalassia* and *Zostera* beds in widely separated locations. In Biscayne Bay (Florida) and the Bahama Banks, mixed gorgonian, sponge, and algal communities can be identified by a granular texture. Areas of sediment usually have a more uniform texture than within rocky communities. Several of the following patterns were useful: bead-line rows of *Thalassia* patches in southern Biscayne Bay identify fossil sinkholes in karstic limestone, "stringy" seagrass beds are often found in wave-washed locations, and reef patches within *Thalassia* beds have the same colour and texture as the beds, but are surrounded by "halos" of sediment where reef fish have grazed them.

Good photointerpretation and mapping depend on careful field techniques, and it is important that the interpreter participates in the field work. A preliminary overflight with a light aircraft, viewing the site from various angles, allows initial identification of major features, such as regions of hard and soft bottom and seagrass beds. These may then be spot-checked by a diver. Regions of interest should then be defined and transects laid out for field identification. To control depth a diver with a diving board may be towed over the transects, stopping as necessary to identify the bottom cover. Very careful navigation and note-taking are important, for, although it may seem as if the transect is carefully located, confusion may result when attempting to locate it on images. Copies of the imagery should be carried in the field. "Ziplock" bags are useful for protection. It is also useful if the diver carries small buoys attached to a monofilament line spooled around weights. These may be dropped when the bottom type changes or when interesting features are seen, so that one may return to them and examine them more closely. Although these techniques may seem tedious, they are essential and not as slow as they might seem. Two of us mapped over 400 km^2 of the Florida Keys in six weeks. Without careful field work, gross misidentification may result. For example, an area in Biscayne Bay was said to have been damaged by effluent from a power plant, when it had actually been buried by sediment deposited when drainage canals were built. In another case, floating debris was attributed to rafts of dead *Thalassia*, but was actually debris from adjacent mangroves. Patch reefs have been misidentified as seagrass beds, as have masses of *Ulva* and *Enteromorpha*.

EXAMPLES OF RESULTS

Rather than describing a wide variety of possible uses of remote sensing with near-shore communities, I will simply describe one example in detail — the factors determining *Thalassia* distribution along the edges of the Florida Straits. Other applications are mentioned below.

Figure 18.1 shows the shallow (1 to 6 m) bottom 3 km south of South Cat Cay. It is south of Bimini, on the west margin of the Florida Straits (north is to the right, and the Gulf Stream lies above the picture; the picture is 4 km wide). Except for the island in the upper left, the area is submerged oolitic sand of recent deposition. This is one of the few sites of modern oolite deposition, as described by Cloud (1962). The area is strongly wave-washed during storms, which results in the spillover

Fig. 18.1. Oolitic sediments and seagrasses on the western edge of the Bahama Banks south of South Cat Cay. See text for details.

lobes of oolite (o) in the figure. The depth of these sand bars is 1 m or less, and the nearly continuous movement of the sand prevents colonization by *Thalassia*. Beds of *Thalassia* (t) have formed in the lee of the bars and island and in channels, although these beds are also dissected by wave-washing during storms. The shaded areas between the arrows at the ends of the lobes were described by Cloud (1962) as the sloped ends of the lobes, lying at the angle of repose. This is not the case, as the shading is caused by sparse vegetation with a striking zonation. Progressing toward the lobes there are zones of mixed *Syringodium* and *Diplanthera*, followed by *Diplanthera* with some mixture of the calcareous algae *Halimeda* and *Penicillus*, and finally a zone of *Halimeda* and *Penicillus* alone. Rhizomes of *Thalassia* colonize the zone at the distal edge of the lobe. There is little variation of depth (10 to 20 cm) over the shaded region. The shaded ends of the lobes appear to be ecotones where *Thalassia* is gradually recolonizing the area, and the *Thalassia, Syringodium, Diplanthera* and *Penicillus–Halimeda* zones may represent successional sequences, with each vegetation type serving increasingly to stabilize the sediment. At depths of 2 to 3 m in the lee (east) of the spillover lobes, and at regions of recolonization immediately adjacent to the ends of the lobes, patchy beds of mixed *Syringodium* and *Thalassia* occur (p). The patchiness is apparent from the texture in the photograph. The sediment in these areas is heavily reworked by arenicolid polychaetes, with cast-mounds of 30 to 60 cm diameter occurring every 1 to 3 m. It may be that the sediment reworking prevents dense development of the seagrasses and causes the patchiness.

It appears that the main factors controlling seagrass distribution in this area are erosion by

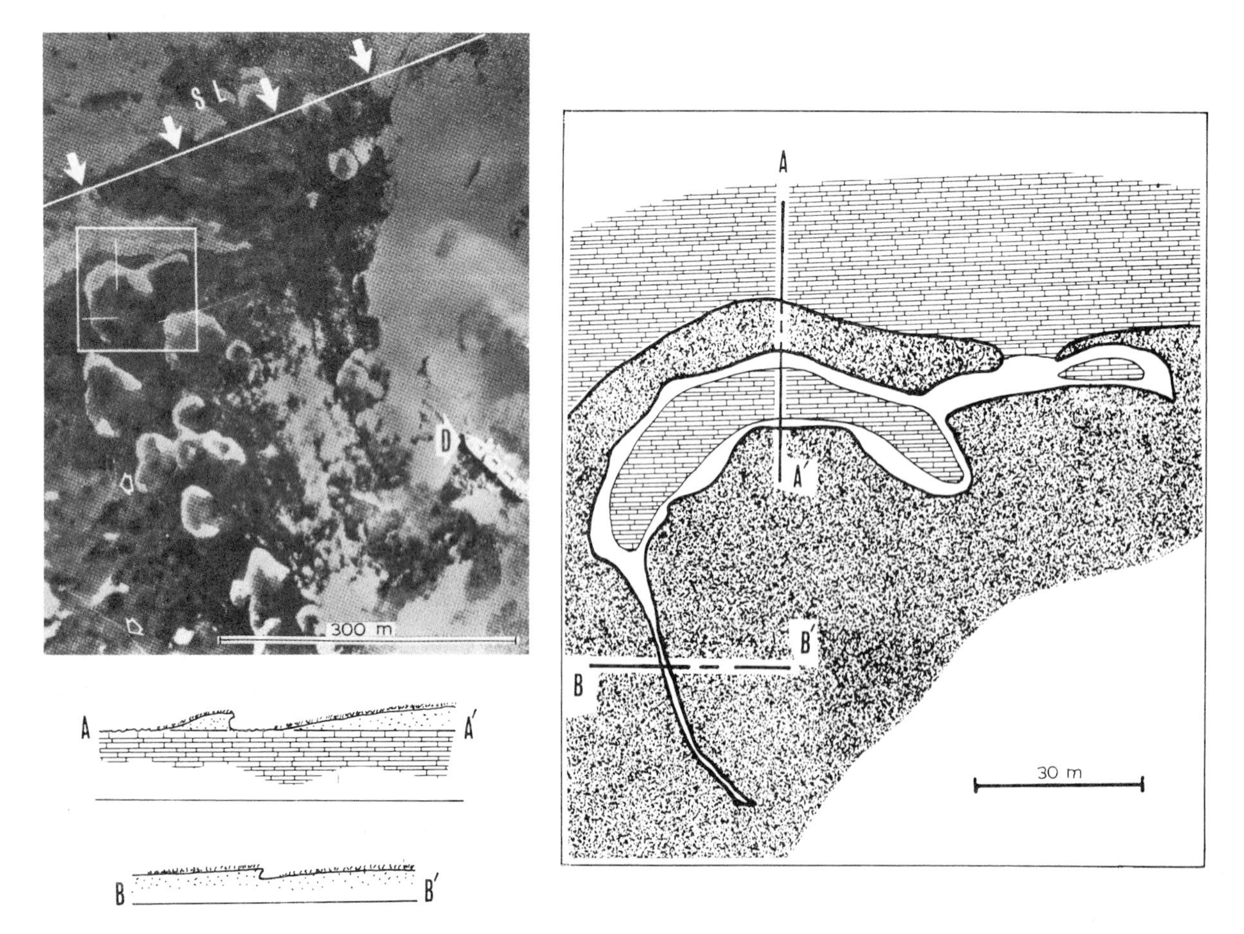

Fig. 18.2. Blowouts from wave-washing in *Thalassia* beds near the area shown in Fig. 18.1. See text for details.

waves and washover by sand. The erosional processes can be seen in Fig. 18.2 (taken from Conrod et al., 1968). The area is 1 km south of South Cat Cay and 4 km north of the area depicted in Fig. 18.1. Depth is 3 to 6 m. The dark areas are *Thalassia* beds (stippled area in the map), the mottled gray areas are coralline limestone (blocked area in the map) with attached algae, sponges, and gorgonians, and the clear areas are clean oolite sediment. The direction of surf action is shown by "SL". The ship at "D" is dredging oolite for use in manufacturing concrete. The erosional process is shown by the crescentric "blowouts" in the *Thalassia* beds. These are commonly found in the seaward edges of *Thalassia* beds along both sides of the Florida Straits. The leading edge of the blowout is an undercut seagrass bed, while the trailing edge is comprised of the same sequence of species as described above for the end of spillover lobes. Examination of photography before and after storms suggests that these blowouts are formed when the bed at the leading edge is torn free, and the bed is then "peeled back" along the horns of the crescent. In some areas the trailing edge is also torn back. It is speculated that the mottled appearance of the beds seen in the lower left of the photograph, caused by variable density of the seagrass, is due to progressive regrowth in old blowouts. Where there is slightly more exposure to waves the blowouts merge, producing striated and patchy beds as shown in Fig. 18.3.

Figure 18.3 is an area on the edge of the Florida reef tract south of Pigeon Key in the Florida Keys. The Gulf Stream is above the photograph, which is to the south-southeast. The width of the photograph is 3 km. The bright object is the nearly emergent upper part of a reef; the reef buttresses are visible above, and a storm-caused rubble bar extends below. The dark areas to the left and right of the rubble bar are *Thalassia* beds. Crescentric blowouts can be seen in the lee of the reef and rubble bar. In more exposed, shallow locations the

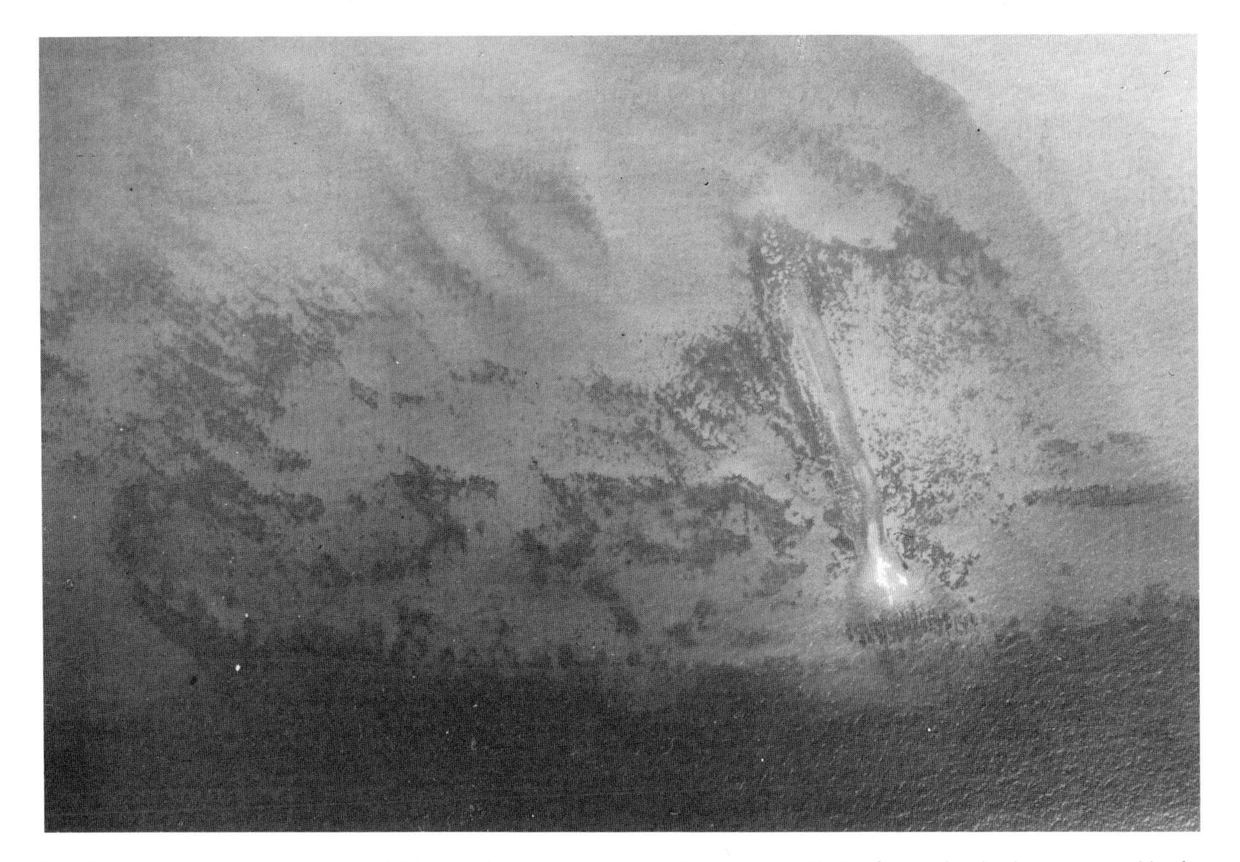

Fig. 18.3. A small reef and wave-washed *Thalassia* beds south of Pigeon Key in the Florida Reef tract, showing how wave-washing has determined the distribution of seagrass beds. See text for details.

blowouts can be seen to merge, forming the striated, patchy distribution of the seagrass beds. The edges of the beds in the patchy areas lie approximately a metre above the surrounding clear sediment and are sharply undercut. Complete beds reappear in deeper water, towards the bottom of the photograph, and gradually become less dense and disappear at 10 to 12 m depth (not visible in the photograph), presumably because seagrasses are light-limited in the channel between the reef tract and the Florida Keys.

It appears that the main factor controlling the distribution of seagrass beds on the edges of the Florida Straits is wave-washing during storms. In more protected locations a dominant factor appears to be the depth of the sediment. This can be

Fig. 18.4. Patches and lines of *Thalassia* beds, the distribution of which is determined by sediment thickness, in southern Biscayne Bay, Florida. See text for details.

seen at almost all locations landward of the Florida Keys. It is particularly striking in southern Biscayne Bay, as shown in Fig. 18.4. This photograph covers an area 4 km wide, with depths ranging to 6 m. The uniformly dark areas around the two islands are *Thalassia* beds growing in sediments more than 20 cm thick. The bead-like strings of dark patches and the dark lines are also beds of *Thalassia* growing in sediments more than 20 cm thick. The surrounding lighter-toned, textured areas are hard limestone bottoms, covered by sediments less than 10 cm thick. The bottom communities are dominated by gorgonians, sponges, and attached algae. Although there is sparse *Thalassia*, it is not dense enough to form beds. The patches and lines are "domed" by about 20 cm, presumably because the seagrass traps and stabilizes sediments, but this is not the cause of the locations of these patterned beds. Probing the center of the beds shows that the sediment is usually several meters thick and deposited in steep-sided depressions resembling sink-holes. Similar depressions are found under mangrove patches in the adjacent landward areas, presumably caused by solution of the limestone under acidic conditions. Zieman (1972) found mangrove peat under seagrass beds in these patches. It appears that *Thalassia* only forms beds in sediments thicker than 20 cm. In Southern Biscayne Bay this limitation causes seagrasses to grow in locations previously occupied by mangroves when the area was above sea level. Similar patterns are found throughout the Florida Keys.

To summarize, examination of aerial photography has enabled us to specify major factors controlling the distribution and patterns of *Thalassia* beds in southern Florida and the western Bahama Banks. In exposed areas, wave-washing during storms is the main control. Either formation and merging of crescentric blowouts removes beds, or they are overlain by sediments. The beds seem to regrow through a predictable sequence of succession, producing a mottled appearance corresponding to the age of regrowth. In less exposed areas the distribution seems to be controlled by sediment reworking (in the Bahamas), light limitation (in the channel between the Florida reef tract and the Florida Keys), and by sediment thickness (in Biscayne Bay and the lee of the Florida Keys).

Although it could be speculated, without using aerial photography, that all of the above factors should be important in controlling the distribution of seagrass beds, it is only through examination of aerial photography combined with extensive field surveys that the importance of the factors and their role in determining pattern and cover could be shown. Such use of remote sensing is not limited to this location or to seagrasses. Although Landsat imagery is available for all the world's coastal areas, it would not have given adequate resolution for the work described here. It is possible to see major areas of seagrass beds in Landsat imagery, but they can only be identified when the imagery is combined with aerial photography and field work. If the aerial photography is available the satellite images become of secondary use, except perhaps in extending observations to other locations. The author strongly urges anyone who uses remote sensing in studying the distribution of coastal communities to rely heavily on aerial photography, which is available for most locations.

The use of remote sensing is of course not limited to the study of seagrasses in clear coastal waters. It was possible to map and study the distribution of assemblages of algae and invertebrates at the sites described above, and to infer some of the factors causing patterns of distribution. *Zostera* patterns and distribution may be mapped and studied in more turbid northern waters (M.G. Kelly, unpubl. results). As light penetration is less (about 5 m under good conditions) the depth of these beds is reduced. For example it could be seen that major growths of *Zostera* in bays south of Long Island, New York, occurred in old overwash channels that were sealed by barrier islands. The upper limit of growth appeared also to be limited by wave washing, but at shallower depths, with less wave energy and finer sediments than for *Thalassia*. The western limit at which *Zostera* was found (as the urban areas of New York City are approached) appears to be limited by epiphyte cover of the seagrasses (i.e., *Cladophora*, *Enteromorpha* and *Ulva*). In areas of heavy nutrient pollution in western Great South Bay, Hempstead Bay, and Jamaica Bay these algae grow massively, forming rafts which are distributed on the bottom in the same areas where *Zostera* would otherwise grow — presumably preventing the growth of seagrasses. Similar growth and distribution of nuisance algae have been seen in aerial photography of many other coastal areas.

The potential use of aerial photography for understanding the distribution of near-shore communities has been little tapped. The above examples should be extended to other environments and species. For example, lagoons and reefs and other tropical habitats have been poorly studied (but see Bour et al., 1986), as well as north-temperate rocky subtidal areas. Factors causing the distribution of terrestrial vegetation have been studied from the earliest days of ecology in nearly all environments, largely because the patterns are readily discernible from the ground. Such studies have contributed greatly to understanding the natural history of dominant terrestrial species. The same approach has not been taken for near-shore communities, probably because most researchers, with the limited perspective of a diver or worker on the surface, have not realized that the patterns were there to be studied. Although aerial photography has been used extensively by governmental agencies to inventory marshes and coastal wetlands (that is, emergent vegetation), shallow submerged communities have largely been ignored. Such communities are important habitats and are frequently destroyed, either directly by dredging or by sedimentation. Aerial photography could be an important tool for the inventory, mapping, and subsequent protection of submerged communities.

REFERENCES

Adams, M.S., Scarpace, F.L., Scherz, J.D. and Woelkerling, W.J., 1977. *Assessment of aquatic environment by remote sensing*. Rep. No. 84, Inst. Env. Studies, Univ. of Wisconsin-Madison, pp. 1–235.

Ball, M.M., Shinn, E.A. and Stockman, K.W., 1967. The geologic effects of Hurricane Donna in South Florida. *J. Geology*, 75: 583–597.

Bour, W., Loubersac, L. and Rual, R., 1986. Thematic mapping of reefs by processing of simulated SPOT satellite data: application to the *Trochus niloticus* biotope on Tetembia Reef (New Caledonia). *Mar. Ecol. Prog. Ser.*, 34: 243–249.

Cloud, P.E., 1962. Environment of calcium carbonate deposition west of Andros Island, Bahamas. *U.S. Dept. Interior Geol. Prof. Pap.*, 350: 1–138.

Colwell, R.N. (Editor), 1983. *Manual of Remote Sensing, Vol. 1. Theory, Instruments and Techniques*. 2nd ed., American Society of Photogrammetry, Falls Church, Va., 1232 pp.

Conrod, A.C., Kelly, M. and Boersma, A., 1968. *Aerial photography for shallow water studies on the west edge of the Bahama Banks*. Rep. No. RE-42. Experimental Astronomy Laboratory, MIT, MA.

Coulson, M.G., Budd, J.T.C., Withers, R.G. and Nicholls, D.J., 1980. Remote sensing and field sampling of mudflat organisms in Langstone and Chichester Harbours, Southern England. In: J.H. Price, D.E.G. Irvine, and W.F. Farnham (Editors), *The Shore Environment. 1. Methods*. Academic Press, London, pp. 241–263.

Kelly, M.G., 1969a. Aerial photography for the study of near shore ocean biology. In: *New Horizons in Color Aerial Photography, Seminar Proc. ASP, SPSE Meetings, June, 1969*. American Society of Photogrammetry, Falls Church, Va., pp. 347–355.

Kelly, M.G., 1969b. *Applications of remote photography to the study of coastal ecology in Biscayne Bay, Florida*. Contract Report, U.S. Naval Oceanographic Office. Contract N-62306–69-C-0032, 52 pp.

Kelly, M.G. and Conrod, A.C., 1969. Aerial photographic studies of shallow water benthic ecology. In: P. Johnson (Editor), *Remote Sensing in Ecology*. Univ. of Georgia Press, pp. 173–183.

Kim, H.H. and Van der Piepen, H., 1986. Sunlight induced 685nm fluorescence imagery. In: M.A. Blizard (Editor), *Ocean Optics 8, Proc. of SPIE — The International Society for Optical Engineering*, 637: 359–364.

Kumpf, H.E. and Randall, H.A., 1961. Charting the marine environments of St. John, U.S. Virgin Islands. *Bull. Mar. Sci.*, 11: 543–551.

Lee, W.T., 1922. *The face of the earth as seen from the air*. Am. Geographical Society, New York, 110 pp.

Meulstee, C., Nienhuis, P.H. and Van Stokkom, H.T.C., 1986. Biomass assessment of estuarine macrophytobenthos using aerial photography. *Mar. Biol.*, 91: 331–335.

Newell, N.D., Imbrie, J., Purdy, E.G. and Thurber, D.L., 1959. Organism communities and bottom facies, Great Bahama Bank. *Bull. Am. Nat. Hist.*, 117: 181–228.

Nichols, M., Kelly, M., Thompson, G. and Castiglione, L., 1972. *Sequential photography for coastal oceanography*. Virginia Institute of Marine Science, Gloucester Point, 142 pp.

Orth, R.J. and Moore, K.A., 1983. Submersed vascular plants: Techniques for analyzing their distribution and abundance. *Mar. Tech. Soc. J.*, 17: 38–52.

Riatala, J. and Lampinen, J., 1985. A Landsat study of the aquatic vegetation of the Lake Luodonjarvi reservoir, Western Finland. *Aquat. Bot.*, 21: 325–346.

Riatala, J., Jantunen, H. and Lampinen, J., 1985. Application of Landsat satellite data for mapping aquatic areas in North-Eastern Finland. *Aquat. Bot.*, 21: 285–294.

Thomas, L.P., Moore, D.R. and Work, R.C., 1961. Effects of Hurricane Donna on the turtle grass beds of Biscayne Bay, Florida. *Bull. Mar. Sci.*, 11: 191–197.

Trees, C.C. and El-Sayed, S.Z., 1986. Remote sensing of chlorophyll concentrations in the northern Gulf of Mexico. In: M.A. Blizard (Editor), *Ocean Optics 8, Proc. of SPIE — The International Society for Optical Engineering, 637*: 328–334.

Venable, D.D., Khatun, S., Poole, L. and Punjabi, A., 1986.

Simulated laser fluoresensor signals from subsurface chlorophyll distributions. In: M.A. Blizard (Editor), *Ocean Optics 8, Proc. of SPIE — The International Society for Optical Engineering, 637*: 365–372.

Zieman, J.C., 1972. Origin of circular beds of *Thalassia* (Spermatophyta: Hydrocharitaceae) in southern Biscayne Bay Florida and their relationship to mangrove hammocks. *Bull. Mar. Sci.*, 22: 559–574.

SYSTEMATIC LIST OF GENERA

INTRODUCTION

The primary source of higher level taxonomy (i.e. to the family level) for this synopsis is S.B. Parker (1982), Synopsis and Classification of Living Organisms, McGraw-Hill Co., N.Y., Vol. 1, 1166 pp., Vol. 2, 1232 pp. Even so, several other sources of information were also utilized, including correspondences from diverse colleagues — see the Preface for details. Generic names within the Systematic List are alphabetically arranged under their appropriate families. Similarly, the various orders are alphabetically designated, with no phylogenetic significance being implied. With respect to generic names in parenthesis a few comments should be made. That is, many names represent "true" generic synonyms. Others, however, either represent differences of spelling for the same taxa (e.g. *Ocypoda* vs *Ocypode*) or previous interpretations/identifications (e.g. *Schottera nicaeensis* [*Petroglossum nicaeensis*]). In the latter case, the generic name *Petroglossum* is still valid and several other species are recognized. However, as no other *Petroglossum* species occur in the volume it was left in parenthesis. with another example like *Ulvaria obscura* (*Monostroma fuscum*) both genera are still valid and different species of both genera are cited in this volume. Hence, the two genera are listed separately. One last comment concerns the names of life history stages, such as *Codiolum* and *Conchocelis*, which are invalid taxa and delineated with quotes. Their "valid" taxonomic partner(s) may or may not be cited in one of the chapters.

PROKARYOTAE

MONERA

BACTERIA

PHOTOTROPHIC BACTERIA
Chlorobiaceae
Chlorobium
Chromatiaceae
Chromatium
Gram-negative aerobic rods & cocci
Pseudomonadaceae
Pseudomonas
Gram-negative anaerobic bacteria
Bacteriodaceae
Desulfovibrio
Endospore - forming rods & cocci
Bacillaceae
Bacillus
Coryneform bacteria
Arthrobacter

CYANOPHYCOTA (CYANOBACTERIA)

CYANOPHYCEAE (MYXOPHYCEAE; SCHIZOPHYCEAE)
Chroococcales
Chroococcaceae
Aphanocapsa
Aphanothece
Dactylococcopsis
Gloeocapsa
Entophysalidaceae
Entophysalis
Nostocales
Oscillatoriaceae
Hydrocoleum (*Hydrocoleus*)
Lyngbya
Microcoleus
Oscillatoria
Phormidium
Schizothrix
Trichodesmium
Rivulariaceae
Calothrix
Dicothrix
Rivularia
Scytonemataceae
Plectonema
Scytonema
Pleurocapsales
Scopulonemataceae
Hyella

EUKARYOTAE

PLANTAE

THALLOBIONTA

RHODOPHYCOTA

RHODOPHYCEAE

BANGIOPHYCIDEAE
Bangiales
Bangiaceae
Bangia
"*Conchocelis*"*
Porphyra
Erythropeltidaceae
Erythrotrichia
Porphyridiales
Goniotrichaceae
Chroodactylon (*Asterocytis*)
Porphyridiaceae
Porphyridium

FLORIDEOPHYCIDEAE
Ahnfeltiales
Ahnfeltiaceae
Ahnfeltia
Bonnemaisoniales
Bonnemaisoniaceae
Asparagopsis
Bonnemaisonia
"*Trailliella*"
Ceramiales
Ceramiaceae
Aglaothamnion
Antithamnionella
Ballia
Callithamnion
Centroceras
Ceramium
Crouania
Griffithsia
Gymnothamnion
Haloplegma
Plumaria
Ptilota
Scagelia
Spyridia
Wrangelia
Dasyaceae
Dasya
Dictyurus
Delesseriaceae
Acrosorium
Botryocarpa
Caloglossa
Cryptopleura
Delesseria
Hymenena
Martensia
Membranoptera
Myriogramme
Pantoneura
Phycodrys
Polyneura
Pseudobranchioglossum
Sarconema
Taenioma
Rhodomelaceae
Acanthophora
Bostrychia (*Bostrichia*)
Chondria
Digenea (*Digenia*)
Herposiphonia
Laurencia
Lophosiphonia
Lophurella
Murrayella
Odonthalia
Polysiphonia
Pterosiphonia
Rhodomela
Tolypiocladia
Corallinales
Corallinaceae
Amphiroa
Bossiella
Cheilosporum
Clathromorphum
Corallina
Dermatolithon
Fosliella
Haliptilon
Hydrolithon
Jania
Leptophytum
Lithophyllum
Lithothamnion
Lithothrix
Melobesia
Mesophyllum
Neogoniolithon
Phymatolithon
Porolithon
Cryptonemiales
Cryptonemiaceae
Aeodes (*Aeodis*)
Cryptonemia

*" " designates a life history stage of another genera

RHODOPHYCOTA *(continued)*

Halymenia
Prionitis
Dumontiaceae
Dilsea (*Neodilsea*)
Dumontia
Endocladiaceae
Endocladia
Gloiosiphoniaceae
"*Cruoriopsis*"
Gloiosiphonia
Kallymeniaceae
Callophyllis
Peyssonneliaceae
Peyssonnelia
Gelidiales
Gelidiaceae
Gelidiella
Gelidium
Pterocladia
Gigartinales
Caulacanthaceae
Catenella
Caulacanthus
Cruoriaceae
Cruoria
Cystocloniaceae
Cystoclonium
Fimbrifolium (*Rhodophyllis*)
Furcellariaceae
Furcellaria
Gigartinaceae
Chondrus
Gigartina
Iridaea
Rhodoglossum
Gymnophlaeaceae
Predaea
Hypneaceae
Hypnea
Petrocelidaceae
Mastocarpus
"*Petrocelis*"
Phyllophoraceae
Ceratocolax
Gymnogongrus
Phyllophora
Schottera (*Petroglossum*)
Plocamiaceae
Plocamium
Solieriaceae
Anatheca
Eucheuma
Solieria
Turnerella
Polyideaceae
Polyides
Gracilariales
Gracilariaceae
Ceratodictyon
Curdiea
Gelidiopsis
Gracilaria
Melanthalia
Hildenbrandiales
Hildenbrandiaceae
Hildenbrandia
Nemaliales
Acrochaetiaceae
Audouinella (*Rhodochorton*)
Galaxauraceae
Actinotrichia
"*Chaetangium*"
Galaxaura
Nothogenia
Helminthocladiaceae
Liagora
Palmariales
Palmariaceae
Devaleraea
Halosaccion
Leptosomia
Palmaria
Rhodophysema
Rhodymeniales
Champiaceae
Champia
Lomentaria
Rhodymeniaceae
Botryocladia
Coelothrix
Epymenia
Rhodymenia

CHROMOPHYCOTA

BACILLARIOPHYCEAE
Centrales
Biddulphiineae
Anaulaceae
Anaulus
Coscinodiscineae
Coscinodiscaceae
Aulacodiscus
Pennales
Araphidineae
Diatomaceae
Fragilaria
Biraphidineae
Cymbellaceae
Amphora
Entomoneidaceae
Entomoneis (*Amphiprora*)
Naviculaceae
Gyrosigma
Navicula
Pleurosigma
Nitzschiaceae
Nitzschia
Monoraphidineae
Achnanthaceae
Cocconeis

DINOPHYCEAE

DINOPHYCIDAE
Gymnodiniales
Gymnodiniaceae
Gymnodinium (*Symbiodinium*)

PHAEOPHYCEAE
Chordariales
Chordariaceae
Chordaria
Eudesme
Corynophlaeaceae
Cylindrocarpus
Leathesia
Elachistaceae
Elachista
Splachnidiaceae
Splachnidium
Cutleriales
Cutleriaceae
"*Aglaozonia*"
Desmarestiales
Desmarestiaceae
Desmarestia
Himantothallus
(*Phaeoglossum* & *Phyllogigas*)
Dictyosiphonales
Dictyosiphonaceae
Coilodesme
Dictyosiphon
Punctariaceae
Adenocystis
Asperococcus
Striariaceae
Isthmoplea
Stictyosiphon
Striaria
Dictyotales
Dictyotaceae
Dictyopteris
Dictyota
Glossophora
Lobophora (*Pocockiella*)
Padina
Stypopodium
Zonaria
Durvillaeales
Durvillaeaceae
Durvillaea
Ectocarpales
Ectocarpaceae
Bachelotia
Ectocarpus
Giffordia
Pilayella
Waerniella
Ralfsiaceae
Basispora
Petroderma
Pseudolithoderma
Ralfsia
Fucales
Cystoseiraceae
Bifurcaria
Cystophora
Cystoseira
Halidrys
Hormophysa
Turbinaria
Fucaceae
Ascophyllum
Fucus
Hesperophycus
Pelvetia
Pelvetiopsis
Xiphophora
Himanthaliaceae
Himanthalia
Hormosiraceae
Hormosira
Phyllosporaceae
Phyllospora
Scytothalia
Sargassaceae
Sargassum
Laminariales
Alariaceae
Alaria
Ecklonia
Egregia
Eisenia
Pterygophora
Undaria
Chordaceae
Chorda
Laminariaceae
Agarum
Costaria
Hedophyllum
Laminaria
Saccorhiza
Lessoniaceae
Lessonia
Lessoniopsis
Macrocystis
Nereocystis
Pelagophycus
Postelsia
Scytosiphonales
Scytosiphonaceae
Colpomenia
Hydroclathrus
Petalonia
Scytosiphon
Sphacelariales
Sphacelariaceae
Sphacelaria
Stypocaulaceae
Halopteris

EUGLENOPHYCOTA

EUGLENOPHYCEAE
Heteronematales
Heteronematina
Peranemaceae
Heteronema

CHLOROPHYCOTA

CHAROPHYCEAE
Charales
Characeae
Chara
Tolypella

CHLOROPHYCEAE
Acrosiphoniales
Acrosiphoniaceae
Spongomorpha (*Acrosiphonia*)
Urospora
Codiolaceae
"*Codiolum*"*
Bryopsidales (Caulerpales)
Bryopsidaceae
Bryopsis
Derbesia
Codiaceae
Codium
Johnson-sea-linkia
Tydemania
Caulerpaceae
Caulerpa
Ostreobiaceae
Ostreobium
Udoteaceae
Avrainvillea
Chlorodesmis
Halimeda
Penicillus
Rhipiliopsis
Udotea

---

*Species of "*Codiolum*" have been shown to be phases in the life histories of members of the order Acrosiphoniales, as well as species of *Ulothrix*, *Monostroma*, *Collinsiella* and *Gomontia*

CHLOROPHYCOTA (*continued*)

Chlorococcales
Oocystaceae
Chlorella
Cladophorales
Anadyomenaceae
Anadyomene
Cladophoraceae
Chaetomorpha
Cladophora
Rhizoclonium
Ctenocladales
Ulvellaceae
Acrochaete (*Endoderma*)
Epicladia (*Entocladia*)
Dasycladales
Dasycladaceae
Batophora
Neomeris
Polyphysa
Prasiolales (Schizogoniales)
Prasiolaceae (Schizogoniaceae)
Prasiola
Rosenvingiella
Siphonocladales
Siphonocladaceae
Boergesenia
Boodlea
Cladophoropsis
Siphonocladus
Struvea
Valoniaceae
Dictyosphaeria
Valonia
Valoniopsis
Ventricaria
Ulotrichales
Ulotrichaceae
Ulothrix
Uronema
Ulvales
Monostromataceae
Blidingia
Monostroma
Ulvaceae
Capsosiphon
Enteromorpha
Ulva
Ulvaria
Zygnematales
Zygnematineae
Zygnemataceae
Mougeotia

EUMYCOTA

ASCOMYCOTINA

LOCULOASCOMYCETES (ASCOMYCETES)

LOCULOANOTEROMYCETIDAE (ASCOMYCETIDAE)
Lecanorales
Lichinaceae
Lichina
Sphaeriales (Verrucariales)
Verrucariaceae
Verrucaria

EMBRYOBIONTA

BRYOPHYTA

BRYOPSIDA

BRYIDAE
Isobryales
Fontinaliaceae
Fontinalis

LYCOPODIOPHYTA

ISOETOPSIDA
Isoetales
Isoetaceae
Isöetes

MAGNOLIOPHYTA (SPERMATOPHYTA)

LILIOPSIDA (LILIATE; MONOCOTYLEDONEAE)

ALISMATIDAE (HELOBIEAE)
Hydrocharitales
Hydrocharitaceae
Elodea
Enhalus
Halophila
Thalassia
Vallisneria
Najadales
Cymodoceaceae
Amphibolis
Cymodocea
Halodule (*Diplanthera*)
Syringodium
Thalassodendron
Najadaceae
Najas
Posidoniaceae
Posidonia
Potamogetonaceae
Potamogeton
Ruppiaceae
Ruppia
Zannichelliaceae
Zannichellia
Zosteraceae
Heterozostera
Phyllospadix
Zostera
Zosterella (=subgenus of *Zostera*)

COMMELINIDAE
Cyperales
Cyperaceae
Scirpus
Poaceae (Gramineae)
Deschampia (*Aira*)
Festuca
Phragmites
Puccinellia
Spartina
Sporobolus

MAGNOLIOPSIDA (DICOTYLEDONEAE; MAGNOLIATAE)

ASTERIDAE
Lamiales
Verbenaceae
Avicennia
Plantaginales
Plantaginaceae
Plantago
Solanales
Convolvulaceae
Ipomoea

CARYOPHYLLIDAE
Caryophyllales
Aizoaceae
Sesuvium
Chenopodiaceae
Salicornia
Sarcocornia

DILLENIIDAE
Primulales
Myrsinaceae
Aegiceras

MAGNOLIIDAE
Ranunculales
Ranunculaceae
Ranunculus

ROSIDAE
Haloragales
Haloragaceae
Myriophyllum
Rhizophorales
Rhizophoraceae
Bruguiera
Ceriops
Rhizophora

ANIMALIA

PROTOZOA

Labyrinthulata
Labyrinthulea
Labyrinthulida
Labyarinthulidae
Labyrinthula (*Labyrinthomyxa*)

SARCOMASTIGOPHORA

MASTIGOPHORA

ZOOMASTIGOPHORA
Retortamonadida
Retortamonadidae
Macrostoma

SARCODINA

RHIZOPODA

GRANULORETICULOSA
Foraminiferida
Miliolina
Milioloidea
Soritidae
Amphisorus
Sorites
Rotaliina
Discorboidea
Asterigerinidae
Amphistegina
Textulariina
Ammodiscoidea
Schizamminidae
Jullienella
Schizammina

CILIOPHORA

KINETOFRAGMINOPHORA

GYMNOSTOMATA
Haptorida
Didiniidae
Askenasia
Karyorelictida
Trachelocercidae
Trachelocerca

OLIGOHYMENOPHORA

HYMENOSTOMATA
Hymenostomatida
Peniculina
Frontoniidae
Frontonia

POLYHYMENOPHORA

SPIROTRICHA
Heterotrichida
Heterotrichina
Stentoridae
Stentor
Hypotrichida
Sporadotrichina
Euplotidae
Discocephalus

PARAZOA

PORIFERA

CALCAREA

CALCINIA
Leucettida
Leucascidae
Leuconia

CALCARONIA
Sycettida
Sycettidae
Sycandra

DEMOSPONGIAE

CERACTINOMORPHA
Dictyoceratida
Spongiidae
Fasciospongia
Halichondrida (Halichondriida)
Halichondridae (Halichondriidae)
Halichondria
Hymeniacidonidae
Hymeniacidon
Haplosclerida
Haliclonidae
Haliclona
Poecilosclerida
Biemnidae
Biemma
Hymedesmiidae
Hymedesia
Myxillidae
Iophon
Myxilla

PARAZOA (*continued*)

TETRACTINOMORPHA
Axinellida
Axinellidae
Acanthella
Phakellia
Hadromerida
Latrunculiidae
Sigmosceptrella
Spirastrellidae
Spirastrella
Suberitidae
Polymastia
Subertechinus
Tethyidae
Tethya
Spirophorida
Tetillidae
Cinachyra

EUMETAZOA

CNIDARIA (COELENTERATA)

ANTHOZOA

ALCYONARIA (OCTOCORALLIA)
Alcyonacea
Alcyoniidae
Alcyonium
Lobophytum
Parerythropodium
Sarcophyton
Sinularia
Xeniidae
Xenia
Gorgonacea
Holaxonia
Gorgoniidae
Lophogorgia
Scleraxonia
Melithaeidae
Acabaria
Pennatulacea
Sessiliflorae
Renillidae
Renilla
Subselliflorae
Virgulariidae
Stylatula
Virgularia
Stolonifera
Clavulariidae
Anthelia
Clavularia
Tubiporidae
Tubipora
Telestacea
Telestidae
Telesto

CERIANTIPATHARIA
Ceriantharia
Cerianthidae
Cerianthus

ZOANTHARIA (HEXACORALLIA)
Actiniaria
Actiniidae
Anthopleura
Bolocera
Bunodactis
Gyrostoma
Macrodactyla
Tealia
Actinodendronidae
Megalactis
Aliciidae
Triactis
Metridiidae
Metridium
Stichodactylidae
Radianthus
Stoichactis
Scleractinia (Madreporaria)
Astrocoeniina
Acroporidae
Acropora
Montipora
Pocilloporidae
Pocillopora
Stylophora
Caryophylliina
Caryophylliidae
Heterocyathus
Faviina
Faviidae
Favia
Favites
Fungiina
Poritidae
Porites
Siderastreidae
Siderastrea
Zoanthinaria (Zoanthiniaria, Zoanthidea)
Zoanthidae
Palythoa
Zoanthus

HYDROZOA
Hydroida
Anthomedusae (Athecata; Gymnoblastea)
Halocordylidae
Halocordyle
Milleporina
Milleporidae
Millepora

SCYPHOZOA
Rhizostomeae
Cassiopeidae
Cassiopea
Semaeostomeae
Cyaneidae
Cyanea

NEMERTEA (NEMERTINEA)

ANOPLA
Heteronemertea (Heteronemertini)
Cerebratulidae (Lineidae)
Cerebratulus

ENOPLA
Hoplonemertea (Hoplonemertini)
Monostilifera
Amphiporidae
Amphiporus
Ototyphlonemertidae
Ototyphlonemertes

GASTROTRICHA
Chaetonotida
Paucitubulatina
Chaetonotidae
Aspidiophorus
Macrodasyida
Macrodasyidae
Macrodasys
Urodasys

ROTIFERA

MONOGONONTA
Ploima
Dicranophoridae
Encentrum

KINORHYNCHA (ECHINDERIDA)
Homalorhagida
Homalorhagae
Pycnophyidae
Pycnophyes

NEMATA (NEMATODA)

ADENOPHOREA

CHROMADORIA
Monhysterida
Monhysterina
Monhysteroidea
Monhysteridae
Monhystera

SECERNENTEA

DIPLOGASTERIA
Tylenchida
Tylenchina
Tylenchoidea
Tylenchidae
Halenchus

MOLLUSCA

BIVALVIA

ANOMALODESMATA
Pandoracea
Pandoridae
Pandora

HETERODONTA
Hippuritoida
Chamacea
Chamidae
Chama
Pseudochama
Myoida
Hiatellacea
Hiatellidae (+Glycimeridae; Saxicavidae)
Hiatella
Myacea
Corbulidae (+Aloididae)
Corbula (*Aloidis*)
Myidae (+Myacidae)
Mya
Pholadacea
Teredinidae
Teredo
Veneroida
Cardiacea
Cardiidae
Cardium
Cerastoderma
Laevicardium
Carditacea
Carditidae
Cardita
Glans
Cyamiacea
Cyamiidae
Kingiella
Lucinacea
Lucinidae
Codakia
Divaricella
Ungulinidae (+Diplodontidae)
Diplodonata
Mactracea
Mactridae (+Tanysiphonidae)
Mactra
Mulinia
Spisula
Mesodesmatidae (+Amphidesmatidae; Paphiidae)
Atactodea
Mesodesma
Solenacea
Cultellidae (Pharellidae)
Cultellus
Ensis
Phaxas
Siliqua
Solenidae
Solen
Tellinacea
Donacidae
Donax
Psammobiidae (+Asaphidae; Garidae; Sanguinolariidae)
Gari
Semelidae
Abra
Theora
Tellinidae
Macoma
Tellina
Tridacnacea
Tridacnidae
Tridacna
Veneracea
Veneridae
Chione
Gemma
Mercenaria
Tivela

PROTOBRANCHIA (+CRYPTODONTA; PALEOTAXODONTA)
Nuculoida
Nuculanacea
Nuculanidae (+Adranidae; Ledidae; Porolepididae; Sareptidae; Zealedidae)
Yoldia

PTERIOMORPHIA
Isofilibranchia
Arcoida
Arcacea
Arcidae (+Anadaridae)
Anadara
Arca

EUMETAZOA *(continued)*

Limopsacea
Glycymerididae
(+Glycymeridae;
Pectunculidae)
Glycymeris
Mytiloida
Mytilacea
Mytilidae
Aulacomya
Austromytilus
Brachidontes
Choromytilus (*Chloromytilus*)
Lithophaga
Modiolus
Mytilus
Perumytilus
Septifer
Eupteriomorphia
Ostreoida
Ostreina
Ostreacea
Ostreidae
Crassostrea
Ostrea
Saccostrea
Pterioida
Pinnina
Pinnacea
Pinnidae
Pinna
Pteriina
Pteriacea
Isognomonidae
(+Isognomontidae; Melinidae;
Pernidae)
Pedalion
Perna
Pteriidae (+Aviculidae)
Pteria
Malleidae (+Eligmidae;
Stefaniniellidae; Vulsellidae)
Malleus

CEPHALOPODA

COLEOIDEA
(DIBRANCHIATA)
Octopoda
Incirrata
Octopodidae
Octopus
Sepioidea
Sepiidae
Sepia
Teuthoidea (Decapoda)
Myopsida
Loliginidae
Loligo
Sepioteuthis

GASTROPODA

OPISTHOBRANCHIA
Acochlidioidea
Philinoglossacea
Philinoglossidae
Sapha
Anaspidea
Aplysiidae
Aplysia
Dolabriferidae
Dolabrifera
Notarchidae
Notarchus
Gymnosomata
Doridoidea (+Doridacea;
Holohepatica)
Anadoridacea
(+Phanerobranchia)
Gymnodorididae
Nembrotha
Eudoridacea
(+Cryptobranchia)
Chromodorididae
Chromodoris
Discodorididae (+Diaululidae)
Discodoris
Hexabranchidae
Hexabranchus
Notaspidea (+Pleurobranchea)
Pleurobranchidae
Berthellina
Nudibranchia
Aeolidoidea
Cleioprocta
Aeolidiidae (+Spurillidae)
Aeolidia
Pleuroprocta
Coryphellidae
Coryphella
Arminoidea
Euarminiacea
Doridomorphidae
(+Doridoeididae)
Onchidoris
PROSOBRANCHIA
Archaeogastropoda
(+Aspidobranchia;
Cyclobranchia; Diotocordia;
Scutibranchia)
Fissurellacea
Fissurellidae (+Emarginulidae;
Hemitomidae)
Diodora
Fissurella
Neritacea
Neritidae (+Septariidae)
Melanerita
Nerita
Smaragdia
Theodoxus
Patellacea (+Cyclobranchia;
Docoglossa; Onychoglossa)
Acmaeidae (+Lottiidae;
Pectinodontidae; Tecturidae)
Collisella (*Acmaea*)
Lottia
Notoacmaea
Patelloida
Patellidae (+Nacellidae;
Trachelobranchia)
Cellana
Nacella
Patella
Pleurotomariacea
(+Zeugobranchia)
Haliotidae
Haliotis
Sanhaliotis
Trochacea
Trochidae (+Calliostomatidae;
Umboniidae)
Austrocochlea
Margarites
Monodonta
Norrisia
Oxystele
Tectus
Tegula
Trochus
Turbinidae (+Liotiidae)
Turbo
Mesogastropoda (Taenioglossa)
Calyptraeacea
Calyptraeidae
Crepidula
Cerithiacea
Cerithiidae (+Bittiidae;
Dialidae; Litiopidae)
Alaba
Bittium
Cerithium
Clypeomorus
Diastoma
Rhinoclavis
Thericium
Modulidae
Modulus
Planaxidae
Planaxis
Potamididae (+Potamidae)
Batillaria
Pirenella
Pyrazus
Turritellidae (+Tenagodidae)
Turritella
Vermetidae
Dendropoma
Petaloconchus
Serpulorbis
Spiroglyphus
Vermetus
Cypraeacea (+Involuta;
Lamellariacea)
Cypraeidae
Cypraea
Eulimacea (+Aglossa;
Melanellacea)
Eulimidae (+Melanellidae)
Monogamus
Robillardia
Heterogastropoda
(Triphoroidea)
Triphoridae
Triphora
Littorinacea
Lacunidae
Lacuna
Littorinidae
Bembicium
Littorina
Littoraria
Melarhaphe
Nodilittorina (*Nodolittorina*)
Peasiella
Risella
Tectarius
Naticacea
Naticidae
Conuber
Polinices
Rissoacea
Assimineidae
(+Falsicingulidae;
Synceratidae; Synceridae)
Assiminea
Hydrobiidae
Hydrobia
Tatea
Strombacea
Strombidae
Lambis
Strombus
Terebellum
Tonnacea (+Canalifera;
Cassidoidea; Cymatioidea;
Doliacea; Dolioidea;
Tonnoidea)
Cassidae (+Cassididae)
Casmaria
Matuta
Cymatiidae (Ranellidae)
Charonia
Tonnidae (+Doliidae;
Oocoritidae; Oocorythidae)
Tonna
Viviparacea
Bithyniidae (+Bulimidae;
Hydroccocidae)
Bithynia
Neogastropoda (+Hamiglossa)
Conacea (+Taxoglossa)
Conidae
Conus
Terebridae (+Perivaciidae)
Impages
Terebra
Turridae (+Clavidae;
Cochlespiridae; Raphitomidae;
Speightiidae; Thatcheriidae)
Drillia
Muricacea (+Buccinacea;
Mitracea; Rachiglossa;
Stenoglossa; Volutacea)
Buccinidae (+Buccinulidae;
Chrysodomidae; Cominellidae;
Neptuneidae; Photidae)
Buccinum
Engina
Kelletia
Columbellidae (+Anachidae,
Beringiidae, Pyrenidae)
Mitrella
Fasciolariidae (+Fusinidae)
Latirus
Melongenidae (+Busyconidae;
Fulguridae; Galeodidae;
Volemidae)
Galeodes (*Volema*)
Pugilina
Mitridae (+Cylindromitridae;
Strigatellacea; Turritidae)
Mitra
Scabricola
Muricidae (+Pseudolividae;
Purpuridae; Rapanidae;
Thaididae; Thaisidae)
Chicoreus
Concholepas
Crassilabrum
Drupa
Morula
Murex
Nucella
Stramonita
Thais (*Mancinella*)
Nassariidae (+Alectryonidae;
Nassidae)
Bullia
Nassa
Nassarius (*Ilyanassa*)
Olivadae (+Olivancilariidae)
Olivella
Vexillidae
Vexillum

EUMETAZOA *(continued)*

PULMONATA
Archaeopulmonata
Ellobiacea (+Actophila)
Ellobiidae (+Auriculidae)
Ophicardelus
Basommatophora
Amphibolacea
Amphibolidae
Salinator
Lymnacea
Lymnaeidae
Lymnaea
Siphonariacea (+Patelliformia; Thalassophila)
Siphonariidae
Siphonaria

POLYPLACOPHORA

NEOLORICATA
Acanthochitonida
Acanthochitonidae
Acanthochitona (*Acanthochiton*)
Ischnochitonida
Callistoplacidae
Nuttallina
Chitonidae
Acanthopleura
Chiton
Ischnochitonidae
Cyanoplax
Ischnochiton
Tonicella
Mopaliidae
Katharina
Mopalia
Plaxiphora

SCAPHOPODA
Dentaliida
Dentaliidae
Dentalium

ANNELIDA

OLIGOCHAETA
Haplotaxida
Tubificina
Enchytraeoidea
Enchytraeidae
Enchytraeus
Michaelsena
Tubificoidea
Tubificidae
Rhyacodrilus
Tubifex

POLYCHAETA
Amphinomida
Amphinomidae
Eurythoe
Hermodice
Euphrosinidae
Euphrosine
Capitellida
Capitellidae
Barantolla
Capitella
Heteromastus
Mediomastus
Maldanidae
Clymenella (*Paraxiotheca*)
Macroclymene
Cirratulida
Cirratulidae
Cirratulus
Cirriformia
Dodecaceria
Eunicida
Eunicacea
Eunicidae
Staurocephalus
Lumbrineridae
Lumbrineris
Onuphidae
Australonuphis
Diopatra
Nothria
Onuphis
Magelonida
Magelonidae
Magelona
Opheliida
Opheliidae
Armandia
Euzonus
Polyophthalmus
Orbiniida
Orbiniidae (Ariciidae)
Haploscoloplos
Scoloplos
Oweniida
Oweniidae
Owenia
Phyllodocida
Aphroditacea
Polynoidae
Harmothoe
Lepidonotus
Sigalionidae
Sthenelais
Pisionidae
Pisionidens
Glyceracea
Glyceridae
Hemipodus
Goniadidae
Glycinde
Goniadides
Nephtyidacea
Nephtyidae
Nephtys
Nereididacea
Hesionidae
Hesionides
Nereididae (Nereidae)
Australonereis
Ceratonereis
Neanthes
Nereis
Perinereis
Phyllodocidacea
Phyllodocidae
Mystides
Polygordiida
Polygordiidae
Polygordius
Protodrilida
Protodrilidae
Protodrilus
Saccocirridae
Saccocirrus
Sabellida
Sabellidae
Branchiomma
Chone
Fabricia
Fabriciola
Myxicola
Oriopsis
Serpulidae
Eupomatus
Galeolaria
Pomatoleios
Spirobranchus
Vermiliopsis
Spirorbidae
Spirorbis
Spionida
Spionidae
Boccardia
Minuspio
Nerine
Polydora
Prionospio
Scolelepis
Streblospio
Terebellida
Pectinariidae
Cistena (*Cistenides*, *Pectinaria*)
Sabellariidae
Gunnarea
Idanthyrsus
Phragmatopoma
Terebellidae
Amphitrite
Thelepus

SIPUNCULA
Phascolosomida
Aspidosiphoniformes
Aspidosiphonidae
Aspidosiphon
Sipunculida
Golfingiaformes
Golfingiidae
Golfingia

ARTHROPODA

CHELICERATA

ARACHNIDA

ACARI (ACARINA; ACARIDA)
Acariformes
Prostigmata (Actinedida; Trombidiformes)
Parasitengona
Hygrobatoidea
Pontarachnidae
Litarachna
Pontarachna
Parasitiformes (Anactinotrichida)
Mesostigmata (Gamasida)
Dermanyssina
Ascoidea
Podocinidae
Andregamasus
Rhodacaridea
Rhodacaridae
Rhodacaroides

MEROSTOMATA
Xiphosura
Xiphosurida
Limulidae
Limulus

CRUSTACEA

BRANCHIOPODA
Sarsostraca
Anostraca
Artemiidae
Artemia
CIRRIPEDIA
Thoracica
Balanomorpha
Balanoidea
Archaeobalanidae
Semibalanus
Balanidae
Austramegabalanus
Balanus
Elminius
Megabalanus
Notomegabalanus
Chthamaloidea
Catophragmidae
Catomerus
Chthamalidae
Chamaesipho
Chthamalus
Jehlius
Octomeris
Tetrachthamalus
Coronuloidea
Tetraclitidae
Tesseropora
Tetraclita
Lepadomorpha
Iblidae
Ibla
Scalpellidae
Pollicipes

COPEPODA
Calanoida
Centropagoidea
Acartiidae
Acartia
Pseudocyclopoidea
Pseudocyclopidae
Pseudocyclops
Ridgewayiidae
Ridgewayia
Cyclopoida
Cyclopidae
Euryte
Cyclopinidae
Cyclopuella
Poecilostomatoida
Myicolidae
Anthessius
Taeniacanthidae
Echinirus
Echinosocius
Harpacticoida
Ameiridae
Nitokra
Argestidae
Dizahavia
Canuellidae
Canuellina
Canthocamptidae
Orthopsyllus
Cletodidae
Cletocamptus
Cletodes
Cylindropsyllidae
Arenopontia
Diosaccidae
Robertsonia

EUMETAZOA *(continued)*

Schizoperoides
Stenhelia
Typhlamphiascus
Laophontidae
Heterolaophonte
Paramesochridae
Kliopsyllus
Tegastidae
Syngastes
Thalestridae
Neodactylopus
Tisbidae
Tisbe

MALACOSTRACA
Eucarida
Decapoda
Dendrobranchiata (Reptantia)
Penaeoidea
Penaeidae
Macropetasma
Penaeus
Pleocyemata
Anomura
Coenobitoidea
Coenobitidae
Coenobita
Diogenidae
Diogenes
Galatheoidea
Porcellanidae
Petrolisthes
Hippoidea
Albuneidae
Blepharipoda
Lepidopa
Hippidae
Emerita
Hippa
Paguroidea
Paguridae
Calcinus
Clibanarius
Pagurus
Pseudopagurus
Astacidea
Nephropoidea
Nephropidae
Homarus
Thalassinidea
Thalassinoidea
Callianassidae
Callianassa
Upogebiidae
Upogebia
Brachyura
Brachyrhyncha
Bellioidea
Belliidae
Acanthocyclus
Bellia
Grapsidoidea
Grapsidae
Grapsus
Metopograpsus
Pachygrapsus
Plagusia
Pseudograpsus
Sesarma
Ocypodoidea
Ocypodidae
Dotilla
Ocypode (*Ocypoda*)
Uca
Portunoidea
Portunidae
Arenaeus
Callinectes
Carcinus
Ovalipes
Xanthoidea
Xanthidae
Eurytium
Lybia
Neopanope
Panopeus (*Eupanopeus*)
Xanthodius (*Xantho*)
Cancridea
Cancroidea
Cancridae
Cancer
Dromiacea
Dromioidea
Dromiidae
Cryptodromia
Dromidia
Oxyrhyncha
Majoidea
Majidae
Hyas
Pugettia
Oxystomata
Leucosioidea
Calappidae
Calappa
Leucosiidae
Ebalia
Caridea
Alpheoidea
Alpheidae
Alpheus
Athanas
Hippolytidae
Calliasmata
Hippolyte
Ogyrididae
Ogyrides
Atyoidea
Atyidae
Antecaridina
Crangonoidea
Crangonidae
Crangon
Palaemonoidea
Gnathophyllidae
Hymenocera
Palaemonidae
Macrobrachium
Palaemon
Palaemonetes
Periclimenes
Palinura
Palinuroidea
Palinuridae
Jasus
Panulirus
Scyllaridae
Scyllarides
Peracarida
Amphipoda
Gammaridea
Ampeliscoidea
Ampeliscidae
Ampelisca
Haploops
Corophioidea
Amphithoidae
Ampithoe
Cymadusa
Corophiidae
Acuminodeutopus
Corophium
Paracorophium
Siphonoecetes
Ischyroceridae
Ischyrocerus
Jassa
Photidae
Gammaropsis
Dexaminoidea
Dexaminidae
Guernea
Eusiroidea
Calliopiidae
Calliopius
Gammarellidae
Gammarellus
Gammaroidea
Gammaridae
Elasmopus
Gammarus
Melita
Oedicerotoidea
Oedicerotidae
Bathyporeiapus
Perioculodes
Phoxocephaloidea
Phoxocephalidae
Grandifoxus
Phoxocephalopsis
Rhepoxynius
Pontoporeioidea
Haustoriidae
Acanthohaustorius
Eohaustorius
Haustorius
Neohaustorius
Pseudohaustorius
Pontoporeiidae
Amphiporeia
Bathyporeia
Talitroidea
Hyalellidae
Parhyalella
Hyalidae
Hyale
Talitridae
Megalorchestia
Orchestia
Orchestoidea
Talorchestia
Cumacea
Bodotriidae
Cumopsis
Cyclaspis
Lampropidae
Hemilamprops
Isopoda
Anthuridea
Paranthuridae
Colanthura
Asellota
Aselloidea
Asellidae
Asellus
Janiroidea
Janiridae
Jaera
Flabellifera
Cirolanidae
Cirolana
Eurydice
Excirolana
Pontogeloides
Sphaeromatidae
Ancinus
Cymodoce
Exosphaeroma
Paracerceis
Syncassidina
Oniscoidea (Oniscidea)
Ligiamorpha
Crinocheta
Atracheata (Oniscoidea)
Halophiloscidae
(Halophilosciidae)
Halophiloscia
Oniscidae
Porcellio
Diplocheta
Ligiidae
Ligia (*Ligyda, Megaligia*)
Tylomorpha
Tylidae
Tylos
Valvifera
Chaetilidae (Chaetiliidae)
Chaetilia
Idoteidae
Chiridotea
Edotea
Erichsonella
Idotea (*Idothea*)
Mesidothea
Paridotea
Mysidacea
Mysida
Mysidae
Archaeomysis
Gastrosaccus
Mesopodopsis
Mysis
Myodocopa
Myodocopida
Myodocopina
Cypridinoidea
Philomedidae
Euphilomedes

PODOCOPA
Podocopina
Bairdioidea
Bairdiidae
Bairdia
Cytheroidea
Cytherideidae
Cyprideis

UNIRAMIA

INSECTA

APTERYGOTA
Collembola
Arthropleona
Hypogastruridae
Anurida

PTERYGOTA

NEOPTERA
Hemipterodea (Oligonephridia)
Hemiptera (Heteroptera)
Gerromorpha
Gerroidea
Gerridae
Halobates
Holometabola

EUMETAZOA (*continued*)

Coleoptera
Polyphaga
Chrysomeloidea (Phytophaga)
Bruchidae
Acanthoscelis
Hydrophiloidea (+Histeroidea)
Hydrophilidae
Anacaena
Paraberosus
Staphylinoidea
Hydraenidae (Limnebiidae)
Ochthebius
Staphylinidae (+Brathinidae; Scaphidiidae)
Bledius
Tenebrionoidea (Heteromera)
Tenebrionidae (+Alleculidae; Cossyphodidae; Lagriidae; Nilionidae; Rhysopaussidae; Tentyriidae)
Pachyphalaria
Phalerisidia
Diptera
Brachycera
Cyclorrhapha
Schizophora
Muscoidea (=Calyptratae)
Anthomyiidae
Fucellia

BRYOZOA (ECTOPROCTA)

GYMNOLAEMATA
Cheilostomata
Anasca

MALACOSTEGA

Cupuladriidae
Discoporella
Electridae
Electra
Membraniporidae
Membranipora

SCRUPARIINA

Scrupariidae
Scruparia (*Plocamionida*)
Ctenostomata
Carnosa
Alcyonidioidea (Alcyonellea; Halcyonellea; Halcyonelloidea)
Alcyonidiidae
Alcyonidium

BRACHIOPODA

ARTICULATA
Terebratulida
Terebratulidina
Terebratulacea
Terebratulidae
Terebratulinae (*Terebratulina*)

ECHINODERMATA

ASTEROZOA

STELLEROIDEA

ASTEROIDEA
Forcipulata (Forcipulatida)
Asteriidae
Asterias
Leptasterias
Pisaster
Heliasteridae
Heliaster
Paxillosida
Astropectinidae
Astropecten
Platyasterida
Luidiidae
Luidia
Spinulosida
Acanthasteridae
Acanthaster
Asterinidae
Patiria
Echinasteridae
Henricia
Poraniidae
Porania
Solasteridae
Crossaster
Solaster
Valvatida
Archasteridae
Archaster
Goniasteridae
Hippasterias
Ophidiasteridae
Linckia
Oreasteridae
Choriaster
Culcita
Oreaster
Protoreaster

OPHIUROIDEA
Ophiurida
Chilophiurina
Ophiocomidae
Ophiocoma
Ophiuridae
Ophiura
Gnathophiurina
Amphiuridae
Amphioplus
Amphipholis
Ophiactidae
Hemipholis
Ophiactis
Ophiopholis

CRINOZA

CRINOIDEA

ARTICULATA
Comatulida
Mariametracea
Himerometridae
Heterometra

ECHINOZOA

ECHINOIDEA

EUECHINOIDEA
Atelostomata
Spatangoida
Amphisternata
Brissidae
Metalia
Loveniidae
Echinocardium
Lovenia
Schizasteridae
Moira
Schizaster
Diadematacea
Diadematoida
Diadematidae
Centrostephanus
Diadema
Echinacea
Arbacioida
Arbaciidae
Arbacia
Tetrapygus
Echinoida
Echinometridae
Echinometra
Heterocentrotus
Echinidae
Echinus
Paracentrotus
Parechinus
Strongylocentrotidae
Strongylocentrotus
Phymosomatoida
Stomechinidae
Stomopneustes
Temnopleuroida
Toxopneustidae
Lytechinus
Nudechinus
Tripneustes
Gnathostomata
Clypeasteroida
Clypeasterina
Clypeasteridae
Clypeaster
Laganina
Laganidae
Laganum
Scutellina
Astriclypeidae
Echinodiscus
Dendrasteridae
Dendraster
Mellitidae
Mellita

HOLOTHUROIDEA

APODACEA
Apodida
Synaptidae
Synapta

ASPIDOCHIROTACEA
Aspidochirotida
Holothuriidae
Actinopyga
Bohadschia
Halodeima
Holothuria (*Brandtothuria*)
Metriatyla
Microthele
Stichopodidae
Stichopus
Thelenota

DENDROCHIROTACEA
Dendrochirotida
Cucumariidae
Cucumaria
Pentacta
Phyllophoridae
Thyone
Psolidae
Psolus

HEMICHORDATA

ENTEROPNEUSTA
Ptychoderidae
Ptychodera

CHORDATA

TUNICATA (UROCHORDATA)

ASCIDIACEA
Aplousobranchia
Polyclinidae
Aplidium
Phlebobranchia
Ascidiidae
Ascidia
Perophoridae
Ecteinascidia
Stolidobranchia
Molgulidae
Molgula
Pyuridae
Boltenia
Pyura
Styelidae
Botrylloides
Dendrodoa

CEPHALOCHORDATA (ACRANIA)
Asymmetrontidae
Asymmetron
Branchiostomidae
Branchiostoma

VERTEBRATA

CHONDRICHTHYES

ELASMOBRANCHII

EUSELACHII

NEOSELACHII
Batoidea
Myliobatiformes
Dasyatoidea
Dasyatidae
Taeniura
Myliobatoidea
Myliobatidae
Aëtobatus
Rhinobatiformes
Rhinobatidae
Rhinobatos
Rhynchobatidae
Rhina
Rhynchobatus

SCRUPARIINA *(continued)*

Torpediniformes
Torpedinoidea
Torpedinidae
Torpedo
Galeomorphii
Carcharhiniformes
Carcharhinidae
Carcharhinus
Galeocerdo
Mustelus
Negaprion
Triaenodon
Sphyrnidae
Sphyrna
Orectolobiformes
Stegostomatidae
(Rhincodontidae)
Stegostoma

OSTEICHTHYES

ACTINOPTERYGII
Anguilliformes
Anguilloidei
Congridae
Gorgasia
Ophichthidae
Callechelys
Antheriniformes
Antherinoidei
Antherinidae
Menidia
Cyprinodontoidei
Cyprinodontidae
Aphanius
Fundulus
Exocoetoidei
Hemirhamphidae
Hemirhamphus
Beryciformes
Berycoidei
Holocentridae
Adioryx
Clupeiformes
Clupeoidei
Clupeidae
Clupea
Engraulididae
Anchoa
Cypriniformes
Cyprinoidei
Cyprinidae
Abramis
Alburnus
Rutilus
Scardinius
Gadidformes
Gadoidei
Gadidae
Gadus
Melanogrammus
Zoarcoidei
Zoarcidae
Macrozoarces
Gasterosteiformes
Gasterosteioidei
Gasterosteidae
Apeltes
Syngnathoidei
Centriscidae
Aeoliscus
Fistulariidae
Fistularia
Syngnathidae
Halicampus
Hippocampus
Siokunichthys
Syngnathus
Gobiesociformes
Gobiesocoidei
Gobiesocidae
Sicyases
Myctophiformes
Synodontidae
Synodus
Trachinocephalus
Perciformes
Acanthuroidei
Acanthuridae
Acanthurus
Ctenochaetus
Naso
Zebrasoma
Siganidae
Siganus
Blennioidei
Anarhichatidae
Anarhichas
Blenniidae
Alticus
Antennablennius
Blennius
Istiblennius
Ophioblennius
Salarias
Xiphasia
Pholididae
Pholis
Stichaeidae
Lumpeninae
Leptoclinus
Stichaeinae
Stichaeus
Ulvaria
Gobioidei
Gobiidae
Acentrogobius
Amblycentrus
Amblyeleotris
Amblygobius
Cryptocentrus
Ctenogobiops
Glanogobius
Glossogobius
Gnatholepis
Lotilia
Monishia
Ptereleotris
Tomiyamichthys
Valenciennea
Vanderhorstia
Yongeichthys
Labroidei
Labridae
Cheilinus
Coris
Hemigymnus
Semicossyphus
Tautogolabrus
Xyrichtys
Scaridae (Callyodontidae)
Pseudoscarus
Scarus
Sparisoma
Percoidei
Apogonidae
Apogon (*Nectamia*)
Cheilodipterus
Siphamia
Malacanthidae
Malacanthus
Carangidae
Chloroscombrus
Trachinotus
Gerreidae
Gerres
Lethrinidae
Lethrinus
Lutjanidae
Lutjanus
Mullidae
Mulloides
Pseudupeneus
Upeneus
Percidae
Perca
Pomacentridae
Abudefduf
Amphiprion
Dascyllus
Microspathodon
Stegastes
Pomatomidae
Pomatomus
Sciaenidae
Bairdiella
Leiostomus
Menticirrhus
Micropogonias
Serranidae
Cephalopholis
Epinephelus (*Epinephalus*)
Sparidae
Acanthopagrus
Diplodus
Lagodon
Pachymetopon
Rhabdosargus
Sphyraenoidei
Sphyraenidae
Sphyraena
Trachinoidei
Creediidae
Limnichthys
Mugiloididae
Parapercis
Trichonotidae
Trichonotus
Pleuronectiformes
Pleuronectoidei
Bothidae
Bothus
Pleuronectidae
Citharichthys
Hippoglossoides
Limanda
Paralichthys
Pleuronectes
Pseudopleuronectes
Soleoidei
Soleidae
Pardachirus
Salmoniformes
Esocoidei
Esocidae
Esox
Salmonoidei
Osmeridae
Mallotus
Salmonidae
Coregonus
Salmo
Cottoidei
Cottidae
Hemitripterus
Myoxocephalus
Triglops
Scorpaeniformes
Platycephaloidei
Platycephalidae
Sorsogona
Scorpaenoidei
Scorpaenidae
Pterois
Synanceiidae
Inimicus
Synanceia
Siluriformes
Ariidae (Tachysuridae)
Arius
Tetraodontiformes
(Plectognathi)
Balistoidei
Balistidae
Balistapus
Pseudobalistes

REPTILIA

ANAPSIDA
Testudines
Casichelydia
Cryptodira
Chelonioidea
Cheloniidae (Cheloniadae, Chelonidae, Chelonioidae, Chelonydae)
Caretta
Chelonia
Dermochelyidae
(Dermatochelydae, Dermochelidae, Dermochelydidae, Sphargidae)
Dermochelys

AVES

NEORNITHES
Neognathae
Anseriformes
Anatidae
Bránta
Bucephala
Clangula
Melanitta
Mérgus
Somatéria
Alcidae
Alca
Fratércula
Úria
Charadriiformes
Dromadidae
Dromas
Haematopodidae
Haemátopus
Laridae
Lárus
Rissa
Scolopacidae
Arenária
Calidris
Limnódromus
Ciconiiformes
Ardeidae
Egretta

SCRUPARIINA (*continued*)

Falconiformes
Acciptridae
Haliaëtus
Pandionidae
Pandion
Pelecaniformes
Phalacrocoracidae
Phalacrocórax

MAMMALIA

THERIA
EUTHERIA
Carnivora
Caniformia
Arctoidea
Otariidae
Otaria
Phocidae
Phoca
Musteloidea
Mustelidae
Enhydra
Lutra
Mysticeta
Balaenopteridae
Balaenoptera
Megaptera
Eschrichtiidae
Eschrictius
Sirenia
Dugongidae
Dugong

AUTHOR INDEX

SYSTEMATIC INDEX

GEOGRAPHICAL INDEX

GENERAL INDEX

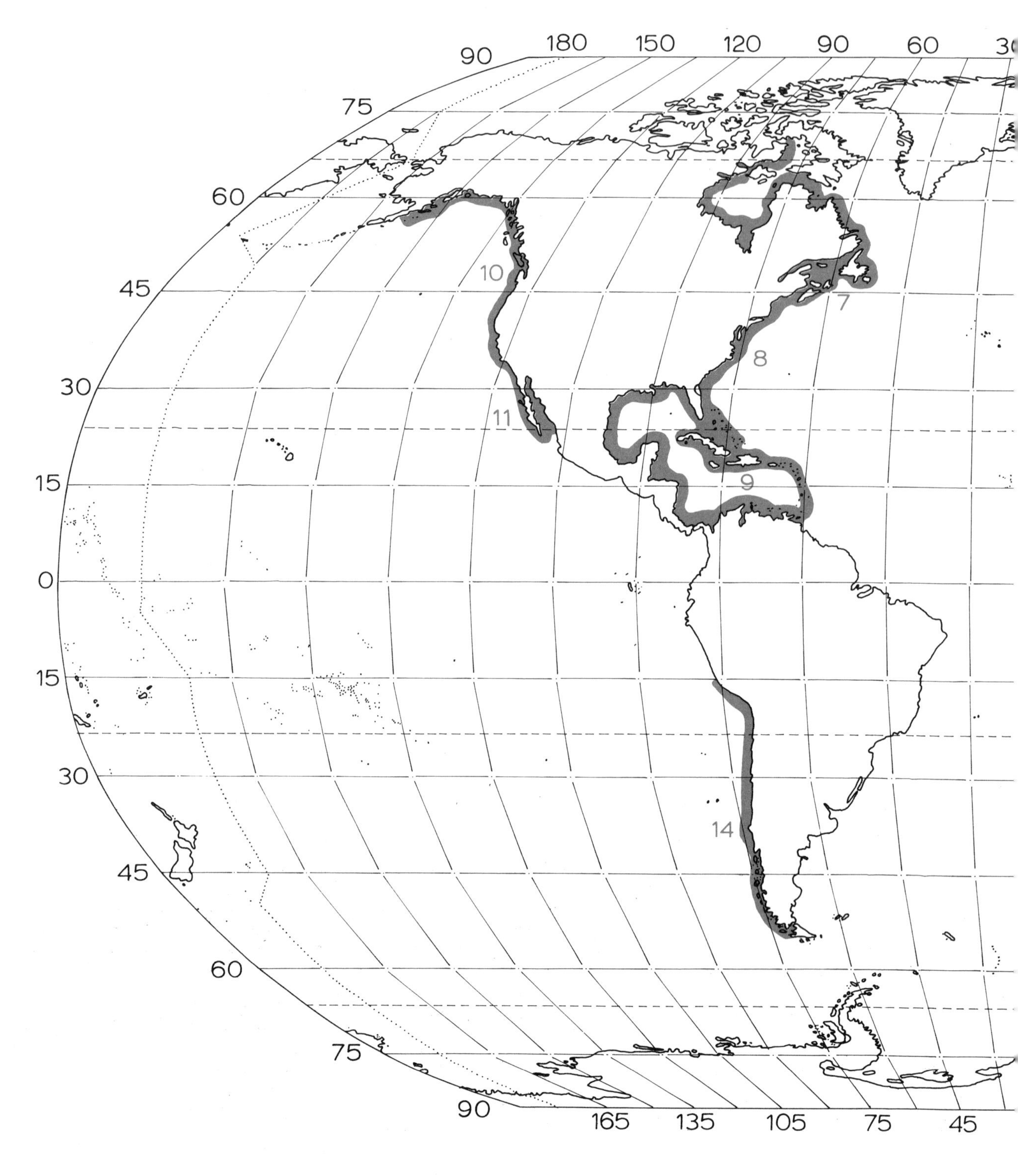

Main areas of intertidal ar
The numbers correspond